Humans and Electricity

Kwang Suk Park

Humans and Electricity

Understanding Body Electricity and Applications

Springer

Kwang Suk Park
Department of Biomedical Engineering
Seoul National University
Seoul, Korea (Republic of)

ISBN 978-3-031-20783-9 ISBN 978-3-031-20784-6 (eBook)
https://doi.org/10.1007/978-3-031-20784-6

This Springer imprint is published by the registered company Springer Nature Switzerland AG
The registered company address is: Gewerbestrasse 11, 6330 Cham, Switzerland

Preface

Humans are electric beings, which is the central concept I hope to explain about the human body by writing this book.

As a biomedical engineer focused on electrical engineering, I have worked on studies and projects developing newer technologies by contacting human bodies with electrical means, intelligently measuring electrical signals, and generating diagnostically useful information with these measured signals, as well as other studies inversely related to efficiently delivering electrical signals to the human body to recover disabled electrical systems. And, always, the first step was understanding how they are working electrically for the tasks within the body. These experiences guided me smoothly to conclude that humans are electric beings. For those who feel this is too provocative, at least humans can be recognized as electrical beings.

Humans are managed by their nervous system, and the signals and their processing in the nervous system are electrical. Humans are "thinking" beings through the brain, and the brain is the central processing unit operating by electrical pulses within the body. Humans are "living" beings by the heart, which is accurately controlled temporally and spatially by its conduction of electrical pulses. Humans are "behavioral" beings by all parts of the muscles. Precisely connected electrical networks control the muscles, whether we are threading a needle or lifting a heavy load. Humans are "sensing" beings through our eyes, ears, nose, tongue, and many kinds of receptors distributed throughout external and internal body surfaces. All these sensing organs and receptors produce electrical signals and transmit them toward the brain to interface humans with internal and external environments. Humans exist by electrical signals and systems that control the body's functions.

Since electrical signals and systems manage humans, it is unsurprising that their behaviors are being monitored electrically outward. Electrical monitoring is the easiest and most appropriate way to find any abnormality in the electrical working body functions. After Einthoven's first electrocardiogram application for medical purposes, attached electrodes measured many electrical signals from the body. They are now widely used in clinics as convenient, noninvasive, cost-effective diagnostic methods. These electrical signals include signals from the heart, brain, muscles, retina, nerves, eyes, ears, and so on. By recording and analyzing these electrical signals, we can immediately monitor how they work and find any abnormalities. These are very similar to the standard approach of an electrical engineer using an oscilloscope to find any malfunctions in electrical circuits or systems. We can consider humans as electrical beings, similar to any electrical circuit system that can be electrically monitored.

Recovering normal function is possible by applying electrical stimulations in some abnormally working organs, as we fix many electrical circuits and systems by adding or blocking electrical signals. Since the successful introduction of cardiac pacemakers as a standard treatment for too slowly beating hearts, electrical stimulation is rising as the efficient therapeutic method for recovering any electrically caused malfunction or disorder. This approach also deals with brain, muscle, and nerve problems. Electrical stimulation is now considered the third therapeutic method in addition to pharmaceutical and surgical treatment. Its minimal side effects are very contrasting to inevitably accompanying adverse effects in many medications. The possibility of modulating human functions externally by electrical stimulations accomplishes humans as complete electrical beings that can communicate with the environment bidirectionally from the body and toward the body. Humans are electric beings managed, monitored, and modulated by electrical signals.

The outer world of the human body has evolved toward the electrically connected world. Data and information are coming and going quickly throughout many electrical units through wired and wireless communication, like computers, smartphones, healthcare devices, and medical instruments. If humans are electrical beings that can be interfaced electrically like other electrical systems, humans can be connected to the environment more seamlessly. Understanding that humans are electric gives readers an extended perspective on integrating humans with the environment for medical purposes or other goals.

This book is composed of three parts. After the introduction overviewing and summarizing related historical progress, Part I deals with how electrical signals manage the human body. This part includes basic units of human electricity, electric potentials at rest and action, receptor potentials, and conduction of electrical potentials. Many of these parts are related to electrophysiology, but I have tried to explain most of the phenomena from an electrical engineering point of view. Part II deals with the kinds of electrical signals the human body monitors and how they are applied. Electrical signals from the heart and the brain are the most representative bioelectrical signals. And there are various kinds of electrical signals originating from nerves and muscles. In addition to the ordinary uses of bioelectrical signals for diagnostic purposes, I have included using these signals for brain-computer interface and human identification and authentication. Part III deals with electrical stimulation methods used in the medical domain and actively being developed for increased performance. Cardiac pacemakers and defibrillators have become indispensable medical therapeutic methods for resuscitating hearts. Deep brain stimulation took a significant position in treating movement disorders and extended its application domain to treating brain-related conditions. Electrical signals are now modulating many kinds of nerves working for specialized functions. Muscles of decreased performance are stimulated directly with electricity to recover their functions. Finally, I have included the health-related effects of electricity and electromagnetic fields, not to dismiss any possible adverse effects of providing electricity to humans.

These three parts and each chapter have their engineering and scientific depth and width. Since this book provides an overall view of the human body and electricity, it is most suitable as a text for graduate students majoring in biomedical engineering and for upper-undergraduate students majoring in electrical engineering. I thank the many people I have met throughout my career when collaborating on studies and projects. My colleagues in the Department of Biomedical Engineering at Seoul National University College of Medicine have motivated and encouraged me to write this book. Many of my students also supported me in writing and preparing materials for this book. Finally, I want to dedicate this book to my lady Mikyoung who shared her whole life with me so that I can develop my career successfully and write this book with all my effort up to this moment.

Seoul, Korea (Republic of) Kwang Suk Park

Contents

1 Introduction 1
1 Electricity and Magnetism 1
1.1 Generation of Electric Field from Electric Charges 1
1.2 Generation of Electric Field by Time-Varying Magnetic Field 1
1.3 Generation of Magnetic Field from the Magnets 2
1.4 Generation of Magnetic Field from the Current 3
1.5 Generation of Magnetic Field by Varying Electric Field 4
1.6 Maxwell Equations 5
2 History of Electricity 5
2.1 Electricity in Ancient Times 5
2.2 Static Electricity 6
2.3 Luigi Galvani 8
2.4 Electrophysiology 12
2.5 Measurement of Bioelectricity 16
2.6 Stimulations with Electricity and Magnetism 17
3 Theoretical Progress in Electricity 20
References 23

Part I Electricity Within the Human Body

2 Nervous System 27
1 Human Body Systems 27
1.1 Systems for Maintaining the Body Structure 27
1.2 Systems for Supplying the Energy 29
1.3 Systems for Controlling the Body Parts 31
1.4 Systems for Survival 32
2 Classification of Nervous System 33
2.1 Functional Classification of the Nervous System 33
2.2 Structural Classification of the Nervous System 34
3 Central Nervous System 36
3.1 Cerebrum 36
3.2 Cerebellum 40
3.3 Diencephalon 42
3.4 Brainstem 43
3.5 Spinal Cord 44
4 Peripheral Nervous System 45
4.1 Cranial Nerves 46
4.2 Spinal Nerves 47
4.3 Sympathetic and Parasympathetic Divisions 49
References 51

3 Neurons at Rest 53
1 The Neuron 53
1.1 Structure of a Neuron 53
1.2 Types of Neurons 54
1.3 Analogy of Neurons to Electrical Circuit Components 55
2 Membrane of the Neuron 57
2.1 Structural Characteristic of the Membrane 57
2.2 Capacitance of Membrane 58
2.3 Resistance of Membrane 59
3 Ion Distribution Across the Membrane 60
3.1 Diffusion Gradients 60
3.2 Potential Gradients 61
4 Membrane Potentials 61
4.1 Nernst's Potential 61
4.2 Goldman Potential 63
5 Ionic Currents in Resting State 66
5.1 Ionic Current by the Difference of the Resting Potential from Its Nernst Potential 66
5.2 Active Transport of Ion Pumps 67
5.3 Affecting Factors for Resting Potential 68
6 Membrane Potentials to Subthreshold Stimulation 69
6.1 Distributed Electrical Equivalent Model 69
6.2 Steady-State Potential of Stimulating Current 70
6.3 Membrane Potential to Step Current Stimulation 72
References 73

4 Action Potential 75
1 Potential Variation During Action Potential 75
1.1 Types of Membrane Potential Variation 75
1.2 Time Course of Action Potential Variation 76
1.3 Recording of Action Potential 77
2 Design of Experiments for the Hodgkin–Huxley Model 77
2.1 Space Clamp 78
2.2 Voltage Clamp 79
2.3 Total Membrane Current During the Voltage Clamp 80
2.4 Separation of Ionic Current 81
3 Quantitative Modeling of Ion Conductances 83
3.1 Modeling of Potassium Ion Conductance 83
3.2 Modeling of Sodium Ion Conductance 87
3.3 Constants of the Model 89
4 Action Potential with the Hodgkin and Huxley Model 90
4.1 Generation of Action Potential 91
4.2 Comparison with the Experimentally Measured Waveform 92
4.3 Refractory Period 92
4.4 Reestablishment of Ion Concentrations 93
5 Voltage-Gated Ion Channels 93
5.1 Types of Ion Channels 94
5.2 Structure of Voltage-Gated Ion Channels 94
6 Interpretation of Action Potential 95
6.1 Positive and Negative Feedback 95
6.2 Threshold for the Generation of Action Potential 96
6.3 Key Points of the Hodgkin and Huxley Model 96
References 97

5 Propagation and Processing of Membrane Potentials 99
1 Propagation of Action Potentials 99
1.1 Passive Propagation and Regeneration 99
1.2 Conduction Speed of Action Potential in the Unmyelinated Fiber 102
1.3 Propagation of Action Potential in the Myelinated Axon 105
1.4 Myelination as an Evolutionary Success 109
2 Synaptic Transmission 111
2.1 Signal Transmission in the Synapse 111
2.2 Neurotransmitters 112
2.3 Metabotropic Receptors 113
2.4 Neuromuscular Junction 113
2.5 Electrical Transmission Between the Neurons 114
2.6 Synapse as an Evolutionary Success 116
3 Processing in the Dendrite and the Soma 116
3.1 Postsynaptic Potentials 117
3.2 Propagation of Postsynaptic Potential Along the Dendrite 117
3.3 Summation of Dendrite Potentials 119
3.4 Dendrite Computation 120
References 121

6 Sensory Receptors 123
1 Properties of Sensory Receptors 123
1.1 Sensory Receptors as Transducers 123
1.2 Types of Sensory Receptors 124
1.3 Characteristics of Sensory Receptors 125
2 Mechanical Receptors 127
2.1 Touch Receptors 128
2.2 Baroreceptors 130
2.3 Auditory Receptors 131
2.4 Vestibular Receptors 132
2.5 Receptors in the Muscles 133
2.6 Kinesthetic Receptor 134
3 Chemical Receptors 135
3.1 Taste Receptors 135
3.2 Olfactory Receptors 137
3.3 Pain Receptors 138
3.4 Chemoreceptors for the Regulation of Ventilation 139
4 Thermal Receptors 140
4.1 Locations of Thermoreceptors 140
4.2 Types of Thermoreceptors 141
4.3 Polymodality of Thermoreceptors 141
4.4 Sensing Mechanism of Temperature 141
4.5 Temperature Regulation in Hypothalamus 142
5 Light Receptors 142
5.1 Rods and Cones 142
5.2 Phototransduction 143
References 145

Part II Electricity from the Human Body

7 **Electrocardiogram** 149
1 History of ECG 149
1.1 The First Human ECG 149
1.2 Discovery of ECG Mechanism 150
1.3 Standard 12-Lead ECG 151
2 Generation of ECG 151
2.1 Evolution of the Human Heart 151
2.2 Conduction of Action Potential in the Heart 153
2.3 Cardiac Pacemaker Potential 154
2.4 Action Potential in Cardiac Muscles 154
3 Recording of ECG 155
3.1 Cardiac Vector Equivalent to Potential Distribution 155
3.2 Standard Limb Leads 156
3.3 Normal ECG Waveform 157
3.4 ECG Recording Leads 158
3.5 ECG Electrodes 161
3.6 Types of ECG Recording 162
4 Diagnosis with ECG 163
5 Hear Rate Variability 164
5.1 HRV and Autonomic Nervous System 165
5.2 HRV Recording Durations 165
5.3 Time-Domain HRV Parameters 165
5.4 Frequency-Domain HRV Parameters 167
5.5 Nonlinear HRV Parameters 168
5.6 HRV Parameters in Applications 171
References 172

8 **Electrical Signal from the Brain** 173
1 History of Brain Electrical Signals 173
1.1 Historical Recording of EEG 173
1.2 Historical Recording of MEG 175
2 Generation of Brain Electrical Signals 176
2.1 Electric Current Generated in the Cerebral Cortex 176
2.2 Spatial Summation of EPSP 176
2.3 Types of Brain Electrical Signals 177
3 Recording of EEG 178
3.1 Electrode Placement System for EEG Recording 178
3.2 EEG Montages 179
3.3 Electrodes Measuring EEG 180
4 Spontaneous EEG 180
4.1 Spectral Bands of EEG Signal 180
4.2 EEG Activities During a Night Sleep 181
5 Evoked and Event-Related Potentials 183
5.1 Evoked Potentials 183
5.2 Event-Related Potentials (ERP) 188
6 Magnetoencephalogram (MEG) 192
6.1 MEG Signal 192
6.2 MEG Sensors 192
6.3 Comparison of MEG and EEG 194

7 Diagnosis and Applications 195
7.1 EEG in Clinical Diagnosis 195
7.2 EEG and MEG as Research Tools 196
References 196

9 Electrical Signals from the Muscles and Nerves 199
1 Electromyogram (EMG) 199
1.1 Types of Muscles 199
1.2 Structure of Skeletal Muscles 200
1.3 Generation of Muscle Contraction 201
1.4 Motor Unit 202
1.5 Types of Muscle Contraction 205
1.6 Recording of EMG 206
1.7 Application of EMG 207
2 Electroneurogram (ENG) 207
2.1 Nerve Conduction Study 208
3 Electrooculogram (EOG) 209
3.1 Generation of EOG 209
3.2 Recording of EOG 210
3.3 Eye Movements 210
4 Electroretinogram (ERG) 213
4.1 History of ERG 213
4.2 ERG Wave Components 213
4.3 Full-field ERG 214
4.4 Multifocal ERG 215
4.5 ERG Electrodes 216
5 Electrogastrogram (EGG) 217
5.1 Gastric Electrical Activity 217
5.2 Recording of EGG 218
6 Electrodermal Activity (EDA) 218
6.1 Sweat Gland 218
6.2 Nervous Control of EDA 219
6.3 Recording of EDA 219
6.4 Tonic and Phasic Components of EDA 220
6.5 Application of EDA 220
References 221

10 Brain-Computer Interface 223
1 Overview of BCI 223
1.1 History of BCI 223
1.2 Need of BCI Systems for the Locked-In Syndrome 225
1.3 Types of BCI Systems 227
2 Invasive BCIs 228
2.1 BCI Using Intracortical Electrodes 228
2.2 BCI Using ECoG 230
3 Noninvasive BCIs Using EEG 231
3.1 Slow Cortical Potential (SCP)-Based BCI 231
3.2 P300-Based BCI 233
3.3 SSVEP-Based BCI 237
3.4 Motor Imagery-Based BCI 240
3.5 Hybrid BCI 243
4 Data Processing in BCI 244
4.1 Preprocessing of Raw Data 244
4.2 Feature Extraction and Selection 244

4.3 Classification . . . 245
4.4 Deep Learning Approaches . . . 245
5 Application of BCI . . . 245
5.1 BCI Systems for the Patients . . . 245
5.2 BCI Systems for the General Population . . . 246
References . . . 246

11 Authentication with Bioelectrical Signals . . . 249
1 Types of Authentication . . . 249
1.1 Knowledge-based Authentication . . . 249
1.2 Possession-based Authentication . . . 250
1.3 Biometric Authentication . . . 250
1.4 Multifactor Authentication . . . 251
2 Required Properties as a Biometric Authentication Item . . . 251
3 Biometric Authentication . . . 252
3.1 Biometric Authentication Using Anatomical Items . . . 253
3.2 Biometric Authentication Using Behavioral Items . . . 256
4 Authentication Using ECG . . . 258
4.1 Characteristics of ECG as an Authentication Item . . . 259
4.2 ECG Leads for Authentication . . . 260
4.3 Structure of Authentication System Using ECG . . . 260
4.4 ECG Features for Authentication . . . 261
4.5 Deep Learning for ECG Authentication . . . 265
4.6 Matching the Features . . . 265
4.7 Performance of Authentication Based on ECG . . . 266
5 Authentication Using EEG . . . 267
5.1 Properties of EEG as an Authentication Item . . . 267
5.2 Tasks for EEG Acquisition . . . 268
5.3 Structure of Authentication System Using EEG . . . 269
5.4 Features Selection for Authentication Using EEG . . . 270
5.5 Matching the Features for Authentication Using EEG . . . 270
5.6 Performance of Authentication Based on EEG . . . 271
6 Multi-factor Authentication Using Biological Signals . . . 272
References . . . 272

Part III Electricity Toward Human Body

12 Electrical Heart Stimulations . . . 277
1 Artificial Cardiac Pacemaker . . . 277
1.1 Brief History of Artificial Cardiac Pacing . . . 278
1.2 Stimulations for Cardiac Pacing . . . 280
1.3 Artificial Pacemaker Device . . . 282
1.4 Modes of Cardiac Pacing . . . 286
1.5 Magnetic Resonance Imaging Compatibility . . . 290
1.6 Leadless Artificial Pacemaker . . . 290
2 Cardiac Defibrillator . . . 291
2.1 Brief History of Defibrillator . . . 291
2.2 Indications for Defibrillation . . . 292
2.3 Generation of Electrical Shock Pulse . . . 294
2.4 Types of Defibrillators . . . 296
References . . . 298

13 Electrical Brain Stimulations 299
1 Brief History of Brain Electrical Stimulation 299
2 Deep Brain Stimulation 301
2.1 Mechanism of Deep Brain Stimulation 301
2.2 Indications for Deep Brain Stimulation 303
2.3 Components of Deep Brain Stimulation Device 307
2.4 Stimulating Pulse Patterns 308
2.5 Battery for DBS 310
2.6 Technologies for Target Placement 311
2.7 Future Directions and Concerns 312
3 Electroconvulsive Therapy 312
3.1 Indications for Electroconvulsive Therapy 312
3.2 Mechanism of ECT 313
3.3 ECT Procedure 314
3.4 Electric Properties of ECT 315
4 Transcranial Direct-Current Stimulation 316
4.1 Indications for tDCS 316
4.2 Mechanism of tDCS 317
4.3 Procedures for tDCS 317
5 Transcranial Magnetic Stimulation 318
5.1 TMS as a Diagnostic Tool 318
5.2 Indications for Therapeutic TMS Treatment 319
5.3 Mechanism of TMS 320
5.4 TMS System 321
5.5 Stimulating Pulse Protocols for rTMS 322
6 Comparison of Brain Stimulation Methods 322
References 323

14 Sensing by Electricity 325
1 Brief History of Sensing by Electricity 325
1.1 History of Artificial Hearing 325
1.2 History of Artificial Vision 326
2 Human Hearing System 327
2.1 Sound That Humans Can Hear 327
2.2 Sound Transmission in the Auditory System 330
2.3 Sound Processing in the Cochlea 331
2.4 Hearing Loss and Indication for Cochlear Implant 333
3 Cochlear Implant System 334
3.1 Functional Architecture of Cochlear Implant 334
3.2 Sound Processor of Cochlear Implant 335
3.3 Electrodes of Cochlear Implant 336
3.4 Pulse Parameters for Loudness Control 337
3.5 Effectiveness of Cochlear Implant 338
3.6 Electroacoustic Stimulation 338
4 Human Visual System 339
4.1 Light that Humans Sense 339
4.2 Retina 341
4.3 Visual Pathway 343
5 Visual Prostheses 344
5.1 Retinal Prosthesis 344
5.2 Indications for Retinal Implants 345
5.3 Visual Prostheses at Other than Retina 346

5.4 Retinal Implant Systems 347
5.5 Spatial Resolution for Prosthetic Vision 347
References 348

15 Electrical Stimulation of Nerves and Muscles 351
1 Functional Electrical Stimulation (FES) 351
1.1 Electrical Activation of Muscles 351
1.2 FES for Foot Drop 352
1.3 FES for Upper Limb Movement 353
1.4 FES for Ambulation 353
1.5 Functional Electrical Stimulation Therapy (FEST) 354
1.6 Electrical Muscle Stimulation 356
2 Electrical Stimulation for the Restoration of Body Function 357
2.1 Phrenic Nerve Stimulation 357
2.2 Gastric Stimulation 358
2.3 Hypoglossal Nerve Stimulation 359
3 Vagus Nerve Stimulation (VNS) 360
3.1 Vagus Nerve 360
3.2 Vagal Nerve Stimulation for the Epilepsy 361
3.3 Vagal Nerve Stimulation for the Depression 362
3.4 Transcutaneous Vagus Nerve Stimulation 363
3.5 VNS and Electroceutical 364
4 Electrical Stimulation for Pain Relief 366
4.1 Spinal Cord Stimulation 366
4.2 Mechanism of Pain Relieving 366
4.3 Finding the Location for Electrical Stimulation 367
4.4 Stimulating Electrodes 368
4.5 Stimulating Parameters 369
4.6 Current Trends and Future Directions 370
5 Transcutaneous Electrical Nerve Stimulation (TENS) 370
5.1 Types of Transcutaneous Electrical Nerve Stimulation 370
5.2 Indication of TENS 372
6 Sacral Nerve Stimulation 372
6.1 Micturition Reflex 373
6.2 Mechanism of Sacral Nerve Stimulation 373
6.3 Sacral Nerve Stimulation Device 374
References 374

16 Biological Effects of Electricity and Electromagnetic Field 377
1 Physiological Effect of Current 377
1.1 Current Perception Threshold 378
1.2 Threshold of Reaction Current 380
1.3 Threshold for Pain Sensation 380
1.4 Let-Go Current 380
1.5 Respiration Paralysis 381
1.6 Ventricular Fibrillation 381
1.7 Electrical Burns 383
1.8 Stun Gun and Taser Gun 383
1.9 Lethality of Electricity 383
2 Biological Effects of Electromagnetic Field 385
2.1 Electromagnetic Field 385
2.2 Ionization Effect in Ionizing EM Waves 386
2.3 Current-Inducing Effect in Low-Frequency EM Waves 387

2.4 Thermal Heating Effect in Radiofrequency and Microwave EM Waves 388
2.5 Specific Absorption Rate (SAR) 389
2.6 Complications with Increased Body Temperature 390
2.7 Photochemical Effect of Light 392
2.8 Biological Effect and Health Effect of EMF 392
3 Nonthermal Biological Effects of Nonionizing EMF 393
3.1 Cancers and Brain Tumors 393
3.2 IARC Classification 395
3.3 Reproduction and Fertility 395
3.4 Neurological Problems in Children 396
3.5 Electromagnetic Hypersensitivity 396
3.6 Beneficial Effect of Electromagnetic Field 398
3.7 Cause-Effect Relation 398
References 399

Index 401

Introduction

1

1 Electricity and Magnetism

Electricity and magnetism have been strange forces for a long period of time throughout history and their relation also was not clearly understood up to the nineteenth century. However, these days we are understanding electricity and magnetism very well, when they are produced, and how they are related to each other. Their relations are beautifully summarized by Maxwell's equations.

1.1 Generation of Electric Field from Electric Charges

Most simply, the electric field is generated if there are any charges nearby. Electric charges produce an electric field, and they exert forces on all other charges in the field. And, other charges within the electric field are being attracted or repelled along the electric field lines depending on their relative polarity. The electric field is defined as the electric force per unit charge.

$$\boldsymbol{E} = \frac{\boldsymbol{F}}{q} \quad (1.1)$$

$\boldsymbol{E}$ is an electric field, $\boldsymbol{F}$ is electric force and q is the amount of charge.

Electric field lines go from positive charges to negative charges and can go to infinity if a single positive or negative charge exists in isolation (Fig. 1.1).

1.1.1 Gauss's Law of Electricity

Gauss's law of electricity describes the relationship between generated electric field and the amount of enclosed charge. If we consider electric field over any closed surface, total electric field flux over a closed surface is equal to the sum of charges enclosed, irrespective of charge polarities and their distribution.

$$\oint \boldsymbol{E} \cdot d\boldsymbol{A} = \frac{Q}{\varepsilon} \quad (1.2)$$

$\boldsymbol{E}$ is the electric field at the surface location, $d\boldsymbol{A}$ is the differential area of the location on a surface, Q is the amount of total charge inside the closed surface, and ε is the permittivity of the space (Fig. 1.2).

Within the human body, there exists a large amount of charge. Most of them are positively or negatively charged ions dissolved in fluids. Sodium, potassium, calcium, and hydrogen ions are positively charged. Chloride, bicarbonate, and large organic ions are negatively charged. Since they have charges, they are generating electric fields by themselves even though they do not move. Potential differences are caused by the uneven distribution of these ionic charges. So, they produce the resting potential across membranes where they are separated by semipermeable membranes. This is the main source of electricity of the human body and the basic existence of life. When we analyze the electrical behavior related to ions, we apply Gauss' law of electricity.

1.2 Generation of Electric Field by Time-Varying Magnetic Field

The electric field is also generated by the time-varying magnetic field, even though there are no electric charges nearby. While a time-varying magnetic field can generate an electric field, a static magnetic field does not generate an electric field. Permanent magnets are generating a magnetic field, but they do not generate an electric field while they are staying static. But if they are moving and causing any temporal differences in magnetic field strength at any position, it produces an electric field depending on the temporal gradient of the magnetic field on the position. This is the principle applied in the electric generator at power plants for the generation of electric current.

K. S. Park, *Humans and Electricity*, https://doi.org/10.1007/978-3-031-20784-6_1

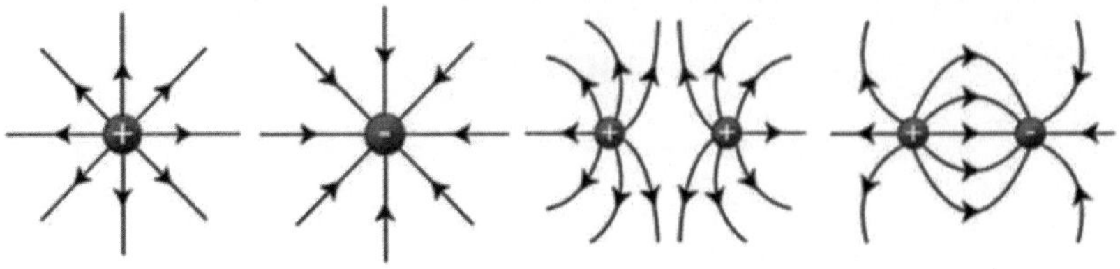

Fig. 1.1 Electric field lines from positive and negative charges

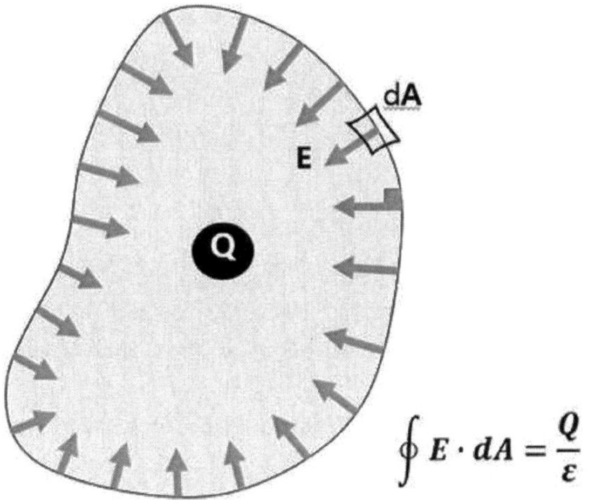

Fig. 1.2 Gauss's law of electricity

1.2.1 Faraday's Law of Induction

Faraday's law of induction describes the relation between electric field generated and varying magnetic fields over time. Electromotive force (EMF), which means electric force produced by a nonelectrical source, around a closed path is equal to the negative temporal change of the magnetic flux enclosed by the path. If the permanent magnet is moving in and out of the internal space enclosed by multiple turns of the coil, the magnetic field within the area enclosed by the coil is varying over time and it produces an electromotive force which can be observed by voltage meter as shown in Fig. 1.3.

$$\oint \boldsymbol{E} \cdot ds = -\frac{\mathrm{d}\Phi_{\mathrm{m}}}{\mathrm{d}t} \tag{1.3}$$

$\boldsymbol{E}$ is an electric field at the location, ds is the differential length, $\Phi_{\mathrm{m}} = \boldsymbol{B}_{\mathbf{ext}}\boldsymbol{A}$, where Φ_{m} is magnetic flux, $\boldsymbol{B}_{\mathbf{ext}}$ is a magnetic field, and $\boldsymbol{A}$ is the area enclosed by the path. A negative sign means that the variation of the magnetic field induced by the generated current is in opposite direction to the applied magnetic field variation.

Faraday's induction law describes that electric current can be generated within the body if there is any temporal variation of the applied magnetic field. The electric field generated from an intrinsic magnetic field within the body is not of big concern. Normally, there is no intrinsic magnetic source within the body and most of the magnetic field within the body is caused by the electric current variation. The strength of this magnetic field is not strong and their temporal variation is also not so fast to induce any influential electric field according to Faraday's law of induction. But there can be external magnetic fields and their strength with fast temporal variation can induce electrical current which can stimulate some neurons and muscles. One of the intended applications of this principle is transcranial magnetic stimulation (TMS). In TMS, they apply a high-frequency magnetic field to induce some electric current in the focused area within the brain. By stimulating neurons located in the focused area they can treat effectively some disorders like depression. Also, humans can be affected by an unintentionally exposed external magnetic field. Magnetic fields are relatively high near the power plants, power transmission cables, relay stations, or at some places of strong electromagnetic wave power. This unintended strong alternating magnetic field can induce a stray current within the body and might result in health-related adverse effects.

1.3 Generation of Magnetic Field from the Magnets

The magnetic field is a vector field like an electric field. It exerts a force on another magnet located within the magnetic field. The static magnetic field is generated by magnets. The earth magnet and small magnets on refrigerator doors are generating a static magnetic field which can be observed by a compass needle. Magnetic field lines go from the north pole to the south pole. Since magnetic poles cannot be isolated, differently to electric charges, magnetic field lines are always closed through the magnet.

1.3.1 Gauss's Law of Magnetism

Gauss' law of magnetism states the relationship of magnetic field intensity with magnet dipole intensity. Net magnetic flux through any closed surface is 0.

$$\oint \boldsymbol{B} \cdot dA = 0 \tag{1.4}$$

$\boldsymbol{B}$ is the magnetic field at the surface location, dA is the differential area of the location.

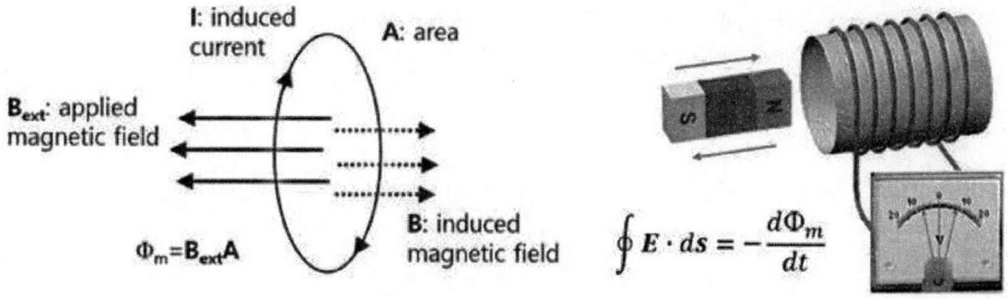

Fig. 1.3 Faraday's law of induction

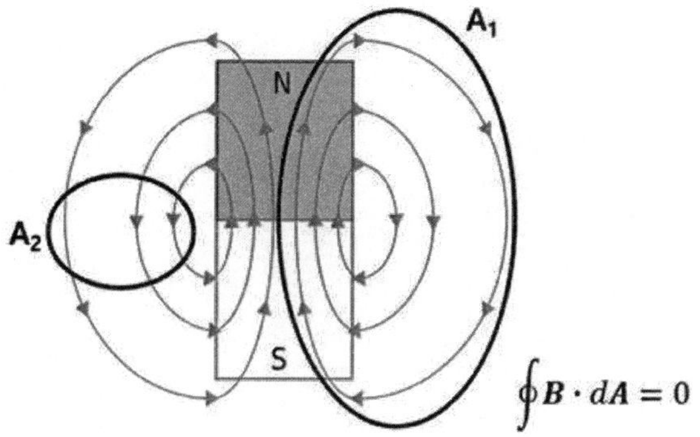

Fig. 1.4 Gauss's law of magnetism

This means that the total amount of magnetic flux directed inward is equal to the total amount of the magnetic flux outward. This is caused by the fact that magnetic monopoles do not exist and the basic entity for magnetism is magnetic dipoles (Fig. 1.4).

Within the human body, there are no noticeable intrinsic magnets or magnetic dipoles. For magnetic resonance imaging (MRI), magnetic dipoles are considered for the explanation of the phenomena, but they don't produce any noticeable magnetic field in the ordinary physiological environment without any strong static external magnetic field. So, the magnetic fields present within the human body are of electric origin. When we analyze magnetic field dynamics within the body, we apply this Gauss's law of magnetism.

1.4 Generation of Magnetic Field from the Current

In addition to that magnetic field is generated by magnets, it also can be generated by flowing current. The magnetic field around the electric current is proportional to the intensity of the electric current flowing. This is the principle that electromagnet is working on. Electric current serves as a source of the magnetic field along the enclosing loop path just as electric charges serve as a source of electric field enclosing the sources.

1.4.1 Ampere-Maxwell's Law for Conduction Current

Ampere-Maxwell's law states the magnetic field generated around the electric current flowing wire. For any closed loop path, the net magnetic field integrated over the path of the closed loop is equal to the permeability times the electric current enclosed by the closed loop.

$$\oint \boldsymbol{B} \cdot ds = \mu_0 I_c \tag{1.5}$$

$\boldsymbol{B}$ is a magnetic field, ds is the differential length segment along the path, I_c is the intensity of conduction current, and μ_0 is the permeability of free space.

When the current is flowing through the wire, an electric field is also generated, since the current is the flow of electrical charges themselves. This means that the electric field and magnetic field are generated at the same time when the current is flowing through. But their field lines are perpendicular to each other as shown in Fig. 1.5. Magnetic field lines are generated concentrically centered to a wire and electric field lines are radiating from a wire.

Within the body, movement of ionic charges are developing current and flowing current also produce magnetic fields. This is the majority of electrical behavior within the body. Current flow through the membrane changes the voltage across the membrane. They are resting potential, action potential, presynaptic, and postsynaptic potentials. These voltages summed over some amount of tissue can be measured externally by electric sensors. At the same time, the developed current produces a magnetic field and it can be measured by magnetic sensors from the outside of the body.

Electric behavior within the body always accompanies magnetic behavior and they can be measured and analyzed electrically or magnetically depending on the convenience and the sensitivity upon application.

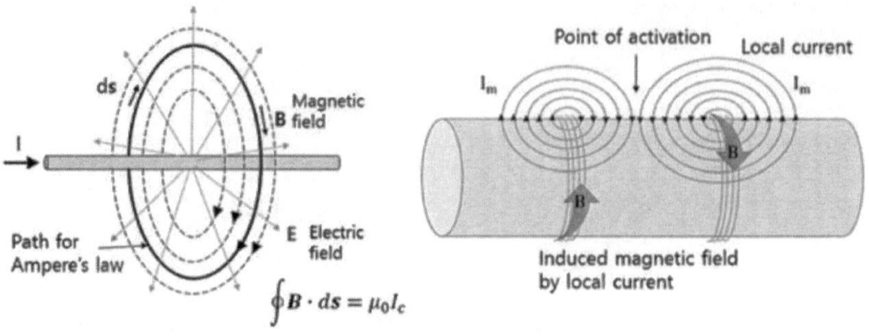

Fig. 1.5 Ampere-Maxwell's law for conduction current

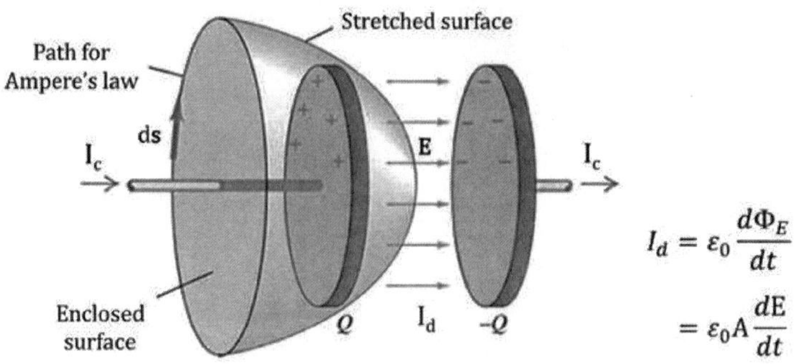

Fig. 1.6 Ampere's law for the varying electric field

1.5 Generation of Magnetic Field by Varying Electric Field

Additionally, the magnetic field can be generated by a varying electric field. The magnetic field generated by a varying electric field is symmetric to an electric field generated by a varying magnetic field. Imagine a wire connected to a charging or discharging capacitor. The surface enclosed by the path for Ampere's law can be stretched into the open region of the capacitor. Since any open surface bounded by the closed path produce the same strength of the magnetic field, the same magnetic field is to be generated by the alternating electric field between the capacitor plates. This induced magnetic field makes it looks like there is current passing through capacitor plates. This is displacement current.

$$I_d = \varepsilon_0 \frac{d\Phi_E}{dt} \tag{1.6}$$

I_d is the displacement current, $\Phi_E = \boldsymbol{E} \cdot \boldsymbol{A}$, where Φ_E is electric flux, $\boldsymbol{E}$ is the electric field, and $\boldsymbol{A}$ is the stretched surface area enclosed by the path. If current flows through a capacitor, displacement current becomes equal to conduction current, that is $I_d = I_c$ (Fig. 1.6).

$$\oint \boldsymbol{B} \cdot d\boldsymbol{s} = \mu_0 I_c = \mu_0 I_d = \mu_0 \varepsilon_0 \frac{d\Phi_E}{dt} \tag{1.7}$$

1.5.1 Ampere-Maxwell's Law

Ampere-Maxwell's law describes the magnetic field generated by both electric currents, displacement current and conduction current. Since displacement current is built by varying an electric field within the capacitor space, this can be considered that magnetic field is generated by a varying electric field.

$$\oint \boldsymbol{B} \cdot d\boldsymbol{s} = \mu_0 I_c + \mu_0 \varepsilon_0 \frac{d\Phi_E}{dt} \tag{1.8}$$

If there are two types of currents enclosed simultaneously within a surface enclosed by the path, they can contribute additionally to the generation of the magnetic field.

Time-varying electric field and magnetic field are inducing each other. If their changing rate is too high, their results can be influential. But, within the body, electrical signals do not change so rapidly. Maximum frequency is in the order of KHz and it is too low compared to the higher frequency in the range of MHz or GHz which might cause electromagnetic hazard. So, the electric and magnetic field induced by the fluctuation of the counter field is minimal, and in most cases, it is ignored for the simplicity of analysis and calculation without causing a large amount of error. But, caution is needed not to forget that Ampere-Maxwell's law is strictly applied in the body, especially if there is coupling with the environment of high-frequency electric or magnetic field.

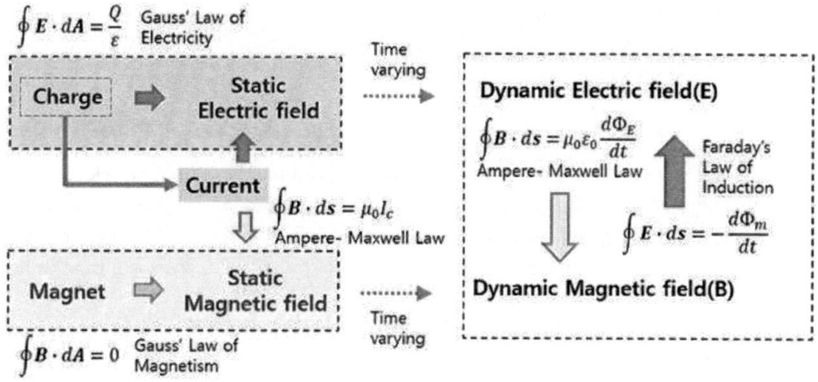

Fig. 1.7 Summary of Maxwell equations

1.6 Maxwell Equations

The generation of an electric field and a magnetic field is summarized by Maxwell equations as shown in Fig. 1.7.

2 History of Electricity

2.1 Electricity in Ancient Times

2.1.1 Electric Fish

The first electricity that human met in history is the electricity from the fish. It was long before we understand the concept of electricity. Its first record is dated back to around 6000 years ago to ancient Egyptian times. The hieroglyphic text in sepulcher fresco describes the electric fish as "Thunderer of the Nile" and the shock they received from them as the "Release of the troop." There were electric sheatfish that generate electricity in some conditions. When they caught such a fish, the fish generated high-voltage electric shocks of more than 450 V, and it forced the fishermen to release the fish.

There are several kinds of electric fish living in the river and the sea. They are mostly found in and around South America and Africa. In addition to electric sheatfish, there are electric eels and electric rays. High voltage is generated by the synchronous activation of electrolytes, which are the modified muscle or nerve cells specialized to produce bio-electric fields. Their electric organ, which is typically located in the tail of the electric fish, is used to discharge a strong electric field around their body with 10–860 V to stun the prey or to defend them from predators. Some electric fishes generate only a weak electric field with less than 1 V to navigate or to communicate. They discharge unique electrical signals in specific types of pulses and waves and this electric discharge carries precise information on their sexes, status, and identification. During the courtship, they use electrical signals to serenade exotic discharge patterns. In contrast to this electrogenic fish, there are electroreceptive fishes that sense electric fields. Most electrogenic fishes are electroreceptive. Many fish such as sharks, rays, and catfishes are electroreceptive, but they are not electrogenic fishes.

Since the electric ray was first used in the treatment of headaches and gouty arthritis in AD 46, it remained as the unique method of retaining electricity for electrotherapeutic treatments until the seventeenth century. The first electricity came from biology and its first application was for medical treatment.

2.1.2 Atmospheric Electricity by Lightning

Lightning is a naturally occurring electrostatic discharge from ancient times. Lightning is a giant spark between the two electrically charged regions in the atmosphere, both in the atmosphere or with one on the ground. In the early stage of its development, air acts as an insulator between the positive and negative charges in the cloud and the ground. When the opposite charges are built up enough, this insulating capacity of the air breaks down and there is a rapid discharge of electricity to generate a spark of lightning. Even though the electricity of lightning exists before humans exist, humans only recognized that lightning is developed from the accumulated charge in recent decades. In ancient times, it was thought of as a message of God or the weapon of Zeus. Spots struck by lightning were regarded as sacred and churches often were erected at these sites in the Greek and Roman ages. Religions such as Christianity used to teach that lightning was God's punishment and God's creation and shouldn't be avoided. During the Napoleonic wars, more than 220 British tall ships were damaged by lightning. Many soldiers have been killed by lightning due to their metal weapons and equipment acting as conductors for lightning. These were all before Benjamin Franklin identify the electric property of the lighting and invent a lightning rod.

2.1.3 Static Electricity from the Amber

Amber is fossilized pine resin that has been appreciated for its color and natural beauty as a gemstone by the Greeks, much as it is today. In about 600 BC, the ancient Greek philosopher

Thales (625–547 BC) of the Greek city Miletus discovered that amber possesses a rather peculiar property. Thales had a particular interest in what materials were made from. So, he carried out many tests to explore the properties of different materials and compare them. In one of his experiments, Thales investigated amber and discovered it develops the ability to attract light objects like feathers when it is rubbed with a piece of animal fur. In another experiment, he noticed that lodestone, which is magnetic iron ore, attracts pieces of iron. Thales compared the behaviors of the rubbed amber and lodestone in attracting other materials and objects. He had actually discovered magnetism and static electricity, but he did not recognize the difference between these two forces at the time and thought they were the same phenomena. At the time, it was generally believed that any sort of movement was due to life, soul, or the gods. He drew a conclusion that lodestone and amber must have a soul and quoted "Communicating life to inanimate things" and made predictions to explain the behavior of the materials he had observed. This is the first historical reference to static electricity.

2.2 Static Electricity

For about 2000 years since Thales, this strange phenomenon was thought to be a unique property of amber and the static electricity and the magnetism remained confused without clearly recognizing their difference. In his book *De Magnete* published in 1600, an English physician William Gilbert (1544–1603) coined the term "Electric," which is a variation of the Greek word for amber, to describe the effect developed by rubbing the amber. He also recognized the difference between static electricity and magnetism, and enforced it with two observations. He observed that a wet surface or a humid atmosphere removes the electricity of amber, which is not the case for the magnet, and that the attractive property of amber, unlike that of the magnet, belongs to a wide variety of substances. In 1733, a French chemist du Fay (1698–1739) discovered the existence of two different types of electricity and named them "resinous" and "vitreous" electricity. When amber is rubbed with fur, it acquires "resinous" type electricity. On the other hand, when glass is rubbed with silk, it acquires "vitreous" type electricity. He also discovered that objects with the same type of electricity would repel each other, and objects with different types of electricity attract. He also noted the difference between conductors and insulators, calling them "Electrics" and "Non-electrics" for their ability to produce contact electrification.

2.2.1 Electroscope

The electroscope is the first electrical measuring instrument used to detect the presence of electric charges on an object. It detects charges by the movement of balls or leaves within it based on the Coulomb electrostatic force and their movements increase with the amount of charge on an object. As it requires hundreds or thousands of volts to operate, it is mostly used with high-voltage sources such as static electricity and electrostatic machines, and it only gives a rough indication of the charge quantity.

The first electroscope was a pivoted needle, called the Versorium, invented by a British physician William Gilbert around 1600. The Versorium is a metallic needle rotating freely on a pedestal and it looks and functions similar to a compass needle, but unmagnetized. The needle is attracted to a charged body brought near it, turning the one end toward the charged object. When a positively or negatively charged object is positioned near the tip of the needle, mobile charges of opposite polarity in the metal are attracted to the tip near the object, and the charges of the same polarity move to the far end of the needle by the electrostatic induction. The attractive force between the opposite charges then rotates the needle to the nearest angle to the charged object and the repulsive force between the charges of the same polarity exerts a force to rotate the other end of the needle to the farthest angle from the charged object. After some swings of the needle, it stops pointing at the charged object. Actually, both ends of the Versorium function in the same manner, and it cannot tell the polarity of the charge of the object, unlike a compass which can tell the "North" and "South" of the magnetic direction. The pith-ball type and the gold-leaf type electroscopes are two classical types of electroscope and they are still used in schools to demonstrate the principles of electrostatics (Fig. 1.8).

2.2.2 Electrostatic Generator

With the knowledge that static electricity can be produced by rubbing the two different materials, electrostatic generators were developed based on this principle. The electrostatic generator is the first electric machine for the generation of electricity. In 1672, German physicist Otto von Guericke (1602–1686) came up with the earliest version of a machine that can generate static electricity. It is composed of a sphere of sulfur, whose size is of an infant's head, and an iron axle mounted in a wooden framework. Static electricity was generated while the sphere was rotating and rubbed.

In 1709, English experimentalist Francis Hauksbee (1660–1713) improved the electrostatic generator using a sphere of glass rotated by a wheel, which produces electricity continuously while the glass is rotated and rubbed (Fig. 1.9). He used his electrostatic generator to demonstrate the electric glow with mercury. While placing a small amount of mercury placed in the glass, he created a mild vacuum by evacuating the glass with an air pump. When he rubbed the glass ball in order to build up a charge, a glow was visible if he placed his

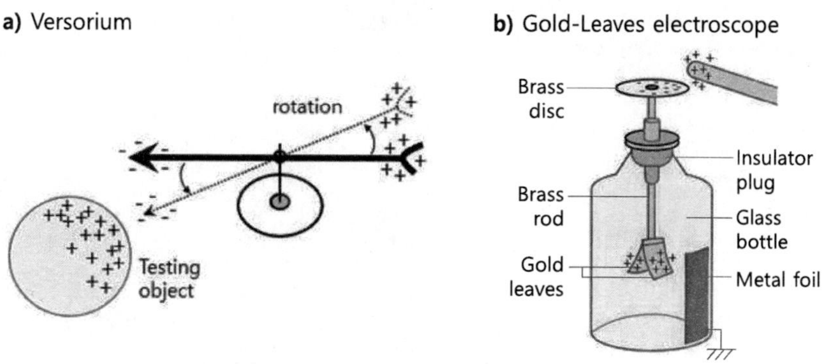

Fig. 1.8 Electroscopes. (**a**) Versorium, the first electroscope, (**b**) Gold-leaves electroscope

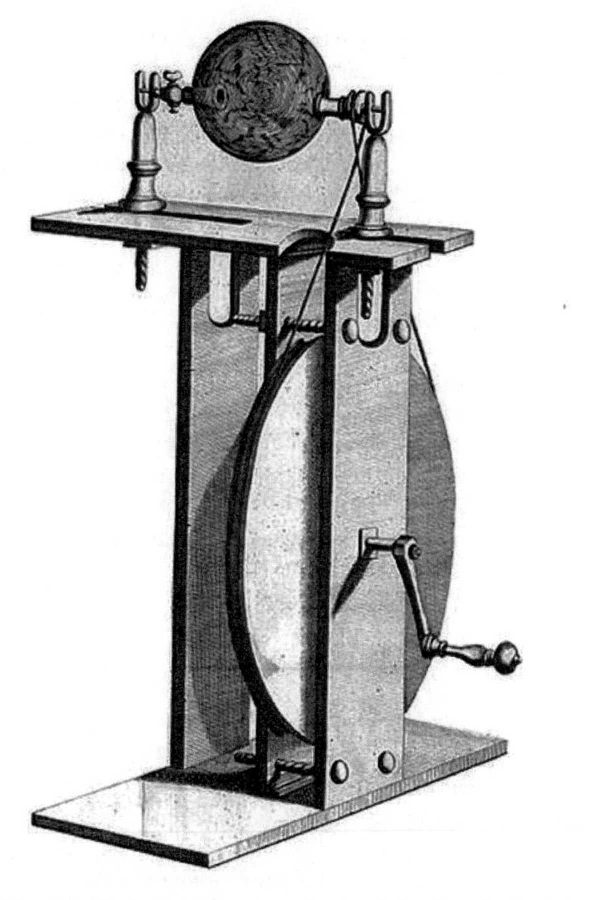

Fig. 1.9 Electrostatic generator developed by Francis Hauksbee in 1709

hand on the outside of the ball. This effect later became the basis of the gas-discharge lamp, which led to neon lighting and mercury vapor lamps.

These early versions of electrostatic generators generated electricity by friction. Even though the generated voltages could be very high, the drawable current was very small and always unipolar static voltage was built. The electricity generated by these electrostatic generators cannot be stored and it should be used at the site and at the time of generation. In those days, generated electricity was used mostly for medical treatment or just for entertainment.

2.2.3 Leiden Jar for Collecting Electricity

A Leyden jar is an electrical device that accumulates and stores electric charges to a high voltage from an external source. It is a prototype of the capacitor. It is composed of two electrical conductors of metal foils on the inside and outside of a glass jar. A metal terminal projecting vertically through the jar lid contacts the inner foil. Leiden jar was invented independently almost the same time in 1745–1746 by Dutch scientist Pieter van Musschenbroek of Leiden and by German cleric Ewald Georg von Kleist. The Leyden jar could be charged to a high voltage of 20,000 to 60,000 Volts and the terminal rod is shaped in a ball to prevent leakage of accumulated charges through the air.

In its earliest form, it was a glass vial, partly filled with water and the orifice of glass was closed by a cork pierced with a wire or nail that dipped into the water. In those days, charges were thought to flow like water. So, they guided generated charges with a conductor to this glass vial and hoped the transferred charges to be accumulated and stored safely in the water within the glass. With this concept, they brought the exposed end of the glass vial into contact with an electrostatic generator. They first thought that charges can be stored in the water. But, the glass did not effectively collect the charges when it was placed normally on the table. However, when they raised the glass by hand, the sparks or shocks were developed by touching the terminal wire. Then, they recognized the outer coating is required connected to the ground. The shock delivered by the Leiden jar was very severe as Musschenbroek wrote "I would not repeat the experiment if offered the whole kingdom of France" and some experimenters have died during the experiment with the Leiden jar.

One of the initial benefits of the Leyden jar was that it could be transported easily, much simpler than the electrostatic generators. Soon after the Leyden jar was invented, the Leyden jar was actively used to conduct many early

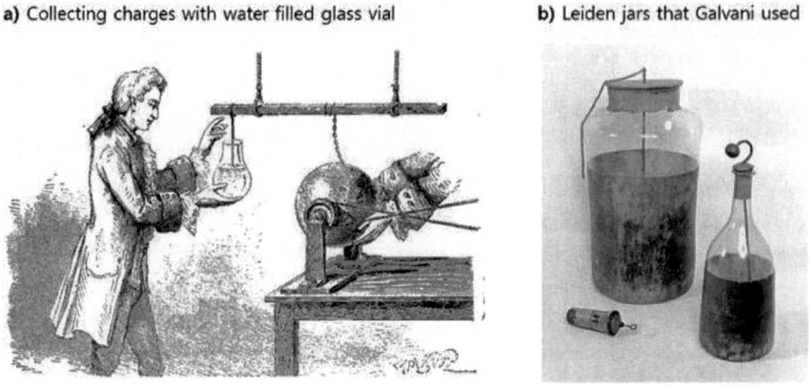

Fig. 1.10 Leiden jar. (**a**) Water filled glass vial as the earliest version of Leiden jar. (**b**) Leiden jars that Luigi Galvani has used in his experiments

experiments in electricity. It was very convenient to treat medical symptoms by providing electrical stimulation and actively used for muscular stimulation and treatment of paralysis. Leiden jar has popularized as a new electrical gadget even for the general public which can provide a kind of strange shock (Fig. 1.10).

2.2.4 Collecting Electricity from Lighting with Leiden Jar

The electrical property of lightning was not fully understood before Benjamin Franklin (1706–1790). By conducting the kite experiment, Franklin proved that lightning is an electrical discharge and verified that it can be discharged safely over a conductor into the ground.

Franklin hypothesized that lightning was an electrical discharge. Before he conducted an experiment with flying a kite, he proposed standing an iron rod into stormy clouds to attract electricity from them. But, he did not perform the iron rod experiment by himself, as he thought the experiment should be done in a higher place to reach close to the clouds. Actually, his idea was performed successfully by the French scientists Delor and Dalibard before his kite experiment. In 1752, without hearing the success of the iron rod experiment in France, he thought that the same experiment can be performed with a kite instead of an iron rod. With the help of his son William, he built a kite with sharp-pointed wire at the top and it was connected to the metal key and Leiden jar. They were remained within the shelter not to be wet and the end of the string was also kept dry to insulate themselves from a possible electric shock. When the storm passed over his kite, the conductor drew electricity into his kite. The kite was not struck by lightning, but the conductor drew negative charges from a charged cloud to the kite, string, metal key, and Leyden jar. When he moved his hand near the key, he received a shock because the negative charge attracted the positive charge in his body.

With the idea of an iron rod experiment for the electricity of the lightning, he first theorized that lightning might be preventable by a lightning rod, which is an elevated iron rod connected to earth to empty static electricity from the clouds. Before his kite experiment, he did many experiments on the electricity using the Leyden Jar and discovered that the charges are in two types and used the term "positive" and "negative" to describe the difference of charges instead of "vitreous" and "resinous."

2.3 Luigi Galvani

2.3.1 Animal Electricity

Luigi Galvani (1737–1798) was an Italian physician, physicist, biologist, and was a professor of anatomy at the University of Bologna. He has discovered animal electricity and is recognized as the pioneer of bioelectricity. He is famous for his experiment of electrical stimulation that induced a twitch in the leg muscle of a dead frog.

Galvani received a doctorate in medicine and his research primarily was concerned with comparative anatomy with a tendency toward physiology. In the late 1770s, with the acquisition of an electrostatic generator and Leyden jars, his interest moved to electrophysiology. He began to experiment with muscular stimulation by electrical means and animal electricity remained his major field of investigation since then. His discovery began with the accidental overlap when the dissected frog was lying on the same table with electric equipment. In 1781, when his assistant touched the femoral nerve of the frog with an anatomical knife, the muscle of the dissected frog contracted violently with sparks simultaneously discharged at the nearby electric equipment. This first experimental scene of bioelectricity was depicted in Fig. 1.11. This can be thought due to the capacitive current occurred in a conductor when the potential in a separated and closely located conductor changes abruptly. It has been said that if Galvani were sufficiently acquainted with the physics of electricity, he might have recognized the external origin of the electricity, would not be surprised by the phenomenon,

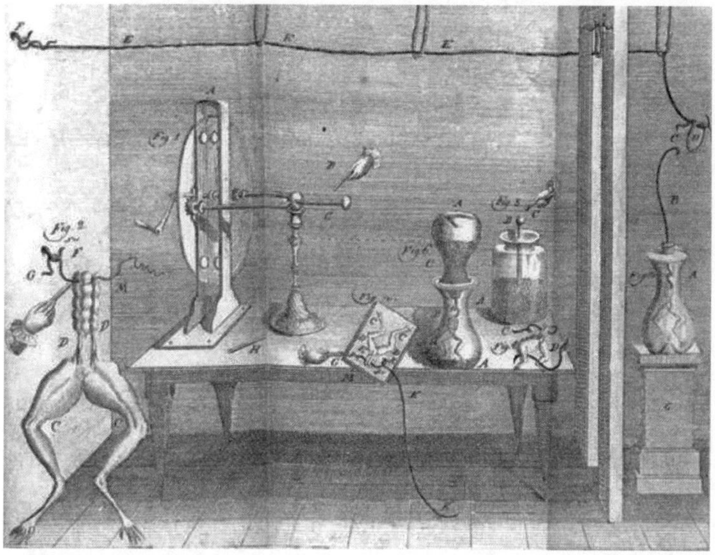

Fig. 1.11 The first experimental scene of bioelectricity of Galvani in 1781. *Fig. 1*, An electrical machine, A: disc, B: an iron cylinder by which a spark is forced out, C: conductor, *Fig. 2*, Frog and experiment prepared, C: legs, D: the sacred nerves, which pass into the leg muscle, E iron wire trajected through the pith, F: a metallic wire that pierces through the foramina of the backbone that passes through the spinal cord, G: Iron cylinder, H: a glass cylinder, K: the conductor of the strings, M: Backbone, *Fig. 3*, A: the glass vial, inside which the frog has been prepared, B Iron wire with a hook of frog, C: the end of a suspended iron wire to which an iron wire B is attached, D: silk trap, E: the iron wire, which in conjunction with the iron thread B makes the conductor of the strings, and the longest pellets, F: hook to which iron wire is attached, *Fig. 4*, C: conductor of nerves, D: conductor of muscles, *Fig. 5*, Leiden jar, A: bowl contacting stimulating content in it, B: ball terminal of conductor, C: a hand inducing spark with B, *Fig. 6*, A: an inverted bowl, inside which contains the stimulating beads, B: a vial, within which a frog is accompanied

and perhaps would not have started his investigations. With further experiments, Galvani assured himself that the twitching of the frog's leg muscle has the involvement of the electrical action.

After assessing the most appropriate conditions for obtaining contractions with various devices capable of producing electricity or of accumulating it, Galvani decided to investigate the effects of natural atmospheric electricity (Piccolino, 1998). He prepared the frog's leg and connected it to a long metallic wire pointing toward the sky in the highest place of his house. To his anticipation, all muscles of the limbs were contracted at the moment of the lightning before the thunders. Then he came to the thought that natural electricity present in the atmosphere may also induce the same contraction as the electricity during lightning. He prepared the frogs in the usual way and suspended them up on the iron railing of the balcony of his house on a clear and calm day and waited. But nothing happened for a long time. However, to his surprise, the contractions appeared while he was manipulating the frogs by pushing and pressing the brass hooks inserted into the frog's spinal cord toward the iron bars of the railing. Soon, he recognized that the contraction does not related to the atmospheric electricity as the contraction could be repeated indoors in a closed room by substituting the balcony railing with an iron plate. With a long series of experiments, he realized that creating a circuit by connecting a nerve structure and the muscle through a bimetallic conductor is sufficient for a contraction (Fig. 1.12). He thought that creating a circuit between the nerves and the leg muscles through a metallic conductor is sufficient to discharge the electricity accumulated in the muscle, like the electricity developed in a Leiden jar resulted in a contraction. Also, he thought that the animal has the capability of storing the electrical fluid and maintaining it in a disequilibrium state, and it is ready to be set in motion using the internal or external conductive arc. Without knowing the mechanism of bimetallic potential, Galvani explained the experimental results that the contraction was caused by the discharging "Animal electricity" existing in the body through connected the bimetallic arch.

As Galvani was also a medical practitioner, he had an interest in medical electricity, the field of therapeutic application of electricity. Medical electricity emerged in the middle of the eighteenth century with the advent of new electrical equipment like an electrostatic generator and Leiden jar to treat some diseases which could not be cured by traditional treatments. He believed that medical therapy should be strictly based on anatomy and physiology and revealing the hidden properties of nerves and muscle will provide safer treatment for the diseases. The prevailing view for the explanation of medical electricity was based on the existence of fluid matter named "Animal spirit." Animal spirit was

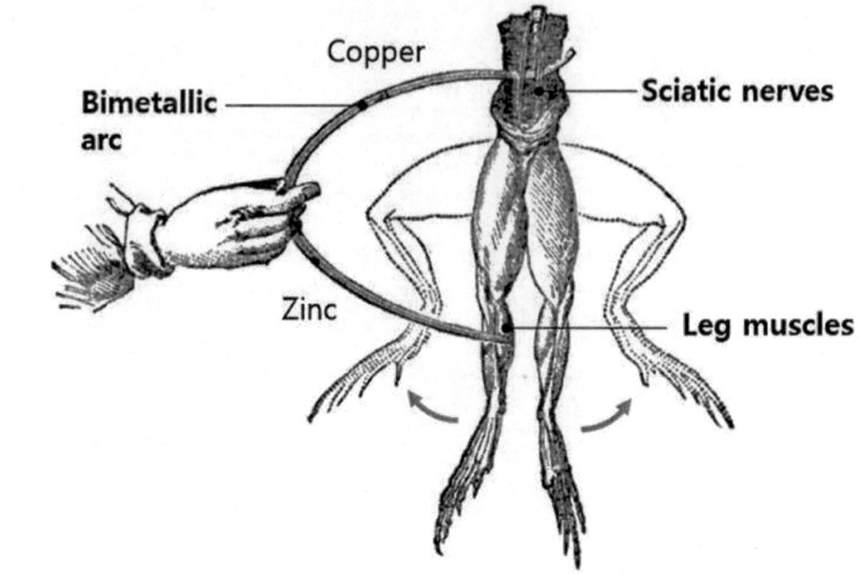

Fig. 1.12 Galvani's experiment on a dead frog's leg with a bimetallic arc. The muscular contraction was resulted by connecting the sciatic nerve and leg muscle with a bimetallic arc composed of two different materials, for example, zinc and copper

thought produced in the brain, flowed from the brain through the nerves to produce muscular contraction and body motion, and flowed oppositely back to the brain by producing sensation (Bresadola, 1998). Even though there was a hypothesis proposing the electrical nature of the "Animal spirit," it was not supported due to the lack of experimental results. Galvani was the first to provide evidence for the electrical nature of the mysterious fluid involved in nerve conduction and muscle contraction. The experimental work of Galvani swept away the mysterious fluid of "Animal spirits" and led to the foundation of new science, electrophysiology and neurosciences. His main work summarized research results over 10 years on the effect of electricity on animal preparations and was published in 1791. It gave a strong impact on the scientific audience comparable to the social and political impact of the French revolution in those same years. Galvani's experiments were perceived as not only the "great and marvelous discovery" of electricity intrinsic to the organism but also the fact that could represent "the most productive of very useful applications to medicine, both practically and theoretically." However, it also triggered the controversy on animal electricity.

2.3.2 Metallic Current

Alessandro Volta (1745–1827) was an Italian physicist and a chemist at the University of Pavia. He is a pioneer of electricity and is the inventor of the electric battery. Actually, his invention of the battery has derived from the controversy of animal electricity with Galvani.

With the acquaintance of Galvani's achievement, Volta repeated and confirmed Galvani's observations and expressed his admiration for the great discovery of animal electricity at first. However, with the progress of his work on this subject in the following 8 years, he recognized some inexplicability of animal electricity and progressively changed his attitude. As Galvani viewed the muscular contraction of a frog's leg was caused by the discharge of the electricity accumulated in the muscle like a Leiden jar through a nerve which behaves as a conductor, contraction caused by connecting the two points of the same nerve through a bimetallic arc without any contact with the muscle was difficult to understand. Disequilibrium of a natural electricity between the two points in the same nerve was strongly not plausible. By repeating the experiments, he was also surprised to find out how a small amount of electricity was capable of eliciting contractions of the frog's muscle. Since it was more sensitive than the sensing instruments, he regraded the frog as an extremely sensitive biological electroscope and accordingly thought that any extremely small external electricity could induce muscle contractions (Piccolino, 1998).

Volta progressively became aware that a bimetallic was more effective in inducing contractions than a monometallic arc and considered the possibility that the electricity was external in origin and derived from the difference of the metals used in a connecting arc. The frog might just react to this metallic current as it reacted to other forms of external electricity. Volta applied the bimetallic arc in other experiments including the living beings. When it was applied to the tongue, a clear acid taste was perceived instead of a muscular contraction, and when it was tried on the eye, the sensation of light has induced. These results matched the concept that different sensations do not come from the different properties of the stimulating energy but from the different types of sensing nerves even though the stimulating energy is the same as the electricity. All with these results, it seemed more correct for Volta to suppose that a bimetallic arc

Fig. 1.13 Voltaic pile. (a) Structure of a Voltaic pile, (b) Voltaic pile of the time

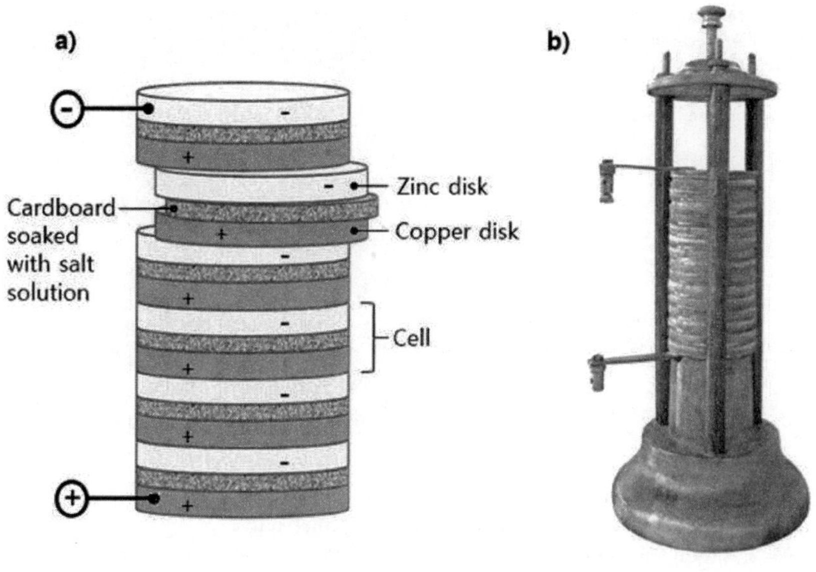

practically produced the disequilibrium in electricity as an electrostatic generator did rather than simply working as a conductor for discharging innate electrical disequilibrium. Then, he confirmed the existence of tiny electricity generated by the contact of the two different materials with the sensitive electroscope associated with a capacitor.

From this moment, Volta's interest moved from electricity in biology to the generation of electricity, which led to the invention of the battery. By building the arrangement of metal plates, he did all experimental efforts to generate a substantial quantity of electricity comparable to that produced by electrostatic generators. He multiplied the generated electrical force by stacking several disks of two dissimilar metals in an alternating fashion interposing the alternating couples of dissimilar metal pieces with disks of salt-soaked cardboard (Fig. 1.13). Volta referred to this "Voltaic pile," which was lately named, as an "Artificial electrical organ" based on the physical similarity to the electrical organ of electrical fish like an electric torpedo and electric eel. The stack-like arrangement of modular elements in the electrical organ of fish indeed provided him the structural concept that guided assembling the elements of the "Voltaic pile." The debate between "Animal current" vs. "Metallic current" in the eighteenth century ended with the apparent triumph on the metallic current's side based on Volta's conceptions, and the hypothesis of animal electricity was abandoned for many years (Piccolino, 1998).

2.3.3 The Most Capital Experiment of Electrophysiology

To the conception of the metallic current of Volta, Galvani did not remain unreacted. Galvani experimentally demonstrated the muscular contraction without using a bimetallic arc. Galvani prepared the two legs of a frog with their respective sciatic nerves sectioned near their emergence from the vertebral canal and placed them somewhat apart. Then, he contacted the nerve from a leg at two different points in the nerve of the other leg by bending a nerve to form a small arc. With the conduction of electricity from the first nerve to the second nerve, muscular contraction appeared in the first leg and, frequently, also in the second one. This experiment was the most capital experiment of electrophysiology that has shown the existence of animal electricity conclusively and unquestionably for the first time. Without any doubt, this experiment showed, in principle, evidence to face any possible criticism from Volta's side, because it did not involve a bimetallic arc or any contact of external electrical origin. It seems to be sufficient to persuade the opponents in controversy because it was intrinsically truthful and obviously self-evident in demonstrating the animal electricity. Nevertheless, Volta refused to see definitive proof of the existence of real animal electricity. Under the immense success of Volta's electrical pile, controversy on animal electricity became the finished one and this capital experiment did not collect scientific interest and remained in the shadow practically unnoticed by the scientific community.

2.3.4 Galvani's Achievement

When Galvani did his experiments, he also considered the possibility of the external origin of electricity. However, the experimental results were difficult to explain with the purely external origin of the electricity. Muscular contractions were more vigorous when the spark was elicited from the electrical machine separated from the frog, than when the frog was directly connected to the charged machine through the

Fig. 1.14 Two pioneers in history of electricity, Luigi Galvani who proposed "Animal current" and Alessandro Volta who proposed "Metallic current"

metallic wire. As direct connection seemed to be clearly favorable to transfer external electricity, its reduced effectiveness in producing contractions was strange. Moreover, even though there seemed to be a relationship between the intensity of the electrical discharge and the strength of muscle contractions, this relationship only worked within a certain range. Increasing the intensity of discharge beyond a certain limit does not result in stronger contractions, and contractions disappeared suddenly when the intensity is reduced progressively. Especially, the smallest quantity of electricity needed for contraction made Galvani believe that the force of contraction is not external and instead internal force was set in motion by the external electrical agent. The amount of electricity for contraction was so small that contraction can be induced with only remained electricity in a Leyden jar almost completely discharged and undetectable with the sensitive electroscope. How could such a tiny electrical force induce a muscle contraction if there were no prepared internal force waiting for the external trigger to set in motion? With all these results, Galvani eventually concluded that this extremely mobile characteristic is electrical, a specific form of electrical force generated in the living organism through its innate life process. Potential generated in the bimetallic arc only depolarized the membrane potential of the nerve. If there were no electricity in the nerve and muscles, the potential generated by the bimetallic arc is not sufficient to induce any muscular contraction. In this respect, the conception of the animal electricity of Galvani is correct. Fortunately and unfortunately, it only came with a bimetallic arc.

If there were no Galvani before him, Volta could not become the inventor of the electrical battery leading him to a pioneer in electricity. And, if there were no controversy with Volta's conception, Galvani might have not proven the existence of animal electricity conclusively before he died some years later in 1798 (Fig. 1.14).

2.4 Electrophysiology

In contrast to the extraordinary progress in the physics of electricity after the invention of the electrical battery by Volta in 1800, there was no substantial progress related to electricity within biology for about three decades after Galvani.

2.4.1 Quantitative Measurement of Biological Current

Carlo Matteucci (1811–1868) was an Italian physicist and neurophysiologist at the University of Pisa. In 1842, he measured a biological current using the galvanometer for the first time and correctly interpreted that the origin of the current is biological. He measured the current from the cut muscular tissue by putting one electrode of a galvanometer on the cut surface and the other electrode on the intact muscular surface. To increase the amount of measured electricity, he connected the thighs of the frog in series like a Voltaic pile by contacting an intact surface of one thigh with a cut surface of the next one. By showing that the deflection of the galvanometer needle is increasing with the number of connected frog's thighs, he clearly demonstrated that the measured electricity is coming from the biological tissue instead of the metallic contacts of the galvanometer electrodes to the biological tissue.

By the objective instrumental measurement, his experiment proved quantitatively the existence of animal electricity, which was proposed qualitatively by Galvani on the basis of biological experiments. The existence of a muscle current also confirmed the electrical disequilibrium in a muscle tissue like a Leiden jar that Galvani proposed. The current measured by Matteucci between the intact and the cut surface of the muscle was indeed derived from the potential difference between the internal and the external compartments of the muscle fibers.

2.4.2 Measurement of Action Potential

Emil du Bois-Reymond (1818–1896) was a German physician and physiologist at the University of Berlin. In 1848, he first recorded the action potential accompanying the excitation of muscle and nerve as a negative variation. Negative variation during the excitation meant that the measured potential between the intact and injured surfaces decreases as well as the potential of the outer membrane surface of the nerve and muscle at the location of excitation becomes negative in reference to the distant intact resting site. Even he did not record the time course of the action potential, his recorded results correctly reflect the variation of membrane potential during the generation of an action potential.

2.4.3 Speed of Propagation of the Nervous Signal

Hermann von Helmholtz (1821–1894) was a German physicist and physician who made significant contributions to several scientific fields. His achievement in electrophysiology is the first measurement of the propagation speed of nerve signals along with the nerve fiber in 1849. He used experimental nerve muscular samples prepared by connecting a recently dissected sciatic nerve of a frog to the calf muscle. He measured the elapsed time interval from the application of an electric stimulus to the nerve until the mechanical response of the muscle, which is the same as the current method of measuring the conduction speed of a nerve fiber. He measured the propagation speed of the nerve signals around 30 m/s instead of several hundreds of km/s. This speed was too slow as many contemporary scientists expected the extremely fast speed approaching the speed of light.

2.4.4 First Measurement of the Time Course of the Action Potential

Even though Helmholtz explicitly measured the speed of nerve signal propagation, too low propagation speed cast a doubt on the electrical nature of the nerve fluid. The measured speed of the nerve signal was too much slower than the propagation speeds of other entities of electrical nature like light or electric field. To re-establish confidence in the electrical identity of nerve activity, demonstration of negative variation propagating along the nerve with measured speed became crucial.

Julius Bernstein (1839–1917) was a German physiologist at the University of Halle. In 1868, he developed a "Differential rheotome" or "Current slicer" to record accurately the time course of an action potential. As he recognized the action potential is a short and fast phenomenon, the galvanometer was too slow to record this variation occurring within several milliseconds. Differential rheotome records potential with sampling and holding at each delayed time with multiple replications of the signal instead of a whole waveform record in a single measurement like a current oscilloscope. Even though its measurement was tedious and time-consuming, it increased the temporal resolution of measurement to the tenth of milliseconds from tens of milliseconds (Carmeliet, 2019). The speed of propagation could be measured by changing the point of stimulation and measuring the delay in the appearance of electrical variation. Recorded action potential clearly showed the transient activity in the order of millisecond consisting of two phases. In the first phase, resting potential disappeared, and during the second phase, the resting potential was rebuilt with a somewhat slow tail. He also measured the propagation speed of this action potential at 28.7 m/s, which is in the same range as the speed measured by Helmholtz.

Even though he recorded overshoot, an increase of membrane potential over zero potential, he did not focus on this recorded peculiarity. Upon his membrane hypothesis, nerve cells have selectively permeable to K^+ ions during the resting period, and negative intracellular potential is developed thermodynamically by the high intracellular and low extracellular K^+ concentration according to the Nernst equation. During the excitation generating an action potential, he thought that membrane loses its selective permeability to K^+ ions and membrane potential only approaches zero value. With his membrane hypothesis, overshot was not predicted. He did not recognize the selective permeability of other ions like Na^+ ions and missed identifying the importance of the overshoot.

2.4.5 Local Current Theory

With these experiments by Bernstein, the electrical nature of nerve signals became clear. However, it was also clear that the propagation of nerve signals does not follow the simple laws of the pure passive propagation of electrical currents, and that there involves a more complex and specific mechanism. The longitudinal resistance of nerve fiber was too high to allow an effective propagation of an electrical signal.

Ludimar Hermann (1838–1914) was a German physiologist and speech scientist and he proposed a local current theory in 1872. He noticed the fundamental fact that the nerve signal propagates along with the negativity of the external surface of the nerve and that at the same time the fiber is stimulated when its external surface becomes more negative. This suggested the possibility that the external surface negativity is produced by the arrival of the negative variation from a region ahead and the region is stimulated by closing a loop capable of propagating the nerve signal. It appeared plausible that the signal propagation was caused by a local excitation of the resting fiber due to a current flow from the active region. To explain the propagation of local currents along with the nerve fiber, he proposed a local circuit theory. This theory was based on the nerve fiber structure consisting of a conductive core separated from an external fluid by a relatively insulating coating. In this structure, electrical disturbance produced in place of the nerve could influence neighboring regions via developing a local current loop that involves the internal core, the insulating coating, and the external fluid.

2.4.6 All-or-Nothing Character

Henry Bowditch (1840–1911) was an American physician and physiologist at the Harvard Medical School. In 1871, he first noticed all-or-nothing character was in an isolated frog's heart. An electrical stimulus applied directly to the apex of the frog heart could induce a contractile response when its intensity was greater than a threshold value. However, the strength of contraction remained constant, even though the intensity of the stimulus was further increased. And, the intensity of stimulus lower than the threshold value did not result in any detectable contraction of the frog's heart. In 1905, Keith Lucas (1879–1916), who was a British scientist and neuroscience at Trinity College, successfully showed that other electrically excitable tissues indeed follow the all-or-nothing law.

2.4.7 Electrical Discharge of a Single Nerve Fiber Under the Physical Stimulus

Edgar Adrian (1889–1977) was an English electrophysiologist and recipient of the 1932 Nobel Prize for Physiology Medicine. He directly demonstrated the all-or-nothing character in the electrical activity of a single nerve fiber.

Using the capillary electrometer and a vacuum tube amplifier, he successfully recorded individual electrical discharges responding to physiological stimulations in various sensory fibers of both frog and cat. In order to isolate the electrical activity of a single fiber, he carefully cut the muscle of the frog and minimized the number of active sensory units. Then he could record a response representing the activity of a single sensory fiber from just one muscle spindle, which was shown in rather a regular train of impulse-like deflections of constant amplitude and duration. When the intensity of the stimulus was increased, the rate of discharged impulse was increased without any change in the shape of discharged impulse in amplitude and duration. This frequency-modulated characteristic of the sensory nerve was later confirmed in a variety of other nervous fibers and led to an understanding of how the nervous system communicates information.

2.4.8 Hodgkin–Huxley Model

Even though the electric nature of the nerve signal was confirmed in many aspects, there was still an unsettled aspect that whether electricity is the fundamental process of nerve signaling, or a simply secondary effect of the underlying more essential process, just like heat production accompanying the chemical reaction. The difficulty in fully accepting the view of the electrical nature of the nerve signal was not without logical grounds. First of all, the conduction speed of the nerve signal was too slow compared to the typical conduction of electrical signals through a metallic wire. Moreover, the conduction speed of nerve signals was profoundly affected by the temperature. Since a high-temperature coefficient was indicated as the involvement of chemical reactions, it made some scientists envision that a nervous impulse is fundamentally a chemical event. Experiments and computational modeling carried out by Hodgkin and Huxley was provided the final explanation of the mechanism underlying the generation and propagation of the nerve signal. They fundamentally verified that electricity is the fundamental characteristic in the generation and propagation of nerve signals and cleared all the suspicious aspects related to animal electricity during the previous 160 years since it is first proposed by Galvani in 1791.

Alan Hodgkin (1914–1998) and Andrew Huxley (1917–2012) were English physiologists and biophysicists at the University of Cambridge. They shared the Nobel Prize in Physiology or Medicine in 1963 "For their discoveries concerning the ionic mechanisms involved in excitation and inhibition in the peripheral and central portions of the nerve cell membrane" (Fig. 1.15).

Fig. 1.15 Alan Lloyd Hodgkin and Andrew Fielding Huxley who jointly developed the Hodgkin–Huxley model and received 1963 Nobel Prize in in Physiology or Medicine

They tried to explain the generation and propagation based on the local circuit theory. They succeeded in demonstrating that the increase in excitability occurs simultaneously with the measurable local current response and the produced local current spreads passively along with the nerve fiber with the spread of increased excitability. This was supported by the change in the conduction speed when the extracellular fluid was changed. The conduction speed of the nerve fiber was decreased when the nerve fiber was immersed in an extracellular medium of lower conductivity like water or oil. This implies that the nerve signal does not propagate solely through the core of the fiber, but rather along the outer membrane of the fiber by producing a local current flowing through the extracellular fluid.

By inserting the electrode inside the giant axon of the squid, they could directly measure the transmembrane potential of the neuronal membrane during resting state and excitation. They confirmed the existence of the polarized membrane potential during the resting state. However, they found that there occurs an overshoot of potential across the zero level during the excitation. This was contradicted the membrane hypothesis, which was suggested by Bernstein, that disappearance of selective permeability of potassium is the base of potential variation during the excitation and potential approaches to zero as the membrane becomes equally permeable to all ions. Then, they found the involvement of a selective permeability for sodium ions in the discharge of the action potential, because changing the extracellular concentration of sodium ions decreased the amplitude of recorded action potential. They thought that the selective increase of membrane permeability for sodium ions during the discharge would result in the change in the polarity of recorded discharge potential from negative to positive. That is, the membrane potential is largely dependent on the selective permeability of potassium ions during the resting state and it is dependent on the selective permeability of sodium ions during the excitation. And the change of permeability was thought could be changed depending on the variation of membrane potential.

To verify their theory, they developed a membrane model based on the potential dependent variation of membrane permeability for major contributing ions, potassium, sodium, and chloride. However, due to the transient and explosive characteristics of the action potential, it was very difficult to obtain accurate quantitative information on the electrical events of the action potential. With the newly developed technique of patch-clamp at the time and the lucky biological material of the giant axon of the squid, they successfully developed the membrane model, which is now known as Hodgkin–Huxley model. This model represents a membrane with parallel variable conductances whose values are dependent on the membrane potential and the variation of conductances relies on a set of differential equations describing the dynamics of the permeability of each ion. Hodgkin–Huxley model describes the variation of membrane potential during the generation of an action potential, propagation of action potential, and refractory period with surprisingly high accuracy and is continuously used thereafter. This model is studied in Chap. 4 of this book. They started the work on their model after the end of World War II, as they have also participated in warfare, and they published the experimental results based on their model in a series of five papers in 1952, which eventually led them to receive the 1963 Nobel Prize in Physiology and Medicine. In addition, their findings led them to hypothesize the existence of ion channels on the cell membrane and the secretory mechanism that pumps sodium and potassium actively against the electrochemical gradients, which was later discovered as the Na-K pump.

2.4.9 Patch-Clamp for Ion Channels

The question left unresolved by the Hodgkin–Huxley study was the practical mechanism of ion permeation through the membrane. While the variation of selective permeability of each ion was clearly described theoretically and experimentally, the mechanism of the moving ions in and out of the membrane and related change of permeability was still vague. Ion movement through the ion channel was clarified with the invention of the patch-clamp technique.

Erwin Neher (1944–) and Bert Sakmann (1942–) are German biophysicists, specializing in the field of cell physiology and they invented a patch-clamp technique for recording the elementary electrical events of the membrane. In 1991, they were awarded the Nobel Prize in Physiology or Medicine for their discoveries concerning the function of single ion channels in cells.

In a patch-clamp, a microelectrode forms a high resistance seal with the cellular membrane, and a small patch of cell membrane containing the ion channel of interest is detached from a cell. An isolated ion channel is independently evaluated by measuring the current with varying potential across the detached patch containing the ion channel. The studies carried out with the patch-clamp technique have clarified the mechanism of each ion involved in membrane permeability and the electrical excitability with a molecular resolution. Studies with the patch-clamp technique have provided definite evidence that membrane currents involved in the electrical behavior of the nerves and muscles are carried by ions passing through ion channels embedded in the membrane. Ion channels have proved to provide hydrophilic paths whereby ions can permeate despite the hydrophobic properties of the lipid constituents of the membrane. The studies of "Animal electricity" started from Galvani have concluded at this moment up to the molecular level.

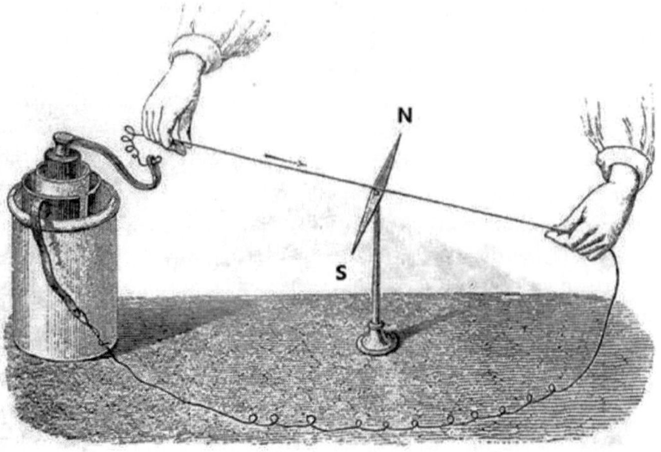

Fig. 1.16 Oersted's experiment. The magnetic needle is located under the conductor during the current flows from left to right

2.5 Measurement of Bioelectricity

Since the first electrical measuring instrument of the electroscope, which was used to detect the presence of electric charges on an object, detecting electricity was developed based on the relation between electricity and magnetism. Christian Oersted (1777–1851) was a Danish physicist and chemist who discovered that electric current in a wire can deflect a magnetized compass needle closely located (Fig. 1.16). It was accidentally discovered during his lecture in 1820. In an experiment to show that the wire heats up when the electric current flows through it, to his surprise, he noticed the rotation of the compass's magnetic needle placed next to the wire. By reversing the direction of the electric current, he could reverse the direction of the needle deflection. Oersted experimented with this phenomenon several times and revealed a close relationship between electrical and magnetic phenomena. Since his discovery, the interaction between electricity and magnetism has been actively studied, and their interacting mechanism has been used in many applications including detecting electricity.

A galvanometer is an electromechanical measuring instrument detecting the weak electric current. It is based on the electromagnetic interaction discovered by Oersted. It was first invented by German physicist chemist Johann Schweigger (1779–1875) in 1821. It was made to deflect the magnetized needle inside of a coil by the current flowing through. Schweigger increased the sensitivity using the multiple loops of wire forming a coil and called his instrument "Multiplikator." In 1825, Leopold Nobili (1784–1835), an Italian physicist, invented the astatic galvanometer. The astatic galvanometer is a galvanometer of increased accuracy by canceling out the earth's magnetic field (Fig. 1.17). To cancel out the earth's magnet, two magnetized needles were employed in as same axis but in an opposite direction. Then, the magnetized needle rotates exclusively by the magnetic field which is generated by the current flowing through the coil. As the magnetized needles are positioned in opposite directions, two needles should be located separately in the reversed magnetic field developed by the figure-8 type coil or only one needle should be located within the generated magnetic field, while the other needle is remained not affected. In 1842, Carlo Matteucci measured a bioelectric current in frog muscle using the astatic galvanometer.

The capillary electrometer was invented by Gabriel Lippmann (1845–1921), a French physicist, in 1873. He was known as a genius on mercury and also received the Nobel Prize for Physics in 1908 for producing the first color photographic plate. While the electromagnetic galvanometer is essentially used to measure electricity, it was a current-measuring device and it was not very convenient and is easily damaged by overload. The device measuring the electricity without draining the current was required. A capillary electrometer detects the potential difference without draining electric current. Lippmann saw a well-known demonstration that involves a drop of mercury covered by dilute sulfuric acid (H_2SO_4). When touched by an iron wire, the drop contracts but regains its original shape on the removal of the wire. He recognized that the effect must be due to a connection between electric polarization and surface tension, and used the phenomenon in the development of the capillary electrometer (Stock, 2004). Capillary electrometer was based on this unique characteristic of mercury that changes its surface tension according to the applied potential. By extending the change in volume in a thin capillary tube, he made the interface surface between the mercury and the sulfuric acid within the capillary move sensitively to the potential difference. He could increase the sensitivity of the measurement to detect the potential difference of 0.03–0.1 mV. This device was used in the recording of the primitive ECG machine invented by Augustus Waller.

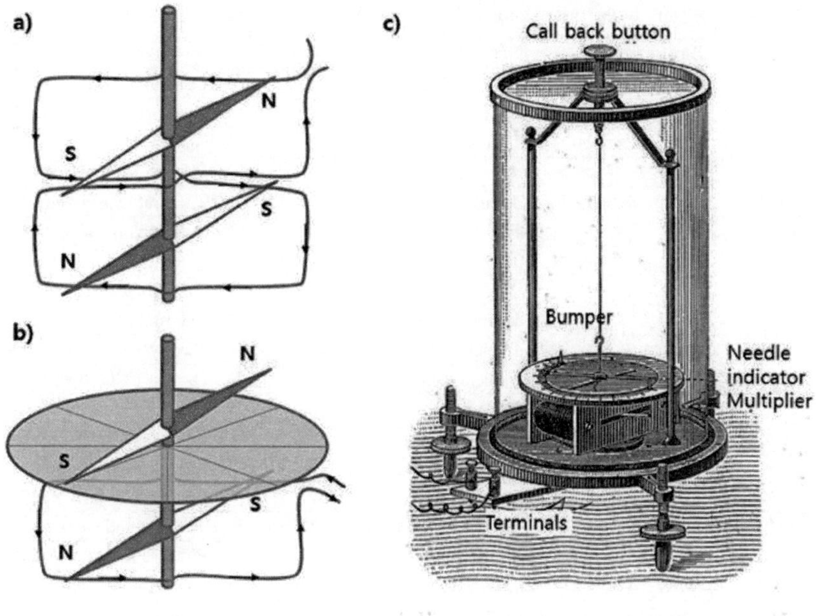

Fig. 1.17 Astatic galvanometer. (**a**) two magnetized needles of opposite direction in the separated magnetic field generated by figure eight type coil, (**b**) only one magnetized needle is used to detect the current while the other magnetized needle is used for indicating as well as canceling out the earth's magnetic field, (**c**) astatic galvanometer developed by Nobili

The string galvanometer is a fast-responding current measuring instrument with increased sensitivity. It was first invented by Clément Adair, a French engineer, in 1872. He replaced the coil with a much fast-moving wire or "string" stretched between poles of a magnet and used it as a telegraph receiver. Willem Einthoven (1860–1927) advanced the form of a string galvanometer. A strong light source was used to illuminate the fine string filament, and the fine movement of the filament was magnified by the optical system to be observed or recorded in photography. He used this string galvanometer to record all wave components of ECG and became a leader in applying the string galvanometer to physiology and medicine. Einthoven was awarded the 1924 Nobel prize in Physiology or Medicine for his work in recording electrocardiograms. The string galvanometer was used as a major recording instrument of electricity until the advent of electronic vacuum-tube amplifiers in the 1920s.

In 1904, John Ambrose Fleming (1849–1945), a British engineer, invented the vacuum tube having two electrodes. And, in 1906, Lee de Forest (1873–1961), an American engineer, invented the three-element vacuum tube, which is the first practical amplification device. Developed vacuum tubes would have far-reaching applications as a standard device in radio receivers, radars, early television sets, and many other forms of electronic communications for about half of a century. Vacuum tubes also allowed the amplification of tiny biological signals and revolutionized the measurement technology in electricity from the human body.

John Bardeen (1908–1991), William Shockley (1910–1989), and Walter Brattain (1902–1987), American engineers and physicists at Bell Labs., invented a point-contact transistor in 1947, and they jointly were awarded the 1956 Nobel Prize in Physics for their research on semiconductors and their discovery of the transistor effect. The transistor revolutionized electronic technology and industry by replacing the vacuum tube and making almost every modern electronic device in compact form. The invention of transistors was the beginning of the semiconductor era toward high-density integrated circuits and an opening of "The information age." Invention and progress of transistor technology made the bioelectrical instruments for measurement and stimulation be miniaturized so that they are portable, implantable, and more reliable in their medical applications. Table 1.1 shows a historical list of the first recordings of major biological signals based on the measurement technologies described above.

2.6 Stimulations with Electricity and Magnetism

Stimulating the part of the body was the primary application of electricity since its first use in ancient times using electric fish. The invention of the Leiden jar, which could collect and store electricity, made the electrical stimulation more accessible.

In 1800, Alessandro Volta (1745–1827), an Italian physicist at the University of Pavia, invented the Voltaic pile, a battery that could produce electric current continuously. With this invention, Volta proved that electricity is not a substance that belongs only to living beings and that electricity could be generated artificially from the chemical interactions of

Table 1.1 History of the first recording and clinical use of various biological signals

Year	Biosignals	Scientist	Remarks
1781	Animal current	Luigi Galvani	Detection of current in frog's muscle
1842	Animal current	Carlo Matteucci	Qunatitative & external measurement of current in frog's muscle
1848	Action potential	Emil du Bois-Reymond	As negative variation accompanying the excitation of muscle and nerve
1865	ERG	Alarik Holmgren	Electric response of eye from amphibian retina
1868	Action potential	Julius Bernstein	The first recording of action potential time course
1875	EEG	Richard Caton	In the brains of rabbits and monkeys
1877	ERG	James Dewar	Electric response from human retina
1887	ECG	Augustus Waller	Human electrical cardiac rhythms
1890	EMG	Étienne-Jules Marey	The first actual recording, also coined the term EMG (Reaz et al., 2006)
1903	ERG	Francis Gotch	Two waves of ERG to a light flash
1908	ECG	Willem Einthoven	Human ECG rhythms of P, Q, R, S, T waves
1921	EGG	Walter Alvarez	Reported rhythmic 3 cpm electrical gastric activity
1924	EEG	Hans Berger	From the scalp of human brain
1951	EOG	Elwin Marg	First described and named EOG
1962	EOG	Geoffrey Arden	First clinical use of EOG
1963	MCG	Gerhard Baule, Richard McFee	The magnetic activity of the heart using a coil magnetometer
1966	EMG	Curtis Hardyck	First clinical use of EMG
1970	MEG	David Cohen	Using the SQUID

ECG Electrocardiograms, *EEG* Electroencephalogram, *EGG* Electrogastrogram, *EMG* Electromyogram, *EOG* Electrooculogram, *ERG* Electroretinogram, *MCG* Magnetocardiogram, *MEG* Magnetoencephalogram

different materials. In contrast to the electricity collected in the Leiden jar, electricity could be extracted continuously for an extended time, and it made the electrical stimulation more conveniently be manipulated.

Giovanni Aldini (1762–1834) was an Italian physician, physicist, and nephew of Galvani. He became a professor of physics at the University of Bologna in 1798, in succession to his uncle Luigi Galvani. He worked as a successor of Alessandro Volta as well as Luigi Galvani. He used the Voltaic pile and water-filled vessel as electrodes to demonstrate the Galvanism for medical applications. Based on Galvani's experiments, he applied electricity to the muscles in different parts of the body and made the muscles contract. He is most famous for the public demonstration performed on the executed criminal in 1803. What happened when the galvanic stimulating process was used on the body was described "On the first application of the process to the face, the jaws of the deceased criminal began to quiver, and the adjoining muscles were horribly contorted, and one eye was actually opened. In the next part of the process the right hand was raised and clenched, and the legs and thighs were set in motion." It made the audience surprised and flee away. As a physician, he also used this method in many therapeutic applications. He used electrical stimulation to treat depression as the first electric stimulation to the brain. He attempted electrical stimulation in several other disorders including paralysis, blindness, deafness, headaches, and rheumatism. He used the method to reanimate the drowned subject, which inspired the electrical stimulation in heart failure leading to the use of cardiac pacemakers and defibrillators. Aldini's experiment is also known that it has motivated the famous novel "Frankenstein" written by English author Mary Shelley in 1818.

In 1872, surgeon Thomas Green used electrical stimulation as a method for cardiorespiratory resuscitation (CPR). For the subjects in cardiac arrest as a side effect of anesthesia with chloroform, he applied electrical stimulation from the battery generating about 300 V between the neck and the lower ribs on the left side to resuscitate. He described six successful resuscitations out of seven attempts. In the same year, Duchenne de Boulogne (1806–1875), a French neurologist, successfully resuscitated a drowned child by rhythmically tapping electrodes between a leg & precordium. John MacWilliam (1857–1937), a Scottish physiologist at the University of Aberdeen, proposed that ventricular fibrillation (VF) was the most common cause of sudden death. While he recognized that most cardiac arrests were hopeless, he considered there might be some situations that could be reversible with one single dramatic shock (Ball & Featherstone, 2019). In 1887, he showed how ventricular fibrillation could be terminated by a serial application of electric shocks to the heart. Studies on the effect of electrical stimulation to resuscitate subjects in the asystolic heart were followed continually since then and their endeavors have resulted in current cardiac pacemakers and defibrillators which are now actively being used as standard methods for resuscitation in medicine. Electrical stimulation methods of cardiac pacemakers and defibrillators are the topic of Chap. 12.

Michael Faraday (1791–1867) was an English scientist who was one of the most influential in the history of electricity. He mainly contributed to the study of electromagnetism by inventing the induction coil. He discovered that electricity and magnetism are not separate substances and just two facets of electromagnetism and proved that electric current can generate magnetic field and magnetic field also generate electric current reversely. He showed that a pulse of electric current passing through a primary coil generates a magnetic field and the generated magnetic field induces an electric current in the second coil. Because the direct current was not effective in inducing current in the secondary coil, he successfully derived induction current by periodically breaking the current flow in the primary coil. The rate of change of this magnetic field determines the magnitude of induced current in a secondary coil. In addition to that his induction mechanism has generated the electric motor which converts electricity into energy, his discovery also built the basis for the electrical stimulation with induced alternating current. Magnetic stimulation can be used to induce a current within the body. By using magnetic stimulation, the brain cortex, peripheral nerves, and muscles could be stimulated without attaching electrodes directly to their surface.

Robert Bartholow (1831–1904) was an American physician who first performed electrical stimulation directly on the human brain cortex. In his study on electrotherapy, he used electricity in his practice and treated subjects with alternating current by the Faradic induction as well as Galvanic direct current. He is best known for his experiment of direct cortical stimulation performed on a 30-year-old female subject in 1874. He did the experiments of applying current directly to the brain cortex through a large hole with a diameter of about two inches in the parietal region, which was caused by the cancerous ulcer in her skull. Using a pair of needle electrodes inserted into the dura and underlying tissues, he applied an electric current at different strengths to different sites of the exposed brain and noticed the corresponding motor responses on the contralateral side to the stimulated cortex. Primarily, he hoped to explore whether the human cortex was excitable, as the scientific community was divided on whether or not the human cortex was electrically excitable at the time of his experiment. Additionally, he wanted to compare the bodily reactions to electrical stimulation on the left and right sides of the brain and wanted to generalize the stimulation effect by comparing the human results with the results that came from the experiments with monkeys. While the low-level current he applied to the brain did not seem to cause any pain, the increased amount of current drove the subject into a distressed, convulsive state and finally into a coma. She revived from the coma three days later, but the following day she had a major seizure and died. His experiment confirmed the excitability of the brain and discovered cortical excitation and localization in the human brain, but also produced strong criticism for possible ethical violations.

Even though Bartholow used alternating current by Faradic induction, he did not stimulate the brain with the magnetic field itself. While the direct physiological effect was not clearly elucidated, magnetic stimulation has been used to treat some symptoms using artificial time-varying electromagnetic fields as well as mineral magnets. Similar to "Animal electricity," "Animal magnetism" was also proposed, which assumed that a healthy person contains a balanced quantity of magnetic fluid and a diseased person is supposed to have a lack this magnetic fluid. Without clear evidence, "Animal magnetism" and treatment with magnetism were treated as quackery, suggestion, or fraud around the eighteenth century. However, as electromagnets could produce stronger magnetic fields than mineral magnets and different magnetic field of time-varying characteristics, it re-encouraged the research on therapy with magnetic stimulation.

Various experiments showing the evidence with magnetic stimulation were first performed by Jacques d'Arsonval (1851–1940), a French physician and physicist. He investigated the consequences of tissue exposure to time-varying electric and magnetic fields. He employed various time-varying magnetic fields with different types of coils of various sizes (Fig. 1.18). In 1893, with a large solenoid cage

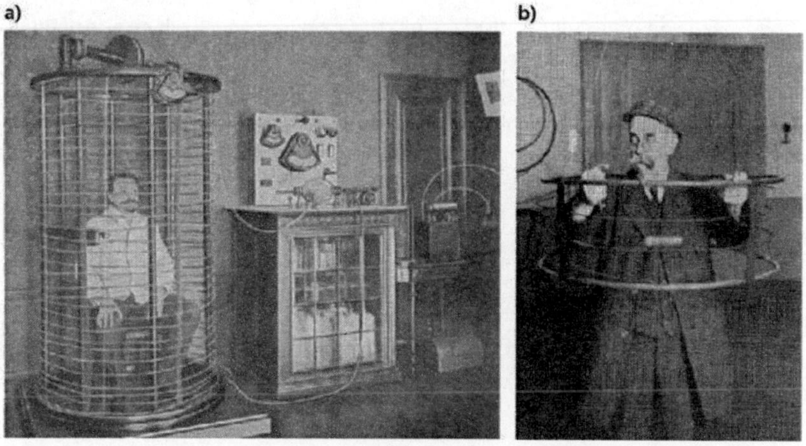

Fig. 1.18 d'Arsonval's experiment of magnetic stimulation. (**a**) Solenoid cage used for magnetic stimulation of full body, (**b**) Magnetic stimulation experiment on a part of his own body. (d'Arsonval, 1897; Reif-Acherman, 2017)

within which a human subject can stand, he exposed high-frequency magnetic fields of at least 10 KHz. An electrical wire ran spirally to induce a magnetic field vertically within a cage. When he placed a mercury thermometer and bulbs without any connection to an external source to prove the presence of a magnetic field, the mercury thermometer rose to 150 °C within seconds and the bulbs burned when the time-varying magnetic field was turned on. Nevertheless, subjects did not experience any direct physiological effects, but according to d'Arsonval, this magnetic field had a positive effect on metabolism as did the time-varying electric field (Martens et al., 2013). In the experiment performed in 1896, he passed a 42 Hz and 30 A current through a coil in which the head of a subject was placed. And he wrote, "There occurs, when one plunges the head into the coil, phosphenes, vertigo, and in some persons, syncope" (d'Arsonval, 1896). A phosphene is defined as visual awareness of light not caused by light. Electrophosphene is a phosphene caused by electricity and magnetophosphene is a phosphene caused by magnetism. He is convinced that this is a magnetophosphene caused by the time-varying magnetic field. He commented that magnetophosphene indicates that either the retina or the visual cortex or both were stimulated by the magnetic field.

Based on d'Arsonval's experimental results of the time-varying magnetic field, magnetic stimulation extended its therapeutic applications. Rheumatoid arthritis and gout were the symptoms most benefited while hysteria and neuralgia did not show any improvement with magnetic stimulation. In 70% of patients with anomalies of the sensory nervous system, such as neuralgia, migraine, sciatica, and irritative forms of neurasthenia with sleeplessness, symptoms improved with magnetic stimulation (Rodari, 1902). Magnetic stimulation was also advised as a sedative treatment not only in all kinds of painful conditions, including neuralgias, lancinating pain in tabes dorsalis, joint and muscle rheumatism, but also for sleeplessness, vasomotor disturbances, angina pectoris, and so forth (Martens et al., 2013).

Like electrotherapy, confidence in the magnetic stimulation waxed and waned due to the lack of clear evidence and only a minority of physicians continued applying electromagnetic therapy in the margins of medicine. Horace Barlow (1921–2020), a British vision scientist, also performed the experiment perceiving the magnetophosphene under a magnetic field up to 900 G produced by 20 A of 60 Hz sinusoidal current through the coil. To illustrate that the phosphenes were due to electrical stimulation of the retina, he recorded an electroretinogram (ERG), by placing an active electrode on the side of the forehead and a reference electrode on the back of the forearm. He reported that phosphenes began to be perceived with 0.3 mA of 60-Hz current and the duration of perception of the phosphenes outlasted the duration of magnetic stimulation in proportion to the stimulation strength. He also stated that the phosphenes were usually colorless, but occasionally they appeared bluish and sometimes slightly yellowish. No phosphenes could be perceived when the core of the coil was placed over the occipital area, the site of the visual cortex (Barlow et al., 1947).

In 1965, Reginald Bickford (1913–1998), an English electroencephalographer, used a pulsed magnetic field instead of sinusoidal alternating currents and observed muscle twitching in frogs, rabbits, and humans (Bickford & Fremming, 1965). In 1985, Anthony Barker (1950–), an English scientist, successfully stimulated the human cortex with electromagnetic stimulation. He demonstrated the possibility of stimulating the human brain through the scalp using transcranial magnetic stimulation (TMS). He placed a circular air-cored coil with a 10-cm outside diameter centrally over the vertex of a normal subject and recorded motor evoked potentials from the abductor's muscle using surface electrodes (Barker et al., 1985). The subject reported no pain or discomfort. This was striking compared to the discomfort and pain associated with transcranial electrical stimulation (ECT) of the motor cortex using electrodes placed on the scalp (Pascual-Leone & Meador, 1998). His demonstration generated an enormous increase of interest in transcranial magnetic stimulation among motor physiologists and clinical neurophysiologists as the years have passed into the twenty-first century.

3 Theoretical Progress in Electricity

There was great progress in understanding electricity with supporting theoretical frameworks during the nineteenth century. Scientific knowledge and developments have transformed electricity from a scientific curiosity to practical uses as applications in ordinary lives. Theoretical understanding of electricity began with the invention of the Voltaic pile, which was publicized since 1800. Voltaic piles, as a primitive type of battery, provided the first tool which can provide the current continuously. Voltaic piles changed the electricity from a transient phenomenon to one that can be adequately managed and studied. The voltaic pile has technically progressed into a battery with increased performance in producing and storing electricity and has enabled many related studies.

In 1820, Christian Oersted discovered that a magnetic field is generated around a wire when a current is flowing by noting the deflection of a compass needle with an electrical current flowing through a wire. Andre-Marie Ampere (1775–1836), a French physicist, quickly picked up Oersted's discovery and developed a mathematical and physical theory that explained the relationship between electricity and magnetism already observed and predicted by many others. He developed the foundation of electrodynamics by showing that two parallel wires carrying electric currents attract or repel each other whether currents flows are in the same direction or in the opposite directions. His principle came to be called Ampère's law, which stated that the magnetic field created by an electric current is proportional to the strength of the electric current flowing through a wire with a proportional constant. It also explained that generated force between the current-carrying wires is proportional to the lengths of wires and to the strength of flowing currents.

In 1827, Georg Ohm (1789–1854), a German physicist and mathematician, discovered the relationship between voltage and current in a conductor. He expressed his law that electromotive force acting between the extremities of any part of a circuit is the product of the strength of the current, and the resistance of that part of the circuit, which is the relationship known as Ohm's law. His work of Ohm's law became the fundamental base for the subject of circuit theory.

Michael Faraday (1791–1867), an English chemist and physicist, built a fundamental framework on electromagnetism. Although he received little formal education, he was one of the most influential scientists in history. In 1831, he established the law of induction, on which the operation of generators, motors, and transformers is based. He discovered that a current also can be generated in a wire reversely from the variation of the magnetic field as well as the magnetic field generated from the current flowing through a wire. And he showed experimentally how the electricity can be transformed into motion by the world's first motor and how the electricity is induced in the second coil (Fig. 1.19). His discovery of mutual induction, a form of electromagnetic induction, became the foundation for many subsequent developments. These advances in electromagnetism were important for the development of the mechanical generation of electricity as well as the force generation with electricity. In 1866, a major step toward effective generation of electricity was developed with the introduction of "self-excitation," which uses the current produced by the generator to energize its own wires by several individuals. In 1883, Nikola Tesla (1856–1943), a Serbian-American inventor and electrical engineer, invented an induction motor using the modern alternating current (AC) electricity, which is based on the principle of Faraday's mutual induction. The development of

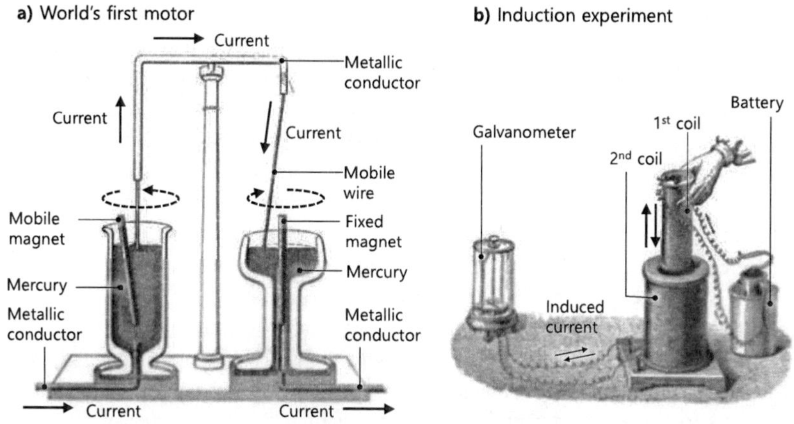

Fig. 1.19 Faraday's experimental setup for the world's first motor and for induction. (**a**) The motors were constructed of a metal wire suspended in a cup of mercury with a permanent magnet protruding up from the bottom of the cup. In the left cup, the magnet was attached to the bottom with a piece of thread and left free to move, while the metal wire was immobile. On the right side, the magnet was held immobile and the suspended metallic wire was free to move. When current from an electric battery was applied to the wire, the circuit was completed via the mercury, which is a good conductor of electricity, and the resulting current flowing through the wire produced a circular magnetic field. The electromagnetic field interacted with the existing magnetic field from the permanent magnet, causing rotation of the magnet on the left, or of the metallic wire on the right (Weise et al., 2011). (**b**) Experiment showing the induction using two coils. A battery provides a current flowing through the first small coil, and it produces a magnetic field. While the coils are stationary, induction does not occur and a current is not induced in the second large coil. However, when the first coil is moved in or out of the second coil, a relative change of the magnetic flux through the second coil induces a current which is detected by a galvanometer

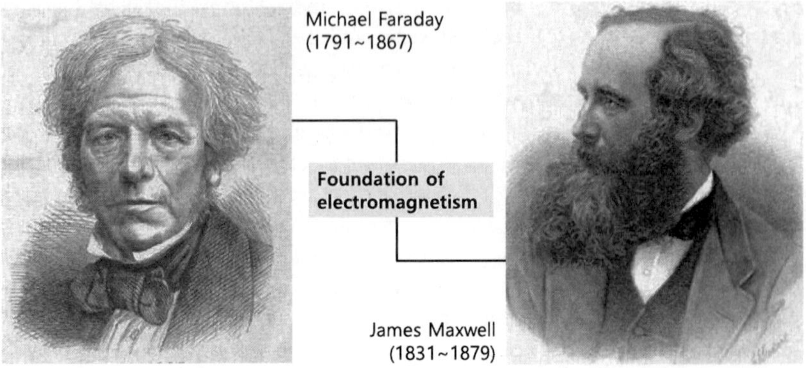

Fig. 1.20 Michael Faraday and James Clerk Maxwell who jointly established the solid foundation of electromagnetism

induction coils also led to the development of the transformer, which could transfer electrical energy from one circuit to another by inductively-coupled conductors. In 1883, a practical transformer was developed by Gaulard and Gibbs, and it allowed the alteration of voltages between circuits using pair of induction coils. It solved several practical problems for the transfer of electricity over long distances and for connecting equipment using different voltages in a single power supply).

James Maxwell (1831–1879), a Scottish mathematician and scientist, formulated the electromagnetic theory completely by unifying the relationship between electricity and magnetism and describing electricity, magnetism, and light as different manifestations of the same phenomenon. Before Maxwell met Faraday, electricity was not his interest. Even though Faraday has demonstrated the relationship between electricity and magnetism, it lacked theoretical solidity. When Faraday postulated the similarity between light and electromagnetism, only with his experimental understanding it was not sufficient to persuade his colleagues. In 1855, when Maxwell of age 24 met Faraday of age 64, Faraday handed his experimental research directly to Maxwell and the quest for the electromagnetic field became the role of Maxwell. Maxwell used his excellent mathematical skill to unify the disparate phenomena in electricity and magnetism. In 1865, he finally formulated the relationships into a coherent and complete theory with mathematic equations. With these equations, he also proved that the propagation speed of the electromagnetic waves is the same as the speed of light and both are different manifestations of electromagnetism. In 1884, Oliver Heaviside (1850–1925), an English mathematician and physicist, reformulated Maxwell's mathematical relations into four unifying equations, which are now known as Maxwell's equations and also called "Four of the most influential equations in all of science." The joint work of Faraday and Maxwell was a paradigmatic example of the modern scientific process. Experimentation and observation yield facts that can then be organized into theory and it finally results in a mathematical formalism that allows the understanding of observed phenomena and the prediction of future phenomena. Faraday and Maxwell jointly established the solid foundation of electricity and magnetism (Fig. 1.20).

Heinrich Hertz (1857–1894), a German physicist, discovered radio waves and experimentally proved that Maxwell's theory of electromagnetism was correct and that light is another facet of electromagnetism. He built an apparatus generating the sparks across a gap in a short metal loop attached to an induction coil and used it as a source of electromagnetic waves. A similar apparatus but without the induction coil was used to detect the electromagnetic wave. When the electromagnetic wave was generated in a spark from the apparatus connected to the induction coil, a weak spark was measured across the gap in the detecting apparatus which was placed close to the generating apparatus. With the various setups, he could confirm the properties of electromagnetic waves and measure their speed. In 1887, he announced his initial discovery, showed that the basic properties of the electromagnetic wave in its vibration and susceptibility to reflection and refraction are the same as those of the light, and proved the correctness of Maxwell's theory. Figure 1.21 is the summarized chronology of human electricity.

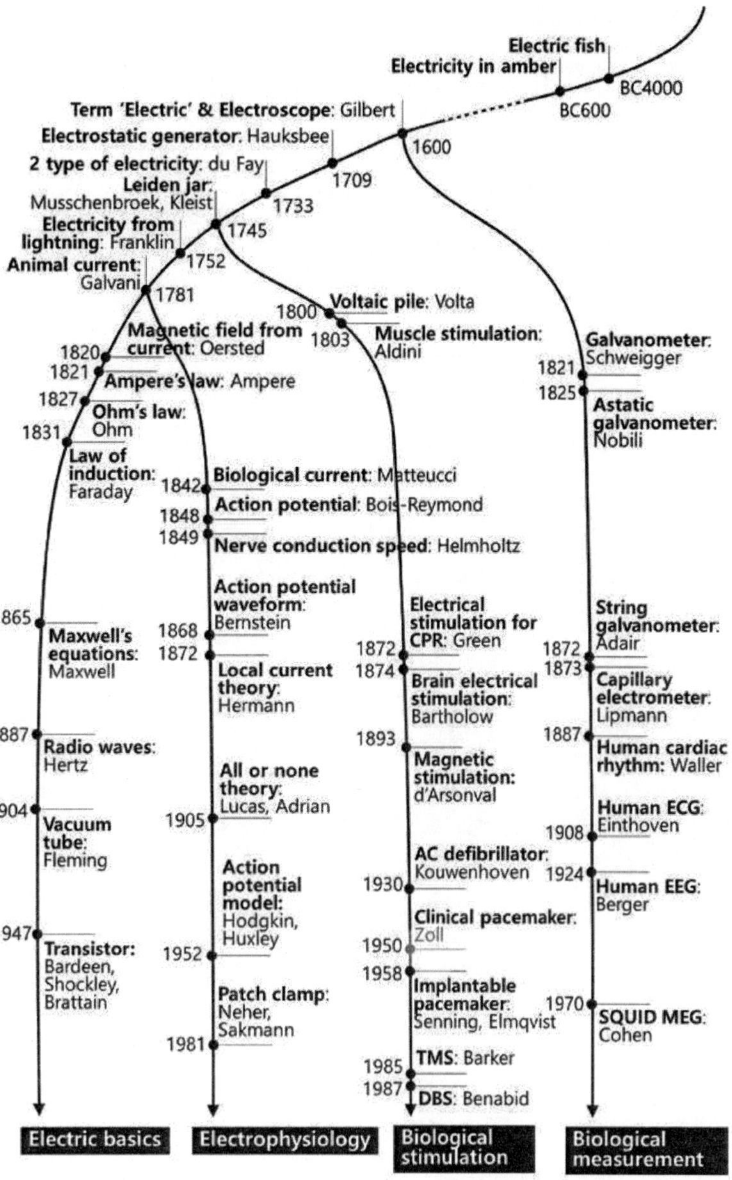

Fig. 1.21 Chronology of human electricity. *CPR* Cardiopulmonary resuscitation, *DBS* Deep brain stimulation, *ECG* Electrocardiogram, *EEG* Electroencephalogram, *MEG* Magnetoencephalogram, *SQUID* Superconducting QUantum Interference Device, *TMS* Transcranial magnetic stimulation

References

Ball, C. M., & Featherstone, P. J. (2019). The early history of cardiac pacing. *Anesthesia and Intensive Care, 47*(4), 320–332.

Barker, A. T., Jalinous, R., & Freeston, I. L. (1985). Non-invasive magnetic stimulation of human motor cortex. *Lancet, 11*(1), 1106–1107.

Barlow, H. B., Kohn, H. L., & Walsh, E. G. (1947). Visual sensations aroused by magnetic fields. *American Journal of Physics, 148*, 372–375.

Bickford R. G., & Fremming B. D. (1965). Neural stimulation by pulsed magnetic fields in animals and man. In *Digest of 6th International Conference on Medical and Biological Engineering*, Tokyo.

Bresadola, M. (1998). History of neuroscience: Medicine and science in the life of Luigi Galvani (1737–1798). *Brain Research Bulletin, 46*(5), 367–380.

Carmeliet, E. (2019). From Bernstein's rheotome to Neher-Sakmann's patch electrode. The action potential. *Physiological Reports, 7*(1), e13861. https://doi.org/10.14814/phy2.13861

d'Arsonval, A. (1896). Dispositifis pour la mesure des courants alternatifs de toutes frequences. *Compt. Rend Soc. Biol., 3*, 450–451.

d'Arsonval, J. A. (1897). Des courants a haute frequence. *Arch. d'Electricite Medicale*, 166–179.

Martens, J. W., Koehler, P. J., & Vijselaar, J. (2013). Magnetic flimmers: 'Light in the electromagnetic darkness'. *Brain, 136*, 971–979. https://doi.org/10.1093/brain/aws185

Pascual-Leone, A., & Meador, K. J. (1998). Is transcranial magnetic stimulation coming of age? *Journal of Clinical Neurophysiology, 15*(4), 285–287.

Piccolino, M. (1998). History of neuroscience: Animal electricity and the birth of electrophysiology, The legacy of Luigi Galvani. *Brain Research Bulletin, 46*(5), 381–407.

Reaz, M. B. I., Hussain, M. S., & Mohd-Yasin, F. (2006). Techniques of EMG signal analysis: Detection, processing, classification and applications (Correction). *Biological Procedures Online, 8*, 163. https://doi.org/10.1251/bpo124

Reif-Acherman, S. (2017). Jacques Arsene d'Arsonval: His life and contributions to electrical instrumentation in physics and medicine. part iii: High-frequency experiences and the beginnings of diathermy. *Proceedings of the IEEE, 105*(2), 394–404. https://doi.org/10.1109/JPROC.2016.2640638

Rodari, P. (1902). Nochmals einige 'elektrotherapeutische' reflexionen. *Korresp Bl Schweiz Arzte, 4*, 103–109.

Stock, J. (2004). Gabriel Lippmann and the capillary electrometer. *Bulletin for the History of Chemistry, 29*(1), 16–20.

Weise, K., Brauer, H., Ziolkowski, M., & Toepfer, H. (2011). Modelling and simulation of a simple homopolar motor of Faraday's type. *Facta Universitatis, 24*(2), 221–242.

Part I

Electricity Within the Human Body

Nervous System

2

1 Human Body Systems

The human body consists of a number of biological systems that do specific functions necessary for active living. The basic unit constituting the body is the cells. Groups of similar cells form tissues, and groups of different tissues build organs. And, groups of organs form an organ system. A system means a set of these tissues and organs working together as parts of an organized mechanism for a common purpose. The human body has 11 specific functional systems, each of which has its own specialized function (Fig. 2.1).

1.1 Systems for Maintaining the Body Structure

The skeletal system, muscular system, and integumentary system are considered to constitute a physical body structure (Fig. 2.2).

Skeletal System

The skeletal system is the basic and central framework of the body. It consists of bones and connective tissue, including cartilage, tendons, and ligaments. The adult human skeleton is made up of 206 bones including the skull, spine, ribs, arms, and legs. It takes about 20% of the body weight. Bones support the body structure, make body parts move, and protect the critical organs of the body. The skeleton maintains the body shape against the gravitational force. Bones are connected with muscles to produce the movement of the body parts. The brain is protected by covering with the fused bones of the cranium and becomes less vulnerable to external injury. Vertebrae keep and protect the spinal cord. Protection of the heart and lungs of the thorax is supported by the bones of the rib cage. The femur bone is the longest bone with an average length of 50.5 cm (19.9 inches) and spans the hip and knee joints maintaining upright posture by supporting the skeleton. The smallest bone in the human body is the stapes located in the middle ear having the size of 3 mm × 2.5 mm. It does the role of transferring the sound from the eardrum to the cochlea. Bones are used to changing shape or remodel in response to variations in mechanical stress.

Active tissues of bones also require a blood connection like other tissues for the supply of oxygen and nutrients and for the removal of waste products in metabolism. Bones contain more calcium than any other organ and work as a repository for calcium. If the calcium level in the blood is lowered than the normal level, calcium in bones is released into the blood to support metabolic demands adequately. On the contrary, if the blood calcium level is increased, the excess amount of calcium is stored in the bone matrix. Three different types of blood cells, red blood cells, white blood cells, and blood platelets, are made in the bone marrow of the bones.

Muscular System

The muscular system is composed of a specialized type of cells, muscle fibers, which can shorten their length upon stimulation by neuronal signals. There are about 700 named muscles attached to the bones of the skeletal system and it takes about 40% of body weight. The primary function of muscle is exerting a force on the tissues attached by contracting itself. Muscles are attached to bones, internal organs, and blood vessels to make them move. All movements within the body are the results of the involvement of muscle contraction.

The human body has three different types of muscle, striated skeletal muscle, smooth muscle, and cardiac muscle. Smooth muscles are involved in involuntary movements and are found in the wall of the stomach, bladder, intestines, and blood vessels. Cardiac muscle is striated but involuntary. It makes up the volume of the heart and does the role of pumping the blood out of the heart with its rhythmic

K. S. Park, *Humans and Electricity*, https://doi.org/10.1007/978-3-031-20784-6_2

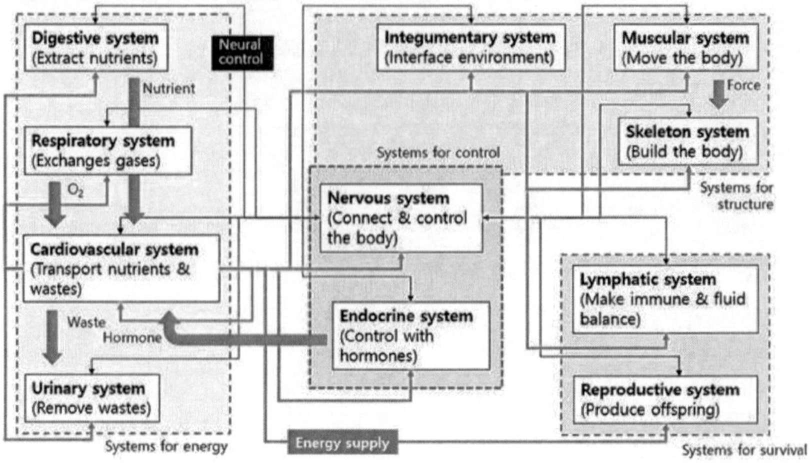

Fig. 2.1 Overview of 11 body systems. While each system does their designated roles, all systems are connected and controlled by the nervous system and powered by the cardiovascular system

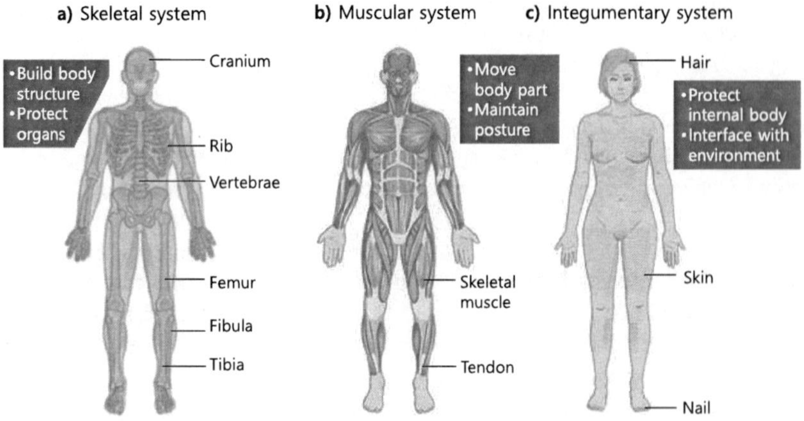

Fig. 2.2 Human body systems for maintaining the body structure. (**a**) Skeletal system, (**b**) Muscular system, (**c**) Integumentary system

contraction. Skeletal muscles, which are mostly attached to the bones, produce a wide range of movements from the obvious movements like walking and running to subtle movements like facial expressions, eye movements, and respiration. In addition to making movements, skeletal muscles also do the role of maintaining the posture and balance, like standing and sitting, by continually making fine muscular contractions to adjust and hold the body in stationary positions. The human muscular system is controlled voluntarily through the nervous system while some muscles like cardiac muscles are completely under involuntary control of the autonomous nervous system.

The muscle also produces heat as a metabolic by-product of muscular contraction. The heat generated from the muscles takes nearly 85% of the heat produced in the body and does a critical role in maintaining or increasing the body temperature.

Integumentary System

The integumentary system is made up of several organs and structures forming the outermost layer of the body. It comprises the skin, hair, nails, glands, and nerves acting as a physical barrier and interface between the external environment and the internal environment. The primary function of the integumentary system is to protect the inside of the body from hazardous elements in the external environment, like bacteria, pollution, and UV rays from the sun. The skin is responsible for excreting various substances, including small amounts of carbon dioxide, sweat, water and waste products like excess sodium chloride and urea.

The skin is the main part of integumentary system and is the largest organ of the body to function as a protective barrier covering the entire outer surface of the body, from the top of the head to the end of the toes. The skin is approximately 2 mm thick, covers a surface area of approximately 2.0 m^2, and its entirety weighs nearly 3.6 kg.

The skin and its associated structures also do many specialized roles in close interactions with other body systems like nervous, cardiovascular and digestive systems. The skin functions to transmit sensations from the environment via its nerve receptors. The skin is populated with receptors which perceive different types of sense including pain, heat, and cold and transmit the impulses to the nervous system. To control the release of heat, the skin works with the cardiovascular system by constricting or dilating the blood vessels. The skin synthesizes vitamin D during exposure to the sun and provides this vital nutrient to the digestive system.

Hair serves to help protect the skin, regulate body temperature, and lend itself to the evaporation and perspiration process. Nail serves to protect the fingers and toes from injury or trauma and assists the sense of touch.

1.2 Systems for Supplying the Energy

All the cells in the body need the energy to survive and to do the defined function. So, the human body is constructed with the detailed energy delivery systems. Cardiovascular system, respiratory system, digestive system, and urinary system can be categorized to belong to this group of system (Fig. 2.3).

Cardiovascular System

The cardiovascular system is composed of the heart, blood vessels, and approximately 5 L of blood. It is a transporting system of the body. Its complex network transports oxygen, nutrients, hormones, and cellular waste products to and from the cells in every corner of the body. Blood flow transporting the materials is powered by the heart, which is the body's hardest-working organ and never stops pumping with only the size of a closed fist. Even in the resting state, the heart continuously pumps out about 5 L of blood per minute. It pumps, on average, 80 beats per minute, 115,200 beats per day, and 42,048,000 beats per year. For a life of 80 years, the heart would beat 3.36 billion times. It is very difficult to find an engineering pump continuously working for 80 years.

The cardiovascular system has two primary circulations, pulmonary circulation, and systemic circulation. Pulmonary circulation is the loop to the lung to transport deoxygenated blood and return oxygenated blood to the heart. As the lung is located close to the heart and at almost the same height, this circulation does not require high pumping power and the pulmonary circulation is performed with maximal arterial pressure of 18–25 mmHg. Systemic circulation transports oxygenated blood from the heart to all of the tissues of the body. On the way back to return, systemic circulation collects carbon dioxide and other wastes from body tissues. Since the blood should be transported to the highest location in the head and be returned from the lowest end of the toe, blood should be pumped with high enough power. Normally, maximal arterial pressure rises up to 120 mmHg. Systemic circulation divided into several branches to reach major organs and neighboring tissues including carotid branch for the brain and the coronary branch for the heart itself.

Blood vessels are the other component constituting the cardiovascular system. They are the body's highways that make blood flow efficiently between the heart and every region of the body. Blood flows through different types of vessels sequentially, aorta, large arteries, small arteries, arterioles, capillaries, venules, and veins. Capillaries are the smallest and thinnest blood vessels running throughout almost every body tissue. They transport the blood close to the surface of the cells of the tissues to exchange gases, nutrients, and waste products most efficiently. Blood includes three different types of cells. In addition to that red blood cells work for transporting the oxygen, blood cells also do the role of protecting the body. White blood cells remove cellular debris and fight against exogenic pathogens. Platelets and red blood cells form scabs to seal wounds and prevent pathogens

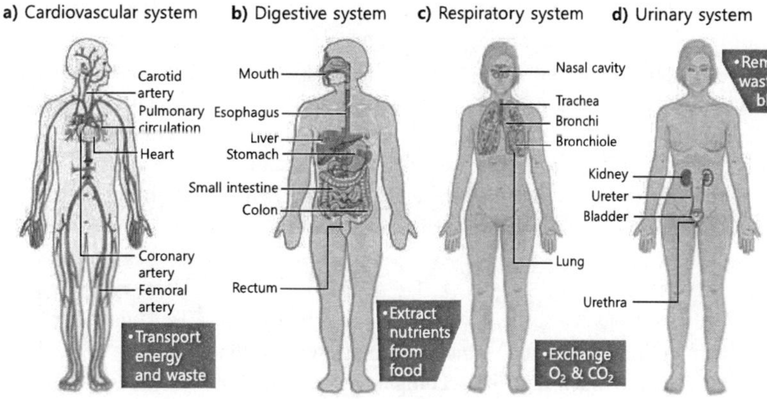

Fig. 2.3 Body systems for supplying the energy. (**a**) Cardiovascular system, (**b**) Digestive system, (**c**) Respiratory system, (**d**) Urinary system

from entering the body and liquids from leaking out. The cardiovascular system also does the role of maintaining homeostatic control of several internal conditions including body temperature. The cardiovascular system maintains a stable body temperature by appropriately controlling the blood flow to the surface of the skin for the dissipation of heat.

Respiratory System

The respiratory system is made up of the organs involved in the exchange of oxygen and carbon dioxide. The respiratory air pathway is composed of two tracts, the upper respiratory tract above the vocal fold and the lower respiratory tract below the vocal fold. When air enters through the nose, it passes the sinuses and nasal cavity of the upper respiratory tract, both of which are behind the nose. They filter out foreign particles and warm up the breathed air. Air can also enter through the mouth and it passes through the pharynx and larynx in the lower portion of the upper respiratory tract. After air enters the larynx, it flows down into the trachea, which belongs to the lower respiratory tract. The trachea is a rigid and hollow windpipe that guides air passage directly to the lungs, and it is divided into two bronchi leading air into each side of the lung. Then, these bronchi continue to branch off into smaller bronchioles, and finally into alveoli, which are air sacks where the exchange of oxygen and carbon dioxide actually takes place with blood vessels. On average, the diameter of alveoli is about 250–300 μm, and there are about 480 million alveoli in an adult human lung. The alveolar wall is covered with a dense network of capillaries. Oxygen in the breathed air passes through the alveoli and diffuses into the blood in capillaries, and it is loaded in the hemoglobin of the blood, and travels to the tissues throughout the body. The diaphragm is a large, dome-shaped muscle located below the lung and contracts rhythmically, continually, and most of the time involuntarily for respiration. For the inhalation, the diaphragm is contracted and flattened to enlarge and produce negative pressure in the chest cavity. The respiration rate of a healthy human is 12–16 times per minute, and the air volume breathed under normal resting circumstances, which is the resting tidal volume of the lung, is about 500 mL while the maximal total lung capacity of holding air is about 6 L.

The respiration system is also involved in vocalization. The air movement across the vocal cord in the larynx, pharynx, and mouth allows humans to speak and phonate.

Digestive System

The digestive system is a series of connected organs leading from the mouth to the anus, which is composed of the gastrointestinal (GI) tract and the accompanying organs for digestion. The gastrointestinal (GI) tract is the largest structure of the digestive system which begins at the lips and ends at the anus covering about nine meters. It includes the mouth, the throat, the esophagus, the stomach; the small intestine, and the large intestine. The large intestine is composed of the cecum, the colon, and the rectum. The digestive system breaks down the food and liquid we eat into smaller components so that its nutrients can be easily absorbed into the bloodstream. Absorbed nutrients include carbohydrates, proteins, fats, vitamins, minerals, and water. The digestive system also includes the structures through which wastes pass for elimination and organs that produce juices necessary for digestion. The glands producing digestive juices include the salivary glands, the gastric glands in the stomach lining, the exocrine glands in the pancreas, and the liver and the gallbladder. The digestive tract and glands jointly contribute to breaking down the ingested food physically as well as chemically and to the elimination of nondigestible components. The digestive process follows three sequential phases, the cephalic phase, the gastric phase, and the intestinal phase.

The first cephalic phase of the digestion process begins with preparatory secretions from gastric glands in the stomach in response to the sight and smell of the food. The mouth, or the oral cavity, does a major role in this phase, by breakdown the food mechanically with teeth and a tongue, and chemically with digestive enzymes within the saliva. Mastication of food in the mouth with enzymes makes the food be swallowed down the esophagus to enter the stomach. The second gastric phase of the digestion process occurs in the stomach. Transferred food is further broken down by mixing with gastric acid until it passes into the duodenum. The third intestine phase begins in the duodenum, where partially digested food is mixed again with enzymes produced in the pancreas. In the small intestine, most of the digestion of food takes place. Some minerals and water are reabsorbed back into the bloodstream in the colon of the large intestine. The remaining waste products of digestion become feces and they are defecated from the rectum via the anus.

The digestive system is supplied by the celiac artery, which is the first major branch of the abdominal aorta. And the gastrointestinal tract is governed by the enteric nervous system (ENS), which is one of the main divisions of the autonomic nervous system (ANS) and consists of a meshlike system of neurons. The enteric nervous system (ENS) controls motor functions, local blood flow, mucosal transport, and secretions of the digestive system.

Urinary System

The function of the urinary system, which is also called the renal system, is to filter blood and produce urine as a by-product of digestion. The urinary system is composed of two kidneys, ureters, bladder, and urethra. The kidney is an essential organ that makes blood clean. To filter the blood, the kidneys have an extensive network of blood vessels connected via the renal arteries and the renal vein. Kidneys eliminate excess water and salt, toxins, and other waste products like urea in the bloodstream, and maintain the chemical balance of potassium and sodium, and water balance. Urea is the metabolic waste produced during the digestion of foods containing protein, like meat, poultry, milk, and beans. Urea carried to the kidneys in the bloodstream is removed in the form of urine along with water and other wastes.

Urine and waste products move from kidney to bladder through the ureters and they are stored in the bladder until they are excreted as urine through the urethra. The normal capacity of the bladder is 0.4–0.6 L, while the average amount of produced urine is 0.8–2.0 L per day. By eliminating waste and excess water, the urinary system also regulates blood volume, blood pressure, concentration of electrolytes, level of metabolites, and blood pH.

1.3 Systems for Controlling the Body Parts

To make the body systems function in harmony, human body should be maintained and regulated fast and accurately under supervised control system. Nervous system and the endocrine system take the part of controlling the body parts collectively for the targeted tasks (Fig. 2.4).

Nervous System
The nervous system is a complex collection of specialized cells known as neurons that transmit signals through its fiber between different parts of the body. It is essentially the body's wired commanding and communication system. It controls all the other systems of the body, connects those systems for various body functions, and integrated them to work as a unified body unit for survival.

The nervous system can be divided into two subsystems, the central nervous system and the peripheral nervous system. The central nervous system is made up of the brain, spinal cord, and nerves and the peripheral nervous system consists of sensory neurons, motor neurons, and interneurons that connect the peripheral part of the body to the central nervous system. Functionally, the peripheral nervous system can be categorized into two parts, the somatic nervous system which can be controlled voluntarily, and the autonomic nervous system which is managed involuntarily. The somatic nervous system consists of nerves that connect the brain with muscles and make the muscles act for the designated purposes, and the nerves that connect the brain and sensory receptors populated in the interface with the external environment. The autonomic nervous system regulates fundamental body processes, like blood pressure, heart rate, and respiration rate, continuously without the subject's conscious effort. The contents of this book are mostly related to the nervous system and each part of the nervous system will be described in detail in the corresponding topics.

Endocrine System
The endocrine system, which is also called the hormone system, is composed of glands located all over the human body, which produce and release hormones into the blood at the appropriate time. Even though hormones reach all parts of the body, only target cells or organs having compatible receptors can respond. This is a kind of wireless communication like broadcasting that releases the message unidirectionally into the open media so that the necessary recipients can receive and respond.

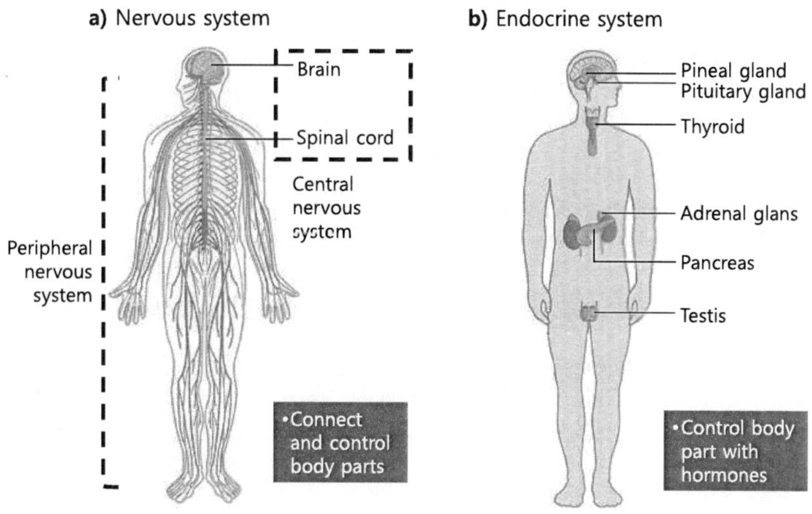

Fig. 2.4 Systems for controlling the body parts. (**a**) Nervous system, (**b**) Endocrine system

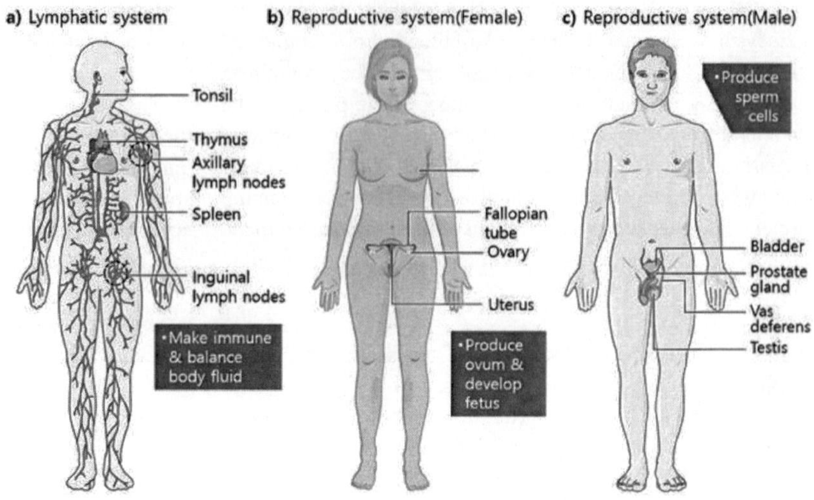

Fig. 2.5 Systems for the survival of individual and species. (**a**) Lymphatic system, (**b**) Female reproductive system, (**c**) Male reproductive system

Hormones are chemicals that carry messages and are released into the blood for the coordination of organs, muscles, skin, and other tissues. Over 50 hormones have been identified in humans and they influence nearly every process in functions, like blood pressure, digestive metabolism, growth and development, mood and emotion, sleep, fertility, and sexual function. So, the imbalance of hormones can cause health problems, like weight gain, high blood pressure, sleep disturbance, and changes in mood and behavior. The balance of the hormone can be disturbed by internal and external causes, like illness, stress, and certain medications.

Glands are located in several places throughout the body, including the brain, the neck, and the reproductive organs in different sizes. Hypothalamus is located in the brain and centrally controls the endocrine system for the management of mood, hunger and thirst, sleep patterns, and sexual function. The pituitary is a small gland about the size of a pea located at the base of the brain. It makes hormones controlling other glands like the thyroid gland, adrenal glands, ovaries, and testicles. The thyroid is a butterfly-shaped gland in the front part of the neck and produces hormones for metabolism. The parathyroid is four tiny glands about the size of a rice grain that control the level of calcium in the body. Two adrenal glands on top of each kidney produce hormones for the control of metabolism, blood pressure, sexual development, and response to stresses. The pineal gland manages the sleep cycle by releasing the melatonin hormone. The pancreas produces a hormone called insulin that controls the level of sugar in the blood in addition to its significant role in the digestive system. Ovaries in women release sex hormones including estrogen and testes in men release the hormone affecting sperm production, muscle strength, and sexual motivation.

1.4 Systems for Survival

To defend the body from any pathogen and continue the life over the generation, specialized systems are required for the survival. These special body systems are lymphatic system and reproductive system (Fig. 2.5).

Lymphatic System

The lymphatic system protects the body from illness-causing invaders as a part of the immune system. It also does the function of maintaining the body's fluid levels, absorbing digestive tract fats, and removing cellular waste. The lymphatic system is composed of lymph, lymph nodes, lymphatic vessels, collecting ducts, and some organs. They are networked together to move a colorless, watery fluid called lymph back into the bloodstream. The lymph is a pale fluid that bathes the tissues of an organism and flows through the lymphatic system. It includes proteins, minerals, fats, nutrients, damaged cells, cancer cells, and foreign invaders as well as lymphocytes doing the role of infection-fighting. Lymph nodes are bean-shaped glands in the middle of lymphatic vessels. These nodes filter out the damaged cells and cancer cells in the lymph and produce lymphocytes and other immune system cells. In the human body, there are about 600 lymph nodes populated throughout the body including the more familiar locations like the neck, armpit, and groin. Lymph in the lymphatic vessels flows through collecting ducts and finally into the subclavian vein and returns to the bloodstream.

The lymphatic system can be thought of as a supplementary one-way drainage system that collects the released fluid during the cardiovascular circulation. When the blood circulates throughout the body, blood plasma releases into tissues across the thin walls of the capillaries. This released blood plasma is called interstitial or extracellular fluid and contains oxygen, glucose, amino acids, and other nutrients needed by tissue cells. While most of this fluid seeps back immediately into the bloodstream, about 15% of this fluid is left behind with included substances. The lymphatic system removes and returns this fluid back to the bloodstream through the lymphatic vessels, and thus prevents a fluid imbalance that would result in the damage of tissues.

The lymphatic organs include the spleen, the thymus, the tonsil and adenoid, the bone marrow, and the appendix. These organs work as a part of the immune system by producing and maturing immune cells like lymphocytes and white blood cells, and by trapping, destroying, and fighting pathogens like bacteria, viruses, parasites, and fungi that entered the body.

Reproductive System

The reproductive system is to continue the survival of the human species over the generation. During the survival of each individual, all body systems work together to maintain a long, healthy, and happy life. However, to continue the survival over the generation, the human body needs a reproductive system to produce offspring. For this biological process, males and females have different reproductive organs and structures. In males, the primary reproductive organs are two testes, which are enveloped in a sack of skin, the scrotum, lying below and outside the abdomen and produce sperm cells. In females, the primary reproductive organ is two ovaries that are situated in the pelvic cavity and produce ova, the female germ cells. Testes and ovaries also produce hormones to mature the reproductive tract, develop sexual characteristics, and regulate the normal physiology of reproduction.

All other organs, ducts, and glands in the reproductive system do the role of transporting and sustaining the gametes and nurturing the developing offspring. Sperm cells produced in the testes are stored and conveyed through the male duct, are contained in the seminal fluid, and ejaculated into the female genital tract through the penis during intercourse. Ovum generated in the ovary is conveyed through the fallopian tube to the uterus during which it can be fertilized by a sperm cell.

Upon successful fertilization, the fertilized ovum travels out of the fallopian tube and into the uterus. In the uterus, the fertilized ovum is implanted in the uterine wall, and gestation for the development of a fetus begins and continues for around nine months. When a fetus has fully developed, gestation is completed and the reproductive process is concluded with childbirth, involving labor. After childbirth, the breasts are involved during the parenting stage of reproduction.

2 Classification of Nervous System

2.1 Functional Classification of the Nervous System

The nervous system is the main control system of the body that collects information from the internal and external environments and coordinates its behavior. Based on its functions, the nervous system can be classified into three parts doing different functions of sensation, processing, and response (Fig. 2.6). Information from the internal environment as well as the external environment is received via sensors populated through the body. This information is processed in various ways by combining and comparing the signals including previously collected and stored information. Upon the processed results, command signals are determined and delivered to each part of the body for the coordinated actions of the body components. While the processing part of the nervous system is centralized in the brain, nerves doing the sensation and responding role are distributed all over the body anatomically intermingled with each other.

Sensation

The first functional role of the nervous system is sensation, to gain information about what is happening inside and outside of the body. The sensation is received with the sensory organs and receptors, and received information is transmitted via peripheral afferent nerves to the central processing part.

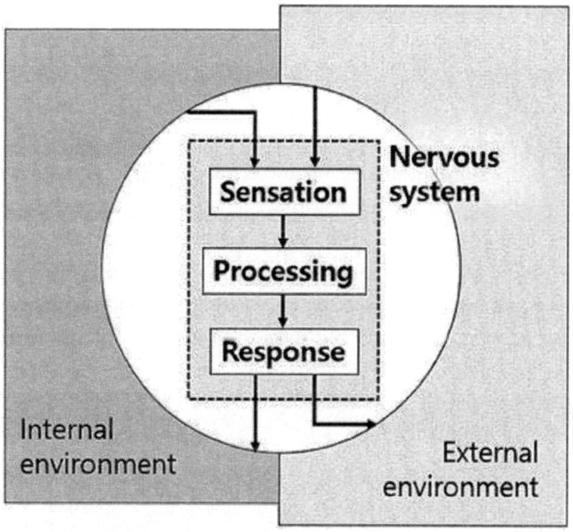

Fig. 2.6 Functional subdivision of the nervous system. Nervous system interacts with internal and external environment by responding based on the processed results using the sensed information

From the outside of the body, human senses the environment with five senses, sight, hearing, smell, taste, and touch. Sight is the sensing of light which is a signal of electromagnetic type. Hearing is the sensing of sound which is a signal of mechanical type. Taste and smell both sense chemical substances like molecules, compounds, and ions, in the food and air. Touch is the sensing of stimuli in physical or thermal types. While the sensation of light, hearing, taste, and smells are localized in sensory organs specialized respectively for their sensations, sensors for touch are distributed over the skin through the body surface.

Also, sensory information is collected from the internal environment. To maintain the body in an optimally balanced state, first of all, knowing the current state and the variation of the states are important. Receptors distributed throughout the internal body parts continuously report the current states of the organs and functional variables. Blood pressure, oxygen concentration, and muscle forces are some of the representative examples of internal information received in the nervous system. In contrast to the information from the external environment, information from the internal environment is collected without any conscious notice.

Processing

All the sensed information is transmitted toward the central part of the nervous system, where the received information is processed. The received information is transmitted to the allocated part of the brain and the received information is integrated by combining and comparing. During the integration, higher cognitive functions such as memories, learning, and emotion are involved. Previously learned, processed and stored information is actively recruited and used. Visual scene is processed in the occipital region of the brain cortex and sound information is primarily processed in the temporal part of the brain cortex. Each part of the brain has its own role in processing and decision-making. They are working parallelly and sequentially most of the time. For the recognition of a person, processed results from the occipital region for a visual image and from the temporal region for a voice are finally integrated and compared with the stored face and voice of each person. Processing in the brain can be a simple one like clicking a mouse or it can be a profound one like to be or not to be. Based on the processed results, the final decision is determined to invoke appropriate responses.

Response

Upon the decisions made from the integrated information, command signals are delivered to each part of the efferent organs of the body to invoke appropriate responses to accomplish a designated task or to maintain the balance of the body state. Most of the responses are performed via muscles in various parts of the body. To invoke responses of each part, decision results are transferred to the corresponding part of the motor cortex, where the motor signals are originated to contract the muscles in the periphery for the desired response. In most cases, muscle contractions are needed for the movement of the body parts. In many cases, the required response can be a speech that needs the movement of the muscles responsible for pronunciation. Sometimes, when subjects might need only facial expressions to represent their emotional response, it also needs the contraction of facial muscles. Even though we do not recognize their actions, all internal organs continuously receive commands from the brain to appropriately respond to the changed circumstance. Heart rate and respiration rate can be increased with the exercise automatically without our intentional notice. Ingested foods are moving from the mouth to the anus via various types of muscular contractions for mastication, peristalsis, mixing, and discharging. Different types of muscle, skeletal, cardiac, and smooth muscle, are involved in voluntary or involuntary movements of body parts. In addition to the responses through muscular contraction, glands can increase the secretion of hormones for the optimal regulation of body state. Systematically, motor responses are the output of the human body to the environment caused by the applied inputs from the environment.

2.2 Structural Classification of the Nervous System

Central Nervous System and Peripheral Nervous System

From a structural point of view, the nervous system can be classified into two major subsystems (Fig. 2.7), the central nervous system (CNS) and the peripheral nervous system (PNS). The central nervous system is the processing center of the body and is composed of the brain and the spinal cord. The peripheral nervous system is all the other parts of the nervous system excluding the CNS and relays information

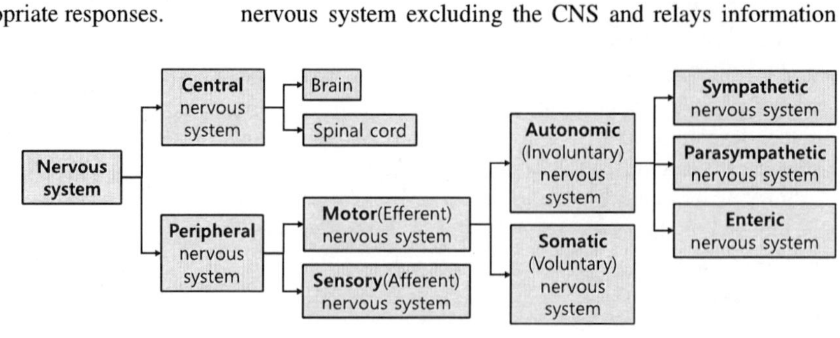

Fig. 2.7 Structural classification of nervous system into different levels of subdivisions

between the CNS and the rest of body. As the brain is contained within the cranial cavity of the skull, and the spinal cord is contained within the vertebral cavity of the backbone, CNS can simply be thought of as the part protected safely within the two bony cavities. And, the peripheral nervous system can be thought of as the part of the nervous system located out of these bony structures mostly in the peripheral part of the body.

Sensory Nervous System and Motor Nervous System
The peripheral nervous system can be further classified into two parts based on the function and the direction of pulse transmission, sensory division, and motor division. Sensory division, which is also called afferent division, transmits neural signals afferently from the sensory organs and the receptors to the central nervous system. These neural signals report the information on the states of the external environment in which the body is interfaced. If there is any stimulus that can be sensed by the sensing organs of the body, it is transformed into a series of electrical pulses and transmitted through afferent sensory nerve fibers. Sensing receptors are also populated in the interior part of the body to sense and report the acting states of the visceral organs continuously. Information from the interior part of the body is also transmitted through afferent division of PNS toward CNS.

The motor division, which is also called efferent division, transmits neural signals in the opposite direction, from the central part to the effector organs in the peripheral part of the body. These neural signals are commands to make each of the body parts respond to the external and internal environments. To click a mouse, a series of electrical impulses are transmitted through the efferent nerve fibers from the brain to the muscle responsible for the contraction of the finger. As there are different types of muscles, the motor division is further can be classified into two parts, the somatic nervous system (SNS) and autonomic nervous system (ANS).

Somatic Nervous System and Autonomic Nervous System
The somatic nervous system is the part of the motor nervous system that can be controlled voluntarily under the conscious notice of the subjects. It involves all the contraction of skeletal muscles and the process of voluntary reflex arcs. The somatic nervous system practically controls most of the muscular movements of the body and has a vital role in human behavior. The basic route of the somatic motor pathway begins in the primary cortex of the cerebrum, where the cell bodies of the primary motor neurons are located. Train of pulses generated in an upper motor neuron is transmitted through its axonal fiber to the brainstem or to the spinal cord, where it is connected to the lower motor neuron by the synapses. Then, the axonal fiber of the lower motor neuron transmits the command signal to each muscle fiber in the motor unit. By inducing contractions of muscle fibers in a motor unit, each motor neuron contributes collectively to produce sufficient enough force for the attraction of a skeletal bone.

While the voluntary muscles are controlled by the signals originating from the brain, during the reflexes, some muscles can be contracted involuntarily without the involvement of the brain. The somatic nervous system includes these reflexes. Reflexes are performed through reflex arcs that consist of sensory nerves carrying signals to the spinal cord, motor neurons, and interneurons which connect sensory nerves and motor nerves in the middle. In a reflex arc, muscle contraction for a reflex is directly triggered by the sensed signal without the processing of the sensed signal in the brain.

The autonomic nervous system (ANS) is the involuntary response of the motor nervous system. It regulates internal body processes autonomously without a subject's conscious effort for the sake of homeostasis, which is maintaining the body in an optimally balanced state. Nerve impulses of the autonomic nervous system go to the smooth and cardiac muscles as well as glandular tissues. The internal organs regulated by the ANS include the blood vessels, stomach, intestine, liver, kidneys, bladder, genitals, lungs, pupils, heart, sweat, salivary, and digestive glands. So, ANS is involved in a variety of involuntary body responses including heart rate, perspiration, respiration, digestion, emotion, and sexual arousal. ANS contains three physiologically distinct divisions, the sympathetic, parasympathetic, and enteric nervous systems.

Sympathetic, Parasympathetic, and Enteric Nervous System
The autonomic nervous system regulates internal body processes through a balanced activation of two antagonistic divisions, sympathetic and parasympathetic activations. They do an instrumental role in the homeostatic mechanisms by activating the internal process in the opposite directions. The sympathetic division activates the fight-or-flight response, and the parasympathetic division drives the body toward a rest and digest state.

The sympathetic division is activated under the condition of stress or danger so that the body becomes more suitable for fight-or-flight response. It functions to produce localized adjustments and reflex adjustments to the cardiovascular system. Sympathetic responses are characterized by the increase in heart rate and cardiac output, vasodilation to increase the blood supply to the skeletal muscles, cutaneous and gastrointestinal vasoconstriction where the blood supply is not so imminent in the stressful condition, the release of large quantities of epinephrine from the adrenal gland, pupillary dilation to allow more light to enter, bronchial dilation to increase breathing, and piloerection. The overall responses make the body prepare for imminent danger.

The parasympathetic nervous system functions to conserve the body's natural activity, and make the subjects relax once the stress or danger has passed. The parasympathetic nervous system produces localized adjustments and cardiovascular reflexes in the opposite direction to the sympathetic activation so that the body can be led to decreased arousal. Parasympathetic activation can be characterized by many responses. Heart rate is decreased, especially during the rest after completing physical exercise or after stressful situations. The bronchi are constricted to slow down breathing to a resting rate. The pupils are constricted to a normal level as increased light is not required in a relaxed condition. Parasympathetic activation also promotes salivation of the mouth to help the digestion of food through the enzymes within saliva. Parallelly, digestion is promoted through the digestive organs by the activation of accompanying muscular movement and secretion of digestive juices.

The enteric nervous system (ENS) is the third division of ANS, which is also controlled autonomously without conscious effort. It is an extensive, web-like structure in the gastrointestinal tract that can function independently of the remainder of the nervous system (Waxenbaum et al., 2022). Its complex reflex circuits detect the physiological condition of the gastrointestinal tract and integrate information about the state of the gastrointestinal tract. It also provides outputs to control gut movement, fluid exchange between the gut and its lumen, and local blood flow (Gershon, 2005). While the ENS has extensive connections with the CNS and works combinedly to control the digestive system to meet the physiological demand of the whole body, its autonomy is extensive so that it is referred to as a second brain.

3 Central Nervous System

The central nervous system (CNS) is composed of the brain and spinal cord (Fig. 2.8). They are protected by the bony frames. That is, the brain is surrounded by the cranium and the spinal cord is embedded in the vertebral column. The meninges, which are three layers of membranes between the bony frames and the soft tissue, cover and additionally protect the brain and spinal cord. Within the meninges, CNS tissues are surrounded and bathed in the cerebrospinal fluid (CSF) which cushions the brain and spinal cord from injury and also delivers nutrients and removes the waste for the CNS. The CNS tissues are composed of gray matter and white matter. Gray matter consists of neural cell bodies, dendrites, and unmyelinated axons. White matter forms the tracts within the CNS and consists of myelinated axon bundles.

The brain is constituted of about 86 billion neurons and its weight is around 1.5 kg. While its weight takes only 2% of total body weight, it is supplied with 20% of total cardiac output due to its tremendously high metabolic rate. It is very similar to the central processing unit (CPU) in the computer which consumes much more electric power compared to its volume and weight. The brain has an extensive vascular structure including the paired internal carotid and vertebral arteries. As the continuous blood supply is crucial, failure of cerebral circulation, for example, as short as 10 s may lead to the death of the brain (Van De Graaff & Fox, 1999). The brain is composed of 4 major regions, cerebrum, cerebellum, diencephalon, and brain stem.

3.1 Cerebrum

The cerebrum is the symbolic region of the brain making up about 82% of the mass of the brain. It contains about

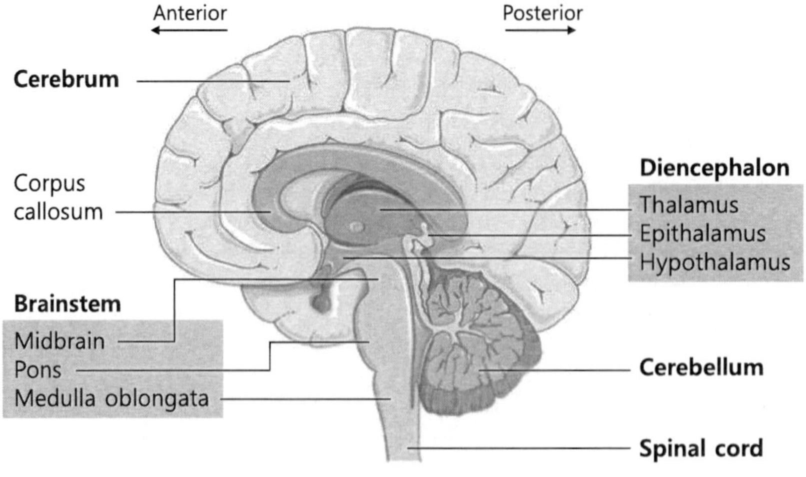

Fig. 2.8 Organization of the brain and spinal cord. The brain is composed of four major regions of cerebrum, cerebellum, diencephalon, and brainstem.

16 billion neurons which is 18% of the total 86 billion neurons in the brain (Herculano-Houzel, 2009). Most of the higher cognitive functions, like thought, memory, analysis, and emotion, are the results of cerebral activations.

Right and Left Hemispheres

By deep longitudinal fissure, the cerebrum is largely separated into two distinct halves, a right and left hemisphere. While they look separated on the cortical surface, both hemispheres are connected beneath the cortical layer by the large nerve fiber bundle called corpus callosum. Based on the possible different functions between two hemispheres, right brain left brain theory was suggested by Roger Sperry, who was awarded the Nobel Prize in 1981 (Gazzaniga, 2014). He studied brain functioning in patients who had their corpus callosum surgically cut to treat refractory epilepsy. The left side of the brain tends to control many aspects of language and logic, and the right side tends to handle spatial information and visual comprehension. So, the left brained person tends to be logical and the right brained person tends to be artistic. However, recent studies including the experiments using MRI do not support the theory in normal person whose two hemispheres are closely connected. Even though some brain functions tend to occur in one side of the hemisphere, for example language on the left and attention on the right, normal subjects don't tend to have a stronger hemisphere dominance. Each subject's personal trait seems to be determined more connection by connection.

Cerebral Cortex

Both hemispheres of the cerebrum are covered by the cerebral cortex, which is a continuous layer of gray matter. This thin and extensive layer of the cerebral cortex is heavily wrinkled with many ridges and grooves called gyrus and sulcus. Extensive folding of the cerebral cortex is considered to pack an increased amount of gray matter into a limited space within the skull. If the wrinkled cerebral cortex were unfolded flat, its surface area would be roughly equal to 1.0 m^2. By the pattern of these folds, the cerebral cortex is divided and indicated into specific regions called lobes (Fig. 2.9). Cerebral functions for the processing of general and special senses, initiation and coordination of movement, and high-level cognitive functions such as judgment, reasoning, problem-solving, and learning are allocated in different parts of the lobes respectively.

The central sulcus transversely divides each hemisphere into the anterior part and posterior part. The precentral gyrus is the ridge just before the central sulcus, and the postcentral gyrus is the gyrus just after the central sulcus. The lateral sulcus separates the temporal lobe from the frontal lobe and parietal lobe. The parieto-occipital sulcus separates parietal lobe and occipital lobe.

Different regions of the cerebral cortex are connected to each other by the axonal tract of the nerves in each region. Myelinated axons of the nerves constitute the white matter of the cerebral cortex with its white appearance by the myelin. These axons are grouped into tracts and connect different parts of the cerebral regions to integrate sensed information and prepare a motor response. These tracts can be grouped into three types, commissural, association, and projection tracts. Commissural tracts integrate the two hemispheres by connecting the left and right regions of the cerebrum. The corpus callosum is the most prominent commissural tract and largest white matter structure. Association tracts connect different cerebral regions within the same hemisphere. Projection tracts are the long extending tracts from the cerebral cortex to the inferior regions of the brain and to the spinal cord (Mazzasette et al., 2021).

Cerebral Lobes for Sensory Information Processing

Brain functions are attributed to the specific region of the cerebral cortex and demonstrate a strong correlation with the structures. Sensory brain functions are attributed to the parietal, temporal, and occipital lobes in the posterior cortical area to the central gyrus. The postcentral gyrus is the most anterior

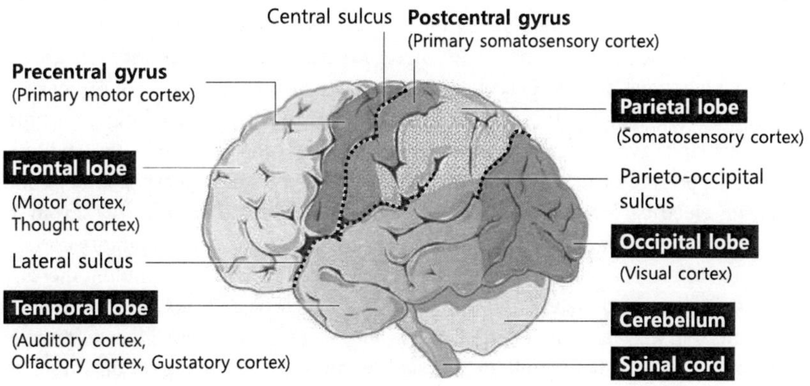

Fig. 2.9 Organization of cerebral cortex. Each lobe of the cerebral cortex is separated by the sulci

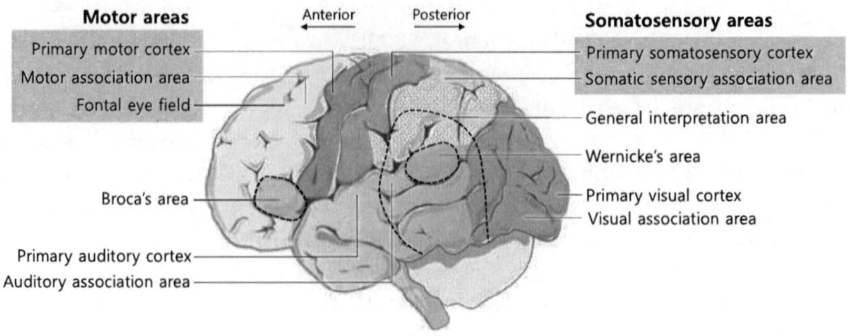

Fig. 2.10 Types of cortical area. Primary cortex and association areas for each sensation and motor response

part of the parietal lobe and composes the primary somatosensory cortex. All the tactile senses are processed in this area, including touch, pressure, tickle, pain, itch, and vibration. This region also processes the sensory information from the other body parts like proprioception and kinesthesia, reporting the body position and movement, respectively. Posterior to the postcentral gyrus in the parietal lobe, there is the sensory association area, which combines information from the primary somatosensory region for higher-level somatosensory recognition.

The occipital lobe is a region of the most post posterior part of the cerebral cortex. The primary visual cortex, which is the most posterior part of the occipital lobe, receives and processes visual information. The visual association area is located just adjacent to the primary visual cortex and does the subsequent visual processing by combing the results from the primary visual cortex. The primary auditory cortex and auditory association area in the temporal lobe sense and further process the auditory information respectively. The primary olfactory cortex which senses and processes smell is in the same temporal lobe. Finally, the primary gustatory cortex which senses and processes taste information is in the insula, which is the deeper location of the lateral sulcus separating the temporal lobe from the parietal lobe.

Cerebral Lobes for Processing the Responses

The cortical regions which process the collected information for the generation of responses are in the frontal lobe. Overall, the frontal lobe is responsible for motor functions, from planning the movements to executing commands to be sent to the peripheral effector organs. The primary motor cortex is in the precentral gyrus, just the anterior part of the central sulcus. Upper motor neurons are in this cortical region and generate electrical pulses to command a contraction of the corresponding motor unit in the muscle. Generated electrical pulses transmit via axons of the nerves to the spinal cord. After the synaptic connection to the lower motor neuron in the brainstem and the spinal cord, electric pulses are finally transmitted to the motor unit in the skeletal muscle. The motor association area or premotor cortex is the anterior area to the precentral gyrus and aids in deciding the final motor response in the primary motor cortex. The frontal eye field (FEF) in the anterior area to the premotor cortex is responsible for eye movements for binocular vision and awareness, as well as for voluntary eye movement. The most anterior portion of the frontal lobe is the prefrontal cortex, and it is associated with personality which influences decision-making on motor responses. Primary cortices are where sensory information is initially received for conscious perception, or where descending commands are sent down to the brainstem or spinal cord to execute movements. The regions around the primary cortices are each referred to as association areas which help the functions of the primary cortices (Fig. 2.10).

Topographical Mapping of the Cerebral Cortex

Neurons in the cerebral cortex are not arranged at random. Each cortical region is topographically mapped to the sensory and motor organs in the body (Fig. 2.11). When sensory neurons and muscles are adjacent to each other in the body, in general, their corresponding sensing and motor areas in the cortex are also adjacent to each other even though their relationship might be somewhat distorted. So, the sensory or responding pattern of neurons changes systematically across the cortical sheet.

The auditory cortex receives sensed information from the cochlea in the ear. When the cochlea detects the frequency characteristics of the input sound, lower frequency activates auditory nerves at the apex and higher frequency activated the auditory nerves at the base of the cochlea. This tonotopic mapping along the auditory nerves is generally maintained across the auditory nerve tract while they arrive at the primary auditory cortex. The auditory cortex is organized in the same way. From the anterior to the posterior direction along the auditory cortex in the temporal lobe, the frequency of best responding increases systematically. The primary visual cortex is also arranged topographically in mapping the visual image even though the relative geometrical relationships are highly distorted.

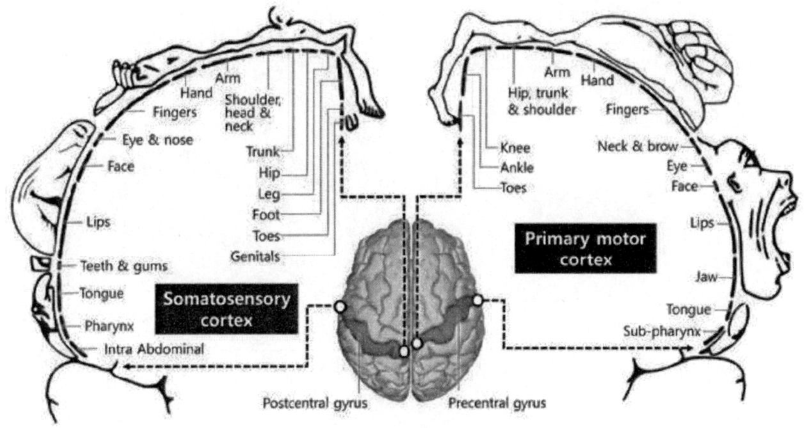

Fig. 2.11 Topographical mapping of somatosensory and primary motor cortex to the body parts

In the somatosensory and motor cortex, topographical mapping is more clearly demonstrated. Different regions of the body are mapped in the cerebral cortex for somatic sensation and movement. The somatosensory receptors across the body are mapped onto the somatosensory cortex. And, the primary motor cortex contains a similar mapping of neurons that control the respective body regions. Motor neurons moving the legs, toes, and knees are mapped into the medial face of the gyrus within the longitudinal fissure, while the motor neuron moving the hands and fingers are in the middle surface of the precentral gyrus. This topographical difference activates neurons in different locations of the motor cortex for each intended movement. As activation occurs in different locations of the motor cortical cortex, the intention of movement can be recognized by identifying the location of the neural activation. This characteristic is actively used in the brain-computer interface to move the external device just with the neural activity of the brain.

Areas of topographically mapped regions in the cerebral cortex do not match the areas or sizes of the corresponding body parts. It rather exaggerates the more sensitive areas of the body for sensation and movement, like the fingers and lips. Less sensitive areas of the body, like the shoulders and back, are mapped to smaller areas on the cortex. When people contact by the lips and hands, it is considered the most effective way to activate the maximal area in the cerebral cortex just by contacting small areas of the body each other.

Cerebral Regions for Higher Cognitive Functions

Planning, reasoning, and deciding are the unique cognitive functions of the cerebral cortex. All these activities can be considered largely as the preliminary process to produce responses and allocated to the prefrontal cortex, which is the most anterior region of the frontal lobe. Planning and reasoning start from this prefrontal cortex without direct connection to the motor function in the earliest phase. As the activation moves from the most anterior part toward the posterior part of the frontal lobe, it passes through the motor association area, the premotor cortex to the primary motor cortex. During this passage, decisions are made as a result of planning and reasoning, responses are prepared, and the motor neurons corresponding to the decided responses are finally activated for muscular contraction. The prefrontal cortex is also involved in developing the subject's mood and personality. When the subjects have disorders associated with their mood and personality, the prefrontal cortex mostly becomes the target region of electrical or magnetic stimulations for the treatment.

The cerebral cortex also functions for the memory in coordination with the structures beneath. Short-term memory, which is also called working memory or active memory, is localized in the prefrontal lobe. The medial temporal lobe is associated with the encoding, storage, and retrieval of long-term memories (Simons & Spiers, 2003). As memory is essentially considered a sensory function, usually the ultimate location of the memory is in the region where the sensory perception was processed. Understanding of language, both written and spoken, is performed in the Wernicke's area, which is located adjacent to the auditory association cortex and just anterior to the visual cortex. Production of the language, by controlling movements responsible for speech, is allocated to Broca's area in the lateral aspect of the frontal lobe anterior to the primary motor cortex associated with the head and neck (Fig. 2.10). Language is the representatively lateralized function, and the functioning areas are usually located only in the left hemisphere. The two regions are connected by white matter tracts connecting the posterior temporal lobe and the lateral aspect of the frontal lobe.

Basal Ganglia

The basal ganglia are a group of subcortical nuclei that belong to the cerebrum by located at the deep base (Fig. 2.12).

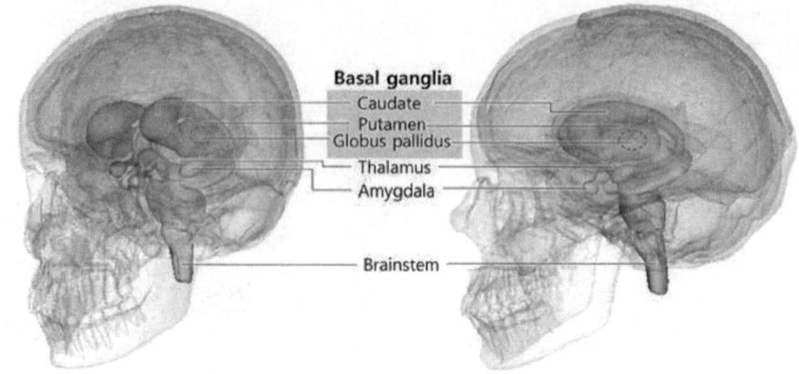

Fig. 2.12 Basal ganglia at the base of the cerebrum are composed of caudate, putamen, and globus pallidus and primarily responsible for motor control

They are responsible primarily for motor control as well as other functions like motor learning, executive functions, and emotional behaviors. The basal ganglia are particularly associated with movement disorders. Damage to these areas induces tremors, involuntary muscle movements, abnormal increase in tone, difficulty in initiating movements, and abnormal posture. So, these areas became the major target for the electrical stimulation for the treatment of movement disorders including Parkinson's disease. And, the basal ganglia play an important role in reward and reinforcement, addictive behaviors, and habit formation.

The major structures of the basal ganglia that control the movement are the caudate, putamen, and globus pallidus. The caudate is a long nucleus that forms the C-shape along the base of the cerebrum from the frontal through the parietal and occipital lobes, and back to the temporal lobe. The putamen and the globus pallidus are in a deeper place than the head part of the caudate in the inferior regions of the frontal and parietal lobes. The globus pallidus is medial to the putamen. The basal ganglia have several loops, where cortical and subcortical projections interact with internal reentry loops forming a complex network. They are ideally designed for selecting one and inhibiting the others among the simultaneously occurring events and signals (Lanciego et al., 2012). The basal ganglia compare cortical processing with the general state of activity in the nervous system to make the movement take place. The basal ganglia are connected below with a few nuclei in the brainstem and they act together as a functional group by forming a motor pathway.

3.2 Cerebellum

The cerebellum, which means the little brain, is located at the back of the brain, underneath the occipital and temporal lobes of the cerebral cortex, and on the posterior surface of the brainstem. The cerebellum takes approximately 10% of the brain's volume, but it contains about 69 billion neurons which are 80% of the total 86 billion neurons in the brain and are more than 4 times larger than the number of neurons in the cerebrum (Herculano-Houzel, 2009). The cerebellum is considered a motor structure. Damages in the cerebellum used to cause impairments in motor control and posture and most of the cerebellum's outputs are connected to the parts of the motor system. Motor commands are initiated in the primary motor cortex in the cerebrum, and the command messages are transferred to the cerebellum. Then, the cerebellum modifies the motor commands of the descending pathways to make movements more adaptive and accurate. The cerebellum is also involved in some cognitive functions like attention, language, and emotional control for fear and pleasure (Wolf et al., 2009).

Organization of the Cerebellum

Like the cerebrum, the cerebellum consists of an outer layer of gray matter and an inner layer of white matter and cerebellar deep nuclei (Fig. 2.13). Its cortical surface is tightly convolved but with finely spaced parallel grooves, in contrast to the broad irregular convolutions of the cerebral cortex. It contains almost all the neurons in the cerebellum. The cerebellum is divided into two cerebellar hemispheres with a narrow middle zone called the vermis based on the functions and connections involved. The vermis connects two cerebellar hemispheres and is involved in regulating the body movements by allowing for error corrections with received proprioceptive information. Both lateral hemispheres are primarily concerned with movement planning with inputs from the cerebral motor cortex and pontine nuclei and return outputs through the thalamic projections back to the premotor and motor cortices. This area also regulates the coordination of muscle activation and is involved in visually guided movements.

Each hemisphere consists of three lobes, the anterior lobe, the posterior lobe, and the flocculonodular lobe. The anterior

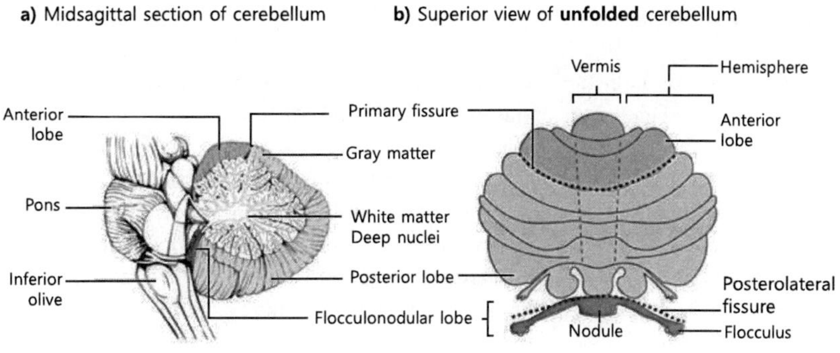

Fig. 2.13 Organization of cerebellum. (**a**) Midsagittal sectional view of cerebellum, (**b**) Superior view of unfolded cerebellum

lobe is responsible for modulating the unconscious proprioception with inputs mainly from the spinal cord. The posterior lobe is much larger than the anterior lobe and separated from the anterior lobe by the primary fissure and from the flocculonodular lobe by the posterolateral fissure. It plays a major role in fine motor coordination, specifically in the inhibition of involuntary movement via inhibitory neurotransmitters, like GABA. The flocculonodular lobe is placed on the anteroinferior surface of the cerebellum and consists of the nodule and the flocculus. This lobe of the cerebellum has connections to the vestibular nuclei and uses information about the head movement to influence eye movement.

The deep cerebellar nuclei are the only output structures of the cerebellum. All their nervous connection with other parts of the brain goes through the pons via cerebellar peduncles. There exist three cerebellar peduncles. The superior cerebellar peduncle (SCP) connects to the thalamocortical pathway and involves the vestibular sense and proprioception. The middle cerebellar peduncle (MCP), which is the largest among the three cerebellar peduncles, conveys impulses from the cerebral cortex to the cerebellum. And, the inferior cerebellar peduncle (ICP) arises from the inferior in the medulla and connects the cerebellum with the spinal cord to integrate proprioceptive sensory input and maintain a posture (Kim et al., 2014).

Coordination of Voluntary Movements

The cerebellum is primarily responsible for coordinated movement. It coordinates a number of different muscle groups to move together in a temporally and spatially arranged fashion, like an alternating movement of the left and right arms and legs during the walking. When the primary motor cortex of the cerebrum generates a command message and sends it down through motor neurons to initiate a walking, this command message is also transferred to the cerebellum. The cerebellum compares this command message with sensory feedback signals coming through the inferior olive from the muscles and joints reporting the proprioceptive information about the movements of walking, and the balance of the body. If walking is disturbed, probably by the rough ground or by a strong wind, then the cerebellum detects the difference and sends out a corrective command to compensate for the difference between the original cortical command and the sensed movement. The output of the cerebellum to correct the messages is sent into the midbrain and then descends to the spinal cord and finally to skeletal muscles.

While the body is not in active movement, the cerebellum does the work of postural adjustments to maintain the balance of the body. Vestibular receptors and proprioceptors provide the information for the postural balance and the cerebellum modulates commands to motor neurons to compensate for any shifts in body position or changes in tension upon muscles.

Procedural Memory

Procedural memory is a type of long-term memory involving how to perform a specific action and skill which involves a certain procedure. In contrast to declarative memory, which stores the events in terms of facts, procedural memories are related to the experiences that make a person remember skills. Riding a bike, tying the knots, and cooking a pizza are all examples of procedural memories. Procedural memory is formed when the synaptic connections are developed between the neurons so that the signals pass through. The more frequently an action is performed, the more often signals are sent through those same synapses. When these synaptic routes become stronger over time, the actions themselves become unconscious and automatic toward procedural memory. As procedural memory is related to coordinating and accurate movement, the cerebellum does an important role in forming procedural memory. So, cerebellar damage produces disorders in motor learning as well as coordinating fine movement, equilibrium, and posture.

3.3 Diencephalon

The diencephalon is in the deeper part of the brain beneath the cerebrum and above the brainstem. It can be described as any region of the brain with "thalamus" in its name including the thalamus, the hypothalamus, and the epithalamus. The word diencephalon originated from the word meaning "through brain." As its name means, at the location beneath the cerebrum, it has an extensive nerve connection between the cerebrum and the rest of the nervous system. Excluding the olfactory system associated with the sense of smell, the other part of the brain, the spinal cord, and the PNS all send information to the cerebrum through the diencephalon, and output from the cerebrum passes also through the diencephalon.

Thalamus

The thalamus is a paired gray matter structure of the diencephalon located near the center of the brain. It is divided into two bulb-shaped masses of around 5.7 cm in length and positioned symmetrically on each side, but connected by a flattened nerve band. As the main structure of the diencephalon, it forms during embryonic development. The primary function of the thalamus is to relay motor and sensory signals to the cerebral cortex (Fig. 2.14). A group of nuclei receive sensory or motor information from the rest of the body and present selected information via nerve fibers to the cerebral cortex through thalamocortical radiation. The connections between the peripheral sensory organs and the thalamus are contralateral, communicated with the opposite side of the body, while the connections between the thalamus and the cerebral cortex are ipsilateral, on the same side of the brain. Additionally, the thalamus is involved in the regulation of consciousness, sleep, alertness, and other sensorimotor associated functions. The thalamus can be grouped into three distinct nuclei, sensory relay nuclei, association nuclei, and nonspecific nuclei.

Sensory relay nuclei relay primary sensations, except for olfactory sensation, by transferring specific sensory information to the corresponding cortical area. Instead of just copying the information for the relay, they do the role of gating by filtering and processing the information. Feedback connections from the primary sensory cortices make the thalamus process sensed information in close connection with those regions. The visual sensory system sends sensed visual information from the retina to the lateral geniculate nucleus (LGN) of the thalamus to route toward the visual cortex of the occipital lobe. The auditory sensory system sends sensed auditory information from the inner ear to the medial geniculate nucleus (MGN) of the thalamus to route toward the auditory cortex of the temporal lobe. The somatosensory information for pain, temperature, and crude touch, coming from the face and other body parts through the spinal is relayed in the ventral nucleus of the thalamus and projected to the somatosensory cortex in the parietal lobe. The gustatory sensation is known to be conveyed through the ventral posterior part of the thalamus.

Association nuclei receive input from specific areas of the cerebral cortex, which is then projected back to the cerebral cortex to somewhat generalized association areas so that they regulate activity. The ventral anterior (VA) nucleus of the thalamus is involved in the induction, execution, and control of voluntary movement. It receives topographically segregated inputs from the basal ganglia of the cerebrum and returns output to the premotor cortex, and the primary motor cortex for the association. The ventral lateral (VL) nucleus is involved in balance and fine motor control.

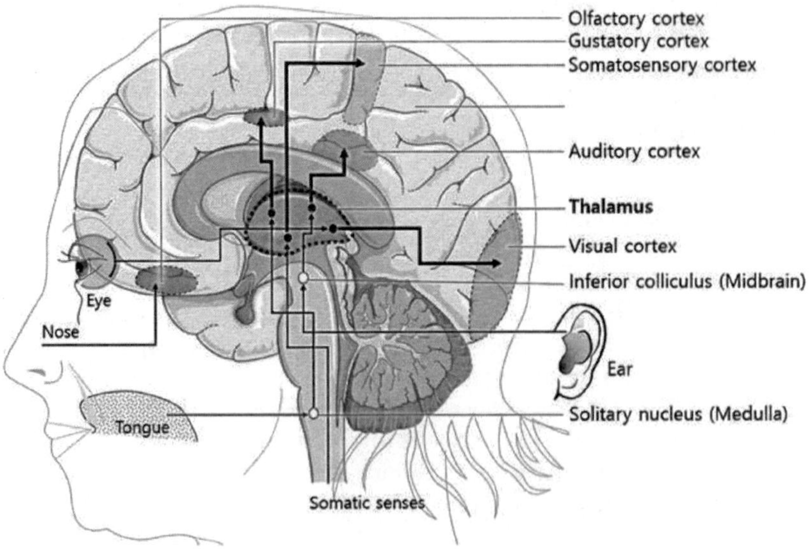

Fig. 2.14 Sensory relay nuclei of the thalamus. All sensory information is gated in the thalamus and projected to the respective cerebral cortices except olfactory sensation

Table 2.1 Major hormones produced in the hypothalamus and their effects

Hypothalamus hormones	Destination	Pituitary hormones	Target organ	Effect
Antidiuretic hormone	Posterior pituitary	–	Kidney	Water reabsorption
Oxytocin		–	Female reproduction system	Uterus contraction, breastfeeding
Gonadotrophin-releasing hormone	Anterior pituitary	Luteinizing hormone	Gonad	Sex hormone
		Follicle-stimulating hormone	Reproductive system	Production sperms and eggs
Thyrotropin-releasing hormone.		Thyrotropin-stimulating hormone	Thyroid gland	Thyroid hormone for metabolism
Prolactin-releasing hormone		Prolactin	Mammary gland	Promote milk production
Corticotropin-releasing hormone		Adrenocorticotropic hormone	Adrenal glands	Regulate stress and metabolism
Growth hormone-releasing hormone		Growth hormone	Liver, bone, muscles	Stimulate growth & high metabolic rate

It receives afferents predominantly from the cerebellum and the brainstem, and project to the cortical motor area.

Nonspecific nuclei receive input from the cerebral cortex and project information diffusely through it. Most of these nuclei interconnect brain activity between different areas of the brain and play a role in general functions like regulating sleep, alerting, and awareness.

Hypothalamus

As a part of the diencephalon, the hypothalamus is located on the inferior and slightly anterior to the thalamus and above the pituitary gland. It is a collection of nuclei linking the nervous system with the endocrine system. The main role of the hypothalamus is to maintain homeostasis, keeping the human body in a constantly balanced condition. The hypothalamus produces releasing and inhibiting hormones, which stimulate and inhibit the production of other hormones throughout the body.

To maintain homeostasis, the hypothalamus responds to a variety of signals from the internal environment including body temperature, hunger and satiety, blood pressure, and levels of hormones in the circulation. It also senses the signals from the external environment including day and night variation, atmospheric temperature, and stress, and regulates body rhythms for its adaptation.

There are two sets of nuclei in producing hormones. One set sends the hormones they produce into the posterior lobe of the pituitary gland where these hormones are released directly into the bloodstream. These hormones are antidiuretic hormone and oxytocin. The other set of nuclei produces stimulating and inhibiting hormones that reach the anterior lobe of the pituitary gland and produces hormones regulating the production of target hormones in the body part including the gonad, thyroid, and adrenal cortex. Table 2.1 shows the list of hormones produced in the hypothalamus and the target hormones regulated by the hormones produced in the hypothalamus.

Epithalamus

The epithalamus is in the dorsal posterior part of the diencephalon. It includes the pineal body and habenular nuclei. The pineal gland is an endocrine gland responsible for the secretion of melatonin which promotes sleep and regulates day-night cycles. The habenula connects the basal ganglia of the cerebrum with the midbrain, and it is involved in reward, emotional responses, and aversion processing via dopamine release and serotonin release.

3.4 Brainstem

The brainstem is the stem-like structure at the base of the brain and the top of the spinal cord. It connects the cerebrum of the brain to the spinal cord and cerebellum. It is divided into three sections, the midbrain, the pons, and the medulla oblongata. Each of the three components has its own unique structure and function. Together, they are responsible for regulating the vital functions of life, including heart rate, breathing, consciousness, blood pressure, and sleep. Even though it is not directly involved in providing the higher intelligence, the brainstem does carry all of the information to and from the areas associated with higher intelligence. Out of 12 cranial nerve pairs, 10 nerve pairs originate from the brainstem (Fig. 2.15).

Midbrain

The midbrain, which is also called the mesencephalon, is the topmost part of the brainstem that connects the pons and cerebellum with the cerebrum. The surface of the upper posterior portion of the midbrain, which is called the tectum meaning "roof," is covered with four bumped structures in two pairs, the superior and inferior colliculi. The superior

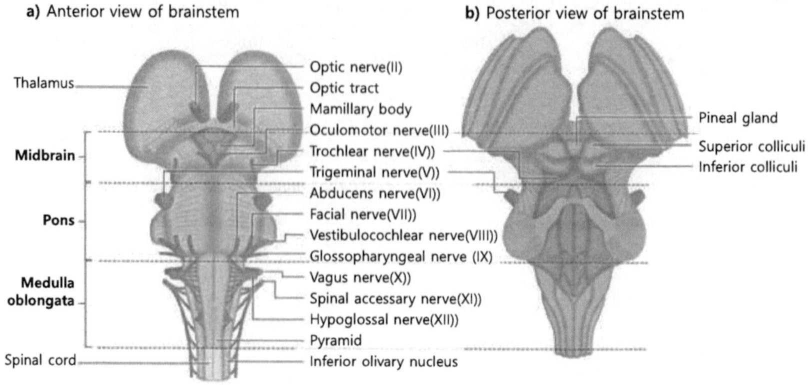

Fig. 2.15 Organization of brainstem. (**a**) Anterior view, (**b**) Posterior view

colliculi are involved in eye movements and visual processing, and the inferior colliculi are involved in auditory processing. The substantia nigra is another important nucleus that appears darker than neighboring areas due to high levels of neuromelanin in dopaminergic neurons. It produces dopamine which affects the many neuronal functions ranging from movement control, cognitive executive functions, and emotional activity. In Parkinson's disease, the hallmark of movement dysfunction comes with the neurodegeneration in the substantia nigra. Among the cranial nerves, the oculomotor nerve (III) and trochlear nerve (IV) have nuclei in the midbrain and emerge from the midbrain.

Pons

The pons, which means a bridge in Latin, is situated in the middle part of the brainstem and in front of the cerebellum. The pons includes neural pathways and tracts that run from the cerebrum down to the medulla, cerebellum, and spinal cord, as well as tracts that travel up into the thalamus. The pons contains autonomic nuclei important for regulating breathing. It also contains the nuclei for cranial nerves from V to VIII, the trigeminal (V), abducens (VI), facial (VII), and vestibulocochlear (VIII), which are involved in sensations of touch, pain, and temperature from the face and head, in motor commands dealing with eye movement, chewing, and facial expressions, and in other functions, including swallowing, tear production, hearing, and maintaining balance.

Medulla Oblongata

The medulla oblongata, also called the medulla, is the lowest part of the brainstem as well as of the brain. It does a critical role in transmitting signals between the spinal cord and the higher parts of the brain. In the anterior region of the medulla oblongata, there shows two longitudinal ridges called pyramids, which contain motor projection tracts. These pyramids cross to the opposite side in the posterior region of the medulla oblongata. By this crossing, motor responses of each part of the body are controlled by the hemisphere of the opposite side. A nucleus called inferior olive, in the inferior to each pyramid, relay information on proprioception coming from the spinal cord and going into the cerebellum. It contains the cardiac, respiratory, vomiting, and vasomotor centers, and is involved in the regulation of the autonomic functions of heart rate, respiration, blood pressure, and sleep-wake cycle. It also contains several nuclei for cranial nerves from IX to XII, the glossopharyngeal (IX), vagus (X), spinal accessory (XI), and hypoglossal (XII) nerves.

3.5 Spinal Cord

The spinal cord is another major organ of the central nervous system. It is continuous with the brain and is the main pathway for transmitting sensory information from the peripheral part of the body to the brain, and motor information from the brain to the periphery. It is also a center for coordination of reflexes, as well as control of reflexes independently from the brain. Along the spinal cord, spinal nerves emerge that carry both sensory and motor information.

Structure of the Spinal Cord

The length of the spinal cord is divided into four regions that correspond to the regions of the vertebral column, cervical, thoracic, lumbar, and sacral regions (Fig. 2.18). The cervical region of the spinal cord is immediately inferior to the brainstem and it contains motor neurons and their axons together with sensory axons which are extending into cervical spinal nerves. The thoracic region is the longest region inferior to the cervical region and it contains neurons for the thoracic spinal nerves. The lumbar region is a short segment inferior to the thoracic region and it contains neurons for the lumbar spinal nerves. The sacral region is the final region inferior to the lumbar region and contains neurons for the sacral nerves.

The spinal cord is contained in a longitudinal cavity of the vertebral column. It is covered and protected by the three layers of the spinal meninges, similar to the brain. Like the brain, cerebrospinal fluid (CSF) encloses the spinal cord within the meninges for protection and supply. The vertebral column is the stack of vertebrae, a spine bone, on top of each other from the pelvic bones to the skull. While the average person is born with 33 vertebrae, some vertebrae at the bottom end of the vertebral column fuse together during growth, and most adults have only 24 vertebrae. There are 7 vertebrae in the cervical column (C1–C7), 12 vertebrae in the thoracic column (T1–T12), and 5 vertebrae in the lumbar column (L1–L5). The spinal cord is not contained in the full length of the vertebral column because the spinal cord does not grow significantly longer after the first or second year, but the skeleton continues to grow. The spinal cord ends between the first and second lumbar vertebrae. The spinal cord and vertebral column together are collectively called the spine.

Cross-Sectional View of the Spinal Cord

The cross-section of the spinal cord is divided into two parts of gray matter and white matter (Fig. 2.16). In contrast to the cerebral cortex of the brain, gray matter resides inside and white matter outside of the spinal cord volume. The gray matter is the part where the most of neural bodies and synapses are populated and the white matter is the part where the myelinated axons of sensory and motor nerves go up and down. The gray matter has an overall shape of "H" and is subdivided into regions that are referred to as horns, anterior, central, and posterior horns.

The anterior horn, also called the ventral horn, is the region of the neurons for motor control and contains the cell bodies of somatic motor neurons that control skeletal muscles. The somatic motor neurons are the largest neurons of the body and their axons cause the contraction of skeletal muscles. Axons of the motor nerves exit the spinal cord through the anterior (ventral) nerve root. The lateral horn is only in the thoracic, upper lumbar, and sacral regions. It contains the cell bodies of autonomic motor neurons which innervate cardiac muscle, smooth muscle, and glands for autonomic nervous control. The posterior horn, also called the dorsal horn, is responsible for sensory processing and contains the axons of sensory neurons, the cell bodies of interneurons, unmyelinated axons, and their synaptic connections. The axons of sensory neurons are coming in a bundle through the posterior (dorsal) nerve root from the posterior root ganglion where the cell bodies of sensory neurons are located. As the anterior and posterior nerve roots fuse together to form the spinal nerve, each spinal nerve contains mixed fibers sending sensory or motor information respectively.

Like gray matter, the white matter of the spinal cord is subdivided into the anterior, lateral, and posterior columns by the "H" shaped gray matter. These columns of white matter extend longitudinally along the length of the spinal cord. Through these columns, ascending tracts of sensory fibers carry sensory information up to the brain, and descending tracts of motor fibers carry motor commands from the brain.

The anterior column is the region bounded by the two anterior horns and the gray matter. The lateral column is the white matter on either side of the spinal cord between the posterior horn and the anterior horn. The posterior column is the region bounded by the two posterior horns and the gray matter. The anterior and lateral columns are composed of many different groups of axons of both ascending and descending tracts while the posterior columns are composed of axons of ascending tracts.

4 Peripheral Nervous System

The peripheral nervous system (PNS) is the other part of the nervous system excluding the central nervous system (CNS) composed of the brain and the spinal cord. PNS includes the nerves which connect the CNS to the rest of the body for transferring information, their roots and branches, peripheral nerves, and neuromuscular junctions. Unlike the CNS, the PNS is not protected by the bonny structure like the skull and vertebral column and by the blood-brain barrier, which makes PNS vulnerable to mechanical injuries and circulating toxins.

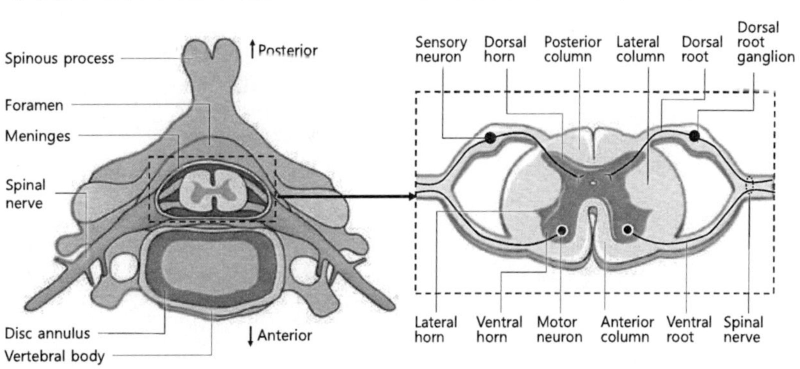

Fig. 2.16 Cross-sectional view of vertebra and spinal cord. (a) Vertebra, (b) Spinal cord

Nerves in the peripheral nervous system can be classified into cranial nerves and spinal nerves by the location where they exit the CNS. Twelve pairs of the cranial nerves exit from the brain and the 31 pairs of the spinal nerves exit from the spinal cord. These cranial nerves and spinal nerves are classified into sensory nerves and motor nerves by the direction of the signal transmission. The sensory nerves are afferent nerves transmitting the sensed signals from the periphery to the brain directly or via the spinal cord. The motor nerves are efferent nerves transmitting the motor commands from the brain to the periphery directly or via the spinal cord. Motor nerves in the PNS can be classified into somatic nerves and autonomic nerves by the involvement of the intention for the action. The somatic nerves are associated with conscious or voluntary activities and carry command messages throughout the body to initiate and control movement. The autonomic nerves are responsible for coordinating involuntary behaviors of the body like heart rate, respiration, and digestion. Autonomic nerves transmit the messages that take place without conscious thought. Autonomic nerves are subclassified into the sympathetic and parasympathetic divisions. The sympathetic division is activated to induce a so-called fight-or-flight response, during times when the body needs to respond to stress and threatening stimuli. The parasympathetic division is activated after threatening stimuli are gone and during the relaxing time to maintain normal bodily functions by decreasing the responses increased during the urgent circumstance.

4.1 Cranial Nerves

The cranial nerves are emerging from the brain and are composed of symmetric 12 pairs (Fig. 2.17). These 12 cranial nerves are designated CN I through CNXII based on the anatomical exiting order in the inferior view of the brain, from anterior to posterior. Except for one, 11 cranial nerves are primarily responsible for the sensory and motor functions of close regions to the brain including the head and neck. The tenth cranial nerve, the vagus nerve, only goes far beyond the head and neck region to the target organs in the thoracic and abdominal cavities as part of the parasympathetic nervous control.

Among 12 cranial nerves, three cranial nerve pairs are totally sensory. Axons of these nerves originate from the sensory ganglia, which is a group of neuron cell bodies, external to the cranium and enter the brain to connect in a nucleus. Five cranial nerve pairs are solely motor nerves with their nuclei within the brainstem and their axons connect to the skeletal muscles of the head or neck. The remaining four cranial nerve pairs are mixed nerves. In a nerve bundle, some nerve fibers convey the sensory information and the other fibers transmit the motor commands.

Among 12 cranial nerve pairs, four of them are involved in the autonomous control of the body. The tenth cranial nerve, the vagus nerve, is totally autonomous nerve primarily targeting the autonomic ganglia in the thoracic and upper abdominal regions for parasympathetic activation. Three cranial nerves are partly autonomous by containing some nerve fibers transmitting signals for autonomous control. The third cranial nerve, the oculomotor nerve, autonomously controls pupillary constriction. The seventh cranial nerve, the facial nerve, and the ninth cranial nerve, the glossopharyngeal nerve, regulate the salivation and lacrimation autonomously (Table 2.2).

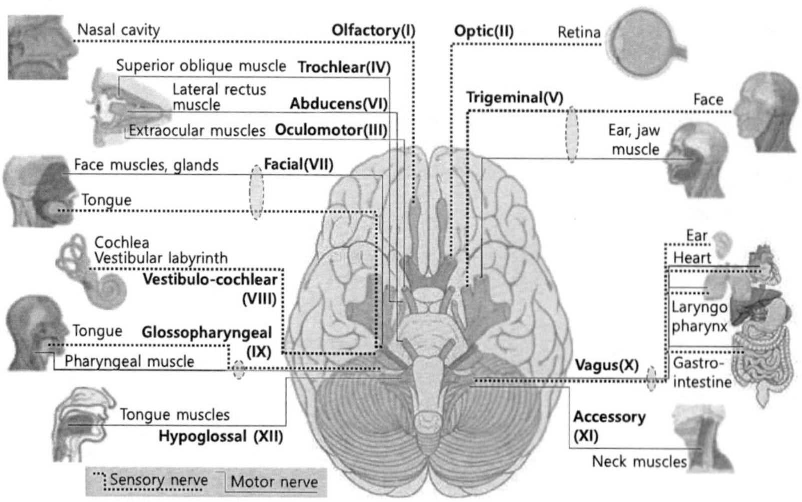

Fig. 2.17 Connections of 12 cranial nerves

Table 2.2 Functions of 12 cranial nerves

No.	Name	Type	Function	Central connection	Peripheral connection
I	Olfactory	Sensory	Sense of smell	Olfactory bulb	Nasal cavity in nose
II	Optic	Sensory	Ability to see	Lateral geniculate body in thalamus	Retina in eyes
III	Oculomotor	Motor, partly autonomous	Move, focus & blink eyes, pupil response	Oculomotor nucleus in midbrain	Extraocular muscles in eyes
IV	Trochlear	Motor	Move eyes up & down, back & forth	Trochlear nucleus in midbrain	Superior oblique muscle in eyes
V	Trigeminal	Sensory & Motor	Sense in face & neck, jaw movement	Trigeminal nuclei in brainstem	Face, jaw & ear muscles
VI	Abducens	Motor	Move eyes to look side	Abducens nucleus in pons	Lateral rectus muscle in eyes
VII	Facial	Sensory & Motor, partly autonomous	Sense taste, facial expression	Facial nucleus in pons	Tongue, facial muscles, glands in head & neck
IIX	Vestibulocochlear	Sensory	Sense hearing & balance	Cochlear nucleus, vestibular nucleus in pons & medulla	Cochlea & vestibular labyrinth
IX	Glossopharyngeal	Sensory & Motor, partly autonomous	Taste & swallow	Solitary nucleus, inferior salivatory nucleus in medulla	Pharyngeal muscles
X	Vagus	Sensory & motor, fully Autonomous	Digestion and heart rate	Medulla	Ear, throat, tongue, thoracic & upper abdominal organs
XI	Accessory	Motor	Neck & shoulder movement	Nucleus ambiguus in medulla	Neck muscles
XII	Hypoglossal	Motor	Tongue movement	Hypoglossal nucleus in medulla	Tongue muscles

4.2 Spinal Nerves

The spinal nerves leave CNS from the spinal cord. All spinal nerves are mixed nerves of sensory nerve and motor nerve. These sensory and motor nerves include autonomic nerve fibers. The axons of the spinal nerves enter spinal cord via separated two roots. The axons of the sensory nerves enter the spinal cord via dorsal root and the axons of the motor nerves exit the spinal cord via ventral root.

There are 31 pairs of spinal nerves, and they are grouped and named according to the location from which each nerve exits the spinal cord (Fig. 2.18). Eight pairs of spinal nerves exiting from the cervical region of the vertebral column are named cervical nerves and designated C1–C8. Twelve pairs of thoracic nerves designated T1–T12 are exiting from thoracic region and five pairs of lumbar nerves exiting from the lumbar region are designated L1–L5. Five pairs of sacral nerves designated S1–S5 exit from the sacrum at the lower end of the vertebral column. The roots of the sacral nerves begin inside the vertebral column at the level of the L1 vertebra, and then descend into the sacrum before exiting the vertebral column. Finally, one pair of coccygeal nerve designated C0 arises from the end of the spinal cord and exits the vertebral column through the sacral canal.

Spinal Plexus

Each of nerve of the spinal nerves and the cranial nerves is not a single nerve fiber originating from a single neuronal cell body. It is actually a bundle of axonal fibers originating from many neuronal cell bodies populated closely and having same direction for the destination. On average, the ventral root of S5 has 906 nerve fibers, which is the smallest number in spinal nerve root, within a diameter of 0.14 mm. And, the dorsal root of C7 contains 39,653 nerve fibers, which is the largest number in spinal nerve root, within a diameter of 4.61 mm (Liu et al., 2015). Even though their directions may be generally same, their detailed final target positions are different. So, each of the axonal fibers should be deviated from the nerve bundle and follow their unique own ways. Nerves are similar to highways where many cars go through together. Each car should exit the highway and follow some street and road, and finally go through the small road leading to its final destination. Highways meet at junctions or interchanges and flow of cars is rearranged depending on their destinations. Nerves also go similar to junctions in plexus.

A plexus is a bundle of interchanging and intertwining nerves with no associated cell bodies. When the spinal nerves from different spinal levels extend outward from the vertebral column to innervate the periphery, they do not go straightly to the periphery. But their composing axonal fibers are rearranged to follow different courses by passing the spinal plexus and form the individual nerves more distally. There are four nerve plexuses along the vertebral column. They are cervical, brachial, lumbar, and sacral plexus.

The cervical plexus is composed of axons from spinal nerves C1 through C5 and branches into individual nerves in the posterior neck and head region including the phrenic nerve, which connects to the diaphragm at the base of the

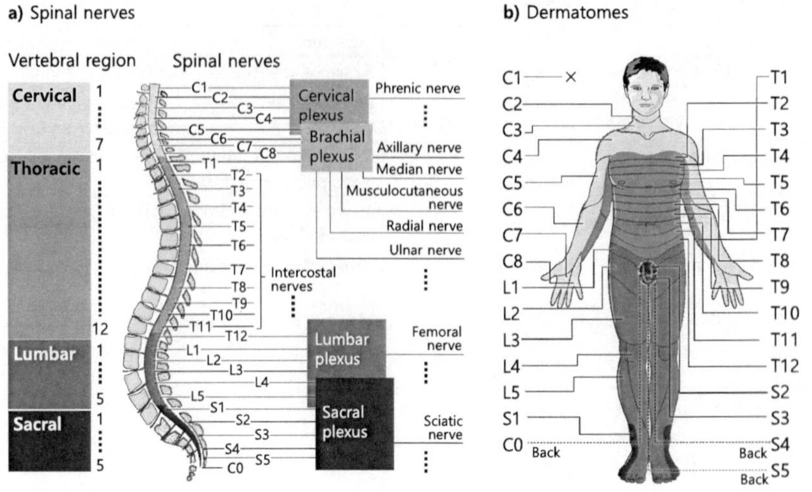

Fig. 2.18 Spinal nerves and their dermatomes. (**a**) Spinal nerves. (**b**) Dermatomes. C1-C8: Cervical nerves, T1–T12: Thoracic nerves, L1–L5: Lumbar nerves, S1–S5: Sacral nerves, C0: Coccygeal nerve

thoracic cavity and regulates respiration. Spinal nerves C5 through T1 reorganize through the brachial plexus to the individual nerves of the arms. As the name implies, this plexus forms the nerves of the arms. With some splitting and converging, finally five systemic nerves, the axillary, median, musculocutaneous, radial and ulnar nerves, branch out the brachial plexus. They innervate separately for muscular contraction and sensations in different region of the upper limb including, arm, forearm, hand and fingers. The lumbar plexus arises from the spinal nerves T12 and L1 to L5 of the lumbar region and forms individual nerves innervating the anterior leg and the pelvic region. The femoral nerve is one of the major individual nerves from the lumbar plexus, which innervates anterior thigh muscles and receives sensation from the thigh. The sacral plexus arises from the spinal nerves L4 and L5 from the lumbar region and S1–S4 from the sacral region of vertebral column. The sciatic nerve is one of the major nerves formed in sacral plexus, which is the one of the largest nerves of the body. It extends across the hip joint and innervates the muscles and receives sensation of the thigh and the leg. Spinal nerves T2 through T11 of the thoracic region do not rearrange the nerves by the plexuses, but they emerge and form the intercostal nerves. The intercostal nerves innervate the intercostal muscles and receive sensation from the skin of the thoracic region. The spinal nerves T7–T12 also innervate the abdominal muscles and receive sensation from the skin overlying the abdominal region.

Dermatomes

The spinal nerve contains sensory nerve fibers and motor nerve fibers. Motor nerve fibers reach to the motor endplates at the end and sensory nerve fibers finally reach to the receptors which sense the stimuli. As most of the receptors are populated in the skin to sense the external stimuli, each skin area is topographically allocated to the spinal nerves for the sensation. A dermatome is an area of skin that is supplied by a single spinal nerve. Dermatomes are the collection of skin areas topographically arranged according to the connected spinal nerves and tell the skin area allocated for each of 30 spinal nerves except the spinal nerve C1 (Fig. 2.18). Dermatomes are important in identifying the potential damage to spinal nerves. By testing the sensation of the dermatome of a specific area, damage to the connected spinal nerve can be evaluated.

Spinal Reflex

The spinal nerves function under the supervision of the brain in most of information processing. All the sensed information is conveyed to the brain and the muscles in the periphery respond by the commands from the brain. However, in a spinal reflex, the information is processed and executed independently within the spinal cord before the commands are delivered from the brain.

The spinal cord is phylogenetically older structure than the brain, and it means that spinal processing of reflex carries more fundamental functions for survival. For example, when a subject touches a hot surface, extracting the hand from the surface is more fundamental for the survival than recognizing and speaking the hotness of the surface. So, many of the fundamental responses for survival are remained within the spinal reflexes.

The spinal reflex is involuntary as it occurs before the brain catch the information. And, the basic unit of the spinal reflex is the reflex arc, which is consisted of a sensory neuron and a motor neuron for effector organ. In a monosynaptic reflex, stimuli sensed in a sensory neuron are conveyed to a motor neuron in the ventral horn of the spinal cord. By a single synaptic connection between the two neurons, information is transferred to the motor neuron and the response can be executed according to the processed result in the single motor neuron. In a polysynaptic reflex, interneurons are intervened between the sensory and motor neurons, and the information is transferred through the multiple synaptic connections between the neurons. Representative example of the monosynaptic spinal reflex is the knee-jerk reflex, also called the patellar reflex. It is a sudden kicking response of the lower leg in response to a sharp tap on the patellar tendon. When the tendon organ of the quadriceps muscle is excessively or suddenly stretched by any stimulation, sensory neuron having the nerve ending in the tendon organ senses the force applied and transfers sensed information to the motor neuron, which is in the lumbar segments of the spinal cord. Motor neuron processes the transferred information and if the contraction of the quadriceps is eminent to avoid injury, it generates and transmits impulses to contact the quadriceps. By simply testing the reflex, the response provides information on the healthiness of the spinal cord including sensory and motor neurons associated.

4.3 Sympathetic and Parasympathetic Divisions

The autonomic nervous system regulates body organs unconsciously and involuntarily in contrast to the somatic nervous system which regulates the body organs voluntarily under awareness. It makes the body respond to external and visceral stimuli reflexively, while the subjects do not aware. The autonomic nervous system innervates smooth and cardiac muscle of the target organs as well as glandular tissues while the somatic nervous system innervates the skeletal muscles of target effectors (Fig. 2.19).

Preganglionic and Ganglionic Neurons

Another difference between the two nervous systems is the difference in the number of neurons involved in the transmission of the signal to the target organs. When the motor signals are transmitted from the brain to the periphery, the upper and the lower motor neurons are involved. The upper motor neurons originate in the cerebral cortex and travel down to the lower motor neurons in the brainstem or spinal cord, and the lower motor neurons extend from the CNS to reach the targets. In the somatic nervous system, a single lower motor neuron of the spinal cord extends from the CNS toward the skeletal muscles. These neurons have myelinated axons of large diameter and release acetylcholine at neuromuscular junctions. In comparison, the autonomic nervous system is composed of a chain of two lower motor neurons. The cell body of the first lower motor neurons resides in the brainstem or spinal cord and is called a preganglionic neuron. The axon of the preganglionic neuron exits the CNS as a nerve fiber within the cranial or spinal nerves. The preganglionic neuron is the neuron before the ganglion, which is a group of neuron cell bodies in the peripheral nervous system. Preganglionic neurons connect to the ganglionic neurons, the second lower motor neuron, in the autonomic ganglion. And, the postganglionic fibers, which are the axons of the ganglionic neuron, extend finally from the ganglion to the target effectors. Preganglionic neurons have myelinated axons of small diameter and release acetylcholine to excite a second lower motor neuron. Ganglionic neurons have unmyelinated axons and release either acetylcholine for excitation or norepinephrine for inhibition depending on the type of receptors on the effectors. As the preganglionic and postganglionic axons are thinly myelinate or unmyelinated, the propagation of

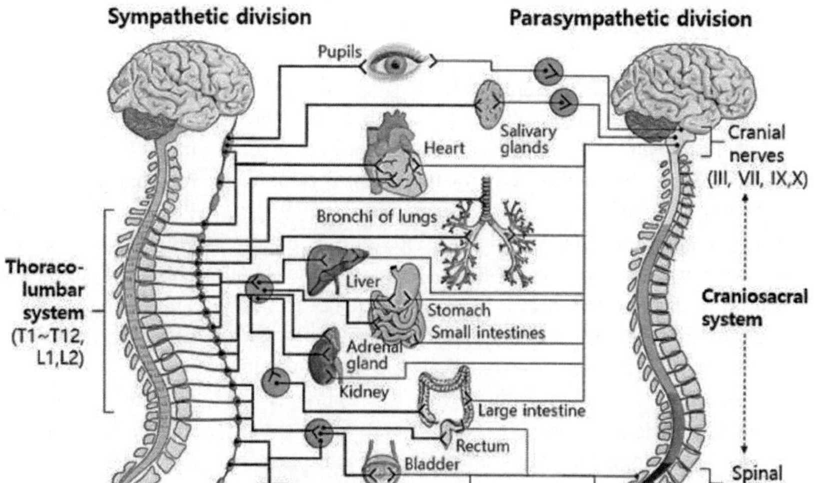

Fig. 2.19 Sympathetic and parasympathetic connection of body organs

autonomic messages is slower compared to the somatic motor commands.

Sympathetic Division

Most of the autonomic target effectors are dually innervated by both divisions of the autonomic nervous system, sympathetic and parasympathetic divisions. In the heart, both divisions innervate their ganglionic neurons to the cardiac muscle. The sympathetic connection increases the heart rate and the parasympathetic connection decreases the heart rate. However, there are anatomical differences between the two divisions in the location of preganglionic neurons, the location of ganglions, and the length of the preganglionic and postganglionic axons.

The sympathetic division of the ANS emerges from the first thoracic (T1) through the second lumbar (L2) spinal segment and it is also called the thoracolumbar system. Cell bodies of the sympathetic preganglionic neurons reside in the lateral horns of the spinal cord and provide axons reaching 23 ganglia on either side of the vertebral column through the ventral roots. The sympathetic ganglia of 23 pairs are vertically connected via the sympathetic trunk of the ganglia. To branch toward 10–20 targets, preganglionic sympathetic fibers make synapses with ganglionic neurons at the same level or in the upper or lower level through the ganglia trunk. After synapses, postganglionic axons travel to their destination in one of the organs. As the sympathetic ganglia are positioned adjacent to the vertebral column, axonal fibers of the preganglionic sympathetic neurons are relatively short, and they are myelinated. In comparison, postganglionic fibers of the sympathetic division are unmyelinated and long to travel the distance from the center to the target effectors in the periphery.

Among the 23 pairs of ganglia, three cervical ganglia innervate structures of the head, neck, smooth muscle of arteries, and heart. Preganglionic sympathetic axons extending from T5–L2 form splanchnic nerves and terminate in three autonomic ganglia called prevertebral ganglia to control the organs in the abdominal cavity, and the enteric nervous system. Preganglionic sympathetic axons from T5–T9 form splanchnic nerves and terminate in the prevertebral celiac ganglia and their axons innervate the stomach, abdominal blood vessels, liver, gallbladder, pancreas, and small intestine. The splanchnic nerves originate from T9–T11 spinal nerves project to prevertebral superior mesenteric ganglia, and their postganglionic axons innervate the large intestine. The splanchnic nerves extending from T12 spinal nerves project to and terminate in the prevertebral renal ganglia and their postganglionic axons innervate the pancreas, the small intestine, the large intestine, the kidneys, and ureters. Lumbar splanchnic nerves extending from L1–L2 spinal nerves terminate to the inferior mesenteric ganglia and their postganglionic axons innervate the large intestine, rectum, kidneys, bladder, gonads, and external genitalia. Small sacral splanchnic nerves originating from the sacral sympathetic ganglia terminate into urinary and reproductive organs.

Parasympathetic Division

The parasympathetic division of the autonomic nervous system has its preganglionic neurons in the brainstem or in the lateral horns of the sacral region of the spinal cord. So, the parasympathetic division is also called the craniosacral system.

Like the sympathetic division, the motor connection is composed of the preganglionic and ganglionic neurons in the parasympathetic division. The axonal fibers of parasympathetic preganglionic neurons from the cranial region travel in cranial nerves, while the axonal fibers from the sacral region travel in spinal nerves like the fibers in the sympathetic division. However, they travel a longer distance to the periphery and terminate in the small terminal ganglia located near the target organs or intramural ganglia that are embedded within the walls of the target organs. These ganglia receiving the preganglionic connection through the cranial nerves are found in the head, neck, thoracic cavity, and upper abdominal cavity. Parasympathetic ganglia in the lower abdomen and pelvic cavity receive preganglionic connection through the sacral portion of the spinal nerves. Compared to the preganglionic nerves in the sympathetic division, preganglionic fibers of the parasympathetic division travel longer distance to reach the terminal ganglia around the target organs, and postganglionic fibers travel only a short distance.

Among the cranial nerves, the oculomotor nerve (III), the facial nerve (VII), the glossopharyngeal nerve (IX), and the vagus nerve (X) are involved in the parasympathetic division. Their preganglionic neurons emerge from the nuclei in the brainstem. The preganglionic fibers in the oculomotor nerve (III) travel to the ciliary ganglion in the posterior orbit of the eyes to control the size of the pupil and the shape of the lens. Parasympathetic preganglionic fibers of the facial nerve (VII) travel to the terminal ganglia to control the secretions of the lacrimal apparatus, nasal epithelium, and salivary glands. Parasympathetic preganglionic fibers of the glossopharyngeal nerve (IX) nerve exits the medulla oblongata and travel to secrete the parotid salivary glands. The majority of parasympathetic preganglionic fibers start from the medulla oblongata and travel through the vagus nerve (X) to the terminal and intramural ganglia of target effectors. Their postganglionic fibers innervate the organs in the thoracic and abdominal cavity including the heart, airways, esophagus, stomach, liver, gallbladder, pancreas, small and large intestine, kidneys, ureters, and gonads. The parasympathetic preganglionic fibers traveling through the spinal nerves emerge from the lateral horns of the sacral

Table 2.3 Sympathetic and parasympathetic effects of body organs

Organ/function	Sympathetic effect	Parasympathetic effect
Overall	Fight of flight	Rest and digest
Pupils	Dilation	Constriction
Lens	Far focus	Near focus
Hairs	Erection	No innervation
Heart	Increase rate & blood pressure	Decrease rate & blood pressure
Arteries to skeletal muscle	Dilation	No innervation
Veins	Variable by neurotransmitters	No innervation
Lung	Dilate respiratory passages	Constrict respiratory passages
Gastrointestinal	Decrease motility Inhibit digestive secretion	Increase motility, stimulate digestive secretion
Glycogen to glucose conversion	Increase	Not involved
Adrenalin gland	Release adrenaline	No innervation
Sweat glands	Increase activity	No innervation
Salivation	Inhibition	Stimulation
Urinary bladder	Relaxation	Contraction
Genitalia	Ejaculation	Erection

S2–S4 segments. These axons form the pelvic splanchnic nerves and project to the terminal or intramural ganglia to activate the distal portion of the large intestine, rectum, urinary bladder, and most of the reproductive organs (Table 2.3).

References

Gazzaniga, M. S. (2014). The split-brain: Rooting consciousness in biology. *Proceedings of the National Academy of Sciences, 111*(51), 18093–18094. https://doi.org/10.1073/pnas.1417892111

Gershon, M. D. (2005). Nerves, reflexes, and the enteric nervous system. *Journal of Clinical Gastroenterology, 38*, S184–S193.

Herculano-Houzel, S. (2009). The human brain in numbers: A linearly scaled-up primate brain. *Frontiers in Human Neuroscience, 3*, 3. https://doi.org/10.3389/neuro.09.031

Kim, M. S., Tak, H. J., & Son, S. M. (2014). Recovery of cerebellar peduncle injury in a patient with a cerebellar tumor: Validation by diffusion tensor tractography. *Neural Regeneration Research, 9*(21), 1929–1932. https://doi.org/10.4103/1673-5374.145364

Lanciego, J. L., Luquin, N., & Obeso, J. A. (2012). Functional neuroanatomy of the basal ganglia. *Perspectives in medicine, 2*(12), a009621.

Liu, Y., Zhou, X., Ma, J., Ge, Y., & Cao, X. (2015). The diameters and number of nerve fibers in spinal nerve roots. *The Journal of Spinal Cord Medicine, 38*(4), 532–537. https://doi.org/10.1179/1079026814Z.000000000273

Mazzasette, W. M., Chiara, J. J., & Nguyen, K. L. (2021). *Brain-Cerebrum*. ASCCC Open Educational Resources Initiative.

Simons, J. S., & Spiers, H. J. (2003). Prefrontal and medial temporal lobe interactions in long-term memory. *Nature Reviews Neuroscience, 4*, 637–638.

Van De Graaff, K. M., & Fox, S. I. (1999). *Concepts of human anatomy & physiology*. WCB/McGraw-Hill.

Waxenbaum, J. A., Reddy, V., & Varacallo, M. (2022). *Anatomy, autonomic nervous system*. StatPearls Publishing.

Wolf, U., Rapoport, M. J., & Schweizer, T. A. (2009). Evaluating the affective component of the cerebellar cognitive affective syndrome. *Journal of Neuropsychiatry and Clinical Neurosciences, 21*(3), 245–253. https://doi.org/10.1176/jnp.2009.21.3.245

3 Neurons at Rest

Humans are electrical beings as well as biological beings. Human are managed by the electricity of the nervous system by themselves. The nervous system is based on the electrical signals in its processing, transmission, and control. As humans are electrical beings, their behaviors and activities can be measured and understood in electrical maneuvers. Practically, electrical signals from the human body, which are called biological signals, are actively used for these purposes mainly to find any abnormalities of the body in the medical domain. Also, as humans are electrical beings, they can be manipulated by the electrical signal and energy from the external side. Currently, there are many methods that apply electricity to change the electrical behavior within the body. Hearts can be paced by the electrical pulses, and the brain can be modulated by the electrical signals to reduce abnormal movements. Humans can be sensed by electricity and modulated by the electricity from the external side. In many aspects, humans can be thought like a computer system that is also controlled by electricity. As all the entities having the electrical touches are connected by the electrical communication, humans also can be part of the electrical connection. Humans can be connected with the outer world smoothly by electricity in both directions bypassing their own interfacing systems of sensory and motor systems. As technologies are advanced, humans are expected more smoothly be connected to the world by electricity. At the base of the human electricity is the nervous system. And, the nervous system is built based on the constructing block of the neuron.

1 The Neuron

The neuron is the basic cell of the nervous system that is electrically excitable. As the transistor is the basic element of the central processing unit (CPU) of a computer, neurons compose the brain and the nervous system. The nervous system includes about 86 billion neurons and an equal number of glial cells. Instead of participating in the neuronal processing directly, the glial cells work to support, nourish, and insulate neurons and remove the waste products of metabolism. However, the total number of cells in the nervous system including the neurons and the glial cells is less than 1.0% of the whole number of cells in the human body, which is estimated to be about 37.2 trillion cells. The relatively small number of neurons in the human body does an important role in information processing, controlling the body, and making the human life human-like.

1.1 Structure of a Neuron

A neuron is composed of three parts, the dendrite, soma, and axon (Fig. 3.1).

Dendrites are the part extruding from the soma like a tree branch and usually receive information from other neurons via synapses. Dendrites generate postsynaptic potentials (PSP) and transmit these electrical signals toward the soma area. While the length of the dendrites is generally shorter than the axon, their lengths are depended on the function of each neuron. In the central nervous system, neurons have long and complex dendrites to receive signals from many other neurons. For instance, the Purkinje cells in the cerebellum have highly developed dendrites to receive signals from thousands of other nerve cells.

The soma is main the body part of the neuron doing the essential function of the neuron. The soma is the part where a cell nucleus and other cellular components are contained just like other types of cells. The soma collects all information received through the dendrites and processes them to generate the output of the neuron. And, the soma generates an action potential if the processed result of potential crosses the threshold at the hillock, which is the position of the soma indicating the entry part to the axon.

The axon is the long cable-like part of the neuron and is also called a nerve fiber. It does the role of transmitting the action potentials generated at the hillock to the terminal part

K. S. Park, *Humans and Electricity*, https://doi.org/10.1007/978-3-031-20784-6_3

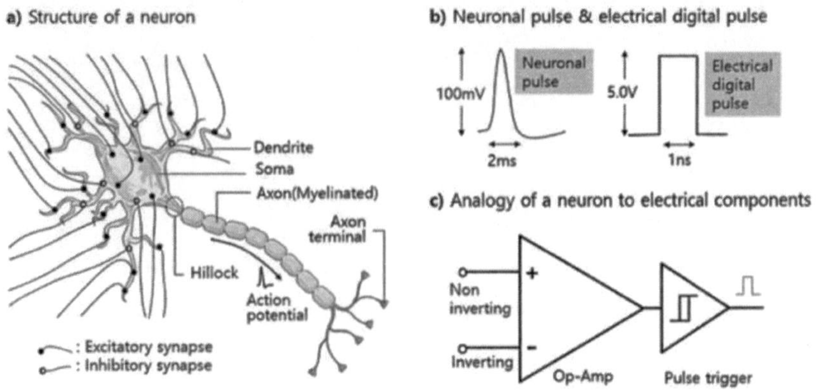

Fig. 3.1 Structure of a neuron and its analogy to electrical circuit components. (**a**) Structure of a neuron, (**b**) Comparison of neuronal pulse and electrical digital pulse in amplitude and in pulse width, (**c**) Analogy of a neuron to electrical components

of the axon, where the signals are conveyed to other neurons via synaptic connections. The length of the axon is variable from 0.1 mm to 1 m. Axons running from the brain to the spinal cord and from the spinal cord to the foot reach almost 1.0 m in their length. There are two types of axons, unmyelinated and myelinated axons. The surface of the unmyelinated axon is covered by the simple membrane as in the other parts of the neuron. However, the surface of the myelinated axon is covered by a myelin sheath throughout its length. Myelin sheath increases the insulation of the axonal membrane and results in increasing the conduction velocity of the electrical signal through the axon. In electrical cable, increasing the insulations of the cable is also most important in increasing the transmission rate of signals through the cable. The axon generally branches off to increase the number of contact areas of synaptic junctions (Fig. 3.1).

1.2 Types of Neurons

Types of neurons can be classified by their functions and their anatomical structure.

1.2.1 Functional Types of Neurons

Based on their function, neurons can be classified into sensory, motor, and interneurons.

Sensory neurons, also called afferent neurons, convert a specific type of stimulus into action potentials and transmit the action potential to the brain. Different types of external or internal stimuli are sensed by the specific receptors for each stimulus and converted into action potentials. In some types of senses, stimuli are sensed first with graded potentials and converted into a series of action potentials in another associated neuron. Sensory neurons of special sensing organs for vision, sound, taste, and smell extend their fibers to the brain through the cranial nerves. Most somatosensory information like pain, touch, temperature, and muscle tension travels afferently from the periphery to the brain via the spinal cord. The sensory neuron transmits this information in a series of action potentials from the periphery to the spinal cord and makes synaptic connections to interneurons which relay the information to the neurons in the brain. Sensory neurons are structurally unipolar neurons, and their bodies are located in the dorsal ganglia of the spinal cord.

Motor neurons, also called efferent neurons, transmit the command messages from the brain to the effector organs including the muscles and glands. The command messages from the brain are transmitted via two types of motor neurons, the upper motor neurons, and the lower motor neurons. Axons of the upper motor neuron extend from the motor cortex of the cerebrum to the brainstem or spinal cord and make synapses with interneurons or directly with the lower motor neuron. Axons of the lower motor neurons are extended from the brainstem or spinal cord to the effector organs in the periphery. Motor neurons are multipolar type neurons. Neuronal bodies of the upper motor neurons are located in the motor cortex and neuronal bodies of the lower motor neuron are located in the brainstem or in the ventral horn of the spinal cord. A single motor neuron innervates a single motor unit which is composed of many muscle fibers. A motor neuron connects to each muscle fiber within a motor unit via neuromuscular junction and transmits many action potentials through the muscle fiber for a single muscle twitch.

Interneurons are the most abundant neurons in the nervous system. Interneurons do not sense the stimuli and are not connected to effector organs. Interneurons connect the neurons in the central nervous system (CNS). Relay interneurons have long axons to relay messages received from the neurons in one region to the neurons in other regions of the CNS. Local interneurons have short axons and make connections with neighboring neurons to process received information. The primary function of local interneurons is integration. When the response to the stimulus becomes complicated, more interneurons are involved in the connection. The interaction between interneurons makes the brain do all higher functions including cognition, learning, and decision making. Most of the interneurons are multipolar of different sizes and have different patterns of fiber branches.

1.2.2 Structural Types of Neurons

Despite the basic similarity of neurons, there is wide variability in details of their shapes and sizes. Their structure can be classified by the number of main processes from the cell body (Fig. 3.2). The unipolar neuron, also called a monopolar neuron, has only a single extension from the cell body. This extension divides into two branches representing the axons and oriented in opposite directions. One branch is the receptive pole, and the other is the output zone. The unipolar neurons are mostly populated as sensory neurons transmitting the somatosensory sensations from the periphery to the brain. For example, unipolar neurons transmit touch information from the surface of a hand to the dorsal horn of the spinal cord.

Bipolar neurons have two extensions from the cell body. One part of the extension is a dendrite of the neuron receiving the signals from the other cells, and the other extension is the axon transmitting the processed signal away from the body to connect to the other neurons. Bipolar cells are mostly found in the receptive layers of the retina of the eyes.

The multipolar neuron is the representative type of the neuron. The majority of human neurons are multipolar having multiple dendritic projections from the cell body and a long axon as well. Multipolar neurons receive signals from multiple neurons via dendrites. The pattern of the dendritic process to collect many synaptic connections is characteristic of the multipolar neuron. The Axon of the multipolar neuron transmits the output of the neuron as far as it reaches. The motor neurons and interneurons belong to a type of multipolar neuron. Multipolar neurons are also commonly found in the cortex of the brain and the spinal cord. In the periphery, multipolar neurons are found in autonomic ganglia.

1.3 Analogy of Neurons to Electrical Circuit Components

The function of a neuron can be analogized to the electrical circuit producing digital pulses. Most simply, a neuron can be analogized to an operational amplifier. The operational amplifier (Op-Amp) accepts inputs by two nodes, one by noninverting node and the other by inverting type. As an input to the noninverting node increases the output, it functions similar to the excitatory input of a neuron. Similarly, as an input to the inverting input decreases the output of the Op-Amp, it functions similar to an inhibitory input of a neuron. The main body of Op-Amp combines all of these inputs based on their connected weights, which are implemented by the value of the resistances connected serially or parallelly.

1.3.1 Dendritic Summation

In a neuron, connected inputs are weighted by the distance from the soma whether they are excitatory or inhibitory. Postsynaptic potentials (PSP) generated at the farthest point of the dendrite should be transmitted a relatively long way to the soma and their amplitudes decrease more heavily than the PSP generated at the base of the dendrites near the soma. That is, in the neuron, the position of the synapse generating PSP is the weight. If the synapse is made close to the soma, input to the synapse has a higher impact on producing the neural output. Generally, inhibitory synapses tend to be made at the base of the dendrites closer to the soma area. So, single inhibitory input can negate the other excitatory inputs generated at the upper part of the dendrite and have propagated down the dendrite with diminished amplitude depending on their propagated distances. If a synapse moves to the farther position, its impact diminishes. Fading of memory is interpreted in this way. When the stored information is not utilized by the recurrent usage, its synaptic

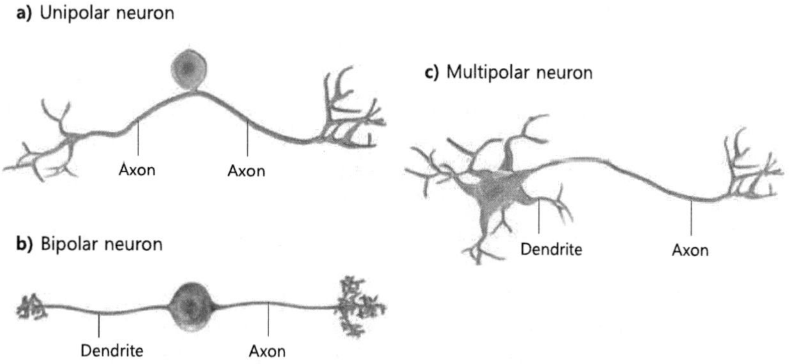

Fig. 3.2 Types of neurons based on the anatomical structure. (**a**) Unipolar neuron, (**b**) Bipolar neuron, (**c**) Multipolar neuron

positions are moved away as the new axonal terminals squeeze into the dendritic surface to make a new synaptic connection.

1.3.2 Analog to Digital Conversion

The propagated PSP are summated and produce a resulting potential in analog form. However, this potential in analog form is not transmitted directly along the axon. This analog form of neuronal signal is converted into a digital form at the hillock. If the resulted voltage is raised higher than a threshold potential, there generated an action potential and this action potential is transmitted away through the axon. The action potential is a kind of digital pulse. According to the "All-or-none" principle, the action potential always appears with a determined shape in amplitude and duration or not occurred. There are no half-tall or shorter action potentials in the same neuron. Occurring of action potential represents a state "1" and not occurring an action potential represents the state "0" exactly like in the digital electrical circuit. In this regard, a simple OP- Amp is not accurately equivalent to a neuron as its output is in the analog form. To make the analogy more accurate, the pulse trigger can be serially connected to the output of the Op-Amp. The pulse trigger produces a digital pulse when the amplitude of the Op-Amp output is higher than the threshold value. So, a single neuron can be represented by a combination of Op-Amp and pulse trigger in electrical components. At the hillock of the neuron, analog to digital conversion is performed. Signal transmission in the axon is in digital form. Transmitted digital signals are converted into analog form by triggering the release of vesicles into a synaptic gap.

When the two digital pulses are compared to each other, they have comparative characteristics. Neuronal pulse has a small amplitude generally less than 100 mV, while the electrical digital pulse has a higher amplitude of 5.0 V. In their pulse width, the action potential has a width of about 2 ms while the electrical digital pulse has a much shorter width of 1 ns order. Compared to each other, the action potential is small and thick and the electrical digital pulse is tall and thin. It can be thought that to save energy in a single pule, an action potential is tried by reducing the amplitude, and an electrical pulse is tried by reducing the pulse width.

1.3.3 The Capacity of a Neuron and the Brain

There are about 86 billion neurons in the brain. The number of transistors in the CPU is increasing rapidly with Moore's law. CPU of the highest performance at the time of writing, which is H100 GPU from NVIDIA, has about 80 billion transistors in it, which is similar to the number of neurons in the brain. If the performance of a neuron is equivalent to a single transistor, the performance of the CPU at the technological edge can be thought of as almost equivalent to the human brain. Even though the performance of the CPUs or GPUs has highly enhanced and made the many specific tasks, that the human brain cannot perform, possible, they are still not equivalent to the brain in overall performance. In a strict sense, the performance of a neuron looks higher than a single transistor. It is more appropriate to analogize a neuron to an Op-Amp, which is composed of 20 transistors on average, than a single transistor. With this analogy, the brain has about 20 times the higher capacity in the number of basic elements than the highest performance CPU. According to Moore's law, the number of transistors in a dense integrated circuit (IC) doubles about every 2 years, and it is expected about 8–10 years before the technology for CPU reaches the level of the brain. Recently, there are studies reporting that each branch of dendrite functions like an Op-Amp. Considering this analogy and that each neuron has 128 basal dendritic segments having up to 40 actual synapses (Hawkins & Ahmad, 2016), it will take additional 14 years that the CPU reaches the brain in performance.

1.3.4 Coding Strategy of Digital Pulses

In the digital circuits, pulses are coded based on the coding or modulation strategy. While the various strategies are possible, with fixed pulse width, pulse code modulation (PCM) is the most popular, which converts information into the pattern of coded pulses. In neural coding, the strategy for coding and decoding neural information is not clearly known, and many theories are suggested to explain the measured results.

The rate coding is a traditional coding scheme, assuming that stimulus information is contained in the firing rate of the pulses. The firing rate increases, generally nonlinearly, with increasing stimulus intensity. While the pulses generally appear at random intervals, information possibly encoded in the temporal structure of the pulse train is not counted. Rate coding is highly robust to noises in the interstimulus intervals for slowly varying sensory stimuli. However, it is inefficient and too simplistic to describe the neural coding of brain activity.

Temporal coding is described to explain the precise spike timing on a millisecond time scale or high-frequency firing-rate fluctuations. Temporal codes employ features of the pulse train other than the firing rate like time to first spike after the stimulus onset, pulse randomness, dynamics of interpulse duration, and specific temporal patterns.

Population coding is a coding strategy to represent stimuli by the joint activities of spatially grouped neurons. In population coding, each neuron responds to some set of inputs with different sensitivity. When the responses of many populated neurons are combined, they collectively provide information in the input. Population coding reduces the uncertainty in the neuronal variability and can represent a number of different stimulus attributes. Population coding can reflect fast changes in the stimulus conditions almost instantaneously. Many strategies have been suggested for describing neural coding and decoding. However, they only cover a part of the coded patterns and cannot explain others. The actual reading and writing of the neural code remain a challenge in understanding human intelligence.

2 Membrane of the Neuron

Electricity within humans starts from the electricity in nerve cells. Neurons have potential across their membrane. These potentials are developed by the distribution difference of charged ions across the membrane and the intrinsic selective permeable characteristic of the membrane. Potential across the membrane occurs in one of these three conditions.

- *Resting state*: The membrane maintains stable and constant potential while there is no stimulation or disturbance. Membrane potential during the resting state is called resting potential and can be defined as a temporarily and spatially stable value across the membrane.
- *Subthreshold fluctuation*: Membrane potential varies by the external stimulus or disturbance, but does not cross the threshold level to fire an action potential. Potential fluctuates under the threshold level and returns to resting level as the stimulus effect fades away.
- *Action potential*: If any stimulus increases membrane potential over the threshold level to fire an action potential, membrane potential generates an action potential. After the onset of an action potential, automatic sequences of potential variations are followed, and it generates always the same shape of the action potential. It is similar to the positive state "1" of the digital pulse in the digital circuit. This action potential propagates along the axon by stimulating neighboring sites by occurring local currents.

Among the three types of membrane potentials, resting potential and membrane potential at the subthreshold level are explained in this chapter.

2.1 Structural Characteristic of the Membrane

The membrane covers the whole volume of nerve cells and separates internal contents and volume of fluids from external fluids. Most electrical phenomena involved in neuronal preservation and functioning are triggered on this membrane. The unique characteristics of the membrane play a key role in developing potentials across. To develop a potential, the membrane should keep the charges effectively on both sides without breaking the potential. That is, the membrane should be a good insulator with minimized thickness and have a high dielectric constant like a good capacitor.

2.1.1 Double Layer Structure

The membrane of neurons is composed of a bilayer of phospholipid molecules. A phospholipid molecule has two parts, phosphoric acid head, and glyceride tails. The head part is hydrophilic and the tail part is hydrophobic. When these molecules are collected and built into a membrane with a double-layer structure in the watery body fluids, their hydrophobic tail parts come inside and hydrophilic head parts arrange outward (Fig. 3.3).

The thickness of the membrane is about 7.5–10 nm, and this is just about 25 times the size of a water molecule. The membrane is very thin, but it is sufficiently strong to keep the materials inside safely. This membrane is not rigid. It is a plasma structure much similar to the membrane of a soap bubble. Even though the membrane is intrinsically permeable to small nonpolar particles like CO_2, N_2, O_2, and uncharged small polar molecules like water, it is not permeable to large polar molecules like glucose and charged particles like sodium, potassium, and chloride ions. Charged particles mostly move across the membrane through specific ion channels.

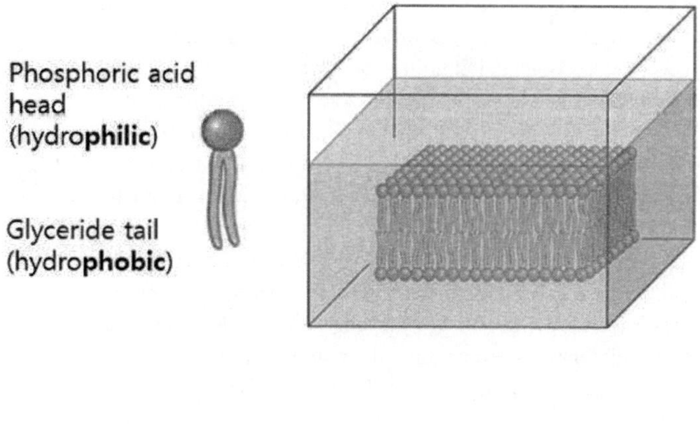

Fig. 3.3 A phospholipid molecule and double-layer structure in the water

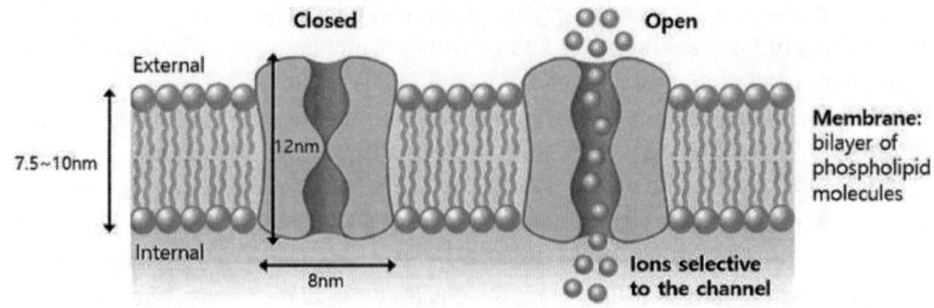

Fig. 3.4 Membrane with ion channels embedded

2.1.2 Embedded Ion Channels

Within this membrane, multiple types of ion channels are embedded (Fig. 3.4). The ion channel, which is also called a transport protein, is the gate for the movement of ions across the membrane. There are many types of ion channels having different selectivity on the ions. Ion channels are taller than membrane thickness and they slowly float around the membrane just like dust particles moving around the surface of soap bubbles. These ion channels control the movement of ions by opening and closing them and these characteristics are represented as the membrane permeability.

2.2 Capacitance of Membrane

Membrane capacitance is the electrical capacitance associated with the membrane of phospholipid layers. The total capacitance of a cell (**C_m**) is dependent on the surface area of the cell, as capacitance is proportional to the area of the surface. The capacitance of the membrane is given by

$$C_m = \frac{\varepsilon_0 K A}{l} \tag{3.1}$$

where C_m is membrane capacitance (F), ε_0 is the permittivity of vacuum (8.854×10^{-12} F/m), K is dielectric constant (relative permittivity), A is the surface area of the membrane (m^2), and l is the thickness of the membrane (m).

Membrane capacitance determines the time constant τ_m of a neuron with the membrane resistance R_m by $\tau_m = R_m C_m$. Time constant plays an important role in the transmission of postsynaptic potential along the dendrite and determines the propagation velocity of an action potential along the axon.

Membrane capacitance can be measured experimentally based on several methods directly or indirectly. Due to the diversity of measurement methods and the complex geometry of cell surface, a wide range of experimental values had been reported (Gentet et al., 2000). When Hodgkin and Huxley published their Nobel Prize-winning experimental results, they have used 1 μF/cm^2 for the simplicity of model calculation while their experimental values showed a wide range of 0.8–1.5 μF/cm^2, with a mean of 0.9 μF/cm^2. Recently, it was reported with an estimation of 0.75 μF/cm^2 in hippocampal pyramidal neurons (Major et al., 1994), 0.9 μF/cm^2 in hippocampal interneurons (Chitwood et al., 1999), and 2.4 μF/cm^2 in ventral horn neurons of the spinal cord (Thurbon et al., 1998). The value most used is 1.0 μF/cm^2 which is reasonable and the simplest. With $C_m = 1.0$ μF for $A = 1$ cm^2 of membrane area, we can calculate the dielectric constant as $K = 8.47$ for the membrane of thickness $l = 7.5$ nm.

In contrast to the dielectric constant, dielectric strength represents the maximum electric field strength above which an insulating material begins to break down and conduct. While a higher dielectric constant collects a larger amount of charges across the same gap thickness, higher dielectric strength enables the capacitor increases capacitance by minimizing the gap between the plates and results in a minimized volume of the capacitor. As a good insulator for a capacitor, it requires a high dielectric constant and high dielectric strength together. When dielectric breakdown of the cell membrane is intended by applying strong electric fields, it occurred around 4×10^8V/m which is very high and comparable with the breakdown voltages for bilayers of most oils (Zimmermann et al., 1998). Table 3.1 shows the dielectric constant and dielectric strength for commonly used materials in comparison with the membrane. In the resting state, a resting potential of −75 mV develops around 10×10^6 V/m across the membrane of thickness $l = 7.5$ nm. The membrane is a good material for capacitors, which can keep a large amount of charges with minimal thickness.

In regard to ion channels, it has been reported that membrane capacitance does not significantly vary as a function of ion channel density (Gentet et al., 2000).

Through the capacitance of the membrane, capacitive current flows when the voltage across the membrane changes. Capacitive current is given by

$$C_m = \frac{Q}{V} \tag{3.2}$$

$$I_C = \frac{dQ}{dt} = C_m \frac{dV}{dt} \tag{3.3}$$

Table 3.1 Comparison of dielectric constant and dielectric strength with commonly used materials

Material	Dielectric constant	Dielectric strength
Vacuum	1.0	Infinity
Air	1.0006	3.2×10^6 V/m
Teflon	2.1	60×10^6 V/m
Polyurethane	2.3	40×10^6 V/m
Silicone oil	2.5	15×10^6 V/m
Paper	3.7	16×10^6 V/m
Mica	4–8	$20–200 \times 10^6$ V/m
Glass	5–12	$12–100 \times 10^6$ V/m
Membrane	*8.47*	[a]*200×10^6 V/m*
Water	80	–
Strontium titanate	233	8×10^6 V/m

[a]This value is the half of the reported electric field causing a dielectric breakdown

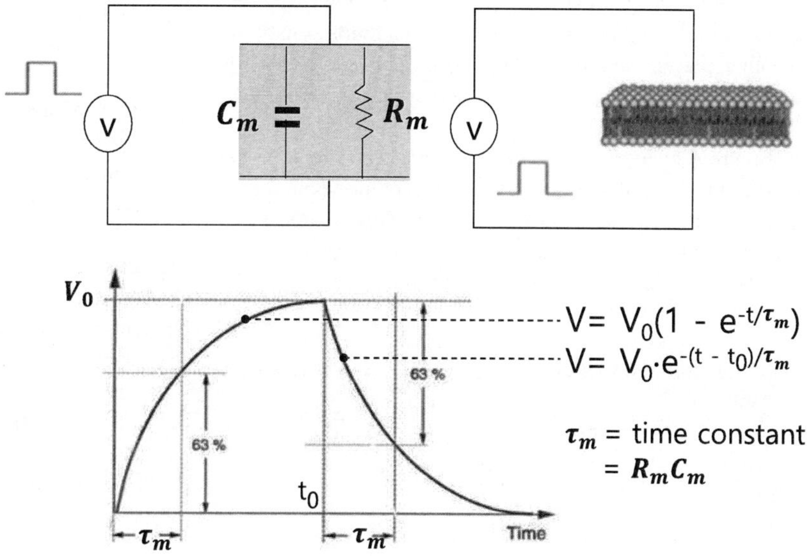

Fig. 3.5 Experiment for the measurement of membrane resistance

2.3 Resistance of Membrane

Membrane resistance means the resistance across the membrane which impedes the flow of current through the membrane. It also represents how effectively the membrane is insulating the internal part from the external fluids. Membrane resistance is dependent on the density of open ion channels. Membrane conductance, which is the reciprocal of membrane resistance, is also used instead to explain membrane potentials by the gating of ion channels. In contrast to capacitance, membrane resistance temporarily varies dependent on the ion channel activities during the firing of an action potential.

Membrane resistance, when it is not changing in a passive state, can be measured experimentally by measuring the time constant. As measuring time constant in a parallel RC circuit, the time constant of the membrane can be measured by applying pulses across the membrane. Figure 3.5 shows the experimental setup and equivalent circuit for the measurement of membrane time constant.

For the measured time constant of $\tau_m = 4$ ms, membrane resistance R_m is calculated by dividing the time constant with the membrane capacitance. Since 1 cm^2 of membrane area has capacitance of $C_m = 1.0$ μF, resistance of 1 cm^2 of membrane area is calculated $R_m = 4$ KΩ. The resistivity of the membrane ρ is calculated from the resulted resistance R_m by the general resistance formula.

$$R = \frac{\rho l}{A} \tag{3.4}$$

where $R = 4$ KΩ is the resistance, ρ is the resistivity, $l = 7.5$ nm is the thickness of the membrane, and $A = 1$ cm^2 is the membrane area for the derivation of membrane resistivity.

$$\rho = \frac{R \cdot A}{l} = \frac{4 \times 10^3 \Omega \cdot 1.0 \times 10^{-4} \mathrm{m}^2}{7.5 \times 10^{-9} \mathrm{m}} = 5.33 \times 10^7 \Omega \mathrm{m} \quad (3.5)$$

The membrane has a resistivity of 5.33×10^7 Ωm. Table 3.2 compares resistivity of the membrane with commonly used materials, and it represents that membrane is a good insulator while it is in a passive state.

3 Ion Distribution Across the Membrane

The membrane separates the internal contents and fluid of nerve cells from the external fluid. Sodium ions (Na^+) and chloride ions (Cl^-) are dominant in external fluids as in blood or other plasma within the body. This fluid constitution is quite similar to seawater and is considered strong support that life came from the sea.

Figure 3.6 gives approximate ion concentrations of the extracellular and the intracellular fluids. While sodium ions and chloride ions take the major parts of the extracellular fluid ion constituents, their intracellular concentrations are 10–30 times lower than their extracellular ones. On the contrary, potassium (K^+) has about 40 times higher intracellular concentration compared to its extracellular concentration. As positive sodium ions are balanced electrically with negative chloride ions in extracellular fluid, positive potassium ions are electrically balanced by negative large organic ions internal side of the nerve cell. But large organic ions are too large to move through ion channels, so they are residing in the internal part of the cell all the time and contribute to biasing the internal part toward negative potential. This difference in ionic concentrations is extremely important to generate electrical potentials and to life.

The membrane of the phospholipid layer works as a barrier for the movement of charged ions in external and internal fluids. Ion channels only allow the movement of ions across the membrane. But one type of ion channel is only selective to a single type of ion. Sodium ion channel only pass Na^+ ions and potassium ion channel only pass K^+ ions. This selectivity of the ion channels embedded in the membrane makes the membrane semipermeable and selective to certain ions, and the density of open ion channels determines the permeability of the membrane.

Charged ions move through the ion channel by the gradient forces applied to them. There are two types of gradients that can cause the movement of the ions. One is the diffusion gradient, and the other is the potential gradient.

Table 3.2 The resistivity of membrane in comparison with commonly used materials

Material	Resistivity	Remark
Silver	1.60×10^{-8} Ωm	Conductors
Copper	1.62×10^{-8} Ωm	
Gold	2.44×10^{-8} Ωm	
Mercury	9.40×10^{-7} Ωm	
Sea water	2.0×10^{-1} Ωm	
Axoplasm	6.0×10^{-1} Ωm	
Ground water	10–100 Ωm	
Membrane	*5.33×10^7 Ωm*	Insulators
Glass	1–10,000 × 10^{11} Ωm	
Air	1.3–3.3 × 10^{14} Ωm	
Dry wood	1–100 × 10^{14} Ωm	
Hard rubber	1–100 × 10^{17} Ωm	

3.1 Diffusion Gradients

Every particle in the fluid moves with Brownian motion and smoothly diffuses from higher concentration to lower concentration side and pushes the whole volume of fluid to be stabilized with the same concentration. This is the phenomenon that makes a beaker of water totally black with one drop of black ink after some time-lapse (Fig. 3.7). Ions within the body fluids also diffuse according to their concentration gradient. If there is a high difference in concentrations, there occurs a strong gradient moving a large number of particles. Since there are strong concentration differences for sodium, potassium, and chloride ions across the

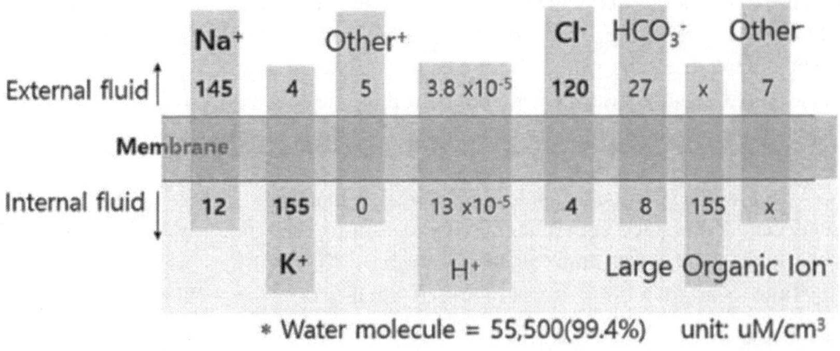

Fig. 3.6 Ion concentrations across the membrane

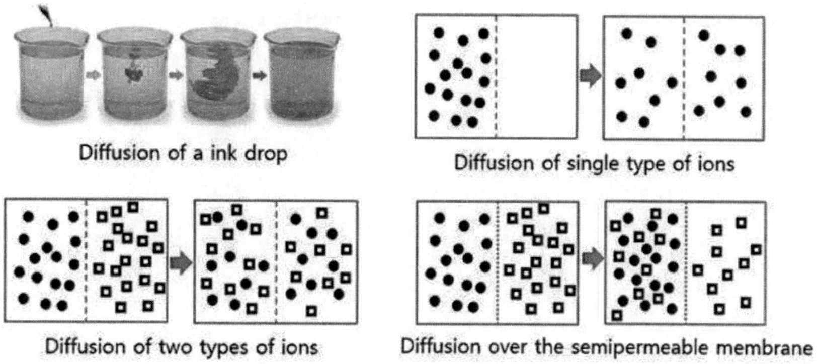

Fig. 3.7 Diffusion of ions over the semipermeable membrane

membrane, there develop concentration gradients. The diffusion of ions is blocked by the membrane and only a small part of them are diffusing through the open ion channels. If the membrane is broken or the gates of ion channels are wide open, there will be a flood of ions crossing the membrane pushed by these strong concentration gradients. The diffusion flow of ion $[A]$ along the concentration gradient is given by

$$J_A(\text{diffusion}) = -D\frac{d[A]}{dx} \tag{3.6}$$

where J_A(diffusion) is the diffusion flow of ions by the concentration gradient, D is the diffusion coefficient, and $[A]$ is a concentration of ion A at location x.

Diffusion of one type of ions works independently of the concentration of other types of ions. The diffusion of sodium ions is irrespective of potassium or chloride ion concentration when their charges are not considered.

If the membrane is not equally permeable to all the particles or ions, it is semipermeable. Semipermeable membrane passes some ions, but does not allow passage of the other ions. For the selected ions, the concentration gradient reduces with time, but for the blocked ions established concentration gradient remains unchanged across the membrane. Since the ions have charges, there also develops a potential gradient across the membrane following the movement of charged ions.

3.2 Potential Gradients

Another gradient is the potential gradient. The potential gradient is the electric field established by the difference in the charged ion concentrations. If the particles do not have charges like ink particles, there will be no potential gradient and the diffusion gradient is the only force moving the particles. Since the potential can be developed by both polarities of charged particles, the potential gradient of both polarities should be combined. If there are multiple types of ions, the potential contributed by the concentration of one type of ions affects the drift of the other ions by the combined potential gradient. This is equivalent to the simple movement of charged particles under the developed electric field. Drift flow of positive ion $[A]$ by potential gradient is given by

$$J_A(\text{drift}) = -\mu\frac{dV}{dx}[A] \tag{3.7}$$

where J_A(drift) is the flow of ions by the potential gradient, μ is the mobility constant, V is the voltage at location x, and $[A]$ is the concentration of ion A at location x. The potential gradient is proportional to the slope of the potential along with the distance and the amount of ion concentration. A negative sign means the direction of the potential gradient is opposite to the direction of potential increasing for the positive charges.

The diffusion coefficient D for the diffusion flow and the mobility μ of drift flow are not independent. Their dynamics are related by the kinetic theory of gases and their relations are equated by Einstein as

$$D = \frac{\mu RT}{zF}, \mu = \frac{DzF}{RT} \tag{3.8}$$

where D is the diffusion coefficient, μ is the mobility, R is the universal gas constant of 8.31 Joule/mol · K, T is the absolute temperature, z is the valence of the ion, and F is the Faraday constant of 9.65×10^4 Coulomb/mol.

4 Membrane Potentials

4.1 Nernst's Potential

To derive a resting potential across the membrane, we first consider the simple case of a single type of ions only working. This will result in Nernst's potential. After that, we will consider all the participating ions working together, and will derive Goldman potential which finally calculates resting potential based on concentrations of multiple ions across the membrane.

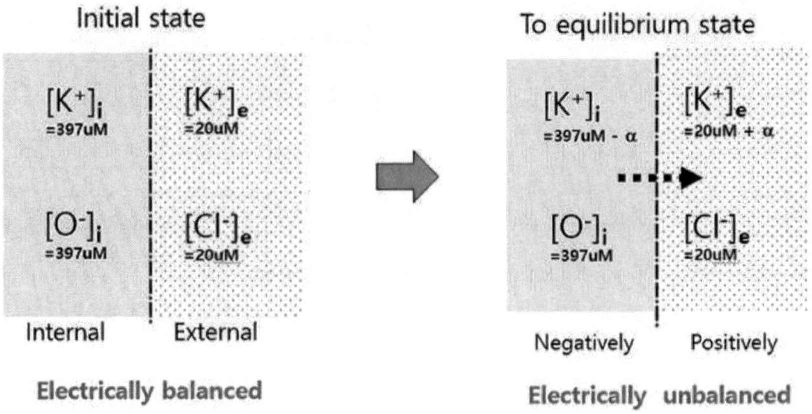

Fig. 3.8 Movement ions over the semipermeable membrane

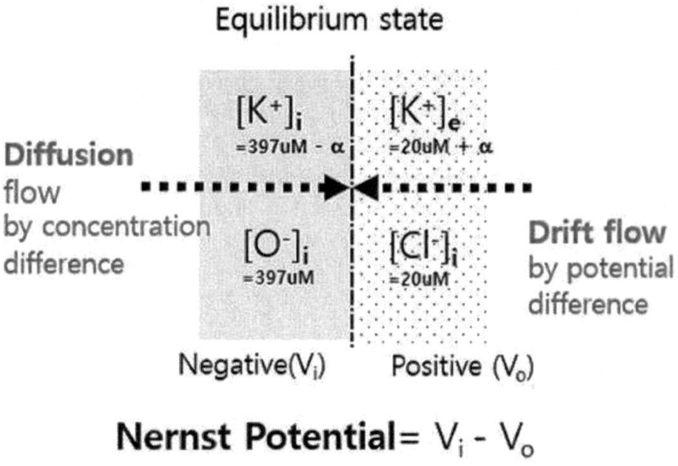

Fig. 3.9 Nernst potential by the balance of diffusion flow and drift flow

Consider the semipermeable membrane which is permeable only to a single type of ions, potassium in this explanation. Figure 3.8 shows how the semipermeable membrane works.

Potassium ion in intracellular fluid is about 20 times higher than external in concentration. Since this membrane is selectively permeable to potassium ions, there established a strong diffusion gradient pushing the potassium ions to the external side across the membrane. And there are other types of ions, chloride ions of negative charge and organic ions of negative charge. These ions are not permeable to the membrane. Since they cannot move across the membrane, their concentration gradients do not have any effect on the movement of potassium ions. But, their negative charges are balancing potassium ions in the internal and external parts by an equal amount of charges. Initially, internal and external parts are electrically balanced and there is no potential gradient across the membrane. Only, the diffusion gradient of potassium pushes potassium ions to the external side of lower concentration.

As the potassium ions are moved out to the external side by the diffusion gradient, the internal concentration of potassium is decreasing and the external concentration is increasing. This decreased concentration difference reduces the outward diffusion flow as time goes by. In addition to this, the internal side becomes negative by the positive charges moved out and the external side becomes positive due to the increased positive ions. Both internal and external parts become electrically unbalanced and there develops a potential gradient across the membrane. The internal side becomes negative to the external side. The negative potential is developed referenced to the external side. This developed potential gradient exerts forces on potassium ions. Since potassium ions have positive charges, negative potential push positive charges back to the internal side. This force is in the opposite direction to the force developed by the diffusion gradient. As the diffusion gradient is decreasing and the potential gradient is increasing, they arrive at an equilibrium state where these two forces are balanced after some time. The number of potassium ions moving out by diffusion gradient becomes equal to the number of potassium ions moving in by the potential gradient. So, there is no net movement of ions across the membrane and the concentration and potential are maintained as it is. Potential developed at this balance is measured with reference to the external side and called Nernst potential (Fig. 3.9).

Nernst potential is derived by equating the total ionic flow, which is the sum of flows caused by diffusion gradient and potential gradient, to 0.

$$J = J_K(\text{diffusion}) + J_K(\text{drift}) = 0 \tag{3.9}$$

$$J = -D\frac{d[\mathrm{K}^+]}{dx} - \mu\frac{dV}{dx}[\mathrm{K}^+] = -D\left(\frac{d[\mathrm{K}^+]}{dx} + \frac{zF}{RT}[\mathrm{K}^+]\frac{dV}{dx}\right) = 0 \tag{3.10}$$

$$\frac{d[\mathrm{K}^+]}{dx} + \frac{zF}{RT}[\mathrm{K}^+]\frac{dV}{dx} = 0 \tag{3.11}$$

Equation (3.11) becomes by dividing with [K$^+$] as

$$\frac{1}{[\mathrm{K}^+]}\frac{d[\mathrm{K}^+]}{dx} + \frac{zF}{RT}\frac{dV}{dx} = 0 \tag{3.12}$$

By integrating from the external ($x = 0$) to the internal ($x = h$) side of the membrane, where h is the thickness of the membrane, the membrane voltage is calculated.

$$[\ln[\mathrm{K}^+]]_{e(x=0)}^{i(x=h)} + \left[\frac{zF}{RT}V\right]_{e(x=0)}^{i(x=h)} = 0 \tag{3.13}$$

$$\ln\frac{[\mathrm{K}^+]_i}{[\mathrm{K}^+]_e} + \frac{zF}{RT}(V_i - V_e) = 0 \tag{3.14}$$

$$V = V_i - V_e = \frac{RT}{zF}\ln\frac{[\mathrm{K}^+]_e}{[\mathrm{K}^+]_i} \tag{3.15}$$

Generally, the Nernst equation for an ion $[A]$ is given as

$$\text{Nernst potential} = V = \frac{RT}{zF}\ln\frac{[A]_e}{[A]_i} \tag{3.16}$$

For the ion concentration of Fig. 3.6, the Nernst potential for each ion is calculated as in Table 3.3.

The diffusion gradient of sodium ions, which has a higher concentration in external fluid, is balanced if the potential gradient is developed that drives positive ions outward. On the contrary, potassium ion has a higher concentration in intracellular fluid, and their diffusion gradient is balanced by the negative Nernst potential which develops the potential gradient pushing positive ions inward. Chloride ion has negative charges and it is pushed in the opposite direction to positive ions by the same potential gradient. So, high concentration in the external fluid is balanced by the negative Nernst potential which pushes chloride ions outward direction.

Nernst potential is the resting potential calculated when a single type of ion is permeable across the membrane. Even in this case, the potential developed is affected by the ions that reside on the internal side and do not move across the membrane. Nernst potential is not occurring in a real circumstance, because there are multiple types of ions across the membrane that are differently selective to the membrane in moving across the membrane. Nernst potential is the theoretical potential for the given ion concentrations, and it is used as the reference voltage for a given membrane potential estimating how strongly the additional potential gradient can be developed for each type of ions.

4.2 Goldman Potential

In real circumstances, there are multiple ions that can move in and out of the membrane. While the diffusion gradients drive each ion independently over the membrane, the potential is developed by the contribution of all ions and the potential gradient established affects all the ions. If the Nernst potentials are all same for all types of ions, this potential will be maintained as resting potential and all types of ions will be calmly and stably resting without producing net ionic movement across the membrane. But, measured ionic concentrations from the body show the differences in the distribution of each type of ions and their Nernst potentials become all different. In this circumstance, resting potential is developed by making the total membrane current crossing the membrane equal to 0, so that membrane resting potential is not changing. While the total membrane current is 0 in this resting state, ionic currents for each type of ion are not 0. Membrane currents of some ions flow inward, membrane currents of some other ions flow outward and they are exactly canceled out resulting in no net current over the membrane and maintaining stable resting state potential.

Figure 3.10 shows the procedure for the derivation of Goldman potential.

Table 3.3 Nernst potential for each ion

Ion	External concentration (μMol/cm^3)	Intracellular concentration (μMol/cm^3)	Permeability (10^{-6}cm/s)	Nernst potential
Sodium (Na$^+$)	145	12	0.07	$V_{\mathrm{Na}} = 25.8\ \mathrm{mV} \cdot \ln\frac{145}{12} = 64.3\ \mathrm{mV}$
Potassium (K$^+$)	4	155	1.8	$V_{\mathrm{K}} = 25.8\ \mathrm{mV} \cdot \ln\frac{4}{155} = -94.4\ \mathrm{mV}$
Chloride (Cl$^-$)	120	4	0.8	$V_{\mathrm{Cl}} = -25.8\ \mathrm{mV} \cdot \ln\frac{120}{4} = -87.8\ \mathrm{mV}$

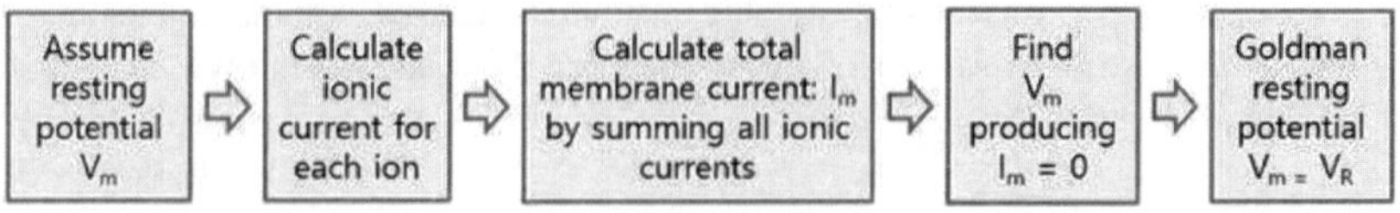

Fig. 3.10 Procedure for the derivation of Goldman potential

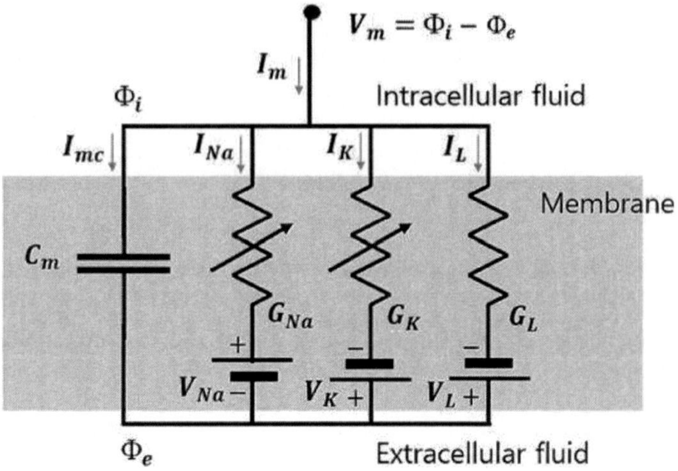

Fig. 3.11 Equivalent electrical circuit model of the membrane

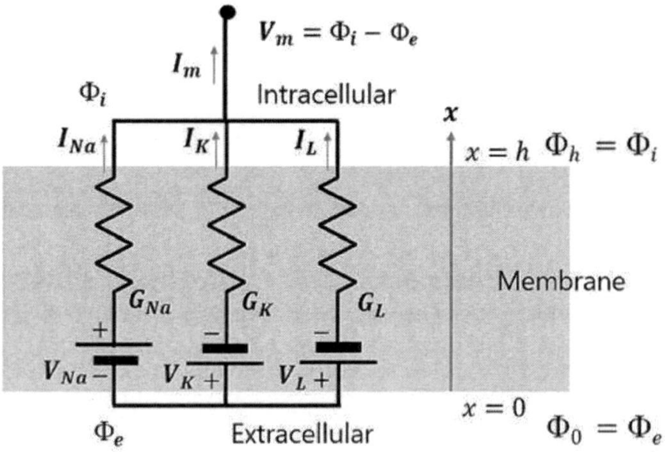

Fig. 3.12 Simplified equivalent circuit model for the derivation of Goldman equation

4.2.1 Equivalent Electrical Circuit Model

Figure 3.11 shows the equivalent circuit model of the membrane. This model can be used for the derivation of Goldman potential and also for the derivation of the action potential. Membrane voltage V_m is defined by the potential of the internal side of the membrane in reference to the external side. That is $V_m = \Phi_i - \Phi_e$.

The membrane is electrically represented by a parallel circuit of capacitance and conductance. Capacitance C_m represents the membrane capacitance caused by the bilayer structure of phospholipid molecules, which is around 1 μF/cm^2 in most of living body. If the membrane potential V_m is not constant and varies over time, there flows a capacitive current I_{mc} through the capacitance. However, when the membrane voltage is not changing as in the steady state of the resting potential, there is no current flowing through the capacitance. Membrane resistance is represented by three resistors, and each of them represents resistance for the sodium, potassium ion flow, and leakage including chloride ion flow. They are represented by their conductance, which is the reciprocal of resistance. While the conductance for leakage current is modeled constant, the conductance of sodium and potassium are modeled as varying. They are actually varying dramatically if the membrane is in an active state of firing action potentials.

For the derivation of resting potential, the equivalent electrical circuit model can be simplified as in Fig. 3.12.

Since it is the case for resting potential, whatever value it is, there is no variation over time causing current flow through capacitance. Current flowing through capacitance C_m can be eliminated. During the resting potential, the membrane is in a passive state and its conductance does not change. So, the current through the membrane is composed of three ionic currents caused by sodium, potassium, and chloride ions. Each ionic current varies upon the membrane potential. They are a function of membrane potential.

$$I_m = I_{\mathrm{Na}} + I_{\mathrm{K}} + I_L + I_{mc} = 0 \tag{3.17}$$

$$I_m = I_{\mathrm{Na}}(V_m) + I_{\mathrm{K}}(V_m) + I_L(V_m) = 0 \tag{3.18}$$

If we find the membrane potential V_m which makes $I_m = 0$, it becomes the resting potential with multiple ions.

For the simplicity of derivation, we assume that the electric field across the membrane is constant. Since the membrane is very thin (h = 7.5 nm), assuming that the electric field for this short distance h is linear and constant is reasonable. The potential gradient along the membrane from the external side to the internal side is approximated as follows.

$$\frac{dV}{dx} = \frac{\Phi_h - \Phi_0}{h} = \frac{\Phi_i - \Phi_e}{h} = \frac{V_m}{h} \tag{3.19}$$

With this simplified potential gradient, we first calculate the ionic particle current J_k for the k−th ion.

$$J_k = J_k(\mathrm{diffusion}) + J_k(\mathrm{drift}) \tag{3.20}$$

$$\begin{aligned} J_k &= -D_k \frac{dC_k}{dx} - \mu_k \frac{dV}{dx} C_k \\ &= -D_k \frac{dC_k}{dx} - \frac{D_k z_k F}{RT} \frac{V_m}{h} C_k \\ &= -D_k \left(\frac{dC_k}{dx} + \frac{z_k F}{RT} \frac{V_m}{h} C_k \right) \end{aligned} \tag{3.21}$$

$$\frac{dC_k}{dx} = -\frac{J_k}{D_k} - \frac{V_m z_k F}{RTh} C_k \tag{3.22}$$

$$\frac{dC_k}{-\frac{J_k}{D_k} - \frac{V_m z_k F}{RTh} C_k} = dx \tag{3.23}$$

Integrating both sides over x from 0 to h,

$$\int_0^h \frac{dC_k}{-\frac{J_k}{D_k} - \frac{V_m z_k F}{RTh} C_k} = \int_0^h dx \tag{3.24}$$

$$\left[-\frac{RTh}{V_m z_k F} \ln\left(\frac{J_k}{D_k} + \frac{V_m z_k F}{RTh} C_k \right) \right]_0^h = h \tag{3.25}$$

$$-\frac{RTh}{V_m z_k F} \ln \frac{\frac{J_k}{D_k} + \frac{V_m z_k F}{RTh} [C_k]^h}{\frac{J_k}{D_k} + \frac{V_m z_k F}{RTh} [C_k]^0} = h \tag{3.26}$$

$[C_k]^0$ and $[C_k]^h$ are concentrations of k−th ion at $x = 0$ and $x = h$, which are the external and the internal concentration C_{ek} and C_{ik} respectively.

$$\ln \frac{\frac{J_k}{D_k} + \frac{V_m z_k F}{RTh} C_{ik}}{\frac{J_k}{D_k} + \frac{V_m z_k F}{RTh} C_{ek}} = -\frac{V_m z_k F}{RT} \tag{3.27}$$

$$\frac{\frac{J_k}{D_k} + \frac{V_m z_k F}{RTh} C_{ik}}{\frac{J_k}{D_k} + \frac{V_m z_k F}{RTh} C_{ek}} = e^{-\frac{V_m z_k F}{RT}} \tag{3.28}$$

$$\frac{J_k}{D_k} + \frac{V_m z_k F}{RTh} C_{ik} = e^{-\frac{V_m z_k F}{RT}} \left(\frac{J_k}{D_k} + \frac{V_m z_k F}{RTh} C_{ek} \right) \tag{3.29}$$

After arranging for ionic current $\boldsymbol{J_k}$, it becomes as follows.

$$J_k = -\frac{D_k V_m z_k F}{RTh} \left(\frac{C_{ik} - C_{ek} e^{-\frac{V_m z_k F}{RT}}}{1 - e^{-\frac{V_m z_k F}{RT}}} \right) \tag{3.30}$$

To derive electric current from ionic particle current, Faraday's constant and valance of ion should be multiplied. And let $D_k/h = P_k$ as permeability for of k−th ion. Then electric current for the k−th ion is given by

$$I_k = z_k F \bullet J_k = -\frac{P_k V_m z_k^2 F^2}{RT} \left(\frac{C_{ik} - C_{ek} e^{-\frac{V_m z_k F}{RT}}}{1 - e^{-\frac{V_m z_k F}{RT}}} \right) \tag{3.31}$$

Now we can have an ionic current for three ions of interest. The valence of sodium (z_{Na}) and potassium (z_{K}) is 1, and the valence of chloride (z_{Cl}) is −1.

$$I_{Na}(V_m) = -\frac{P_{\mathrm{Na}} V_m F^2}{RT} \left(\frac{C_{i\mathrm{Na}} - C_{e\mathrm{Na}} e^{-\frac{V_m F}{RT}}}{1 - e^{-\frac{V_m F}{RT}}} \right) \tag{3.32}$$

$$I_K(V_m) = -\frac{P_{\mathrm{K}} V_m F^2}{RT} \left(\frac{C_{i\mathrm{K}} - C_{e\mathrm{K}} e^{-\frac{V_m F}{RT}}}{1 - e^{-\frac{V_m F}{RT}}} \right) \tag{3.33}$$

$$I_{Cl}(V_m) = -\frac{P_{\mathrm{Cl}} V_m F^2}{RT} \left(\frac{C_{e\mathrm{Cl}} - C_{i\mathrm{Cl}} e^{-\frac{V_m F}{RT}}}{1 - e^{-\frac{V_m F}{RT}}} \right) \tag{3.34}$$

For resting potential, membrane current $\boldsymbol{I_m}$ becomes 0.

$$I_m = I_{Na}(V_m) + I_K(V_m) + I_{Cl}(V_m) = 0 \quad (3.35)$$

$$-\frac{P_{Na}V_mF^2}{RT}\left(\frac{C_{iNa} - C_{eNa}e^{-\frac{V_mF}{RT}}}{1-e^{-\frac{V_mF}{RT}}}\right) - \frac{P_K V_m F^2}{RT} \times \left(\frac{C_{iK} - C_{eK}e^{-\frac{V_mF}{RT}}}{1-e^{-\frac{V_mF}{RT}}}\right) - \frac{P_{Cl}V_mF^2}{RT} \times \left(\frac{C_{eCl} - C_{iCl}e^{-\frac{V_mF}{RT}}}{1-e^{-\frac{V_mF}{RT}}}\right) = 0 \quad (3.36)$$

After dividing common terms, the equation is given by

$$P_{Na}\left(C_{iNa} - C_{eNa}e^{-\frac{V_mF}{RT}}\right) + P_K\left(C_{iK} - C_{eK}e^{-\frac{V_mF}{RT}}\right) + P_{Cl}\left(C_{eCl} - C_{iCl}e^{-\frac{V_mF}{RT}}\right) = 0 \quad (3.37)$$

$$e^{-\frac{V_mF}{RT}}(P_{Na}C_{eNa} + P_KC_{eK} + P_{Cl}C_{iCl}) = P_{Na}C_{iNa} + P_KC_{iK} + P_{Cl}C_{eCl} \quad (3.38)$$

$$e^{\frac{V_mF}{RT}} = \frac{P_{Na}C_{eNa} + P_KC_{eK} + P_{Cl}C_{iCl}}{P_{Na}C_{iNa} + P_KC_{iK} + P_{Cl}C_{eCl}} \quad (3.39)$$

Finally resting potential V_R is derived as

$$V_R = V_m = \frac{RT}{F}\ln\left(\frac{P_{Na}C_{eNa} + P_KC_{eK} + P_{Cl}C_{iCl}}{P_{Na}C_{iNa} + P_KC_{iK} + P_{Cl}C_{eCl}}\right) \quad (3.40)$$

This equation is called as Goldman–Hodgkin–Katz equation.

In contrast to the Nernst potential, the permeability of the ions is involved in the Goldman potential. It means even though their concentration gradients are the same, ion of a high permeability has a greater influence on the development of resting potential. For example, in the squid axon, the permeability of the potassium is almost 25 times higher than the permeability of the sodium. This difference in the permeability results in the resting potential much closer to the Nernst potential of potassium ion than the Nernst potential of sodium ion.

If all other ions are excluded except one type of ion, Goldman potential returns to the Nernst potential. In this case, permeability is also canceled out.

For the ion distribution in Table 3.3, resting potential is given by

$$V_R = 25.8 \cdot \ln\left(\frac{0.07 \cdot 145 + 1.8 \cdot 4 + 0.8 \cdot 4}{0.07 \cdot 12 + 1.8 \cdot 155 + 0.8 \cdot 120}\right) = -75 \text{ mV} \quad (3.41)$$

In most mammalian cells, the resting membrane potential varies from cell to cell. It is measured inside negative up to −100 mV. In typical neurons, the resting membrane potential is around −70 mV, and in typical muscle cells, it is around −90 mV. In nonexcitable cells, such as the epithelial cells and the red blood cells, the resting membrane potential is generally lower than excitable cells. It is around −50 mV for the epithelial cells and −5~−8 mV for the red blood cells and does not change appreciably over time (Hoffman, 1974). In the sympathetic neuron of frogs, it was measured with almost similar value of −75 mV (Jones, 1989).

5 Ionic Currents in Resting State

5.1 Ionic Current by the Difference of the Resting Potential from Its Nernst Potential

According to Eqs. (3.32, 3.33, and 3.34), ionic currents during the resting potential of −75 mV become I_{Na} = 357 nA/cm^2, I_K=−285 nA/cm^2, and I_{Cl}=−72 nA/cm^2. Even though during the resting state, the membrane potential is constant with Goldman potential and there is no net current across the membrane, there still are ionic currents flowing through the membrane. This is caused by the difference between the Goldman potential and their Nernst potential. For sodium ions, there generated −139 mV of equivalent electric force pushing sodium ions inward and resulting in 357 nA/cm^2 ionic currents. For potassium ions, the developed resting potential is closer to its Nernst potential and it causes 19.3 mV of equivalent electrical force. It pushes potassium ions outward and results in a resting ionic current of −285 nA/cm^2. Potassium current flows in opposite direction to sodium current and cancels a large part of sodium current. For chloride ions, there generated 12.8 mV of equivalent electric force pushing chloride ions inward and resulting in −72 nA/cm^2 ionic currents. The sum of potassium and chloride currents exactly cancels out inward potassium current and makes net membrane current totally null.

The equivalent membrane conductance of each ion is calculated for the example case as in Fig. 3.13. Sodium has the lowest conductance of 2.56 μS/cm^2, and potassium has the highest conductance of 14.76 μS/cm^2, which is about 6 times higher than sodium. While sodium current is developed by a large electric force multiplied by low conductance, a similar amount of potassium current is developed by a small electric force multiplied by high conductance. The low electric force of chloride and its medium level of conductance cause a small chloride current which just fills the gap between sodium and potassium currents.

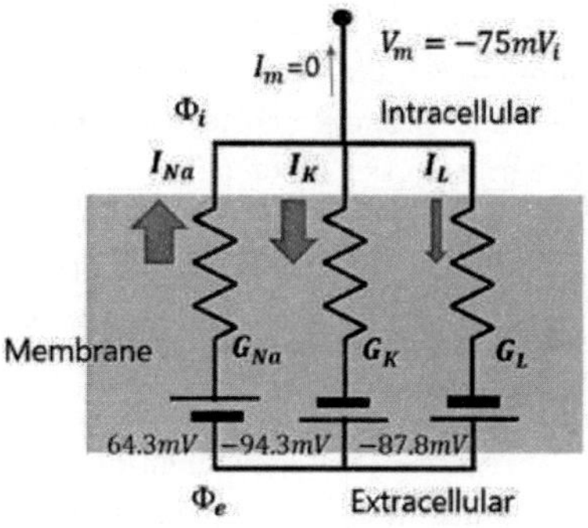

	Sodium	Potassium	Chloride
Resting potential (V_R)	$-75mV$	$-75mV$	$-75mV$
Nernst potential	$64.3mV$	$-94.3mV$	$-87.8mV$
$V_R - V_{Nernst}$	$-139.3mV$	$19.3mV$	$12.8mV$
Conductance ($\mu S/cm^2$)*	2.56	14.76	5.63
Ionic current (nA/cm^2)	357	−285	−72

*S: Siemens = (R: Resistance)$^{-1}$

Fig. 3.13 Current of each ion during the resting state

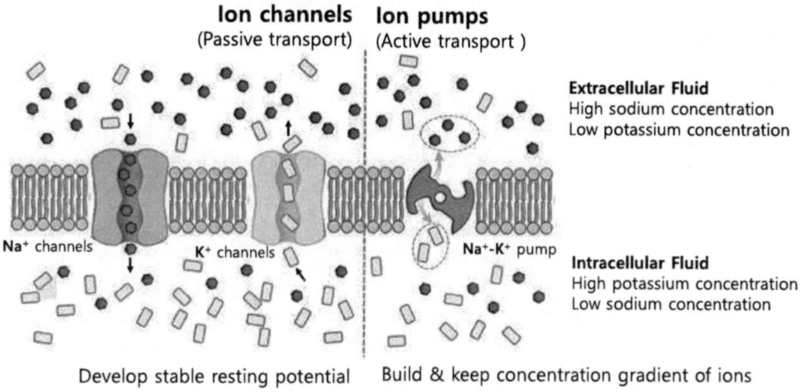

Fig. 3.14 Ion channels and ion pumps for the development of resting potential among multiple ions

5.2 Active Transport of Ion Pumps

Up to this point, the development of resting potential was explained within the scope of passive transport. Resting potential is produced by the balance of ionic flows which are developed by the diffusion gradient and potential gradient applied to each ion. This movement is passive. It does not require the use of any energy. But, if the resting state continues with passive transport only, ion concentrations of the resting state are not maintained. Sodium inflow will increase the internal concentration of sodium ions above the balanced level and internal potassium concentration will be lowered by the outflow of potassium ions as the resting state is continued. And the ion concentrations we have applied for the derivation of Goldman potential are not maintained and the derived Goldman potential will be no more correct.

But there is active transport. Active transport is the transport of ions across the membrane overcoming the gradients by using the energy. It is just like a pump that raises water by expending energy. This is done by the special proteins called ion pumps. They are also embedded in the membrane as ion channels and actively transport ions against gradients falling on each type of ion. The most relevant ion pump to the generation of neuronal potential is the sodium–potassium (Na^+-K^+) pump. They are known as exchanging three sodium ions outward with two potassium ions inward using the energy derived from a single ATP molecule. This 3:2 ratio is almost equivalent to their ionic current ratio during resting state, which is 357 nA/cm^2 and 285 nA/cm^2, respectively, in the example case. Sodium ions moved in during the resting state by the developed passive force are moved out by the sodium–potassium pump. Similarly, potassium ions moved out during the resting state are moved in by this sodium–potassium pump. The original balance of ion concentrations is maintained by this active transport using the energy and the resting potential continues. Using energy in a resting state is the one reason that we become hungry even though we do nothing and rest all day.

This can be considered in the reverse direction. Live cells continuously pump ions actively in and out of the membrane with the consumption of energy. As a result, concentration differences are built up in charged particles, and the potential is generated to balance out this diffusion gradient. If the active transport of the ion pump is stopped, the developed potential will be decreased to ground level, and there will be no more neuronal potential and no more life (Fig. 3.14).

Sodium–potassium pump was discovered in 1957 by the Danish scientist Jens Christian Skou, who was awarded a Nobel Prize in chemistry in 1997 for this contribution. Before Skou, even though the existence of concentration gradients for Na^+ and K^+ across the plasma membrane was well-established, it was difficult to explain the maintenance of resting potential just with passive transport based on the selective permeability of the membrane. To explain this paradox, he proposed that the membrane should contain a structure which could pump out the Na^+ and pump in the K^+. Then, he isolated a special enzyme from a nerve cell and verified that this enzyme is involved in the active transport of Na^+ and K^+ across the cell membrane (Clausen & Persson, 2002).

5.3 Affecting Factors for Resting Potential

In resting potential represented by the Goldman equation as in Eq. (3.40), there are two types of variables affecting the calculated resting potential. These are permeabilities and ion concentrations in and out of the membrane and their difference.

Membrane permeability does a key role in developing resting potential. Importance feature is its selectivity on the ions. The permeability of each ion is different. In the resting state, potassium has a much higher permeability of more than 20 times than sodium ion. So, potassium ions move relatively easily across the membrane and their concentration gradient has more effect on developing a balancing potential gradient than sodium ions. By permeability dominance, resulted resting potential is close to the Nernst potential of potassium ion only. On the contrary, sodium ion does not contribute significantly to resting potential development since its permeability is much lower than potassium ion. It is rather similar that the membrane totally blocks sodium movement across the membrane. For sodium ions, resting potential is developed in the opposite direction by the potassium ion, which has the reverse concentration gradient.

If permeabilities are the same for sodium and potassium ions, there is no difference in the development of resting potential since both ions have a single positive charge. High extracellular sodium concentration will cancel out high potassium concentration in intracellular fluid. When two ions are combined, the internal side is only $1.12(=(12 + 155)/(145 + 4))$ times higher than the external side. High concentration ratio difference of 12 : 1 for sodium ions and 38 : 1 for potassium ions disappear. If the average permeability of 0.935 is used, Goldman potential of Eq. (3.40) becomes −14.7 mV. Most of this value is developed by chloride ions. When chloride ion is neglected, resting potential developed by sodium and potassium ion is only −2.94 mV. With this level of resting potential, nerve cells may remain stable continuously. But it is not sufficient for the development of action potential, which requires a large potential gradient to support a rapid influx of sodium ions in the beginning phase of an action potential.

Change in permeability regulates the flow of ions across the membrane. If permeability for the sodium is increased significantly, there will be a large amount of sodium ion influx, and the increase of potassium ion permeability increases the outflux of potassium ions. Even though permeability is not changing during the stable resting state of the constant resting potential, it is dependent on membrane potential. Permeabilities are changing dramatically, during the generation of the action potential, when the membrane potential increased over some threshold potential.

Another important affecting factor is active transport. Within the body, there are several kinds of active transport for each type of molecules and ions. Active transport is the movement of particles against their concentration gradients with the consumption of energy. Sodium–potassium pump is the most known active transport and explains the most active transport in neuronal potential generation. Life seems to start with the active transport of ions. Active transport builds up an ion concentration gradient across the membrane. Gradients of ion concentrations are balanced by the potential gradient to maintain a stable resting state in the environment of multiple ions. During the resting state, there is no net current flowing across the membrane with the determined resting potential. But developed resting potential does not make all ions stay in their resting state. Because developed resting potential is different from its own Nernst potential for each type of ions, there exists a potential gradient to move the ions by this difference. This potential gradient inevitably produces ionic currents over the membrane. If moved ions are not returned, concentration differences will reduce continuously, and resting potential also disappears. Sodium–potassium pump returns the ions moved in and out of the membrane back to the higher concentration side by using energy so that they can move passively again. Even though the membrane has selective permeability, membrane potential cannot be maintained if there were no active transport. Even though active transport builds up the concentration gradient of each ion, it is not possible to develop membrane potential without selectivity of membrane permeability. Permeability and active transport are the two key players for the generation of resting potential, and also for the generation of the action potential (Fig. 3.15).

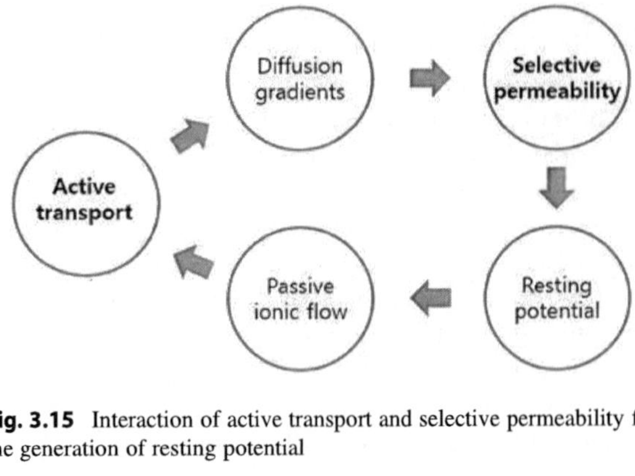

Fig. 3.15 Interaction of active transport and selective permeability for the generation of resting potential

6 Membrane Potentials to Subthreshold Stimulation

The membrane does not maintain the resting potential for a long time. While the nerves are alive, there are many inputs from the other neurons and various types of stimulations from the environment. These inputs and stimulations produce additional electric current within the cell and through the membrane. Membrane potential can be no more stable and the resting potential used to be disrupted.

If the stimulation is strong enough to induce membrane potential over the threshold, the membrane changes its dynamics, and an action potential is generated. If the stimulation is not strong and the induced membrane voltage is lower than the threshold, the action potential does not occur and the disturbed membrane potential returns back to resting potentials smoothly. Even though the stimulation is not strong enough to generate an action potential, it disturbs the resting potential and membrane potential varies in time and space. Let's see how the membrane potential changes over time and along the axis of the axon when there is some stimulus of electric current.

Figure 3.16 shows the segment of the axon of unmyelinated fiber during the stimulation with a stimulating current I_s. When the stimulating current is delivered to the internal part of the axon, current I_i flows longitudinally along the internal part of the axon since the axoplasm which fills the interior part of the axon is conductive. As the current flows along the axon, some part of the current also leaks through the membrane, since the membrane is not a perfect insulator. This leaky current constitutes membrane current I_m. Leak current through the membrane flows through the extra fluid and collected extra fluid current I_e returns to the reference electrode. At a certain point of distance from the stimulating site, the internal axial current I_i and the external fluid current I_e flow in opposite direction, and their magnitudes are the same, while there is no inputted current at the point.

$$I_e = -I_i \tag{3.42}$$

In contrast to the resting state where there is no current flowing along the axon, the internal current flowing along the axon produces a voltage drop and the potential diminishes along with the axial distance. Generally, membrane potentials decrease along the axon. The membrane potential also does not follow the time course of the stimulation current. When stimulating current is applied in a step function, membrane potentials do not follow the rapid increase of current as it is. Membrane potential increases gradually toward the maximum value as the stimulating current continues.

6.1 Distributed Electrical Equivalent Model

Figure 3.17 is the extended equivalent circuit model of Fig. 3.11 for the axon to explain the potential distribution along with the distance. The minute axon segment is

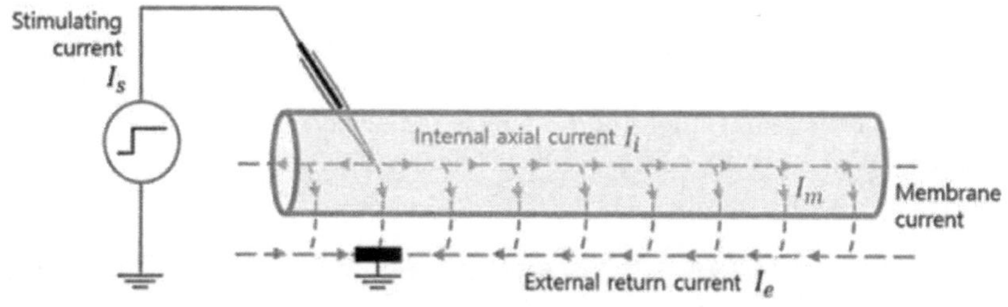

Fig. 3.16 Currents in axons segment during the stimulation with current

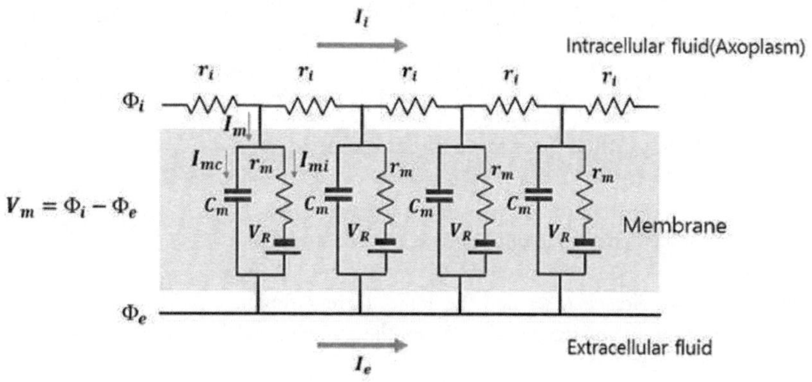

Fig. 3.17 Equivalent circuit model of an axon with distributed resistances and capacitances

represented with lumped elements of internal resistance (r_i), membrane resistance (r_m),and membrane capacitance (C_m). The model can include external resistance representing resistance when the current flow through extracellular fluid. However, because the cross-section of the external fluid space is much larger than the cross area of the internal axon, generally external resistance is much lower than internal resistance and is neglected in the model for simplicity.

If there is no stimulating current, there is no axial current, no external current, and no membrane current as in the resting state.

$$I_s = 0 \rightarrow \; I_i = I_e = I_m = 0 \tag{3.43}$$

The stable membrane potential of the resting state is maintained along the axial direction and there is no voltage drop along the axonal direction.

$$V_m = V_R, \;\; V' = V_m - V_R = 0 \tag{3.44}$$

$$\frac{dV_m}{dx} = \frac{dV_R}{dx} = \frac{dV'}{dx} = 0 \tag{3.45}$$

But, if a stimulating current is applied, it will flow along the axonal direction producing an internal axial current. While the current flows along the axonal direction, a small amount of current flows through the membrane since the membrane is not a perfect insulator. On the external side, transmembrane currents are collected and returned to the external fluid current which is equal in magnitude but flows in opposite direction to the internal axial current.

$$I_s \neq 0 \;\; \rightarrow \; I_i \neq 0, \;\; I_m \neq 0, \;\; I_e \neq 0 \tag{3.46}$$

Membrane potential also deviates from the resting potential and there appears voltage gradient along the axial direction.

$$V_m \neq V_R, \;\; V' = V_m - V_R \neq 0 \tag{3.47}$$

$$\frac{dV'}{dx} = \frac{dV_m}{dx} - \frac{dV_R}{dx} = \frac{dV_m}{dx} \neq 0 \tag{3.48}$$

6.2 Steady-State Potential of Stimulating Current

Transmembrane current (I_m) during the electrical stimulation consists of two parts, the capacitive current flowing through membrane capacitance (I_{mc}) and the ionic current flowing through membrane resistance (I_{mi}). When the stimulation is applied in a step function, the capacitive membrane current appears only at the starting point, and it soon disappears during a stimulating current is applied constantly. In the steady state, three is no capacitive current component and only the ionic current component remains in the membrane current.

$$I_m = I_{mc} + I_{mi} = I_{mi} \tag{3.49}$$

After membrane potential reaches a steady state, it does not vary over time.

$$\frac{dV'}{dt} = \frac{dV_m}{dt} - \frac{dV_R}{dt} = \frac{dV_m}{dt} = \frac{d\Phi_i}{dt} = 0 \tag{3.50}$$

According to Ohm's law, voltage drops along with the distance.

$$\frac{d\Phi_i}{dx} = -r_i I_i \tag{3.51}$$

Membrane current is the decreased part of the axial current.

$$I_m = -\frac{dI_i}{dx} = \frac{1}{r_i}\frac{d^2\Phi_i}{dx^2} \tag{3.52}$$

Varied membrane potential from resting potential also changes along with the axial distance, and its rate of change is equal to the changing rate of internal potential.

$$\frac{dV'}{dx} = \frac{dV_m}{dx} - \frac{dV_R}{dx} = \frac{dV_m}{dx} = \frac{d\Phi_i}{dx} - \frac{d\Phi_e}{dx} = \frac{d\Phi_i}{dx} \tag{3.53}$$

Differentiating the varying membrane potential twice is given as

$$\frac{d^2V'}{dx^2} = \frac{d}{dx}\left(\frac{d\Phi_i}{dx}\right) = \frac{d}{dx}(-r_i I_i) = -r_i\frac{dI_i}{dx} = r_i I_m \tag{3.54}$$

Since there is only ionic current in the steady state, ionic membrane current flowing through membrane resistance r_m is calculated by dividing the voltage over resting potential by the membrane resistance.

$$I_m = I_{mi} = \frac{V'}{r_m} \tag{3.55}$$

Substituting Eq. (3.55) into (3.54) becomes

$$\frac{d^2V'}{dx^2} = r_i I_m = r_i\frac{V'}{r_m} = \frac{r_i}{r_m}V' \tag{3.56}$$

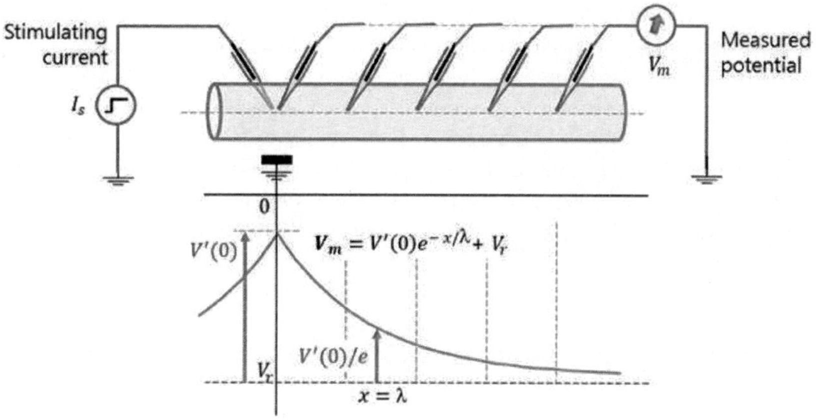

Fig. 3.18 Membrane potential distribution in the steady state along with the axial distance to a consistent current stimulation

Finally, the equation describing the raised membrane potential along the axial distance has been derived as follows.

$$\frac{d^2V'}{dx^2} = \frac{r_i}{r_m}V' \tag{3.57}$$

This second order differential equation is easily solved as

$$V'(x) = V'(0)e^{-x/\lambda} \tag{3.58}$$

where $\lambda = \sqrt{r_m/r_i}$ is a characteristic length constant.

As shown in Fig. 3.18, consistent electrical stimulation depolarizes membrane potential. Continuous inward electrical current raises the intracellular potential above the resting potential. A major part of the stimulating current flows along the axon and increases the potential of the neighboring site above the resting potential. Since there is a minor current flowing out through the membrane, the current flowing along the axial direction decreases gradually and the amount of raised potential also decreases. Raised potential decreases exponentially along the axial direction with characteristic length constant λ which is determined by the ratio of membrane resistance to internal axial resistance. Spatial distribution of membrane potential is maintained while the stimulating current is applied consistently and maximal raised potential is still under the threshold potential of the membrane to generate an action potential.

The characteristic length constant λ represents how the stimulated current can be transmitted away from the stimulation site within the subthreshold voltage range. For the example case of squid axon of diameter 0.5 mm, the characteristic length constant is expressed as follows.

Membrane resistance r_m per unit axon length is given by

$$r_m = \frac{\rho_m l_m}{\pi a(1)} = \frac{(5.33\times10^7\Omega\text{m})\cdot(7.5\times10^{-9}\text{m})}{\pi(0.5\times10^{-3}\text{m})} = 2.54\times10^4\Omega\text{cm} \tag{3.59}$$

Intracellular axial resistance r_i per unit axon length is given by

$$r_i = \frac{\rho_i(1)}{A_i} = \frac{\rho_i}{\pi\left(\frac{a}{2}\right)^2} = \frac{(0.6\ \Omega\text{m})}{\pi(0.5\times10^{-3}\text{m})^2/4} = 3.06\times10^4\Omega/\text{cm} \tag{3.60}$$

Then characteristic length constant of squid axon is calculated by

$$\lambda = \sqrt{r_m/r_i} = \sqrt{2.54\times10^4\Omega\text{cm}/(3.06\times(10^4\Omega)/\text{cm})} = 0.91\text{cm} \tag{3.61}$$

Axon with a diameter of 0.5 mm transmits potential only 0.91 cm away along the axial direction from the stimulating site. It means that raised potential will drop to 36.7% level just after it transmits 0.91 cm, which is too short to transmit signals within the body.

How the characteristic length will change if the internal fluid is simply replaced with the copper as in the electrical wires having the resistivity as in Table 3.2.

$$\lambda = \sqrt{r_m/r_i} = \sqrt{\frac{(5.33\times10^7\Omega\text{m})\cdot(7.5\times10^{-9}\text{m})}{\pi(0.5\times10^{-3}\text{m})}\Big/\frac{(1.62\times10^{-8}\Omega\text{m})}{\frac{\pi(0.5\times10^{-3}\text{m})^2}{4}}} = 55.5\text{m} \tag{3.62}$$

Characteristic length is increased about 6000 times from 0.91 cm to 55.5 m. It is long enough within the human body range to transmit electrical signals by a single neuron with the membrane in a passive state. Then electrical signals also can be transmitted much faster like in electrical systems. Why human and biological systems did not evolve in this direction of efficient transmission based on high conductivity metal as a core component of a neuronal element? Copper is heavier than water by 8.96 times. If intracellular fluid were replaced with copper, our body becomes heavier by 8.96 times. Having a lighter bodyweight with rather a slow transmission might have caused better survival than a heavier body with faster signal transmission.

6.3 Membrane Potential to Step Current Stimulation

During the steady state, there is no current flowing through the capacitance, since the membrane potential does not vary over time. However, this steady state is not common, and most of the time there are stimulation and membrane voltages vary over time. So, we cannot neglect the membrane current in most cases.

When we input the equation $I_m = I_{mi} + I_{mc}$ into the Eq. (3.54), the raised membrane potential is represented by

$$\frac{d^2V'}{dx^2} = r_i I_m = r_i(I_{mi} + I_{mc}) = r_i\left(\frac{V'}{r_m} + c_m\frac{dV'}{dt}\right) = \frac{r_i}{r_m}V' + r_i c_m\frac{dV'}{dt} \tag{3.63}$$

$$-\lambda^2\frac{d^2V'}{dx^2} + \tau\frac{dV'}{dt} + V' = 0 \tag{3.64}$$

, where $\lambda = \sqrt{r_m/r_i}$ is a characteristic length constant and $\tau = r_m c_m$ is a time constant of the membrane.

In nondimensional form with normalization to characteristic length constant $X = x/\lambda$ and time constant $T = t/\tau$, Eq. (3.64) becomes

$$\frac{dV'}{dT} = \frac{d^2V'}{dX^2} - V' \tag{3.65}$$

This is the cable equation that applies in describing potential over electrical cable as well as in membrane potential. For an ideal axon cable extending to $\pm\infty$in both directions from the stimulating point, a stimulating current I_s is applied at a middle point $x = 0$. The solution for this cable equation obtained directly for these conditions is (Holmes, 2014)

$$V'(X,T) = \frac{r_i I_s \lambda}{4} \times \left\{e^{-X} erfc\left(\frac{X}{2\sqrt{T}} - \sqrt{T}\right) - e^{X} erfc\left(\frac{X}{2\sqrt{T}} + \sqrt{T}\right)\right\} \tag{3.66}$$

, where *erfc* is complementary error function.

For special case of stimulating site of $x = 0$, membrane potential in given by

$$V'(0,T) = \frac{r_i I_s \lambda}{4}\left\{erfc\left(-\sqrt{T}\right) - erfc\left(\sqrt{T}\right)\right\} = \frac{r_i I_s \lambda}{4}\left\{2 - 2erfc\left(\sqrt{T}\right)\right\} = \frac{r_i I_s \lambda}{2} erf\left(\sqrt{T}\right) \tag{3.67}$$

Figure 3.19 shows how the membrane potential is varying along with the distance at a specific time after stimulation onset and along the time at a specific location. At a stimulating location, resting potential approaches 84% of its steady-state value within a time of one time constant, which is faster than the general exponential curve approaching 67% of its maximum during the same time interval.

We have applied the cable equation for the explanation of subthreshold potential. This equation is derived with the assumption that the membrane conductance does not vary during the stimulation. These are the cases when postsynaptic potentials are transmitted along the dendrite after they are generated in the location of synapses. Many types of receptors do not generate an action potential. They generate receptor potentials at the subthreshold level and transmit the potentials to the neurons generating action potentials and transmitting them afferently.

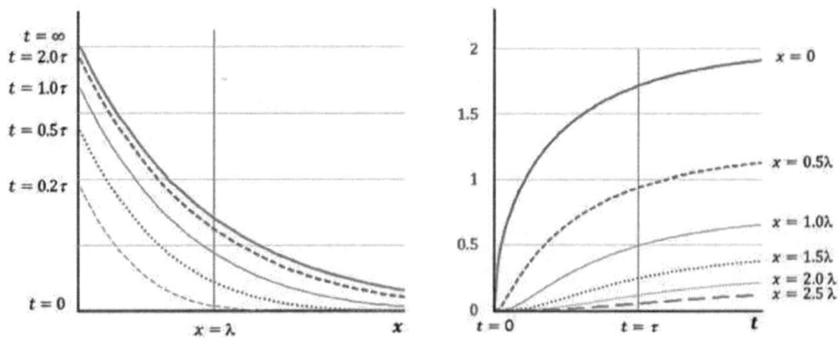

Fig. 3.19 Distribution of the membrane potential along the axon at some discrete time and along the time at some distance from stimulating site after the onset of electrical stimulation

References

Chitwood, R. A., Hubbard, A., & Jaffe, D. B. (1999). Passive electrotonic properties of rat hippocampal CA3 interneurons. *The Journal of Physiology, 515*, 743–756.

Clausen, T., & Persson, A. E. G. (2002). Jens Christian Skou awarded the Nobel prize in chemistry for the identification of the Na+, K+-pump. *Acta Physiologica Scandinavica*. https://doi.org/10.1046/j.1365-201x.1998.00367.x

Gentet, L. J., Stuart, G. J., & Clements, J. D. (2000). Direct measurement of specific membrane capacitance in neurons. *Biophysical Journal, 79*(1), 314–320.

Hawkins, J., & Ahmad, S. (2016). Why neurons have thousands of synapses, a theory of sequence memory in neocortex. *Frontiers in Neural Circuits, 30*, 23. https://doi.org/10.3389/fncir.2016.00023

Hoffman, J. F. (1974). Determination of membrane potentials in human and amphiuma red cells by means of a fluorescent probe. *The Journal of Physiology, 239*, 519–552.

Holmes, W. R. (2014). Cable equation. In *Encyclopedia of computational neuroscience*. Springer Science+Business Media. https://doi.org/10.1007/978-1-4614-7320-6_478-1

Jones, S. W. (1989). On the resting potential of isolated frog sympathetic neurons. *Neuron, 3*(2), 153–161.

Major, G., Larkman, A. U., Jonas, P., Sakmann, B., & Jack, J. J. (1994). Detailed passive cable models of whole-cell recorded CA3 pyramidal neurons in rat hippocampal slices. *The Journal of Neuroscience, 14*, 4613–4638.

Thurbon, D., Luscher, H. R., Hofstetter, T., & Redman, S. J. (1998). Passive electrical properties of ventral horn neurons in rat spinal cord slices. *Journal of Neurophysiology, 79*, 2485–2502.

Zimmermann, U., Pilwat, G., & Riemann, F. (1998). Dielectric breakdown of cell membranes. *Biophysical Journal, 14*(11), 881–899.

Action Potential

4

A binary system, which represents the states only by "0"and "1", is the most essential system to represent data and information. Like the current advanced electrical systems are based on this digital system, the neural network system of our body is also based on this binary system. While the resting potential represents the nonactive ground state "0", the action potential is considered representing the active state "1". As 3.3 V or 5.0 V of a rectangular pulse represents the active state in an electrical digital circuit, 100 mV of a determined shape of action potential represents the active state in the human neural circuit.

Until the end of the eighteenth century, the connection between electricity and nervous function was not explored and most contemporary researchers still believed that the nervous system is operating via the movement of nervous fluids or ethers. The connection between nervous function and electricity was first verified by the Luigi Galvani in 1791 based on the countless experiments with frog muscles. He also demonstrated that animal electricity is free from metallic electricity artifacts by showing that a muscle can be stimulated by bringing its nerve into contact with a skinned part of the muscle (Schuetze, 1983). Even though the following studies on animal electricity were not active by late 1930, researchers had recognized several important things about the neuronal potentials. They knew that signaling within the neuron is electrical in nature in contrast to the synaptic signaling between the neurons, that it occurs in bursts of activity called action potentials, and that action potentials are stimulated by the increase of membrane permeability on sodium ions. Alan Hodgkin and Andrew Huxley discovered the full details of membrane permeability changes during the action potential in their Noble prize-winning work published serially in 1952. In this chapter, the generation of an action potential is explained based on the Hodgkin–Huxley model.

1 Potential Variation During Action Potential

1.1 Types of Membrane Potential Variation

Membrane potentials are varying in different directions from one state to another. These are the types of membrane potential variation upon their directions (Fig. 4.1).

- *Polarization*: In membrane potential, polarization means the generation of the polar characteristics of potential deviated from "0" potential. Resting potentials of any negative potential are in a polarized state. Polarization generally means that the membrane is in a resting state with its negative resting potential.
- *Depolarization*: This means the escaping of potential from the polarized state. Since the polarization state is the state of negative potential, depolarization means the disappearance of a negative membrane potential toward "0" potential. When there is an overshoot in potential variation crossing the potential toward reverse polarized direction by a small amount, we still use the term "depolarization." It is because most of the variation is devoted to escaping from the polarized state. Inward current increases the membrane potential and depolarizes the membrane and occurs in the early phase of action potential generation.
- *Repolarization*: Repolarization means the polarization occurred again. So, repolarization is the same as polarization from an electrical point of view, and generally means polarization when it occurred just after the depolarization. Outward current decreases the membrane potential and repolarizes the membrane and occurs in the later phase of action potential generation.

K. S. Park, *Humans and Electricity*, https://doi.org/10.1007/978-3-031-20784-6_4

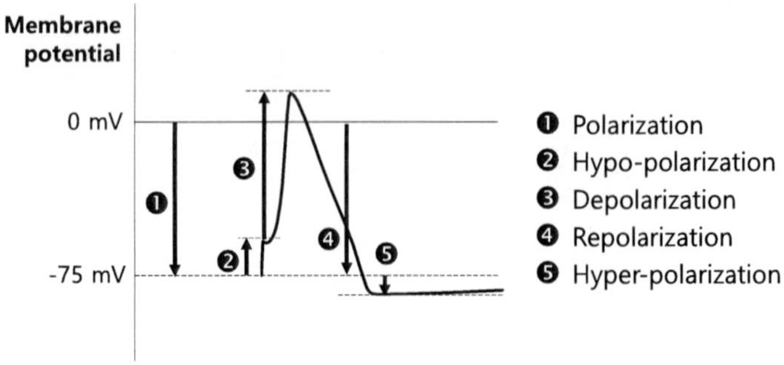

Fig. 4.1 Potential variation during the generation of an action potential

- *Hyperpolarization*: Hyperpolarization means increased polarization. When the resting potential is −75 mV, if the potential is lowered more than the resting potential to −80 mV, it is said that the membrane is hyperpolarized. The efflux of positive charges, which is an outward current, decreases the internal potential and hyperpolarizes the membrane.
- *Hypopolarization*: Hypopolarization is the reverse direction of potential variation to hyperpolarization, and means the decreased polarization. When the resting potential of −75 mV is increased to −60 mV, it is said that the membrane is hypopolarized. The influx of positive charges, which is an inward current, increases the internal potential and hypopolarizes the membrane. The direction of hypopolarization is the same as depolarization, but the amount of potential variation is not so big to approach "0" potential.

1.2 Time Course of Action Potential Variation

The action potential is generated from the neuron at a resting state. Neurons are polarized in the resting state into negative potential, for example, of −75 mV. If there is any disturbance that increases the membrane potential, the membrane is hypopolarized to a lesser polarized state. If the disturbance is not strong and the hypopolarized potential is lower than the threshold potential to fire action potential, membrane potential smoothly returns to the polarized resting-state potential as the stimulus occurred stayed in the subthreshold level. However, when the occurred stimulus is strong enough to hypopolarize the membrane potential over the threshold potential, there fires an action potential. Membrane potential rapidly depolarizes toward "0" potential and crosses "0" potential and reaches some positive potential, for example, +15 mV. After it arrived at the peak potential, membrane potential starts to repolarize back to its resting potential. Instead of the membrane potential stopping immediately as it approaches its resting potential level, there is an undershoot of the membrane potential. This hyperpolarized membrane smoothly returns to its resting potential with time. Finally, membrane potential returns to its resting potential and finishes one cycle of an action potential generation routine.

How the action potential follows the determined routine automatically was a great concern in the study of neural electricity. Resting potential is established based on the selective characteristics of the membrane to the movement of ions. The high permeability of potassium ions practically determines the resting potential, while the practically blocked low permeability of sodium ions does not contribute effectively to establishing the resting potential. For sodium ion, established resting potential is in opposite direction to its Nernst potential. In this circumstance of resting potential state, depolarization of membrane potential toward the Nernst potential of sodium ions seems to be clear as the result of increased permeability to sodium ions. The most plausible explanation for the increased permeability of sodium ions seemed to be the loss of membrane selectivity on the ions. Loss of selectivity of ions means totally opening all ion channels and allowing all charged particles to move freely across the membrane. In this case, the permeability of sodium increases greatly, since it was almost blocked during the resting state, and it explains the depolarization of membrane potential toward the Nernst potential of the sodium ions. And, when the depolarizing membrane potential looks to approach "0" potential in earlier recordings, loss of membrane selectivity seemed to be the appropriate explanation for the establishment of the action potential.

But, when Hodgkin and Huxley first recorded the action potential intracellularly, they recognized that there is an overshoot of membrane potential and the action potential is rising across the "0" potential to a positive peak value, for example, of +15 mV. This recording cannot be explained by the total loss of membrane selectivity. It seems to be clear that there is another mechanism changing the permeability of sodium ions more actively. A kind of positive feedback seems to be involved in a rapid increase of membrane potential in the early phase of action potential establishment, and a kind of negative feedback might explain the returning of action potential to its resting potential in the later phase.

1.3 Recording of Action Potential

Membrane potential can be measured intracellularly or extracellularly (Fig. 4.2).

The intracellular recording is used to record the membrane potential at a specific location of nerve or muscle with a reference electrode located in the extracellular fluid. The intracellular recording allows a very accurate assessment of the electrical activity at the location, but there is a considerable chance of damage to the membrane during the insertion of the electrode. Also, it is very difficult to do in human nerve cells and in most vertebrate cells which are in the range of 10–100 μm in diameter.

The extracellular recording places one electrode on the external membrane of the nerve cell or in close proximity to the excitable cell with the reference electrode at some location in the extracellular fluid. This arrangement records potential changes at the membrane surface rather than across the membrane, which is a far less demanding technique for the accurate analysis of membrane potentials. Action potentials recorded extracellularly differ from those recorded intracellularly in some aspects. The size of the recorded action potentials is generally reduced and the waveform for action potentials shows a biphasic or triphasic shape rather than a monophasic shape intracellularly recorded. Extracellular recordings are therefore better suited where one only wants to know the occurrence of an action potential or to record the activity of an entire population of nerve cells.

So, the extracellular recording is not appropriate to analyze the membrane potentials and the variation of its conductance along the time course. Though the intracellular recording is preferred for the analytic studies, human nerve cells and axons with a diameter less than 100 μm are too small to insert an intracellular electrode successfully. Here, the axon of squid has appeared as a very appropriate experimental subject. Squid has a giant axon whose diameter is in the range of 500 μm–1.0 mm, which is 10–100 times bigger than the mammalian axons. This axon controls the part of the water jet propulsion in squid and has evolved by increasing its diameter to improve the speed of escaping response for survival. The bigger size of axon diameter provides great advantages in intracellular electrode insertion. It is the giant axon of squid that made Hodgkin and Huxley record the first intracellular action potential (Fig. 4.3). After the measurement of the intracellular action potential, they did all the following experiments for the study of membrane electrical behavior during the action potential with this giant axon of a squid. They have stayed at Plymouth Marine Laboratory during their experiments where they can get fresh samples of squid axons more easily. The axon of squid is considered that it has done one of the key roles in the success of the Hodgkin–Huxley model development. It is the giant axon of a normal-sized squid, not the axon of the giant squid.

2 Design of Experiments for the Hodgkin–Huxley Model

When Hodgkin and Huxley started to develop a model, their object was to find equations that describe the conductances with reasonable accuracy and are sufficiently simple for theoretical calculation of the action potential and refractory period. We are going to follow the steps of experiments and analysis done by Hodgkin and Huxley. The first part is to

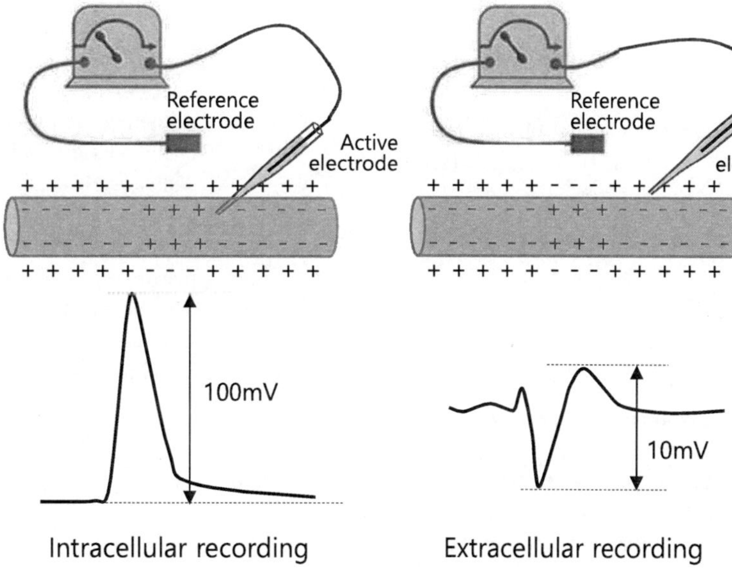

Fig. 4.2 Intracellular and extracellular recording of membrane potential

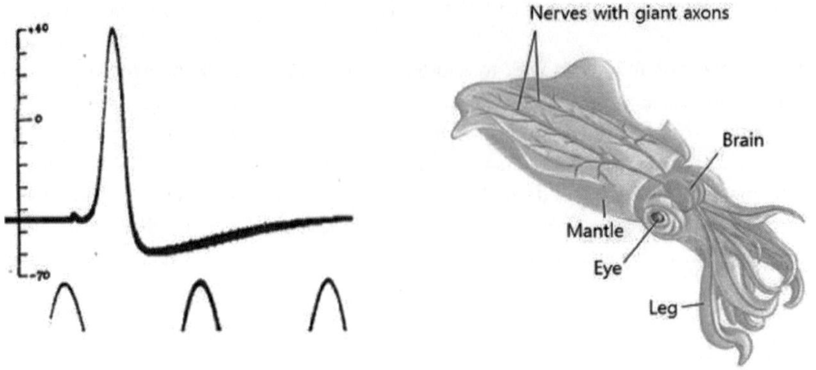

Fig. 4.3 First intracellular recording of an action potential by Hodgkin and Huxley with the giant axon of a squid. The vertical scale indicates the potential of the internal electrode in millivolts. The time marker is a waveform of 500 cycles/s. (Hodgkin & Huxley, 1939)

understand how they designed the experiments. When the membrane potential is hypopolarized from the resting potential and increases over the threshold level, the action potential starts to fire. And the membrane potential varies along the time and also along with the distance when action potential propagates along the axon. Membrane current also varies along the time and the axial distance as membrane voltage varies. Membrane current consists of the ionic conductive current and the capacitive current through the membrane capacitance. Ionic conductive current is contributed by sodium ionic current, potassium ionic current, and leakage current which also includes the contribution of chloride ion. Each ionic conductive current is represented by the multiplication of its ion-dependent conductance and electric driving force. Membrane conductance also varies along the time and axial distances. Membrane conductance is regarded as dependent on transmembrane voltage since permeability appears to depend on the membrane potential and not on the membrane current. The electric driving force for a particular ion is clearly zero at the Nernst potential for that ion. The electrical driving force may therefore be measured as the difference between the membrane potential and its Nernst potential (Hodgkin et al., 1952). The membrane current is described

$$I_m(x,t) = G_{\mathrm{Na}}(x,t,V)(V(x,t) - V_{\mathrm{Na}}) + G_{\mathrm{K}}(x,t,V) \times (V(x,t) - V_{\mathrm{K}}) + G_L(x,t,V)(V(x,t) - V_L) + C_m \frac{dV(x,t)}{dt} \tag{4.1}$$

where

$I_m(x, t)$: membrane current,
$V_m(x, t)$: membrane voltage measured intracellularly to the extracellular reference,
$G_{\mathrm{Na}}(x, t, V)$: sodium conductance depending on the membrane voltage,
$G_{\mathrm{K}}(x, t, V)$: potassium conductance depending on the membrane voltage,
$G_L(x, t, V)$: conductance for the leakage current including the chloride ion depending on the membrane voltage,
V_{Na}: Nernst potential of sodium ion,
V_{K}: Nernst potential of potassium ion,
V_L: Nernst potential for the leakage current including the chloride ion,
C_m: membrane conductance.

In Eq. (4.1), $V(x, t)$ and $I_m(x, t)$ are variables, V_{Na}, V_{K}, and V_{Cl} are constants representing Nernst potential of sodium, potassium, and chloride respectively for the given experiment set-up, and C_m is the membrane capacitance which is reasonably assumed not varying upon time and location. $G_{\mathrm{Na}}(x, t, V)$, $G_{\mathrm{K}}(x, t, V)$, and $G_{\mathrm{Cl}}(x, t, V)$ are the conductance parameters whose dependency on the transmembrane voltage needs to be analyzed. In this form, these parameters are too complex to analyze since they are varying along the time, location, and voltage simultaneously. Some efficient steps to simplify the problem are required.

2.1 Space Clamp

Space clamp is the first step to simplifying the problem and designing the experiment at a manageable level. The space clamp eliminates the space gradient along with the axial distance. This can be done by threading a highly conductive metallic wire as an electrode along the interior of the axon to short-circuit the whole longitudinal resistance of the axoplasm (Fig. 4.4). As the potential along the axial distance becomes equal, all space gradients of variables along with the axial distance x disappear. Equation (4.1) can be simplified by eliminating a space variable x.

$$I_m(t) = G_{\mathrm{Na}}(t,V)(V(t) - V_{\mathrm{Na}}) + G_{\mathrm{K}}(t,V) \times (V(t) - V_{\mathrm{K}}) + G_L(t,V)(V(t) - V_L) + C_m \frac{dV(t)}{dt} \tag{4.2}$$

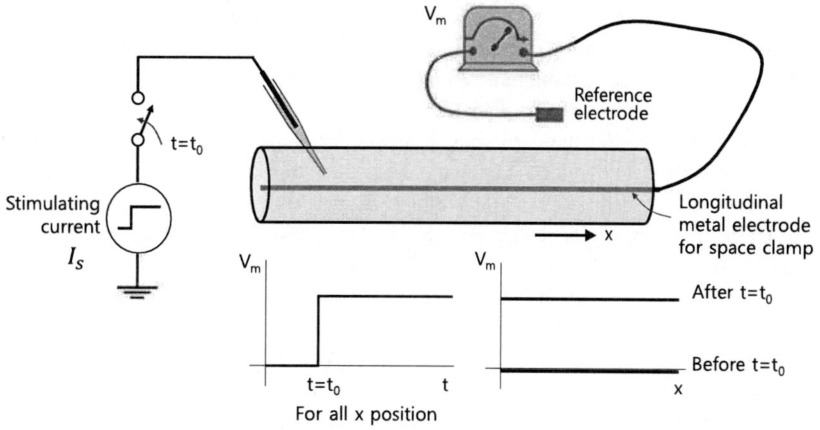

Fig. 4.4 Space clamp by the threading of a metallic wire electrode. Membrane potential changes uniformly throughout the whole length of an axon

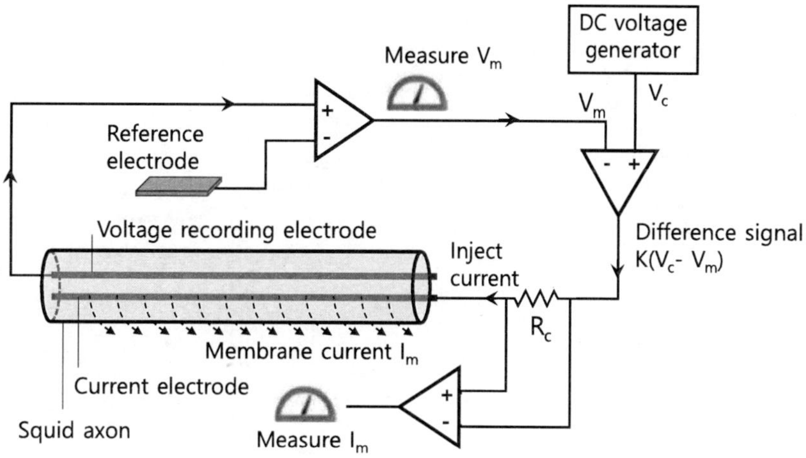

Fig. 4.5 Voltage clamp circuit

2.2 Voltage Clamp

Now the $G_{Na}(t, V)$, $G_K(t, V)$, are $G_L(t, V)$ are simplified by being independent to a space variable x. But they are still dependent on membrane voltage. While the membrane conductance is dependent on varying membrane potential during the action potential, it is very difficult to analyze its dependency on rapidly varying membrane voltage. If we set the membrane potential to some specific value and maintain this constant potential throughout the experiment, membrane conductance is no more dependent on voltage. Then, time t is the only independent variable and the whole characteristics of membrane conductance are recorded solely along the time. In addition, capacitive current vanishes as there is no more voltage gradient along the time. Solely, conductive current by the three different types of ions becomes the contributing component of membrane current. The voltage clamp is an experimental setup to fix the membrane potential to some predetermined voltage V_c throughout each of the experiments. Equation (4.2) is simplified again.

$$I_m(t) = G_{Na}(t, V_c)(V_c - V_{Na}) + G_K(t, V_c)(V_c - V_K) + G_L(t, V_c)(V_c - V_L) \tag{4.3}$$

Figure 4.5 describes the configuration of the voltage clamp circuit. Two longitudinal electrodes are inserted while maintaining the condition for the space clamp. One electrode is to record intracellular membrane potential, and the other one is to inject current so that the membrane potential is clamped to a targeted voltage V_c. The first amplifier measures membrane potential V_m in reference to the electrode located in the extracellular fluid. The second amplifier compares the measured voltage V_m to the targeted membrane voltage V_c and generates an electric current proportional to the difference in voltage ($V_c - V_m$). If the membrane voltage is lower than the clamping voltage ($V_c > V_m$), ($V_c - V_m$) becomes greater than zero and the current flows inward to increase the membrane potential toward clamping potential. Oppositely, if the membrane voltage is higher than the clamping voltage ($V_c < V_m$), ($V_c - V_m$) becomes less than zero and the current flows outward to decrease the membrane potential toward clamping potential. In both cases, a current is fed through the current injecting electrode so that the membrane potential V_m approaches to a targeted clamping voltage V_c. When the gain K is set to a very high value of 10^5–10^6, the membrane potential immediately follows the clamping voltage. Injected current flows through the measuring resistor R_c and the voltage drop across the measuring resistor R_c is measured by the third amplifier. From this signal, current injecting into the current electrode and then flowing through the membrane is measured.

Hodgkin and Huxley designed their own electrode unit with two silver wire electrodes in a single glass capillary shank as shown in Fig. 4.6. The external diameter of the electrode unit is 120 μm, and it is composed of 2 silver wire electrodes (20 μm in diameter) helically circulating over a glass capillary shank (70 μm in diameter). This electrode unit is inserted longitudinally 20 – 30 mm into the inside of the squid axon. Only the central portion of the wire is exposed, and the remaining outer portion is insulated. The current injecting wire is exposed along the whole length, while the voltage sensing wire only is exposed over a smaller portion in the center partition. This is to ensure that the current is flowing in a straight line while the voltage is just being measured.

2.3 Total Membrane Current During the Voltage Clamp

Figure 4.7 shows the current waveform measured with different clamping voltage settings. Each current curve is derived by switching the clamping voltage from the resting

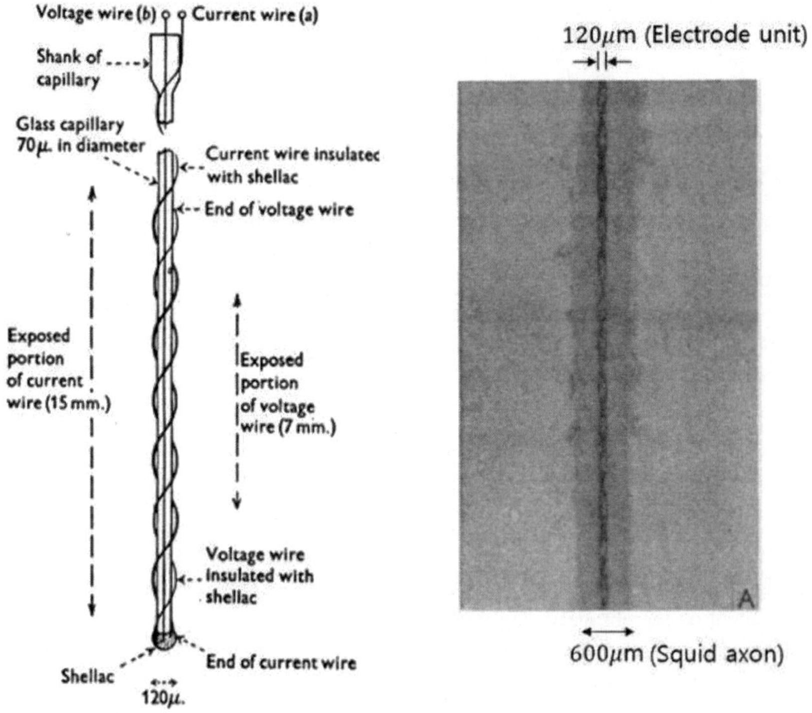

Fig. 4.6 Structure of the internal electrode unit Hodgkin and Huxley have used for their experiments and the photomicrographs of the electrode unit inserted into the squid axon of diameter 600 μm. (Hodgkin & Huxley, 1952a)

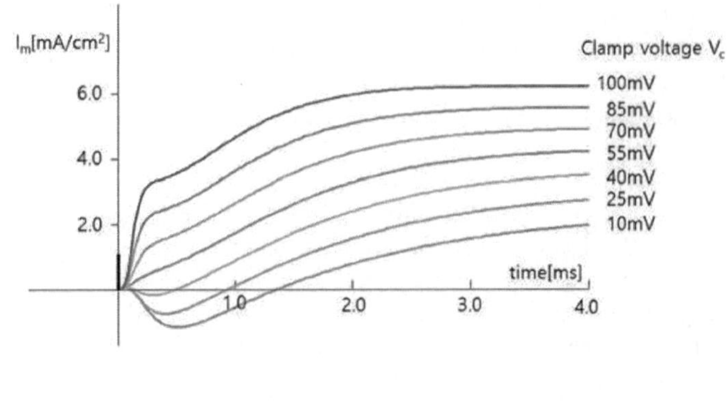

Fig. 4.7 Measured membrane currents with different clamping voltages

potential V_R of −65 mV to some clamping voltage V_c. Since the direction of clamping voltage is in increasing the membrane potential, changing to clamping voltage hypopolarizes and depolarizes the membrane. At the timing of switching, there is a pulse of capacitive current momentarily. But it disappears soon while the membrane voltage is clamped to a fixed constant voltage. Patterns of membrane current are different upon the clamping voltage. For lower clamping voltage less than 50 mV, the negative current appears first and then it changes to positive current later with different time delay. Here, negative current means the inflow current and it means that inward current is dominant in the early phase of the depolarization. After some time, the direction of the current flow changes, the outward current starts to increase, and it seems to approach some saturating level. Inward current decreases as the clamping voltages increases, while the clamping voltage is lower than 50 mV. Later asymptotic potential increases monotonically as the clamping voltage increases.

On the contrary, if the clamping voltage is higher than 50 mV, there appears no inward current at all. Instead, there appears a small hump in the beginning part. Membrane current flows totally outward from the beginning and seems approaching to some higher value of saturation. And, later asymptotic potential increases monotonically as the clamping voltage increases.

Since the voltage across the membrane is clamped, a change in membrane current can be considered caused by the change in membrane conductance. Conductance change can occur in any of or all of the ions contributing to membrane current. In this experimental setup, membrane current includes all conductive currents generated by sodium, potassium, and chloride ions. Since the concentrations of the sodium ions and potassium ions are differently dominant on the external and internal sides respectively, it can be guessed that this pattern is caused by the different roles of these ions. But, while their currents are summed in measured membrane current, the role of each ion cannot be differentiated. An additional strategy is required to separate each ionic component and evaluate its characteristics depending on voltage.

2.4 Separation of Ionic Current

In Eq. (4.3), there are three types of ionic current, $I_{Na} = G_{Na}(V_c - V_{Na})$, $I_K = G_K(V_c - V_K)$, and $I_L = G_L(V_c - V_L)$, sum together to membrane current I_m.

To measure current contributed by one ion only, current contributed by other ions except for one ion of interest need to be eliminated. In this case, the measured current becomes the ionic current solely developed by the ion of interest and the variation of the current curve directly depicts the conductance variation of the ion. To measure the current of potassium ion only, the ionic current established by the sodium ion and the leakage should be eliminated. This can be done by canceling out electrical driving force ($V_c - V_{Na}$) and ($V_c - V_L$) to zero. It makes sodium conductance and leakage conductance not effective. And then the measured current is solely the current established by the potassium ion.

$$V_{Na} = V_c,\quad V_c - V_{Na} = 0 \tag{4.4}$$

$$V_L = V_c,\quad V_c - V_L = 0 \tag{4.5}$$

$$I_m(t) = G_K(t, V_c)(V_c - V_K) \tag{4.6}$$

Nernst potential is determined by the ratio of ion concentration between the internal and external sides of the membrane. If we find a concentration ratio producing some targeted Nernst potential, we can obtain targeted Nernst potential by changing the internal or external concentration of the ion. For sodium ions, to set the Nernst potential equal to some targeted potential V'_{Na}, the ion concentration ratio is given by

$$V'_{Na} = \frac{RT}{zF}\ln\frac{[Na]'_e}{[Na]'_i} \rightarrow \frac{[Na]'_e}{[Na]'_i} = e^{\frac{zFV'_{Na}}{RT}} \tag{4.7}$$

Change of concentration ratio can be achieved by changing the external concentration while maintaining the internal concentration as it is. Compared to the original concentration of sodium ion, the external concentration of sodium ion needs to be increased k times.

$$\begin{aligned} k &= \frac{[Na]'_e}{[Na]_e} = \frac{[Na]'_e}{[Na]_i} \Big/ \frac{[Na]_e}{[Na]_i} \\ &= e^{\frac{zFV'_{Na}}{RT}} \Big/ e^{\frac{zFV_{Na}}{RT}} = e^{\frac{zF\left(V'_{Na} - V_{Na}\right)}{RT}} = e^{\frac{V'_{Na} - V_{Na}}{25.8}} \end{aligned} \tag{4.8}$$

where

V_{Na}: Nernst potential of the sodium ion with original ion concentration (mV),

V'_{Na} : targeted Nernst potential of the sodium ion with altered ion concentration (mV)

$[Na]_i$, $[Na]_e$: the original internal and external concentrations of the sodium ion (mM)

$[Na]'_i$, $[Na]'_e$: the altered internal and external concentration of the sodium ion (mM)

If we want to change the Nernst potential of 50 mV to new targeted Nernst potential of 24.2 mV, which is equal to the clamp voltage, according to Eq. (4.8), increase of external

Table 4.1 Derivation of new concentration according to the targeted Nernst potential

	Original	Targeted	
Internal concentration ($[Na]_i$)	66 mM	66 mM	Internal concentration remains unchanged
External concentration ($[Na]_e$)	460 mM	169 mM	External concentration increased by $k = 1/e$ times
Concentration ratio $\left(\frac{[Na]_e}{[Na]_i}\right)$	6.945	2.56	Concentration ratio increased by $k = 1/e$ times
Nernst potential (V_{Na})	50 mV	24.2 mV	Nernst potential is changed from 50 to 24.2 mV

concentration of $k = 1/e$ times is needed as in Table 4.1. This can be done by lowering the external concentration of sodium ions from 460 to 169 mM.

Alteration of concentration ratio also can be achieved by changing the internal ionic concentration while maintaining the external concentration unchanged. The required changing ratio of the internal concentration can be calculated similarly to Eq. (4.8). In a later study, a clever technique for changing the internal ionic concentration was developed (Baker et al., 1962). In this approach, normal internal fluid of axoplasm is squeezed out, and it is replaced with perfusion fluid of altered ion concentration.

To make the membrane current only for potassium ion, the Nernst potential of the leakage current including chloride ion also needs to be modified to targeted clamping voltage. This also can be done in the same manner as in the case of sodium ions. But in their experiment, Hodgkin and Huxley regraded the leakage current component as constant and not depending on the membrane voltage, and its contribution to the total membrane was minimal. Then, the variation of total membrane current is contributed only by the variation of sodium and potassium conductance depending on membrane voltage.

During the process of equalizing the Nernst potential of some ion to a clamping voltage, there is a chance that it modifies the resting potential. Since ions have charges, altered external concentration inevitably causes an alternation of voltage across the membrane if altered charges are not replaced by another ion. But the resting potential should be kept constant through the series of experiments since it is the referential potential that clamping voltage is switched from. To keep the resting potential constant, they have used the choline ion which has one positive charge like the sodium ion and is impermeable to the membrane due to its large size. When eliminated part of sodium ion is replaced with choline ion, resting potential remains unchanged because the number of charges across the membrane is not changed.

In their experiment, Hodgkin and Huxley used seawater as the external fluid and build another sodium-free solution by replacing total sodium ions in seawater with the same number of choline ions remaining all other ions unchanged. And they made a certain percentage of seawater, for example, 10%, 30%, and 50%, by mixing the two solutions at an appropriate rate. For 30% of seawater, sodium concentration is reduced to 3/10 compared to 100% seawater solution and it will decrease the Nernst potential of new solution by 25.8 $\ln\left(\frac{10}{3}\right) = 31$ mV. Nernst potential for a new solution is changed from 50 mV to 19 mV. If we set the clamping voltage to 19 mV, sodium ions across the membrane do not cause any ionic current across the membrane, because the clamping voltage is set to its Nernst potential. So, the variation of measured membrane current is considered solely contributed by the potassium ion and the variation of its conductance.

If we repeat the experiment with a different solution of altered sodium ion concentration and with the clamping voltage equivalent to the Nernst potential of altered concentration, the dynamics of potassium ion conductance can be measured by the membrane current when membrane voltage is switched from the resting voltage to this clamping voltage. Figure 4.8 shows an example of measured potassium ion currents with different settings of clamping voltage.

The conductance of sodium also can be derived in a similar way. Alternatively, since potassium current was derived for each clamping voltage, sodium current can be derived by subtracting potassium current from the measured total membrane current. In this case, the total membrane current is measured with the ion concentration as it is, there is no need to change the potassium concentration of the external or internal fluids. Figure 4.9 shows the measured conductance curves for the sodium ions when the membrane potential is switched from resting potential of −65 mV to each clamping voltage V_c. In this figure, displacement voltage, which is clamping voltage in reference to resting voltage ($V_d = V_c - V_r$), is expressed together with V_c.

When the measured sodium conductance curve is compared to the potassium conductance, there are two significant differences in their shapes. First, the final values of the sodium conductance are different from the potassium conductance. For potassium ions, conductance increases monotonically to some saturation level and the final saturation value increase as the displacement voltage increases. In contrast, the conductance of sodium ions returns to zero value after it reaches a peak value in less than 1 ms. These peak values of sodium conductance increase as the displacement voltage increases. The slope of conductance increase also gets steeper and also reaches its peak value earlier as the displacement voltage increases. Second, their responding speeds are different. The conductance of sodium ions responds fast and rises up to its peak value within a millisecond. And it returns back to zero value during the potassium conductance starts to rise. In contrast, potassium ions respond

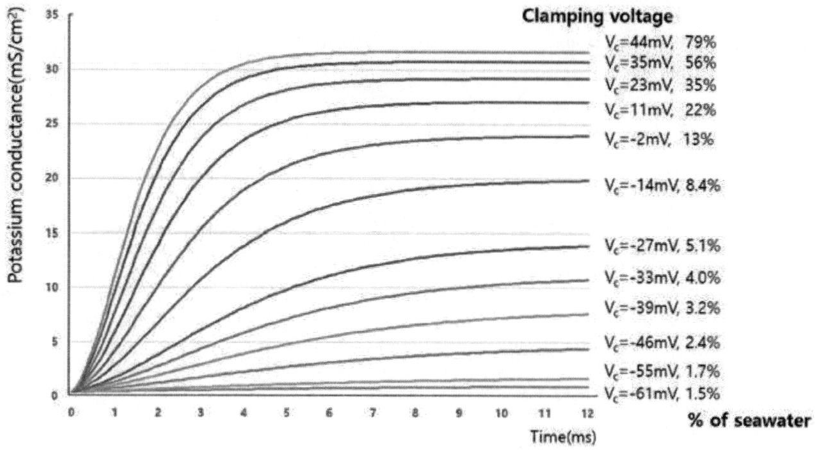

Fig. 4.8 Curves of measured potassium conductance when the membrane potential is switched from resting potential ($V_r = -65$ mV) to a different clamping voltage (V_c) corresponding to the altered Nernst potential developed by a different percentage of seawater

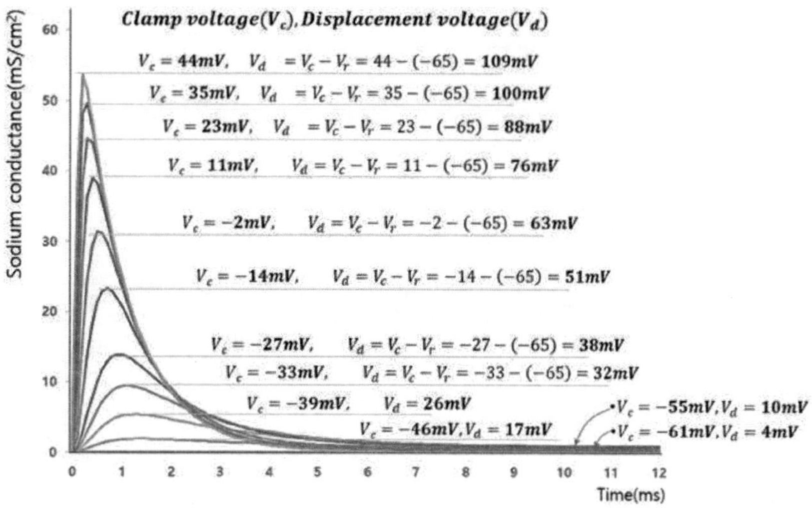

Fig. 4.9 Curves of measured sodium conductance when the membrane potential is displaced by switching from resting potential ($V_r = -65$ mV) to each of different clamping voltage (V_c)

slowly. It takes more than 4 ms to approach its saturation level. Based on these responding properties of potassium and sodium conductances, sodium conductance can be regarded as responsible for the early phase and potassium is responsible for the later phase of the action potential generation process.

In contrast to separating ionic current electrically based on the manipulation of ion concentration by canceling out one ionic current, a pharmacological method is also possible. In the pharmacological method, ion channel-specific agents are used. For example, tetrodotoxin (TTX) blocks the sodium ion channels selectively without affecting the conductance of potassium ions (Narahashi et al., 1964). The potassium ion channel is selectively blocked by the tetraethylammonium (TEA) (Armstrong & Hille, 1972). Currently, various types of ion channel blockers are known and some of them are also used for medical purposes, for example, to improve neuropathic pain treatments. Sodium channel blocker is also used as an insecticide. It acts on voltage-gated sodium channels and irreversibly blocks the depolarization of the membrane potential (Silver et al., 2010) (Fig. 4.10).

3 Quantitative Modeling of Ion Conductances

The next step is to derive a quantitative model for voltage dependency of the sodium and potassium ions conductances. First, let's see how the model for the conductance of potassium ions, which seems rather easier, has been derived.

3.1 Modeling of Potassium Ion Conductance

Measured conductance curves of potassium ion look familiar at first glance. It is generally quite similar to the exponentially increasing curve of $y = 1 - e^{-t}$ form. But, with a closer look, the starting part is quite different from a simple exponentially

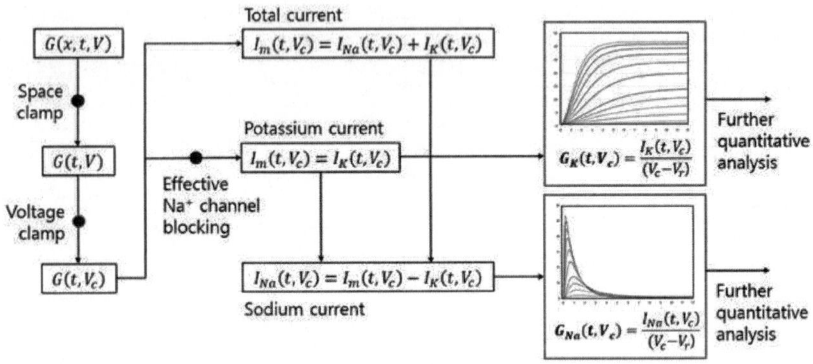

Fig. 4.10 Overview of the experimental design to measure sodium and potassium ion conductance separately

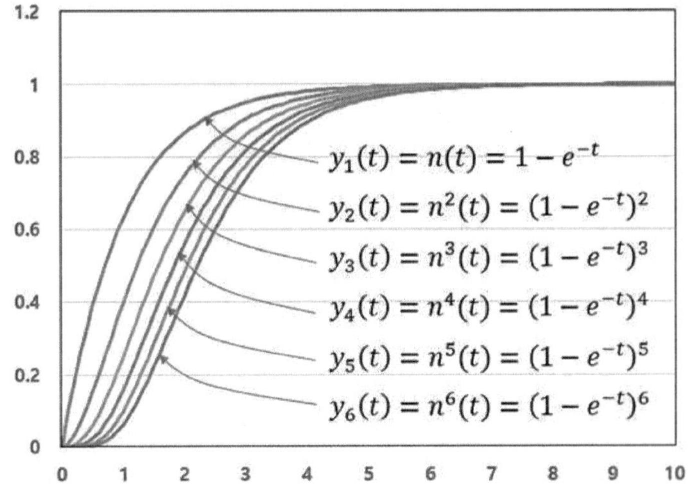

Fig. 4.11 Variation of exponential curves as the exponent increases

increasing curve. While the simple exponential curve starts with a rapid increase, the measured conductance curves have a rather slow increase in this part. However, if the exponential curve is multiplied multiple times, the curve changes by lowering its slope in the beginning part. As shown in Fig. 4.11, when the exponent quotient increases, the curve has a lower slope in starting phase and starts to rise lately. Hodgkin and Huxley found that the power 4 fit best numerically with their measured potassium conductance data. Even though the general form of the curves fits the power 4 of the exponentially rising curve, their rising slope, that is the time constant of the curve, is not the same. When the displacement voltage is higher, it approaches the saturation level earlier than the curves measured with lower displacement voltage. That is, the rising rate seems to be dependent on the displacement voltage from its resting membrane voltage. Since each exponentially rising curve is a solution of a first ordinary differential equation, potassium conductance might be explained by the quadruple action having the same dynamics as a first-order differential equation.

Conductance change is dependent on the membrane voltage. If the membrane voltage is displaced from the resting potential, conductance starts to increase and approaches its saturation value which is dependent on the strength of depolarization. In addition to the final saturation level, its rising time constant also looks different depending on the strength of depolarization, since the conductance curve approaches its saturation level in the earlier phase as the displacement voltage increases. It seems there are at least two independent parameters representing each property of potassium conductance variation.

As actuators working for the opening of the potassium ion channel, Hodgkin and Huxley proposed the "n" particle. The "n" particle can be considered as a particle whose activity is dependent on the membrane voltage. It physically can be a charged particle or a dipole moment within the membrane which is affected by the electric field produced by the potential across the membrane. If the "n" particle is in the active state, it works for the opening of the ion channel, and if it is not in the active state or in the inactive state, it is not contributing to the opening of the ion channel. That is, this "n" particle is sometimes in the active state and other times it is in the inactive state. It is transferred between these two states randomly. When the transferring rates between two states are determined by the voltage across the membrane, we can say that the behavior of "n" particles is dependent on the membrane potential and the transferring rates can represent the dynamics of the "n" particles depending on the membrane voltage. And, this particle does not act singly for the opening of an ion channel, but these particles are working in foursome as it was recognized by fitting the experimental data with power 4.

The dynamics of a "n" particle can be described with a first-order ordinary differential equation with two parameters.

$$\frac{dn}{dt} = \alpha_n(1-n) - \beta_n(n) \quad (4.9)$$

where

n: the proportion of "n" particles in the active state (dimensionless)
$1 - n$: the proportion of "n" particles in the inactive state
α_n: the activating transfer coefficient representing the transferring rate of "n" particles from the inactive state to the active state (1/ms)
β_n: the inactivating transfer coefficient representing the transferring rate of "n" particles from the active state to the inactive state (1/ms).

Equation (4.9) represents the increasing rate of "n" particle in the active state. It increases with the activating transfer coefficient α_n applied to "n" particles in the inactive state, and decreases with the inactivating transfer coefficient β_n applied to "n" particles in the active state. Since the value of "n" is within the range of 0–1, it also can be considered as the probability that a "n" particle is in the active state. Solving Eq. (4.9), "n" particle is given

$$n(t) = n_\infty - (n_\infty - n_0)e^{-t/\tau_n} \quad (4.10)$$

where

$$n_\infty = \frac{\alpha_n}{\alpha_n + \beta_n}$$

$$\tau_n = \frac{1}{\alpha_n + \beta_n}$$

Equation (4.10) explains that when the membrane potential is switched from resting potential to a certain clamping voltage, the proportion of "n" particle increases exponentially from an initial value of n_0 to a final value of n_∞ with a time constant τ_n. And these values are determined by two transfer rates α_n and β_n as above. For different pairs of transfer coefficients α_n and β_n, potassium conductance curves increase with a different time constant and approaches a different final saturation level. Even while the transfer coefficients α_n and β_n are constant, proportion of "n" particles changes over the time. The real ones who are dependent on the membrane voltage are these transfer coefficients. They are dependent on the change of membrane voltage. When the experiments are done with different clamping voltages, these two coefficients are determined with different values for each clamping voltage. If the coefficients α_n and β_n are determined for the displaced membrane voltage, the proportion of "n" particle varies with time as in experimental data, while the membrane voltage is maintained with a set clamping voltage. The difference between the behaviors of the "n" particle and the coefficients α_n and β_n is that proportion of "n" particle in active state varies with time while the coefficient α_n and β_n are not a function of time and constant unless the membrane voltage is changed.

For each of the potassium conductance curves derived from experimental data, we can estimate α_n and β_n by measuring the time constant and final value of each curve using Eq. (4.10) pertaining to the set clamping voltage.

$$\alpha_n = \frac{n_\infty}{\tau_n} \quad (4.11)$$

$$\beta_n = \frac{1 - n_\infty}{\tau_n} \quad (4.12)$$

By repeating the experiment with a clamping voltage set to different values, multiple data points of α_n and β_n can be derived for different clamping voltages. Figure 4.12 shows the collected set of α_n and β_n values for different values of clamping voltage. In Fig. 4.12, the horizontal axis is represented by displacement voltage with the reference to resting potential in seawater.

$$V_d = V_c - V_r \quad (4.13)$$

where

V_d: the displacement voltage from resting potential
V_c: the clamping voltage for the experiment
V_r: the resting potential of the axon membrane in seawater

Collected data of transfer coefficients α_n and β_n for potassium ion were fit to the equations given

$$\alpha_n = \frac{0.01(10 - V_d)}{e^{0.1(10 - V_d)} - 1} \quad (4.14)$$

$$\beta_n = 0.125e^{-V_d/80} \quad (4.15)$$

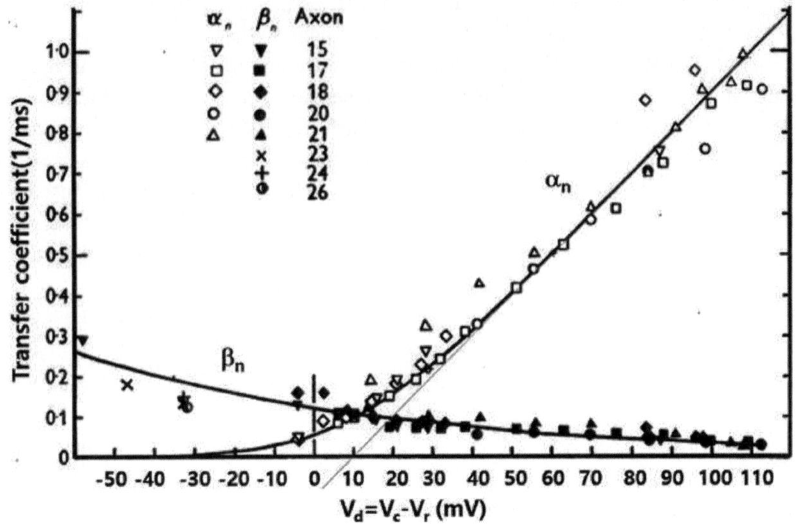

Fig. 4.12 Obtained transfer coefficient data. The horizontal axis represents the displaced membrane potential referenced to resting potential in seawater. Data are a collection of experimental results obtained from different axons numbered from 15 to 26. Data for $V_d < 0$ were obtained from the decline of potassium conductance associated with a decrease of the membrane potential or from the repolarization to the resting potential in choline seawater. The smooth curves were drawn from Eqs. (4.14) and (4.15). (Hodgkin & Huxley, 1952b)

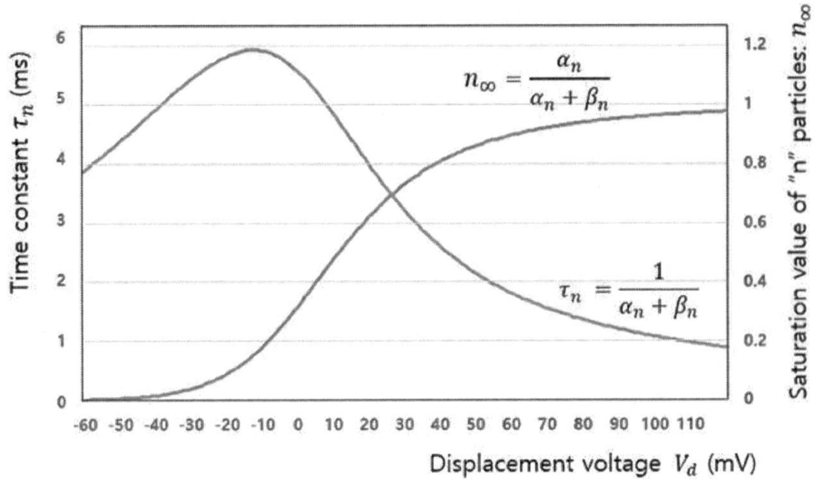

Fig. 4.13 Variation of time constant and the final saturated proportion of "n" particles along with the variation of displacement voltage

Activating transfer coefficient α_n described in Eq. (4.14) shows a pattern that is converging to a linear line of 0.01 $(V_d - 10)$ for a large positive value of V_d, converging to *zero* value for a negative value, and smoothly transiting between these two lines around the 10 mV of displacement voltage. This form of an equation is physically relevant to the Boltzmann equation describing the probability of particles in the opening state (Dubois et al., 2009). Inactivating transfer coefficient β_n is easily described by the form of exponentially decaying.

Figure 4.13 shows the variation of the time constant τ_n and the final saturated proportion of "n" particles n_∞ along the variation of displacement voltage, which are calculated directly from the estimated transfer coefficients α_n and β_n. The time constant τ_n decreases from 6 ms level to less than 1 ms level as the depolarizing voltage increases from 0 to 120 mV. This shorter time constant exactly describes the faster increase of potassium conductance according to the increased displacement voltage. This curve also shows that the time constant is decreasing from the peak value of 5.8 ms at $V_d = -12$ mV as the membrane potential is lowered than the resting potential. Gradual increase of the finally saturated proportion of "n" particles n_∞ also properly describes the general increasing behavior of final potassium conductance shown in Fig. 4.8. Finally, setting the maximum value of potassium conductance $G_{K,max}$ to the experimentally measured value, potassium conductance is given by

$$G_K(t) = G_{K,\max}\, n^4(t) \tag{4.16}$$

where

$G_K(t)$: the potassium conductance varying upon time (mS/cm^2),

$G_{K,\,max}$: the maximal potassium conductance when the membrane is fully permeable to potassium ions, 36 mS/cm^2,
$n(t)$: the time-dependent proportion of "n" particles in the active state.

3.2 Modeling of Sodium Ion Conductance

Two methods are possible for describing the transient changes of sodium conductance. The first method is to describe the sodium conductance with a single variable that obeys a second-order differential equation. The second method is to describe sodium conductance by two variables, each of which obeys the first-order equation. The second method was chosen since it is simpler to apply to the experimental results and to explain the procedure similarly to the case of potassium conductance.

So, modeling of the sodium conductance also can be done by following a similar procedure applied for the derivation of equations describing the potassium conductance. As shown in Fig. 4.9, the dynamics of sodium ion conductance are quite different from the potassium ion case. In the beginning part, sodium conductance is rising as in the case of potassium conductance. Even though it rises much faster than potassium conductance, this type of rising pattern can be similarly modeled with the concept of the charged particle "m" as "n" particle that worked for the modeling of potassium ion conductance. But with this particle only, the latter part of the curves cannot be described, since the sodium ion conductance differently falls back to zero level after reaching the peak value instead of approaching to certain saturation level. For the explanation of the latter part, the extra particle "h" is added which works rather differently from the former "m" particle. While "m" particles work for increasing the sodium conductance, "h" particles work for the decreasing the sodium conductance. In this concept, "m" particle is called the activation particle and "h" particle is called the inactivation particle. If they are considered as charged particles or dipole moments, they might have opposite charges or opposite directions of each other.

The beginning part of sodium conductance, which is rising faster than potassium conductance, also can be thought of as following the general form of the exponential curve $y = 1 - e^{-t}$ as in potassium ion conductance. As in potassium ion conductance, a single particle is not enough to explain the delayed start of the curve. In contrast to power 4 in potassium ion case, power 3 is enough to describe the starting delay and the earlier part of the curve. Three actuators having the same dynamics are working together for the opening of sodium ion channels.

The dynamics of "m" particle can be described similarly to the first-order ordinary differential equation with two parameters.

$$\frac{dm}{dt} = \alpha_m(1-m) - \beta_m(m) \tag{4.17}$$

where

m: the proportion of "m" particles in the active state (dimensionless)
$1 - m$: the proportion of "m" particles in the inactive state
α_m: the activating transfer coefficient representing the transferring rate of "m" particles from the inactive state to the active state (1/ms)
β_m: the inactivating transfer coefficient representing the transferring rate of "m" particles from the active state to the inactive state (1/ms).

Equation (4.17) represents the increasing rate of "m" particle in the active state. It increases with the activating transfer coefficient α_m applied to the particles in the inactive state and decreases with the inactivating transfer coefficient β_m applied to the particles in the active state. Similarly solving Eq. (4.17), "m" particle is given

$$m(t) = m_\infty - (m_\infty - m_0)e^{-t/\tau_m} \tag{4.18}$$

where

$$m_\infty = \frac{\alpha_m}{\alpha_m + \beta_m}$$

$$\tau_m = \frac{1}{\alpha_m + \beta_m}$$

The latter part of sodium conductance, which is smoothly falling back toward zero value, looks following the general decaying routine of the exponential curve of $y = e^{-t}$. It also can be modeled as a first-order differential equation of "h" particles.

The dynamics of "h" particle is also described similarly with the first-order ordinary differential equation with two parameters.

$$\frac{dh}{dt} = \alpha_h(1-h) - \beta_h(h) \tag{4.19}$$

where

h: the proportion of "h" particles not in the inactive state, that is in the active state (dimensionless),
$1 - h$: the proportion of "h" particles in the inactive state

α_h: the activating transfer coefficient representing the transferring rate of "h" particles from the inactive state to the not inactive state (1/ms)

β_h: the inactivating transfer coefficient representing the transferring rate of "h" particles from the not inactive state to the inactive state (1/ms)

Equation (4.19) represents the increasing rate of "h" particle not in the inactive state. It increases with the activating transfer rate α_h applied to particles in the inactive state and decreases with the inactivating transfer rate β_h applied to particles not in the inactive state. Similarly solving Eq. (4.19), "h" particle is given

$$h(t) = (h_0 - h_\infty)e^{-t/\tau_h} + h_\infty \tag{4.20}$$

where

$$h_\infty = \frac{\alpha_h}{\alpha_h + \beta_h}$$

$$\tau_h = \frac{1}{\alpha_h + \beta_h}$$

Then the sodium conductance can be modeled as

$$G_{\mathrm{Na}}(t) = G_{\mathrm{Na,\,max}}\, m^3(t) h(t) \tag{4.21}$$

Figure 4.14 shows the concept of sodium conductance modeling. As an example, for one conductance curve in Fig. 4.9 measured with clamping voltage $V_c = -33$ mV, the curve was fit by two exponential curves graphically. The beginning part is fit by a 3rd power of exponentially rising curve and the latter part is fit by a simple exponentially decaying curve. It looks like they fit quite well for each part of the curve. The middle part which is not matched to any of these curves is expected to be fit by the multiplication of two curves as a smooth transition between two curves.

For the estimation of four parameters α_m, β_m, α_h, and β_h from a single measured conductance curve, the equation is simplified for the easiness of derivation. Since sodium conductance is very small during the resting potential compared with the value attained during a large depolarization, m_0 is neglected when the displacement voltage is greater than 30 mV. Further, inactivation is nearly completed if the displacement voltage is more than 3 mV, h_∞ may also be neglected. With this simplification Eq. (4.21) can be given

$$\begin{aligned} G_{\mathrm{Na}}(t) &= G_{\mathrm{Na,\,max}}\left(m_\infty\left(1 - e^{-\frac{t}{\tau_m}}\right)\right)^3\left(h_0 e^{-\frac{t}{\tau_h}}\right) \\ &= G_{\mathrm{Na,\,max}}\, m_\infty^3 h_0 \left(1 - e^{-t/\tau_m}\right)^3\left(e^{-t/\tau_h}\right) \end{aligned} \tag{4.22}$$

where

$G'_{\mathrm{Na,\,max}} = G_{\mathrm{Na,\,max}}\, m_\infty^3 h_0$ is the value that the sodium conductance would attain if h remained at its resting level h_0.

Equation (4.22) is fit to each of the experimental data with the time constant (τ_m) estimated from the slope in the rising part of the curve. m_∞ is estimated by matching the peak values of the curve to Eq. (4.22). From the estimated τ_m and m_∞, transfer coefficients α_m and β_m are calculated

$$\alpha_m = \frac{m_\infty}{\tau_m} \tag{4.23}$$

$$\beta_m = \frac{1 - m_\infty}{\tau_m} \tag{4.24}$$

After applying the same procedure for each of the measured conductance curves, multiple data points of α_m

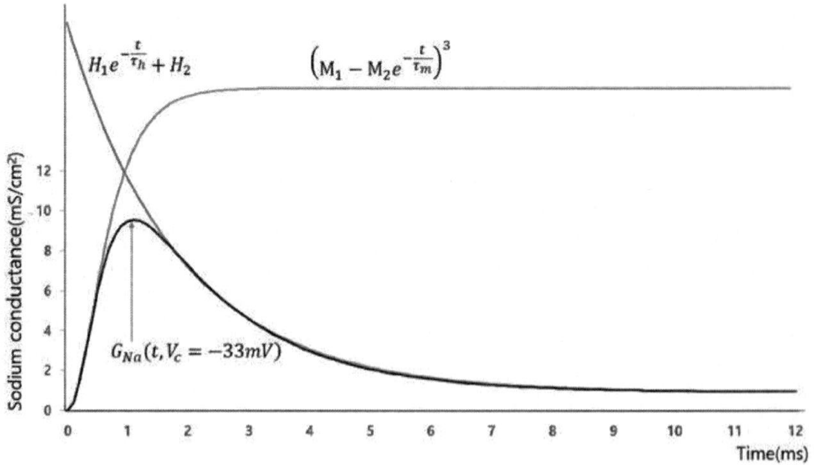

Fig. 4.14 Concept of sodium conductance modeling. The beginning part is graphically fit by the 3rd power of the exponentially increasing curve and the latter part is fit graphically by the simple exponentially decaying curve

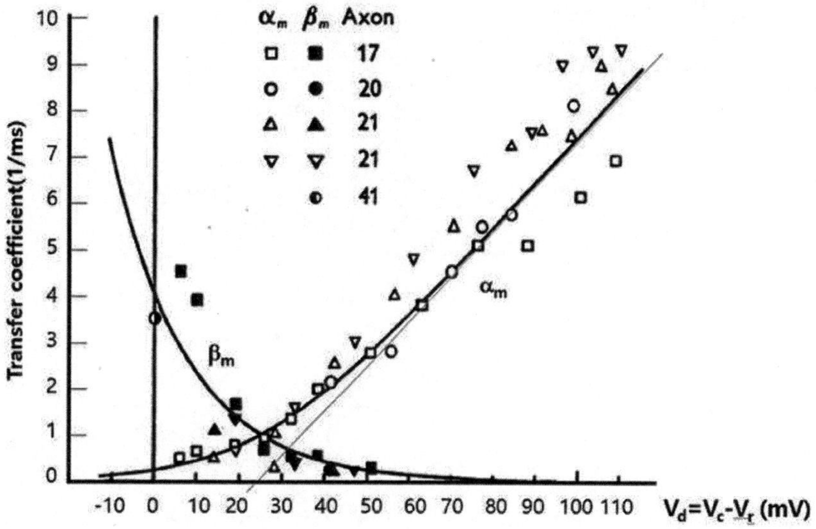

Fig. 4.15 Collected data for the transfer coefficient of "*m*" particle and the fit curve of the estimated model equation. (Hodgkin & Huxley, 1952b)

and β_m can be derived for different clamping voltages. Figure 4.15 shows the collected set of α_m and β_m values for different value of displacement voltage. In Fig. 4.15, the horizontal axis is represented by displacement voltage with the reference to resting potential in seawater.

Collected data of transfer coefficients α_m and β_m was fit to the equation given

$$\alpha_m = \frac{0.1(25 - V_d)}{e^{0.1(25 - V_d)} - 1} \tag{4.25}$$

$$\beta_m = 4e^{-V_d/18} \tag{4.26}$$

Activating transfer coefficient α_m described in Eq. (4.25) shows a pattern that it is converging to a linear line of 0.1 (V_d − 25) for a large displacement voltage, converging to *zero* value for a negative value and smoothly transiting between these two lines in the middle range of displacement voltage. Inactivating transfer coefficient β_m is easily described by the form of exponentially decaying as in the case of β_n.

For the transfer coefficients of "*h*" particles, time constant (τ_h) estimated from the falling slope in the latter part of the curve and h_∞ is estimated from the final values of the sodium conductance curve. In the steady state approaching zero value, "*m*" particle is nearly at the saturation level and small values of sodium conductance can be considered as the contribution of the "*h*" particle. From the estimated τ_h and h_∞, transfer coefficients for "*h*" particle α_h and β_h are calculated in the same way

$$\alpha_h = \frac{h_\infty}{\tau_h} \tag{4.27}$$

$$\beta_h = \frac{1 - h_\infty}{\tau_h} \tag{4.28}$$

After applying the same procedure for each of the measured conductance curves, multiple data points of α_h and β_h can be derived for different membrane voltages. Figure 4.16 shows the collected set of α_h and β_h values for different value of displacement voltage.

Collected data of transfer coefficients α_h and β_h was fit to the equation given

$$\alpha_h = 0.07e^{-V_d/20} \tag{4.29}$$

$$\beta_h = \frac{1}{e^{0.1(30 - V_d)} + 1} \tag{4.30}$$

In contrast to the previous activating transfer coefficients of "*n*" and "*m*" particles, activating coefficient α_h follows the pattern of inactivating transfer coefficient of the previous "*n*" and "*m*" particles. Activating transfer coefficient α_h described in Eq. (4.29) shows a pattern that is simply decaying exponentially as in the case of β_n and β_m. β_h shows a pattern of sigmoid waveshape which is antisymmetric with the center at V_d = 30 mV.

3.3 Constants of the Model

In addition to the variables for describing potassium and sodium conductance, there are some constants in the model. Constants used in the model are given as follows. Table 4.2 shows equations and constants derived from the Hodgkin-Huxley model.

C_m = 1 μF/cm^2, the membrane capacitance

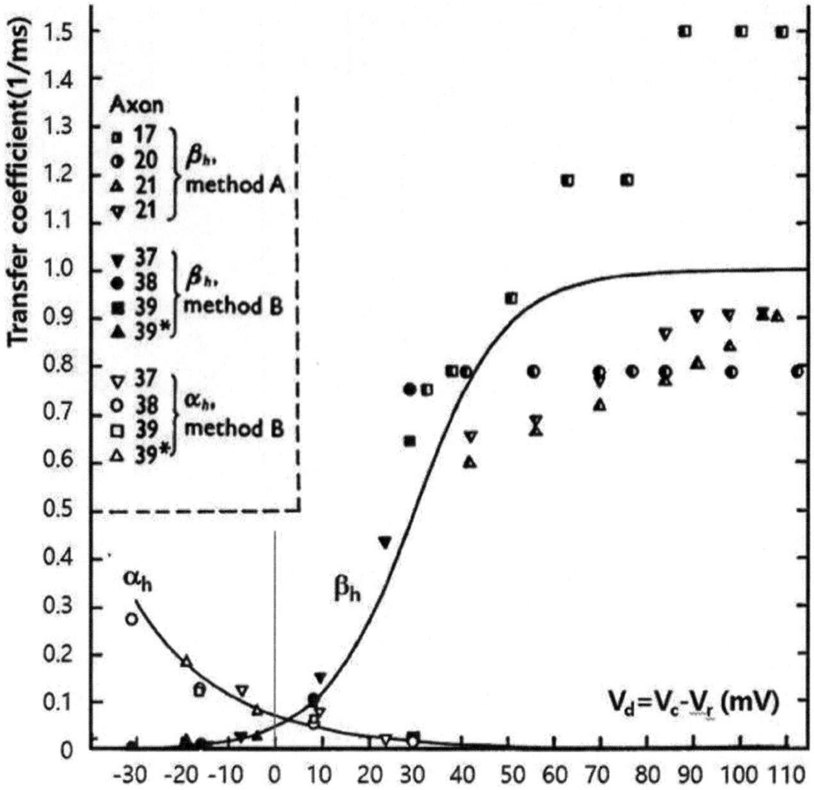

Fig. 4.16 Collected data for the transfer coefficient of "h" particle and the fit curve of the estimated model equation. (Hodgkin & Huxley, 1952b)

$G_{K,\ Max} = 36\ \text{mS/cm}^2$, the maximum conductance of the potassium conductance

$G_{Na,\ Max} = 120\ \text{mS/cm}^2$, the maximum conductance of the sodium conductance

$G_L = 0.3\ \text{mS/cm}^2$, a constant conductance for the leakage current and the chloride ion

$E_K = V_K - V_r = -12$ mV, Nernst potential of the potassium ion (−77 mV) in reference to the resting potential (−65 mV)

$E_{Na} = V_{Na} - V_r = 115$ mV, Nernst potential of the sodium ion (50 mV) in reference to the resting potential (−65 mV)

$E_L = V_L - V_r = 10.6$ mV, Nernst potential for the leakage current and the chloride ion (−54.4 mV) in reference to resting potential (−65 mV).

Figure 4.17 shows the overall flow of variable calculation for the model. When the membrane voltage is set to a certain voltage, displacement voltage is calculated by the membrane potential in reference to the resting potential. Then activating and inactivating transfer coefficients α_n, β_n, α_m, β_m, α_h, and β_h are calculated from the changed displacement voltage using the modeled equations. The proportion of "m", "n, " and "h" particels in their active state are updated with the updated transfer coefficients. These updated "m", "n, " and "h" particels update sodium and potassium conductance G_{Na} and G_K. Then the updated conductance and the displaced membrane voltage determine each ionic current. Ionic membrane current summed with capacitive current builds up total membrane current. If the space is clamped and there is no axial current along with the distance, the total membrane current becomes zero and the total ionic current is balanced by the capacitive current. This causes a change in membrane voltage. Change in membrane voltage again causes the alteration in transfer coefficients of acting particles and the total loop repeats again. And, the membrane by the Hodgkin and Huxley model is summarized with the equivalent circuit in Fig. 4.18.

4 Action Potential with the Hodgkin and Huxley Model

Even though we have assumed the constant membrane voltage for the derivation of model equations, membrane voltage does not remain at a constant level and varies dynamically during the action potential generation. Modeled equations can be applied in the generation of action potential and the model represents the real behavior very closely. It provides the insight into how the membrane is working for the generation of the action potential.

Fig. 4.17 Overall flow of variable calculations for the Hodgkin–Huxley model

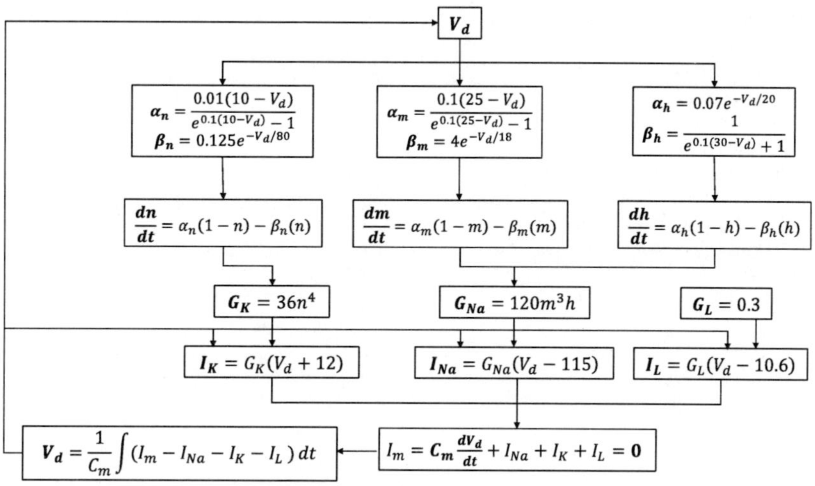

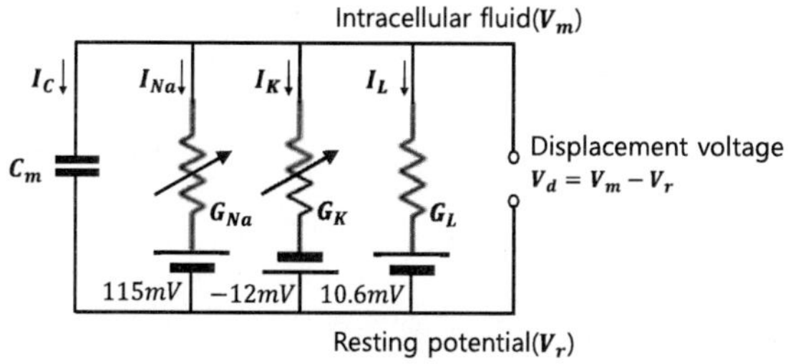

Fig. 4.18 Equivalent circuit for the Hodgkin and Huxley model

4.1 Generation of Action Potential

Figure 4.19 shows the simulated waveforms during the generation of action potential in a nonpropagating axon. Figure 4.19a shows the waveform of the membrane voltage represented in displacement from the resting potential of −65 mV. Displacement voltage is switched from 0 to 20 mV at $t = 0$, which is sufficiently higher than the threshold voltage to fire an action potential. The voltage waveform rapidly rises to the peak value of 107 mV within 1 ms. After passing the peak, membrane voltage decreases back to the resting potential and reaches the resting level almost in 3 ms. Instead of returning to the resting level directly, there is hyperpolarization and the membrane potential decreases lower than the resting level. Finally, hyperpolarized membrane potential smoothly returns back to the resting membrane potential with time, and a routine of action potential finishes.

Figure 4.19b shows how the sodium and potassium conductance is varying during the course of action potential generation. The sodium conductance increases rapidly before the potassium conductance increases. This increased sodium conductance allows a large number of sodium ions to flow inward driven strongly by the voltage gradient and concentration gradient. The rapid inflow of positive sodium ions explains the rapid rise of membrane potential in the initial phase of the action potential. Potassium ion conductance starts to increase lately while sodium conductance is approaching its peak level. Increased potassium conductance causes increased outward flow of positive potassium ions mostly driven by the concentration gradient. This outward current is in the reverse direction to the initial inward current caused by the sodium inflow and it cancels out the current caused by the sodium ion. It decreases the membrane potential and does the role of returning membrane potential back to the resting state. Even if the conductance is increased for both sodium and potassium ions, the peak value of the potassium conductance is less than half of the peak sodium conductance. Their increasing pattern is different in time and in magnitude scale and this difference practically shapes the waveform of an action potential. Both ion conductances completely decrease and return back to the resting state level finally.

Figure 4.19c shows the current waveform caused by the sodium ion and the potassium ion. Inward sodium current is in the negative direction, and it appears earlier than potassium ion current. Sodium current is bigger in magnitude than potassium current before the potential reaches its maximum, and it does the role of increasing the membrane potential. Potassium current is bigger than sodium current after the membrane potential passes the peak value, and it does the role of decreasing membrane potential. The influence of the current induced by the leakage and the chloride ion is minute, and its peak is less than 30 μA/cm^2. Figure 4.19d shows the variation of intermediate particles that have appeared to explain the voltage dependency of potassium and sodium conductance. Since the voltage is not fixed as in the case of voltage clamping experiments, transfer coefficients of each particle also vary dynamically according to the instantaneously varying membrane potential. "n" particle clearly represents the variation of conductance variation of

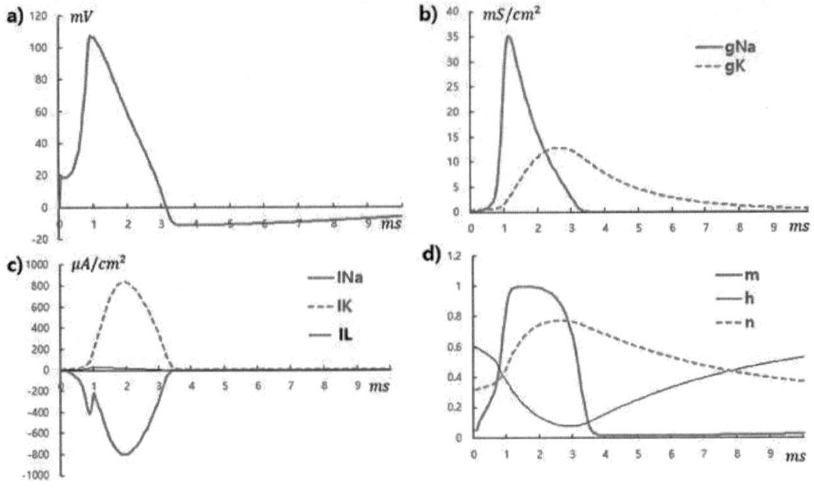

Fig. 4.19 Simulated waveforms of nonpropagating action potential with the Hodgkin–Huxley model. (**a**) Action potential (mV), (**b**) Membrane conductance for the sodium and the potassium ions (mS/cm^2), (**c**) Membrane currents for the sodium, potassium, and leakage ions ($\mu V/cm^2$), ions, (**d**) Proportion of particles describing the sodium and the potassium conductance

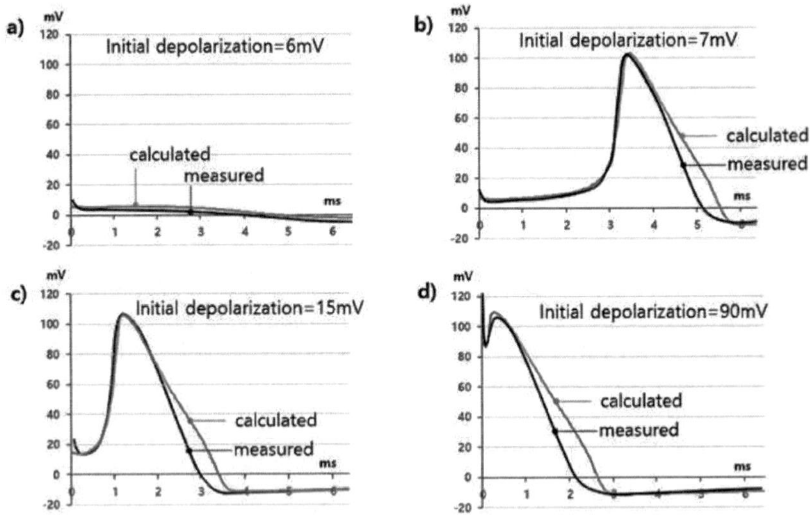

Fig. 4.20 Comparison of the measured and the calculated action potential waveforms with various levels of initial depolarization stimulus. (**a**) Responses to stimulus of a subthreshold level of 6 mV, (**b**) Responses to stimulus just above the threshold level of 7 mV (**c**) Responses to an increased level of 15 mV (**d**) Responses to a strong level of 90 mV, black lines: experimentally measured membrane potential waveforms, gray lines: calculated membrane potential waveforms using the model equations. (Hodgkin and Huxley, 1952b)

potassium ion, since the potassium ion conductance is proportional to the 4th power of "n" particle fraction in the active state. The rapid increase of "m" particle explains the rapid rise of sodium conductance before "h" particle starts to decrease. The later falling phase of sodium ion conductance looks mostly driven by the "h" particle, while the "m" particle is staying almost at the saturation level. Level of all of these particles also returns to resting state levels as the action potential finishes.

4.2 Comparison with the Experimentally Measured Waveform

Figure 4.20 shows the calculated waveforms by the developed model equations in comparison with waveforms experimentally measured with different levels of initial stimulus intensities. Calculated waveforms are remarkably similar to the real waveforms measured with experiments. With the low level of stimulus intensity under the threshold level, membrane potential returns smoothly to its resting potential in both waveforms. When the stimulus intensity increased just above the threshold level, an action potential is generated with some delay in both cases with similar waveforms. As the stimulus intensity is increased, calculated waveforms are following the changing pattern of the real response of shortened peak time while maintaining the overall waveform unchanged. One noticeable difference in this comparison is the small downward deflection in the later part of the repolarization period.

4.3 Refractory Period

The object of the model was to find the equations which describe the conductance variation of the action potential and refractory period with reasonable accuracy and sufficient simplicity for theoretical understanding. As the model explains the variation of conductance during the generation of an action potential, it also explains the refractory period following an action potential.

The refractory period mentions the period during which firing of the following action potential is not possible or more difficult than in the state of resting potential (Fig. 4.21). This refractory period is the depolarized and hyperpolarized phases of an action potential. For some period just after the onset of an action potential which is called the absolute refractory period, it is impossible to start another action potential regardless of the amplitude of the stimulus. The absolute refractory period extends almost to the time that the potential reaches its resting potential level before it is hyperpolarized. As the refractoriness is weakened, second action potential can be generated with relative difficulty. In this relative refractory period, the membrane is in the state where its threshold level to fire action potential is increased and a higher stimulating amplitude is required to generate a second action potential.

The Hodgkin–Huxley model shows a satisfactory degree of agreement for the threshold variation during the refractory period. There are two changes resulting from the depolarization during the generation of an action potential which make the membrane unable to respond to another stimulus until a certain time has elapsed. One is the "inactivation" of "*h*" particle, which is lowered so that the sodium conductance can be raised by a depolarization. As shown in Fig. 4.18, "*h*" particles start to decline just after the onset of an action potential, stay at the lower values during the action potential period, and only return to resting level near the end of the whole action potential period. Another change is the delayed rise in potassium conductance, which is represented by the behavior of "*n*" particle, and it tends to hold the membrane potential near to the equilibrium value for potassium ions. Delayed recovery of these "*h*" and "*n*" particles up to the end of a long hyperpolarized phase is quite contrast to the "*m*" particle which immediately returns to its normal level when the membrane voltage is returned to its resting level prior to the hyperpolarization phase. The model shows the occurrence and the strength of stimulus to fire second action potential which are matching quite well with the experimentally measured results.

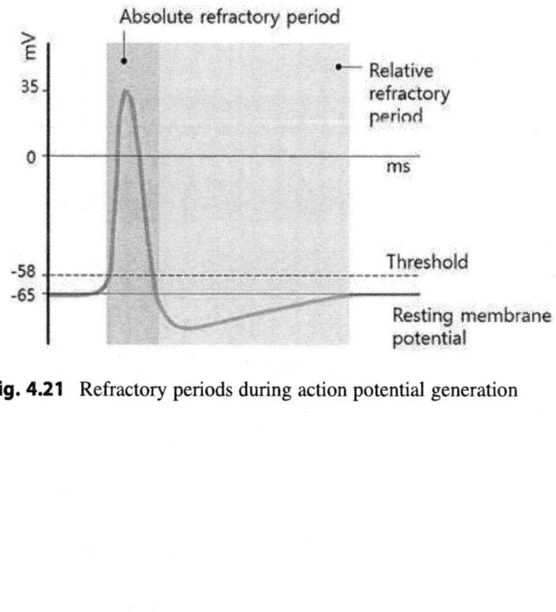

Fig. 4.21 Refractory periods during action potential generation

In addition to these major features of the model, the model could reproduce and explain a remarkable range of data from the squid axon, including the shape and propagation of the action potential, its sharp threshold, anode-break excitation, accommodation, and subthreshold oscillations (Häusser, 2000).

4.4 Reestablishment of Ion Concentrations

As a result of an action potential, ion concentrations across the membrane are disturbed from the resting state. During the generation of an action potential, sodium ion influx increases the internal concentration of sodium ions, and the outflux of potassium decreases the internal concentration of potassium ions. The absolute and relative amount of moving ions is dependent on the size of the axon diameter. As the diameter of the axon increases, the membrane surface for a unit length also increases linearly and an absolute number of ions moved also increases proportionally to the increased surface. So, the absolute number of moved charges increases in proportion to the axon diameter. But the relative number of moved ions decreases as the diameter increases since the volume of the internal axon of unit length is increasing with the 2nd power of the axon diameter. So, the relative portion of the moving ions is decreasing as the axon diameter increases.

Even though the relative portion of moving ions as a result of action potential generation is minute, ion concentrations need to be reestablished to fire action potentials repetitively. This is done by the sodium-potassium pump in the same way as already mentioned in the maintenance of resting potential. Extra sodium ions moved in are pumped out and the deficit potassium ions moved out are pumped in by this sodium-potassium pump and ion concentrations are reestablishing toward the ones of resting potential. It uses metabolic energy derived from adenosine triphosphate (ATP). The degree of activity of the sodium-potassium pump is known to be stimulated by the sodium ion concentration accumulated inside the cell membrane.

5 Voltage-Gated Ion Channels

Membrane voltages are changed by the inflow or the outflow of the current, which is actually the transport of the charged ions across the membrane. In the model, varying characteristics of the membrane are represented by the varying conductances of each ion depending on the membrane voltage. Since ions are moving in and out of the cell through the ion channels, variation of ionic conductance is considered as the macroscopic representation of ion channels of each ion depending on the voltage. When Hodgkin and Huxley developed the model, there was no clear understanding of ion channels. But, the model based on voltage-dependent conductance and the concept of actuating particles to explain conductance are matching very nicely with the results proven on ion channels by the advanced technologies in later times.

5.1 Types of Ion Channels

There are several types of ion channels (Fig. 4.22). The simplest type is the ion channel always open regardless of the membrane condition. Even though they are always open, they are maintaining their own selectivity to specific ions, and provide the basic selective permeability need to develop a resting potential. Second type is the voltage-gated ion channel. It switches between open and closed states depending on the membrane voltage. Voltage-gated ion channel is the ion channel that contributes to the voltage-dependent conductance variation and the generation of action potential. Third type is the ligand-gated ion channel. It gates channel pore depending on the acceptance of special chemical component on its receptor site. This type of ion channel is most popular in synaptic junctions. When propagated neurotransmitter is accepted on the receptor site of the ion channel, it opens and allows the movement of ions across the membrane. Finally, there are other types of ion channels which are depending on different types of physical or mechanical variables other than voltage or specific chemical components. This type ion channels are related to the sensing of different type of physical quantities like temperature, pressure or light and found mostly in various types of sensory receptors.

5.2 Structure of Voltage-Gated Ion Channels

The transmembrane structure of voltage-gated ion channels has some common recognizable features. For the case of voltage-gated potassium ion channel, whose behavior is represented by the varying potassium conductance in the model, many studies have been done. So, potassium ion channel is most extensively recognized for its structure and characteristics. Figure 4.23 shows the molecular structure of the voltage-gated potassium ion channel. Surprisingly, it consists of four subunits that act almost independently in opening the channel. It is accurately matching with the foursome characteristic of "n" particle which has appeared to explain potassium conductance. Each subunit is consisting

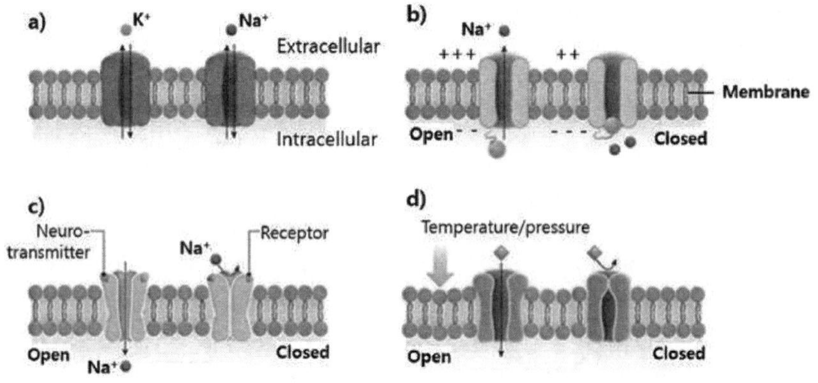

Fig. 4.22 Different types of ion channels. (**a**) Ion channels always in open state, (**b**) Voltage-gated ion channels, (**c**) Ligand-gated ion channels, (**d**) Ion channel gated by other physical or mechanical variables like temperature, pressure, and light

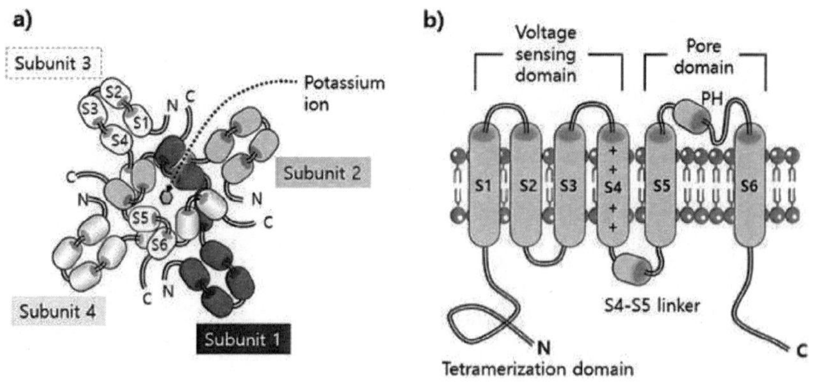

Fig. 4.23 Molecular structure of voltage-gated potassium ion channel. (**a**) Extracellular side view of a potassium ion channel composed of four homogeneous subunits. (**b**) Schematic structure of a single subunit of potassium ion channel composed of 6 transmembrane domains termed S1–S6. (Tombola et al., 2006)

of 6 helical transmembrane domains termed S1–S6. Among these 6 domains, four domains of S1–S4 are voltage-sensing domains and do the role of sensing the membrane voltage and conformationally changing the structure of the ion channel to open or close the gate. Domain S4 is considered the location where the positive charges are located and respond to the change of membrane voltage. This finding also matches with the assumption that "n" particles are composed of charged particles or dipoles which do not act directly as carriers, but which allow potassium ions to pass through the membrane by altering their positions in the membrane. And two domains of S5–S6 are pore domains and constitute the ion channel pore portion by combining four sets of them from each of four subunits. The pore domain is connected to the voltage-sensing domain by the linker between S4 and S5 domains, and conformational change caused by the voltage-sensing domain is transferred to the pore domain through the linker.

6 Interpretation of Action Potential

6.1 Positive and Negative Feedback

The process of generation of an action potential can be viewed with the feedback control theory (Fig. 4.24). In this case, two kinds of feedback, positive feedback and negative feedback, are involved simultaneously for the generation of an action potential. The cycle depolarizing the membrane is the positive feedback process. The initial stimulus over the threshold triggers positive feedback by increasing the membrane potential. Increased membrane potential increases sodium conductance rapidly, and this increased sodium conductance is followed by the increased inward sodium current. Inward current increases membrane potential further and depolarize the membrane. This process of feedback occurs in a very fast mode due to the high sensitivity of sodium conductance to increased membrane potential. This cycle does the work only for the depolarization of the membrane.

The cycle for the repolarization of the membrane is a negative feedback process. Depolarized membrane potential increases the potassium conductance and the increased outward potassium current is following. Increased outward potassium current decreases membrane potential in contrast to the increased sodium current. As the cycles are repeated, the membrane potential is decreasing further and the potential returns to the resting potential. This repolarization cycle is slow compared to the fast depolarization cycle. So, the negative feedback starts much later than the onset of positive feedback and takes the role of repolarization in the later part of the action potential generation. The difference in reaction speed is caused by the difference in sensitivity of the transfer coefficients between the "m" particle for sodium ion and the "n" particle for potassium ion. The transfer coefficient α_m is almost 10 times higher than α_n as shown in Table 4.2. In the depolarizing phase, sodium conductance increases at a much

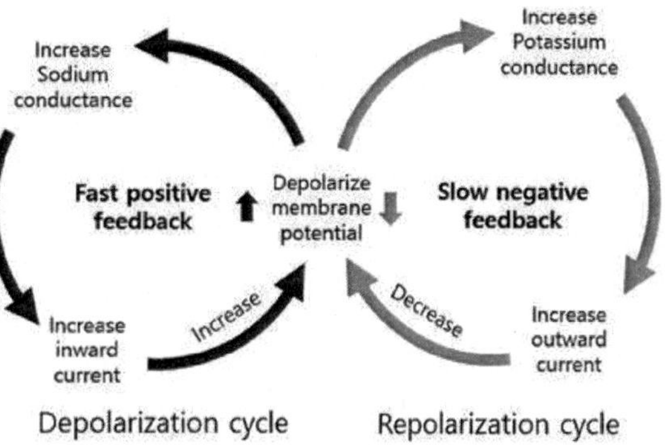

Fig. 4.24 Understanding of action potential generation with control concept of a positive feedback and a negative feedback

Table 4.2 Summary of equations for the Hodgkin–Huxley model

Membrane variables		
Membrane current	$I_m = C_m \frac{dV_d}{dt} + G_{Na}(V_d - 115) + G_K(V_d + 12) + G_{Cl}(V_d - 10.6)$	μA/cm²
Displacement voltage	$V_d = V_m - V_r$	mV
Ion conductance		
Sodium ion	$G_{Na} = 120m^3h$	mS/cm²
	$\frac{dm}{dt} = \alpha_m(1-m) - \beta_m(m)$, $\frac{dh}{dt} = \alpha_h(1-h) - \beta_h(h)$	
Potassium ion	$G_K = 36n^4$	mS/cm²
	$\frac{dn}{dt} = \alpha_n(1-n) - \beta_n(n)$	
Leakage and chloride ion	$G_{Cl} = 0.3$	mS/cm²
Transfer coefficients		
Sodium ion	$\alpha_m = \frac{0.1(25 - V_d)}{e^{0.1(25 - V_d)} - 1}$, $\beta_m = 4e^{-V_d/18}$	1/ms
	$\alpha_h = 0.07e^{-V_d/20}$, $\beta_h = \frac{1}{e^{0.1(30 - V_d)} + 1}$	1/ms
Potassium ion	$\alpha_n = \frac{0.01(10 - V_d)}{e^{0.1(10 - V_d)} - 1}$, $\beta_n = 0.125e^{-V_d/80}$	1/ms

faster speed, and the positive feedback dominates the initial part of action potential generation.

6.2 Threshold for the Generation of Action Potential

There exists a threshold for the generation of an action potential. The quantitative model reflects clearly an abrupt difference around the threshold level for firing an action potential. Which factor does a key role in the decision on generating action potential? For the generation of an action potential, there should be a rapid increase in sodium conductance to depolarize the membrane progressively. It is shown by the rapid increase of the portion of "*m*" particles in the active state as in Fig. 4.19d. To initiate positive feedback of increasing membrane potential, increased inward current caused by the rapid increase in sodium conductance should surpass the accompanying increase of outward potassium current. But, it is not easily achieved since the potassium conductance is much higher than sodium conductance near the resting potential level. Figure 4.25 compares the variation of sodium conductance ratio to potassium conductance when the membrane is stimulated by the voltage just above the threshold (6.6 mV) and just below the threshold level (6.4 mV). This conductance ratio is less than 0.1 during the resting state. When an action potential is generated, this ratio increases over 1.0, meaning that sodium conductance becomes higher than potassium conductance, and progressively increases up to the level of 10–30 depending on the initial stimulating intensity. But when the stimulation of lower than the threshold intensity fails to fire an action potential, sodium conductance remains less than 20% of potassium conductance and never goes higher above that level. Finally, in this range, the membrane fails to initiate positive feedback to rapidly increase sodium conductance. The trial to increase the sodium conductance is blocked by the accompanying increase of potassium current supported by the much higher conductance around this resting state. Sufficient stimulation intensity of increased initial depolarization voltage is required to start the rapid increase of sodium conductance and to induce following positive feedback capable of overcoming the accompanying increase of potassium conductance.

6.3 Key Points of the Hodgkin and Huxley Model

The Hodgkin and Huxley model for the explanation of membrane potential is one of the most successful models in science. This Nobel Prize-winning achievement also provides us the lessons on modeling biological phenomena which are quite difficult by the innate variability of biological lives. Here are summarized the key points related to the success of the model.

First of all, they pointed out the core issue correctly that the action potential is associated with a large increase in membrane conductance and that it is dependent on the membrane voltage.

Second, they found the appropriate experimental subject, the giant axon from the squid. It made the first intracellular recording of an action potential and the following voltage-clamp experiments possible with its large enough diameter for inserting an electrode unit.

Third, they developed a voltage-clamp circuit to enable the quantitative measurement of ionic currents from the squid axons. As many new achievements come with the advent of a new device or new method, their success is also strongly in debt to the new method of voltage clamp.

Fourth, they have developed the model with the right level of simplicity. They have simplified the experimental setup by eliminating the variability along with the space and along with the voltage. Also, they were successful to simplify the measured current by separating a current component caused by a single ion type. In their quantitative model, they derived

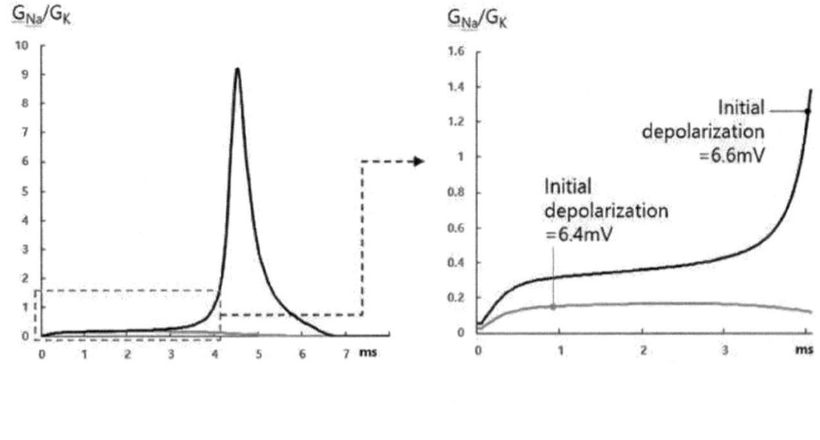

Fig. 4.25 Comparison of the conductance variation. Black line: conductance variation with the action potential firing, gray line: conductance variation without the action potential firing

the model equations with reasonable simplicity. They achieved a good balance between the accuracy coming from the detailed model equations and the simplicity to show the essence of the behavior.

Fifth, they showed the importance and usefulness of quantitative modeling in the biological phenomena and the role of computers.

Finally, their collaboration exemplified a good balance between the experiment and the theory that is difficult to be appropriately matched. Right experimental preparation is needed to support a theory, and adequate theoretical analysis is required to explain the experimental results correctly.

References

Armstrong, C. W., & Hille, B. (1972). The inner quaternary ammonium ion receptor in potassium channels of the node of Ranvier. *The Journal of General Physiology, 59*, 388–400.

Baker, P. F., Hodgkin, A. L., & Shaw, T. I. (1962). The effects of changes in internal ionic concentration on the electric properties of perfused giant axons. *The Journal of Physiology, 164*, 355–374.

Dubois, J. M., Ouanounou, G., & Dubois, B. R. (2009). The Boltzmann equation in molecular biology. *Progress in Biophysics and Molecular Biology, 99*, 87–93.

Häusser, M. (2000). The Hodgkin-Huxley theory of the action potential. *Nature Neuroscience, 3*, 1165.

Hodgkin, A. L., & Huxley, A. F. (1939). Action potentials recorded from inside a nerve fibre. *Nature, 144*, 710–711.

Hodgkin, A. L., & Huxley, A. F. (1952a). Current carried by sodium and potassium ions through the membrane of the giant axon of Loligo. *The Journal of Physiology, 116*, 449–472.

Hodgkin, A. L., & Huxley, A. F. (1952b). A quantitative description of membrane current and its application to conduction and excitation in nerve. *The Journal of Physiology, 117*, 500–544.

Hodgkin, A. L., Huxley, A. F., & Katz, B. (1952). Measurement of current-voltage relations in the membrane of the giant axon of Loligo. *The Journal of Physiology, 116*, 424–448.

Narahashi, T., Moore, J. W., & Scott, W. R. (1964). Tetrodotoxin blockage of sodium conductance increase in lobster giant axons. *The Journal of General Physiology, 47*, 965–974.

Schuetze, S. M. (1983). The discovery of the action potential. *Perspectives, 6*, 164–168.

Silver, K. S., Song, W., Nomura, Y., Salgado, V. L., & Dong, K. (2010). Mechanism of action of sodium channel blocker insecticides (SCBIs) on insect sodium channels. *Pesticide Biochemistry and Physiology, 97*, 87–92.

Tombola, F., Pathak, M. M., & Isacoff, E. Y. (2006). How does voltage open an ion channel? *Annual Review of Cell and Developmental Biology, 22*, 23–52.

5 Propagation and Processing of Membrane Potentials

Generated action potentials are propagating along the axon to the end terminal of the neuron. And, the propagated action potentials are conveyed to the next neurons or muscles through the connection of synapses or neuromuscular junctions. Then postsynaptic potentials are generated in the connected neurons. These postsynaptic potentials transmitted from the hundreds of presynaptic neurons are integrated via the processing of the soma. And, a new action potential is generated if the processed result crosses the threshold. In this way, the action potentials are generated, propagated, and processed repeatedly and parallelly through the nervous system within the body (Fig. 5.1). In this chapter, we are going to study how the generated action potentials are propagated, transmitted, and processed for the integrated function of neural activity.

1 Propagation of Action Potentials

1.1 Passive Propagation and Regeneration

1.1.1 Types of Nerve Fibers

Generated action potentials propagate along with the axon fiber of the neuron. It is just like electrical voltages are transmitted via a current flowing through the electrical wires. As there are various types of electrical wires, there are also different types of neuronal fibers that have different diameters and different internal resistances. Axon fibers are divided into unmyelinated fibers and myelinated fibers depending on the existence of myelinated sheath enveloping the membrane layer (Fig. 5.2).

The unmyelinated fibers are covered only by the cellular membrane without an additional layer of the myelin sheath. They are generally small in diameter, short in length, and look gray in color. Their diameters are less than 1.2 μm, and the small cross-sectional area due to their small diameter generates high resistance for the flowing of electrical current along the axial direction. Since the voltage drops rapidly along with the distance due to the high internal resistance, without a special mechanism of regeneration, action potentials cannot propagate further down to the axonal terminal. Action potentials are continuously regenerated during the propagation, and the conduction speed of action potential is low limited to around 1.5 m/s. Due to their low conduction speed, they are conducting information that is not in a hurry like reporting mild pain, temperature, itch, and related to housekeeping chores.

The myelinated fibers are generally larger in diameter, longer in length, and look white in color. Their diameters are up to 22 μm. As defined, myelinated fibers are covered with a thick myelin sheath. This myelinated layer increases the insulation of the membrane by decreasing membrane capacitance by about 50-folds, increasing the membrane resistance by about 200-folds, and decreasing the ion flow as much as 5000-folds. The myelin sheath is deposited by wrapping around the axon many times with the membrane of the Schwann cells. These myelin sheath layers are interrupted by the nodes of Ranvier where the ions can flow easily and the action potentials can occur. Since action potentials are generated only in the place of the nodes of Ranvier, action potentials are conducted in a saltatory pattern in contrast to the continuous pattern of unmyelinated fibers. Due to the larger diameter, increased insulation, and saltatory conduction, the conduction speed of the myelinated fibers is higher and reaches up to 130 m/s (Table 5.1). They are found in the peripheral nervous system reporting time-critical information, spinal and white matter of the brain.

1.1.2 Passive Propagation of Action Potential Along the Axon Fiber

Propagation of an action potential is not passive as in the case of voltage propagation in the electrical wires. Propagation of an action potential is the active process of repetitive regeneration that requires the consumption of energy.

If we consider the axon fiber as a passive cable, the voltage cannot be transmitted a long way down to the end

K. S. Park, *Humans and Electricity*, https://doi.org/10.1007/978-3-031-20784-6_5

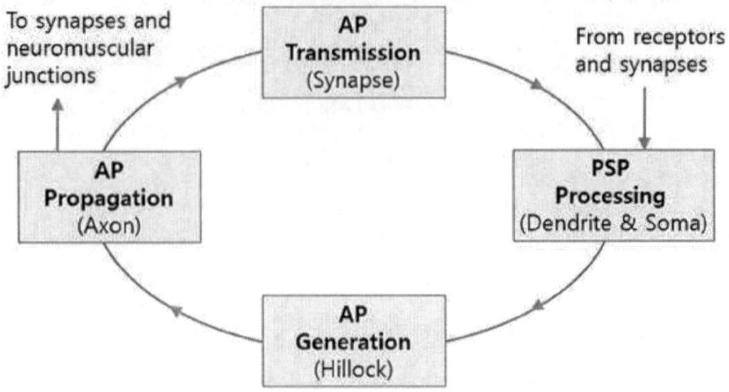

Fig. 5.1 Processing cycle of the neuronal potentials in the nervous system. *AP* Action potential, *PSP* Postsynaptic potential

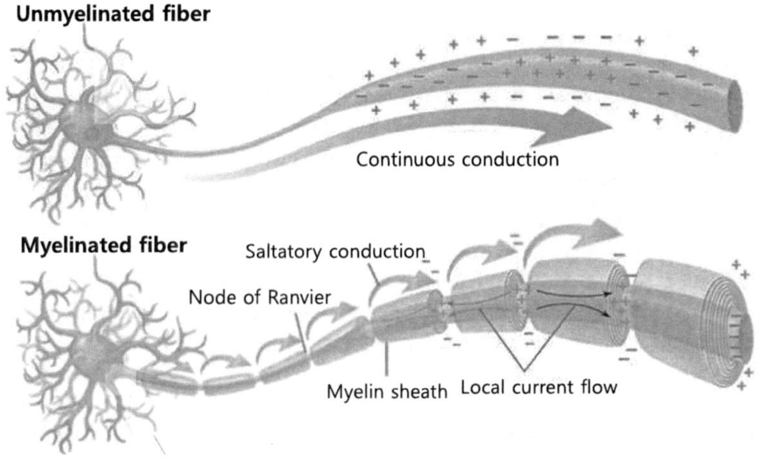

Fig. 5.2 Different types of conduction in unmyelinated and myelinated fibers (Encyclopedia Britannica, 2002)

Table 5.1 Comparison of the unmyelinated and myelinated fibers

Type	Unmyelinated fiber	Myelinated fiber
Myelin sheath	No	Yes
Nodes of Ranvier	No	Yes
Diameter	Small (0.2–1.2 μm)	Large (~22 μm)
Conduction speed	Slow (~1.5 m/s)	Fast (~130 m/s)
Conduction type	Continuous	Saltatory
Length	Short	Long
Population	2/3 in number	1/3 in number
Role	Chores reporting pain, temperature, itch	Speed critical tasks
Color	Gray	White
Location	Small axon neurons in CNS In ANS	Most peripheral nerves White matter in brain, spinal cord,

CNS Central Nervous System, *ANS* Autonomic Nervous System

of the axon. When the action potential is generated at the location, which is mostly the hillock of the neuron, the number of positive ions within the inside of the axon is increased by the influx of sodium ions. This increased number of ions builds the potential gradient as well as the concentration gradient along the axial direction. These developed gradients move the ions and develop local current and increase the potential of the neighboring site toward the threshold voltage to generate an action potential. When we see the potential distribution along the axonal direction, potential drops exponentially as shown in Eq. (3.58) in Chap. 3 with characteristic length λ. The distance that the action potential can be transmitted passively is recognized by this characteristic length λ of the axon fiber, which is represented as the square root of resistance ratio of membrane resistance to the internal resistance of axoplasm as in Eq. (5.2).

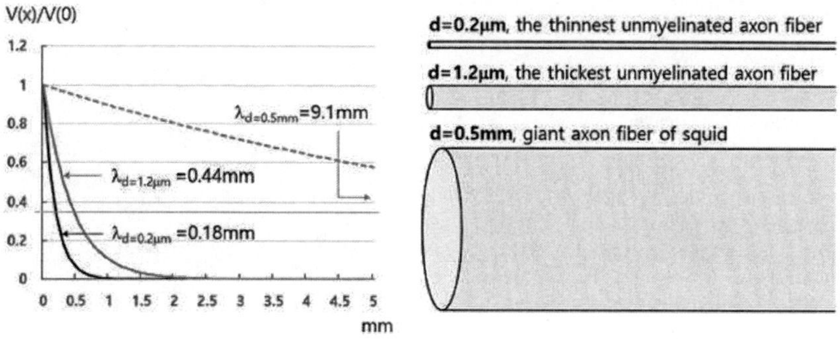

Fig. 5.3 Voltage drops along the unmyelinated fiber axon depending on the diameter

$$V'(x) = V'(0)e^{-x/\lambda} \tag{5.1}$$

$$\lambda = \sqrt{r_m/r_i} \tag{5.2}$$

The membrane resistance r_m per unit axon length of unmyelinated fiber of diameter 0.2 μm, which is considered the thinnest found in the human body, is given by

$$r_m = \frac{\rho_m l_m}{\pi d(1)} = \frac{(5.33\times10^7\Omega m)\cdot(7.5\times10^{-9}m)}{\pi(0.2\times10^{-6}m)} \\ = 6.37\times10^5\Omega m \tag{5.3}$$

Intracellular axial resistance r_i per unit axon length of the unmyelinated fiber of diameter 0.2 μm is given by

$$r_i = \frac{\rho_i(1)}{A_i} = \frac{\rho_i}{\pi\left(\frac{d}{2}\right)^2} = \frac{(0.6\ \Omega m)}{\pi(0.2\times10^{-6}m)^2/4} \\ = 1.91\times10^{13}\Omega/m \tag{5.4}$$

Then characteristic length constant of unmyelinated fiber is calculated by

$$\lambda = \sqrt{\frac{r_m}{r_i}} = \sqrt{\frac{\rho_m l_m d}{4\rho_i}} \\ = \sqrt{\frac{(5.33\times10^7\Omega m)\cdot(7.5\times10^{-9}m)d}{4(0.6\Omega m)}} \\ = 0.408\sqrt{d} \tag{5.5}$$

For an unmyelinated axon fiber of diameter 0.2 μm, it becomes

$$\lambda = 0.408\sqrt{(0.2\times10^{-6}\mathrm{m})} \\ = \sqrt{6.37\times10^5\ \Omega\mathrm{m}/(1.91\times10^{13}\ \Omega/\mathrm{m})} = 0.18\ \mathrm{mm}$$

The characteristic length of 0.18 mm means that the voltage drops to 36.7% when the current is flowing through this distance. For the thickest unmyelinated axon fiber of 1.2 μm, it is calculated to be 0.44 mm in the same way. This is too short to transmit action potential successfully down to the axon terminal more than hundreds of micrometers away. Figure 5.3 shows the potential distribution along the axial direction of different size of unmyelinated axon fibers.

A commonly used analogy for the propagation of these local currents is a leaky garden hose. Water pushing down the hose flows following the path of minimal resistance. If the diameter of the hose is narrow and there are many holes on the surface, water leaks out and the propagating distance is limited. There are two solutions for increasing the propagating distance of water further down the hose. One is increasing the diameter and thereby reducing the resistance (equivalent to reducing the axial resistance r_i) against the water flow through the core of the hose, and the other solution is increasing the difficulty for water leakage (equivalent to increasing the membrane resistance r_m) through the surface of the hose by patching up the holes (Barnett & Larkman, 2007).

1.1.3 Regeneration

Action potential propagates along with the axon fiber by the regeneration mechanism. Instead of passively transmitting current, each location of the membrane along the axial distance regenerates action potential of its own if the membrane voltage of the location is increased over the threshold. When the action potential is started, sodium current flows in. This influx current again flows down along the axial distance through the internal axoplasm and increases the potential of the neighboring sites. While this current increases the membrane potential of the neighboring sites, it partly leaks to the outside through the membrane and constitutes a local current loop. When the membrane potential increased over the threshold level, the process of generating an action potential is started at this neighboring site and the whole sequence of

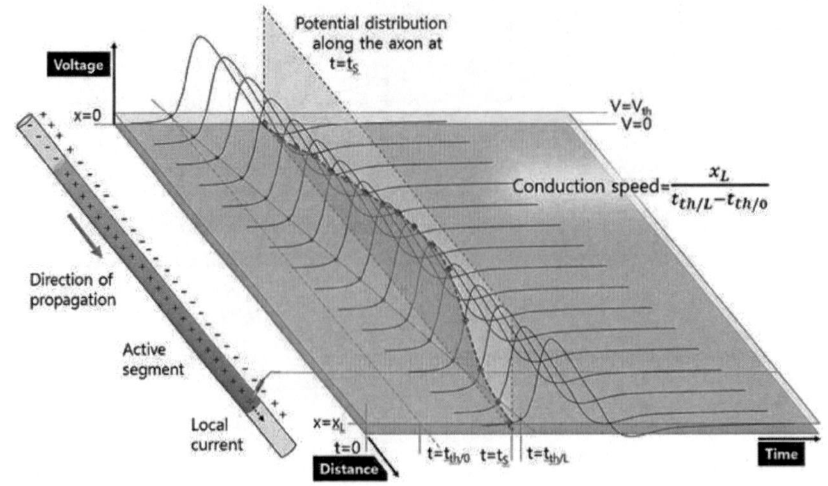

Fig. 5.4 Propagation of an action potential in the unmyelinated axon fiber. V_{th}: threshold voltage to start action potential generation, $t_{th/O}$: time that the membrane potential reaches a threshold voltage V_{th} at $x = 0$ position, $t_{th/L}$: time that the membrane potential reaches a threshold voltage V_{th} at $x = L$ position, t_S: time at which the potential distribution along the axonal distance sampled

action potential generation follows. As the inflow current increases with the progress of action potential generation, this current increases the membrane potential of this neighboring site and also spreads to the next neighboring sites again by increasing the membrane potential of that location. This process is repeated from site to next site up to the final endpoint of the axon terminal. Since this process occurs smoothly and continuously along with the axon fiber, it seems overall that action potential is propagating smoothly along with the nerve fiber. Figure 5.4 shows the propagation of action potential along with the unmyelinated axonal fiber. As the membrane potential increases by the local current, it generates an action potential at the location when the membrane potential reaches a threshold voltage V_{th}. The conduction speed of the action potential can be calculated by finding the time at which the potential reaches the threshold voltage V_{th} at some position x_L departed away by distance L. Figure 5.4 also shows the potential distribution along with the axonal distance at some sampled time t_S. It shows the spatial potential distribution shape is quite similar to the action potential in the time axis, and this potential distribution pattern moves down smoothly along the axial distance with the propagation speed of an action potential.

1.2 Conduction Speed of Action Potential in the Unmyelinated Fiber

As shown in the previous section, action potentials are propagated by the development of local currents. Local current is the passive transport of ions along the axial direction by the developed potential and concentration gradients of ions. When the voltage of the neighboring site reaches and crosses the threshold voltage by the flow of local currents, the action potential sequence is followed automatically. So, the conduction speed of an action potential is practically determined by how fast the potential of the neighboring site is raised up to the threshold voltage. That is, the conduction speed of the wavefront of this threshold voltage level V_{th} represents the conduction speed of the action potential. The conduction speed of an action potential can be estimated by the conduction speed of this wavefront. And, the propagation of action potential by the local currents to raise the potential of the neighboring site is totally passive phenomenon where the active membrane is not involved before it reaches the threshold.

With this perspective, we can use the passive RC cable theory for the derivation of the conduction speed of the unmyelinated fiber. Figure 5.5 shows the simplified *RC* cable model for the derivation of conduction speed in the unmyelinated fiber. This model is in contrast to the precise model we have derived in Chap. 4 for the generation of an action potential. In this case, we are only interested in the passive conduction of the wavefront along the axial direction of an axon fiber. We do not need to include the voltage-dependent active conductance components across the membrane. A simplified model is consisting of internal axial resistance R and the membrane capacitance C which are evenly distributed along the axonal direction.

Axial resistance R, which is the same as r_i in Eq. (5.4), and membrane capacitance C for the unit length of axon fiber of diameter d are described by

$$R = \rho\frac{l}{A} = \rho\frac{l}{\pi\left(\frac{d}{2}\right)^2} = (0.6)\frac{4(1)}{\pi d^2} = \frac{0.764}{d^2} \tag{5.6}$$

$$C = \epsilon_0 K\frac{A_m}{l_m} = 8.854\times 10^{-12}\times 8.47\frac{\pi d(1)}{75\times 10^{-10}} = 0.0314d \tag{5.7}$$

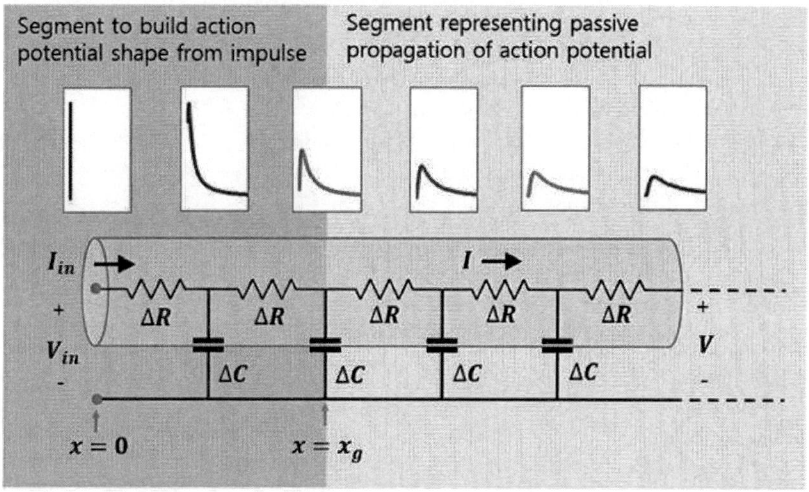

Fig. 5.5 Simplified ***RC*** cable model for the derivation of conduction speed of the unmyelinated fiber

$$RC = \frac{0.764}{d^2} 0.0314d = \frac{0.024}{d} \tag{5.8}$$

When the current flows through the longitudinal internal resistance, there is a voltage drop along with the distance. The equation representing the voltage drop in the time domain, and its transform into the Laplace domain become as follows.

$$\frac{dV(t)}{dx} = -R \bullet I(t) \rightarrow \frac{dV(s)}{dx} = -R \bullet I(s) \tag{5.9}$$

Since during the passive phase of not producing action potential, the resistance across the membrane is very high compared to the resistance in the longitudinal direction, a current flowing through the membrane is represented mostly by the capacitive current.

$$\frac{dI(t)}{dx} = -C \bullet \frac{dV(t)}{dt} \rightarrow \frac{dI(s)}{dx} = -sC \bullet V(s) \tag{5.10}$$

Combining these two equations, we have a single equation to solve.

$$\frac{d^2V(s)}{d^2x} = sRC \bullet V(s) \tag{5.11}$$

The solution of the second-order differential equation in the Laplace domain is easily derived with the appropriate boundary condition.

$$V(s) = V_{in}(s)e^{-x\sqrt{sRC}} \tag{5.12}$$

When an impulse of area A_0 $V \cdot s$ is applied as $V_{\text{in}}(s)$ at $x = 0$ position, $V(s)$ becomes

$$V(s) = A_0 e^{-x\sqrt{sRC}} \tag{5.13}$$

Since this is the solution in the Laplace domain, the solution in the time domain can be found by the inverse transforming the solution back to the time domain.

According to the Laplace transform table

$$\frac{k}{2\sqrt{\pi t^3}} e^{-k^2/4t} \leftrightarrow e^{-k\sqrt{s}} \tag{5.14}$$

And replacing k with $x\sqrt{RC}$, we can obtain time-domain solution.

$$V(x,t) = \frac{A_0 x}{2t^{3/2}} \sqrt{\frac{RC}{\pi}} e^{-\frac{x^2 RC}{4t}} \tag{5.15}$$

This is the equation representing the voltage waveform along with the time and distance. When we want to see the waveform at a specific distance, we can fix the distance to specific value $x = x_1$ and find the variation of the waveform along the time axis. Also, by fixing the time to specific value $t = t_1$, we can find the voltage variation along with the distance at a specific time. When we find the waveform at some distance, their shapes are different from distance to distance. As shown in Fig. 5.5, when we applied an impulse at the leftmost position of $x = 0$, its waveform changes from an impulse to broadened exponential waveshapes as the voltage propagates along the cable. Even though the earlier waveforms are not similar to the general action potential shape, at some distance its shape is quite similar to an action potential. We call this distance "the action potential generating distance" x_g, and action potential is regarded as generated from here for the following analysis. The segment before the distance x_g is only devoted to generating the action potential shape from the impulse shape, but it will not be used for the analysis and will be forgotten.

To numerically fit the waveform to the general shape of the action potential of 100 mV amplitude and 2 ms width at the firing location x_g, we can apply these conditions to Eq. (5.15). The time for the peak amplitude of 100 mV can be found by differentiating Eq. (5.15).

$$\frac{dV(x=x_g,t)}{dt}=0 \rightarrow t_p=\frac{x_g^2RC}{6} \tag{5.16}$$

$$V\left(x_g,t_p=\frac{x_g^2RC}{6}\right)=100\ mV=\frac{A_0x_g}{2\left(x_g^2RC/6\right)^{3/2}} \times \sqrt{\frac{RC}{\pi}}e^{-\frac{x^2RC}{4\left(x_g^2RC/6\right)}} \tag{5.17}$$

$$A_0=(100\ mV)\frac{x_g^2RC}{3}\sqrt{\frac{\pi}{6}}e^{3/2} \tag{5.18}$$

To apply the condition of waveform width, for example of 2 ms wide action potential, we need the definition of the waveform width. Since the shape of the equation of (5.15) has a long tail, we approximate the width of this action potential waveform to $t_{\text{width}}=x_g^2RC$, which is 6 times of the peak occurring time $t_p=\frac{x_g^2RC}{6}$. Even though it is a crude approximation intended for the calculation simplicity, it is quite reasonable since the voltage at this time is 23.7% of peak amplitude (Fig. 5.6). So, the width can be considered as the width of the waveform from the start to the time when its amplitude decreased up to 23.7% of the peak value. At this point, the amplitude of the waveform is given by

$$V\left(x_g,t=x_g^2RC\right)=\frac{A_0x_g}{2\left(x_g^2RC\right)^{3/2}}\sqrt{\frac{RC}{\pi}}e^{-\frac{x^2RC}{4\left(x_g^2RC\right)}} \tag{5.19}$$

When we apply Eq. (5.18) to Eq. (5.19), the amplitude at $t=t_{\text{width}}$ becomes

$$V\left(x_g,t=t_{width}\right)=\left(100\ mV\frac{x_g^2RC}{3}\sqrt{\frac{\pi}{6}}e^{\frac{3}{2}}\right)\frac{x_g}{2\left(x_g^2RC\right)^{\frac{3}{2}}} \times \sqrt{\frac{RC}{\pi}}e^{-\frac{1}{4}}=100\ mV\frac{1}{6\sqrt{6}}e^{\frac{5}{4}}=23.7\ mV \tag{5.20}$$

Similarly, we can find the time t_{th} at which the potential reaches the threshold voltage of 8 mV. Numerically, we can find that at $t_{th}=0.04072 \bullet x_g^2RC$, voltage approximately approach the threshold voltage of 8 mV. We are going to use this threshold time to calculate the conduction speed of an action potential since the speed of action potential is considered as the conduction speed of the wavefront crossing the threshold level V_{th}. When we apply the 2 ms of action potential width, x_g^2RC equals to 2 ms, peak occurs at $t_p=0.33$ ms, and the waveform reaches a firing threshold of an action potential at $t_{th}=0.08144$ ms.

To find the conduction speed of propagating action potential, it is useful to derive an algebraic expression of the conduction speed dx/dt of propagating waveform. This expression is derived by differentiating Eq. (5.15) with the voltage held constant.

$$speed=s_{um}=\frac{dx}{dt}=\frac{x}{2t}\left(\frac{x^2RC-6t}{x^2RC-2t}\right) \tag{5.21}$$

The waveform has the shape of an action potential at $x=x_g$ and we are evaluating the conduction speed of action potential by the conduction speed of the threshold wavefront which occurs at $t=t_{th}$. By applying $x=x_g$ and t_{th}=0.04072 $\bullet$ x_g^2RC, we can obtain the conduction speed of unmyelinated axon fiber as follows.

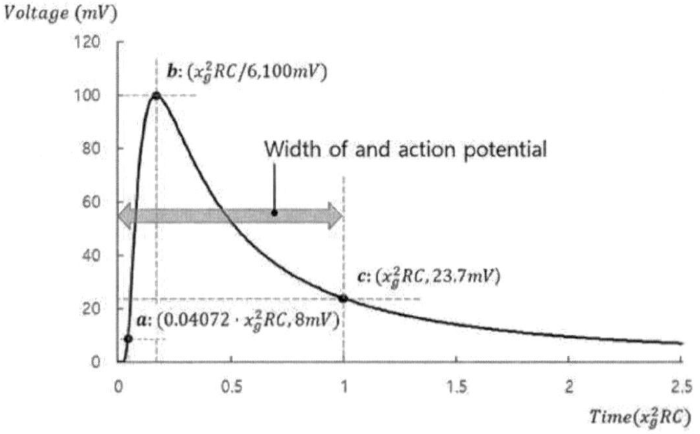

Fig. 5.6 Waveform and critical points of action potential in RC cable model at the action potential generating distance $\boldsymbol{x_g}$. Point ***a*** represents the time at which the voltage crosses the threshold voltage of 8 mV for generating an action potential. Point ***b*** represents the time at which the voltage shows the maximum amplitude of **100 mV**. Point ***c*** represents the time to define the width of the waveform

$$s_{um} = \frac{dx}{dt} = \frac{x_g}{2\left(0.04072 \bullet x_g^2 RC\right)} \times \left(\frac{x_g^2 RC - 6\left(0.04072 \bullet x_g^2 RC\right)}{x_g^2 RC - 2\left(0.04072 \bullet x_g^2 RC\right)} \right) \times \cong \frac{10}{x_g RC} = \frac{10 x_g}{t_{width}} \tag{5.22}$$

Action potential generating distance x_g is dependent on the values of R and C which are a function of the diameter of axon fiber. When the width of an action potential is set to 2 ms, x_g is represented by $x_g^2 RC = 0.002$ and Eq. (5.8) as

$$x_g = \sqrt{\frac{t_{width}}{RC}} = \sqrt{\frac{0.002}{0.024/d}} = 0.289\sqrt{d} \tag{5.23}$$

The conduction speed of the unmyelinated axon fiber is finally represented by

$$s_{um} = \frac{10 x_g}{t_{width}} = \frac{10}{\sqrt{t_{width} RC}} = \frac{2.89\sqrt{d}}{0.002} = 1445\sqrt{d} \tag{5.24}$$

The conduction speed of the unmyelinated axon fiber is increasing with the increase of the axon diameter. But it only increases in proportion to the square root of the diameter. To increase the conduction speed 2 times faster, we should have a 4 times the thicker unmyelinated fiber. For the unmyelinated axon fiber of the minimal diameter of 0.2 μm, its conduction speed becomes 0.65 m/s (=2.3 km/h), and for the thickest diameter of 1.2 μm, its conduction speed becomes 1.58 m/s (= 5.7 km/h), which is the level of the walking speed of a person. This level of speed is too slow to propagate time-critical information for survival. The unmyelinated fibers are found in nerve fibers mostly with thin diameters, short lengths, and for reporting less critical messages.

Even though we have applied the structural and numerical approximations for the derivation of waveforms and conduction speed, more detailed studies also show the same results. The conduction speed of unmyelinated nerve fibers is analyzed by considering the propagation as a direct consequence of the coexistence of two structurally distinct regions, active and resting (Ichiji, 2002, 2004). Akaishi et al. employed a new conductive model focusing on the distribution of voltage-gated sodium ion channels and Coulomb force between electrolytes (Akaishi, 2017). They theoretically have shown the same results that the conduction speed of unmyelinated nerves is proportionate to the square root of axonal diameter.

1.3 Propagation of Action Potential in the Myelinated Axon

In contrast to the unmyelinated axon fibers, axons of larger diameter and longer length in central and peripheral nervous systems are wrapped with a myelin sheath. The spirally-wrapped myelin sheath is an evolutionary specialization of vertebrates and it is produced by oligodendrocytes and Schwann cells. Myelin sheath covers most of the axonal membrane surface except the nodes of Ranvier. The schematic myelinated axon looks like a string of sausage with an exaggerated diameter. In reality, internodal length is about 100 times the outer diameter of the myelinated axon fiber (Hursh, 1939). Myelin sheath increases the insulation of axon fibers, reduces the leaky current across the membrane while action potential is propagating along with the axonal fiber, and results in a substantial increase in the conduction speed.

Myelin sheath reduces the capacitance of the membrane greatly. Reduced membrane capacitance causes a reduced population of ions along the inside of the membrane. The density of the ions is sparse along the internodal segment while the most of ions are concentrated at the node of Ranvier where the capacitance is much higher due to not having the myelin layer. With this sparse ion distribution in the internodal segment, ions can diffuse much more easily to the next node site which results in an increase in conduction speed. Also, as the ions are highly populated at the nodal site due to the increased local capacitance, the process of generating action potential is accelerated and the duration of generated action potential becomes shorter. It can be shortened up to 1.0 ms instead of 2.0 ms of the unmyelinated axon fiber.

A thick myelin sheath layer blocks the transportation of ions across the membrane. The action potential is not generated across the myelin sheath layer between the nodes of Ranvier. Action potentials are generated only on the sites of the node of Ranvier. Since the action potential is not generated during the internodal segments of the thick myelin sheath, the potential is propagating along the axial direction totally passively based on the concentration and voltage gradients. Even though the potential crosses the threshold voltage to generate an action potential, the active process of generating action potential is not occurring at this internodal segment. Instead, the voltage at this segment just follows the voltage of the previous node with minimized leaky current by the increased myelinated insulation.

Action potential propagates through the internodal segment of myelin to the next node. When the potential at the next node crosses the threshold of generating an action potential, there starts a generation of an action potential at

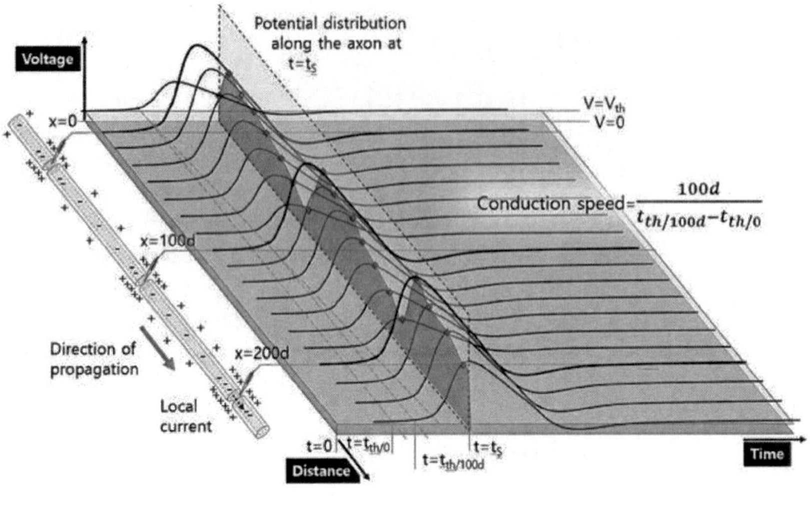

Fig. 5.7 Propagation of action potential in the myelinated fiber. V_{th}: the threshold voltage to start an action potential generation, $t_{th/0}$: time that membrane potential reaches a threshold voltage V_{th} at $x = 0$ position, $t_{th/100d}$: time that membrane potential reaches a threshold voltage V_{th} at next node of $x = 100$d, t_S: time at which the potential distribution along the axonal distance sampled

the node of Ranvier. As the action potential is started to actively generate an action potential, voltage increases rapidly over the voltage propagated from the previous node. Figure 5.7 shows the propagation of action potential along with the myelinated axon fiber. It shows that the potential waveform at the location between the nodes follows the action potential of the previous node with propagation delay. And the amplitude of the waveform decreases away from the action potential generating node due to the leaky current across the membrane. Voltage distribution at some sampling time shows a weird pattern in contrast to the potential distribution of unmyelinated axon fiber. In the unmyelinated axon fibers, voltage distribution along the axial direction at some sampling time is the same shape as the action potential along the time axis as shown in Fig. 5.4. This pattern is caused because the action potential is generated continuously along the axial direction. However, for the myelinated axon fiber, voltage is decreasing during the internodal segment, since it is only propagating from the previous node by the diffusion. But the voltage increases actively and rapidly at the nodal site regardless of the propagated potential amplitude once the voltage crosses the threshold of generating an action potential. There appears a discrete abrupt increase of amplitude at the nodal site and the overall shape of the voltage distribution along with the axial direction results in a sawtooth-like pattern.

1.3.1 Conduction Speed of the Myelinated Axon

As we have seen in the previous section, the conduction speed of the unmyelinated axon fiber is dependent on the axon parameters. Since it is based on the RC cable model, conduction speed is dependent on the R and C values. Here we have assumed that the characteristic properties of the membrane and the axoplasm are the same for all sizes of unmyelinated fibers and the R and C values are only dependent on axon diameter. For the derivation of the conduction speed of the myelinated axon fiber, it is useful to analyze the relation of the R and C values on the conduction speed. For the myelinated axon fibers, propagation of an action potential can be interpreted in the same way as the unmyelinated axon fibers in the first stage. Propagation of potential from the site of a node of Ranvier to the next node is totally passive and driven by the concentration gradient and potential gradient along the axial direction. Even though the axon fiber is covered with the myelin sheath, the mechanism of increasing the potential of the neighboring site is the same up to the site of the next node. So, the conduction speed of the myelinated fibers is also dependent on the R and C values of the axon fiber. The difference is their relation to the diameter of the axon. For the pulse width of 2.0 ms,the relation between the conduction speed of the unmyelinated axon fiber and the R and C values is given by

$$s_{um} = \frac{10}{\sqrt{t_{width}RC}} = \frac{10}{\sqrt{0.002RC}} = \frac{223.6}{\sqrt{RC}} \tag{5.25}$$

With this equation, we can find that conduction speed is increased as the R and C values are decreased. When they cannot be decreased simultaneously, RC values should be decreased in their combination. Figure 5.8 shows the cross-sectional diagram of the myelinated axon fiber. For the fixed diameter size d of the myelinated fiber, the cross-sectional area is represented by the concentric circles with inner axonal diameter d_a, and g-ratio for their relative size.

$$g = \frac{inner\ axon\ diamter}{outer\ fiber\ diameter} = \frac{d_a}{d} \tag{5.26}$$

With these diameters, R and C values of the myelinated fiber are given by

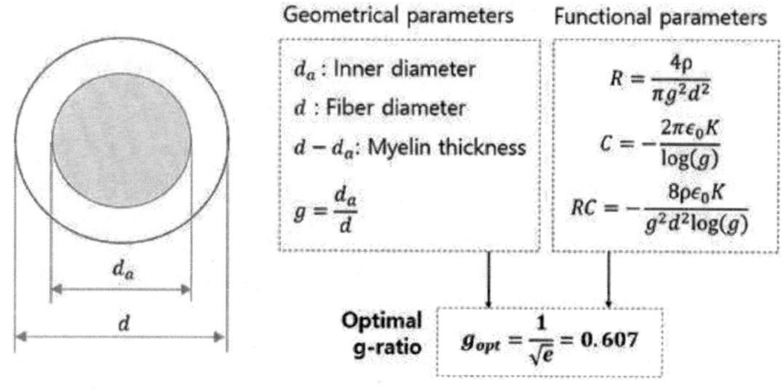

Fig. 5.8 Schematic cross section of the myelinated fiber for the derivation of optimal *g*-ratio

$$R = \frac{4\rho}{\pi d_a^2} = \frac{4\rho}{\pi g^2 d^2} \tag{5.27}$$

$$C = -\frac{2\pi\epsilon_0 K}{\log(g)} \tag{5.28}$$

$$RC = -\frac{8\rho\epsilon_0 K}{g^2 d^2 \log(g)} \tag{5.29}$$

Resistance along the axial direction decreases as the *g*-ratio increases. To achieve an increased conduction speed, higher *g*-ratio is required from the view of the internal resistance. To reduce the capacitance, a thicker myelinated sheath is required, which is equivalent to the decreased *g*-ratio. That is, increased *g*-ratio decreases resistance but it increases the capacitance. The optimal *g*-ratio which maximizes the conduction speed can be found by deriving the optimal *g*-ratio value which minimizes the *RC* product.

By differentiating Eq. (5.29), optimal value of *g*-ratio is derived as

$$g_{opt} = \frac{1}{\sqrt{e}} = 0.607 \tag{5.30}$$

And the minimized *RC* value becomes

$$RC_{min} = \frac{16e\rho\epsilon_0 K}{d^2 \log(e)} = \frac{16 \times 2.72 \times 0.6 \times 8.854 \times 10^{-12} \times 8.47}{d^2} = \frac{1.96 \times 10^{-9}}{d^2} \tag{5.31}$$

This theoretical *g*-ratio nicely fits the experimental data for the myelinated axon fibers obtained in the peripheral nerve fibers. But it appears lower for myelinated axons found in CNS. It is considered that the myelinated fibers in CNS must achieve multiple goals including reducing conduction delays, promoting conduction fidelity, lowering energy costs, and saving space. For the central nervous system where axons tend to be smaller and their myelin sheath thinner, optimal *g*-ratio can be increased up to 0.77 considering the additional constraints (Chomiak & Hu, 2009).

If we apply this value to Eq. (5.25), the conduction speed of the myelinated fiber having the optimal myelin sheath thickness becomes

$$s_m = \frac{10}{\sqrt{t_{width} RC}} = \frac{10}{\sqrt{0.001RC}} = \frac{316.2}{\sqrt{\frac{1.96 \times 10^{-9}}{d^2}}} = 7.14 \times 10^6 d \tag{5.32}$$

This value was derived with the assumption that the width of action potential is 1.0 ms at the node of Ranvier due to the accelerated action potential generation process and the action potential is propagating by continuously firing as in the case of unmyelinated fibers, which is not appropriate in the case of myelinated axon fibers. Without the active process of action potential generation, the potential at this internodal segment is lower than the case of potential with the process of action potential and the diffusion of ions is somewhat slower due to the less steep concentration gradient and potential gradient toward the next node. This results in the delayed rise of membrane potential at the next node and the lowered conduction speed. Figure 5.9 shows the action potentials generated at two nodes. When the action potential is generated at node of $x = x_g$, the membrane potential approaches the threshold level (point ***a*** in Fig. 5.9) 8 mV at $t = 0.04072 x_g RC$ as the same way in the unmyelinated case of Fig. 5.6. Membrane potential at the next node separated by one nodal segment of $100d$ can be calculated with $x = x_g + 100d$ by Eqs. (5.15) and (5.18). Potential at the next node rises passively up to the threshold of 8 mV (point ***b*** in Fig. 5.9) by the propagation of action potential from the $x = x_g$. Once the membrane potential crosses the threshold level, the action potential generation process starts and the potential at this node follows the shape of an action potential.

The relationship between x_g and the diameter d with *RC* value of Eq. (5.31) is given by

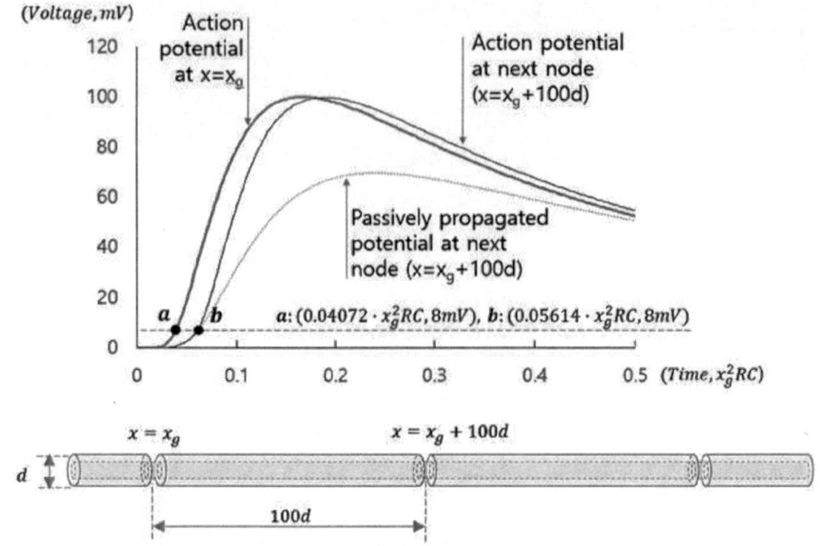

Fig. 5.9 Action potential of the myelinated axon fiber at action potential generating node and at the next node of Ranvier

$$x_g = \sqrt{\frac{t_{width}}{RC}} = \sqrt{\frac{t_{width}}{\frac{1.96\times10^{-9}}{d^2}}} = d\sqrt{\frac{0.001}{1.96\times10^{-9}}}$$
$$= 714d \tag{5.33}$$

$$d = \frac{x_g}{714} \rightarrow 100d = \frac{100x_g}{714} = 0.14x_g \tag{5.34}$$

The voltage waveform at $x = x_g + 100d$ can be derived with Eq. (5.15)

$$V(x,t) = V(x_g + 100d, t) = \frac{A_0(x_g + 100d)}{2t^{3/2}} \times \sqrt{\frac{RC}{\pi}} e^{-\frac{(x_g+100d)^2RC}{4t}} \tag{5.35}$$

With $t = k \bullet x_g^2 RC$ and $100d = 0.14x_g$, Eq. (5.35) is simplified to

$$V(x_g + 100d, t) = V\left(x_g + 0.14x_g, kx_g^2RC\right)$$
$$= \frac{A_0(x_g + 0.14x_g)}{2\left(k \bullet x_g^2 RC\right)^{3/2}} \sqrt{\frac{RC}{\pi}} e^{-\frac{(x_g+0.14x_g)^2RC}{4k\bullet x_g^2 RC}}$$
$$= \left(100mV\frac{e^{3/2}}{6\sqrt{6}}\right)\frac{1.14}{k^{3/2}} e^{-\frac{1.3}{4k}} \tag{5.36}$$

Equation (5.36) reaches 8 mV at $k = 0.05614$, that is at $t = 0.05614x_g^2RC$. For 1.0 ms width of action potential, it is 56.1 μs.

The conduction speed of the myelinated axon fiber can be calculated most simply by dividing one internodal distance of 100d with the time elapsed between the threshold crossing of point ***a*** and ***b*** in Fig. 5.9.

$$s_m = \frac{100d}{\left(0.0561x_g^2RC - 0.0407x_{lg}^2RC\right)}$$
$$= \frac{100d}{0.0154x_g^2RC} = \frac{100d}{0.0154 \bullet 0.001} = 6.48\times10^6 d \tag{5.37}$$

According to this result, the myelinated axon fiber of 10 μm propagates an action potential with the speed of 64.8 m/s (223 km/h). Table 5.2 shows the types of peripheral nerve fibers classified according to the conduction speed. Considering the simplicity of the model and the wide variance of the nerve types, the derived equation for the conduction speed of the myelinated fiber is considered nicely represent the relation to the diameter.

1.3.2 Comparison of Conduction Speed

Based on the derived equations of conduction speed for the unmyelinated fiber and the myelinated fiber, we can compare the conduction speed of different type axon fibers as shown in Table 5.3. For the same size of axon diameter, the myelinated fiber propagates action potentials faster than the unmyelinated fiber. For the smallest size of 0.2 μm of axon diameter, conduction speed increases 2.0 times from 0.65 to 1.30 m/s for the myelinated axon fiber compared to the unmyelinated fiber. The thicker the axon diameter is, the higher the increasing rate is. For the axon of the largest diameter of 22 μm, conduction speed increases by 21.0 times from 6.78 to 142 m/s. If this conduction speed is realized by the unmyelinated fiber, its diameter should increase by 441 times to 9.7 mm. Its cross-sectional area should be increased by 441^2 times, and then it covers the cross-sectional area equivalent to

Table 5.2 Classification of the peripheral nerve fibers according to the conduction speed (Hert et al., 2014)

Fiber type	Sensory classification	Myelin	Diameter (μm)	Speed (m/s)	Function	Speed/diameter ratio (×10^6)
Aα		Yes	12–20	70–120	Extrafusal muscle fiber	5.83–6.0
Aα	Ia	Yes	12–20	70–120	Annulospiral endings	5.83–6.0
Aα	Ib	Yes	12–30	70–120	Golgi tendon organ	4.0–5.83
Aβ	II	Yes	5–12	30–75	Flower-spray endings	6.0–6.25
Aδ	III	Yes	2–5	12–30	Pain, touch, temperature	6.0
Aγ		Yes	3–6	15–30	Intrafusal muscle fiber	5.0
B		Some	<3	3–14	Preganglionic	1.0–4.67
C (Dorsal root)	IV	No	0.4–1.2	0.5–2.0	Pain, touch, temperature	1.25–1.67
C (Sympathetic)		No	0.3–1.3	0.7–2.3	Postganglionic sympathetic fibers	1.77–2.33

Table 5.3 Comparison of conduction speeds of the unmyelinated and the myelinated axon fibers

Diameter	Conduction speed of unmyelinated axon	Conduction speed of myelinated axon	Increase of speed (folds)
0.2 μm	0.65	1.30	2.0
1 μm	1.45	6.48	4.48
5 μm	3.23	32.4	10.0
10 μm	4.57	64.8	14.2
22 μm	6.78	142.6	21.0

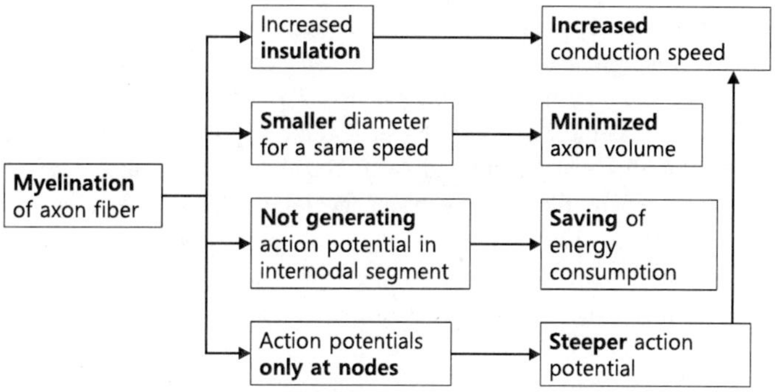

Fig. 5.10 Advantages by the introduction of myelination in axon fibers

the cross section of about 200,000 myelinated fibers having the same conduction speed. Definitely, myelination is the intelligent evolutionary solution to save the body space while maintaining the conduction speed.

1.4 Myelination as an Evolutionary Success

The myelin sheath is a crucial acquisition of vertebrates to increase the propagation speed of action impulses. Even though some invertebrate axons are wrapped with sheaths by glial cells, they do not have a compact myelin. As a result, action potentials in invertebrate axons propagate with a conduction speed of less than 1.0 m/s. For small invertebrate animals between 0.1 and 30 cm, this speed is sufficient for their survival. Larger size invertebrates like squids have evolved by increasing their axonal diameters up to 1 mm to increase the speed of action potential propagation for their survival. Their increased axonal diameter also provided good experimental samples for the intracellular recording of action potential and their analysis (Zalc, 2006).

In contrast, while vertebrates have a strong need of increasing the conduction speed due to their larger body size, they also have physical constraints imposed by the skull and the vertebrae. As an alternative solution, they evolved by introducing the myelin sheaths and achieved the conduction speed of 50–100 m/s successfully without increasing the diameter of their axons.

Introducing the myelination in axon fibers provides advantages in the following viewpoints, as summarized in Fig. 5.10.

- Myelination increased the conduction speed dramatically compared to the unmyelinated fiber of the same diameter by increasing the insulation of the membrane. It is an efficient method to increase the conduction speed of the axon fiber when the space is constrained for the volume of the axon fiber.

- Myelination can save space by minimizing the diameter of the axon fiber to achieve the targeted conduction speed. As the conduction speed is achieved with a minimized diameter of axon fiber, it results in minimized space, volume, and weight which is optimal for escaping as well as attacking for survival.
- Myelination saves energy consumption. There is no regeneration of action potential during the internodal segment of myelination. There is also no ATP consumption to rebuild the ion concentration by the Na-K pumping. Since the potentials are propagating passively over the long segment, which is about 100 times longer than its diameter, it provides a great amount of energy saving, which might have been used otherwise, by not regenerating action potential. If there were action potentials generated during the internodal segment with myelinated sheaths, conduction speed is increased only 10% as in Eq. (5.32) compared to the real case. So, myelination is considered the optimal selection with the minimal loss of conduction speed and the maximal saving of energy consumption. Even though considering the additional energy consumption by the glial cells wrapping the axions, overall energy saving is substantial by the myelination.
- Myelination allows ion transportation across the membrane only at the nodes of Ranvier. The density of ion channels is increased in this area and the population density of ions is also increased due to the higher capacitance compared to the lower capacitance internodal segment area. This environmental situation accelerates the action potential generation process and the width of action potentials generated at these nodes is shorter than the ones generated in unmyelinated axon fiber. Through the modeling analysis, we can find that shorter width action potential increases the diffusion rate by the higher spatial gradient along the axial direction. This results in an increased passive propagation speed of action potential and finally the increased conduction speed of action potential between the nodes.

The concept of regeneration and myelination is quite similar to the relay or repeater in the electrical or optical transmission of signals in telephone cables. Signals are transmitted passively as far as the signals can be recognized with confidence. Before the amplitude of the transmitted signal is decreased to an unrecognizable level, the repeater is used to restore the amplitude of the signal to the level of the sending site. This is exactly the same with the regeneration of action potential at nodal sites. As the maximal length of passive transmission of the signal is determined to minimize the number of repeaters, the length of internodal myelination sheaths is determined optimally to minimize the number of nodes to save energy consumption. But, the length of the internodal segment should be appropriate to have some safety margin so that action potential is continued over some nonregenerating nodes. Even though one or two nodes are not firing action potentials due to some defects, action potential still should be able to propagate over these nodes by passive propagation. While the propagated maximum voltage is higher than the threshold, an action potential can be regenerated at the next active node over some damaged nodes with some time delay.

1.4.1 Demyelination of Axon

There are some diseases like multiple sclerosis which disrupt the layer of myelin sheaths (Fig. 5.11). If the total layer of myelin is disrupted by the disease, myelinated axon fiber becomes the unmyelinated axonal fiber of a smaller diameter and there is a great reduction in conduction speed. For example, consider the myelinated axon fiber of diameter 10 μm. As shown in Table 5.3, it propagates action potential with the conduction speed of 64.8 m/s. If the demyelination occurs, the whole thickness of myelination is disrupted and the only axonal fiber part remains with a diameter of 6.07 μm. Conduction speed of the unmyelinated axon fiber of diameter 6.07 μm becomes 3.56 m/s by Eq. (5.24). There is a great reduction of conduction speed from 64.8 to 3.56 m/s almost by 18-folds. This reduced conduction speed causes significant problems in communication between the brain and the peripheral body parts. The existence and the progress of myelin disruption can be evaluated by the clinical measurement method of nerve conduction speed.

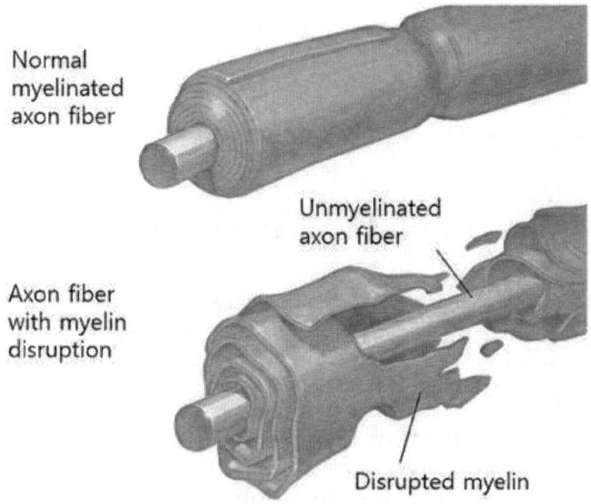

Fig. 5.11 Disruption of myelin sheath in demyelination disease of multiple sclerosis

2 Synaptic Transmission

Neurons are connected with the other neurons or muscles by the junctions called synapses. After action potential is propagated up to the end part of the axon, it spreads to multiple terminal knobs following the branches of the axon. Each terminal knob is landing on other neurons in their dendrites or soma area by building the connection of synapses. The propagated action potential is transmitted to the next neurons by these synaptic junctions. The neuron before the synapse and sending the action potential is called *presynaptic neuron* and the neuron after synapse and receiving the action potential is called *postsynaptic neuron*. One presynaptic neuron has synapses with multiple postsynaptic neurons and one postsynaptic neuron receives signals from the multiple presynaptic neurons. This is the basic unit of connecting neurons and building the neural network.

While 80–95% of synapses are landing on the dendrite area, 5–20% of them are landing directly on the soma area. The number of synapses per neuron varies considerably. Many of the neurons found in the cerebellum have only a handful of synapses, while Purkinje cells in the cortex have up to 200, 000 synapses. On average, each neuron has around 1000 synapses. Since each synapse is a connection between one presynaptic neuron and one postsynaptic neuron, there are equal number of presynaptic neurons and postsynaptic neurons. While each neuron receives signals via about 1000 synapses, its axon is considered also have an equal amount of about 1000 terminal knobs on average. There are about 10^{11} neurons in the brain, and the total number of synapses in the brain raises up to 10^{14}. This number is beyond the 1.2×10^{10} bits of information that can be coded in the human genes, which can be stored in 6×10^{9} base pairs of diploid genes by four base types of A, C, G, T. This number supports that the human neuronal network, intelligence, and memory ability are difficult to be encoded totally in genes, and they can be thought to be developed mostly after birth.

2.1 Signal Transmission in the Synapse

Synaptic transmission of signals starts from the action potential propagated along with the axon fiber up to the terminal part of the axon. At the axon terminal, the axon branches into hundreds of finer paths, and the very end of each path is bulged to increase the contacting area for synaptic contact. This bulged contacting part of the presynaptic neuron is called synaptic *bouton* or synaptic *knob*. The diameter of the synaptic knob is around 1 μm with some variance depending on the diameter of the axon.

When the action potential is propagated up to this synaptic knob, there occurs a series of microprocedures to transmit a signal from the presynaptic side to the postsynaptic side through the synapse as follows (Fig. 5.12).

1. Propagated action potential arrives at the axon terminal.
2. Increased membrane potential opens voltage-gated calcium ion channels and increases the permeability of calcium ions transiently.
3. Calcium ions flow in and increase the concentration of calcium ions at the localized zone of the synaptic knob of the presynaptic neuron.
4. Calcium ions trigger the movement of vesicles, which contain the neurotransmitters such as acetylcholine, toward the synaptic membrane.
5. Vesicles move to the synaptic membrane, bind to the membrane by the SNARE proteins, and fuse with the membrane.
6. Neurotransmitters in synaptic vesicles are released into the synaptic cleft by the exocytosis and the released

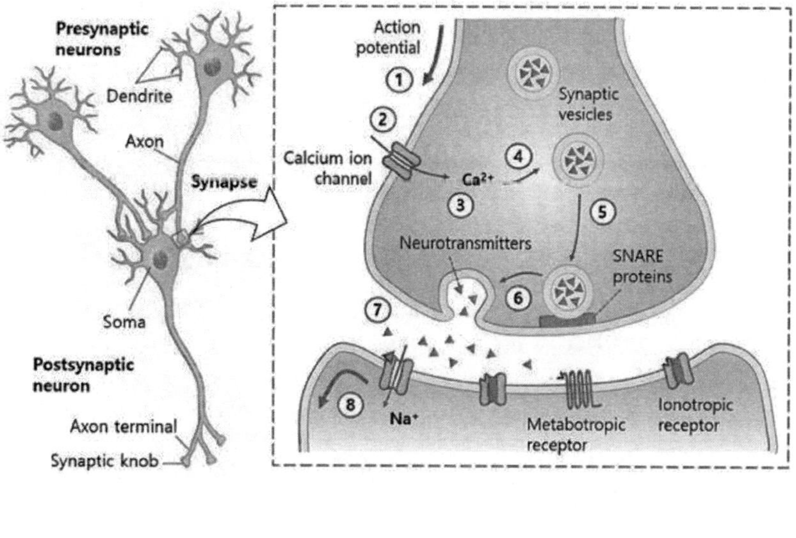

Fig. 5.12 Signal transmission in the synapse

neurotransmitters march across the synaptic cleft, which is about 20–40 nm in distance.

7. Neurotransmitters crossed the synaptic cleft arrive at the synaptic membrane of the postsynaptic neuron and attach to the ligand-gated ion channels. Ion channels coupled with the neurotransmitter(ligand) are open and allow the movement of channel-specific ions across the membrane. Sodium, calcium, potassium, and chloride ions have their specific ion channel and move upon the opening of the corresponding channels.
8. Movement of ions across the synaptic membrane increases (or decreases) the local ion concentration and develops a local postsynaptic potential and the postsynaptic potential propagates along the dendrite toward the soma.

2.2 Neurotransmitters

During the transmission through the synapse, the electrical signal of the action potential is converted to a chemical signal of neurotransmitter and back to an electrical signal again on the postsynaptic side. While the signal is transmitted through the neural network composed of a series of neuronal connections, this conversion of signals from electrical to chemical and chemical to electrical occurs multiple times. This is the biggest difference between neural circuits and ordinary electrical circuits where there is no chemical intervention. What are the benefits or disadvantages of chemical intervention on electrical neural signals?

The first interest is the time that is consumed passing through the small gap of the synapse. Synaptic delay, which is the time required for a neurotransmitter to be released from a presynaptic membrane, diffuse across the synaptic cleft, and bind to a receptor site on the postsynaptic membrane, is in the range of 0.3 – 0.5 ms for traveling about 1 μm at most. This is equivalent to the conduction speed of 2 mm/s which is much lower than the conduction speed of axons of several m/s. When the signal is transmitted through the peripheral axon fiber with the 100 m/s velocity, it will take 10 ms to travel 1 m body distance. If there are two synaptic connections during the pathway, it will increase the conduction time by 1 m/s. If there are 4 synaptic connections, conduction is delayed by 2 ms, and 5 ms if there are 10 synaptic connections. Conduction delay of 20–50% by the chemical intervention seems to be a great disadvantage for the transmission of time-critical information. This disadvantage can be minimized by limiting the number of synaptic connections by increasing the length of axonal conduction as shown in the many peripheral nerves. Nevertheless, the biological system adopted the synaptic connection to get more benefits than a paid disadvantage.

We have got a great diversity of processing of information by the introduction of the synapse. Ion channels in the postsynaptic membrane are ligand-gated types instead of voltage-gated type ion channels that appeared in the action potential generation. This ligand-gated ion channel, which is also called ionotropic receptors, opens its gate when the specific neurotransmitter is attached to the docking site at the external side of the channel protein. This is just like a lock is opened by the insertion of a matching key. When they are open, they allow the movement of channel-specific ions across the membrane and alter the local potential of the membrane. As there are many types of neurotransmitters, there are also many different types of ligand-gated channels. While neurotransmitters are specific to the types of presynaptic neurons, the distribution of ionotropic receptors is specific to the roles of postsynaptic neurons. Even a single transmission of action potential from a presynaptic neuron, diverse responses are possible among the postsynaptic neurons by the diverse combination of neurotransmitters on the presynaptic side and the responding ligand types of ionotropic receptor on the postsynaptic side. Types of released neurotransmitters are related to the roles of presynaptic neurons regulating behavioral or cognitive functions like motor function, arousal, memory, learning, mood and emotion, reward, and so on. Postsynaptic neurons can respond selectively to the released transmitters by possessing ionotropic receptor types. This diverse and selective connection in a neural network provides much more benefit than can be achieved by the regulation of voltage only.

It is known that there are more than 40 types of neurotransmitters. Table 5.4 shows the representative neurotransmitters acting on synaptic junctions. Glutamate is the most common excitatory neurotransmitter in the central nervous system. Acetylcholine, norepinephrine, epinephrine, and histamine are neurotransmitters acting as excitatory types. Gamma-aminobutyric acid (GABA) is the most powerful inhibitory neurotransmitter acting as a brake to the excitatory neurotransmitters. Serotonin is another inhibitory type neurotransmitter. Dopamine is a neurotransmitter working as an excitatory type or inhibitory type depending on the counteracting ionotropic receptors. When it is binding to the ionotropic receptors gating sodium or calcium ions, it acts as an excitatory type neurotransmitter. And it acts as an inhibitory neurotransmitter when it binds to the ionotropic receptors gating potassium or chloride ions.

Since the synaptic cleft is open in the extracellular space, additional chemical messengers originating other than presynaptic neurons can be involved in the modulation of the signal transmission. Many types of drugs and alcohol are acting on this site and modify the excitability of the postsynaptic neurons.

Table 5.4 Important neurotransmitters for synaptic connections

Neurotransmitter	Type	Releasing locations	Functions	Related disorders
Glutamate (Glu)	Excitatory	Sensory neurons, Cerebral cortex,	CNS excitability, learning, memory	Epilepsy, cognitive & affective disorder
Acetylcholine (Ach)	Excitatory	Motor neurons, basal ganglia, ANS neurons	Muscle contraction Sleep cycle	Muscle paralysis
Norepinephrine (Ne) [Noradrenaline]	Excitatory	Brainstem, hypothalamus, adrenal glands	Increase alertness & wakefulness,	Depression, Anxiety disorder, Impaired sleep cycle
Epinephrine (Epi) [Adrenaline]	Excitatory	Medulla, Adrenal glands	Prepare for dangers	Hypoglycemia, myocardial infarction
Histamine	Excitatory	Hypothalamus, stomach mucosa, mast cells	Wakefulness, blood pressure, pain, sexual behavior, inflammatory response	Immune system diseases, Allergies
Dopamine [DA]	Excitatory & Inhibitory	Substantia nigra	Inhibit unnecessary movement, Growth hormone secretion	Parkinson's disease, psychotic disorder, schizophrenia
Serotonin (5-HT)	Inhibitory	Brainstem, gastrointestinal tract, Platelets	Emotion & mood, body temperature, pain perception, Sleep cycle	Decrease immune system function, emotional disorder
Gamma-aminobutyric acid (GABA)	Inhibitory	Spinal cord, cerebellum, cerebral cortex	Reducing neural excitability, Mood & emotion	Anxiety disorder, schizophrenia

2.3 Metabotropic Receptors

In addition to the ionotropic receptors acting on the generation of postsynaptic membrane potential, there is another type of receptor called a metabotropic receptor. It is not acting as an ion channel and is not directly related to the transmission of the signal to the postsynaptic side. Instead, it works as a "Second messenger" activator. Upon the binding of the transmitted neurotransmitters, it activates one or more substances inside the postsynaptic neuron which serves as the second messengers and changes a specific cellular function (Guyton & Hall, 1996).

Once the neurotransmitters are released into the synaptic cleft, they are rapidly removed from the synaptic cleft space by degradation, diffusion, or neuronal uptake for the recycling. This is to prevent repetitive and unwanted stimulation of the postsynaptic cell. So, the responses of ionotropic receptors are fast and last only milliseconds. It is not suitable for many functions, such as the process of memory, requiring prolonged change for seconds to months after the synaptic activation. Prolonged neuronal actions are achieved by the activation of the second messengers, which trigger a series of intracellular events. Among the several types of the second messenger, G-protein is one of the most prevailing types. Inside the cytoplasm of the postsynaptic neuron, it performs specific functions depending on the characteristics of each neuron. G-protein can trigger the opening of extra ion channels for a prolonged time, activate a process altering the long-term excitability of the neuron, activate intracellular enzymes and cause-specific chemical functions, and activate gene transcription which can change the cell structure related to the long-term memory process. Metabotropic glutamate (mGlu) receptors, which exhibit a widespread distribution in the CNS, play a major role in various neuronal processes including synaptic plasticity and memory formation (Mukherjee & Manahan-Vaughan, 2013).

The storage of information is an additional function of the synapse. Each time the sensory signal is transmitted through the synapses, these synapses are facilitated to be more capable to transmit the same signal again. After the signals have been transmitted through the synapses many times, the synapses become so facilitated that transmission of signals through a sequence of the synapses can be occurred by the brain itself without any sensory stimulations. This gives the same perception as the original sensation. They are memories of the sensation that are developed in the synapses. However, still, we do not know much about the precise mechanism of the roles of synapses in memory formation.

2.4 Neuromuscular Junction

A neuromuscular junction is a chemical synapse between the motor neuron and the muscle fiber. Overall steps of transmitting action potential from the presynaptic motor neuron to the postsynaptic membrane of muscle fiber are the same as the chemical synapse between the neurons. The differences come from the receiving side of the muscle fiber.

The receiving side of the neuromuscular junction is called a motor endplate. The motor endplate is folded into peaks and troughs as shown in Fig. 5.13. The peaks are densely packed with acetylcholine receptors, while the troughs contain the enzyme acetylcholinesterase which catalyzes the breakdown of acetylcholine. At each motor endplate, about 10^6 acetylcholine receptors are populated. As acetylcholine is the neurotransmitter in the neuromuscular junction, it is always

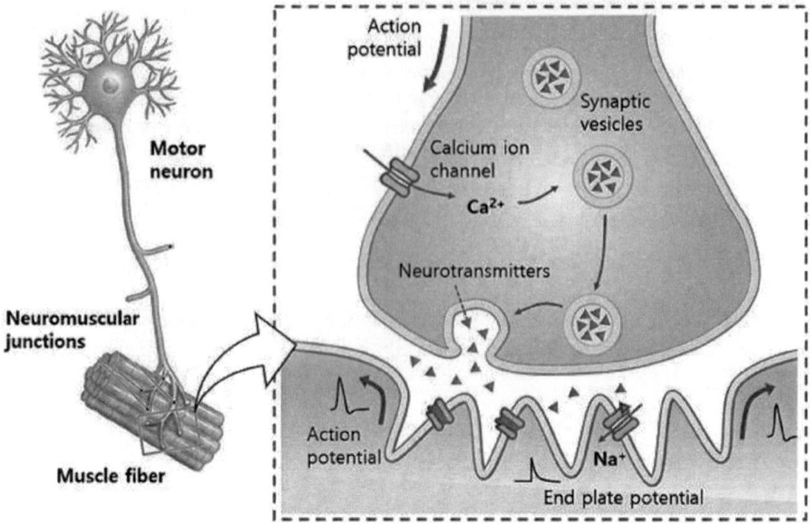

Fig. 5.13 Neuromuscular junction

excitatory. There is no inhibitory connection through the neuromuscular junction. When there is no motor signal, muscles are already in a relaxed and flaccid state, and there is no lower state that muscles can go to by the inhibition.

In contrast to the synapses between the neurons, there is only one neuromuscular junction per muscle fiber. Each muscle fiber is solely controlled by a single motor neuron through its neuromuscular junction. There is no summation of potentials generated from different synaptic transmissions as in postsynaptic neurons. The potential generated by the activation of the ligand-gated ion channel in the endplate is called endplate potential. Each vesicle is contributing to the generation of the endplate potential of 0.4 mV. This voltage is much lower than the threshold to generate action potential which will propagate along with the muscle fiber and cause muscle contraction. Almost 100 vesicles are required to make the potential reach the threshold and generate an action potential.

2.5 Electrical Transmission Between the Neurons

Although synaptic transmission between the nerve cells is performed by chemical intervention, electrical transmissions between the excitable cells also occur. An electrical type synapse is an electrically conductive gap junction between the two neighboring neurons through which the spread of electrical currents can occur. The diameter of the gap junction is about 1.2 to 2.0 nm, which is wide enough to pass the ions and some medium-size molecules and to connect the cytoplasm of the two neurons. As the ions carrying the positive charges move through the gap from the presynaptic side to the postsynaptic side, neurons on the postsynaptic side can be depolarized. Electrical gap junctions are most commonly operating as ohmic resistors, providing bidirectional communication for electrical signals between the two neurons. Currents generated by the action potentials in the presynaptic neuron can directly flow through the gap junction to the postsynaptic neuron. Not only the currents generated by action potentials but also any graded signals of depolarizing or hyperpolarizing can spread to the postsynaptic neuron to develop a coupling potential. The amplitude of the coupling potential is determined by the conductance of the gap junction channels and the input resistance of the postsynaptic neuron.

The intercellular gap junction is formed by the docking of two apposed individual channels, called *connexons* or *hemichannels*, from each of the coupled neurons. Connexons forming the electrical gap junction can be made of the same or different types of proteins. Electrical gap junctions formed by the same type of connexons are called *homotypic*, whereas gap junctions formed by the different types of connexons are called *heterotypic*. Structural differences of the coupled connexons affect the pattern of electrical transmission (Fig. 5.14). While the gap junctions formed by the homotypic connexons allow the bidirectional transmission of electrical signal regardless of the polarity of the potential, gap junctions of the heterotypic connexons rectify the signal, like a diode in electrical circuits, depending on the direction and the polarity. This rectification pattern was first observed in the synapse of the crayfish giant fiber (Furshpan & Potter, 1959; Giaume et al., 1987). Depolarized potentials travel from the presynaptic to the postsynaptic side but not in the opposite direction, and hyperpolarized potentials can travel only from the postsynaptic to the presynaptic side but not in the reverse direction. The rectified transmission feature depending on the polarity of the potential suggests the existence of a voltage-sensitive mechanism among the connexons (Faber & Pereda, 2018).

Compared to the chemical synapse, electrical gap junctions have different characteristics.

- Since there is no synaptic cleft and no chemical intervention, the signal is transmitted rapidly without delay. Mostly electrical synapses are found in neural circuits that require the fastest possible response, such as the escape or defensive reflexes.
- The transmission of the signal is bidirectional. As gates are open between the neurons, ions also can move reversely from the postsynaptic side to the presynaptic side as well as from the presynaptic side to the postsynaptic side.
- As there is no involvement of neurotransmitters, electrical transmission is not diverse in their ways of communication as chemical synapses.
- Instead of voltage gain, potentials in the postsynaptic side are smaller than the presynaptic side due to the resistance of the gap channel. As their polarity cannot be altered, the hyperpolarized postsynaptic potential is not developed by the depolarized presynaptic potential. That is, the inhibitory connection is not possible.

Electrical synapses are present throughout the nervous system, including many areas in the brain, and are crucial to the functioning of the cardiac myocytes and smooth muscles. They are found abundantly during nervous system development.

Another type of electrical transmission occurs as a consequence of the extracellular electrical fields generated by the neurons during the electrical signaling. Current generated during the action potential composes current loops by flowing through the membrane and extracellular space. It can be predicted that this extracellular current will build an electric field around and may influence the neighboring neurons. Depending on the spatial orientation and the conductivity of the extracellular space, a single neuron or a group of synchronously active neurons provide enough extracellular current to modify the excitability of neighboring neurons. Currently, most examples of electrical interactions by electric fields are inhibitory. However, it is known that its actions are localized on critical subcellular locations, such as the neuron's initial segment. This feature allows synapses mediated by electric fields to convey exquisite timing information to decision-making neurons within a network. Figure 5.15 shows that presynaptic electric fields can exert both excitatory and inhibitory actions on a postsynaptic

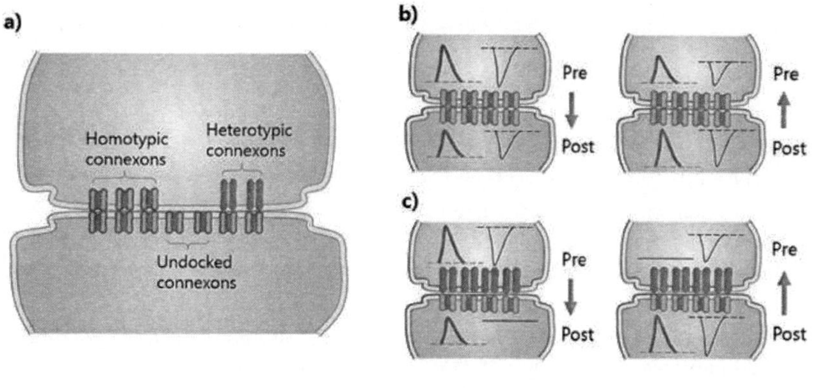

Fig. 5.14 Electrical gap junction. (**a**) Gap junction formed by the apposed individual channels, (**b**) Bidirectional transmission in homotypic connexons, (**c**) Rectified transmission in heterotypic connexons. (Faber & Pereda, 2018)

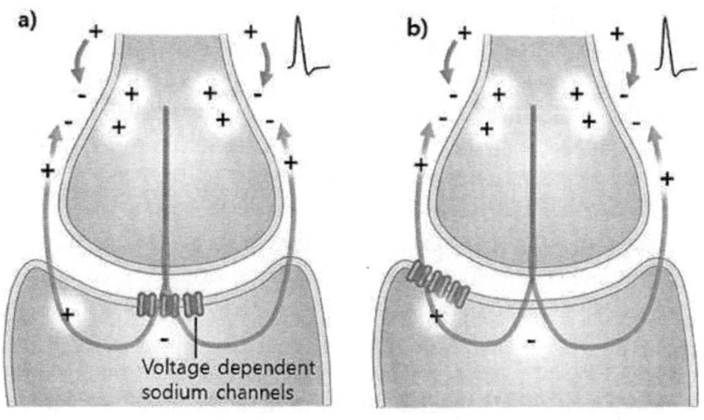

Fig. 5.15 Spatial-dependent membrane excitation by the electric current field. (**a**) Inward postsynaptic current builds inhibitory potential at central site, (**b**) Outward postsynaptic current builds excitatory potential at laterally shifted sites. (Faber & Pereda, 2018)

neuron. Whether a developed ephaptic current is excitatory or inhibitory depends on both the direction of current flow and the properties of the postsynaptic membrane (Faber & Pereda, 2018). Due to the fewer examples of electrical communication mediated by the electric fields, it is generally less appreciated, and less is known of the underlying mechanism.

2.6 Synapse as an Evolutionary Success

In the nervous system, synapses do a great role in the efficient signal transmission and the intelligent processing of information. It can be summarized as follows.

- Synapse is a unidirectional connection. Signals transmit only from the presynaptic side to the postsynaptic side. Even though the electrical synapses are bidirectional for some specialized purposes, fixing the direction of neural pathways is considered the efficient way for the larger amount of information traffic. When it is compared to the electrical circuits, it shows some contrast. In analog electric circuits, input and output lines are separated, while in digital circuits, data lines are shared for reading and writing. Rather, it looks similar to one-way streets which are better to increase the overall speed and traffic volume of vehicles.
- Conversion of an electrical action potential to the concentration of neurotransmitters and finally to the dendrite potentials of variable amplitudes in postsynaptic neurons can be considered as the process of converting signals from digital to analog format. Since the processing is not possible in the digital form of the action potential, it needs to be converted to an analog form which can be added and subtracted with weighted strength even with thousands of inputs. In neurons, the digital form is adopted for efficient propagation, but the analog form is adopted for the processing of signals in the dendrite and soma of the neuron. A/D conversion is occurring in the hillock, the starting point of the axon, and D/A conversion is occurring in the synapses.
- In addition, synapses provide the diversity of manipulating the signals. For a presynaptic action potential, the amplitude of the postsynaptic potential can be multiplied by increasing the number of synapses on the same postsynaptic neuron. Synapses closer to the soma have a greater impact than synapses further away at the end part of the dendrite, since postsynaptic potential decreases as it propagates along the dendrite. By the selection of neurotransmitters, it can transmit the action potential as an excitatory or inhibitory message. Depending on the ionotropic receptors on the postsynaptic side, the same action potential can be recognized selectively based on the specialty of the postsynaptic neurons. Synapses provide the ability to transform a presynaptic message into a variety of spatial and temporal shapes. This adaptive property contributes greatly to the diversity and the intelligence of processing in the brain. Considering the reduced conduction speed by the chemical synapses in the peripheral nervous system, the synapse is considered that it has evolved with an emphasis on the processing ability in the central nervous system. Under this evolving trend, nerves in the peripheral system have adapted by minimizing the number of synapses and maximizing their axon lengths.
- The purpose of the synapse is more than the immediate transmission and processing of the signal. Released neurotransmitters also bind to metabotropic receptors in postsynaptic neurons. This metabotropic receptor activates second messengers and triggers a series of prolonged intracellular events. Second messengers can cause specific chemical functions, change cell structure related to the long-term memory and provide the plasticity of the neuron. Synapses play a major role in various neuronal processes other than transmitting the signals. While the role of the ionotropic receptor is related to the survival and interfacing with the environment, the role derived from the metabotropic receptor is related to the growth and increasing intelligence. As the roles related to the metabotropic receptors are not smaller than the roles of the ionotropic receptors, developing chemical synapses must be the inevitable evolutionary direction.
- In addition to the chemical synapses, humans also possess electrical synapses for the neural connection without chemical delay. Electrical synapse is considered as appeared later than chemical synapses in evolutionary history. It is mostly for the specialized functions or delicate controls like the synchronized activation of the neurons or muscles. Its need has appeared as the biological systems are getting bigger and more complex. Different types of synapses are much beneficial in making the nervous system work faster and act intelligently. Chemical synapses and electric synapses may compete sometimes for the transmission of the signal, but they are complementing each other with their own characteristics for the optimal maintenance of the body.

3 Processing in the Dendrite and the Soma

Transmitted signals through the synapses produce dendrite potentials in postsynaptic neurons. Hundreds and thousands of these dendrite potentials are collected, propagated, and processed in the dendrite and soma areas of the neuron. Depending on the resulting potential at the hillock, the neuron fires a timely action potential as a result of its complex processing.

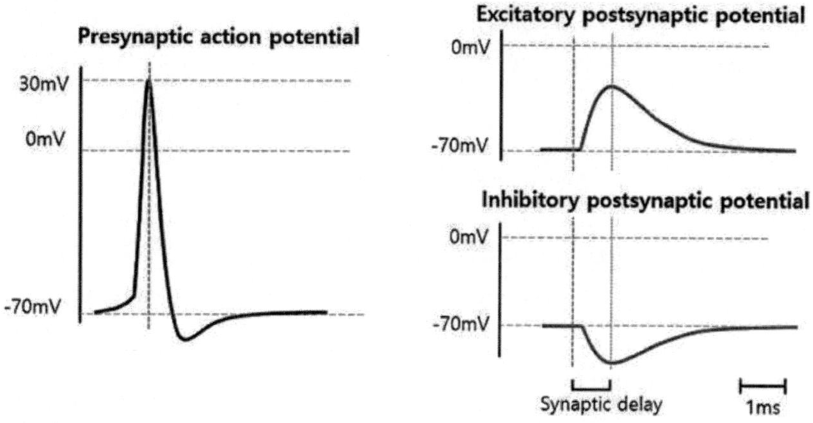

Fig. 5.16 Presynaptic and postsynaptic potentials

3.1 Postsynaptic Potentials

The postsynaptic potential is the membrane potential developed over the synaptic membrane of the postsynaptic neuron. When the ligand-gated ion channels (ionotropic receptors) are open by the attachment of the neurotransmitters of excitatory type, positive sodium ions flow in by the voltage gradient and the concentration gradient across the membrane. This influx of sodium ions is similar to the influx of sodium ions in the initial phase of the action potential generation process. It increases the membrane potential and depolarizes the membrane potential. This type of postsynaptic potential is called *excitatory postsynaptic potential* (*EPSP*). The difference is that these ion channels are not sensitive to the membrane voltage. There is no positive feedback increasing the sodium conductance and no rapid increase of membrane potential. Membrane potential is increasing while the ligand-gated ion channels are open, which is controlled by the attachment and detachment of neurotransmitters to the binding site. The postsynaptic potential does not rise to the level of an action potential. Its amplitude is smaller than action potential and graded upon the strength of neurotransmitter density (Fig. 5.16).

Instead of excitatory ion channels, inhibitory ion channels can be activated by inhibitory type neurotransmitters. If the ion channels allow the passage of positive potassium ions, the membrane potential is decreasing and is hyperpolarized by the outflow of positive potassium. Also, when chloride ion channels are activated, the inflow of negatively charged chloride ions decreases the membrane potential. These ion channels are open while the neurotransmitters are binding and the resulting hyperpolarized potential is graded according to the strength of the inhibitory type neurotransmitters. This postsynaptic potential is called *inhibitory post synaptic potential* (*IPSP*). Compared to the short duration ($2ms$) of presynaptic potential, the duration of postsynaptic potentials is lengthened to $4ms$ for example, because there is no active positive feedback increasing the sodium conductance rapidly as in the action potential generation.

3.2 Propagation of Postsynaptic Potential Along the Dendrite

Generated postsynaptic potentials at the synapse of dendrite propagate along with the dendrite fiber toward the soma. Dendrite fiber is unmyelinated and does not regenerate the action potential. Generated postsynaptic potential propagates passively along the dendrite as in the case of unmyelinated axon fibers. Since there is no regeneration of action potential, potential decreases as it propagates away toward the soma area. Passive dendrite fiber can be modeled as passive cable as in the model of unmyelinated axon fiber. But in this case, as the potential propagates the whole length passively without regeneration, the leakage of the current through the membrane should be included to explain the decrease of potential along with the propagation. In the unmyelinated model, leakage was neglected because the potential is regenerated at the right next site and the passively propagating distance is negligibly short. Figure 5.17 shows the cable mode for dendrite fiber including the leakage over the membrane.

We can derive the membrane potential with this cable model similarly to the membrane potential in the unmyelinated model.

When the current flows through the longitudinal internal resistance, there is a voltage drop along with the distance. The equation representing the voltage drop in time domain, and its transform into the Laplace domain become as follows.

$$\frac{dV(t)}{dx} = -R_i \bullet I(t) \rightarrow \frac{dV(s)}{dx} = -R_i \bullet I(s) \qquad (5.38)$$

But, due to the leakage current flowing through the membrane, the conductive current through the membrane resistance is added to the capacitive current.

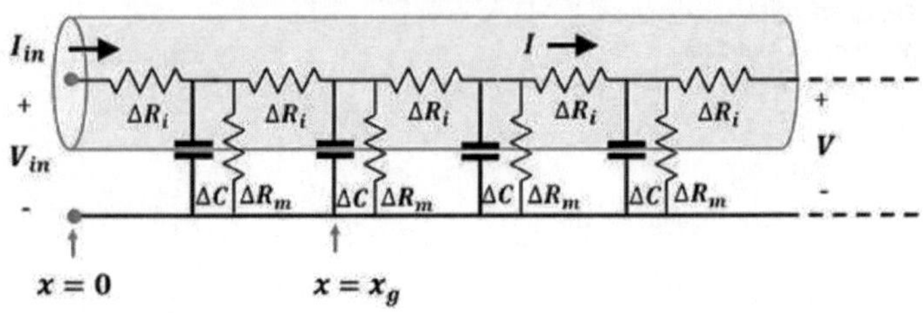

Fig. 5.17 Cable model for the dendrite with leakage through the membrane

$$\frac{dI(t)}{dx} = -\left(\frac{1}{R_m}\right) \bullet V(t) - C \bullet \frac{dV(t)}{dt} \rightarrow \frac{dI(s)}{dx} = -\left(\frac{1}{R_m} + sC\right) \bullet V(s) \quad (5.39)$$

Combining these two equations, we have a single equation to solve.

$$\frac{d^2V(s)}{d^2x} = R_iC\left(\frac{1}{R_mC} + s\right) \bullet V(s) \quad (5.40)$$

The solution of the second-order differential equation in the Laplace domain is easily derived with the appropriate boundary condition as in the case of Eq. (5.12)

$$V(s) = V_{in}(s)e^{-x\sqrt{R_iC\left(\frac{1}{R_mC}+s\right)}} \quad (5.41)$$

When an impulse of area A_0 $V \cdot s$ is applied as $V_{in}(s)$ at $x = 0$ position, $V(s)$ becomes

$$V(s) = A_0e^{-x\sqrt{R_iC\left(\frac{1}{R_mC}+s\right)}} \quad (5.42)$$

Since this is the solution in the Laplace domain, the solution in the time domain can be found by the inverse transforming the solution back to the time domain.

According to the Laplace transform table

$$\frac{k}{2\sqrt{\pi t^3}}e^{-\frac{k^2}{4t}} \leftrightarrow e^{-k\sqrt{s}}, \quad e^{-at}f(t) \leftrightarrow F(s+a) \quad (5.43)$$

And by replacing k with $x\sqrt{R_iC}$ and a with $1/R_mC$, we can obtain time-domain solution.

$$V(x,t) = \frac{A_0x}{2t^{3/2}}\sqrt{\frac{R_iC}{\pi}}e^{-\frac{x^2R_iC}{4t}}e^{-\frac{t}{R_mC}} = \frac{A_0x}{2t^{3/2}}\sqrt{\frac{R_iC}{\pi}}e^{-\left(\frac{x^2R_iC}{4t}+\frac{t}{R_mC}\right)} \quad (5.44)$$

Compared to Eq. (5.15), added part of $e^{-(t/R_mC)}$ represents the effect of the leakage through the membrane.

The characteristic length of the cable is the length of the cable before the potential drops to $1/e$ of the DC input. It is calculated by letting $C = 0$ in Eq. (5.42).

$$V(s) = A_0e^{-x\sqrt{\frac{R_i}{R_m}}} = A_0e^{-x/\lambda} \rightarrow \lambda = \sqrt{r_m/r_i} = 0.408\sqrt{d} \quad (5.45)$$

This is the same as Eq. (5.5), and the characteristic length is dependent on the diameter of the dendrite. As the diameter of the dendrite is increasing as the position moves to the soma direction, the characteristic length is increasing in proportion to the square root of the diameter. Because the increase of diameter is not so steep and the rate of increase is also dependent on the square root, it is assumed that the diameter is constant throughout the dendrite for the simplicity of the model. Since the distance will be normalized to the characteristic length, the increased diameter can be interpreted with the accompanying increase in characteristic length.

The distance is x is normalized to the characteristic length λ

$$z = x/\lambda = x/\sqrt{r_m/r_i} \quad (5.46)$$

And, let the time constant of the membrane r_mC, which is the time that membrane potential drops to $1/e$ level on the DC input, equal to $4ms$ as in other part of the membrane.

$$r_mC = 4ms \quad (5.47)$$

Then Eq. (5.44) is simplified to

$$V(z,t) = \frac{A_0x}{2t^{3/2}}\sqrt{\frac{R_iC}{\pi}}e^{-\frac{x^2R_iC}{4t}}e^{-\frac{t}{R_mC}} = \frac{A_0z}{\sqrt{\pi}t^{3/2}}e^{-\left(\frac{z^2}{t}+\frac{t}{4}\right)} \quad (5.48)$$

where units for voltage and time are mV and ms.

This equation shows the potential at distance z and at time t. To find the constant A_0 with 50 mV of peak postsynaptic potential amplitude, peak occurring time and the postsynaptic potential generating distance are decided by

$$t_p = \sqrt{4z^2 + 9} - 3 \quad (5.49)$$

When we set the postsynaptic potential generating distance z to 2, peak occurs at $t_p = 2$ ms.

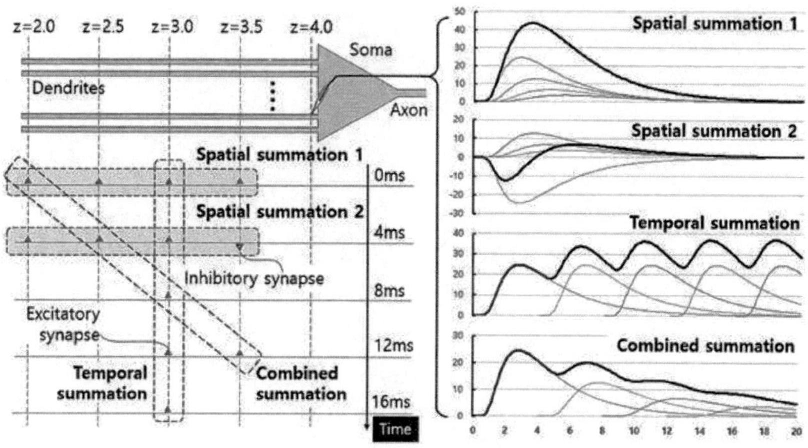

Fig. 5.18 Postsynaptic potentials and their attenuation as they propagate toward the soma

With these values in Eq. (5.48), constant A_0 is calculated as 1527 mV/ms, and the postsynaptic potential with a peak amplitude of 50 mV is represented as follows.

$$V(z,t) = \frac{862z}{t^{3/2}} e^{-\left(\frac{z^2}{t}+\frac{t}{4}\right)} \tag{5.50}$$

Figure 5.18 shows how the dendrite potential is attenuating as it moves from the site of its generation toward the soma. Its amplitude is attenuated almost to half level for each half of the characteristic length. Also, the peaks are delayed and the duration of waveforms is lengthened as it moves toward the soma. When it arrived at the soma, its amplitude is attenuated greatly and is lower than the threshold level to generate an action potential. Since a single synaptic potential is generally not enough to generate an action potential, a number of postsynaptic potentials and summation or processing are required to generate an action potential.

3.3 Summation of Dendrite Potentials

To generate an action potential, postsynaptic potentials from multiple synapses and multiple presynaptic neurons should be integrated. As they are arriving at different locations of the dendrite and at different times in excitatory or inhibitory type, integrating potentials occurs over space and time.

Spatial summation is a summation of dendrite potentials coming from the different sites of synapses that occurred at the same time. Synapses come from different presynaptic neurons and there can be multiple synapses from a single presynaptic neuron via multiple synapses over the space. As the synapses are closer to the soma, their propagating distance is shortened and the attenuation becomes smaller. Synapses positioned closer to the soma have greater influence in increasing the summed potential and generating an action potential. In spatial summation, positions are their weight coefficients. Dendrite potentials are either excitatory or inhibitory. Excitatory dendrite potentials are additively summed and increase the excitability of the postsynaptic neuron to generate an action potential. Contrary, inhibitory dendrite potentials are decreasing the summed potential and deactivate the postsynaptic neurons not to generate an action potential. When inhibitory synapses are involved in a dendrite branch, it used to nullify the contribution of other excitatory synapses behind it.

Temporal summation is a summation of potentials from a single synapse. If the second postsynaptic potential (PSP) arrives before the first postsynaptic potential disappears, the remaining amplitude of the first PSP can be additive to the second PSP. As the presynaptic neuron elicits action potentials at a higher frequency, the effect of temporal summation is increasing in the postsynaptic neurons. If the second PSP is generated after the first PSP has totally disappeared, there is no effect of temporal summation. This temporal summation occurs simultaneously with spatial summation at each time. When the two stimuli from different presynaptic neurons are arrived at different times, resulting postsynaptic potential are not synchronized and they are summed in and out of phase state based on their spatially weighted amplitude. Figure 5.19 shows some examples of spatial, temporal, and combined summations. In this figure, the summed potential is recorded at the nearest location to soma. In the "stimulation 2" example, a single inhibitory postsynaptic potential blocks the sum of the other three excitatory potentials and deactivates the neuron not to reach the threshold level. The temporal summation example shows that summed potential can increase higher than the maximum amplitude of a single postsynaptic potential by its rapid repetition. A combined summation example shows that the combination of spatial and temporal summation from many synapses can produce an arbitrary pattern that activates or deactivates the postsynaptic neuron.

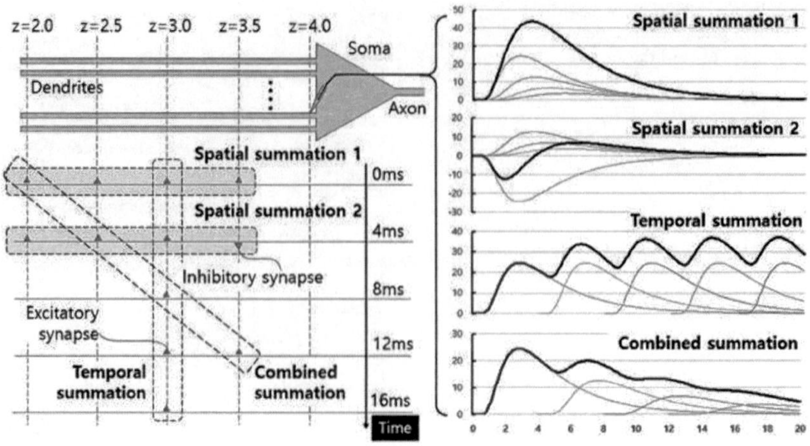

Fig. 5.19 Spatial and temporal summation of dendrite potentials

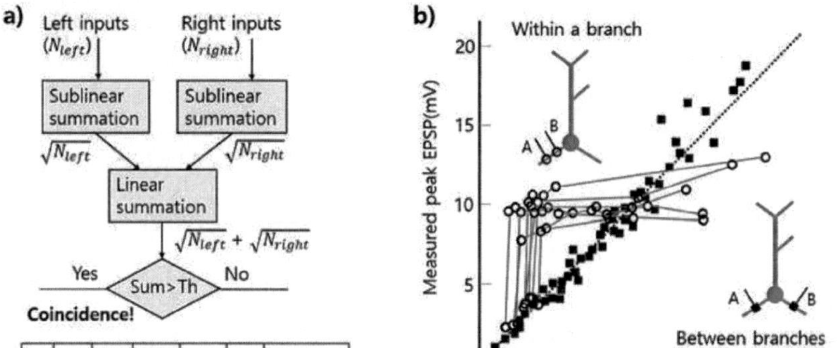

N_{left}	N_{right}	$\sqrt{N_{left}}$	$\sqrt{N_{right}}$	Sum	Th	Coincidence
2	2	$\sqrt{2}$	$\sqrt{2}$	$2\sqrt{2}$	2.5	Yes
4	0	2	0	2	2.5	No

Fig. 5.20 Nonlinear computations in dendrites. (**a**) Sublinear summation (London & Hausser, 2005), (**b**) Supralinear summation. Squares: stimulation synapses in different dendrite branches, Open circles: stimulation synapses in a same branch with inhibitory ion channels blocked. (Poirazi, 2014)

3.4 Dendrite Computation

Dendrites are tree-like elaborate processes extending from the cell body and propagating the received messages with linear or nonlinear integration. The traditional simple view of dendrites is that dendrites are means of increasing the surface area for receiving synaptic inputs, whereby all processing is done at the soma. A single neuron is treated as a simple summation unit of generating an axon potential if the weighted sum of received signals reaches a certain threshold. Recently many experimental studies support the prediction of modeling studies that dendrite's role is much more than the simple propagation and summation of received signals (Poirazi & Papoutsi, 2020). Dendrites have a computational ability that individual dendritic branches can perform elementary computations on incoming signals. These computations include addition, subtraction, multiplication, and division and can also take the form of logical operations (Poirazi, 2014).

One example is the sublinear summation which produces a response lower than the linear summation. When two excitatory inputs arrive together within a short distance, depolarization of the membrane by one input reduces the driving force of the other input, and thus the sum of responses to the simultaneous inputs becomes smaller than the linear sum of the individual responses (Rall, 1967). Due to the passive properties of the membrane, combined activation of close inputs would result in a smaller depolarization than they were placed further apart. This mechanism is found in the coincidence detection in the auditory system (London & Hausser, 2005). As shown in Fig. 5.20a, inputs to each ear are summed sublinearly, and their responses from both ears are then summed linearly and compared to the threshold. When inputs from both ears arrive simultaneously, the summed response exceeds the threshold and recognizes the coincidence, whereas if the inputs of the same amount come only from one ear, the summed response is lower than the threshold (London & Hausser, 2005).

Another example is the supralinear summation which produces a response higher than the linear summation of responses to each received input. Supralinear summation is an active process where the regenerative dendrite spikes are involved. In dendrites of most neurons, many types of voltage-gated ion channels, in addition to ligand-gated ion channels, are found along the morphologically complex dendrite branches. Activation of voltage-gated ion channels alters the synaptic integration nonlinearly and generates localized regenerative dendrite spikes. Dendrite spikes increase the response supralinear when multiple inputs arrive. Figure 5.20b shows the experimental results verifying the supralinear mechanism in dendrites. Single-pulse stimulation of synapses in different branches results in linear summation at the cell body. When the inhibitory ion channels are blocked, PSP to the within-branch synaptic stimulations is summed supralinear with a shape of sigmoidal nonlinearity. This property can be activated as a feature detection capability. It can produce an enhanced summed response only when some selected synapses are anchored on the same branch, while not responding to their spread connections over branches.

Dendrites act as passive low-pass filters and filtering characteristics are affected by the local diameter. Additional active dendrite spikes provide dendrites with a range of computational capabilities traditionally thought possible only at the level of neuronal networks. With these capabilities, it is not appropriate to regard a neuron as a simple unit of summation and triggering an action potential with a threshold. Each single dendrite branch can be regarded as a processing unit. The whole individual neuron is a more intelligent and complex unit that integrates and processes dendrite signals. A single neuron is similar to artificial neural networks of at least two or more layers and can play a critical role in information coding and memorizing the pattern.

References

Akaishi, T. (2017). New theoretical model of nerve conduction in unmyelinated nerves. *Frontiers in Physiology, 8*, Article 798.

Barnett, M. W., & Larkman, P. M. (2007). The action potential. *Practical Neurology, 7*, 192–197.

Chomiak, T., & Hu, B. (2009). What is the optimal value of the g-ratio for myelinated fibers in the rat CNS? A theoretical approach. *PLoS One, 4*(11), e7754. https://doi.org/10.1371/journal.pone.0007754

Faber, D. S., & Pereda, A. E. (2018). Two forms of electrical transmission between neurons. *Frontiers in Molecular Neuroscience, 1*. https://doi.org/10.3389/fnmol.2018.00427

Furshpan, E. J., & Potter, D. D. (1959). Transmission at the giant motor synapses of the crayfish. *The Journal of Physiology, 145*, 289–325. https://doi.org/10.1113/jphysiol.1959.sp006143

Giaume, C., Kado, R. T., & Korn, H. (1987). Voltage-clamp analysis of a crayfish rectifying synapse. *The Journal of Physiology, 386*, 91–112. https://doi.org/10.1113/jphysiol.1987.sp016524

Guyton, A. C., & Hall, J. E. (1996). *Textbook of medical physiology* (pp. 565–582). W.B. Saunders Company.

Hert, S. D., Baerdemaeker, L. C., & Maeseneer, M. D. (2014). What the phlebologist should know about local anesthesics. *Phlebology, 29*(7), 428–441.

Hursh, J. B. (1939). Conduction velocity and diameter of nerve fibers. *The American Journal of Physiology, 127*, 131–139.

Ichiji, T. (2002). On the cable theory of nerve conduction. *Bulletin of Mathematical Biology, 64*, 1069–1082.

Ichiji, T. (2004). On the conduction velocity of nonmyelinated nerve fibers. *Journal of Integrative Neuroscience, 3*(2), 115–124.

London, M., & Hausser, M. (2005). Dendritic computation. *Annual Review of Neuroscience, 28*, 503–532.

Mukherjee, S., & Manahan-Vaughan, D. (2013). Role of metabotropic glutamate receptors in persistent forms of hippocampal plasticity and learning. *Neuropharmacology, 66*, 65–81.

Poirazi, P. (2014). Dendrite computation. *Encyclopedia of Computational Neuroscience*. https://doi.org/10.1007/978-1-4614-7320-6_125-1

Poirazi, P., & Papoutsi, A. (2020). Illuminating dendritic function with computational models. *Nature Reviews Neuroscience, 21*, 303–321.

Rall, W. (1967). Distinguishing theoretical synaptic potentials computed for different soma-dendritic distributions of synaptic input. *Journal of Neurophysiology, 30*(5), 1138–1168.

Zalc, B. (2006). The acquisition of myelin: A success story. *Novartis Foundation Symposium, 27*, 15–21. https://doi.org/10.1002/9780470032244

Sensory Receptors

6

Humans are living in the environment. The first requirement to live embedded in the environment is to have the means that recognize the environmental contents and any of its variations. As the means to recognize the environment, humans are endowed with the many types of sensory units called *sensory receptors*. As there are various types of physical quantities, like heat, light, sound, and pressure, we need to have various types of sensory receptors to recognize each of the different contents. In addition to the outer environment, we also need to recognize the inner environment of the body. So, we are also endowed with the sensory receptors that sense the internal states of the body and report how the body parts are working. Sensory receptors sense and report external and internal environmental states and make the brain perceive and recognize the environment for optimal decisions.

1 Properties of Sensory Receptors

1.1 Sensory Receptors as Transducers

All types of sensory receptors are alike in their concept. Each type of receptor detects a physical stimulus and converts them into electrical signals that give rise to behavioral responses or conscious perceptions. Sensory receptors are regarded as transducers, which translate signals in one form of energy to signals in another form of energy. Sensory receptors convert signals of the external and internal environment in various forms of energy into action potentials of electrical form within the body (Fig. 6.1).

There are two roles in the transducers. The first role, which is the role of the *physical sensing element*, is to sense the signals in their specific form of physical energy. The second role, which is the role of the *variable conversion element*, is to convert the energy form from one form to another form. Physically these roles are implemented by separate elements or can be integrated into a single element.

Similarly, sensory receptors also need a primary sensing component and a variable conversion element. Environmental signals are sensed by the primary sensing capability of sensory receptors. When sensory receptors sense the environmental stimuli, they convert them into the electrical form of signal, *receptor potential*. In sensory receptors, the primary sensing and the variable conversion components are combined into a transmembrane receptor protein and occur simultaneously within a single cell. Different types of senses are dependent on the differences in transmembrane receptor proteins. Many transmembrane receptor proteins are ion channels. They directly open the ion channels and increase the ion flow passing through the gates. As a result, the membrane potential of the sensory receptor is depolarized and receptor potential is developed. Some ion channel receptors are activated by the chemical ligands as in the gustatory receptors, and other ion channel proteins are sensitive to changes in physical quantities like force and temperature for touch and warmness perception. But, some other transmembrane receptor proteins do not possess the ion channels within and do respond indirectly. When they sense the applied stimuli, they release G-protein connected to the receptor protein from the inside. Released G-protein triggers a series of enzymic and chemical processes and finally ion channels around are open and the receptor potential is developed. In this case, all the procedures occur within a single receptor cell, and the roles of primary sensing and variable conversion are separated. Receptor proteins do the role of the primary sensing element and internally gated ion channels do the role of variable conversion. Receptors of this type are found in the sensory receptors for transduction of smell, light, and some types of pain.

Transduced receptor potentials are in the form of graded potential. Their amplitudes represent the intensity of the sensed input stimuli. But, they are not appropriate to be transmitted over long distances from the periphery to the central brain via the nerve axonal fibers. They need to be converted to a series of action potentials for the successful

K. S. Park, *Humans and Electricity*, https://doi.org/10.1007/978-3-031-20784-6_6

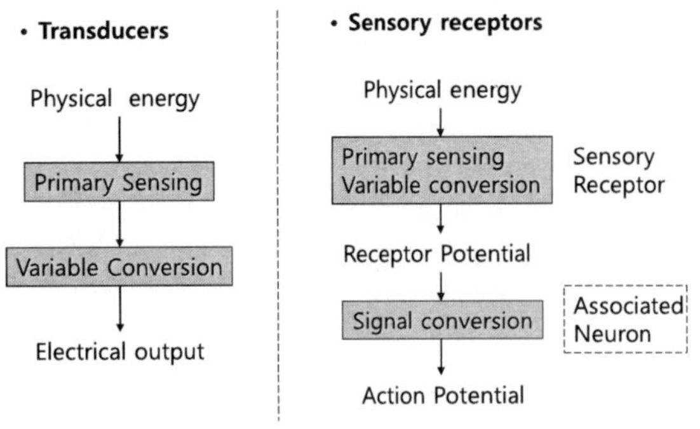

Fig. 6.1 Role of sensory receptors in comparison with transducers

transmission over the axonal fibers by the regeneration mechanism. The signal conversion can occur within a single cell where the receptor potential is transduced. Or, associate neurons, which are connected to receptor cells by the synapse, do the role of signal conversion from receptor potential to a series of action potentials. The signal conversion process from the information in the amplitude of receptor potential to the information in the rate of action potentials is a kind of *frequency modulation* (FM).

1.2 Types of Sensory Receptors

There are many different types of sensory receptors. They are so diverse in their sensitive physical quantity and in their acting locations within the body parts. For the overall understanding of sensory receptors in the human body, they are classified depending on some criteria.

Sensory receptors can be classified into external receptors and internal receptors depending on their locations relative to the stimuli. External receptors are responding to the stimuli from the external part of the body and are also called *exteroceptors*. They are the five senses of the human, including vision, hearing, smelling, tasting, and somatic sensing. When the external stimuli are sensed via exteroceptors, we normally perceive the application of external stimuli. Internal receptors are responding to the stimuli generated within the internal body part and also called *interoceptors*. Interoceptors monitor the changes inside the body. There are many types of physical quantity within the body to be monitored, including blood pressure, O_2 and CO_2 concentration, muscle length, temperature, etc. In contrast to the exteroceptors, we generally do not perceive the sensed signals by the interoceptors.

Sensory receptors can be classified on the structures of the primary sensing part of the cells (Fig. 6.2). The first type is the receptor neurons having *free nerve ending*. Free nerve endings are the dendrites of the afferent neurons. Transmembrane receptor proteins populated over the dendrite membrane directly respond to the physical quantities of the stimuli and produce receptor potentials. The pain and temperature receptors in the dermis of the skin are examples of these types of receptors having free nerve endings. The second type is the receptor neurons having *encapsulated ending*. In contrast to free nerve ending dendrites, their receptive dendrites are encapsulated with additional connective tissue, which is to increase the sensitivity of the receptor to the stimuli. Nerve endings with lamellated corpuscles found in the dermis of the skin, which are sensitive to tactile touch and pressure, are good examples of this type. The third type is the *specialized receptor* neurons having distinct structural components to sense a specific type of stimulus. The photoreceptors in the retina and the hair cells in the cochlea are examples of this type specialized to sense light or sound respectively.

Since each sensory receptor converts signals of different types of energy into a signal of electrical form, they can be classified depending on their physical types of energy. *Mechanoreceptors* sense and convert the mechanical type of stimuli like force, pressure, length, etc. Many tactile receptors populated in the skin are belonging to this type of receptor. They sense the touches and the pressures applied to the skin. In the muscles, there are receptors that sense the length of each part of the muscle component to monitor the exact contraction of the muscle. Hearing is also sensed by the movement of the hair cells in the cochlea. The vestibular receptor in the ear senses mechanically the position and movement of the body. Baroreceptors distributed over the major body area sense the blood pressure to monitor the supplied level of blood. *Chemoreceptors* sense various types of chemical molecules in gases, fluids, and food with their own specificity. Tastes and smells are recognized by their constituent molecules and sensed by the corresponding chemoreceptors located in the tongue and the nose respectively. A number of gases in the blood, like oxygen and carbon dioxide, are sensed by their molecular characteristics. And, there are many chemical contents within the body fluid like glucose, amino acids, and fatty acids. The concentrations of these chemical contents are measured by the chemical type receptors. *Nociceptor*, which senses the pain caused by tissue damage, also can be classified into a chemical type receptor, because the pain is sensed by the chemical contents released by the cell damage. *Thermoreceptors* sense the thermal energy. There are two different types of thermal receptors, cold receptors and warm receptors. Even though the temperature of coldness or warmness can be measured by a single thermal sensor in an out-of-body world, the human body senses coldness and warmness by different types of receptors just like different types of the physical quantity. The last type is *electroreceptors* which senses electromagnetic type quantities. Among the whole spectrum of electromagnetic wave energy, visible light is the only band that human receptors can sense. There are two types of receptors for vision. Rods are for sensing black and white, dim and broad

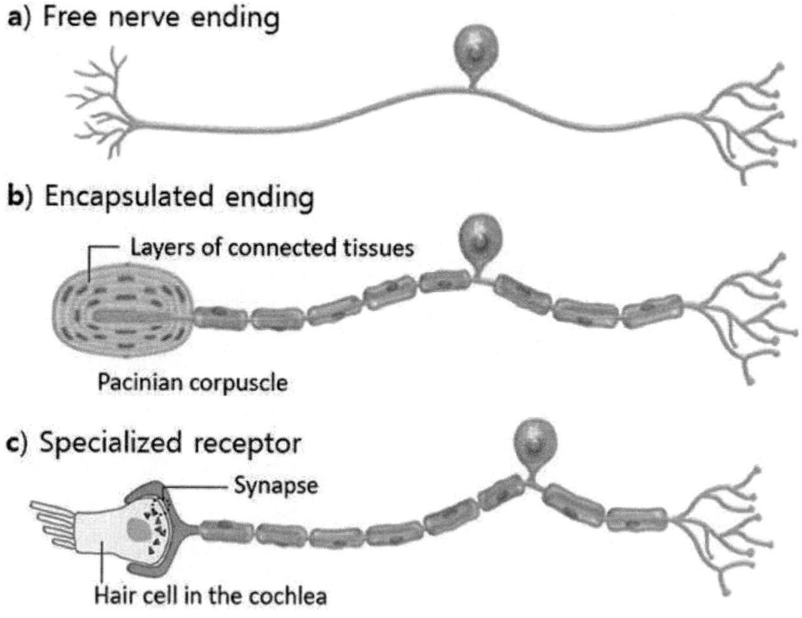

Fig. 6.2 Structural classification of receptor types: (**a**) free nerve ending, (**b**) encapsulated ending, (**c**) specialized receptor

vision, and cones are for sensing colored, bright, and focused vision. Some other organisms have electroreceptors that humans lack, such as snakes having infrared sensors for sensing heat, bees having ultraviolet sensors, and migratory birds having navigation magnetic receptors (Table 6.1).

1.3 Characteristics of Sensory Receptors

1.3.1 Labelled Line Principle

Each sensory receptor has its own specificity in transducing input stimuli and transmits the message along the pathway to the specific area in the brain just as the transmitting nerve pathway is labeled only for that sensory receptor. Any messages coming through the specific neural pathway are recognized as they originated from the labeled sensory receptor (Fig. 6.3).

There are many different types of sensory receptors each of which transduces different stimuli. Even though there are so many sensory receptors and the neural network looks complex by the intermingled connection of their neural fibers, each transmitting line is specialized to transmit only one type of message predetermined to the sensory receptor. This is based on the premise that each individual sensory receptor is responding only to the predetermined specific type of stimulus and the allocated afferent neural pathway is only for that single sensory receptor and not shared with any other sensory receptors. This principle applies to the sensory system generally including the somatosensory system. Photoreceptors in the retina are specialized to sense the light, and signals transmitting through the fibers from any photoreceptor are interpreted as a light-related message. If the photoreceptor generates some potentials by a sudden blow of force instead of light stimulus, the connected occipital cortex by the "labeled line principle" interprets the transmitted potential as some amount of received light instead of applied force. Another example is the *phantom limb*, which sense the amputated or missing limb is still attached. This occurs after the amputation of a limb. While the afferent nerves are deactivated by the loss of the sensory receptors on the limb side, the brain cortex interprets any signal transmitted through the remaining fibers originating from the labeled sensory receptor as before the deafferentation.

1.3.2 Sensory Neural Coding Strategies

Various types of stimuli applied to sensory receptors are transformed into a series of action potentials transmitting over the afferent neural fibers. Analog types of input stimuli are converted to a digital form of action potentials. What are the encoding strategies? In electrical circuits and computers, pulse-coded modulation is the most popular which converts the analog amplitude to certain codes of digital impulses. But, in the physiological system, this is not the strategy taken.

There have been extensive studies to understand the coding strategy of human sensory systems and many theories are suggested to explain the underlying fundamentals of sensory neural encoding. A basic and traditional coding strategy is *rate coding* which modulates the rate of action potentials according to the strength of the stimulus. It is a kind of pulse frequency modulation. Rate coding has been experimentally verified in most sensory systems, such as the visual cortex and motor cortex. However, the rate coding strategy

Table 6.1 Classification of sensory receptor types

Receptors	Functional type	Locational type	Structural type	Transducer
Touch receptor	Mechanoreceptor (Stretch)	Exteroceptor (Skin)	Free nerve ending Encapsulated ending	Touch pad
Auditory receptor	Mechanoreceptor (Sound)	Exteroceptor (Ear)	Specialized receptor	Microphone
Vestibular receptor	Mechanoreceptor (Acceleration)	Exteroceptor (Ear)	Specialized receptor	Accelerometer, Gyroscope
Baroreceptor	Mechanoreceptor (Pressure)	Interoceptor (Artery)	Free nerve ending	Pressure sensor
Kinesthetic receptor	Mechanoreceptor (Angle)	Interoceptor (Joint)	Free nerve ending	Goniometer
Muscle spindle	Mechanoreceptor (Length)	Interoceptor (Muscle)	Free nerve ending	Strain gauge
Tendon organ	Mechanoreceptor (Tension)	Interoceptor (Muscle)	Free nerve ending	Pressure sensor
Pain receptor	Chemoreceptor [Nociceptor]	Exteroceptor (Skin)	Free nerve ending	Biosensor
Taste receptors	Chemoreceptor (Taste)	Exteroceptor (Tongue)	Specialized receptor	Biosensor
Olfactory receptors	Chemoreceptor (Smell)	Exteroceptor (Nose)	Specialized receptor	Biosensor
Gas receptors	Chemoreceptor (O_2, CO_2)	Interoceptor (Artery)	Specialized receptor	Biosensor
Chemical receptors	Chemoreceptor (Glucose, acids)	Interoceptor (Artery)	Specialized receptor	Biosensor
Cold receptor	Thermoreceptor (Temperature)	Exteroceptor (Skin)	Free nerve ending	Thermistor
Warmth receptor	Thermoreceptor (Temperature)	Exteroceptor (Skin)	Free nerve ending	Thermistor
Temperature receptor	Thermoreceptor (Temperature)	Interoceptor (Brain)	Specialized receptor	Thermometer
Visual receptors (Rods, Cones)	Electroreceptor (Vision)	Exteroceptor (Eye)	Specialized receptor	Camera

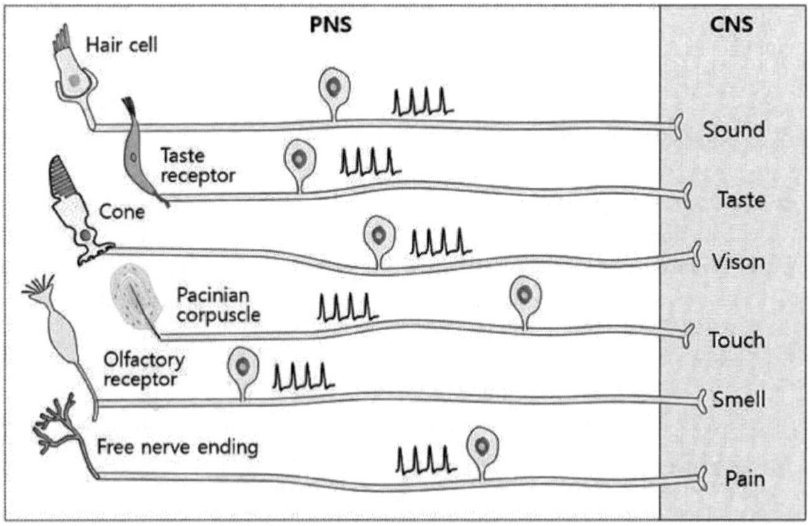

Fig. 6.3 Labelled line principle

requires a series of impulses to represent a stimulus and cannot help being slow in stimuli conversion and transmission. Since humans are responding faster than expected based on this coding strategy, many supplementing theories are emerged to explain the limitation of rating coding. *Temporal coding* suggests that information is carried by the precise timing of action potential occurrence. Different forms of temporal coding are suggested including burst coding,

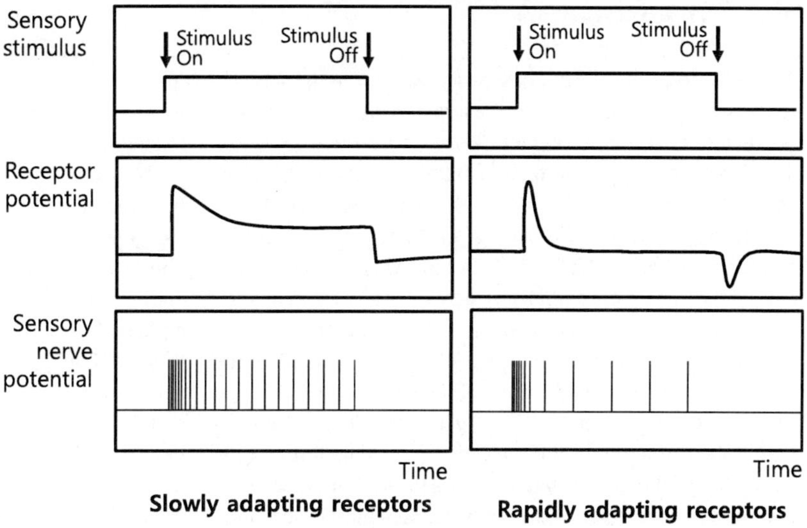

Fig. 6.4 Adaptation of receptors

phase coding, and timing of the first spike. *Burst coding* explains impulse transmissions widely observed in a burst of spikes in various parts including the auditory system. This coding strategy increases the contents of the information carried, as the information can be differentiated by the different temporal bursting patterns even with the same number of impulses. *Phase coding* carries information in the phase of action potential occurrence in reference to the internally generated global oscillating rhythm. Impulses coded with phase coding have been experimentally observed in the olfactory system (Laurent, 1996). Phase coding provides faster-coded pulses compared to simple rate coding. In this coding strategy, information is carried on the time between the stimulus and the first action potential. Since it does not require a series of impulses, the brain can recognize the strength of the stimulus upon the arrival of the first action potential. The significance of the time to first spike strategy has been verified by experimental studies over various sensory systems including visual, auditory, and somatosensory systems (Ponulak & Kasinski, 2011).

1.3.3 Adaptation to Stimuli

Since the generation and transmission of action potentials are accompanying energy consumption, the continuous generation of action potentials for constant sensory stimuli is not efficient. So, the adaptation is occurring in the very first stage of sensory receptors (Fig. 6.4). Some receptors are slowly adapting, which are also called tonic receptors, and show little adaptation in response to a prolonged stimulus. Others are rapidly adapting, which are also called phasic receptors, and decline responses rapidly. Slowly adapting receptors are proper to report static information. Rapidly adapting receptors act like differentiators by transducing the difference instead of the absolute intensity, and they are proper for reporting dynamic stimuli. For example, the Pacinian corpuscle, which is a rapidly adapting receptor, responds briefly at the beginning and at the end of a constant mechanical stimulus. We only recognize the touch of a chair when we sit on and rise from a chair and do not recognize the touch of a chair during sitting. But, rapidly adapting receptors respond continuously to dynamic stimuli like vibrations. Most receptors are actually somewhere between the extremes of slowly and rapidly adapting. The response may be exaggerated at the beginning and at the end of a stimulus but maintained to some level during the stimulus.

2 Mechanical Receptors

Mechanical receptors are called mechanoreceptors. Mechanoreceptors respond to the mechanical type stimuli occurred by the relative movement of the body to the environment or applied forces. There are many types of mechanoreceptors around the body. *Touch receptors* populated mostly in the skin are mechanoreceptors. *Baroreceptors* sensing the blood pressure are also mechanoreceptors. *Hair cells* found in the inner ear for the sensing of sound in the auditory system and acceleration in the vestibular system are also mechanoreceptors. There is a special type of mechanoreceptors *in the muscle* to sense the contracted distance and the applied tension of the muscles. In the joints, there are mechanoreceptors called proprioceptors which sense the relative positions of neighboring body parts.

2.1 Touch Receptors

Touch receptors are the major sensory receptor in the somatosensory system and the skin is the largest sensory organ most populated with touch receptors (Fig. 6.5). Touch receptors are also found on the surface of internal organs. There are several types of touch receptors. They have different structures, different responding characteristics, and are located at different depths of the skin. Most of them have encapsulated nerve endings to increase the receptive efficiency, which is the nerve ending of the myelinated afferent neurons. Mostly, applied mechanical force is sensed by the stretch of the nerve-ending membrane and sensed by the stretch-gated ion channels. They generate action potentials within the cell and transmit the action potentials through the axon fiber of the myelinated afferent neuron. Some receptors are slowly adapting when the applied stimulation is sustained and these receptors are appropriate to sense the indentation distance. Some other receptors are rapidly responding to sustained stimuli and suitable to sense the difference or the velocity of the applied stimuli. Figure 6.6 shows the structure of different types of touch receptors.

Merkel cells are usually found in both hairy and glabrous skin and are located in the basal layer of the epidermis. Merkel cell is a special type of cell contacting by synapse to the Merkel disk, which is the terminal ending of an afferent neuron. Merkel cell is a modified epithelial cell and is tightly coupled with the surrounding tissue by finger-like protrusions. Merkel cell contains synaptic vesicles and released neuropeptides modulate the postsynaptic potential at the Merkel disk. As force is applied to the skin coupled with the Merkel cell, it deforms the Merkel cell and stimulates the release of a neuropeptide at its synaptic junctions. The afferent nerve of the Merkel cell receptor is different from other touch receptors in that the major signal is sensed by ligand-gated ion channels rather than stretch-gated ion channels. Each afferent neuron is innervated only to a few Merkel cells in a discrete patch of skin and it has a very small receptive field of touch. As Merkel cells are tightly coupled to the surrounding tissue, they remain distorted while the force is applied. Merkel cells are considered to be fine tactile receptors and slowly adapt to sustained stimuli.

Meissner's corpuscles are distributed mostly over glabrous skin areas, especially areas sensitive to light touches like fingers and lips. They are located beneath the epidermis within the dermal papillae mostly in glabrous skin. Its nerve ending is elongated and encapsulated with a stack of flattened epithelial cells. Disk-like laminar layers are innervated with terminal endings of an afferent nerve. Force applied to the skin causes the laminar layers in the Meissner corpuscle to slide apart. This movement distorts the terminal ending membrane of the afferent nerve located between the laminar layers. During the force is sustained, the laminar cells remain in a displaced position, but the stretching force on the terminals' membranes rapidly diminishes. As a result, Meissner's corpuscle senses with rapid adaptation to sustained stimuli and responds transiently to the change of applied force. Meissner's corpuscle is suited to transduce and transmit time-varying information of vibrating or moving mechanical stimuli. They are tuned to low-frequency vibrations under 40 Hz, which is appropriate to detect the slip of an object in the hand.

Pacinian corpuscles are found deep in the dermis of the skin and also in the connective tissues of bone, body wall, and body cavity. Pacinian corpuscles are oval in shape of 1 mm in length and contain layered lamellae. In cross-section, the Pacinian corpuscle looks like an onion slice, with a single afferent terminal fiber located in its center. The outer layers of laminar cells contain fluid that is displaced when a force is applied to the corpuscle. When a force is applied, it initially pushes laminar cell layers and distorts the membrane of the terminal nerve ending at the center. While the applied force is sustained, the force is distributed through the fluid in all directions even to the

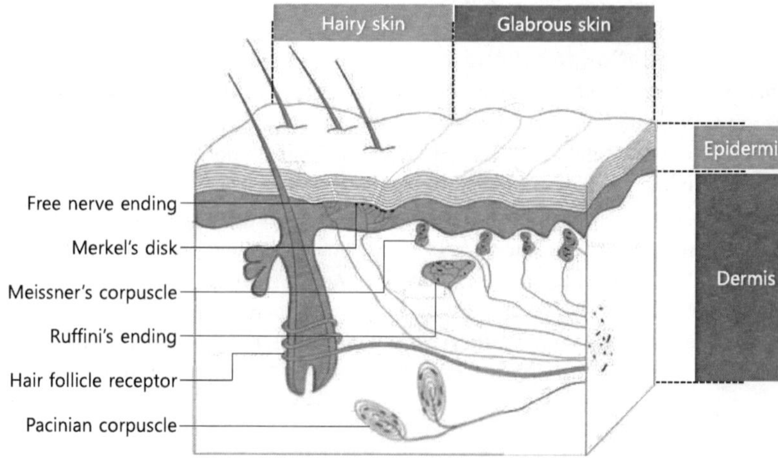

Fig. 6.5 Touch receptors on the skin

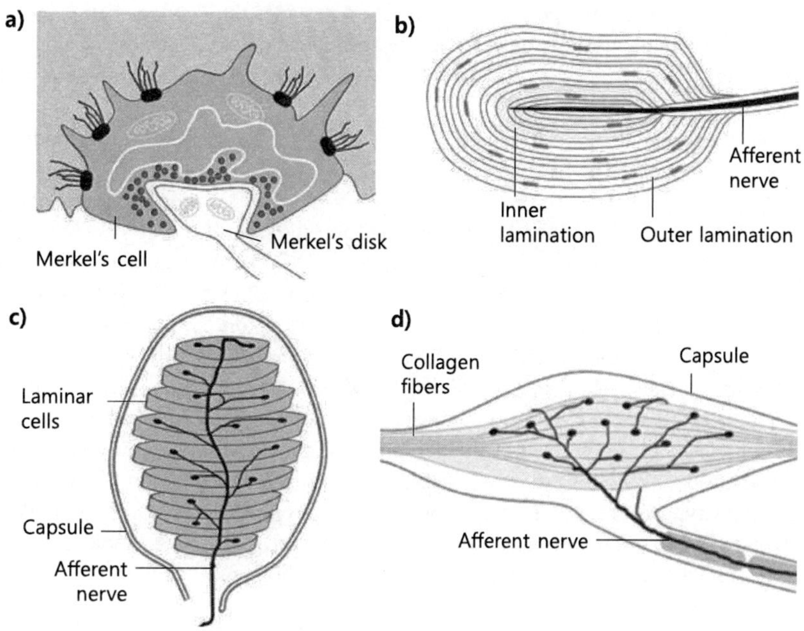

Fig. 6.6 Structures of touch receptors: (**a**) Merkel's cell, (**b**) Pacinian corpuscle, (**c**) Meissner's corpuscle, (**d**) Ruffini's corpuscle

opposite side. Soon later, the nerve ending receives force from all directions with equal strength and the receptor ceases to respond. Outer lamellar cells of Pacinian corpuscle and surrounding fluid act as a high-pass filter. They decrease lower frequency mechanical stimuli and allow higher frequency vibration to reach the terminal ending of the afferent nerve through interconnected collagen fibers.

So, Pacinian corpuscles are also sensitive to high-frequency vibration and respond with optimal activation around 200 Hz, which is appropriate to sense the vibration of hand-held objects. Recent studies report that inner core lamellar cells have glutamate and gamma-aminobutyric acid (GABA) receptors and suggest that they are even transducing force directly in addition to modulating the applied force (Zimmerman et al., 2014). Pacinian corpuscles are regarded as broad tactile receptors and rapidly adapt to sustained stimuli.

Ruffini corpuscles are found deep in the dermis of the skin and particularly sensitive to skin stretches as their longitudinal axes are positioned parallel to the surface of the skin. They are also found in joint ligaments and function as proprioceptive receptors reporting the position of joints. Ruffini corpuscles have a cigar-like shape and are encapsulated. Within a capsule, they contain strands of longitudinal collagen fibers that are smoothly connected with the skin or joint tissue. Nerve endings of afferent nerve fiber branches in many directions and their branches are interwound with the collagen fibers. When the stretching force is applied to the skin, collagen fibers in the Ruffini's corpuscle stretch together. Stretched strands of collagen fiber distort the nerve endings membrane of interwound branches and open stretch-gated ion channels. Collagen fibers remain stretched while the force is sustained and the ion channels remain open. Ruffini corpuscle is a slowly adapting receptor.

In hairy skin, there are specialized touch receptors forming nerve terminals associated with *hair follicles*. The terminal ending of the afferent nerve winds spirally around the base of the hair follicle. The hair follicle receptor is not encapsulated and responds to the hair deflection even without a touch to the skin surface. Some animals, like rodents and cats, have very sensitive hair follicle receptors with their whiskers. Hair follicle receptors are rapidly adapting types and sensitive to low-frequency vibration. Deflection of hair produces a transient response at the onset of deflection and rapidly reduces responses as the stimuli are sustained. As Meissner corpuscles are not present in hairy skin, hair follicle receptors are considered the discriminative movement-sensing receptors in the hairy skin areas.

Free nerve endings are found in the skin and in many other body parts. There are several types of free nerve endings in the skin. Some are nociceptive and stimulated by tissue damage. Some are sensing the temperature by responding to cooling or to warming of the skin. Also, there are some free nerve endings responding to touch. Even though all these free nerve endings are very similar in their morphology, they have different sensing characteristics based on the type of transducing ion channels in the sensing membrane of the nerve endings.

Each type of touch receptor encodes a specific feature of the mild force applied to the skin. The collection of encoded messages via different types of touch receptors provides the brain with enough information to recognize the applied touch accurately.

2.2 Baroreceptors

Baroreceptors sense the blood pressure and report the information required for the baroreflex, negative feedback regulating the blood pressure within the normal range (Fig. 6.7). Baroreceptors are types of mechanoreceptors with free nerve endings and do not have any specialized transducing cells. They are found in the walls of large blood vessels and are extremely abundant in the wall of the carotid sinus and in the wall of the aortic arch. Transduced signals of baroreceptors are transmitted through the afferent glossopharyngeal nerves (ninth cranial nerve) from the carotid sinus and through vagus nerves (tenth cranial nerve) from the aortic arch to the cardiovascular center in the pons and medulla of the brain.

Baroreceptors are extremely effective in preventing rapid changes in arterial blood pressure. Increased blood pressure stretches the vessel wall which accompanies the stretch of nerve endings embedded in the vessel wall. The stretch of the vessel wall is sensed by the baroreceptors, converted to a series of action potentials, and transmitted through the afferent nerve fibers. In the cardiovascular center, it gives rise to decrease sympathetic outflow by increasing the inhibition and ultimately leads to decreased blood pressure by reducing heart rate, peripheral resistance, myocardial contractility, and cardiac output. In contrast, when the blood pressure is decreased, baroreceptors send the signal in decreased strength. Decreased baroreceptor activity leads to increased sympathetic outflow by decreasing inhibition. It also leads to decreased parasympathetic activity and finally increases the blood pressure back to the normal range. Baroreceptors are mainly for short-term regulation of blood pressure from moment to moment or hour to hour. When the increased or decreased blood pressure is sustained longer time, for example, more than 2 or 3 days, adaptation occurs and the normal range is newly set. The baroreflex system is not suitable to regulate blood pressure in the long-term range.

Baroreceptors are found also in large systemic veins, pulmonary vessels, and the atrial and ventricular wall of the right heart. Since blood pressures at these locations are not high, they are differentiated as low-pressure baroreceptors. They are involved mostly in the regulation of blood volume than blood pressure. These baroreceptors mediate the secretion of hormones and intake of salt and water in the kidney as well as the circulatory effects. Even though this reflex process is not rapid, reflex via renal regulation provides the long-term stability of blood pressure.

Even though the basic understanding of baroreceptor reflex was known back to 19C, molecular identities of baroreceptors are not clearly verified. Several types of mechanoreceptor proteins are proposed as substrates of baroreceptors. Epithelial sodium channel (ENaC), acid-sensing ion channel 2 (ASIC2), and transient receptor potential channel 5 (TRPC5) proteins have impaired the baroreceptor reflex when they are genetically deleted. But these channels lack in providing the physiological sensitivity to relevant mechanical stimuli. Recently, piezo channels (Piezo1 and Piezo2) and the Tentonin 3 (TTN3) ion channel are identified as major sensors verifying the required characteristics. Piezo channels mediate rapid adapting kinetics and TTN3 channels mediate slow adapting kinetics to

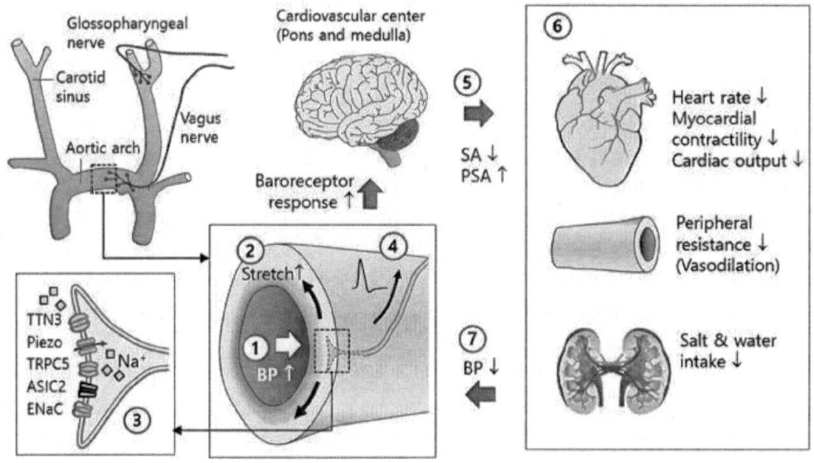

Fig. 6.7 Baroreceptors and baroreflex. *BP* Blood pressure, *SA* Sympathetic activity, *PSA* Parasympathetic activity

blood pressure change (Gu & Berkowitz, 2020). While multiple mechanoreceptors are considered to function as baroreceptors, their competitive and complementary roles are remained to be further studied at the molecular level.

2.3 Auditory Receptors

Humans have an extraordinary auditory system. Each segment of sound is represented by the combination of different strengths of its frequency components. The cochlea in the inner ear distinguishes frequency components and transduces electrical signals for each frequency component. The dynamic range of sound amplitude that humans can hear is very large from 0 dB to 140 dB. The human can hear the sound of frequency from 20 Hz to 20KHz, which is almost 10 octaves.

The human auditory system is briefly divided into three parts of the outer ear, middle ear, and inner ear (Fig. 6.8). *Outer ear* consists of the ear auricle (or pinna) which collects and funnels sound waves from the environment and the ear canal which guides sound waves to the tympanic membrane (or eardrum). Sound waves are conveyed to the vibration of the tympanic membrane in the outer ear. *Middle ear* is the air-filled space between the out ear and the inner ear and connected to the nasal cavity by the eustachian tube. This provides equal atmospheric pressure to both sides of the eardrum to maximize the hearing sensitivity. The eardrum is connected to the ossicles, which is the chain of three delicate and the smallest bone in the body. Ossicles do the role like a lever that transfers and amplifies the vibrational amplitude of the tympanic membrane to the opening of the oval window in the cochlea. Ossicles also do the role of impedance matching in transmitting sound from the air to the fluid in the inner ear. *Inner ear* is a labyrinth structure consisting of the cochlea, vestibule, and semicircular canals. The cochlea is the auditory part of the inner ear. It has a tubular structure encased in rigid bone, is about 35 mm long, and coiled 2.5 times into a snail shape. The cochlea is connected to ossicles by the oval window, which is a unique opening covered by the membrane in the bony cochlea structure. Through the oval window, the vibration of sound is transmitted from the ossicles to the internal fluid of the cochlea. To maximize the fluctuation of the incompressible fluid in the cochlear, there is another opening covered by a flexible membrane called the round window. When we see the cross-sectional view of the cochlea, two membranes, the basilar membrane, and the Reissner's membrane, separate the cochlea into three parts of scala tympani, scala media, and scala vestibuli. Scala vestibuli and scala tympani are filled with the perilymph fluid, which has a high Na^+ ion concentration and a lower K^+ concentration similar to extracellular fluid and cerebrospinal fluid. In contrast, scala media is filled with endolymph which has a high K^+ ion concentration and a lower Na^+ concentration. Vibrational force fluctuates the internal fluid and fluid fluctuation causes the movement of the basilar membrane, where the specialized sensory epithelium, the organ of Corti, that allows the transduction of sound vibrations into neural signals, resides.

The organ of Corti is the active site where sound transduction occurs. There are four rows of hair cells, one row of inner hair cells, and three rows of outer hair cells. Each row contains about 3500 hair cells which are lined up along the whole length of the cochlea. The tectorial membrane overall covers the organ of Corti and hair cells. Each hair cell is connected to the auditory nerve to transmit the transduced electrical signal to the auditory center in the brain. Sensory transduction is mostly achieved in the single row of inner hair cells and 95% of afferent auditory nerves have synapsed to the inner hair cells. Only 5% of afferent auditory nerves are

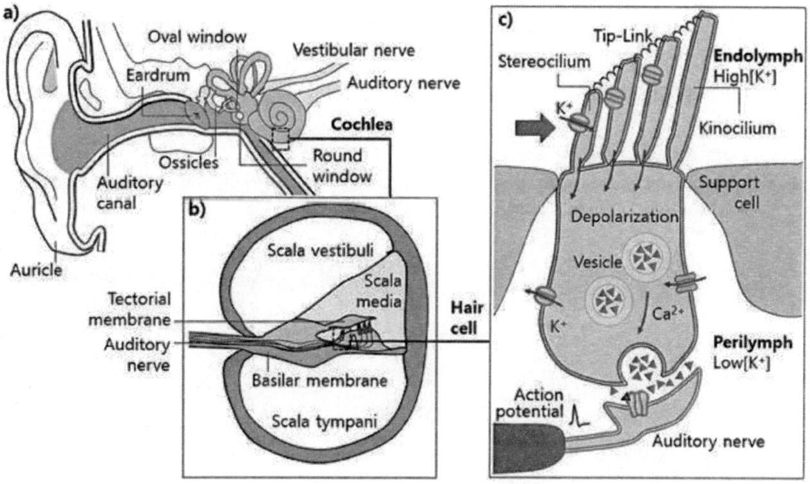

Fig. 6.8 Auditory receptors transducing the sound. (**a**) Structure of the ear, (**b**) Cross-sectional view of the cochlea, (**c**) Mechanoreception of sound in the hair cell

innervated to outer hair cells. Instead, outer hair cells are endowed with many efferent neurons and mainly work to amplify the sound rather than sense the sound.

Hair cells are mechanoreceptors (LeMasurier & Gillespie, 2005). Each hair cell has 50~150 shorter hairs called stereocilia and the tallest hair called kinocilium and their tips are connected by tip-links like spring coils. Hairs of outer hair cells are connected to the tectorial membrane. Upon the movement of the basilar membrane and the following movement of the tectorial membrane and the endolymph in the scala media, stereocilia deflect toward the kinocilium and produce tension via tip links among the membrane of stereocilia. Applied tension to each stereocilium opens the mechanoreceptive ion channels. Since hair cells are protruding their hairs into the endolymph, the K^+ ion, which has a higher concentration on the outer side, moves in when the cation ion channels are open. With the influx of K^+ ions, depolarization starts to occur and increased receptor potential is developed. As the membrane potential increases, voltage-gated calcium ion channels start to open and the influx of calcium ions promotes the vesicle to release neurotransmitters. Hair cells release neurotransmitters, glutamate, into the synaptic gap, and the postsynaptic potential is developed in the afferent auditory nerves.

Cochlea also has a unique feature in way of differentiating the frequency components of the input sound. The vibrating pattern of the basilar membrane in the cochlea is dependent on the frequency constituent of the sound. When the high frequency of sound is applied, maximal oscillation of the basilar membrane occurs at the basal part of the membrane. In contrast, for lower frequency sound, maximal oscillation of the basilar membrane occurs at the apical part of the basilar membrane. Along with the two-and-a-half turns of the cochlea, the basilar membrane has its location-specific resonant frequencies like a piano keypad. That is, the basilar membrane transforms sound in the time domain signal into the sound in the frequency domain signal as we do with Fourier transform in signal analysis.

2.4 Vestibular Receptors

The vestibular system senses the balance of the body and transmits information on the direction and the movement of the head to the brain. This information is essential to properly coordinating the body parts responding to the targeted or external movements. The vestibular system is located in the bony structure of the labyrinth in the inner ear which is composed of cochlea, vestibule, and three semicircular canals. Within the bony structure, the vestibular system is integrated with the auditory system by the sets of interconnected tubes and chambers called the membranous labyrinth. The vestibular parts are composed of three *semicircular canals* and the *otolithic organ* in the vestibule (Fig. 6.9). Three semicircular canals sense the angular acceleration and the otolithic organ senses the linear acceleration. These two parts do the role of six-axis motion sensors in electrical systems which measure three axes of linear accelerations and three angular velocities with gyroscopes.

Linear accelerations are sensed by the otolithic organ which is located in the central part of the labyrinth. The otolithic organ consists of the *utricle* and the *saccule* which are a pair of relatively large chambers. The utricle and the saccule respectively have a sensory epithelium called the *macula*. The macula in the utricle is oriented close to the horizontal plane and the macula in the saccule is oriented close to the vertical plane to sense the accelerations in different directions. The macula contains a cluster of hair cells among the bed of supporting cells. Hair cells protrude their hairs, kinocilium, and stereocilia into the gelatinous fluid in the otolithic membrane. The otolithic membrane contains *otoconia* (or otolith), which are the crystals of calcium carbonate and make the otolithic membrane heavier than the surrounding endolymph fluid. With an applied acceleration, the inertia of the otoconia causes the fluid movement within the otolithic membrane and the fluid movement deflects the embedded kinocilium and stereocilia of the hair cells.

Fig. 6.9 Vestibular receptors

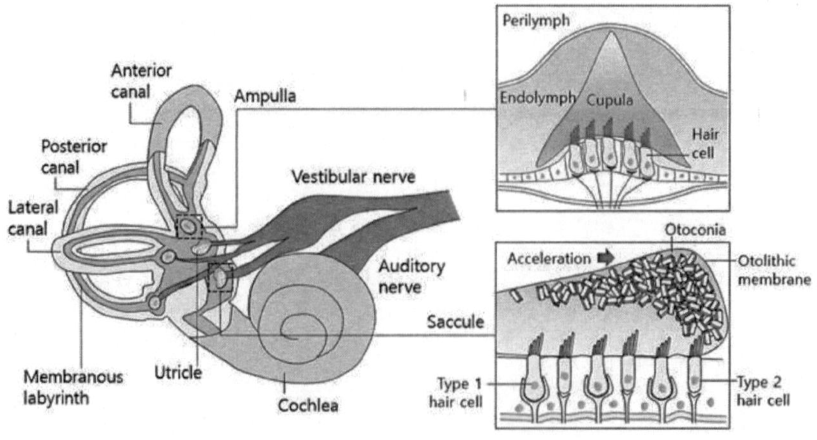

The functioning of hair cells is quite similar to the hair cells in the organ of Corti for sound transduction. The endolymph that kinocilium and stereocilia are embedded in is K^+ ion dominant as in the endolymph of the cochlea. Tension applied to the stereocilia opens the cation ion channels and the inward flow of K^+ ion develops the depolarization of the receptor potential. And, the influx of calcium ions, release of vesicles, synaptic transmission, postsynaptic potential, and transmitting action potential in afferent nerves are followed. There are two types of hair cells. Type I hair cells have flask shape and synapses to a large area of calyx-like nerve endings. Type II hair cells have a cylindrical shape and synapses to a small area of button-like nerve endings. Functional differences are reported between the types like a different sensitivity to the same movement of head rotation.

Angular acceleration generated by the head rotations is sensed by the three semicircular canals. Three semicircular canals, anterior canal, posterior canal, and lateral canal, are positioned approximately in the orthogonal planes to sense the angular accelerations centered on three perpendicular axes. Sensory parts of the semicircular canals are located in the *ampulla*, the bulged parts at the end of each canal. Within each ampulla, there is a sensory epithelium that contains a cluster of hair cells. Hairs of hair cells, kinocilium, and stereocilia are projected into the gelatinous fluid within the dome-shaped *cupula* which covers the whole lumen of the ampulla. Cupula does not have otoconia and is not sensitive to linear acceleration. However, when there occurs a sudden rotation around the specific axis, stagnant endolymph within the ampulla tends to stay and applies inertia force to the movable cupula. This force shifts the cupula and a tension is applied to the embedded kinocilium and stereocilia of the hair cells. Depending on the direction of angular acceleration, hair cells respond by increasing or decreasing the receptor potentials and the release of neurotransmitters. Same routines are followed as in the transmission of linear acceleration through the vestibular nerve fibers. Since each canal responds most sensitively to the angular rotation along a specific axis, a combination of three canals provides a good representation of any angular movement in the 3D space. This information is further ensured by the paired semicircular canals on the contralateral side of the head, which reside in the same planes and respond along the same axis.

2.5 Receptors in the Muscles

The muscle is an effective organ rather than a sensing organ. But for an accurate contraction of each muscle, its contractile length and strength should be monitored. Receptors in the muscles do the role of monitoring the ordered work to muscles. Each motor unit is innervated by a motor neuron and it contracts upon the action potential transmitted via a motor neuron. The strength of muscle contraction to exert a force is modulated by the rate of the action potential and the number of recruited motor units. To exert an accurate amount of contraction, open control by the efferent motor neurons only is not sufficient. While the rate of action potentials and the number of recruited muscle fibers can be decided by the brain, it is not possible to confirm that the exact force is exerted as commanded via the efferent motor neurons. Actual forces and the contracted muscle distances can be different depending on the heaviness of the load and the degree of muscle fatigue. So, feedback is required to evaluate the accurate contraction compared to the ordered signals.

There are two types of muscle contractions. *Isotonic contraction* is the contraction that muscle is shortened while maintaining constant tension in the muscle. To evaluate the isotonic contraction, sensory receptors that can sense the length or shortened distance of the muscle are required. In contrast, *isometric contraction* is the exertion of force without shortening the muscle length. This type of contraction occurs mostly when subjects are maintaining body balance. To evaluate the isometric contraction, sensory receptors that can sense the exerted force are required.

Within the motor unit, there are two sensory organs. *Muscle spindle* is parallelly in the middle of muscle fiber to monitor the shortening of the muscle fiber. *Tendon organ* is located serially in the junction connecting muscle with tendon and monitors the force exerted.

Figure 6.10 shows the sensory receptors in the muscle. In the muscle spindle, there are *intrafusal muscle fibers*, which contract according to the gamma efferent neuron. Intrafusal muscle fiber does not exert any force on the external loads. This is a great contrast to *extrafusal muscle fibers* which exert forces on external loads and do the main role of muscle contraction according to the command received through the innervated alpha motor neuron.

There are two kinds of receptors within the spindle. The primary sensing receptor is *annulospiral receptor* that wrap spirally around the central part of intrafusal muscle fibers and senses the stretch of the muscle fiber. As it is located in the central part, it measures the small change and dynamic length of the muscle. Annulospiral receptors are Group Ia(Aa) fibers that transmit the message with 80–120 m/s. *Flower spray receptor* is the secondary receptor that is located in the distal part of the intrafusal muscle fibers and also senses the stretch of intrafusal muscle fibers. As it is located in the distal parts, it measures the static length of the muscle. The conduction speed of flower spray receptors is slower than the annulospiral receptors and these receptors are connected to Group II(Ab) fiber. Receptors within the spindle increase the firing rate of action potential when the muscle is stretching, not contracting. When the muscle is contracted, there can occur some slack between the receptor and the intrafusal muscle fiber surface and the sensitivity of the receptor can be decreased. In this circumstance, intrafusal muscle fiber is contracted to match the shortened length with the extrafusal

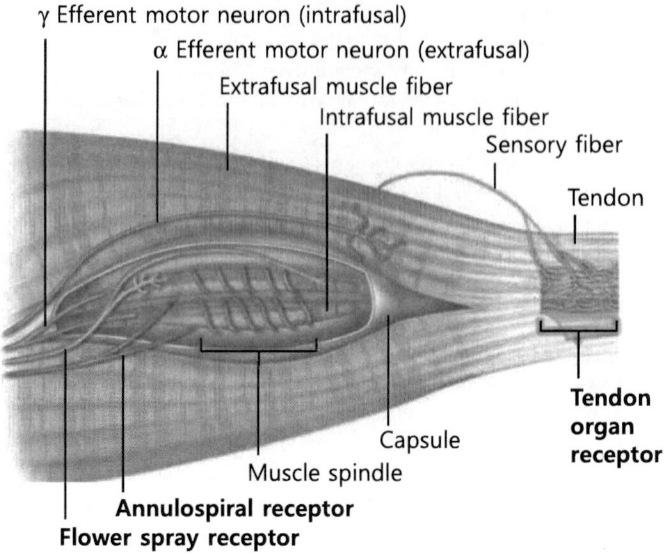

Fig. 6.10 Mechanoreceptors in the muscle

muscles and to maintain the optimal sensitivity of receptors on its surface without losing sensitivity due to the slack between them. By the feedback of two kinds of receptors, muscle length is accurately controlled by appropriate feedback signals.

The tendon organ is a specialized receptor to sense the tension applied to the muscle. The tendon organ is in the shape of a capsule containing numerous collagen fibers innervated by primary afferents fiber. This afferent fiber belongs to Group Ib(Aa) fibers and transmits the information with a speed of 33–75 m/sec for immediate responses. When the force is exerted, the tendon organ senses the strength of the force and sends the feedback message to increase or decrease the force toward a targeted level. The tendon organ also does the protective role of preventing the detachment of a muscle from the bone. However, we still see some cases of the broken tendon which is caused by the improper function or delay of tendon organ feedback.

Ion channels in the muscle receptors are stretch-activated ion channels. They are composed of transmembrane proteins and their structures are deformed to open the gates by the applied tension. Then, ions move rapidly across the channels, membrane potential increases immediately by the influx of positive ions, and the degree of stretch is sensed by the increased potential.

2.6 Kinesthetic Receptor

Kinesthetic receptors sense the information regarding the positions and movements of the body parts in space. Different kinds of sensory receptors provide kinesthetic information.

The sensory receptors in the muscles sense the contraction of the muscles and provide information about connected body parts. The primary nerve ending, the annulospiral receptors, in the muscle spindles are sensitive to the velocity and extent of muscle stretch, and the secondary nerve ending, the flower spray receptors, in the muscle spindles sense the static length of the muscles. The Golgi tendon is specialized to sense the tension applied to the muscle and provides the cue for the range of possible movement.

Vestibular receptors also provide kinesthetic information by sensing the linear and angular acceleration during the body movement which are important for accurate coordination of related body parts. Visual information reported from the visual receptors is also necessary for the accurate coordination of the body and movement. We cannot move and maintain standing with our eyes closed. Skin receptors provide additional information related to the motion by the accompanying stretch of the skin.

Proprioceptors provide information on the relative position of the neighboring body parts and the strength of effort employed for the movement. Broadly, proprioceptors are synonymous with kinesthetic receptors. But, sometimes they refer restrictedly to the receptors reporting the information on the position and motion of the joints. With proprioceptors, we can know what the body part is doing and touch a finger to the nose accurately.

Joint proprioceptors are located in the joints, where bones are connected to each other (Fig. 6.11). Different types of receptors are found in the ligaments and joint capsules. Mechanoreceptors of free nerve endings followed by Ruffini endings are the most common type of proprioceptors overall (Cabuk & Cabuk, 2016). These

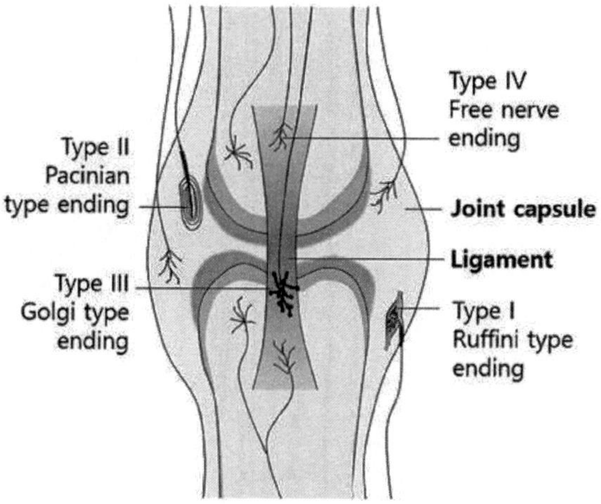

Fig. 6.11 Proprioceptors in the joints

receptors synapse to very small myelinated or unmyelinated nerves (Taylor, 2013).

Joint proprioceptors are generally classified into four types. Type I receptors are like Ruffini endings and are mostly found in the outer layers of the fibrous joint capsule. These mechanoreceptors are slowly adapting with a low threshold. Their nerve endings are interwound with the collagen fibers in the joint capsule and respond to the stretch of the joint capsule. They are generally synapsed to small myelinated afferents nerves and increase firing rate at both ends of the movable range. Type II receptors have the shape of flattened Pacinian corpuscles and are found in the deeper layers of the joint capsules. These receptors are a type of rapid adapting with a low threshold and are inactive when the joints are stationary. Connected afferent nerves transmit action potentials at the onset or termination of movement or in response to sudden changes of mechanical stress. They are also sensitive to compression of the joint rather than stretch along the plane of the capsule (Taylor, 2013). Type III receptors are the shape of the Golgi tendon organ and are embedded in the ligaments and in terminal regions of the tendons near the joint capsule. They synapse to large myelinated afferent nerves. These receptors adapt slowly with a high threshold. Type IV receptors are free nerve endings located globally through ligaments and joint capsules. These receptors have a high mechanical threshold and also contribute to sensing the pain.

The majority of joint afferent nerves have high firing rates at both ends of the joint range and others fire only at flexion or extension. But some joint afferent nerves increase the rate of action potentials in proportion to the joint angle.

3 Chemical Receptors

Chemical receptors are to sense the chemical contents in the environment or in the internal body parts. Chemical receptors allow humans to find favorite foods, avoid harm, and regulate internal body physiological processes. Chemoreception is the oldest form of sensory transduction that single-celled bacteria also detects and responds to the chemical contents in the environment.

Among the five major sensory systems of humans, gustation and olfaction systems are based on chemical receptors. External chemical contents reach the human by oral ingestion or nasal inhalation and they are continuously monitored by the diverse channels of chemosensory receptors. Also, humans have widespread chemoreceptors throughout the body. Chemoreceptors in the gut warn against irritating substances and chemoreceptors in the carotid arteries monitor the gas concentrations and the acidity of the blood.

Both gustatory and olfactory systems sense the concentration of dissolved molecules and transmit the sensed information to the brain. They are closely linked in the brain, and the two systems function in parallel during food taking. The flavors of most foods are strongly dependent on smell as well as taste. While their functions are similar in transducing the electrical signals from the chemical concentrations, their ways of transduction are different. While olfactory receptors are transducing the signal within a single olfactory nerve cell, gustatory reception is composed of specialized receptor cells which sense specifically one type of taste, and the associated ganglionic nerves transmit the action potentials.

3.1 Taste Receptors

Taste receptors are distributed over the surface of the tongue and located within the ridges like small papillae (Fig. 6.12). Within each papilla, numerous taste buds are embedded. Each taste bud contains 50–150 receptor cells sensing the taste. There are also basal cells and supporting cells around the receptor cells. Nerve endings of sensory afferent nerves are connected to the receptor cells at the base of the taste buds. Humans have 2000–5000 taste buds usually.

Even humans can discriminate more than 4000 tastes, each taste is recognized by the patterned combination of five primary taste components as color is recognized by the patterned combination of three primary colors. These five primary tastes are sour, salty, sweet, bitter, and savory tastes. Savory taste, which is also called *umami* meaning delicious in Japanese, is the lately discovered and senses the amino acids like monosodium glutamate (MSG) which is popular as a seasoning adding to the foods.

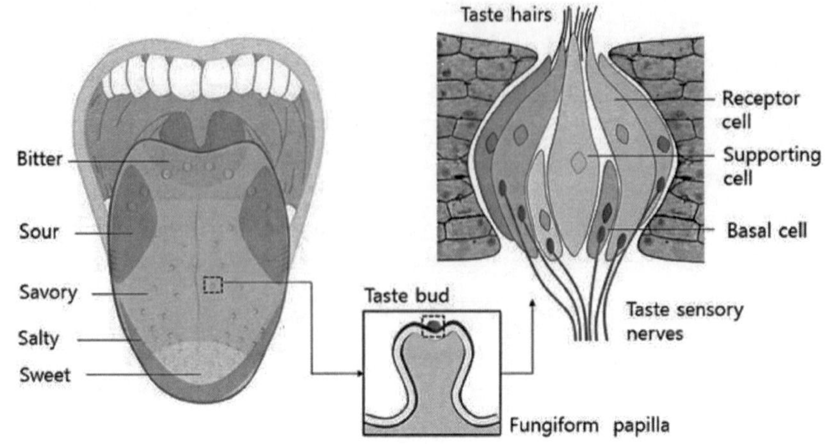

Fig. 6.12 Distribution and structure of taste buds

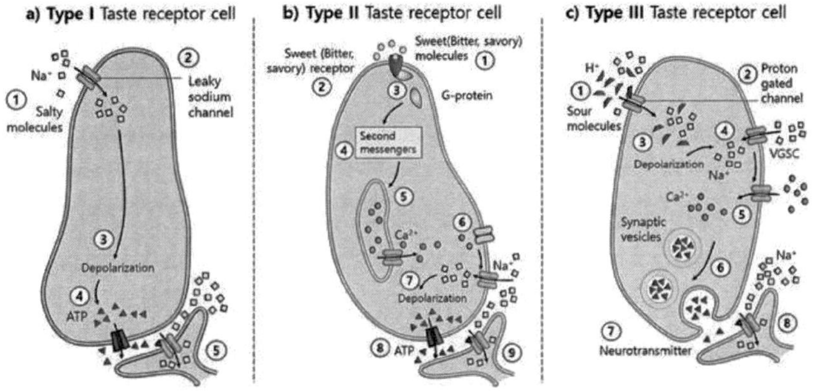

Fig. 6.13 Transduction of tastes in 3 different types of taste receptor cells. (**a**) Type I taste receptor cell for salty taste, (**b**) Type II taste receptor cell for sweet, bitter, and savory taste, (**c**) Type III taste receptor cell for sour taste. *VGSC* voltage-gated sodium channel

Multiple receptor cells packed into a single taste bud protrude taste hairs through the exposed pore to the surface of the tongue. The role of receptor cells is transducing the chemical contents into the release of the transmitters, which in turn depolarize the postsynaptic afferent neuron to generate the action potentials. While all five types of gustatory receptor cells are responding by releasing the transmitters and not transmitting action potentials directly, they are different in their ways of transducing the chemical concentrations.

Transduction of tastes is performed by 3 different types of taste receptor cells as shown in Fig. 6.13. Salty taste is sensed by Type I taste receptor cells. Salty taste is represented by the concentration of sodium ion Na^+. When the concentration of the sodium ion is increased, there is an increased influx of sodium ions into the salty receptor cell through the sodium selective ion channels (ENaC) which are leaky ion channels staying open in a resting state. An increased influx of sodium ions depolarizes the membrane potential. When the membrane is depolarized, Type I taste receptor cells release ATPs instead of neurotransmitters. Gustatory afferent nerve endings have ionotropic receptors responding to the released ATPs (Spector & Travers, 2021).

Bitter, sweet, and savory tastes are transduced by Type II taste receptor cells which open ion channels indirectly. Each of these tastes is sensed when molecules bind to specific receptor sites on the corresponding transmembrane receptor proteins. But, these membrane receptor proteins do not have gates and only activate the second messengers. These receptors are G-protein-coupled receptors (GPCR) which activate G-proteins as second messengers. Activated G-proteins trigger a series of chemical and enzymatic actions and finally result in the opening of cation-gating ion channels. The influx of sodium ions through the open cation channels depolarizes the membrane potential, and Type II taste receptor cells also release ATPs as a transmitter. Transduction mechanisms are the same for the three tastes and the differences are in the receptor proteins responding to the different chemical molecules at the external site of receptor proteins. Based on the labeled line principle, messages of each receptor cell are transmitted to the corresponding site of the brain through the hard-wired pathway.

Sour taste is transduced by the Type III taste receptor cells. Sour taste is represented by the concentration of hydrogen ion H^+ as the acidity is increased with the concentration of hydrogen ion. Increased hydrogen ion concentration directly makes many proton-gated ion channels open and allows sodium ions Na^+ and hydrogen ions H^+ themselves to flow into the receptor cells (Tu et al., 2018). Membrane potential is depolarized and voltage-gated calcium ion channels open. Type III taste receptor cells release neurotransmitters by the increased calcium ion concentration and are connected to afferent nerve endings by the synapses.

Many of the taste transducing is not clear. Taste receptor cells produce the electrical signal and release ATPs and neurotransmitters. Some of these messages are conveyed to basal cells around and the basal cells also have synapses to afferent sensory neurons. So, it is suggested that there is some kind of basic information processing like the retina.

3.2 Olfactory Receptors

The smell is transduced by the olfactory receptor cells located in a thin layer of the olfactory epithelium at the upper part of the nasal cavity (Fig. 6.14). Each olfactory receptor cell has an external process extending up to the epithelial surface and terminated with a number of long slender cilia. It also sends the axonal branch to the olfactory bulb through the cribriform plate, which is the bony structure perforated by many tiny holes like a sieve at the roof of the nasal cavity. In the epithelium layer, there are also supporting cells that help produce mucus and basal cells which progress later into new olfactory receptor cells. Olfactory receptor cells are of a special type that are regularly regenerated within a period of 1~2 months. The surface area of the human epithelium layer is about 10cm^2 and contains five million receptor cells (Li & Lui, 2021). With this relatively small area compared to the dog, the human olfactory system can discriminate against more than 10,000 substances which is much more developed than the discrimination ability in the gustatory system. Since more than 80% of the odorants are unpleasant, the olfactory sensory system is considered more weighted on the protective function to escape away from any harmful substance in the earliest time.

Even though the discriminating ability in detecting inhaled odorant molecules is much better than the gustatory sensory system, its way of transaction is all same by the G-protein-coupled receptors (GPCR). Olfactory G-protein-coupled receptors belong to the same family as Type II taste receptor cells and phototransduction receptors. Following are the procedures by which action potential is transduced from the inhaled odorant molecules.

Step 1: Inhaled odorant molecule binds to the specific olfactory receptor cell in its external binding site of the receptor protein.
Step 2: Reception of odorant molecule triggers structural changes and stimulates olfactory G-protein internally attached to the receptor protein.
Step 3: Olfactory G-protein activates adenylyl cyclase.
Step 4: Adenylyl cyclase converts ATP into cAMP (cyclic adenosine monophosphate).
Step 5: cAMP binds to active cAMP-gated cation channels.
Step 6: Positive ions Na^+ and Ca^{2+} move in and increase (depolarize) the receptor potential.
Step 7: Increased calcium concentration activates the calcium-gated chloride ion channels.
Step 8: Internal chloride ions move out through the open chloride ion channels.
Step 9: Receptor potential is further increased and crosses the threshold to generate an action potential.
Step 10: Action potential is generated and propagates along the axon to the synapse in the olfactory bulb.

How are the odorants differentiated by the same transduction mechanism? The difference is in the receptor proteins. Olfactory receptor cells express a large number of receptor proteins. Richard Axel and Linda B. Buck won the Noble

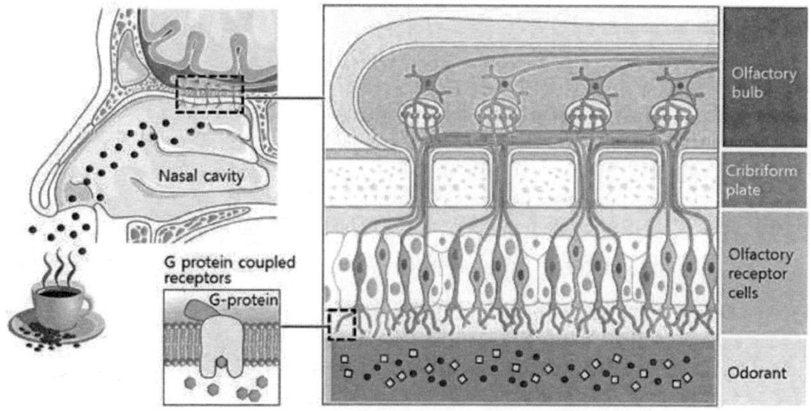

Fig. 6.14 Olfactory receptor cells

Prize in 2004 for their achievement in the discovery of odorant receptors and the organization of the olfactory system. They discovered a large gene family comprised of around 1000 different genes in mammals. Humans also have around 400 functional olfactory receptor genes (Gilad & Lancet, 2003; Verbeurgt et al., 2014). Since each olfactory receptor gene expresses only one type of receptor protein, this number gives rise to an equal number of olfactory receptor cell types. Even though each olfactory receptor cell expresses only one type of receptor protein, each receptor protein does not respond to only one odorant molecule. Each receptor protein responds to several different odorant molecules of a similar structure while the sensitivities are different among the detectable molecules. Also, one type of molecule can be detected by multiple receptor cells whose affinities are different among the responding receptor cells. Most odors are composed of multiple odorant molecules, and constituent molecules stimulate multiple olfactory receptor cells with varying responses.

Axonal branches of olfactory receptors reach the olfactory bulb where around 2000 glomeruli are populated. Glomeruli are the spherical structure where synaptic networks are formed between the axonal terminals of the olfactory receptor cells and the dendrites of mitral cells which transfer sensed olfactory information to the several areas in the brain. Receptor cells carrying the same type of receptor protein converge to the same glomerulus and glomeruli exhibit remarkable specificity. Each mitral cell is activated only by one glomerulus, and the specificity in detected information is maintained up to the brain (Buck & Axel, 1991).

Different responding intensities from the hundreds of receptors lead to a unique olfactory response pattern for each inhaled odor and this pattern is recognized as a specific odor in the brain. Based on this patterned recognition of odor, humans can interpret more than 10,000 different odors and recollect the old memories triggered by the odor.

3.3 Pain Receptors

Pain receptor, which is also called nociceptor, transduces electrical signals from the harmful stimuli that are causing actual or potential damage to the body tissues. Transduced electrical signals are transmitted through the spinal cord to the brain and recognized as the pain we cannot ignore.

There are different kinds of external or internal stimuli that are recognized as painful. As there are multiple types of painful stimuli, there are as many types of nociceptors (Fig. 6.15). Different types of pain can be categorized into three kinds of *nociceptive pain*, *inflammatory pain*, and *neuropathic pain* (Scholz & Woolf, 2002).

Nociceptive pain is caused by the harmful stimuli applied to various parts of the body including the skin, muscles, bones, or other tissues. Nociceptive pain is the most common type of pain and is sensed by the nociceptors around the body. The nociceptors detect mechanical, thermal, and chemical stimuli causing pain. Mechanical nociceptors transduce action potentials responding to the strong mechanical pressure, for example by pushing the sharp objects, before the cell damage. Thermal nociceptors transduce action potentials based on the temperature of the input stimuli. When the thermal nociceptors respond to the temperature stimuli of noxious heat over 43 °C or unhealthy cold, the transmitted message is recognized as painful in the brain (Craig et al., 2001). There are many chemical constituents whose appliance is recognized as painful. Chemical nociceptors respond to a variety of agents, including K^+ ion, extremes of pH, and irritants from the environment. Many of these nociceptors are polymodal which respond to more than one type of noxious stimuli.

Inflammatory pain is caused by damage to the tissue cells and the following inflammation. Cell damage can be caused by external stimuli, such as cuts, strong mechanical forces,

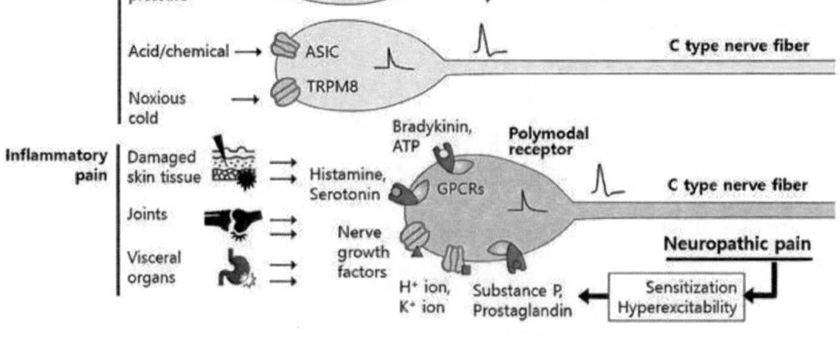

Fig. 6.15 Types of pain receptors

heat, etc. Damaged tissue cells, co-populated blood cells, and nerve endings release a variety of chemical substances into the external fluid at the injured site. These substances include bradykinin, prostaglandins, ATP, serotonin, substance P, K^+ ion, H^+ ion, and many others. They trigger local inflammation. Blood vessels become leakier and cause tissue swelling and redness. Mast cells release the chemical histamine. Action potentials generated in nociceptors propagate along their axons from the site of damage to neighboring branches of the same axon and release substance P which increases capillary permeability and promotes inflammation. These substances have counteracting receptor proteins in the membrane of nociceptor receptors. When each chemical substance is anchored at the outer segments of receptor proteins, ion channels are open directly or indirectly and the receptor potential is developed.

Neuropathic pain, which is also referred to as nerve pain, describes the pain that develops when the somatosensory nervous system is damaged or affected by diseases or injuries. It is different from nociceptive pain because it does not develop in response to any specific circumstance or external stimulus. As the nerve damages cause the hyperexcitability or sensitization of the peripheral nociceptors, mostly it accompanies abnormally high sensitivity to stimuli or pain caused by the never-painful stimuli. Neuropathic pain is usually chronic and includes the sensation of burning or coldness, needle-pricking sensations, and numbness. Many abnormal conditions and diseases cause neuropathic pain including carpal tunnel syndrome, spinal cord injury, and thalamic stroke.

Cell bodies of nociceptor neurons are located in the dorsal root ganglia of the spinal cord. And their axons are extended to the site of sensing location with sensitive free nerve endings. Generated action potentials are transmitted through these axon fibers to the spinal cord. After the synapse to the second afferent neurons, action potentials are transmitted to the thalamus of the brain. Depending upon the conduction speed of the propagating axon fiber, pain is described by first pain or second pain. First pain is the pain delivered fast through the myelinated Aδ type fiber which has a diameter of 2~5 mm and a fast conduction speed of 5~40 m/s. Nociceptors reporting stabbing mechanical pain and burning heat pain belong to this type and transmit a painful message within the fastest time. Second pain is the pain delivered through the unmyelinated C-type nerve fiber which has a diameter of 0.4–1.2 mm and a slow conduction speed of 0.5~2 m/s. Polymodal nociceptors responding to cold stimuli and the chemical stimuli belong to this type and report painful stimuli with the delayed time. After first pain is recognized, second pain is recognized some seconds later.

3.4 Chemoreceptors for the Regulation of Ventilation

For a sufficient supply of oxygen to the brain and the cells of the vital organs, maintaining a normal oxygen level is very important. Inhaled oxygens are dissolved into the blood, carried by the hemoglobin molecules, and transported to the tissues in peripheral organs. When the oxygen is not supplied sufficiently, for example by reduced ventilation, it should be recovered to an appropriate level immediately. This ventilatory reflex is maintained by the sensing role of chemoreceptors.

Hypoxia, a state of lowered oxygen supply in the tissues, is mainly caused by the reduced supply of oxygen in the blood. For medical purposes, oxygen supply is monitored by SaO_2, the oxygen saturation in the arterial blood which is measured by pulse oximetry. Hypoxemia is the low oxygen concentration in the blood. Insufficient ventilation is followed by hypoxemia in the blood, decreased SaO_2 in hemoglobin, and hypoxia in the tissue sequentially.

Hypoxemia is represented by decreased pO_2, partial pressure of oxygen in the blood, accompanied by the increased pCO_2, partial pressure of carbon dioxide, and increased H^+ ion concentration. Increased H^+ ion concentration, which is represented also by decreased pH, is the result of increased pCO_2 by the chemical reaction:

$$CO_2 + H_2O \leftrightarrow H_2CO_3 \leftrightarrow HCO_3^- + H^+ \qquad (6.1)$$

Peripheral chemoreceptors to sense the blood gases are located in carotid arteries and aortic arch. These are unique organs to sense pO_2, pCO_2, and pH to maintain the homeostasis of the body. The carotid body is located in the bifurcation of the common carotid artery and contains primary chemoreceptors provoking the cardiorespiratory, autonomic, and humoral responses to hypoxemia (Fig. 6.16). Chemoreceptors in the aortic body do a similar work supplementing the carotid body.

The carotid body is the cluster of capillary blood vessels, two types of cells, and the afferent and efferent nerve fibers. Type 1 cells are neuron-like glomus cells that transduce the rate of action potentials depending on the oxygen level in the blood and contain multiple types of neurotransmitter vesicles. Type 2 cells do a glia-like supporting role by enveloping the type 1 cells and receiving inputs through efferent nerves. Type 1 glomus cells are synapsed to the afferent glossopharyngeal nerve transmitting the action potentials to the medullary respiratory center in the brainstem. Transmitted signals activate reflective feedback including hyperventilation, tachycardia, and vasoconstriction ensuring sufficient blood and oxygen supply to the brain and vital organs.

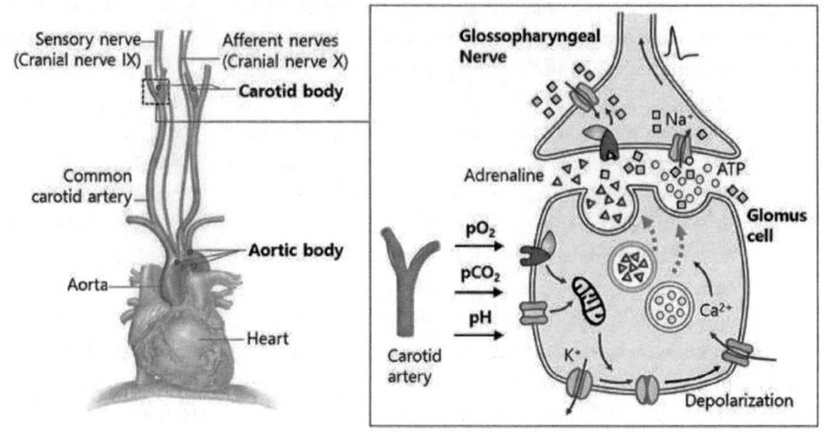

Fig. 6.16 Chemoreceptors for the ventilation reflex

Glomus cells sense the degrees of hypoxemia by the altered level in pO_2, pCO_2, and pH and modulate the internal processes involving the mitochondria and enzyme activation. As a result, potassium ion K^+ channels are closed and blocked outward potassium ion K^+ current depolarizes the membrane. Developed depolarization is followed by a series of action potentials and the opening of the calcium ion channels. Increased calcium ions move the vesicles to the membrane surface and release the transmitters into the synaptic gap. Released transmitters from the glomus cells include ATP, adrenaline, dopamine, and serotonin, and most of them are sensed by G-protein-coupled receptors in the membrane of the afferent nerves (Aldossary et al., 2020). While the consensus is developed that peripheral chemoreceptors sense the hypoxemia by closing the potassium ion channels without additional specific receptor proteins, the molecular mechanism gating the potassium ion K^+ channel from the sensing of oxygen tension is not clear yet (López-Barne et al., 2008).

Central chemoreceptors are traditionally found in the medulla and at some other sites in the brain away from the medullae recently (Nattie & Li, 2012). They detect changes in the arterial pCO_2., and send action potentials to the respiratory center in the brainstem as well. The mechanism detecting changes in arterial pCO_2 in central chemoreceptors is complex and is related to changes in the pH of the cerebral spinal fluid (CSF). Central chemoreceptors do the role of monitoring brain blood flow and acid-base balance in addition to the alveolar ventilation. But, central chemoreceptors respond more sensitively in regulating normal gas exchange than in regulating the acid-base balance.

4 Thermal Receptors

Temperature is an important cue to regulate many essential processes ranging from enzymatic reactions to behavioral responses. Due to the significant impact of temperature, humans have evolved sophisticated temperature-sensing mechanisms based on thermoreceptors. At a first glance, one type of sensory receptor seems to be enough for sensing the temperature of the environment or internal body parts. But, humans have multiple types of thermoreceptors sensing different bands of the temperature range. It can be thought similarly that we are using multiple thermistors to increase the sensitivity of the temperature change for each specialized task instead of a single thermistor covering the whole band of temperature. Thermoreceptors are divided into two groups, warm receptors and cold receptors. It is the division based on the flow of the heat. When heat moves from the environment to the body, we sense the warmth by the warm thermoreceptors. Contrarily, when heat moves away from the body, we feel the coldness sensed by the cold thermoreceptors.

A special group of *transient receptor potential* (TRP) ion channels do the role of transducing the electrical voltage from the temperature change by the heat flow. These ion channels change their structures depending on the applied temperature and gate the movement of ions across the channel. The sensing range of thermoreceptors is determined by the characteristics of these TRP ion channels.

4.1 Locations of Thermoreceptors

Thermoreceptors to sense the temperatures of the environment are populated mostly in the *skins*. Thermoreceptors distributed in the skin have a structural type of the free-nerve ending in all layers of the skin near blood vessels and hair follicles. Cold receptors are more densely populated than warm receptors in the skin. In addition, thermoreceptors are also innervated various tissues within *internal body organs* including the cornea, tongue, and urinary bladder. Thermoreceptors innervated internal organs have their own tasks. For example, cold receptors innervated in the tongue deliver coldness that modulates the degree of taste. Some foods taste better when they are cold, while others do not.

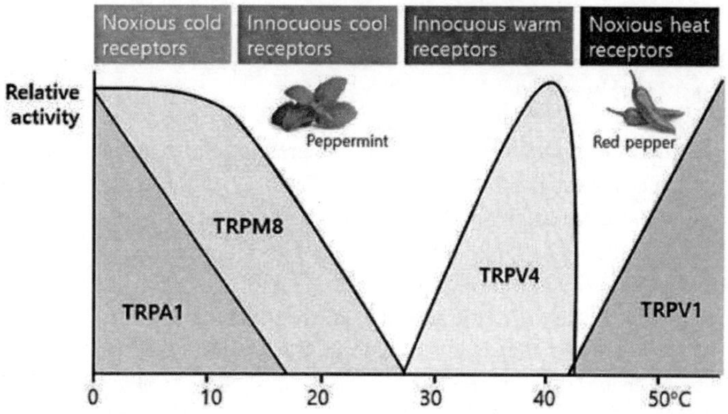

Fig. 6.17 Types of thermoreceptors and their relative responses over temperature range. (Patapoutian et al., 2003)

Warm receptors are distributed with higher density than cold receptors among internal organs. *Hypothalamus* is the special area innervated with thermoreceptors to sense and regulate core body temperature.

4.2 Types of Thermoreceptors

Thermoreceptors can be divided into warm receptors and cold receptors by the flowing direction of the heat. Warm receptors generally increase their activities as the temperature increases while cold receptors increase their activity when the temperature decreases reversely. Many types of thermoreceptors are found to respond sensitively to the changes in the temperature. Some of them sense noxious temperatures and some other thermoreceptors sense the temperature in an innocuous range (Xiao & Xu, 2021). Figure 6.17 shows the responsive characteristics of four representative thermoreceptors and their relative sensing activities over a temperature range.

Noxious heat is the temperature range noxious to the body and immediate responses should be followed. Thermoreceptor *TRPV*1 (transient receptor potential cation channel subfamily V member 1) is the representative receptor sensing noxious heat. It senses temperature higher than 42 °C and also senses pain. *Innocuous warmth* receptors increase the activity according to the temperature without accompanying pain reception. *TRPV*4 (transient receptor potential cation channel subfamily V member 4) is representative one and senses temperature between 27°and 42 °C. It transmits the channel activity through the unmyelinated C-fiber. *Innocuous coolness* receptors sense the coolness and increase the receptor activity as the temperature decreases without accompanying pain reception. *TRPM*8 (transient receptor potential cation channel subfamily M member 8) is the representative one and it senses the temperature below 27 °C. It transmits the receptor activity through unmyelinated C-fibers and thinly myelinated A-delta fibers. *TRPA*1 (transient receptor potential cation channel subfamily A member 1) is a representative thermoreceptor sensing the *noxious coldness*. It senses temperature range below 15 °C and accompanies pain reception.

There are additional multiple types of thermoreceptors other than these, and their sensing ranges are highly overlapped. When one type of thermoreceptor is blocked, even though total activities are decreased, there are still responses to the same temperature caused by other types of thermoreceptors having overlapped temperature spectrum.

4.3 Polymodality of Thermoreceptors

Noxious thermoreceptors sense pain as well as heat or coldness. In this concept, they are nociceptors as well as thermoreceptors. In addition, some temperature receptors simultaneously act as chemotype receptors and respond to special chemical molecules. They do not work as a single modality sensor. They are polymodality-type sensors. TRPV1, a representative noxious heat sensor, also responds to capsaicin, which is an active chemical commonly found in red chili peppers. This is the reason why we are sweating and feel burning hot when we eat red pepper as we are being at the right next to the hot stove. TRPM8, which is a representative innocuous coolness sensor, responds to menthol or other cooling agents. We feel the coolness when we taste the peppermint which has a high menthol content.

4.4 Sensing Mechanism of Temperature

An important question on the thermal sensory mechanism is whether the thermal system detects absolute temperature or just detects temperature change. Recent experiments revealed that these two mechanisms are coexisting among warmth and coldness thermoreceptors. Heat-sensing thermoreceptors encode absolute skin temperature during heat sensation, and some coldness thermoreceptors are often activated by the cooling process rather than absolute values of temperature (Xiao & Xu, 2021).

Sensed temperature signals are transmitted to and collected in the central nervous system, where the final determination of the temperature cue is made. There are two theories on temperature coding in the thermoregulatory circuits in the brain. Both are supported by experimental evidence. One is the *specificity theory* based on the well-known labeled line principle. Warmth or coldness is specifically encoded in the brain circuits by the signals transmitted through the labeled lines from the corresponding warmth or coldness thermoreceptors. This theory is supported by the ablation experiments of one type of thermoreceptor. When warmth thermoreceptors are ablated, warmth sensitivity is largely diminished while the coldness sensitivity is normal. Another coding strategy is *pattern theory*. The pattern theory explains the final perception of temperature is decided by processing the patterned inputs from multiple channels of thermoreceptors. According to the pattern theory, reported one channel signal can be combined with other channels and the perceived temperature may be different from any of the signals transmitted through the labeled lines. While the patterned processing provides more accurate and adaptive temperature perception, a temperature stimulus may be misinterpreted leading to temperature illusion. A good example of the pattern theory is the synthetic heat or thermal grill phenomenon, which is a paradoxical sensation of heat and pain occurring from the simultaneous application of adjacent warm and cold stimuli to the skin.

How can we accept the conflicting two theories? Since pattern theory is based on the more complex neural circuits in higher species like humans, this coding strategy is considered more evolved lately than the labeled line mechanism. Also, the labeled line strategy might be more efficiently used in the peripheral nervous system, while the patterned coding strategy is more properly utilized in the central nervous system for the integration of a large number of input messages.

4.5 Temperature Regulation in Hypothalamus

Maintaining an appropriate body temperature is very important for survival. Thermoregulation is a process that controls the core body temperature within the narrow temperature range of around 37 °C. Even though the human body has some flexibility, driving the core temperature out of this range decreases body function significantly and the damages to organs follow. If the core temperature is decreased to lower than 35 °C, which is the hypothermic state, this condition can potentially lead to cardiac arrest, brain damage, or even death. Also, if the core temperature is raised as high as 42 °C, which is the hyperthermal state, it can cause brain damage or even death.

To avoid critical body conditions, the thermoregulation center in the hypothalamus does the work for maintaining the normal core body temperature. Based on the transmitted temperature information from the thermoreceptors over the body parts and finally decided temperature cue, the hypothalamus delivers a signal to each body part for an appropriate reaction to alleviate the distorted condition and drive the core temperature back to the normal range.

As the core temperatures is decreasing, there occurs a series of feedback responses to minimize heat loss and increase heat production. Hormone epinephrine is secreted into the blood and facilitates vasoconstriction. Vasoconstriction restricts the blood flow into the skin area, stops sweating, and makes the skin fat act as a blanket. By contrast, blood flow into the muscles is increased to facilitate muscle contraction and produce additional heat by muscular work. Shivering, involuntary vibration of muscles follows. Skin hairs stand up by the activation of piloerection muscles and trap air to build the air-insulating layer. Endocrine tissues encourage an increased metabolic rate. Especially, brown adipose tissue is consumed which produces heat without shivering. The body is curled up to reduce the surface area and the behavioral response of moving to the warmer place follows.

When the core temperature is increasing, thermoregulatory feedback occurs in the reverse direction to increase heat loss and minimize heat production. Vasodilation occurs and it increases the blood flow into the skin area and the heat loss by radiation and conduction. Increased sweating is the effective response to consuming a large amount of heat by evaporation. To reduce heat production, metabolic rate is decreased, muscles are stretched, the body is extended to increase the surface, and the brain orders to move to a cooler place.

5 Light Receptors

Vison is the most informative sensory domain in humans. Among the wide range of electromagnetic wave spectrum, humans can sense the electromagnetic waves only in the visible light range. This can be explained by the spectral sensitivity of human photoreceptors. Rods and cones are receptors transducing electrical signals from the light. As a sensory receptor, the transduction of photoreceptors has its own specific features.

5.1 Rods and Cones

There are two types of photoreceptors in the retina, rods, and cones. Their names represent the shape of their outer segments. Rods have a comparatively long cylindrical

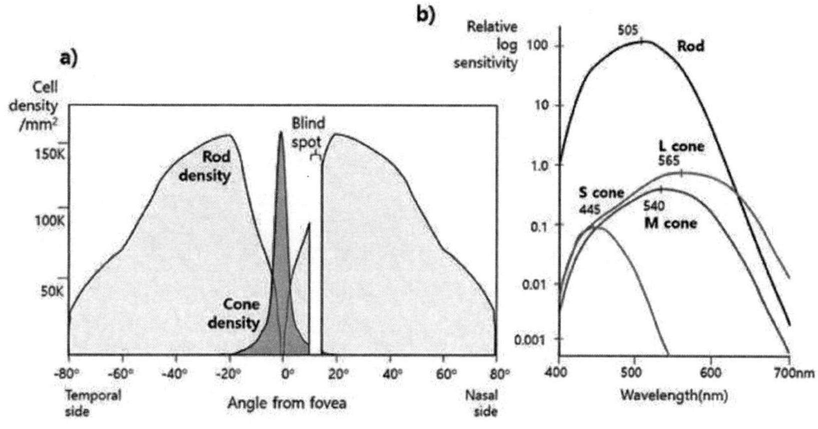

Fig. 6.18 Distribution and sensitivity of photoreceptors. (**a**) Distribution of rods and cones along with the angles from the fovea. (**b**) Relative sensitivity of rod and 3 different types of cones

shape in their outer segments and cones have shorter cone-shaped outer segments.

Rods are distributed through the retina and there are almost a 60 million rod cells in the retina. The density of rods is highest at a distance of 3–5 mm from the foveola with a mean of about 72,000 cells/mm^2 (Fig. 6.18a). They can detect minute changes in light intensity and are accustomed to dim light vision. This vision in the dim light, or low-light vision, is called *scotopic vison*. Rods are about 100 times more sensitive than cones in their maximal sensitive wavelength. While the rods are very sensitive to recognizing the brightness, they do not contribute to recognizing the colors. So, the rods are most actively working in the night vision. Rods are poor in spatial resolution. Many rod cells are connected to a single bipolar cell. In addition to the lower spatial density of rod cells distributed all over the retina area, averaging the multiple rod cells reduces the spatial resolution further. This is the tradeoff between the brightness resolution and the spatial resolution. It is the reason that we have difficulty differentiating the sharp edges of the object during the night. Rods are also poor in temporal resolution. When the lights are flashed on and off at some frequency, they sense the flickering light as continuous light at a lower frequency. We see the movies in a darkened room to minimize the flickering caused by the finite number of image frames per second.

Cones are distributed only in a small central part of the retina called *fovea*, whose diameter is 1.5 mm. There are about 3.2 million cone cells and most of them are populated in the fovea area. Their density in the fovea is 150,000~180,000 cells/mm^2 (Jonas et al., 1992). With this higher cell density, they provide higher spatial acuity when the images are formed in this small area of the retina. This is the reason that we focus an image on this small area when we try to find the detail of the object. They are accustomed to sensing the bright lights and this vision is called *photopic vision*, vision of the eye under well-lit conditions. With the bright light, cones generate signals for recognizing the color of the input lights. So, cones are mainly working for daytime vision. In contrast to rod cells, each cone has its synapse on a single bipolar cell and this connection provides the enhanced visual acuity in daylight vision. Temporal resolution is also high in the cone cells.

There are three types of cone cells that have different spectral sensitivity (Fig. 6.18b). Each type of cone cell has a specific type of photoreceptor pigment, *photopsin*. Depending on their sensitivity over spectral ranges of light, they are called separately as *S cones*, *M cones*, and *L cones*. S cones are most sensitively responding to the light of short wavelengths and have a peak sensitivity at around 445 nm of blue-violet color. M cones are most sensitively responding to the light of middle wavelengths and have a peak sensitivity at around 540 nm of green color. L cones are most sensitively responding to the light of long wavelengths and have a peak sensitivity at around 565 nm of greenish-yellow color. These three types of cones are not equally distributed in number. L cones are the most popular and 63% of cones are L cones. S cones are the least popular and only 6% of cones are S cones. M cones take up 31% population. If any of these cone types are missing or malfunctioning, people cannot discriminate color accurately and become some type of color blindness.

5.2 Phototransduction

Photoreceptors transduce electrical potentials from the triggered light (Fig. 6.19). Rods and cones are the same in their transducing mechanism. When the light is applied to photoreceptor cells, it does not function directly to the ion channels gating the ion movements. Light is firstly sensed by the photosensitive pigment called rhodopsin in the rods and photopsin, a different type of rhodopsin, in the cones. Rhodopsin is densely deposited in the folded layers of the outer segment of the photoreceptors. These pigments determine the type of photoreceptor by how the photoreceptor responds to light.

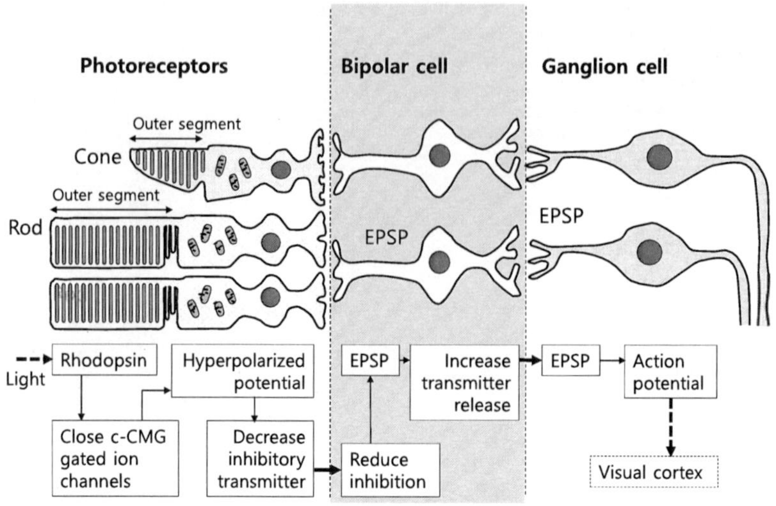

Fig. 6.19 Phototransduction in the retina

To change the membrane potential of the photoreceptor cell, the ion channel should do the work of gating the ion flows upon the reception of the light. This work is done in photoreceptors by the *cyclic nucleotide – gated (CNG) ion channels* which gate the transfer of positive ions like potassium and calcium across the membrane. This ion channel becomes active and opens the channel when the cyclic nucleotide, cyclic guanosine monophosphate (cGMP), is attached to the internal side of the ion channel protein. It does not respond to the light directly and is activated by the attachment of the chemical, which is quite similar to ligand-gated ion channels. But, cyclic nucleotides are internal contents and attached from the internal side of the membrane differently from the ligand-gated ion channels where the ligands are attached to the outer side of the membrane. When there are no light stimuli as in darkness, these CNG channels are usually activated and open by the increased concentration of cGMP inside the membrane. By the inflow of the positive sodium ions Na^+ and calcium ions Ca^{2+} through the open gate, the internal membrane potential is increased, that is depolarized potential is developed across the membrane, about -40 mV for example.

When the light is applied and received by the rhodopsin, there starts a series of chemical and enzyme-activated processes and it finally changes the concentration of cGMP. cGMP molecules are hydrolyzed into non-activating 5′-GMP and the concentration of cGMP is reduced upon the reception of the light stimuli. As the concentration of cGMP is reduced, many of the cGMP molecules attached to the ion channels are detached and many of the channel gates are closed. Inward current during the darkness, which is also called *dark current*, is reduced and membrane potential is decreased, that is hyperpolarization is developed. In all the other sensory receptors, the membrane potential is depolarized when the stimuli are applied and this potential is used to be conveyed to the generation of an action potential within a single cell. But in photoreceptors, strangely, the membrane potential is developed in the reverse direction and action potential cannot be generated within a single cell by this lowered potential. If the darkness is a stimulus instead of light, it can be explained partly in the general view. In the early stage of evolutional history, the darkness of shadows by the predators might be a life-threatening stimulus. This hyperpolarized potential is similar to the inhibitory postsynaptic potential (IPSP) which occurs when inhibitory messages are transmitted over the synapses.

The developed graded receptor potential is propagated along the rod fibers to the end part of the receptor cells, where the synapse is built to the bipolar cell. Propagated hyperpolarized potential reduces the release of neurotransmitters to the synapse, which is acting as an inhibitory ligand to the postsynaptic bipolar cells. As a result, when light is triggered, reduced inhibition is developed at the postsynaptic bipolar cell which is equivalent to the increased excitatory postsynaptic potential (EPSP). This depolarized potential is not directly converted to action potentials. The EPSP maintains the form of graded potential to be integrated with the inputs from the neighboring rod cells. The integrated graded potential propagates along with the bipolar cell fiber to the end part where the synapse is built to the ganglion cell. Presynaptic bipolar cell increases the release of neurotransmitters acting as excitatory stimuli and the EPSP is developed in the postsynaptic ganglion cell. Then, ganglion cells act as associated neurons converting graded potentials into a series of action potentials. Generated action potentials are transmitted to the visual cortex along with the optical fiber and visual pathway. Consequently, light stimulus decreases the dark current, and the action potential is generated by the depolarization developed through the reduced inhibition.

References

Aldossary, H. S., Alzahrani, A. A., Nathanael, D., Alhuthail, E. A., Ray, C. J., Batis, N., Kumar, P., Coney, A. M., Holmes, A. P., & A.P. (2020). G-protein-coupled receptor (gpcr) signaling in the carotid body: Roles in hypoxia and cardiovascular and respiratory disease. *International Journal of Molecular Sciences, 21*, 6012. https://doi.org/10.3390/ijms21176012

Buck, L., & Axel, R. (1991). A novel multigene family may encode odorant receptors: A molecular basis for odor recognition. *Cell, 65*(1), 175–187.

Cabuk, H., & Cabuk, F. K. (2016). Mechanoreceptors of the ligaments and tendons around the knee. *Clinical Anatomy, 29*, 789–795.

Craig, A. D., Krou, K., & Andrew, D. (2001). Quantitative response characteristics of thermoreceptive and nociceptive lamina i spinothalamic neurons in the cat. *Journal of Neurophysiology, 86*(3), 1459–1480. https://doi.org/10.1152/jn.2001.86.3.1459

Gilad, Y., & Lancet, D. (2003). Population differences in the human functional olfactory repertoire. *Molecular Biology and Evolution, 20*(3), 307–314. https://doi.org/10.1093/molbev/msg013

Gu, J. G., & Berkowitz, D. E. (2020). Tentonin 3 as a baroreceptor mechanosensor: Not a stretch. *The Journal of Clinical Investigation, 130*(7), 3412–3415. https://doi.org/10.1172/JCI138120

Jonas, J. B., Schneider, U., & Naumann, G. O. (1992). Count and density of human retinal photoreceptors. *Graefe's Archive for Clinical and Experimental Ophthalmology, 230*(6), 505–510. https://doi.org/10.1007/BF00181769

Laurent, G. (1996). Dynamical representation of odors by oscillating and evolving neural assemblies. *Trends in Neurosciences, 19*, 489–496. https://doi.org/10.1016/S0166-2236(96)10054-0

LeMasurier, M., & Gillespie, P. G. (2005). Hair-cell mechanotransduction and cochlear amplification. *Neuron, 48*(3), 403–415.

Li, X., & Lui, F. (2021). *Anosmia*. StatPearls.

López-Barne, J., Ortega-Sáenz, P., Parda, R., Pascual, A., & Piruat, J. I. (2008). Carotid body oxygen sensing. *European Respiratory Journal, 32*, 1386–1398. https://doi.org/10.1183/09031936.00056408

Nattie, E., & Li, A. (2012). Central chemoreceptors: Locations and functions. *Comprehensive Physiology, 2*(1), 221–254.

Patapoutian, A., Peier, A., Story, G., et al. (2003). ThermoTRP channels and beyond: Mechanisms of temperature sensation. *Nature Reviews. Neuroscience, 4*, 529–539. https://doi.org/10.1038/nrn1141

Ponulak, F., & Kasinski, A. (2011). Introduction to spiking neural networks: Information processing, learning and applications. *Acta Neurobiologiae Experimentalis, 71*, 409–433.

Scholz, J., & Woolf, C. J. (2002). Can we conquer pain. *Nature Neuroscience, 5*, 1062–1067. https://doi.org/10.1038/nn942

Spector, A. C., & Travers, S. P. (2021). *Taste buds and gustatory transduction: A functional perspective*. Neuroscience, Oxford Research Encyclopedias, 23 February 2021. https://doi.org/10.1093/acrefore/9780190264086.013.100

Taylor, J. L. (2013). Kinesthetic inputs. In D. W. Pfaff (Ed.), *Neuroscience in the 21st Century* (pp. 931–964). Springer. https://doi.org/10.1007/978-1-4614-1997-6_31

Tu, Y., Cooper, A. J., Teng, B., Chang, R. B., Artiga, D. J., Turner, H. N., & Liman, E. R. (2018). An evolutionarily conserved gene family encodes proton-selective ion channels. *Science, 359*, 1047–1050. https://doi.org/10.1126/science.aao3264

Verbeurgt, C., Wilkin, F., Tarabichi, M., Gregoire, F., Dumont, J. E., & Chatelain, P. (2014). Profiling of olfactory receptor gene ex-pression in whole human olfactory mucosa. *PLoS One, 9*(5), e96333.

Xiao, R., & Xu, X. Z. D. (2021). Temperature sensation: From molecular thermosensors to neural circuits and coding principles. *Annual Review of Physiology, 8*, 205–230. https://doi.org/10.1146/annurev-physiol-031220-095215

Zimmerman, A., Bai, L., & Gint, D. D. (2014). The gentle touch receptors of mammalian skin. *Science, 346*, 950–995. https://doi.org/10.1126/science.1254229

Part II

Electricity from the Human Body

Electrocardiogram

7

An electrocardiogram or electrocardiograph is the record of electrical heart rhythm to check for any abnormality in the heart. The electrocardiogram is abbreviated as ECG or EKG. Electrocardiography is a method or procedure of recording an electrocardiogram. ECG is the oldest and most commonly using electrical signal recorded from the human body. ECG is measured simply by attaching the electrodes to the body surface within a short time, is very informative on the heart function, is not invasive, and is not expensive to use. ECG with 12 leads is universally and routinely used as an essential tool to evaluate heart diseases in current clinical diagnosis since its first use 120 years ago.

1 History of ECG

The first biological electricity was discovered by Luigi Galvani, an Italian physician, and physicist at the University of Bologna, in 1781. He verified the electrical intervention in biological tissues by showing that the muscle of a dead frog's leg is twitching by an electrical spark. In 1842, Carlo Matteucci, an Italian physicist and neurophysiologist at the University of Pisa, first observed that the beating heart of a dissected frog was a source of electrical current, and he built a sort of generator with stressed muscles of pigeon hearts (Maddio, 2020).

1.1 The First Human ECG

In 1887, Augustus Waller (1856–1922), a British physiologist at the St Mary's Medical School, first recorded a human electrocardiogram using a capillary electrometer. He also introduced the term electrocardiogram. He is the pioneer of ECG development. Like the former researchers, he started his study on the heart rhythm with the exposed hearts of animals. One day, he happened to think that limbs can be used as electrodes and then heart rhythms can be recorded away from the heart without directly touching the surface of the exposed heart. He thought that if this were possible, heart rhythms could be recorded even in the intact and normal hearts and humans could be easily the best candidates. He dipped his right hand and left foot into a couple of basins of salt solution, which were connected to the two poles of the electrometer. At once, he had the pleasure of seeing the mercury column pulsating according to the beating of his heart. He made the first demonstration in St Mary's laboratory in 1887 to audiences, and his friend Willem Einthoven was among them.

Waller also recognized that the whole muscles of the heart are not contracting totally at one time. He mentioned that contraction of the ventricles is not simultaneous throughout the mass, but traverses it as a wave. He noticed that potential distribution is not equal throughout the mass and the inequalities of potential are established at the beginning as well as at the end of each systole. He postulated that this is the evidence for the passage of excitation waves, which is the conduction of electrical pulses from the sinoatrial (SA) node to ventricles in the current view. Based on the spatial inequalities, he thought that potential distribution could be represented as a cardiac vector and the overall behavior could be interpreted by the movement of this vector.

Waller proposed a cardinal lead system to record ECG with 5 electrodes, electrodes at the right arm, left arm, right foot, left foot, and mouth. Among the 10 possible leads, he selected mouth to left arm and mouth to left leg as the orthogonal leads with additional lead from front to back. With this lead system, he made a complete survey of all sorts of leads from the hands, feet, and mouth (Fig. 7.1). In 1921, one year before his death, he published his study of over 3000 traces of heart conditions. He lectured and demonstrated the ECG by circulating in Europe and America and is famous for having used his dog, Jimmy, in his ECG recording demonstration. He did achieve major progress in ECG development. However, from the beginning, he did not expect the potential of ECG in medical usage. Unfortunately,

K. S. Park, *Humans and Electricity*, https://doi.org/10.1007/978-3-031-20784-6_7

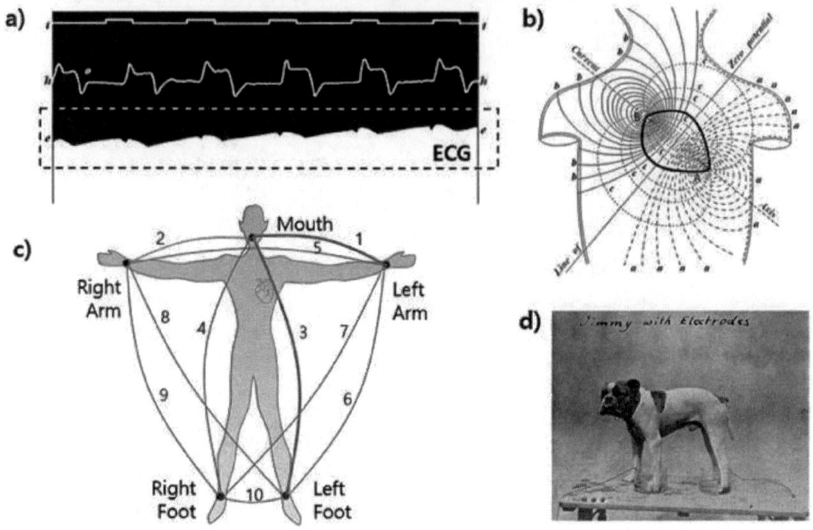

Fig. 7.1 Augustus Waller's achievements. (**a**) The first record of human ECG. ECG is in the third row. The first row is a time marker, the second row is a pulse wave recorded by a sphygmometer. (**b**) Inequalities of potential distribution generating ECG. (**c**) Waller's cardinal leads with 5 electrodes, (**d**) Jimmy in ECG recording demonstration

he did not become a joint awardee when Willem Einthoven won the Nobel prize in 1924 since he died 2 years before in 1922.

In 1891, William Bayliss and Edward Starling, British physiologists at the University College London, demonstrated triphasic cardiac electrical activity in each beat with an improved capillary electrometer (AlGhatrif & Lindsay, 2012).

1.2 Discovery of ECG Mechanism

Up to this time, ECG was recorded based on the capillary electrometer. The capillary electrometer was devised by the mercurial genius Gabriel Lipmann in 1873. It is based on the specific characteristics of the mercury that it changes its volume according to the applied voltage. Minute volume change of mercury is extended into the capillary of the smallest diameter to maximize the sensitivity, and the movement of surface-level interfacing to sulfuric acid (H_2SO_4) was recorded as representing the change of applied voltage. This device has the sensitivity to discriminate the voltage difference of 0.03–0.1 mV. But due to the high inertia of mercury, it distorted the response and its slow response was not able to record high-frequency components. Even though it can record the general variation and a major peak of ECG, it is not sensitive enough to record the details and faster components of ECG for diagnostic use.

Willem Einthoven (1860–1927), a Dutch physiologist at the University of Leiden, is the founder of modern ECG and has recognized the great potential of the ECG as a diagnostic tool in the early stage. Einthoven, greatly inspired by the work of Waller, started to study ECG with a capillary electrometer like Waller. After he recognized the innate distortion of the capillary electrometer, he devoted a lot of effort to correcting the distortion mathematically. He was able to register a fairly good representation of the ECG which is not much different from the ECG obtained with the string galvanometer. During these early years, he identified 5 wave deflections from the corrected ECG and assigned the letters P, Q, R, S, and T to each deflection to differentiate them from the waves A, B, C, and D of uncorrected ECG (Fig. 7.2a). Thus, the basic waveforms of the ECG were characterized by Einthoven before the advent of the string galvanometer (Barold, 2003). Based on early experience, he concluded that the ECG of the diseased heart might deviate from the normal. Then he tried to improve the sensitivity of the recording instrument based on galvanometry and in 1901, he successfully developed a new string galvanometer. It has highly increased sensitivity based on a different mechanism. It did not use mercury. It is based on electromagnetic interaction. In the sensing part of the equipment, a very thin string wire is stretched vertically between the poles of a very strong electromagnet. When the current flows through the string, force is developed in the horizontal direction by the interaction with the established magnetic field between the poles of the electromagnet. Shadow of the string by the focused shining light was designed to be recorded continuously on the moving film roll. The string galvanometer was originally invented by Clément Adair, a French engineer, before Einthoven. Adair replaced the coil in the galvanometer with fine wire to increase the sensitivity and he used it for telegraph receivers. Einthoven just rebuilt the string galvanometer for his purpose and nicely recorded the details of ECG. This instrument required water cooling for the strong electromagnets and could not move around due to its heavyweight of 270 kg (Fig. 7.2b). Large buckets of saline were used as electrodes with the subject's limbs immersed in. Due

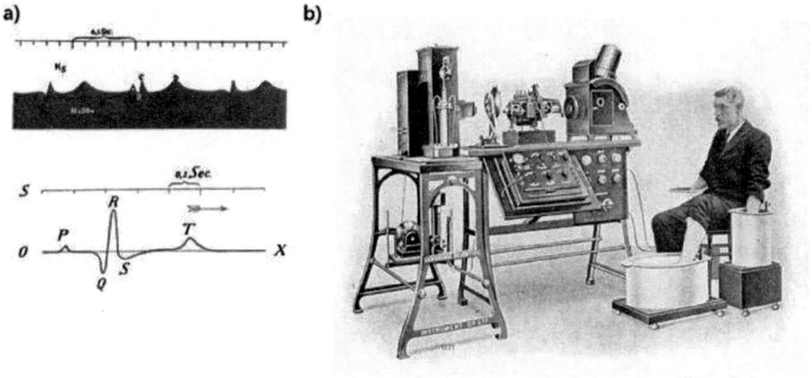

Fig. 7.2 Willem Einthoven's achievement. (**a**) ECG record from the improved capillary electrometer (upper) and reconstructed ECG with mathematical correction (lower). The lower tracing shows a close resemblance to subsequent ECG tracing obtained with the string galvanometer. (**b**) ECG equipment based on a string galvanometer

to the large size and heavyweight of the equipment, many of the ECG recordings were performed through the telephone line between his laboratory and the hospital located a mile away. The invention of the string galvanometer for ECG was a breakthrough in ECG and his successful achievement is mainly based on it. He also developed a standardization for ECG recording. He reduced the number of electrodes by selecting only three electrodes, the left arm, right arm, and left foot. Einthoven triangle is the triangle composed of leads I, II, and III, which provide the recording of potential between the two electrodes out of these three electrode positions respectively. Leads I, II, and III of Einthoven became the standard leads in current ECG recording. Willem Einthoven won the Nobel Prize in physiology or medicine in 1924 "for his discovery of the mechanism of the electrocardiogram".

1.3 Standard 12-Lead ECG

Thomas Lewis (1881–1945), a British cardiologist at the University College Hospital in London, has used ECG in exploring many disorders of the heart especially the mechanisms of arrhythmias and cardiac electrophysiology. In 1909, he identified atrial fibrillation and suggested an occurring mechanism. He did more than anyone else to establish the clinical value of ECG.

His successor Frank Wilson (1890–1952), an American cardiologist at the University of Michigan, made major contributions to the diagnostic usefulness of the ECG including the identification of bundle branch block, ventricular hypertrophy, and myocardial infarction. He improved the accuracy of ECG diagnosis by introducing the central terminal concept. Wilson's central terminal is established by connecting three electrodes with resistors to a single terminal and making the potential at this terminal equivalent to the average of potentials of three limb electrodes. By considering this central terminal as a theoretical reference electrode within the center of the thorax like a ground electrode of potential zero, the potential of any point of the body part could be measured. It could increase the detecting sensitivity on pathology just below the electrode. He added three unipolar leads, which measure the potential at three electrodes with reference to the central terminal. While Einthoven's 3-lead ECG covered the frontal plane with 60° separations, the addition of unipolar leads made the recorded ECG cover the frontal plane with 30° separations and increased the possibility to record formerly silent myocardial pathology.

In 1938, the American Heart Association and the Cardiac Society of Great Britain published their recommendation for six precordial leads. Six precordial electrodes are placed around the precordium with 30° separations in the horizontal plane and the leads recording the potential with reference to the central terminal were named V1–V6.

In 1942, Emanuel Goldberger, an American cardiologist at the Lincoln Hospital in New York, designed the method to augment the unipolar leads of small amplitudes. He increased the sensitivity by 50% by recording the potential of one electrode, for example, LA, with reference to the average of the other two electrodes, for example, RA and LF instead of referencing the central terminal. He named these leads aVL, aVR, and aVF which augment previous unipolar leads VL, VR, and VF respectively.

In 1954, the committee of ECG in the American Heart Association reported a recommendation for the standardization of 12-lead ECG (Wilson et al., 1954).

2 Generation of ECG

2.1 Evolution of the Human Heart

The human heart is the most vital of all organs. It has the size of one's own fist and weighs about 250~300 g. It is a large muscular organ composed of four chambers and four valves with interconnected nerves and vessels. The main role of the

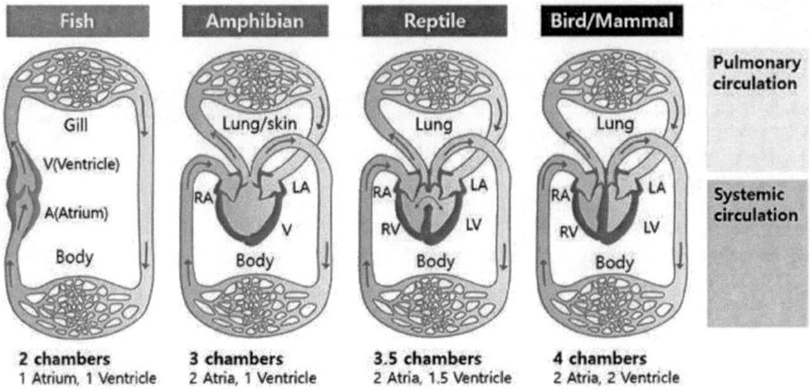

Fig. 7.3 Evolution of four-chamber heart

heart is to provide enough power for pumping the blood and to make the blood circulate all around the human body. It is an engine to accomplish delicate functions required in the human circulatory system and it is updated through the billions of evolutional years. Observing the heart function in other vertebrates might help to understand the current human heart (Fig. 7.3).

Heart and blood vessels started to appear in the vertebrates in evolutional history. Of the vertebrates, fishes have the simplest form of heart and it is considered the first type of closed circulatory system. Fishes have only two chambers, one atrium and one ventricle. A fish's heart is directly connected to the gills by one large vessel to supply oxygen to the blood. And oxygenated blood is transported by the same flow around the fish's body. Collecting the blood in the atrium is considered a more urgent need for effective circulation than having the separate circulation between the body and the lung.

Amphibians like the frogs are the evolutional link between the water-dwelling animals and the land-dwelling animals. Frogs have two atria and still one big ventricle. The frog's heart collects blood separately from lung circulation and body circulation. Oxygenated blood from the lung circulation is mixed with deoxygenated blood returning from the body circulation as they enter the same ventricular chamber. The ventricle pumps out partly oxygenated blood into the vessels toward the lung and toward body circulation. Lung oxygenates the blood which is partly oxygenated, and oxygens are supplied to body tissues by the partly oxygenated blood.

Reptiles like turtles have little advanced hearts. They have two atria and the ventricle which are partly divided. Their hearts are sort of three-and-half chambered. The small septum only separates the ventricle incompletely and the blood can be mixed through the still remained opening. But the mixing cannot be complete by the restricted size of the opening and the time allowed. Blood toward the body circulation is slightly deoxygenated by the partial mixing in the ventricle chambers.

The human heart has evolved into four chambers with completely separated ventricles like birds and the rest of the mammals. Atria and ventricles are separated by a fully formed septum. The right atrium receives and accumulates deoxygenated blood returning from the body's circulation. Then, the accumulated volume of blood is squeezed to the right ventricle. The right ventricle pumps the blood to the lungs through the pulmonary artery for oxygenation. Oxygenated blood returns from the lung through the pulmonary veins and accumulated in the left atrium. Accumulated blood on the left atrium is transferred to the left ventricle within a short period of time. The oxygenated blood is then pumped out toward body circulation through the aorta, the largest artery in the body. This is the most efficient and evolved way to oxygenate blood and supply blood to all around the body parts.

A single human heart works for two circulations. The heart can be considered as a complex of two mechanical pumps but controlled by a single electrical circuit. One pump is for the pulmonary circulation of a short distance and the other is for the systemic circulation of a long and wide circulation. For the systemic circulation, which takes 45 s on average, the left ventricle pumps blood with a high blood pressure of 80~120 mmHg to overcome the high vessel resistances. In contrast, for pulmonary circulation, which takes 7 s on average, the right ventricle pumps blood with a much lower blood pressure of 8~25 mmHg overcoming one-tenth of vessel resistance compared to the systemic circulation. However, the ejected volumes of each beat are the same for both ventricles. So, the two pumps are not of the same characteristics. The left ventricle requires much larger pumping power which is supported by the enlarged muscular structure around the left ventricle.

Considering that the atrium appeared earlier than the separated circulations in the evolutionary phase, the accumulation of returning blood for a long time into an antechamber and transferring the whole volume into the ventricle within a brief instance looks more important for the efficient function of the heart. Enough time should be guaranteed so that the whole blood is completely transferred to the ventricles before the start of the ventricular contraction. This needs some temporal delay between the myocardial contraction of the atria and ventricles.

To maximize the pumping ability of ventricles, especially for systemic circulation, efficient timing sequences are required instead of contracting all the ventricular muscles at a single time. Squeezing the ventricles from the apex, which is the farthest part from the exit valves of the blood, is efficient. This also needs accurate time mapping among the ventricular muscle cells. All these complex and accurate control of myocardial contractions are only possible by the conduction of contraction command of the action potential through the accurately designed conduction pathways.

2.2 Conduction of Action Potential in the Heart

The heart is contracted synchronously to the conducted impulses at the same rate. The impulses are generated at the sinoatrial (SA) node and conducted to every cardiac muscle with an accurately controlled time delay (Fig. 7.4). Action potentials generated at the SA node propagate through three radial pathways stimulating neighboring atrial muscles and arriving at the atrioventricular (AV) node. With some delay in the AV node, action potentials are transverse to the ventricle region following the His bundle in the ventricle septum. His bundle is branched into the right bundle branch and left bundle branch and action potentials are conducted through these branches to each of the ventricles. Bundle branches are connected to Purkinje fibers and action potentials spread to the myocardial cells in the apical end of the ventricles. Each myocardial cell contracts according to the timing of received action potential and the orchestra of integrated muscle contraction are performed.

Sinoatrial (SA) node is located on the roof of the right atrium and serves as the pace-making source of the heart contraction. Even though the rate of action potentials is modulated by the sympathetic and parasympathetic tones, the SA node generates action potential automatically without external stimuli from the brain.

Action potentials generated from the SA node propagate through both atria. Action potentials generally propagate through the myocardial cells in the atria by exciting neighboring cells. Also, there are three pathways conducting action potentials in a rapid manner to the left atrium and the atrioventricular (AV) node. General propagation between myocardial cells is connected by the gap junctions instead of synapses and the action potentials spread from cells to cells throughout the entire atrial myocardium. Generation of action potential on the atrial side and stimulating atrial myocardium before sending the impulses to the ventricle side is to contract the atria earlier and to forward the collected blood to the ventricles before the start of ventricular contraction.

The primary role of the atrioventricular (AV) node is to relay the conduction of action potentials from the atria to the ventricles. AV node is located in the interatrial septum at the base of the right atrium. When the AV node is activated, the action potential is conducted via a slow pathway to His bundle, which is indicated by the long-time delay between the action potentials in atria and His bundle. This delay is important because it makes the atria completely empty the blood into the ventricles. A simultaneous contraction would cause inefficient filling and backflow. Since atria and ventricles are separated by the fibrous tissues, direct propagation through the cell-to-cell conduction is not possible. In the normal circumstance, the conduction via the AV node and His bundle is the unique conduction pathway from the atria to the ventricles.

After passing the His bundle, the spread of action potentials is divided into right and left bundle branches. Following the branched pathways, action potentials are

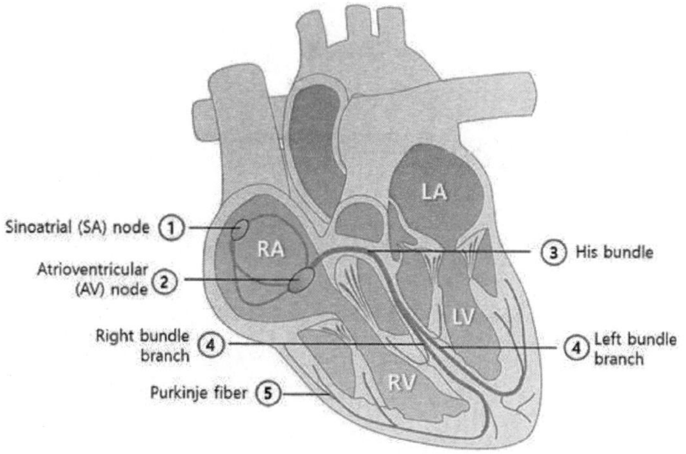

Fig. 7.4 Conduction pathways of cardiac action potentials

conducted rapidly to the apical regions of the left and right ventricles. Finally, the action potentials spread broadly through the Purkinje fibers to the remainder of the ventricular myocardium. Purkinje fibers are extended from either the right or left bundle branches and conduct action potential rapidly to coordinate a contracting pattern among the various parts of the ventricular myocardium. For example, rapid conduction of impulses to the apical area of the ventricles makes the ventricles squeeze from the apical areas toward the exits to optimize the contraction efficiency. The cells in the His bundle, both left and right bundle branches, and Purkinje fibers are specialized to conduct action potentials differently from nonspecialized contractile myocardial cells.

Any dysfunction of the conduction system can cause irregular and abnormal heart rhythms and it used to be recorded in ECG by deviating patterns from the normal.

2.3 Cardiac Pacemaker Potential

The sinoatrial (SA) node is the site that does the role of pacemaking for the synchronous contraction of heart muscles. Pacemaker cells in the SA node are specialized to generate action potentials automatically without any external stimulus. The rate of the action potentials determines the interval of heart contraction. Action potential generated in the SA node has differences in shape and properties compared to the ordinary action potentials. While Na^+ and K^+ play essential roles in the generation of ordinary action potentials, Ca^{2+} is actively involved in the generation of action potentials in pacemaker cells.

Generation of pacemaker action potential is composed of three phases (Fig. 7.5a). During the first phase, *prepotential* is developed. Unlike neurons and skeletal muscles, pacemaker cells do not have a resting period of constant resting potential. As the membrane potential is lowered to −60 mV at the end of the repolarization phase of the previous cycle, the *hyperpolarization activated cyclic nucleotide*-gated (HCN) channels are open. These channels are activated automatically by the lowered membrane potential and the channels are permeable to both Na^+ and K^+. While both ions are allowable to move through the channels, a higher voltage gradient for Na^+ produces a small excessive inward current exceeding the K^+ efflux. Membrane potential rises slowly from an initial value of −60 mV to −40 mV. When the membrane potential is reached −40 mV, the second phase of *depolarization* is started. During this phase, transient Ca^{2+} channels are open resulting in a rapid inward current by the calcium ions. Since the concentration of calcium ions in external fluid (1.2 mmol/L) is much higher than the internal concentration (<0.5 mmol/L), the opening of calcium ion channels causes the influx of calcium ions and the membrane is rapidly depolarized to a level of 0 mV and a little higher above 0 mV. From the peak point of an action potential, the third phase of *repolarization* begins. Calcium ion channels are inactivated and open potassium channels increase the outflow of potassium ions. The membrane is repolarized and the potential is lowered to the minimum level of −60 mV. When the membrane potential is reached −60 mV, potassium ion channels are closed and HCN channels are open again and a new cycle of pacemaker potential begins. Generating pacemaker potentials without external stimuli is the autorhythmic property of cardiac pacemaker cells.

2.4 Action Potential in Cardiac Muscles

Cardiac muscles are contracted depending on the generated and transmitted action potentials across their membranes. The shape of the myocardial action potential accurately controls the timing for activation and relaxation of cardiac muscle cells. In contrast to the ordinary action potentials in neuronal cells and other skeletal muscles, myocardial action potential has an extended plateau to maintain enough time for pumping out and refilling the blood in the ventricles. An extended plateau of the action potential is established by the

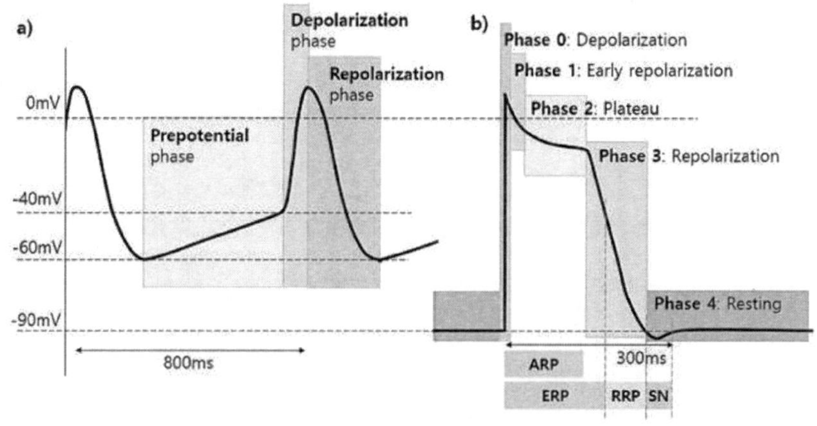

Fig. 7.5 Cardiac action potentials. (**a**) Cardiac pacemaker potentials, (**b**) Myocardial action potentials. *ARP* Absolute refractory period, *ERP* Effective refractory period, *RRP* Relative refractory period, *SN* Supranormal period

intervention of calcium ions Ca^{2+}, which is also doing a major role in the contraction of muscle fibers. Generation of a myocardial action potential is explained by the five sequential phases (Fig. 7.5b).

In the resting state, −90 mV of resting membrane potential is developed in cardiac muscle cells by the ion distributions across the membrane which are not much different from other neurons or skeletal muscle cells. Different shape of the myocardial action potential is caused by the existence of calcium ion channels in the membrane.

Phase 0 is the *depolarization* phase. As in normal nerve cells, the action potential starts with the phase of depolarization. An action potential transmitted from the neighboring muscle cells starts to deviate and increase the membrane potential from the resting potential of −90 mV. Sodium-ion (Na+) channels start to open and the inflow of sodium ions raises the membrane potential further. As the membrane potential approaches −70 mV, the threshold for the generation of action potential in cardiac muscle cells, fast-acting sodium ion (Na^+) channels open their gates very rapidly. The explosive inflow of a large Na^+ current rapidly depolarizes the membrane potential to 0 mV and slightly above 0 mV in an instant. Soon, fast-acting sodium ion (Na^+) channels are closed by their time-dependent inactivation property. When the membrane potential is increased above −40 mV, long opening (L-type) calcium ion (Ca^{2+}) channels start to open, and a small but steady calcium ion current starts to flow in.

Phase 1 is the *early repolarization* phase before the plateau of the potential. Depolarized membrane potential slightly above 0 mV returns to approximately 0 mV by the excessive outflow of potassium ions through the voltage-gated potassium ion channels.

Phase 2 is the *plateau* phase specific to the myocardial action potential. Steady inflow current by the calcium ions through the long opening calcium ion channel electrically balances out the outward current by the potassium ions and maintains the membrane potential in the depolarized plateau just below 0 mV throughout this phase. This extended depolarized potential is important to maintain the ventricle contraction enough to provide maximal stroke volume.

Phase 3 is the *repolarization*. During this phase, calcium ion channels become gradually inactivated. Persistent potassium outward current exceeds calcium current and the membrane potential slowly returns to resting potential as in the case of the ordinary action potential.

Phase 4 is the *resting* phase. The resting potential of −90 mV is maintained mostly by the leaky potassium ion channels. Disturbed ionic concentration during the process of an action potential is restored. Flowed in sodium ions and calcium ions are pumped out to extracellular fluid and outflowed potassium ions are returned to intracellular fluid. This exchange of ions is performed by the mechanism of ATP-dependent Ca^{2+}pump, Na^+/K^+ pump, and Na^+/Ca^{2+} exchanger.

Cardiac muscle cells have a longer refractory period than other skeletal muscle cells related to the depolarized plateau in phase 2 of the action potential. Extended refractory periods in cardiac muscles constrain the occurrence of the next contraction to provide the ventricles with enough time to empty the space maximally before the next contraction. *Absolute refractory* period (ARP) is the duration that the cell is completely unexcitable by a new stimulus. Phases 0, 1, and 2 are the duration of absolutely refractory. *Effective refractory* period (ERP) includes ARP and the period during which a stimulus may cause a minimal depolarization within the cell, but not propagate (effectively of no use) to the neighboring cells. Short segment of phase 3 of myocardial action potential is added to effective refractory period. *Relative refractory* period (RRP) allows a generation of effective action potential if the strength of the stimulus is sufficiently higher than normal. More than half of the later segment of phase 3 is relative refractory. *Supranormal* period is a hyper-excitable period with hyperpolarized membrane potential. During this period, a stimulus lower than normal intensity can depolarize the membrane and derive an action potential.

3 Recording of ECG

3.1 Cardiac Vector Equivalent to Potential Distribution

ECG is the recording of the electrical activity of the heart. It is recorded by attaching the electrodes to the predetermined body surface locations and measuring the potentials between some pairs of electrodes. Recorded activities are dependent on the combinations of electrode pairs and the potential of each lead shows the activity of the cardiac vector at different angles. ECG electrodes are physical conductive pads through which the currents are flowing and sometimes they also indicate the places where the physical electrodes are attached. A lead is composed of a pair of electrodes and the measured potential between a pair of electrodes is called the potential of the lead. For example, lead II is composed of electrodes at the right arm (RA) and left foot (LF), and the potential measured between these electrodes is called lead II potential. The cardinal lead system proposed by Waller has used 5 electrodes and recorded all possible 10 lead potentials. The current standard ECG lead system uses only 12 representative leads by attaching 9 active electrodes.

Heart muscle cells are contracted when their membranes are depolarized by the conduction of action potentials. Action potentials are conducted through the conduction pathway and by the gap junctions between the myocardial cells. When myocadiac cells go through a process of depolarization and repolarization, currents flow in and out across the membrane. These membrane currents develop a local current and the local current flows through the volume conductor up to where they can reach. Neighboring heart muscle cells work as passive volume conductors when they are not actively generating local current and neighboring tissues to the heart also work as volume conductors. The local current generated within the heart can reach the body surface through volume conduction. Electrodes attached to the body surface can sense the minute currents caused by the electrical activation of heart muscle cells through the volume conductors.

For theoretical simplicity, the heart is considered immersed in a homogenous volume conductor. Practically, the conductivities of body tissues are all different. Blood has the highest conductivity and muscles also have relatively high conductivity but are dependent on the directions of muscle fibers. In contrast, lung and fat have lower conductivity and the bone is almost nonconductive to current. Even with these complexities of the structural difference in volume conductance, explaining heart activity as a single vector in a homogamous medium provide a good overall understanding of electrical activities. Distortions and complexities that cannot be explained by the simple view are complemented by the use of multiple redundant leads.

Complex electrical activity of the heart can be represented as an equivalent cardiac vector representing the potential distribution at the time in three-dimensional space. As the potential distribution changes according to the time course of heart activation, the equivalent cardiac vector also varies its direction and magnitude temporally. If the movement of this cardiac vector is properly monitored, we can recognize the overall activity of the heart by the dynamics of this cardiac vector. Since the cardiac vector resides in the 3D space, monitoring it in a single plane is not enough. We observe the electrical activity of the cardiac vector in two planes, in the frontal plane and the horizontal plane.

3.2 Standard Limb Leads

The ability to observe the cardiac vector in a certain plane is dependent on the angle of the lead to the major axis of cardiac vector variation. If we were to record only one lead potential, the lead should be the one that shows the maximal sensitivity to the variation of cardiac vector during the contraction of the heart. But when we are allowed to use multiple leads, a minimal number of leads are required to provide complete behavior of the cardiac vector combinedly by complementing each other.

Standard limb leads were established by Einthoven who observed the cardiac vector components in the frontal plane (Fig. 7.6). Among the 5 electrodes and 10 leads proposed by Waller, he selected three electrodes and three leads which are clinically effective with minimized redundancy. Electrodes at the left arm (L), right arm (R), and left foot (F) have been selected, and each lead potential is measured as the potential between a pair of electrodes.

Lead I potential is measured between electrodes R and L with electrode R as the reference. The lead vector of lead I is horizontal. Lead II potential is measured between electrodes R and F with electrode R as the reference. The lead vector of lead II points the direction 60° rotated from the lead I vector. Lead III potential is measured between electrodes L and F with electrode L as the reference. The lead vector of lead III points the direction 120° rotated from the lead I vector and 60° rotated from the lead II vector. These three vectors constitute an equilateral triangle and this triangle is called Einthoven's triangle. Recorded potential in each lead is the difference of the potentials measured at each electrode and represents the cardiac vector components projected to the corresponding lead vector.

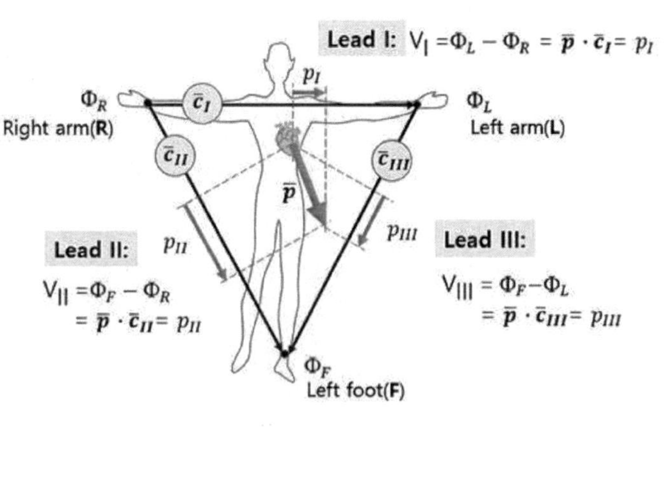

Fig. 7.6 Standard limb leads and Einthoven's triangle

$$\text{Lead I}: V_I = \Phi_L - \Phi_R = \overline{p} \cdot \overline{c}_I = p_I \quad (7.1)$$

$$\text{Lead II}: V_{II} = \Phi_F - \Phi_R = \overline{p} \cdot \overline{c}_{II} = p_{II} \quad (7.2)$$

$$\text{Lead II}: V_{III} = \Phi_F - \Phi_L = \overline{p} \cdot \overline{c}_{III} = p_{III} \quad (7.3)$$

where V_I, V_{II}, V_{III} are the voltages of lead I, II, and III,

Φ_L, Φ_R, Φ_F are the potentials at electrodes L, R, and F
$\overline{p}$ is a cardiac vector
$\overline{c}_I$, $\overline{c}_{II}$, $\overline{c}_{III}$ are lead vectors of lead I, II, and III
p_I, p_{II}, p_{III} are components of a cardiac vector projected to lead I, II, and III

Theoretically lead III potential is redundant and it can be calculated by subtracting lead I potential from lead II potential. But, the measured potential of lead III is not exactly the same as the potential calculated by lead II and lead I potentials. This discrepancy is caused by the inhomogeneous property of volume conductors.

Since the cardiac contraction starts from the atria and progresses to the ventricles, variation of cardiac vector magnitude is maximal along the axis connecting the center of the atria and the center of the ventricles. As the lead II vector is most parallel to this axis, normally maximum amplitude in recorded ECG appears in the lead II potential.

3.3 Normal ECG Waveform

Waveforms of recorded ECG are variant depending on the recorded leads. Figure 7.7 shows the most representative waveform of lead II potential. Positive or negative deflections of ECG waveform are named P, Q, R, S, T, and U waves sequentially and each of them represents the special phase of heart contraction.

P wave is a small positive deflection in lead II recording which is generally smooth and not peaked. It represents the depolarization of the atria. Atrial depolarization spreads from the SA node toward the AV node, and from the right atrium to the left atrium. Its duration is less than or equal to 110 ms. P wave of a longer duration may represent atrial enlargement.

PR interval is the duration from the beginning of the P wave to the beginning of the QRS complex. This interval reflects the time the electrical impulse takes to travel from the SA node to the AV node. Normally it takes between 120 ms and 200 ms. It is used to assess the normality of conduction from the atria to the ventricles. A shorter interval suggests an unusual pathway bypassing the AV node. An extended interval longer than 200 ms is diagnosed as a first-degree atrioventricular block.

PR segment is the isoelectric potential duration between the end of the P wave and the beginning of the QRS complex. It reflects the delayed conduction through the AV node. The potential level of the PR segment serves as the baseline potential.

QRS complex is the combination of negative deflection of the Q wave, peaked positive deflection of the R wave, and following negative deflection of the S wave. It represents the rapid depolarization of the right and left ventricles. As the ventricles, especially the left ventricle, have a larger volume of myocardial mass compared to the atria, the QRS complex usually has a much larger amplitude than the P wave. In the lead II, amplitude of the QRS complex is around 1 mV and its duration is less than or equal to 120 ms. A shorter QRS duration is desirable as it proves the rapid and synchronous contraction of ventricles. The extended duration of the QRS complex suggests a disruption of the impulse conduction

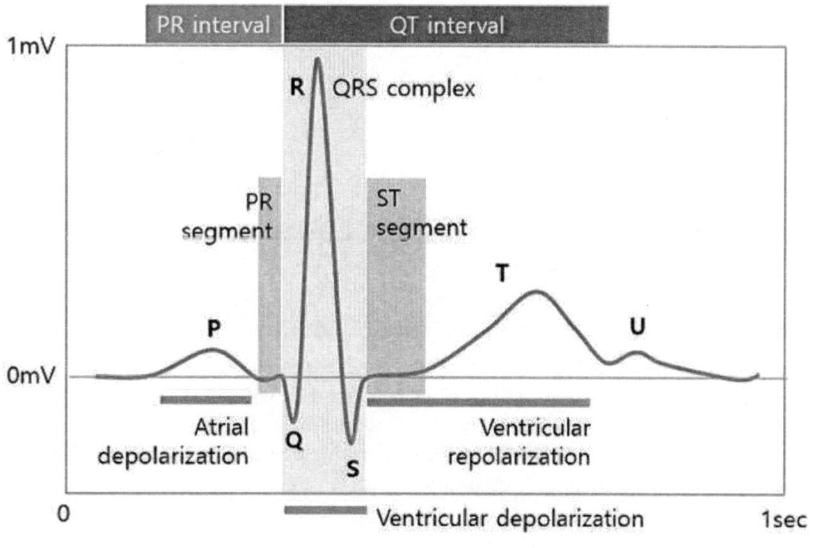

Fig. 7.7 Waveform of normal ECG lead II potential

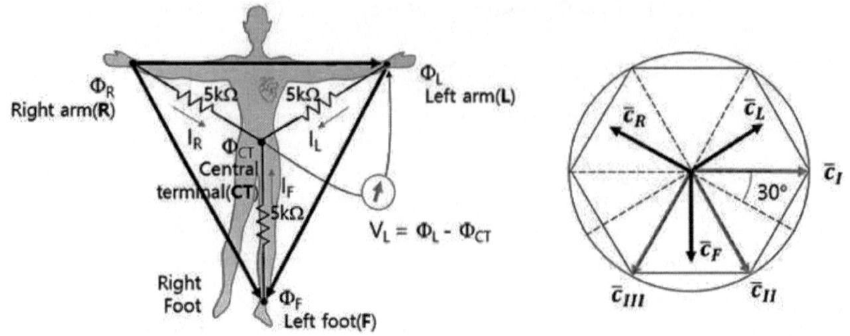

Fig. 7.8 Wilson's central terminal and three unipolar limb leads

system. Left bundle branch block (LBBB), right bundle branch block (RBBB), and pathological ventricular rhythm such as ventricular tachycardia are possible causes of the extended QRS duration.

ST segment is the duration of isoelectric potential slightly slanting upward toward the T wave between the S wave and T wave. It corresponds to the plateau of myocardial action potential (phase 2). Deviation of the ST segment from the baseline has been particularly informative in the diagnosis of acute myocardial ischemia.

T wave is the positive deflection in lead II potential. It represents the rapid repolarization of the ventricles (phase 3). It is normally rounded and asymmetrical, more gradual ascent than descent.

QT interval is the duration from the beginning of the QRS complex to the end of the T wave. It reflects the consumed time of ventricular contraction by depolarization and repolarization. It strongly varies depending on the heart rate. The QT interval increases at lower heart rates and decreases at higher rates. The QT interval should be assessed with the adjustment of the heart rate. Extended QT intervals increase the risk of ventricular arrhythmias.

U wave is normally a low amplitude positive deflection after T wave and often is completely absent. It is caused by the repolarization of the interventricular septum region. Its amplitude is generally one-fourth of the T wave.

3.4 ECG Recording Leads

3.4.1 Central Terminal and Unipolar Leads

The concept of the central terminal was developed in the process of investigating unipolar recording. Theoretically, the reference for the unpopular recording should be the point indifferent to the electrical activity of the heart. As an electrode at infinity or far away is impractical as a reference, an effort to find the equivalent and practical reference within the body was pursued. Wilson suggested the use of a central terminal as the reference. This central terminal is formed by connecting the three limb electrodes to a common point called the central terminal through resistors. He considered this terminal as in the center of the body with zero potential. With this central terminal, any potential measured referencing to this terminal can be thought of as unipolar reflecting the activity of the recording site more directly. In reality, Wilson's central terminal is not indifferent to the electrical activity of the heart. It is the average of three limb potentials. Wilson's central terminal is an approximately constructed virtual reference potential for the recording of unipolar leads. With tolerable deviation, the central terminal provides a practically useful reference. In standard 12-lead systems, 6 precordial lead potentials are recorded with a reference to this central terminal.

Figure 7.8 shows the construction of the central terminal. Three limb electrodes are connected to the central terminal through 5KΩ resistors respectively. As the sum of current coming from three electrodes becomes zero, the potential of the central terminal is derived as the average of the potentials measured at three electrodes.

$$I_R + I_L + I_F = \frac{\Phi_R - \Phi_{CT}}{5000} + \frac{\Phi_L - \Phi_{CT}}{5000} + \frac{\Phi_F - \Phi_{CT}}{5000} = 0 \tag{7.4}$$

$$\Phi_{CT} = \frac{\Phi_R + \Phi_L + \Phi_F}{3} \tag{7.5}$$

Potentials of unipolar lead V_R, V_L, and V_F are recorded with a reference to the central terminal.

$$\begin{aligned} V_R &= \Phi_R - \Phi_{CT} = \Phi_R - \frac{\Phi_R + \Phi_L + \Phi_F}{3} \\ &= \frac{-(\Phi_L - \Phi_R) - (\Phi_F - \Phi_R)}{3} = -\frac{V_I + V_{II}}{3} \end{aligned} \tag{7.6}$$

$$\begin{aligned} V_L &= \Phi_L - \Phi_{CT} = \Phi_L - \frac{\Phi_R + \Phi_L + \Phi_F}{3} \\ &= \frac{(\Phi_L - \Phi_R) - (\Phi_F - \Phi_L)}{3} = \frac{V_I - V_{III}}{3} \end{aligned} \tag{7.7}$$

$$V_F = \Phi_F - \Phi_{CT} = \Phi_F - \frac{\Phi_R + \Phi_L + \Phi_F}{3} = \frac{(\Phi_F - \Phi_R) + (\Phi_F - \Phi_L)}{3} = \frac{V_{II} + V_{III}}{3} \quad (7.8)$$

Lead vectors of unipolar leads are pointing the direction between the lead vectors of standard limb leads. These 6 lead vectors compose a vector set whose directions are equally separated by 30° and each lead makes the potential to be recorded from different angles 30° apart incrementally.

3.4.2 Precordial Leads

Precordial electrodes have been used from the beginning of ECG recording as an effort to record heart activity as close to the heart as possible. Since then researchers have applied precordial electrodes to diagnose the heart pathology and they made many important discoveries in electrocardiography and cardiology. Soon, it added the value of ECG in the diagnosis of infarction, ventricular hypertrophy, bundle branch block, and many abnormal cardiac states. With the introduction of Wilson's central terminal, potentials of precordial leads are standardized to be recorded with a reference to the central terminal.

Precordial leads are to record the cardiac vector components in the horizontal plane which cannot be observed with leads in the frontal plane. Like the lead vectors in the frontal plane, 6 precordial lead vectors are positioned to record the horizontal components from different angles 30° apart incrementally. Electrode positions are standardized over the anterior part of the chest with references to the sternum and intercostal spaces and the leads are named as V1, V2, V3, V4, V5, and V6 sequentially (Fig. 7.9).

Electrode for V1 lead is placed vertically in the fourth intercostal space and horizontally right border of the sternum. Electrode for V2 lead is placed vertically in the fourth intercostal space and left border of the sternum horizontally. Lead V1 and V2 are also called septal leads and primarily observe ventricular septum activity and occasionally display ECG changes originating from the right ventricle. Electrode for V3 lead is placed in halfway between the electrodes for lead V2 and V4 after placing the electrode for lead V4. Electrode for V4 lead is placed vertically in the fifth intercostal space and horizontally at the nipple line or at the midclavicular line. For women, placing under the breast is recommended. Lead V3 and V4 are called anterior leads and observe the anterior wall of the left ventricle. Electrode for V5 lead is placed between the electrodes for lead V4 and V6 after placing the electrode for lead V6. Electrode for V6 lead is placed vertically in same height as V4 and horizontally at midaxillary line, which is an imaginary landmark line separating the body into front and back halves. Lead V5 and V6 are called anterolateral leads and observe the lateral wall of the left ventricle.

3.4.3 Augmented Limb Leads

Three unipolar limb leads, V_R, V_L, and V_F record potentials between each limb electrode and Wilson's central terminal. The resulting potentials are related to the standard limb lead potentials as shown in eqs. (7.6)~(7.8). For example, unipolar lead potential V_F is equivalent to the sum of lead potentials VII and VIII divided by 3, which is 33.3% lower in magnitude than the average of these two potentials. Goldberger observed that these unipolar limb potentials can be augmented in magnitude by removing the resistor connecting the central terminal and the recording limb electrode. Then, the potential at the reference terminal becomes the average of potentials at the other two electrodes except the active limb electrode. In this way, three unipolar limb leads are replaced with a new set of leads that are augmenting the unipolar lead potentials by 50% (Fig. 7.10). These leads are named aVR, aVL, and aVF, where "a" means augmentation.

$$aV_R = \Phi_R - \frac{\Phi_L + \Phi_F}{2} = \frac{-(\Phi_L - \Phi_R) - (\Phi_F - \Phi_R)}{2} = -\frac{V_I + V_{II}}{2} = 1.5V_R \quad (7.9)$$

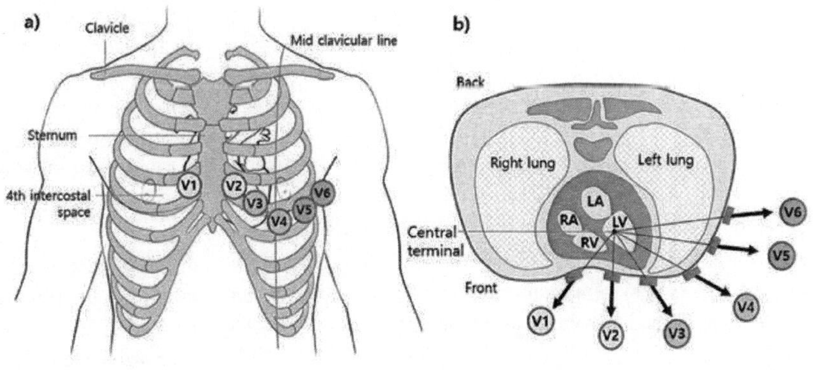

Fig. 7.9 Electrode locations and lead vectors of precordial leads. (**a**) Locations of precordial electrodes, (**b**) Lead vector for precordial leads

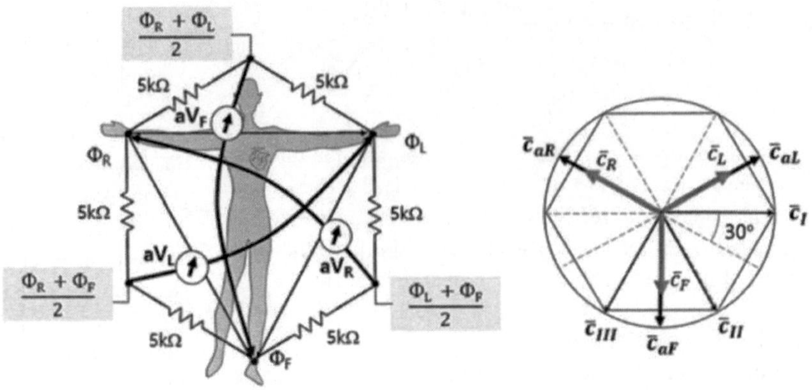

Fig. 7.10 Construction and lead vectors of augmented limb leads

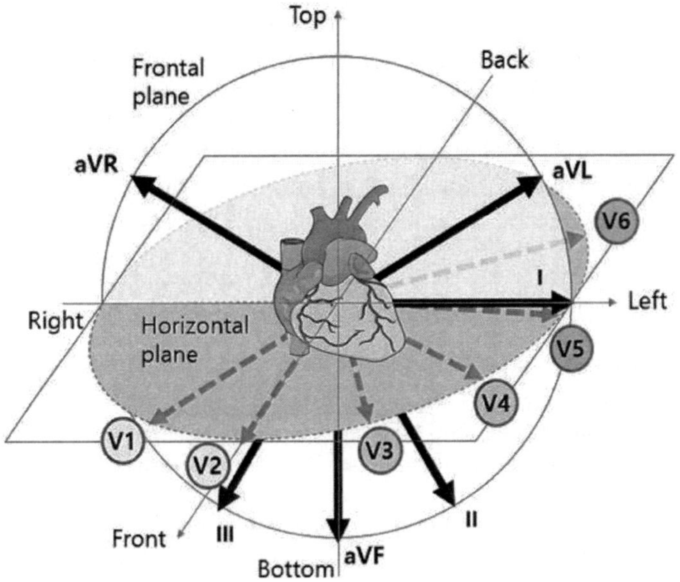

Fig. 7.11 Lead vectors of standard 12 leads in 3D space

$$aV_L = \Phi_L - \frac{\Phi_R + \Phi_F}{2} = \frac{(\Phi_L - \Phi_R) - (\Phi_F - \Phi_L)}{2} = \frac{V_I - V_{III}}{2} = 1.5V_L \tag{7.10}$$

$$aV_F = \Phi_F - \frac{\Phi_R + \Phi_L}{2} = \frac{(\Phi_F - \Phi_R) + (\Phi_F - \Phi_L)}{2} = \frac{V_{II} + V_{III}}{2} = 1.5V_F \tag{7.11}$$

3.4.4 Standard 12-Lead System

After the establishment of augmented limb leads, the ECG recording system has standardized to 12-lead systems. The standard 12-lead system includes 6 leads in the frontal plane and 6 leads in the horizontal plane (Fig. 7.11). Standard leads in the frontal plane are leads I, II, III, aVR, aVL, and aVF. These 6 leads record the activity of the cardiac vector in the frontal plane from different angles with 30° angular resolution. Standard leads in the horizontal plane are 6 precordial leads V1, V2, V3, V4, V5, and V6 referenced to the central terminal. These 6 precordial leads record the activity of the cardiac vector in the horizontal plane from different angles with almost 30° angular resolution. These 12 leads record the activity of the cardiac vector during the contraction of the heart from different planes and from different angles.

Why do we have 12 leads to record a single dipole of cardiac vector? Theoretically, we do not need 12 leads to record the activity of a single vector in 3D space. Three leads seem to be enough. Among the 12 leads, we can select three orthogonal leads. The lead vector of lead I directs horizontally from the right to left, the lead vector for augmented limb lead aVF directs vertically from top to bottom, and the precordial lead V2 directs horizontally from back to front.

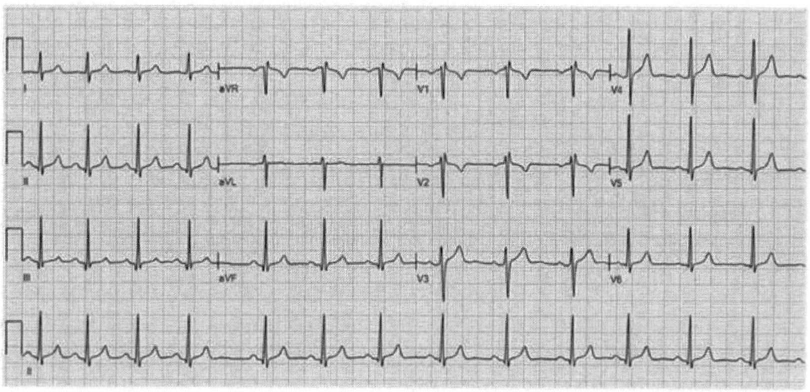

Fig. 7.12 Example of 12-lead ECG recording. The upper three lines are the 12-lead recordings of 2.5 s respectively to show the morphology of each wave deflection. The bottom line is the 10 s recording of the lead II to show the rhythmic pattern of the heart. Each large box in the recording sheet represents 0.2 s horizontally and 0.5 mV vertically

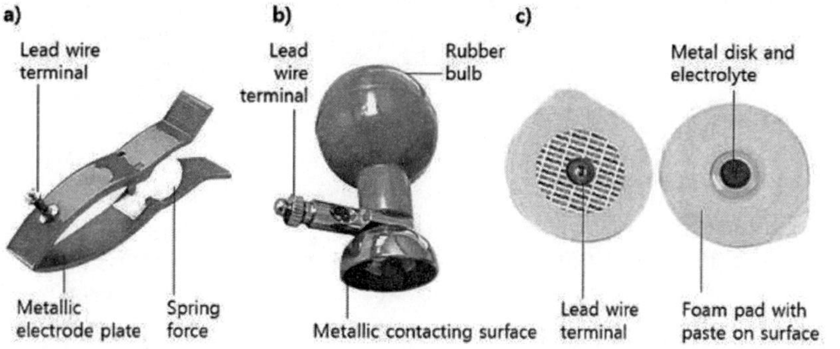

Fig. 7.13 Types of ECG electrodes. (**a**) Metallic plate electrode, (**b**) Suction electrode, (**c**) Ag-AgCl electrode

These three leads a compose orthogonal lead vector set perpendicular to each other. Potentials recorded in these orthogonal leads can perfectly define the cardiac vector in space and all other lead potentials can be calculated from three potentials. Additional 9 leads seem of no use theoretically.

However, practical circumstances are different. First, the cardiac vector is not an ideal dipole and the volume conductors are not homogenous. Regional distortion caused by inhomogeneity makes the recorded potentials different from the computed potentials. Also, potentials recorded on the chest surface close to the heart reflect cardiac activity more sensitively by bypassing a large mass of volume conductors. Additionally, all 12-lead potentials used to be recorded without calculation and without the use of digital circuits in the ECG recording device. Of course, most of the current ECG recording devices are CPU-based and have an automatic interpretation function too. Figure 7.12 shows an example of 12 lead ECG recordings.

3.5 ECG Electrodes

For the recording of an electrocardiogram, several types of electrodes are used (Fig. 7.13). The type of electrode is selected by the type of recording. For short-term recording as in the outpatient clinics, usually reusable metallic electrodes are used. For long-term recording as in the intensive care unit or for the increased stability during the recording, disposable silver-silver chloride electrodes are widely used.

Metal plated electrodes, which are also called dry electrodes, are usually made of an alloy of copper, zinc, and nickel. The electrode metallic part is silver coated to allow a safe and clear signal transmission. These electrodes are designed to record the biopotentials from the skin surface repeatedly and are suitable for short-term ECG recording. These electrodes are used mostly for limb electrodes and attached to the limbs by elastic straps or in clamp-type electrodes. A thin layer of gel or electrolyte must be applied between the metal electrodes and the skin to build a good electrical contact during the ECG recording. Applied electrolytes are solutions containing additional ionic salts such as sodium chloride (NaCl) or potassium chloride (KCl).

Suction electrodes are specially designed metallic electrodes to attach easily to the chest. These electors are composed of a metallic cup and a connected rubber bulb. Electrodes are attached to the skin by creating a vacuum inside the hollow metallic cup by squeezing the rubber bulb. These electrodes are convenient to locate around the chest according to the predefined precordial electrode positions.

Silver – silver chloride (Ag-AgCl) electrodes, which are also called wet electrodes, are the most common ECG electrodes and are used for long-term recording designed for one-time use. A thin electrochemical layer of silver chloride layer is formed on the silver electrode. They are non-polarizable electrodes, which allows the charges to cross the electrode-electrolyte interface, unlike the metallic electrodes. Nonpolarizable biopotential recording generates lower electrode-skin interface impedance and is less sensitive to noises. Electrolyte soaked in the foam paste also provides a good contact to the skin and reduces the motion artifacts which could be generated during the exercise testing.

3.6 Types of ECG Recording

Several types of ECG recordings are used depending on the clinical and health monitoring purposes (Fig. 7.14).

Resting 12 – lead ECG is the standard type of ECG recording examining the basic electrical function of the heart. The examination is performed with the subject lying still and records electrical potentials based on the standard 12 ECG lead system. This ECG examination is used as a routine checkup to screen for heart conditions before any signs or symptoms are recognized.

Exercise ECG examination, also called a stress test, is to record the heart function during the stress of the increased physical loads. During the examination, the subject walks on a treadmill or pedals a stationary bicycle with ECG electrodes attached. Often blood pressure and breathing functions are recorded together. This test is used to detect coronary artery diseases and to determine the safe levels of exercise for patients having a heart attack or heart surgery.

Holter monitoring is the continuous ECG recording for a period of 24 h or longer. This recording is to find any abnormal rhythms that are not detected during the short-time recording in the clinics. Patch-type electrodes are placed at certain locations on the chest. Data are recorded directly in the built-in memory or transmitted to be stored in the mother unit. Stored ECG recordings are further interpreted with the help of automatic screening.

Recently, the ECG recording function is started to be included in the *smart watches* as one of the health monitoring signals. Lead I potential is usually recorded by touching the other hand to the electrode on the surface of the smartwatch while the one electrode at the back of the watch is already touching the wrist of one arm.

Capacitive recording is new technology to record biopotentials without direct touches to the skin surface. Capacitive coupling is formed between the body surface and the recording electrodes even though they are contacted over the wearing of clothes. Potentials of the body surfaces are picked up through the capacitance and the waveform of the recorded potential is a kind of high pass filtered one of the ordinary ECG. When the capacitive electrodes are installed on the back of chairs or on top of beds, ECGs of sitting or sleeping subjects can be recorded without directly attaching the electrodes to the body surface (Lim et al., 2011).

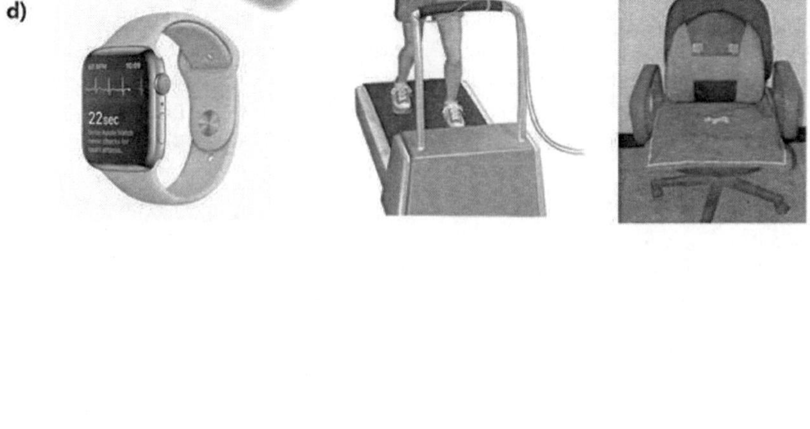

Fig. 7.14 Various types of ECG recording. (**a**) Standard ECG recording device, (**b**) Exercise testing, (**c**) Patch type ECG recording unit, (**d**) ECG in a smartwatch, (**e**) Chair installed with capacitive ECG electrodes

4 Diagnosis with ECG

Standard 12-lead ECG is recorded on the 10 s strip of scaled graph paper. The large box and small box on the graph help to find the duration and amplitude of each interesting wave deflection. Each large box represents 0.20 s horizontally and 0.5 mV vertically. Each of the five small boxes in a large box represents 0.04 s horizontally and 0.1 mV vertically. 12-lead potentials of 2.5 s are recorded in the upper three lines and the bottom line records 10 s of lead II potential to show the rhythmic pattern (Fig. 7.12).

First, the *heart rate* can be easily evaluated by counting the number of beats within 10 s interval of the bottom line. Heart rate is calculated in beats per minute (bpm) by multiplying the number of beats within 10 s by 6. Normal heart rate is between 60 and 100 bpm. If the heart rate is higher than 100 bpm, it is tachycardia, a fast heart. If it is lower than 60 bpm, it is bradycardia, a slow heart.

The second procedure is to check the *heart rhythm* (Fig. 7.15). Any abnormality in the generation or conduction of cardiac action potential can be recognized and differentiated by its recorded rhythmic pattern. Abnormal rhythm can be recorded in a regular pattern or in irregular patterns depending on the causing problem. One type of abnormal pattern is the type caused by abnormal sources other than the SA node. These abnormal sources can occur temporarily or consistently, can be single or multiple sources, and can occur in the atria or in the ventricles. Another type of abnormal pattern comes from the abnormal conduction of cardiac action potentials. Cardiac action potentials are used to be delayed or blocked in abnormal patterns. There are many abnormal heart rhythms by the combination of these problem characteristics. Atrial fibrillation, atrial flutter, ectopic atrial rhythm, ectopic ventricular rhythm, first degree AV block, second degree AV block, third degree AV block, and ventricular tachycardia are representative examples of abnormal heart rhythms.

The *axis* of ECG can give a clue to the physical and pathologic states. By comparing the amplitudes of the recorded lead potentials, the axis of cardiac impulse conduction is evaluated. Hearts with normal axis record lead II potential most positively compared to lead I and lead III potentials. If the lead III potential has the most positive deflection and lead I is negative, then the cardiac axis deviates to the right side and it may be associated with ventricular hypertrophy. If lead I has the most positive deflection and lead II and III are negatively deflected, the cardiac axis deviates to the left direction and it can be the result of the conduction abnormalities.

Most complex part is to interpret the *morphological distortion* of wave deflections that are deviating from the normal shapes. Interpretation of wave distortions is performed by checking the polarity, and absolute or relative levels of lead potentials. Temporal duration and rising or falling slopes of

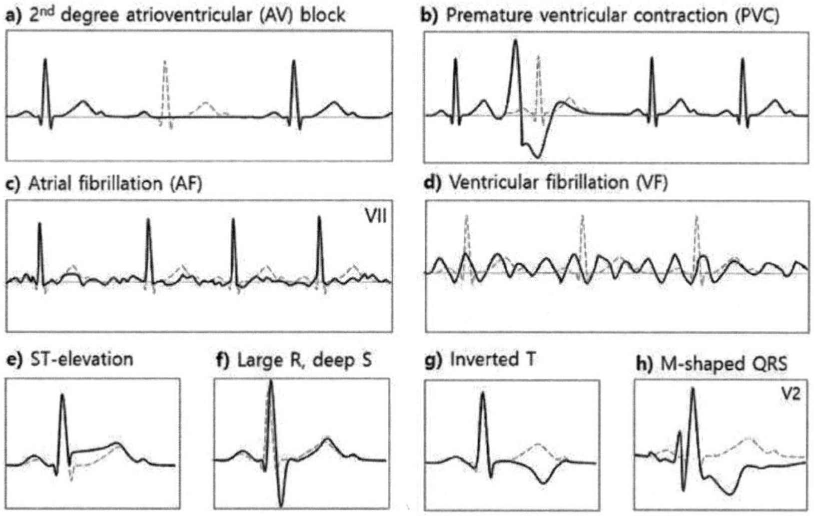

Fig. 7.15 Examples of abnormal ECG waveform. (**a**) second-degree atrioventricular block. Ventricular beats are skipped intermittently. (**b**) Premature ventricular contraction. Ventricles contract earlier than the normal beat and the normal beat is blocked. (**c**) Atrial fibrillation. Atrial muscle fibers contract randomly without synchronization to the normal sinus rhythm. But, the ventricles contract synchronous to the earliest rhythm conducted. (**d**) Ventricular fibrillation. Ventricle muscle fibers contract randomly without synchronization to the normal sinus rhythm. Enough contraction to pump the whole volume of blood is not possible. (**e**) Elevated ST segment above the baseline associated with the myocardial infarction. (**f**) ECG waveform with large R and deep S deflection indicative of the enlargement of the heart. (**g**) Inverted T-wave in association with acute ischemia. (**h**) M-shaped QRS wave pattern in V2 lead associated with right bundle branch block (RBBB)

each wave are also considered. The distortions of ECG waves are caused by many reasons other than the abnormality in the conduction pathway of electrical pulses. One representative example is the elevation of the ST segment where the ST segment is abnormally high above the baseline. It appears associated with a myocardial infarction caused by the occlusion of coronary arteries. The problems that are associated with morphological distortion include enlargement of the heart, congenital heart defects, damage to the heart by coronary occlusion, the poor blood supply to the heart, abnormal position of the heart, heart inflammation, cardiac arrest, and imbalances in the electrolytes that control heart activity. Usually, the ECG examination does not provide the complete diagnostic information of heart problems and additional examinations, like chest X-ray, echocardiography, magnetic resonance imaging, blood test, and cardiac catheterization, are followed to clarify the problems more accurately.

5 Hear Rate Variability

In contrast to our simple thought that hearts are beating periodically, heart rhythms are not periodical in the strict sense (Fig. 7.16). Beating interval of heart rhythm is very varied and its irregularity is rather dominant in healthier people. Healthy heart rhythm used to be described by the chaotic fluctuation mathematically (Goldberger, 1991). The beating rhythm of healthy people is constantly changing and it allows the cardiovascular system to rapidly adapt and respond to sudden external physical and psychological stimuli. While healthy hearts show complexity and irregularity in their rhythms, deviation from the normality can induce loss of irregularity and complexity (Vaillancourt & Newell, 2002) (Fig. 7.17). Since heart rhythm is affected by various interactions with other organs within the body, a detailed analysis of heart rhythm variability provides valuable information on the body's status.

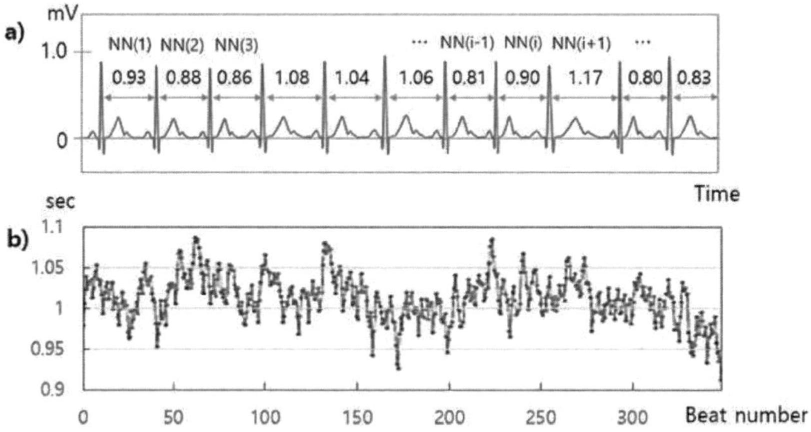

Fig. 7.16 Derivation of interbeat interval signal from raw ECG signal. (**a**) Raw ECG signal and interbeat intervals between consecutive heartbeats, (**b**) Interbeat time series showing the complexity of the heartbeat fluctuation

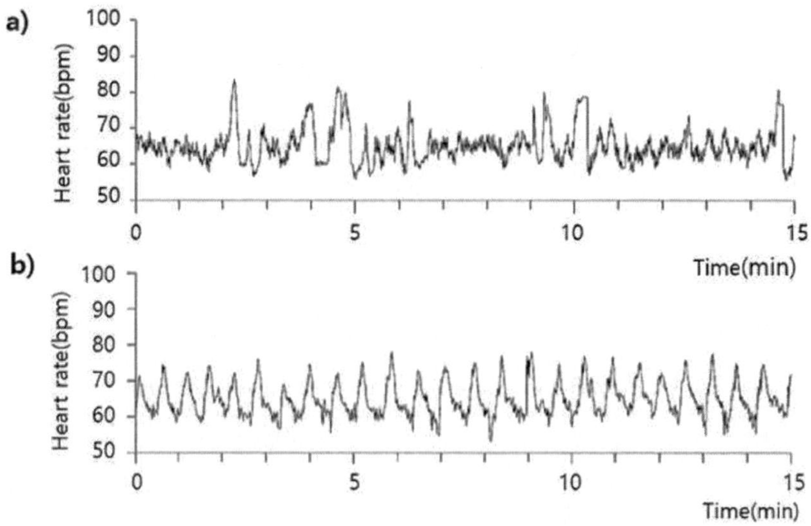

Fig. 7.17 Examples of heart rate variation for a healthy and an abnormal subject having equal level statistical mean and standard deviation. (**a**) Healthy subject with mean = 64.98 beats/min and SD = 4.79beats/min, (**b**) Subject during obstructive sleep apnea with mean = 64.95 beats/min and SD = 4.68beats/min (Goldberger, 2006)

Heart rate variability (HRV) represents the fluctuating characteristics of heart rate or beating interval of the heart within some time duration, typically 5 min. As it is based on measuring the beating interval of the heart, only the dominant R-wave of ECG is used for the evaluation and other ECG wave deflections are not used in the derivation. Other signals from the heart, like photoplethysmogram (PPG) and ballistocardiogram (BCG) are also possible to measure the beating intervals of the heart and quantify HRV. But, sharp peaks of ECG R-wave provide increased temporal resolution compared to the PPG and BBG which vary with only smoother peaks. As ECG R-peaks can be measured easily with simplified ECG recording units like patch-type and watch-type wearable devices, HRV is of great concern and actively used in monitoring illness and evaluating wellness in extended healthcare environments.

5.1 HRV and Autonomic Nervous System

The human autonomic nervous system regulates body processes to react optimally to physical and psychological conditions at each moment. The sympathetic and parasympathetic nervous systems are the two axes controlling the body's processes. Sympathetic rhythm drives body processes so that the organs are to be prepared for the "fight" or "flight" condition and parasympathetic tone drives body processes into the "rest" or "recover" conditions. Sympathetic rhythm increases a heart rate, raises blood pressure, drives blood flow to the muscles, inhibits digestion, reduces appetite, dilates pupils, and does many more controls adaptable to the "fight" or "flight" conditions. Parasympathetic rhythm acts in a reverse way to the sympathetic rhythm. It slows heart rate, decreases blood pressure, dilates blood vessels, promotes energy storage, promotes digestion, increases appetite, and does many more controls adaptable to maintain the body in the "rest" and "restore" conditions. Human body organs are endlessly controlled by these two rhythmic activations and the prepared body reacts in the optimized condition. As abnormal activation of sympathetic or parasympathetic activation drives the body into abnormal status, many physical and psychological disorders are closely related to the unbalanced activation of these rhythms.

Heart rate is one of the most actively controlled variables by the autonomic nervous system. Without any intervention of the autonomic nervous system, the human heart rate is about 60~100 beats/min. But, real heart rate is an endless reflection of autonomic nervous system activation. Sympathetic nerves accelerate the heart rate and parasympathetic nerves try to slow it down. Active involvement of the autonomic nervous system in heart rate makes the monitoring of the heart rate variation provide indirect insight into the activation of the autonomic nervous system. Well-established analyzing parameters on heart rate variability can be used as biological markers representing the sympathetic and parasympathetic activations and their balance. The impairment of the autonomic balance is applicable to assess the cardiovascular risk and predict several neurological disorders.

5.2 HRV Recording Durations

HRV parameters can be derived from the short-term recording of 5 min, from the ultra-short-term recording of fewer than 5 min, or from the long-term recording of 24 h. While they are analyzed with the same HRV parameters, the results of one recording duration cannot substitute the results of the other recording durations. They should be understood with different physiological meanings. HRV parameters from the short-term recording reflect the complex and dynamic involvement of the sympathetic and parasympathetic nervous systems. It includes the regulatory mechanisms that control the heart rate through the baroreflex and respiratory sinus arrhythmia. HRV parameters from the long-term recording of 24 h reflect the effects of circadian rhythms, core body temperature, metabolism, and the sleep cycle. For 24 h recording, the HRV parameter of ULF, representing the power in the ultra-low frequency range lower than 0.003 Hz, is additionally applicable. This recording provides greater predictive power than the results of short-term recording and it is considered the gold standard for clinical HRV assessment (Shaffer & Ginsberg, 2017). HRV parameters being used can be classified into the time domain, frequency domain, and nonlinear parameters (Table 7.1).

5.3 Time-Domain HRV Parameters

Time-domain HRV parameters are just simply calculated from the measured beat interval data during the analyzing period, for example, of 5 min. Many parameters can be calculated from the inter-beat time interval data. Among those parameters, mNN, mHR, SDNN, SDRR, SDANN, RMSSD, NN50, pNN50, HTI, and TINN are the most actively used ones.

mNN is the average of the normative heartbeat interval. Normative heartbeat means that intervals intervened by the abnormal beats like ectopic beats are not included in the calculation of the parameter. If the average is calculated including all abnormal beats, the derived mean is differentiated by *mRR* instead of *mNN*.

Table 7.1 Summary of HRV parameters

Domain	Parameter	Description	Unit	Normal range[a]
Time	mNN	Mean interbeat interval	ms	906 ± 117*2
	mHR	Mean heart rate	Beats/min	66.2 ± 8.7*2
	SDNN	Standard deviation of all NN intervals	ms	83.7 ± 32*2
	SADNN	Standard deviation of the averages of NN intervals in all 5-min segments of the entire recording	ms	33.5 ± 17.7*2
	RMSSD	Square root of the mean of the sum of the squares of differences between adjacent NN intervals	ms	55.9 ± 32.0*2
	NN50	Number of pairs of adjacent NN intervals differing by more than 50 ms	Beats	42.5±27.5*2
	pNN50	NN50 count divided by the total number of consecutive NN interval pairs	%	12.8 ± 8.3*2
	HTI	Area of histogram divided by the height of the histogram	ms	37±15*1
	TINN	Baseline width of the fitted triangle to the histogram of all NN intervals	ms	289*5
Frequency	VLF	Power in VLF range(<0.04 Hz)	ms^2	60,441 ± 43561*2
	LF	Power in LF range(0.04~0.15 Hz)	ms^2	1728 ± 1522*2
	HF	Power in HF range(0.15~0.4 Hz)	ms^2	1508 ± 1876*2
	nLF	LF power in normalized units	nu	0.54 ± 0.04*2
	nHF	HF power in normalized units	nu	0.29 ± 0.03*2
	LF/HF	Ratio of LF to HF	nu	1.5~2.0*1
Nonlinear	SD1	Width of scatter plot in Poincare map	ms	42.1 ± 22.0*2
	SD2	Length of scatter plot in Poincare map	ms	111.1 ± 39. *2
	SD2/SD1	Ratio of SD2 to SD1	nu	2.8 ± 0.71*2
	ApEn	Approximate entropy		0.76 ± 0.07*4 ($m = 2$, $r = 0.2$)
	SampEn	Sampled entropy		1.05 ± 0.10*4 ($m = 2$, $r = 0.2$)
	α1	Short-term correlation in DFA analysis		0.958 ± 0.034*3
	α2	Long-term correlation in DFA analysis		
	D2	Correlation dimension		4.1 ± 0.28*4

[a]Normal ranges are variable in the studies. They are included to provide an overall range of the HRV parameters. *1: (Task Force Report, 1996), *2: (Corrales et al., 2012), *3: (Yeh et al., 2006), *4: (Bolea et al., 2018),*5: (Tarvainen et al., 2017)

$$mNN = \frac{1}{N}\sum_{i=1}^{N} NN(i) \tag{7.12}$$

mHR is the mean heart rate, represented by beats per minute and calculated as the inverse of the mean interbeat interval multiplied by 60.

$$mHR = \frac{60}{mNN} \tag{7.13}$$

SDNN is the standard deviation of normative heartbeat intervals and measured in ms. If the standard deviation is calculated including all abnormal beats, the derived parameter is differentiated by *SDRR* instead of SDNN. SDNN is primarily affected by respiratory sinus arrhythmia (RSA), which refers to the naturally occurring variability in heartbeat intervals coordinated with the respiratory cycle. Like RSA, SDNN reflects the parasympathetic influence. Patients can be classified into three groups based on their SDNN monitored during 24 h, unhealthy group if SDNN is less than 50 ms, compromised health group if SDNN is 50–100 ms, and healthy group if SDNN is above 100 ms. Subjects with SDNN above 100 ms have a 5.3 times lower mortal risk at follow-up than those with SDNN below 50 ms (Kleiger et al., 1987).

$$SDNN = \sqrt{\frac{1}{N-1}\sum_{i=1}^{N}(NN(i) - mNN)^2} \tag{7.14}$$

SDANN is the standard deviation calculated for the 5 min average of normative beat intervals during the explored time, usually 24 h, and also measured in ms as SDNN.

$$SDANN = \sqrt{\frac{1}{M-1}\sum_{i=1}^{M}\left(mNN(i) - \frac{1}{M}\sum_{i=0}^{M} mNN(i)\right)^2} \tag{7.15}$$

NN50 is the number of differences above 50 ms between successive beat intervals during the analyzing duration of 5 min. pNN50 is the percentage of NN50 over the whole number of successive differences. The pNN50 is influenced by parasympathetic nervous system (PNS) activation.

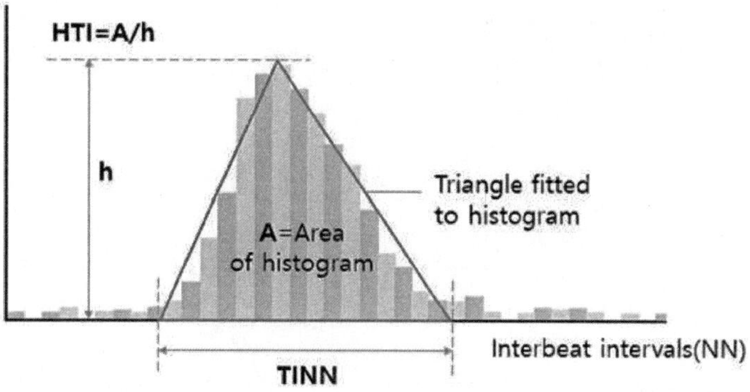

Fig. 7.18 HRV triangular index (HTI) and triangular interpolation of the NN interval histogram (TINN)

$$pNN50 = \frac{number\ of\{|NN(i+1) - NN(i)| > 50ms)\}}{N-1} \tag{7.16}$$

RMSSD is the root mean square of interbeat interval differences. The RMSSD reflects the beat-to-beat variance in heartbeat intervals. RMSSD is influenced by the PNS more strongly than SDNN and is widely used to evaluate the activation of PNS.

$$RMSNN = \sqrt{\frac{1}{N-1}\sum_{i=1}^{N-1}(NN(i+1) - NN(i))^2} \tag{7.17}$$

HTI is the HRV triangular index and is calculated from the histogram of interbeat intervals (Fig. 7.18). HTI is calculated by dividing the total number of interbeat intervals by the height of the histogram, the number of the maximally occurred interbeat interval. TINN is a triangular interpolation of the NN interval histogram and is represented by the baseline of the root mean square fit triangle to the interbeat histogram.

5.4 Frequency-Domain HRV Parameters

Frequency-domain parameters are derived after obtaining the power spectrum by transforming a time series of interbeat intervals. Since the interbeat interval is not a periodically sampled signal, there should be an interpolation procedure before transforming the data which makes the signal of equal sampling distance. After interpolating the instantaneous interbeat interval signals, signals are usually resampled with a 4 Hz sampling rate. Then power spectrum is calculated by a Fourier transform or parametric autoregressive modeling methods. Frequency domain parameters are derived by dividing the power spectrum into three spectral bands (Fig. 7.19).

VLF is the power of a very-low-frequency band below 0.04 Hz. As the period of the maximal frequency of 0.04 Hz is 25 s, VLF power represents the strength of very slowly

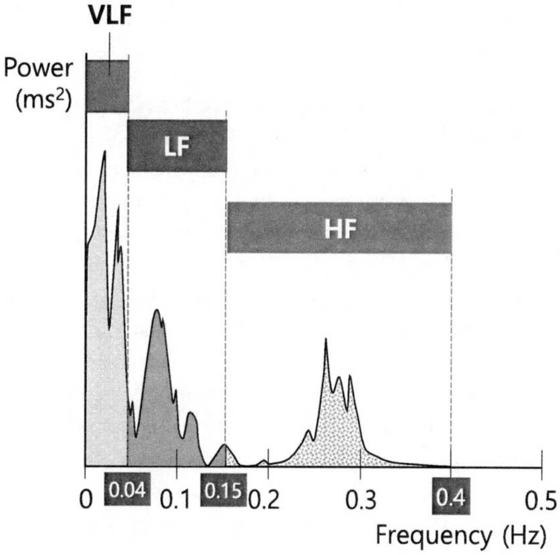

Fig. 7.19 Frequency-domain HRV parameters

fluctuating components with a period longer than 25 s. VLF power is strongly correlated to the mortality of many diseases and is regarded as the fundamental component of health. The lower VLF component has been shown to be associated with arrhythmic death, post-traumatic stress disorder (PTSD), high inflammation, and lower level of testosterone (Shaffer & Ginsberg, 2017).

PNS activation is considered to contribute to VLF power since parasympathetic blockade almost completely abolishes it (Taylor et al., 1998). Unaffected VLF power in tetraplegics, whose sympathetic nervous system (SNS) innervation of the heart and lungs is disrupted, suggests that SNS activation is not contributing VLF power as well (Task force report, 1996). VLF is represented in absolute form with the unit of ms^2.

LF is the power in a low-frequency band of 0.04~0.15 Hz. LF reflects the strength of fluctuating components in heart rate having periods of 6.7~25 s. In resting conditions, LF power is influenced by the baroreflex activation. LF power is correlated to SNS as well as PNS activation. Among the

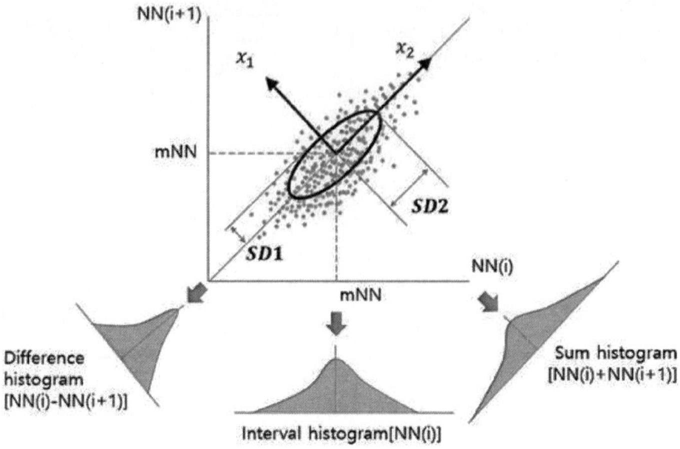

Fig. 7.20 Nonlinear parameters SD1 and SD2 from the Poincare map

HRV parameters, LF is most sensitive to SNS activation. LF is represented in absolute form with the unit of ms^2 and the relative LF is represented by *nLF* in normalized unit.

HF is the spectral power in a high-frequency band of 0.15~0.4 Hz. HF reflects the strength of fast fluctuating components having periods of 2.5~6.7 s. HF power reflects parasympathetic activity clearly and reflects heart rate variation caused by the respiratory phases. Heart rate increases during the inspiratory phase and decreases in the expiratory phase. The cardiovascular center controls the outflow of parasympathetic tone during the respiratory phases. In the inspiratory phase, it inhibits the parasympathetic tone and results increasing in heart rate. In the expiratory phase, it restores parasympathetic tone and decreases the heart rate. If the parasympathetic tone is totally blocked, the HF component is eliminated or greatly reduced in strength (Shaffer et al., 2014). HF power increases during the night and decreases during the day. Lowered HF is recognized as the indication of stress, panic, anxiety, or worry. HF is represented in absolute form with the unit of ms^2 and the relative HF is represented by *nHF* in normalized unit.

LF/HF is the ratio of low-frequency band power to high-frequency band power. This parameter is considered to reflect the sympathovagal balance. It increases as the sympathetic tone is activated and decreases as the parasympathetic tone is activated.

5.5 Nonlinear HRV Parameters

Nonlinear parameters are used to quantify the complexity and unpredictability of the signal which are not clearly quantified in linear parameters. Nonlinear HRV parameters reflect the complexity of the mechanism regulating the heart rate and are more sensitive than time-domain and frequency-domain parameters in some stressors and diseases.

Poincaré plot or return map is the graphical plot to evaluate the nonlinear parameters. It is constructed by plotting every interbeat interval against the previous interval in 2D space by creating a scattered plot. It provides a characteristic visual pattern buried in successive measurements. It is irrespective of the temporal or spectral trend of the heart rate fluctuation. Nonlinear HRV parameters are derived by fitting the scattered pattern into an ellipse (Fig. 7.20). The data points of the scattered plot are gathered around a line of slope 1 passing through the origin. The center point of the scattered data points represents the average of interbeat intervals. By matching the center of an ellipse to the center of scattered points, the long axis of the ellipse is aligned along the line of the slope of 1, and the short axis of the ellipse is positioned passing the center of scattered points crossing the long axis perpendicularly. The length of the long axis is defined as the parameter SD2 and it is calculated as the standard deviation of data points along the long axis. SD1 is defined similarly as the length of the short axis and calculated by the standard deviation of data points along the short axis. SD1 is identical to RMSSD as both are calculated by the same formula from the differences of consecutive interbeat intervals. The ratio of SD2/SD1 measures the unpredictability of the interbeat intervals and it is used to measure the autonomic balance similarly to LF/HF ratio.

$$x_1(i) = \frac{(NN(i) - mNN) - (NN(i+1) - mNN)}{\sqrt{2}} = \frac{NN(i) - NN(i+1)}{\sqrt{2}} \tag{7.18}$$

$$x_2(i) = \frac{(NN(i) - mNN) + (NN(i+1) - mNN)}{\sqrt{2}} = \frac{NN(i) + NN(i+1) - 2mNN}{\sqrt{2}} \tag{7.19}$$

$$\mathrm{SD1} = \sqrt{variance[x_1]} \tag{7.20}$$

$$\mathrm{SD2} = \sqrt{variance[x_2]} \tag{7.21}$$

Approximate entropy (ApEn) is a nonlinear parameter quantifying the unpredictability of intricately fluctuating time series. It can be applied without any assumption on the underlying dynamics and is applicable even in some noisy circumstances. Large ApEn values indicate a high unpredictability in predicting the successive interbeat interval fluctuations and small ApEn values mean that heartbeat intervals are varying in a relatively regular and predictable manner. ApEn is calculated by embedding data into m-dimensional space and by counting the portion of points that are closer than distance r. A point in m-dimensional space is represented by the *m* dimensional vector constructed by the consecutive *m* data points.

$$x(i) = (NN(i), NN(i+1), \ldots, NN(i+m-1)), \quad i = 1, 2, \ldots, N-m+1 \tag{7.22}$$

Distance between the points is defined by

$$d(x(i), x(j)) = \max\{|NN(i+k) - NN(j+k)| k = 0, 1, 2, \ldots m-1)\} \tag{7.23}$$

Next, for each $x(i)$ the relative number of points closer than r is calculated.

$$C_r^m(i) = \frac{number\ of\{x(j)|d(x(i), x(j)) < r)\}}{N-m+1} \forall j \tag{7.24}$$

Then, take the natural logarithm of each $C_r^m(i)$ and average over all points.

$$\Phi_r^m = \frac{1}{N-m+1} \sum_{i=1}^{N-m+1} \ln\left(C_r^m(i)\right) \tag{7.25}$$

ApEn is finally calculated as the difference of Φ_r^m between m dimensional space and m + 1 dimensional space.

$$ApEn(m, r, N) = \Phi_r^m - \Phi_r^{m+1} \tag{7.26}$$

ApEn is dependent on three parameters, the dimensional space m, the tolerance distance r, and the number of data points N.

Sample entropy (SampEn) is similar to ApEn, but less biased by not counting the point itself in the calculation of relative numbers closer than distance r. It is a more reliable measure of signal regularity and complexity.

$$C_r^m(i) = \frac{number\ of\{x(j)|d(x(i), x(j)) < r)\}}{N-m+1} \forall j \neq i \tag{7.27}$$

$$C_r^m = \frac{1}{N-m+1} \sum_{i=1}^{N-m+1} C_r^m(i) \tag{7.28}$$

$$SampEn(m, r, N) = \ln\left(\frac{C_r^m}{C_r^{m+1}}\right) \tag{7.29}$$

Nonlinear parameters $\alpha 1$ and $\alpha 2$ are calculated from the detrended fluctuation analysis. *Detrended functional analysis* (DFA) quantifies the correlations present in the successive interbeat intervals over varying time scales. Parameters α1 and α2 are estimated by the increasing slopes of detrended fluctuations upon the increase of calculating segment length (Fig. 7.21). The followings are the steps to derive parameters α1 and α2.

1. Detrended fluctuation is calculated on the integrated data of the cumulative sum after subtracting the average of the total mean.

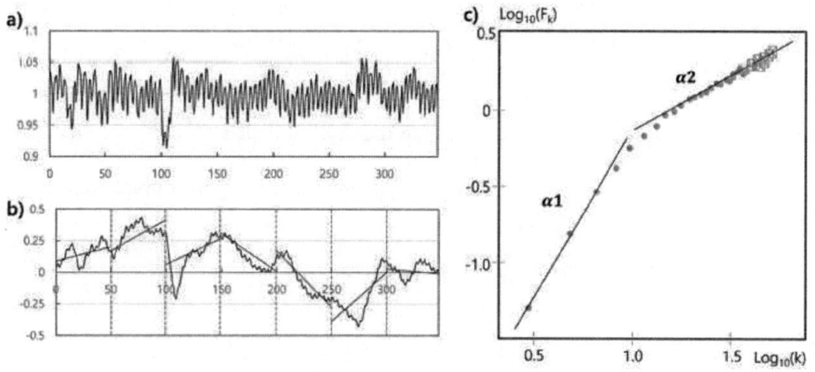

Fig. 7.21 Calculation of α1 and α2 in detrended fluctuation analysis. (**a**) Raw interbeat interval signal recorded during the sleep, (**b**) Integrated interbeat interval signal and locally regressed lines for computation of detrended fluctuation for k = 50. (**c**) α1 and α2 by the slopes in linear parts of the double logarithmic plot of F_k over segment length k

$$x(i) = \sum_{l=1}^{i} (NN(l) - mNN) \tag{7.30}$$

2. Data samples are divided into the segments of length k, which varies from 4 to 64.
3. For each k-length segment, a linear regression line is estimated by minimizing the squared error, and data is detrended by subtracting the linearly trended line. And, the root-mean-square fluctuation of this integrated and locally detrended time series is calculated by

$$F_k(n) = \sqrt{\frac{1}{k} \sum_{i=1}^{k} (x_{nk}(i) - y_{nk}(i))^2} \tag{7.31}$$

$$F_k = \frac{k}{N} \sum_{n=1}^{N/k} F_k(n)$$

where $x_{nk}(i)$ is integrated data of n-th segment of length k, $y_{nk}(i)$ is the local regression line for n-th segment, $F_k(n)$ is detrended error for the n-th segment of length k and F_k is the averaged detrended fluctuation error with segment length k.

4. Computation of F_k is repeated over different segment lengths and produces series of data F_k as a function of k. F_k values are plotted against segment length k on the double log map. A linear relationship on a double log plot can be characterized by scaling exponents.
5. Slopes of the plot are characterized by two parts. Short-term fluctuations are characterized by the slope α1 for the data obtained with smaller data segments, for example, k = 4~12. Correspondingly, long-term fluctuations are characterized by the slope α2 for the data obtained with longer data segments, for example, k = 13~64. Characteristics of the signal can be interpreted with the estimated alpha values (Table 7.2).

Table 7.2 Interpretation of alpha values

Alpha values	Characteristics
α = 1.5	Brownian noise, random walk
1.0<α < 1.5	Different kinds of noises
α = 1.0	1/f noise
0.5<α < 1.5	Large values are likely to be followed by large value and vice versa
α = 0.5	White noise
α < 0.5	Large value is likely to be followed by small value and vice versa

Correlation dimension (CD, D_2) is another nonlinear parameter measuring the complexity or strangeness of the time series. It gives information on the minimum number of variables required to model the dynamics underlying the heart rate variation. As more variables are required to depict the more complex dynamics, the correlation dimension represents the complexity of the time series. Dimension is interpreted as the power of volume increasing with the scale increase. For example, a 2D object increases its volume (area in 2D) by k^2 times and a 3D object increases its volume by k^3 times as the scale is increased by k times. If we can represent the time series data in sufficiently high dimensional space, we can find the dimensional characteristics of the signal by finding the increasing power as the measuring scale is increased. To calculate the correlation dimension, data are needed to be represented in sufficiently high m dimensional space. This is done by constructing the m dimensional vectors by consecutive m data points.

$$x(i) = (NN(i), NN(i+1), \ldots, NN(i+m-1)), \quad i = 1, 2, \ldots, N-m+1 \tag{7.32}$$

Then $x(i)$ represents a point in the m-dimensional space and its position is described by the consecutive m heartbeat intervals. Then, the number of points that are closer than the distance scale r are counted.

$$C_r^m(i) = \frac{number\ of \{x(j) | d(x(i), x(j)) < r\}}{N-m+1} \quad \forall j \tag{7.33}$$

$$d(x(i), x(j)) = \sqrt{\sum_{k=1}^{m} (x_k(i) - x_k(j))^2} \tag{7.34}$$

$$C_r^m = \frac{1}{N-m+1} \sum_{i=1}^{N-m+1} C_r^m(i) \tag{7.35}$$

where $C_r^m(i)$ is the portion of points closer than distance r, $d(x(i), x(j))$ is the distance between i-th and j-th point in the m-dimensional space, and C_r^m is the average of $C_r^m(i)$ over all points in the space, which is called correlation integral.

The correlation dimension is defined as the limit value by approaching the scale r to zero and the number of data point N to infinity. In practice, the correlation dimension is calculated by finding the consistent linear portion in the plotted graph of correlation integral C_r^m against the scale r (Fig. 7.22).

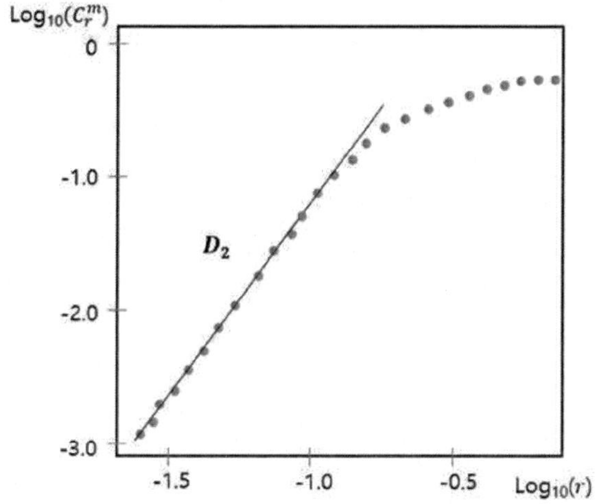

Fig. 7.22 Correlation integral plot for the calculation of correlation dimension

$$D_2^m = \lim_{r \to 0} \lim_{N \to \infty} \frac{log_{10}(C_r^m)}{log_{10}(r)} \tag{7.36}$$

5.6 HRV Parameters in Applications

HRV analysis has become a popular method for the studies of physiologic mechanisms in which the autonomic nervous system appears to play a primary role. Its application is extended from clinical applications to sports medicine, health monitoring, and wellness care. Table 7.3 shows the summarized results of studies showing how HRV parameters are changing to the conditions. HRV parameters are used to show the differences between the selected groups when their autonomic influences are different. As there are many factors related to the autonomic balance, the results of HRV analysis should be understood carefully assuming the possibility that any other factors might have affected the results. The popularization of HRV analysis is based on the advent of portable, wearable, and patch-type ECG recording units which make heart rhythm monitoring easier even in ordinary daily life.

Table 7.3 Example of HRV variation depending on the conditions

Conditions	HRV parameter changes	
As age increases	↓SDNN, ↓RMSSD, ↓pNN50,↓LF, ↓HF, ↑LF/HF, ↓SD1, ↓SD2	*1
As heart rate increases	↓SDNN, ↓RMSSD, ↓pNN50,↓LF, ↓HF, ↓SD1, ↓SD2	*1
As blood pressure increases	↓HF,↑nLF, ↓nHF,↑LF/HF, ↓SD2	*1
As sleep efficiency increases	↑RMSSD, ↑pNN50,	*2
With high cognitive performance	↑LF, ↑HF, ↓LF/HF, ↑SDNN, ↑RMSSD,	*3
As memory indices increase	↑mNN,↑LF, ↑HF, ↓LF/HF	*4
With mental work stress	↓RMSSD, ↓TINN, ↓SD1	*5
In trained athletes	↑mNN,↑SDNN, ↑RMSSD, ↑pNN50	*6
In fatigue	↓mNN, ↓RMSSD, ↓HF, ↑LF/HF,	*7
In obese subjects	↓SDNN, ↓RMSSD, ↓pNN50, ↓HF, ↑LF/HF, ↓SD1	*1
In sleep apnea subjects	↑SDNN, ↑HTI, ↑VLF, ↑LF, ↑LF/HF	*8
Faller in hypertension subjects	↓SDNN, ↓RMSSD, ↓pNN50,↓LF, ↓HF, ↑LF/HF, ↓SD1, ↓SD2, ↑ApEn,↑SampEn, ↑α1, ↑α2, ↓D2	*9
With myocardial infarction	↓SDNN, ↑VLF,	*10
With diabetic neuropathy	↓LF, ↓HF,	*10
In diabetes patient	↓mNN, ↓SDNN,↓RMSSD, ↓LF, ↓HF, ↑SampEn,	*11
With major depressive disorder	↓HF	*12
With schizophrenia	↓HF, ↓SDNN, ↓RMSSD,	*12
With posttraumatic stress disorder	↓HF	*12
As inflammatory markers increase	↓SDNN, ↓VLF, ↓LF, ↓HF	*13
With air pollutant (PM25)	↓SDNN, ↓RMSSD, ↓HF	*14

*1: (Yadav et al., 2017), *2: (Gouin et al., 2013), *3: (Hilgarter et al., 2021), *4: (Morandi et al., 2019), *5: (Orsila et al., 2008), *6: (Aubert et al., 2003), *7: (Leti & Bricout, 2013), *8: (Kim et al., 2015), *9: (Castaldo et al., 2017), *10: (Task Force of the European Society of Cardiology, 1996), *11: (Javorka et al., 2008), *12: (Jung et al., 2019), *13: (Haensel et al., 2008), *14: (Guak et al., 2015)

References

AlGhatrif, M., & Lindsay, J. (2012). A brief review: History to understand fundamentals of electrocardiography. *Journal of Community Hospital Internal Medicine Perspectives, 2*(1). https://doi.org/10.3402/jchimp.v2i1.14383

Aubert, A. E., Seps, B., & Beckers, F. (2003). Heart rate variability in athletes. *Sports Medicine, 33*, 889–919. https://doi.org/10.2165/00007256-200333120-00003

Barold, S. S. (2003). Willem Einthoven and the birth of clinical electrocardiography a hundred years ago. *Cardiac Electrophysiology Review, 7*, 99–104.

Bolea, J., Bailón, R., & Pueyo, E. (2018). On the standardization of approximate entropy: Multidimensional approximate entropy index evaluated on short-term hrv time series. *Complexity, 2018*, 4953273. https://doi.org/10.1155/2018/4953273

Castaldo, R., Melillo, P., Izzo, R., De Luca, N., & Pecchia, L. (2017). Fall prediction in hypertensive patients via short-term hrv analysis. *IEEE Journal of Biomedical and Health Informatics, 21*(2), 399–406. https://doi.org/10.1109/JBHI.2016.2543960

Corrales, M. M., Torres, B. C., Esquive, A. G., Salazar, M. A. G., & Orellana, J. N. (2012). Normal values of heart rate variability at rest in a young, healthy and active Mexican population. *Health, 4*(7), 377–385.

Goldberger, A. L. (1991). Is the normal heartbeat chaotic or homeostatic? *News in Physiological Sciences, 6*, 87–91.

Goldberger, A. (2006). Giles F. Filley Lecture. Complex systems. *Proceedings of the American Thoracic Society, 3*, 467–471. https://doi.org/10.1513/pats.200603-028MS

Gouin, J., Wenzel, K., Deschene, S., & Dang, V. T. (2013). Heart rate variability predicts sleep efficiency. *Sleep Medicine, 14*, e142. https://doi.org/10.1016/j.sleep.2013.11.321

Guak, S., Lim, C., Lee, K., & Park, J. Y. (2015). Review of association between air pollution and heart rate variability. *Journal of Environmental Health Science, 41*(4), 223–230.

Haensel, A., Mills, P. J., Nelesen, R. A., Ziegler, M. G., & Dimsdale, J. E. (2008). The relationship between heart rate variability and inflammatory markers in cardiovascular diseases. *Psychoneuroendocrinology, 33*(10), 1305–1312. https://doi.org/10.1016/j.psyneuen.2008.08.007

Hilgarter, K., Schmid-Zalaudek, K., Csanády-Leitner, R., Mörtl, M., Rössler, A., & Lackner, H. K. (2021). Phasic heart rate variability and the association with cognitive performance: A cross-sectional study in a healthy population setting. *PLoS One, 16*, e0246968. https://doi.org/10.1371/journal.pone.0246968

Javorka, M., Trunkvalterova, Z., Tonhajzerova, I., Javorkova, J., Javorka, K., & Baumert, M. (2008). Short-term heart rate complexity is reduced in patients with type 1 diabetes mellitus. *Clinical Neurophysiology, 19*(5), 1071–1081. https://doi.org/10.1016/j.clinph.2007.12.017

Jung, W., Jang, K., & Lee, S. H. (2019). Heart and brain interaction of psychiatric illness: A review focused on heart rate variability, cognitive function, and quantitative electroencephalography. *Clinical Psychopharmacology and Neuroscience, 17*(4), 459–474. https://doi.org/10.9758/cpn.2019.17.4.459

Kim, Y. S., Kim, S. Y., Park, D. Y., Wu, H. W., Hwang, G. S., & Kim, H. J. (2015). Clinical implication of heart rate variability in obstructive sleep apnea syndrome patients. *The Journal of Craniofacial Surgery, 26*(5), 1592–1595.

Kleiger, R. E., Miller, J. P., Bigger, J. T., Jr., & Moss, A. J. (1987). Decreased heart rate variability and its association with increased mortality after acute myocardial infarction. *The American Journal of Cardiology, 59*, 256–262. https://doi.org/10.1016/0002-9149(87)90795-8

Leti, T., & Bricout, V. A. (2013). Interest of analyses of heart rate variability in the prevention of fatigue states in senior runners. *Autonomic Neuroscience, 173*(1), 14–21. https://doi.org/10.1016/j.autneu.2012.10.007

Lim, Y. G., Hong, G. H., et al. (2011). Monitoring physiological signals using nonintrusive sensors installed in daily life equipment. *Biomedical Engineering Letters, 1*(1), 11–20.

Maddio, S. (2020). The electrostatic foundation of the electrocardiogram. *The Radio Science Bulletin, 373*, 78–82.

Morandi, G. N., Lin, S. H., Lin, C. W., et al. (2019). Heart rate variability is associated with memory in females. *Applied Psychophysiology and Biofeedback, 44*, 117–122. https://doi.org/10.1007/s10484-018-9425-1

Orsila, R., Virtanen, M., Luukkaala, T., Tarvainen, M., Karjalainen, P., Viik, J., et al. (2008). Perceived mental stress and reactions in heart rate variability – A pilot study among employees of an electronics company. *International Journal of Occupational Safety and Ergonomics, 14*, 275–283.

Shaffer, F., & Ginsberg, J. P. (2017). An overview of heart rate variability metrics and norms. *Frontiers in Public Health, 28(5)*, 258. https://doi.org/10.3389/fpubh.2017.00258

Shaffer, F., McCraty, R., & Zerr, C. L. (2014). A healthy heart is not a metronome: An integrative review of the heart's anatomy and heart rate variability. *Frontiers in Psychology, 5*, 1040. https://doi.org/10.3389/fpsyg.2014.01040

Tarvainen, M. P., Lipponen, J., Niskanen, J. P., & Ranta-Aho, P. (2017). *Kubios HRV Version 3 – User's Guide*. University of Eastern Finland.

Task Force of the European Society of Cardiology the North American Society of Pacing Electrophysiology. (1996). Standards of measurement, physiological interpretation, and clinical use. *Circulation, 93*, 1043–1065. https://doi.org/10.1161/01.CIR.93.5.1043

Task Force Report. (1996). Heart rate variability: Standards of measurement, physiological interpretation, and clinical use. *Circulation, 93*, 1043–1065. https://doi.org/10.1161/01.CIR.93.5.1043

Taylor, J. A., Carr, D. L., Myers, C. W., & Eckberg, D. L. (1998). Mechanisms underlying very-low-frequency RR-interval oscillations in humans. *Circulation, 98*, 547–555. https://doi.org/10.1161/01.CIR.98.6.547

Vaillancourt, D. E., & Newell, K. M. (2002). Changing complexity in human behavior and physiology through aging and disease. *Neurobiology of Aging, 23*, 1–11. https://doi.org/10.1016/S0197-4580(01)00247-0

Wilson, F. N., Kossmann, C. E., Burch, G. E., Goldberger, E., Graybiel, A., Hecht, H. H., et al. (1954). Recommendations for standardization of electrocardiographic and vectorcardiographic leads. *Circulation, 10*(4), 564–573.

Yadav, R. L., Yadav, P. K., Yadav, L. K., Agrawal, K., Sah, S. K., & Islam, M. N. (2017). Association between obesity and heart rate variability indices: An intuition toward cardiac autonomic alteration – A risk of CVD. *Diabetes, Metabolic Syndrome and Obesity, 10*, 57–64. https://doi.org/10.2147/DMSO.S123935

Yeh, R. G., Shieh, J. S., Han, Y. Y., Wang, Y. J., & Tseng, S. C. (2006). Detrended fluctuation analyses of short-term heart rate variability in surgical intensive care units. *Biomedical Engineering: Applications, 18*(2), 68–72.

8 Electrical Signal from the Brain

The brain is the organ where most of the human neurons are concentrated and all the electrical processing of the neurons occurs. There are about 86 billion neurons in the brain. They all act electrically based on the neural membrane potentials and the transmission of action potentials. When the local current is generated during the activation of neurons, the electrical and accompanying magnetic field is also produced around. Even though the measurement of each neural activity is not possible, generated local current and electromagnetic field can be recorded as the collective activities of the brain tissues.

Neural activities of the brain are recorded in several types. Electroencephalogram (EEG), commonly described as brainwaves, is the noninvasive record of brain electrical activities with electrodes attached to the scalp surface (Fig. 8.1). It represents the macroscopic activity of the cerebral cortex just underneath the electrode. Clinically, EEG is used to evaluate any functional abnormality of the brain activity before the structural changes appear. EEG is most actively used in the diagnosis of epilepsy and to evaluate sleep disorders. Magnetoencephalogram (MEG) is the record of magnetic fields accompanying the brain's electrical activities. MEG is recorded with a superconducting quantum interference device (SQUID) which senses the minute magnetic field changes within the brain very sensitively. Electrocorticogram (ECoG) is the record of brain electrical activity right on the surface of the cerebral cortex and local field potential (LFP) is recorded with the electrodes inserted into the brain cortical tissues.

1 History of Brain Electrical Signals

1.1 Historical Recording of EEG

1.1.1 First Record of Brain Electrical Current

In 1875, about 90 years after the first discovery of biological electricity by Galvani in 1781, *Richard Caton* (1842–1926), an English physician in Liverpool, reported that he had successfully observed the electrical impulses from the surfaces of living brains in the rabbit and the monkey. Before his discovery of functional electrical current in the brain, there were experiments on cerebral cortical localization, which observes the corresponding focal movement of body parts to the faradic stimulations on the brain cortex. As the body parts are working by the applied electric current, Caton deduced that there might be intrinsic electric currents in the resting state to make the body parts move. Caton placed electrodes on the exposed surface of the cerebral cortex or on the grey matter and measured the currents by the optical magnification of the meniscus movement in his galvanometer. He recorded the corresponding variations of currents upon the induced movements of animals' heads and the light exposure to the eyes. Recorded currents changed as the animals fall asleep and fluctuation of baseline unrelated to cardiac or respiratory rhythms was also observed. This current easily changed its strength depending on the oxygen level and disappeared with the animal's death. He is also recognized as the first person discovered cerebral evoked potential by sensory stimulation of light. Electrical currents during the brain activities in animals were confirmed and related studies were further advanced by the following physiologists, Adolf Beck, Vladimir Pravdich-Neminsky, and Napoleon Cybulski.

1.1.2 First Record of Human EEG

In 1924, Hans Berger (1873–1941), a German neuropsychiatrist at the University of Jena, first recorded human brain activity (Fig. 8.2a). And he has introduced the term electroencephalogram (EEG). The discovery of EEG was a historical breakthrough providing a new neurologic and psychiatric diagnostic tool at the time, especially for patients with seizures.

Different from other pioneers in electrophysiology, Berger was not a physiologist. He was a psychiatrist and did not have a good background in electricity. One day at the age of 18, Hans Berger experienced a near-fatal accident. At that

K. S. Park, *Humans and Electricity*, https://doi.org/10.1007/978-3-031-20784-6_8

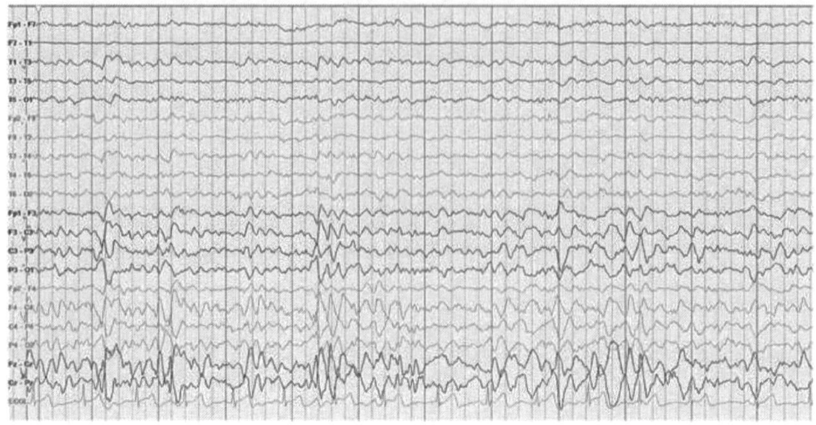

Fig. 8.1 Example of 20-channel bipolar EEG record with a longitudinal montage

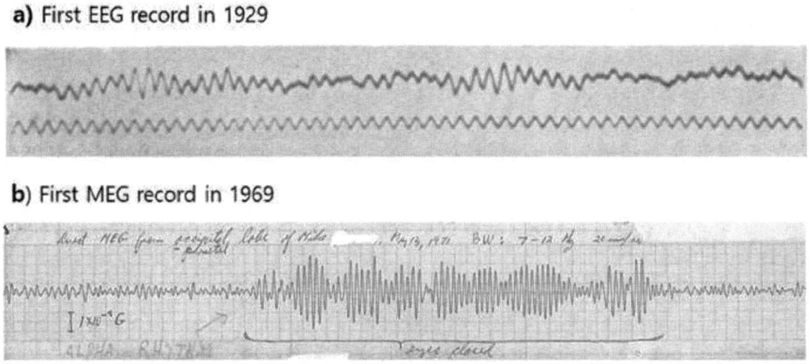

Fig. 8.2 First record of EEG and MEG. (**a**) EEG recorded by Hans Berger in 1929. The upper line shows the fluctuation of alpha wave upon the closing and opening of the eyes and the bottom line is the time marker of 10 Hz. (**b**) MEG recorded by David Cohen in 1969 in a good shield room with a SQUID sensor. Alpha waves are clearly seen in the middle part of the record when the eyes are closed

time, his sister, hundreds of miles away, told their father that Hans was in grave danger and the father sent him a telegram. Since then, Berger became a strong believer in psychic phenomena or telepathy, and he thought that the psychic energy of the human would be released by metabolic energy in the form of localized heat or electrical currents. This experience and belief have driven him to study brain electrical activity in his following career.

Berger made the first EEG recording on the scalp of a 17-year-old boy with skull defects during a neurosurgical operation. He recorded the responses with a string galvanometer having a sensitivity of 1 mV/cm for the first time. With the increased sensitivity of the double coil galvanometer having a sensitivity of 130 uV/cm, he began to use low impedance surface electrodes, experimented with many different surfaces, and improved the results consistently from the patients having skull defects to subjects without any skull defects, especially in bald males. He published his works on human EEG in 1929. In this report, he also first described the waves appearing with the eye closing as the "alpha waves" and low-amplitude high-frequency waves appearing with the eye opening as "beta waves". In 1930, Berger recorded the first human electrocorticogram (ECoG) from the adjacent white matter to brain tumor in a patient undergoing neurosurgery. He also noted the differences in EEG among the groups of infants, young children, and adults. Berger described paroxysmal sharp waves appearing between the epileptic seizures and recognized them as the precursor to a seizure (Stone & Hughes, 2013). However, his discoveries did not receive the right credit from contemporary physiologists. Some of them considered Berger's discoveries as being misguided and he is a non-physiologist with limited expertise in electricity and technical support. Others believed the blocking of a prominent alpha rhythm with the eyes opening is contrary to the existing theory of brain response to light stimulation.

Berger's achievements were fully credited by Edgar Adrian (1889–1977), an English neurophysiologist at the University of Cambridge, who is the 1932 Nobel Prize laureate in Medicine or Physiology with fellow Charles Sherrington for their discoveries regarding the functions of neurons. Using the independently designed differential input amplifiers with increased sensitivity, Adrian was able to record the EEG and confirmed Berger's observations, including the alpha rhythm and the blocking effect appearing with the eyes opening. In 1934, Adrian demonstrated their EEG findings at a meeting of the Physiological Society in Cambridge, England, and gave full credit to Hans Berger for this discovery. Adrian and the other two nominators nominated Hans Berger for the 1940 Nobel Prize laureate in Medicine or Physiology, but unfortunately, no Nobel Prize was awarded this year due to World War II. Hans Berger finished his life the next year in 1941 by suicide.

Since the confirmation of Adrian on Berger's work, EEG gained widespread recognition by eminent researchers in the field leading to its practical use in diagnosis. Remarkable progress has followed in using EEG as a clinical tool by many investigators in Europe and in America. EEG is recognized as an efficient tool in the diagnosis of epilepsy and altered states of consciousness. In 1946, EEG investigators gathered and formed the American Clinical Neurophysiology Society. Currently, EEG remains an indispensable method for the study of brain functional states from a clinical point of view complementing the other neuroimaging diagnostic tools each other.

1.2 Historical Recording of MEG

1.2.1 First Record of MEG

Magnetoencephalogram (MEG) is the record of the magnetic field generated by the electrical activity of the neurons in the brain. The magnetic field generated from the brain is extremely low in intensity. It is about one-billionth of the static earth magnetic field and very difficult to measure and record. In 1969, David Cohen (1930–), an American physicist at MIT, first recorded the MEG using the newly developed superconducting quantum interference device (SQUID) at the time (Fig. 8.2b).

Before Cohen has an interest in the biological magnetic field, he was an accelerator physicist specializing the strong magnetic fields. Then, he began to think about the measurement of the minimal level of magnetic fields, and that the human body must be a source of the weakest magnetic fields. Before MEG was recorded, magnetocardiogram (MCG) has recorded earlier in 1963 with a coil magnetometer. As MCG is related to the electrical activity of the heart, he thought that MEG related to the electrical activities of the brain can be recorded. The issue was its much smaller intensity than MCG. To increase the signal-to-noise ratio of the measurement, he firstly built a magnetically shielded room to reduce the effect of a much larger external noisy magnetic field. Within the shielded room and using coiled sensors of increased sensitivity, he could obtain a much clearer MCG signal. He also measured much weaker MEG, but the recorded MEG was not clear as EEG. Even though the low-quality records were partly caused by the imperfect shielding of the background field, a clear recording of MEG did not seem possible with the intrinsic noise of the coil magnetometer.

In 1969, James Zimmerman, at Ford Motor Company, invented an extremely sensitive magnetic detector with almost no intrinsic noise, called a SQUID. With this SQUID sensor and the rebuilt good shield room in MIT, he successfully recorded MEG as well as MCG which were as clear as conventional EEG and ECG. Since his first MEG record, interest in MEG has aroused rapidly in the medical and research community. The current MEG system engages about 300 SQUID sensors within a helmet-type unit covering most of the head and records human brain activity in many clinical and research examinations.

1.2.2 MEG Using Optically Pumped Magnetometer (OPM)

In 2018, Matthew Brooke, an English physicist at the University of Nottingham, developed a new technology detecting magnetic fields for MEG with similar sensitivity as SQUID without using cryogens. Even though the SQUID sensor provides high sensitivity, the MEG system should maintain the superconductivity by cooling the sensing part to -269 °C. A thermal insulating gap to prevent injury to the patient's head increases the distance between the sensing part and the head surface and decreases the measuring sensitivity. One fixed heavy helmet unit is not movable and does not suitably fit different sizes of the head. Head movement during the recording decreases the quality of recorded signals greatly. And SQUID-based MEGs are expensive to purchase and run (Coleman & Brookes, 2021).

A newly developed optically pumped magnetometer (OPM) is based on quantum technology. When the polarized optical laser beam is pumped into the glass cell enclosed the vapor of rubidium-87 atoms, spin-polarized atomic vapor, which is highly sensitive to external magnetic fields, attenuates the intensity of light passing through the cell by absorption (Boto et al., 2018). This smaller sensor can build the sensor array allowing the movement of the head during the measurement. The flexibility of OPM placement makes the sensing array adapt to any head shape and size. Especially, the lack of complex cryogenics is expected to provide OPM-based MEG at a much cheaper price and to evolve into a more practical and clinically acceptable diagnostic tool.

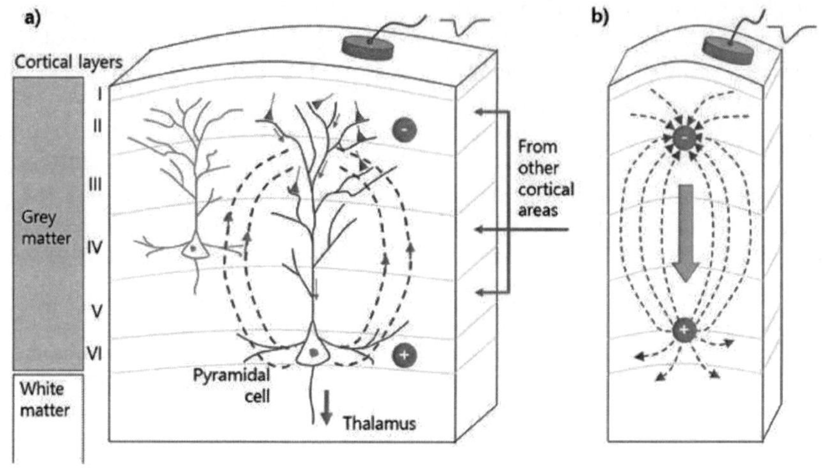

Fig. 8.3 Generation of brain electrical potential. (**a**) Electrical potential recorded in the scalp surface is generated by the synchronous activation of perpendicular pyramidal cells. (**b**) Equivalent electrical dipole model for the recorded brain electrical potential

2 Generation of Brain Electrical Signals

2.1 Electric Current Generated in the Cerebral Cortex

Brain electrical potentials are originating from the cerebral cortex, the outmost neural tissue layer of the cerebrum (Fig. 8.3). The cerebral cortex is distinguished by grey matter and white matter. The grey matter contains numerous cell bodies and relatively few myelinated axons, and the white matter contains relatively few cell bodies and is composed of mostly myelinated axons, which develops the color difference of the region to whiteness. Grey matter is organized into 6 different layers based on the composing cell types. Pyramidal cells and stellate cells are the most representative cells in the cerebral cortex. Among the cells in the cerebral cortex, pyramidal cells do the main role in generating the potential recorded on the brain surface. Pyramidal cells are positioned perpendicular to the brain surface and have long dendrites spanning multiple layers.

When a pyramidal cell receives inputs from the neuronal cells in other cortical regions via synapses, excitatory postsynaptic potentials (EPSPs) are developed in the apical part of the pyramidal cell. It depolarizes the membrane and increases the membrane potential from the resting state. The inner potential of this part becomes positive compared to the soma part of the pyramidal cell, and internally current flows from the apical dendrites to the soma area. When the internal current is developed from the apical dendrites to the soma area, current in the extracellular fluid is developed in the reverse direction, from the soma area to the apical area, to close the local current loop. It is equivalent to the development of an electrical dipole, which directs perpendicular to the surface with a negative charge closer to the surface. A surface electrode just above the cortical area detects negative deflection in this case.

2.2 Spatial Summation of EPSP

Even though each pyramidal cell produces an electrical current, the current from a single neuron is too small to be detected in the electrode on the surface. Thousands of simultaneous activations of pyramidal neurons are necessary to generate a detectable potential from the electrodes on the surface. Electrical potential recorded on the surface of the cortex can be considered the result of fluctuating electric fields generated by the large populations of cortical pyramidal cells. Detectable extracellular current flows are built by the synchronous activations of many cells and their spatial summation at each time. The potential producing the detectable electric current is mostly the excitatory postsynaptic potential (EPSP) generated in the postsynaptic pyramidal neurons. A multitude of pyramidal neurons having plenty of dendritic organization in the cortex, running parallel with each other and perpendicular to the cortical surface, can generate easily additive current when they are activated synchronously.

Generally, action potentials generated in the cortical neurons have a larger amplitude of electrical potentials than EPSP. But, the contribution of action potentials in the generation of detectable brain electrical activity is minimal. Firstly, action potentials are increasing and decreasing rapidly. Their durations are about 1~3 ms. This short time duration makes it very difficult to have thousands of action potentials synchronously at the given time. In contrast, EPSP has a longer duration of 10~40 ms and it greatly increases the probability that multitudes of neurons are maintaining the increased ESPS at a given time (Brienza & Mecarelli, 2019). Secondly, the single presynaptic neuron has about 1000 synapses on average. That is, for a single action potential, 1000 postsynaptic potentials can be followed. Even though multiple synapses can be connected to each of the postsynaptic neurons, the number of generated postsynaptic potentials increases greatly and most EPSPs are synchronous to the presynaptic action potential. This synchronous activation of multiple EPSPs greatly increases their contribution to the generation of detectable potentials.

As the potentials recorded from the surface of the cortex, or on the scalp, are the result of the thousands of synchronously activated neurons, it is not possible to discriminate the activity of each neuron from that of neighboring cells. The brain electrical potentials recorded at the cortical surfaces are the summary of the collective activation of many neurons.

2.3 Types of Brain Electrical Signals

There are three types of electrical potentials recorded from the brain cortex depending on the location of electrodes (Fig. 8.4).

Local field potentials (LFPs) are recorded by the needle-type microelectrodes inserted into the cortical tissues. Inserted microelectrode records the composite local electrical current generated by the multiple nearby neurons. Local field potential is the recording of the electric field in the extracellular fluid which is developed by the ionic charges separated according to the activation of neighboring neurons. Its amplitude is less than 1 mV as it is the measurement in extracellular space. It is much smaller in amplitude than the action potential or resting membrane potential measured inside the neural cell. LFP is the micro-scale potential reflecting the activity of nearby cells within a range of 0.1~1.0 mm radius from the electrodes. Due to its invasiveness in recording, LFP is not routinely used for clinical purposes or human applications.

Electrocorticogram (*ECoG*) is recorded with the intracranial electrodes placed over the exposed surface of the brain cortex. Multiple electrodes in an array, for example, 10~256 electrodes, separated about 10~20 mm distance from each other and arranged in the connecting sheet are used for ECoG recording. Each electrode records the collective activities of cortical neurons just beneath the electrode. ECoG is a mesoscale electrical potential summating the neuronal activities within a distance of 2~5 mm from each electrode. Compared to the EEG, it provides potentials with increased amplitude and exceptionally high SNR. It has a higher spatial resolution than EEG and provides detailed information generated from the smaller volume of neuronal cells. As it requires an incision of the skull to locate the electrodes, clinically it is used during the planned neuronal surgery to find and confirm the location of the pathological lesions.

Electroencephalogram (*EEG*) is recorded with the electrodes positioned on the scalp. It is the representative electrical activity of the brain and it is called generally a brain wave. Due to its distance from the cerebral cortex, about 12~18 mm, and the non-conductive skull within conducting volume, recorded electrical potential is greatly reduced and maximum amplitude is in the range of 200 μV. It is a macroscopic scale electrical signal reflecting the activity of the neurons at a distance of 50 mm from the electrode (Table 8.1). Due to its noninvasiveness, EEG is actively used for clinical purposes to diagnose many neurological disorders. Epilepsy is most often diagnosed by abnormalities in EEG readings. Clinically, multiple channels of EEG are recorded simultaneously using the multiple electrodes placed over the whole brain hemispheres. To evaluate the brain

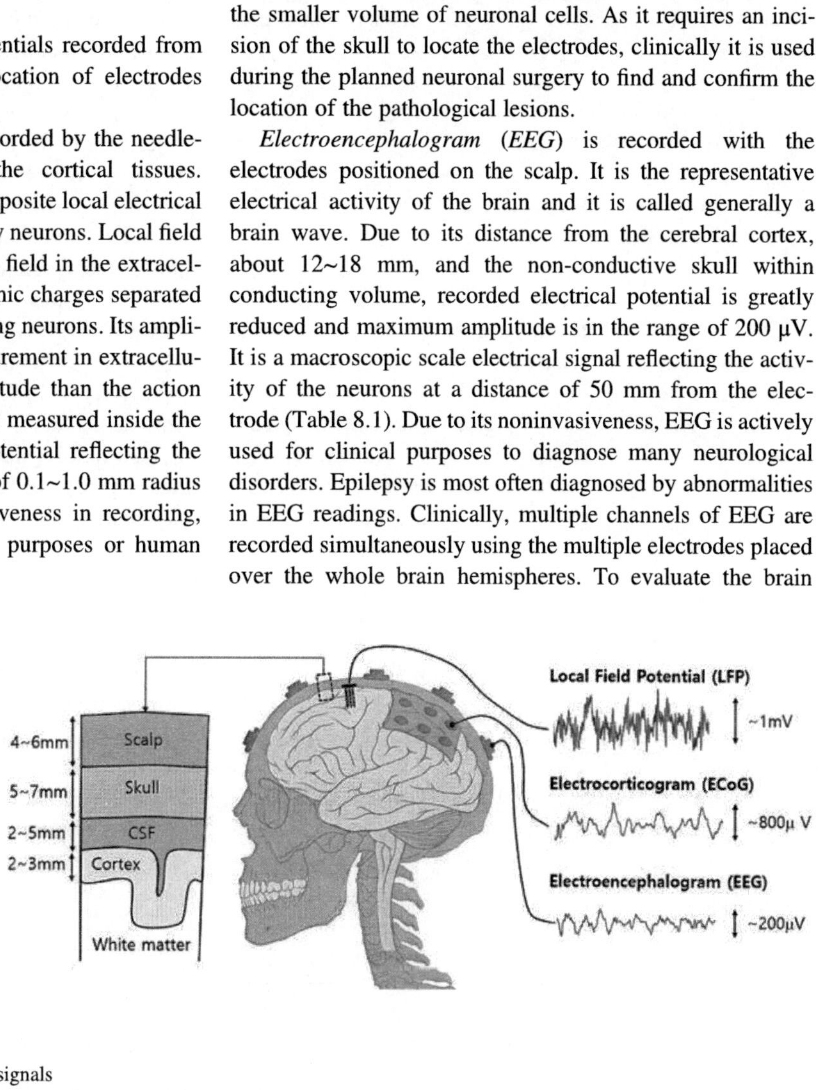

Fig. 8.4 Types of electrical potentials recorded from the brain cortex

Table 8.1 Types of brain-recorded electrical signals

Signal type	Electrode position	Spatial scale	Measurement range	Signal-collecting tissue volume	Brain activity
LFP	Within the cortical tissue	Micro–	0.1~1.0 mm	10^{-3}~1.0 mm^3	Extracellular electric field generated by neighboring neurons
ECoG	Surface of the cortex	Meso–	2.0~5.0 mm	1.0~20 mm^3	Substantial portion of local cortex
EEG	On the scalp	Macro–	~50 mm	10^3~10^4 mm^3	Large portion of underlying cortical tissues

activity to the physical stimulation or contextual information, EEG can be recorded with a sequence of programmed stimulations. EEG is also actively used in sleep medicine as an essential signal to discriminate sleep stages.

With the advent of neuroanatomical imaging systems of high spatial resolution like magnetic resonance imaging (MRI), computerized tomography (CT), and positron emission tomography (PET), the role of EEG as a diagnostic tool for neurological disorders has decreased relatively. But EEG has a much higher temporal resolution which is not possible in neuroanatomical imaging systems. EEG consistently maintains a role as a diagnostic tool and is widely accepted as the easiest tool for exploring the brain in many non-clinical applications.

3 Recording of EEG

3.1 Electrode Placement System for EEG Recording

The 10/20 EEG electrode placement system shown in Fig. 8.5 is the internationally recognized system to locate EEG electrodes on the scalp for ordinary EEG examinations. This system provides a standard for EEG examination so that the subject's EEG can be interpreted across the examination studies. The 10/20 system is based on the anatomical reference landmark points and the relative distances between the points. Names of the location of electrodes are reflecting the underlying cerebral cortex. The "10" and "20" refer to the relative distances between adjacent electrodes in percentage to the whole skull distances of front-back or left-right.

To standardize the electrode placement, anatomical landmarks are used as the reference points for electrode placement. An anatomical landmark in the front part of the skull is the *nasion*. The nasion is the midline bony depression between the eyes, just above the bridge of the nose, where the frontal and two nasal bones meet. An anatomical landmark in the back of the skull is *inion*. The inion is the tip of the external occipital hump, the midline bony prominence in the occipital bone, which can be located with mild palpation. Anatomical landmarks on the left and right sides of the skull are preauricular points on each side. The preauricular point is in front of each ear, in front of the upper end of the tragus which is a small backward-pointing eminence situated in front of the ear canal. It can be easily located with palpation, and more easily with the slight opening of the subject's mouth.

Locations for electrode positioning are determined by dividing the distances between these reference points. To decide the locations along the line connecting the nasion and the inion from the front to the back side of the skull, the whole circumferential distance along the scalp surface is divided into 6 segments of 10%, 20%, 20%, 20%, 20%, and 10% of the whole distance. Five intermediate points from the frontal side to the backside are named FP_Z, F_Z, C_Z, P_Z, and O_Z respectively. Electrode locations from the left side to the right side are determined similarly along the line connecting the left preauricular point to the right preauricular point through the determined central point of C_Z. By dividing this line into 6 segments of 10%, 20%, 20%, 20%, 20%, and 10% of the whole left-right distance, five intermediate points from the left side to the right side are named T_3, C_3, C_Z, C_4, and T_4 respectively. Additional electrodes are located along the circumferential line connecting the FP_Z, T_3, O_Z, T_4, and back to FP_Z. By dividing the left circumferential line from FP_Z to O_Z into 6 segments of 10%, 20%, 20%, 20%, 20%, and 10%, five intermediate points from the front side to the backside are named FP_1, F_7, T_3, T_5, and O_1 respectively. Similarly, by dividing the right circumferential line from FP_Z to O_Z into 6 segments of 10%, 20%, 20%, 20%, 20%, and 10%, five intermediate points from the front side to the backside are named FP_2, F_8, T_4, T_6, and O_2 respectively. Four additional positions for the electrodes F_3, F_4, P_3, and P_4 are determined as the points located at equal distances to neighboring electrodes in both directions. For example, the position of electrode F_3 is determined so that the distance from F_3 to FP_1 is equal to the distance from F_3 to C_3 as well as the distance

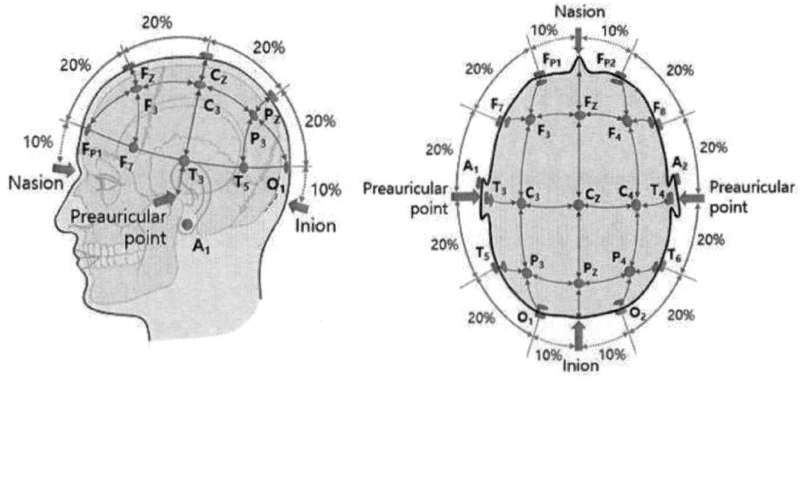

Fig. 8.5 10/20 international EEG electrode placement system

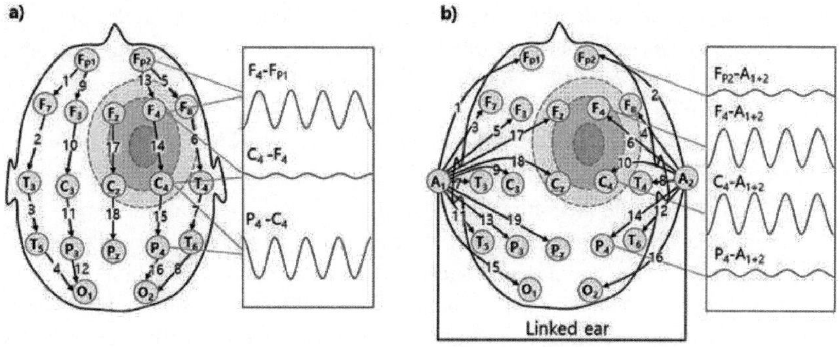

Fig. 8.6 Example of EEG montages. (**a**) Longitudinal bipolar montage called double-banana montage, (**b**) Referential montage using linked ear (A1 + A2) as the common reference

from F_3 to F_7 is equal to the distance from F_3 to F_Z. During the EEG recording, one reference electrode is used. After recording all channels of EEG with one referential electrode, potentials can be recalculated with another reference point in digital EEG recording systems. During the recording, the reference electrode is selected as the maximally neutral point around the skull, which is the most irrelevant point to the brain activity as well as not affected by other biological signals like ECG and EMG. Earlobe electrodes A_1/A_2 or mastoid electrodes M_1/M_2 are mostly selected as the reference during the recording. The mastoid process is the bony prominence behind the ear. In some monopolar EEG recordings not considering re-referencing, wire-linked electrodes of A_1 and A_2 or M_1 and M_2 are used as the reference. Totally 21 electrodes including two referential electrodes are placed over the scalp surface to cover the whole active cerebral cortical areas by this 10/20 EEG electrode placement system.

The first letter of each electrode placement site is to identify the lobe or area of the brain, pre-frontal (FP), frontal (F), temporal (T), parietal (P), occipital (O), and central (C). The odd-numbered electrodes refer to electrodes placed on the left hemisphere and the even-numbered electrodes refer to the electrodes placed on the right hemisphere. Letter "Z" (zero) refers to the electrodes placed along the midline dividing the brain into two hemispheres.

To increase the spatial resolution of EEG recording with an increased number of electrodes, additional electrodes are placed between the electrode positions determined in the 10/20 electrode system. The 10/10 EEG electrode placement system determines the positions of the 74 electrodes with a 10% distance between the electrodes. And 10/5 electrode positioning system describes the positions of electrodes to extend the number of electrodes up to 345 electrodes.

3.2 EEG Montages

EEG montage is an essential group of derived recording channels. In routine EEG examinations, 16~20 channel montages are most often used. Among a large number of possible montages, only a few montages are used in clinical practice (Acharya & Acharya, 2019). EEG montage is used to display activity over the entire head so that important activity cannot be missed. By displaying the entire activity, activities on the left and right hemispheres can be compared, and abnormal activity of a specific disorder can be localized. In patients with focal epilepsy and especially if surgery is planned, localization of an epileptogenic lesion is particularly informative (Fig. 8.6).

In bipolar montages, each channel records the potential difference between two adjacent active electrodes. Recorded potential can be considered as an approximated derivative of the potential at the recording site along the direction of two electrodes.

$$V_{Cz-Fz}(t) = P_{Cz}(t) - P_{Fz}(t) \tag{8.1}$$

The channels are usually arranged in longitudinal or transverse directions. One of the commonly used longitudinal bipolar montages is the so-called "double-banana" montage, which provides spatial information in the sagittal plane (Fig. 8.6a). Transverse bipolar montages are arranged in a similar manner and it provides spatial information in the coronal plane. Bipolar montages are useful for detecting the highly localized discharges, appearing as a steep potential gradient between adjacent channels. Compared to the referential montages, generally, amplitudes are smaller and artifacts are lower. Potentials generated in a broad region may be canceled, as no waveform is recorded if the potentials at two electrodes are varying in the same pattern.

Referential montages, which is also called monopolar montage, use one or common electrodes as a reference input for all channel recordings of each hemisphere (Fig. 8.6b). Ideally, the reference electrode is an indifferent passive electrode that does not pick up any brain activity, so that each channel records the activity of the location at which the active electrode is positioned. Channels are arranged similarly to bipolar montage, longitudinally and transversely. The ipsilateral ear (A_1, A_2) or mastoid process (M_1, M_2) is recommended as a standard reference electrode. The midline reference electrode (C_Z) is also suggested as a reference electrode as it does not amplify the signal in one hemisphere. In some clinical examinations, other references such as linked ears or a non-cephalic reference are used. The advantage of referential montages is the maximal preservation of the shape and amplitude of local potential. Often reference electrodes can be contaminated by the noises or artifacts, and then it contaminates the recording of all channels.

Common average montage is using the average of all the amplifier outputs as a common reference. The advantage of a common average montage is that it can minimize the effect of prominent activity or artifact at a single electrode. However, as the average reference includes the activity of the recording electrode of interest, the amplitude of the potential of interest will be slightly reduced.

Laplacian montage is using a combination of multiple electrodes as the reference. However, the reference is not the same for all channels. The average of the amplifier output of the neighboring 4 electrodes is used as the reference. The potential at each active electrode is compared with the locally averaged potentials of surrounding electrodes. For electrode C_Z, Laplacian reference is determined as the average of potentials recorded at 4 neighboring electrodes F_Z, P_Z, C_3, and C_4. Recorded potential can be considered as the sum of 2nd derivatives in a longitudinal direction and transverse directions.

$$
\begin{aligned}
V_{Cz}(t) &= P_{Cz}(t) - \frac{1}{4}(P_{Fz}(t) + P_{Pz}(t) + P_{C3}(t) + P_{C4}(t)) \\
&= \frac{1}{4}[((P_{Cz}(t) - P_{Fz}(t)) - (P_{Pz}(t) - P_{Cz}(t))) \\
&\quad +((P_{Cz}(t) - P_{C3}(t)) - (P_{C4}(t) - P_{Cz}(t)))]
\end{aligned}
\tag{8.2}
$$

Laplacian montage is helpful for detecting the highly localized activities by filtering out the regionally common activities.

When analog paper recording EEG is used, prewired montage patterns are selected by the switches during the recording. With a digital EEG system, potentials at all electrode sites are recorded with a predetermined reference montage and stored after digitization. Then, the recording of any montage can be reconstructed mathematically from the stored recording with the referential montage.

3.3 Electrodes Measuring EEG

EEG is recorded with the silver/silver chloride (Ag/AgCl) electrodes most commonly in clinical practices. Ag/AgCl electrodes, which are also called wet electrodes, are attached to the scalp surface using a paste or electrolyte gel. A paste or electrolyte gel decreases the electrical contacting impedance along the conductive path between the skin and the electrodes. Decreasing the contacting impedance between skin and electrodes is an important factor. It determines the electrical properties of the electrode-skin interface and provides better quality in recorded EEG signals. Decreased contacting impedance also reduces the interference from the power line and the artifacts caused by the movements of subjects. Ag/AgCl electrodes are currently the gold standard in clinical practice, however, they have some disadvantages. Attaching wet electrodes using electrolyte gel or paste needs some preparation time and the time increases with the number of electrodes. As the examination time increases, electrolyte gel dries up, the contacting impedance increases, and the recorded signal deteriorates. Thus, wet electrodes are not suitable for long-term EEG recording. As the distances between the electrodes are decreased with the increased number of electrodes, two electrodes can be electrically shorted unintentionally by the electrolyte gel. Skin abrasion to reduce the contacting impedance to an acceptable level of 5~20 KΩ and using a paste for electrode attatchment used to make subjects inconvenient.

Dry electrodes are proposed as an alternative to the wet electrodes to diminish the problems raised on wet electrodes. They are basically used without laborious skin preparation and using a paste or electrolyte gel. Dry electrodes also do not show signal deterioration over time and they are suitable for long-time applications. A most common type of dry electrode is a metallic array of spikes that directly contact the scalp. Instead of a direct contacting electrode, a capacitive coupling electrode can be used as a dry electrode without direct conductive contact with the scalp surface. However, dry electrodes are prone to noises due to the increased contacting impedance by eliminating the layer of the electrolyte gel. To reduce the effect of power line interference and movement artifact, dry electrodes are shielded and the shielded cables are used from the electrodes to the amplifier unit. Active electrodes, which amplify the signal at the back of the electrode, decrease the effect of the environmental noises.

4 Spontaneous EEG

4.1 Spectral Bands of EEG Signal

Spontaneous EEG is the EEG recorded without any physical or contextual stimulation. It represents the activations of the brain in its various states. Spontaneous EEG is typically described by the rhythmic activity based on its frequency bands. Typically, frequency components of EEG signal appear up to 50 Hz and most of the spectral power is concentrated below 20 Hz (Fig. 8.7).

Alpha wave is the spectral component in the frequency range of 8~13 Hz. Its name is coined by Hans Berger when he first identified the rhythmic activity of the brain. It is the most

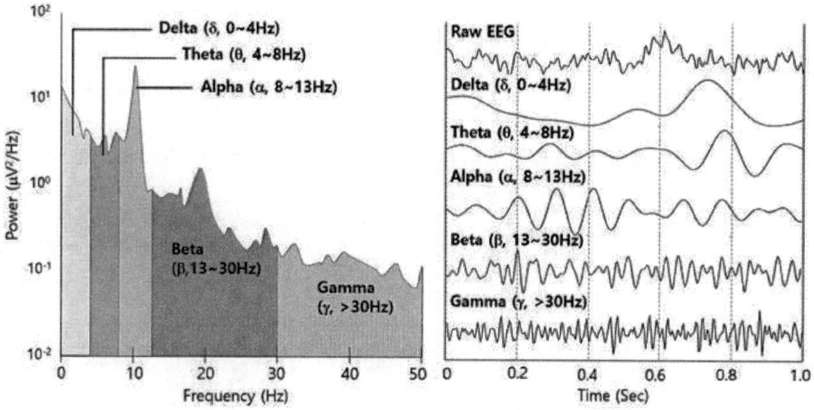

Fig. 8.7 Spectral bands of spontaneous EEG signal

prominent normal adult brain rhythm. It is likely originating from the synchronous and coherent electrical activity of cerebral neural cells. Alpha rhythm appears when the eyes are closed in a mentally inactive relaxed state and disappears with the eyes opening. Its maximum amplitude is in the range of 20~60 μV and dominant at occipital and parietal electrodes in the posterior region. The envelope of alpha rhythm often waxes and wanes over 1~2 s intervals. For each subject, the central frequency of alpha rhythm is generally quite stable across recording sessions and hemispheres.

Beta wave is the spectral component in the frequency range of 13~30 Hz. This rhythm is also named by Han Berger when he identified the rhythm of higher frequency and smaller amplitude replacing the alpha rhythms with the eyes opening. Beta rhythm reflects desynchronized active brain activities. It is generally associated with normal waking consciousness representing active, busy, or anxious thinking and active concentration. Beta rhythms are most often recorded in frontal and central areas. Beta rhythm maximally appears over the sensory-motor strip in central areas. Beta rhythm appearing in frontal lobes does not exhibit clear maximums in the beta frequency band and its maximum amplitude is less than 20 μV.

Theta wave is the spectral component in the frequency range of 4~8 Hz. Theta rhythms are frequently observed in normal young children. In normal adults, theta rhythm is not observed in an ordinary waking state. It mostly appears during meditation, prayer, spiritual awareness, hypnotic state, and drowsy state before falling asleep. It reflects the subconscious mind between wakefulness and sleep state. Theta rhythm is often recorded in association with several types of brain pathology.

Delta wave is the lowest spectral component in the frequency range of 0~4 Hz. The Delta wave is the slowest wave in EEG and its maximum amplitude is in the range of 50~200 μV. The Delta wave is the dominant rhythm in infants during their first year. In adults, the delta wave is most prominently observed in normal deep sleep, which is also called slow-wave sleep. The delta waves are also associated with different brain functions other than deep sleep, like cortical plasticity and cognitive processing, especially in event-related studies.

Gamma wave is the highest spectral component in the frequency range of 30~50 Hz. Gamma wave is the fastest brain waves with a low amplitude of less than 10~20 μV. It is hard to measure accurately and the spectral range used to be extended beyond 50 Hz. Gamma rhythm is attributed to sensory perception integrating different areas. It tends to appear when a subject is intensely focused or actively engaged in a task. Altered gamma rhythm is observed in some neurological disorders such as Alzheimer's disease, epilepsy, and schizophrenia (Oostenvelda & Praamstrac, 2001).

4.2 EEG Activities During a Night Sleep

EEG is the record of the electrical activities of the brain. During the waking state, EEG reflects the various activation of the brain related to concentration, emotion, memory, and so on. When a subject falls asleep, EEG reflects the brain states differently acting upon the sleep stages.

Sleep is the natural and periodic suspension of consciousness with a slowed metabolism. In contrast to the state of wakefulness, sleep is clearly differentiated by immobility and the restricted responsiveness to environmental stimuli. While the physical and mental restorative effect on body organs is the main function of sleep, sleep also provides many beneficial functions including energy conservation and memory consolidation.

Sleep is not a single homogenous state of loss of consciousness. Sleep is classified into several stages of brain activity (Fig. 8.8). Basically, sleep is divided into two macro stages, non-rapid eye movement (NREM) sleep and rapid eye movement (REM) sleep. And NREM sleep is subclassified into three sleep stages, Stage N1, N2, and N3,

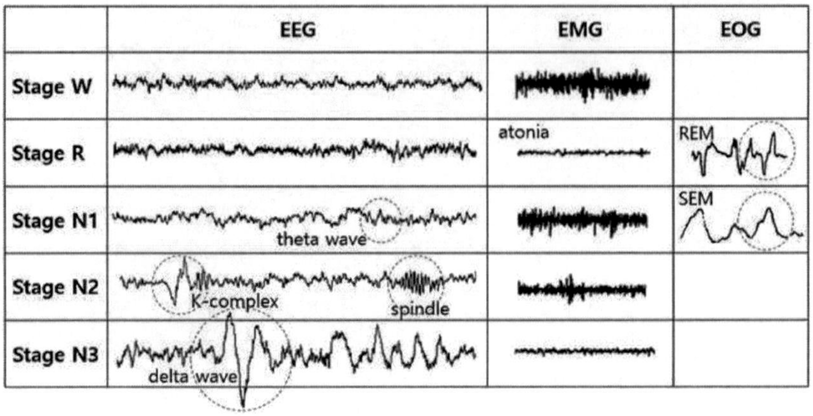

Fig. 8.8 Characteristic signal patterns of EEG, EMG, and EOG for each sleep stage. *EEG* Electroencephalogram, *EMG* Electromyogram, *EOG* Electrooculogram, *REM* Rapid eye movement, *SEM* Slow eye movement

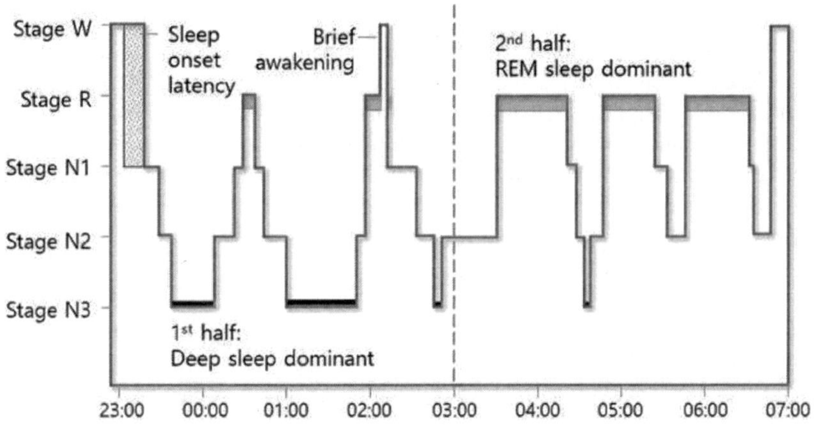

Fig. 8.9 Example of hypnogram showing the change of sleep stages during a whole night sleep

on the depth of the sleep. When a subject falls into the sleep from the wakefulness state and goes into a deeper sleep, EEG signals change from the low-voltage fast activity of beta rhythm to the high-amplitude slower activity of delta rhythm. Each sleep stage is characterized by special patterns of EEG rhythms and the accompanying changes in electromyogram (EMG) and electrooculogram (EOG) signal patterns. NREM sleep is interrupted by periods of REM sleep and these sleep stages are repeated in several cycles during the one-night sleep. In normal adults, 4–6 sleep cycles occur during a normal night's sleep and each sleep cycle lasts for 90~120 min. NREM sleep predominates the first half of the sleep and REM sleep occurs actively in the latter half (Fig. 8.9).

Stage W is the awake state before the sleep and brief arousal during the sleep. In the earlier phase before entering sleep, subjects are still relatively awake and alert. This stage is characterized by the presence of a predominant beta rhythm over the anterior regions and a posterior dominant alpha rhythm over the occipital regions. As the wakefulness transits to drowsiness, the alpha rhythm starts to dominate with intermittent beta rhythm, and occasionally slow eye movements (SEM) are appearing in EOG. Stage W takes less than 5% of total sleep time.

Stage N1 is the lightest sleep stage and is characterized typically by the disappearance of the alpha rhythm and the appearance of slow eye movements. At this stage, the brain is still active and the EEG shows medium amplitude, theta band activity of 4~7 Hz. There also occur bursts of slow waves and vertex sharp transients (VST). During the N1 stage, heartbeat and respiration slow down and reduced muscle activity is recognized by the decreased amplitude of EMG. Usually, N1 sleep takes about 5~10% of total sleep time.

Stage N2 is a state of increased sleep depth from the N1 sleep and is characterized by the presence of theta activities accompanied by the sleep spindles and K-complexes. During the N2 stage, subjects become less aware of their surroundings, and body temperature drops. Eye movements stop and the heartbeat and respiration become more regular. Sleep spindles are defined as a burst of distinct 12~14 Hz wave patterns with a duration of greater than 0.5 s. Sleep spindles are thought to be a feature of memory consolidation appearing when the brain gathers, processes, and filters new memories acquired the previous day (Andrillon et al., 2011). K complexes are defined by a wave complex of a negative sharp peak immediately followed by a positive wave standing out from the background EEG and lasting about 0.5 s. While this stage is continued, body organs slow down in preparation for NREM stage N3 sleep and REM sleep. N2 sleep takes a major portion, about 50%, of the total sleep time. N1 sleep and N2 sleep are collectively called light sleep.

Stage N3 is the deepest sleep state where any noises or activity in the environment may fail to wake a sleeping subject. This stage is characterized by the high amplitude and slow waves in the delta frequency band. So, N3 sleep is also called deep sleep, delta sleep, or slow-wave sleep (SWS). During the N3 sleep, delta waves in the range of 0.5~2.0 Hz with amplitudes of higher than 75 μV are predominantly recorded, and K-complexes and sleep spindles may be present. In this deepest sleep stage, muscles are completely relaxed, blood pressure drops and breathing slows. Body organs start their physical restorations and subjects feel refreshed the next day. In addition, during the N3 sleep, the brain consolidates declarative memories that have been learned before the sleep, like general knowledge, facts or statistics, and personal experiences. In normal adults, N3 sleep occurs most frequently during the first half of the night, and typically N3 sleep takes about 20% of total sleep time.

Stage REM (R) is the dreaming sleep and EEG records show low amplitude fast activity like the W stage. However, the REM stage is clearly characterized by the occurrence of rapid eye movements (REM) in EOG and muscle atonia, which is recognized by the disappeared muscle activity tone in EMG. While the brain is aroused with mental activities during REM sleep, voluntary muscles of the body become immobilized. A temporarily paralyzed body prevents subjects from acting out their dreams. Like the N3 stage, memory consolidation also occurs during REM sleep. In contrast to N3 sleep, REM sleep is thought to process and store emotions and emotional memories. During the REM sleep, the threshold of auditory stimulation level to arouse from the sleep is highest. The REM stage takes about 20% of total sleep time.

Polysomnogram is the record of the biological signals during sleep which includes EEG, EMG, EOG, and respiratory signals. Each epoch of 30 s segment is scored into one of the sleep stages and the variation of sleep stages during a whole night is summarized by a hypnogram (Fig. 8.9). The hypnogram provides valuable descriptions of the sleep structure and architectural abnormalities in sleep disorders. EEG does the main role in sleep stage scoring.

5 Evoked and Event-Related Potentials

5.1 Evoked Potentials

The evoked potential (EP) is the recorded brain electrical activity in response to the specific types of stimuli applied to the subject. The electrode for recording the evoked potential is placed above the cortical area corresponding to the stimulation modality. Stimulation type can be visual, auditory, or somatosensory depending on the purposes of the examination. Different types of evoked potentials result in corresponding to the different modalities of stimuli. Evoked potential represents the normality related to the targeted sensory pathway. If there occurs a problem in conducting the action potentials from the sensory organ to the receiving cortical area, amplitude peaks responding to the stimuli are delayed or attenuated. The evoked potential is different from the spontaneous EEG since it is done with stimuli application. Also, the evoked potential is differentiated from the event-related potential (ERP), which is the response in association with higher cognitive processing to the applied event stimulations. Clinically, evoked potentials are useful in the diagnosis of neurological disorders and drug-related sensory dysfunctions. During the surgical operation, the evoked potential is recorded on the cortical surface to infer the functional integrity of the sensory neural tracts of concern.

The amplitude of evoked potentials responding to the stimuli is less than several μ V and much smaller than the spontaneously generated background EEG having an amplitude of 50~100 μV. Even though the evoked responses occur in response to the applied stimulation, they are buried in the simultaneously occurring spontaneous EEG activities and cannot be identified simply in a single trial of stimulation. To resolve the low-amplitude evoked components buried in the recorded EEG signal, the signal averaging method is required. The signal-averaging method simply averages the recorded signals synchronized to the applied stimulations. While the signal component of the evoked potential is appearing consistently time-locked to the stimulations, the spontaneous EEG component is fluctuating randomly. Sometimes it is in a positive value and sometimes it is in a negative value, and it is canceled out as the number of stimulating repetitions is increased. The required number of repetitions is depending on the ratio of the evoked response to background noise. Sometimes more than one thousand repetitions are required and total recording examination can take more than an hour.

From the engineering point of view, the evoked potential is equivalent to the impulse response of the sensory system. As the unknown system is fully identified by the resulting output responding to the applied impulse input, the sensory organ system is identified by the resulting output responding to the applied sensory input stimulation.

Even though any sensory modality can be tested with stimulation, clinically most encountered evoked potentials are limited to visual evoked potential (VEP), brainstem auditory evoked potential (BAEP), and somatosensory evoked potential (SEP).

5.1.1 Visual Evoked Potential (VEP)

Visual evoked potential (VEP) refers to the evoked electrical potential in response to visual stimulation. It is recorded with electrodes placed on the scalp overlying the visual cortex in the occipital lobe. VEP is used to quantify the functional integrity of the visual pathway from the eyes to the visual cortex, which includes the retina, optic nerve, optic chiasm, lateral geniculate body (LGN), optic radiations, and occipital cortex (Fig. 8.10). Abnormalities affecting the conduction of nerve impulses through the visual pathway can distort the recorded evoked potential away from the normative ones. If the conduction is slowed down by some neurological disorders like multiple sclerosis, peaks of evoked response appear with varied delays depending on the severity of the disorders. By selectively stimulating half of the visual fields of each eye, each visual pathway from the retina to the occipital lobe can be examined separately (Fig. 8.11).

For the generation of VEP, two kinds of visual stimuli are used. The first type is the *flashing light* stimulation. It is the brief flashes of light and not preferred due to the high variability within and across subjects. But it is beneficial to use when the patterned stimulation is not appropriate due to the poor optical media, lack of subject cooperation, or diminished vision of the subject. The second type of stimulation is *patterned stimulation*. A checkerboard pattern of an equal number of black and white squares is recommended. Different sizes of the squared pattern are used and they are defined by the subtended visual angle of the square size. Grating patterns composed of light and dark stripes are another type of patterned stimulation. Patterned stimulation is applied in on/off mode or in a pattern reversal mode using a video

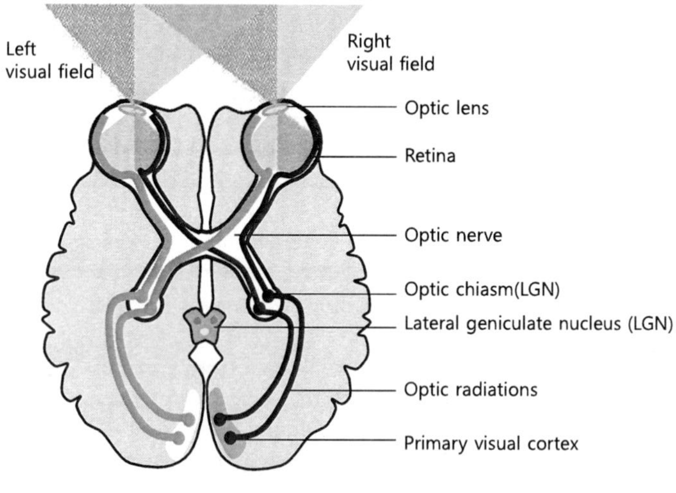

Fig. 8.10 Visual pathway conducting the electrical signals for the generation of VEP

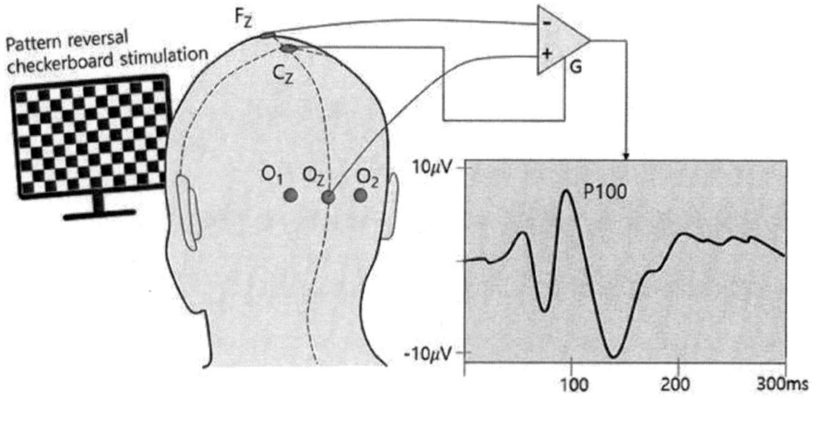

Fig. 8.11 Recording of visual evoked potential

monitor. In on/off patterned stimulation, patterned stimulation is applied for 200 ms followed by 400 ms of the diffuse gray background. In pattern reversal stimulation mode, white and black squares are reversed for each stimulation without change in the overall luminance of the screen at a specific reversal rate. Pattern reversal VEP waveforms are more consistent in morphology and timing compared to VEPs produced by other stimulations (Kothari et al., 2016).

To record the one channel of VEP in response to the patterned stimulation, an active electrode is placed at O_Z in the occipital cortex, a reference electrode at F_Z in the forehead, and ground electrodes at C_Z in the vertex according to the 10/20 EEG placement system. For a more detailed analysis, two additional channels of VEP are recorded with the electrodes placed at O_1 and O_2 positions.

A major component of clinical importance is the P100 wave, which appears as a peak in a positive direction with a time delay of about 100 ms from the stimulation onset. Its amplitude is usually in the range of 5~20 μV. P100 wave is attenuated or delayed upon the disordered state. For example, patients with acute severe optic neuritis often lose the P100 response or have highly attenuated peaks. Clinical recovery and visual improvement come with restoration of P100 amplitude but with an abnormally increased latency that continues indefinitely (Aminoff et al., 2001).

There are many variant forms of VEP with different types of stimulation protocols. *Steady state visually evoked potential* (SSVEP) is the recording of the evoked potential with continuously applied visual stimulation. While the visual stimulation of a specific frequency in the range of 5~75 Hz is applied, recorded evoked potential has spectral peaks at the same frequency and multiples of the frequency of the visual stimulation. Due to its high signal-to-noise ratio, it can be used without averaging process and it is widely accepted in EEG-based research including brain-compute interface. VEP is also used to evaluate the *visual acuity* objectively and of the infants. Visual acuity is commonly assessed with eye charts (Ex: Snellen chart), which use optotypes of different sizes. This technique is subjective and relies on the subjects' cooperation and cannot be done in infants. By applying the visual stimulation of different square sizes, visual acuity can be evaluated objectively. If a subject can see the pattern of visual stimulation, peaks appear in VEP recording and if a subject cannot see the fine visual pattern, peaks do not appear in VEP recording.

5.1.2 Signal Averaging

Visual evoked potential (VEP) is buried in simultaneously generating spontaneous EEG and is not resolved with a single stimulation. VEP cannot be separated by the spectral filtering methods in this case since the frequency bands of the VEP and spontaneous EEG overlap. To separate VEP from spontaneous EEG, VEP separation should be based on the different characteristics of the two signals. The difference between these two signals is their time-locking property. While the VEP is generated time-locked to the visual stimulation, spontaneous EEG is not time-locked to the stimulation onset and is generated randomly. At a certain time point from the stimulation onset, it fluctuates up and down randomly for each trial. When the average is performed in synchronization with the stimulating time, fluctuating spontaneous EEG components statistically cancel out as the number of repetitions for averaging is increased. Signal-to-noise ratio (SNR) increases as the number of repetitions for averaging increases.

Recorded EEG signal to the i-th visual stimulation, $x_i(t)$, is the sum of the signal component, $s_i(t)$, which is the VEP, and the noise component, $n_i(t)$, which is the spontaneous EEG in this case.

$$x_i(t) = s_i(t) + n_i(t) \quad (8.3)$$

The mean of each noise component is assumed to be 0 for all iterations and the variances of signal and noise are assumed not to change during the repetitional stimulation.

$$\begin{aligned}\sigma_s^2 &= E\left\{(s_1(t) - E\{s_1(t)\})^2\right\} \\ &= E\left\{(s_2(t) - E\{s_2(t)\})^2\right\} = \ldots \\ &= E\left\{(s_N(t) - E\{s_N(t)\})^2\right\}\end{aligned} \quad (8.4)$$

$$\sigma_n^2 = E\left\{n_1(t)^2\right\} = E\left\{n_2(t)^2\right\} = \ldots = E\left\{n_N(t)^2\right\} \quad (8.5)$$

When the average of N repetition is performed, the resulting signal is represented as

$$x(t) = \frac{1}{N}\sum_{i=1}^{N} x_i(t) = \frac{1}{N}\sum_{i=1}^{N} s_i(t) + \frac{1}{N}\sum_{i=1}^{N} n_i(t) \quad (8.6)$$

The signal component is consistent during the N repetition of stimulation.

$$s_1(t) = s_2(t) = \ldots = s_N(t) \quad (8.7)$$

$$\bar{s}(t) = \frac{1}{N}\sum_{i=1}^{N} s_i(t) = \frac{1}{N}(s_1(t) + s_2(t) + \ldots + s_N(t)) = s_1(t) \quad (8.8)$$

$$\begin{aligned}\sigma_{\bar{s}}^2 &= E\left\{(\bar{s}(t) - E\{\bar{s}(t)\})^2\right\} \\ &= E\left\{(s_1(t) - E\{s_1(t)\})^2\right\} = \sigma_s^2\end{aligned} \quad (8.9)$$

As the signal is not correlated with noise, and i-th noise is not correlated with j-th noise except when i is equal to j, cross variances between these signals become 0.

$$E\{(s_i(t) - E\{s_i(t)\})(n_j(t))\} = 0 \text{ for all } i,j \tag{8.10}$$

$$E\{n_i(t)n_j(t)\} = 0 \text{ for all } i,j \text{ except } i = j \tag{8.11}$$

$$\overline{n}(t) = \frac{1}{N}\sum_{i=1}^{N} n_i(t) = \frac{1}{N}(n_1(t) + n_2(t) + \ldots + n_N(t)) \tag{8.12}$$

$$\sigma_{\overline{n}}^2 = E\left\{\left(\frac{1}{N}(n_1(t) + n_2(t) + \ldots + n_N(t))\right)^2\right\}$$
$$= \frac{1}{N^2}\sum_{i=1}^{N} E\left\{n_i(t)^2\right\} = \frac{N\sigma_n^2}{N^2} = \frac{\sigma_n^2}{N} \tag{8.13}$$

After averaging the responses to N repetitive stimulations, the variance of the noise component is decreased to one N-th of the variance of single noise while the variance of the signal component is the same without any decrement.

When the signal-to-noise ratio is calculated after signal averaging, it increases $\sqrt{N}$ times.

$$\text{SNR}_1 = \frac{\sigma_s}{\sigma_n} \tag{8.14}$$

$$\text{SNR}_N = \frac{\sigma_{\overline{s}}}{\sigma_{\overline{n}}} = \frac{\sigma_s}{\sigma_n/\sqrt{N}} = \sqrt{N}\cdot\frac{\sigma_s}{\sigma_n} = \sqrt{N}\cdot\text{SNR}_1 \tag{8.15}$$

For example, the average is performed over 4 repetitive stimulations, SNR increases 2 times, and SNR increases 32 times with 1024 repetitive stimulations.

Figure 8.12 shows the example of signal averaging for simulated and real VEP recordings. The left part of the figure shows the averaging results with the simulated signal of the triangular waveform when it is buried in the noise of the spontaneous EEG. The signal component becomes clearer as the number of averaging repetitions is increased. The right part of the figure is the recording of VEP with the increased number of averaging. In this recording, VEP peaks are almost clearly recognized with the repetition number 128 in amplitude and in time delay. Even though the SNR increases with the repetition number, recording time also increases with the repetition number and it can decrease the concentration of the subject. Since the assumption that the signal and noise characteristics are consistent cannot be preserved with a large repetition number, optimal repetition for the signal averaging should be determined carefully.

5.1.3 Auditory Evoked Potential (AEP)

Auditory evoked potentials (AEP) is another type of sensory-evoked potential in response to the auditory stimulation and it is recorded with the active electrode placed at C_Z in the vertex. AEP is used to quantify the functional integrity of the ascending auditory pathway from the cochlea in the inner ear to the primary auditory cortex. After the sound is sensed by the hair cells at the organ of Corti in the cochlea, action potentials are transmitted through ganglion cells whose axons constitute auditory nerves, via the cochlear nucleus in the medulla, superior olivary complex in the pons, lateral lemniscus, inferior colliculus in the midbrain, medial geniculate body in the thalamus, finally to the auditory cortex in the cerebrum (Fig. 8.13).

Components of auditory evoked potential can be grouped into three parts depending on the appearing latency (Fig. 8.14). Early response before 10 ms is called brainstem auditory evoked potential (BAEP) or auditory brainstem response (ABR). The middle latency part recorded with 10~50 ms latency is called middle latency auditory evoked potential (MLAEP). The mismatch negative

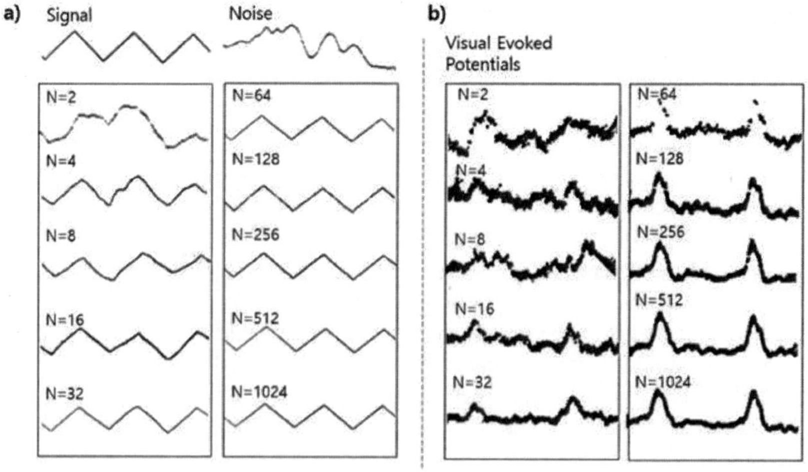

Fig. 8.12 Signal averaging for the visual evoked potentials. Signal-to-noise ratio (SNR) increases as the number of stimulating repetition 'N' increases. (**a**) Simulation of averaging effect with triangular wave buried in spontaneous EEG. (**b**) Recorded VEP in response to the increasing number of stimulations

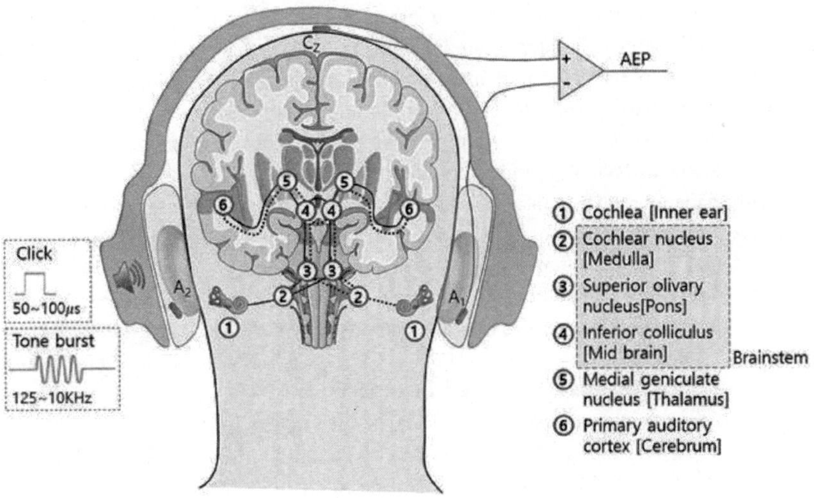

Fig. 8.13 Auditory pathway conducting sound response to the auditory cortex

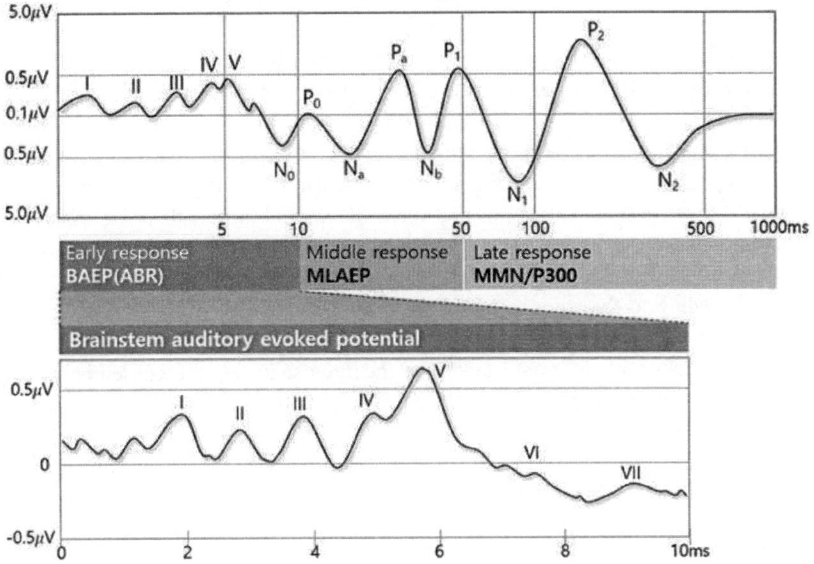

Fig. 8.14 The waveform of auditory evoked potential and brainstem auditory evoked potential. *BAEP* Brainstem auditory evoked potential, *ABR* Auditory brainstem response, *MLAEP* Middle latency auditory evoked potential, *MMN* Mismatch negative

component (MMN) and P300 wave, which appear later than 100 ms, comprise the late auditory evoked potential. Each group of AEP is applied for different purposes and recorded with a different recording paradigm.

The early portion of AEP within 10 ms of the recording is recognized as brainstem auditory evoked potential (BAEP). To evoke a BAEP, acoustic stimulation is applied via a headset or plugged-in earphones. Click sound of 100 ms pulse duration and 80~120 dB sound pressure level is used for stimulation and applied at a rate of 10 stimulations/sec. Usually, the active electrode is located at C_Z and the reference electrode is located selectively at A_1 or A_2 according to the stimulating side. BAEP examination can be performed even if the subjects are asleep as the response appears during the sleep. Normal BAEP has several peaks labeled using Roman numerals I-VII. As each peak of BAEP is related to the integrity of each conducting segment, the delay between the peaks and the attenuation of peak amplitudes are measured and used to localize the abnormality of the disorder. As the amplitude of BAEP is very small in amplitude and buried in spontaneous EEG, it also needs average processing, with 1000~4000 repetitive acoustic stimulation, to resolve the evoked components. BAEP is used in the diagnosis and localization of disorders affecting brainstem pathways. BAEP is also widely used in testing auditory function, and especially as a diagnostic tool for the screening of infant hearing (Boston, 1981).

AEP in the middle latency of 10~50 ms is elicited when a subject does not perceive the direction of attention to the auditory stimulus. AEP in middle latency is used in the objective assessment of auditory functions, in the depth monitoring of anesthesia, localization of thalamocortical lesions, and in assessing the effectiveness of electrical stimulation for cochlear implantation (Wilkinson & Morlock, 1967). AEP in the later part after 100 ms comprises the mismatch negativity (MMN) component, which is recorded when the flow of identical standard pure-tone sounds is followed by a deviant sound (Näätänen et al., 2007). The MMN phenomenon is valuable in investigating the subjects whose cooperative feedback responses are unavailable or unreliable, such as infants. The MMN is a component of brain event-related potential rather than an evoked potential since the cognitive processing of the brain cortex is involved.

5.1.4 Somatosensory Evoked Potential (SEP)

Somatosensory evoked potential (SEP) is the electrical activity recorded over the brain cortex evoked by the stimulation applied to the skin. SEP examination is used to evaluate the functional integrity of the somatosensory pathway from the periphery to the cortex. The somatosensory pathway consists of the peripheral nerve, dorsal columns of the spinal cord, medial lemniscus, ventroposterior lateral thalamus, and primary sensory cortex. Stimulations to evoked SEP are applied in mechanical stroke, by air puff, and most commonly by bipolar transcutaneous electrical stimulation over the trajectory of peripheral nerves. Stimulation on the median nerve at the wrist and posterior tibial nerve at the ankle are the most commonly used sites. The active electrode for SEP recording is placed at C_3, C_Z, or C_4, on the somatosensory cortex corresponding to the stimulating site with a reference electrode at F_Z. Recorded SEPs have a series of positive and negative peaks with different latencies as the impulse travels along the somatosensory pathway (Fig. 8.15).

SEP is divided into the short-latency SEP (SL-SEP) and long-latency SEP(LL-SEP). SL-SEP indicates the components appearing before 40 ms after the stimulation onset and the LL-SEP indicates the components appearing after 40 ms. While SL-SEP is actively used in assessing the integrity of the somatosensory pathway, LL-SEP is used to evaluate the following activation at the cortical level to the somatosensory stimulation. For example, if the lesion is involved in the primary thalamic relay, all SEP components are affected, particularly the short-latency components. But if the lesion is involved in other thalamic areas, particularly the long-latency components can be affected (Chu, 1986). Anatomical relations of SEP peak components to each of conducting pathway segments assist in the localization of lesions along the somatosensory pathway.

5.2 Event-Related Potentials (ERP)

Event-related potential (ERP) is the averaged EEG over the scalp generated in the brain structures in response to the specific events. Discrimination between ERP and evoked potential (EP) is not clear. Both are recorded over the scalp in response to the time-locked stimulation or events. If the stimulations are part of the events, then EP is a part of ERP. Reversely, if events are considered as the different types of stimulations, then ERP becomes a subset of EP. However, there are clearly different points in responding signals. When a stimulation or event is applied, responses coming with early latency, for example before 100 ms, are caused by the physiological integrity of conducting pathway and the physical nature of the stimulation or event. These responses are recorded before the cognitive processing is performed in the brain. The responses are not related to the meaning or content of the stimulation. These responses are termed sensory or exogenous responses. These potentials recorded in response to the physical stimulations are appropriate as evoked

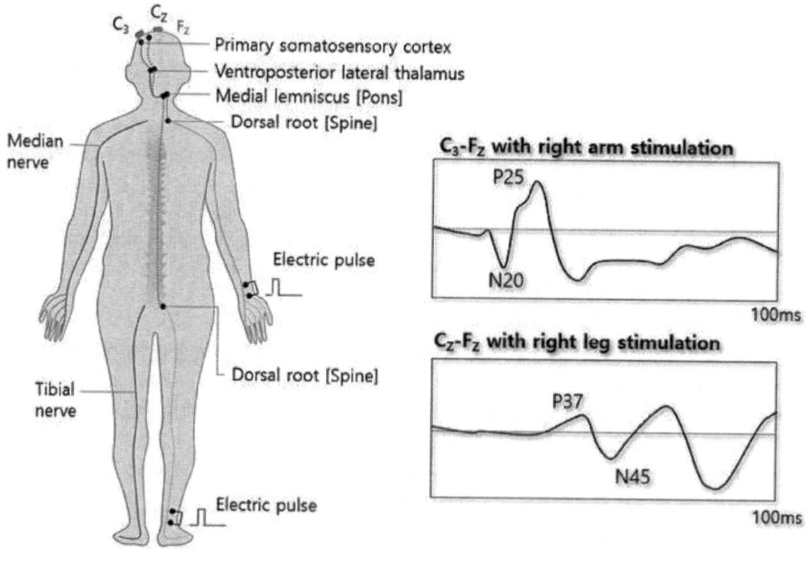

Fig. 8.15 Somatosensory evoked potential

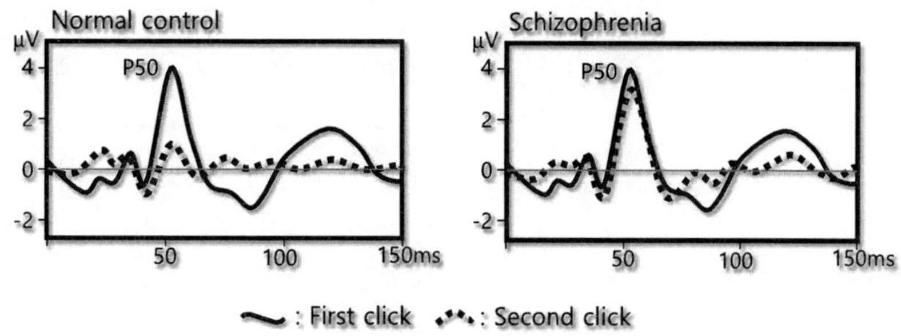

Fig. 8.16 P50 waves in sensory gating examination with normal control and schizophrenia subject

potentials. Visual evoked potential (VEP), brainstem auditory evoked potential (BAEP), and somatosensory evoked potential (SEP) all belong to this category of the evoked potential.

In contrast, the responding signals with longer latency, for example after 100 ms, generally reflect the intervention of the cognitive processing of the brain. Responses are dependent on the content of the event and how the subject evaluates the event rather than the physical nature of the event. In some paradigms, different physical stimulations of visual and auditory nature generate similar responses according to the same content those stimulations possess. These responses are also termed cognitive or endogenous responses. Potential recorded in response to the content of the event with the involvement of cognitive processing is appropriate as event-related potential. There are many wave components belonging to this category of ERP described according to the latency and the polarity of the peaks.

5.2.1 P50 Wave

P50 wave is the responding component appearing around 50 ms, between 40 and 75 ms, in a positive direction after the onset of stimulation (Fig. 8.16). P50 wave is used for sensory gating test. Sensory gating describes a neurological process of filtering out redundant or unnecessary stimuli in the brain. It is the mechanism of the brain to prevent an overload of redundant information in the higher cortical centers. For example, when a pair of clicks with a time gap of 500 ms is applied to a subject, the normal subject gates out the second click by perceiving it as a redundant one. The integrity of the sensory gating mechanism can be confirmed by the P50 wave. While the P50 wave to the 1st click has a high amplitude, response to the 2nd click is occurring with 80~90% reduced amplitude compared to the 1st one. It represents that the inhibitory pathway is properly working to inhibit second click stimulation. In contrast, schizophrenia patients reduce the P50 wave to 2nd click only by 10~20%. Sensory gating is important in selective attending to salient stimuli and ignoring the redundant, repetitive or trivial information, to prevent the central brain from information overload (Light & Braff, 2003).

5.2.2 N100 Wave

N100 or N1 wave is a negative-going component with a peak between 80 and 120 ms after the stimulus onset. It is recorded maximally in the frontocentral region, between F_Z and C_Z electrode positions, of the cortex. N100 wave is thought to record early perceptual processing, for orienting or matching, in response to the auditory stimuli even in the absence of task demands. Individuals with schizophrenia demonstrates markedly decreased N100 amplitude suggesting that the subject has difficulty in allocating attention to auditory stimulation. As N100 may be intact in patients with bipolar disorder, N100 might be used as a biological index for differentiating bipolar disorder from schizophrenia. In children with a high risk of developing psychosis, decreased N100 waves potentially predict the transition to psychosis. While N100 waves are mostly recorded with auditory stimulations, they also can occur in other sensory stimulations including visual and olfactory stimulations.

5.2.3 P200 Wave

P200 or P2 wave refers to the positive deflection peaking around 100~250 ms following the event stimulation onset. Most commonly P200 wave is recorded with auditory stimulation, but it can be also obtained following somatosensory and visual events. P200 appears to be modulated by a diverse number of cognitive tasks and represents some aspects of higher-order perceptual processing modulated by attention. Latency and amplitude of P200 waves vary according to selective attention or event encoding. Latency reflects the speed of event classification and amplitude is proportional to the amount of attentional engagement of cortical regions. Latency and amplitude of the P200 wave may be examined with tasks using "standard" and "oddball" events. In combination with the N100 wave, the N1/P2 complex is used to evaluate the sensation-seeking behavior of an individual, that compares sensory inputs with stored memory (Freunberger et al., 2007).

5.2.4 N200 Wave

N200 or N2 wave is a negative-going deflection peaking between 200 and 350 ms after the onset of event stimulation.

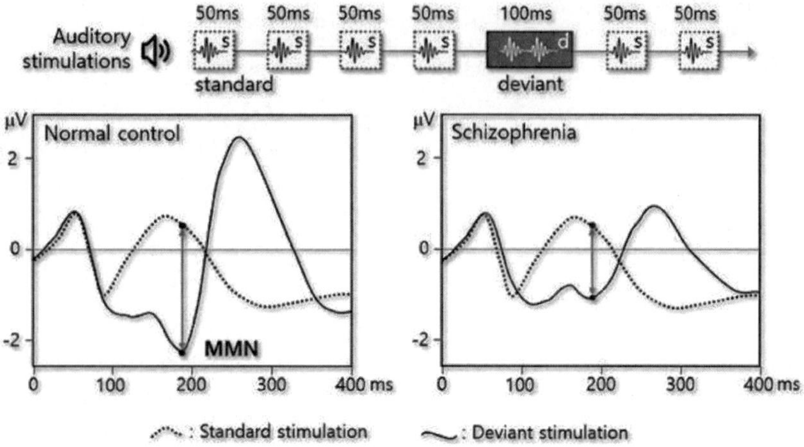

Fig. 8.17 Mismatch negative (MMN) appearing as a N200 deflection

Mismatch negativity (*MMN*) is representative N200 wave (Fig. 8.17). MMN is recorded in response to an odd deviant event in a sequence of event stimulation. MMN is most frequently examined with hearing and vision stimulations. For example, a rare "deviant" sound can be irregularly interspersed among a series of frequent "standard" sound stimulations. The deviant sound is different from the standard sound in one or more perceptual features such as frequency, loudness, or duration. MMN can be elicited relatively independently to the subject's attention level. MMN reflects the function of sensory memory by the automatic process that detects a difference between an incoming event stimulation and the memory trace of preceding sensory stimulations. In subjects with schizophrenia, the attenuated MMN amplitude is recognized as a robust neurophysiological abnormality. MMN can be used as a specific and useful biomarker for auditory disturbances in schizophrenia because MMN is not altered in other psychiatric disorders such as unipolar and bipolar depression (Umbrichta et al., 2003). N200 wave was also elicited in the go/no-go task examinations. In the go/no-go task, subjects are requested to react only to the "go" event stimulation that occurred after the cue sign. Event simulations are mostly applied in auditory or visual event stimulations. This task tests the subject's capacity not to respond to the stimulation as the no-go trials require inhibition of attentively prepared information. N200 wave appears only in the no-go trials. N200 is thought to reflect a cognitive control function, specifically an inhibitory response control mechanism or conflict monitoring.

5.2.5 P300 Wave

P300 or P3 wave is the positive deflection peaking between 250 and 500 ms after the onset of event stimulation. It is thought to reflect processes for the evaluation and discrimination of presented event stimulations, in association with attention and memory mechanisms. P300 has been the major component of ERP studies in clinics and laboratories with its reproducibility and ubiquity. Among the various paradigms to elicit P300 waves, the "oddball" paradigm is the most utilized. In the oddball paradigm, auditory or visual event stimulations are presented in a series such that oddball events, which used to be the target event stimulations to be reacted, occur infrequently interspersed among the frequent standard event stimulations. The subject is instructed to respond to the oddball event stimulations and not to the frequent standard event stimulations. P300 wave is commonly elicited when a subject detects a target event in a regular train of standard events. P300 wave occurs only if the subject is actively involved in the task of detecting the target events.

P300 wave is typically measured most strongly with the electrodes at the parietal lobe. The latency of the P300 peak is usually interpreted as the time consumed in the discrimination of applied event stimulation from the other. Shorter latencies can be interpreted as superior mental capability or easiness of discriminating the target events as the latency increases with the difficulty in discriminating the target from standard events. The amplitude of the P300 wave varies with the improbability of the target event and increases with the attention level (Sur & Sinha, 2009). P300 wave comprised of two sub-components, P3a, and P3b waves. P3a component is elicited earlier with peak latency in the range of 250~280 ms. P3a is associated with brain activity related to early attention and the processing of novelty. P3b component is recorded with a peak at around 300 ms, and the peak latency varies between 250 and 500 ms depending upon the task and the subject. It occurs when the attention-driven event signal is transmitted to temporal and parietal structures. P300 amplitude is used as an indicator of the broad neurobiological vulnerability that underlies disorders. In patients with decreased cognitive ability, P300 waves are recorded with smaller amplitudes and increased latencies than age-matched control subjects.

Brain fingerprinting is a technology to determine whether specific information is stored in a subject's brain based on the

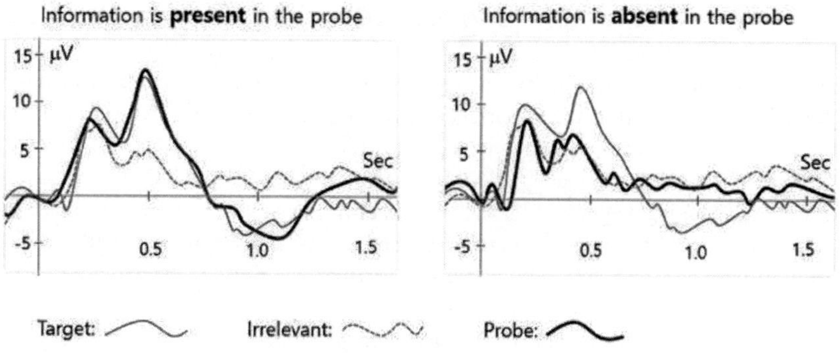

Fig. 8.18 Brain fingerprint based on the P300 wave. When the information is present in the probe event stimulation, P300 is elicited to the probe event like to target event stimulation

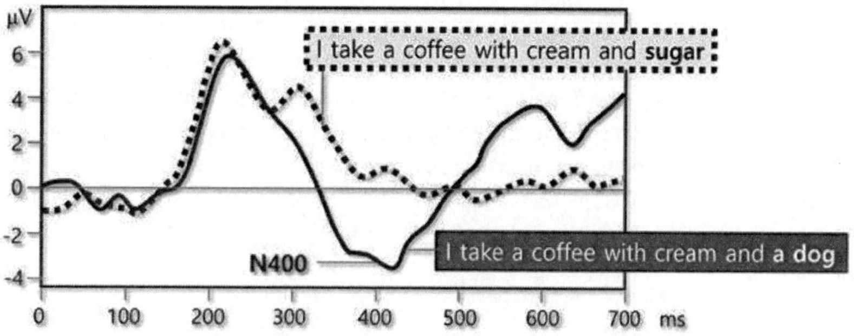

Fig. 8.19 N400 wave in response to semantically unexpected final word

elicitation of a reproducible P300 wave (Fig. 8.18). Probe event stimulation is prepared, which includes the information to identify whether the subject has any meaningful memory of it or not. Oddball paradigm is applied with rare target event stimulation and irrelevant frequent event stimulations. And prepared probe events are presented infrequently interspersed among a series of event stimulations. If the subject recognizes the information in the probe event, the P300 wave is elicited as in the case of target stimulation. However, if the subject does not recognize the information included in the probe event as of any special meaning, the P300 wave is not elicited in the recorded EEG and the recorded potential is close to the response to irrelevant event stimulations. Since the P300 wave has unconscious and uncontrollable aspects, brain fingerprinting is used as a lie detector indicating familiarity with the details of the crime and legal evidence of the criminal investigation just like a fingerprint. The technique is based on the idea of a "memory and encoding related multifaceted electroencephalographic response" (MERMER) developed by Dr. Lawrence Farwell.

5.2.6 N400 Wave

N400 is a negative-going deflection with a peak around 400 ms between 300 and 600 ms after the event stimulation onset (Fig. 8.19). N400 is the brain response to words and other meaningful eventual stimulation and is typically maximal with electrodes over centroparietal sites. N400 is first described during the experiment examining the response to unexpected words in reading sentences. Sentences with anomalous endings with unexpected words, like "I take a coffee with cream and a dog," are presented as an event stimulation for N400. In contrast to the anticipated P300 waves responding to unexpected final words, these anomalous endings elicited a large negativity wave with around 400 ms latency compared to the response to the sentences with expected endings, like "I take a coffee with cream and sugar." N400 wave is related to semantic disagreement and is not just a response to unexpected words. N400 is inversely related to the expectancy of a given word to end a sentence.

5.2.7 P600 Wave

P600 wave is a positive-going deflection with its peak around 600 milliseconds after the presentation of stimulation. P600 wave is a language-sensitive response like the N400 wave. While the N400 wave is sensitive to the semantic violation, P600 occurs when the sentences contain a syntactic violation, have a nonpreferred syntactic structure, or have a complex syntactic structure (Osterhout & Holcomb, 1992). For example, P600 is recorded with the stimulation like "The dog will eating" compared to the correct sentence "The dog will eat." It is used as a common experimental method in investigating sentence processing of the human brain.

6 Magnetoencephalogram (MEG)

Magnetoencephalogram (MEG) is the record of magnetic fields generated during the electrical activity of the brain using very sensitive magnetometers. As static current produces a static magnetic field and changing current produces changing magnetic field as formulated in Maxwell's equation, a current produced during the activation of the brain produces a magnetic field around the acting neuronal cells. MEG records the same electrical activities of the brain as EEG does, but MEG senses accompanying magnetic fields while EEG senses electrical fields.

The magnetic field generated during brain activity is extremely weak in its strength. The strength of the magnetic field is determined by the number of neuronal cells synchronously activated for the generation of brain rhythms. The strength of cortical activity is in the order of 10^{-14} Tesla and alpha rhythm produces a magnetic field of strength in the order of 10^{-12} Tesla (Fig. 8.20). This strength of MEG is much smaller than the earth's static magnetic field of 10^{-4} Tesla and smaller than environmental magnetic noise in the order of 10^{-7} Tesla. To avoid excessive contamination from environmental magnetic noises, basically, MEG systems are operated in a magnetically shielded room. Within a body, there are also other magnetic field signals generated from other organs, like the heart, eyes, and muscles. Their amplitudes, if they are generated, have generally stronger than the brain activity and interfere with brain activity to be measured clearly. So, the essential issue of MEG recording is to devise the detecting method to sense the weakest signal competing with stronger environmental and biological noises.

6.1 MEG Signal

When the magnetic field is generated from the activation of the neuronal cells in the cortex, the measured signal is dependent on the direction of the generated current. At the cortex in the gyrus region, pyramidal cells are positioned perpendicular to the scalp, and the generated current is also directed perpendicular to the scalp. Current flowing perpendicular to the scalp builds a magnetic field perpendicular to the flow of the current, which is parallel to the scalp surface. In this configuration, it is difficult to sense the generated magnetic field with the coil sensors located over the scalp. In contrast, the current generated in the sulcus region is parallel to the scalp and the accompanying magnetic field is built in the plane perpendicular to the scalp. In this configuration, generated magnetic field is easily extended to the scalp and it can be sensed by the pick-up coils located over the scalp. That is, MEG is most sensitive to cortical currents that are oriented tangentially to the scalp and skull, located in the cortical sulcus and fissure region. If the current is tilted with respect to the skull surface, the MEG signal is developed in proportion to the tangential component. This directional preference is in great contrast to the EEG which is more sensitive to a current directed perpendicular to the scalp and skull (Fig. 8.21).

6.2 MEG Sensors

6.2.1 Superconducting Quantum Interference Device (SQUID)

A superconducting quantum interference device (SQUID) is the sensor currently used to sense the magnetic field in the

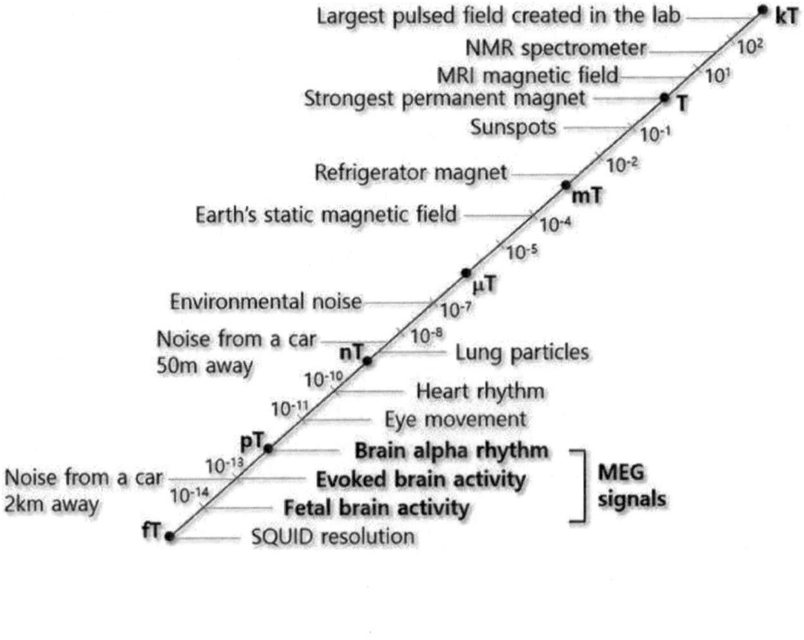

Fig. 8.20 Strength of magnetic fields of brain activities compared to the other magnetic fields

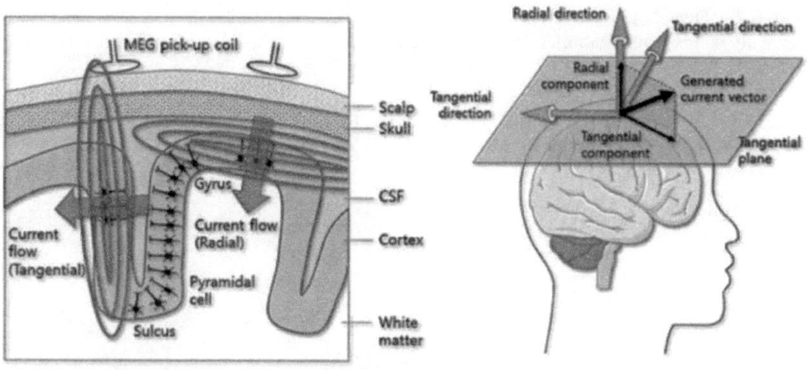

Fig. 8.21 MEG mostly picks up magnetic field generated from the pyramidal cells directed tangentially to the scalp in the sulcus region

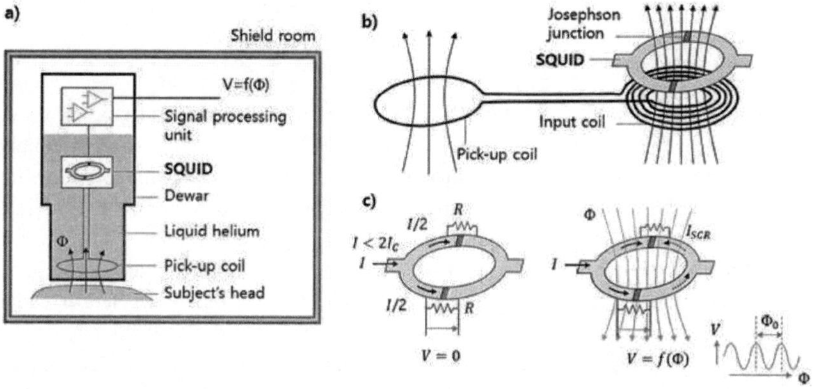

Fig. 8.22 SQUID for MEG. (**a**) MEG system based on the SQUID sensor detects magnetic field from the subject's head in a shielded room. (**b**) Structure of SQUID sensor and macrostructure to multiply the magnetic flux. The pick-up coil detects the magnetic flux of the site and multiple turns of the input coil provides the multiplied magnetic flux to SQUID to increase the sensitivity. (**c**) Working principle of SQUID with and without applied magnetic flux. I bias current, I_C critical current of superconductor, I_{SCR} screening current caused by the applied magnetic flux. Φ_0 one flux quantum, Φ applied magnetic flux

MEG system. SQUID can measure an extremely small magnetic field with a sensitivity that approaches the quantum limit of 10^{-15} Tesla. It was invented in 1964 at Ford Research Labs after Brian Josephson postulated the Josephson effect in 1962.

MEG system using SQUID sensor is used in a shielded room. The recording unit is composed of a pick-up coil, input coil, SQUID sensor, and signal processing unit (Fig. 8.22a). Liquid helium fills the dewar to keep the extremely low temperature for maintaining the superconductor characteristics of the sensing part. The pick-up coil picks up the magnetic field of the sensing site with a larger coil sensing area. Input coil multiples the sensed magnetic flux with an increased number of turns and focuses magnetic flux to a smaller sensing area of the SQUID (Fig. 8.22b). The pickup coil and input coil are also made of superconductive material. And, the applied flux is sensed by the SQUID.

SQUID is based on two effects appearing in the superconductor loop, flux quantization, and superconducting tunneling. The basic structure of SQUID is a superconducting loop interrupted by two Josephson junctions with parallelly connected resistors to eliminate hysteresis in their current-voltage characteristics (Fig. 8.22c). With no magnetic field, applied bias current I flows through two branches splitting into equal halves $(1/2)I$ that also flow through the two Josephson junctions by the tunneling effect. The critical current of superconductor, I_C, is the maximum current that a superconductor can carry with zero resistance and a current greater than I_C will cause resistance in a flow of current. As the biased current is lower than the critical current I_C, no resistance develops along the path and no voltage develops across the junction. If a magnetic field, Φ, is applied through the SQUID loop, a screening current, I_{SCR}, is developed and circulates the loop. Developed screening current is added to the biased current. Due to the relative direction of both currents, currents flowing through the junctions become $\left(\frac{1}{2}\right)I + I_{SCR}$ at one junction and $\left(\frac{1}{2}\right)I - I_{SCR}$ at the other junction. If the summed current $\left(\frac{1}{2}\right)I + I_{SCR}$ becomes bigger than the critical current I_C, there develops a resistance and the voltage is measured across the parallel resistor which is the

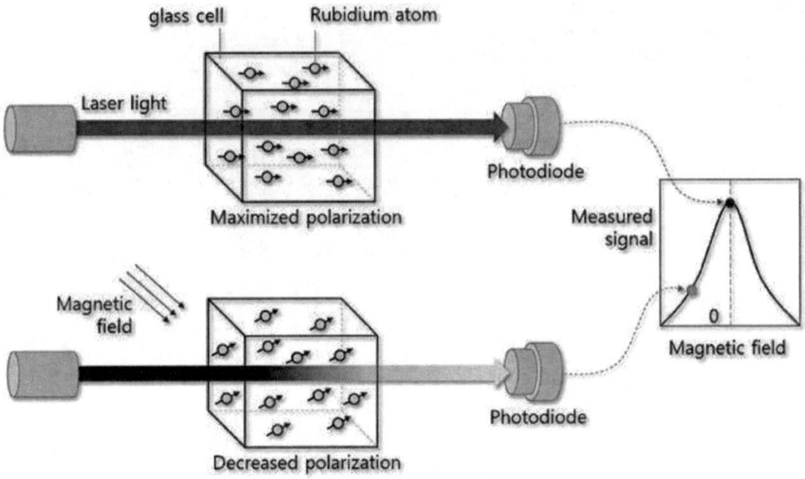

Fig. 8.23 Operating principle of optically pumped magnetometer (OPM)

nonlinear periodic function of the applied magnetic field. The following electrical circuit picks up this voltage and the magnitude of the applied magnetic field is detected. A flux-locked-loop (FLL) circuit is often used, in which the output voltage is converted into a current by a resistor and fed back into the SQUID as a flux, nulling the input magnetic flux and the voltage across the feedback resistor is proportional to the input magnetic flux. Thus, the SQUID works as a flux-to-voltage transducer, which converts a small change in magnetic flux into voltage and has an extremely low magnetic flux noise.

6.2.2 Optically Pumped Magnetometer (OPM)

Recently, new sensor technology has emerged to measure MEG signals, called an optically pumped magnetometer (OPM). OPM measures the intensity of the laser light, that has passed through the gas-filled glass cell, using photodiodes (Fig. 8.23). The glass cell encloses the vapor of rubidium-87 atoms. The wavelength of the transmitting laser is precisely tuned to the resonance frequency of gas atoms in the glass cell (Boto et al., 2018). A resonant laser beam "pumps" the Rubidium atoms into a quantum state in which their angular momentum is aligned with the beam. The pumped atomic vapor has a net magnetization and it is highly sensitive to external magnetic fields. With a zero magnetic field, all atoms are in the same state induced by polarization. In this state, there is no further absorption of photons and the intensity of light passing through the cell is maximized. However, if an external magnetic field affects the glass cell, it makes developed net magnetization process and results in a drop of polarization. A drop in polarization increases the absorption of light and the signal at the photodetector decreases. If the magnetic field affecting the glass cell is changed, it results in a corresponding change in the transmitted intensity of laser light and it is precisely detected by the photodiode (Drung, 2016).

Currently, the sensitivity of OPMs is at a similar level to the SQUID sensors. However, in contrast to SQUID, OPMs do not engage the expensive cryogenic system and can be operated at room temperature providing great benefits in several aspects. Excluding the cryogenic system allowed the OPM to be implemented with a greatly reduced compact size. Therefore, the OPM sensor can be integrated into a mobile system and placed directly over the head in a headset. The mobile system in the headset unit allows the sensor array to move with the head and makes the MEG measurement less susceptible to subject motion. The smaller size of the detector and removing the thermally insulating gap reduced the distance between detectors and the brain cortex. It can significantly increase the power of the measured signal by a factor of 5~7 compared to the SQUID (Iivanainen et al., 2017). While the background noises make the measurement still difficult, another issue to be solved is the effect of static Earth's magnetic field. It has no impact on MEG measurement based on SQUIDs implemented in the fixed headset. However, with OPMs, when subjects move freely during a scan, OPMs also move relative to the Earth's static magnetic field and artifactual changes in the magnetic field are generated, which should be eliminated or compensated.

6.3 Comparison of MEG and EEG

There has been a continuous controversy on the role of the MEG. Does the MEG really provide information that cannot be derived from the EEG? It is basically the question of whether biomagnetic measurements are independent or inter-dependent to bioelectrical measurements.

Theoretically, electrical behavior and the accompanying magnetic behavior are completely dependent as described in Maxwell's equations. If we were able to measure all the magnetic signals accurately at all points of the space, we

Table 8.2 Comparison of EEG and MEG

	EEG	MEG
Signal type	Electrical voltage	Magnetic flux
Signal strength	0.1~200 μV	1~1000 fT
Primary source	Extracellular current	Intracellular current
Detected current sources	Radial and tangential current	Tangential current
Skull effect	Greatly blocks and distorts current flow	Almost transparent to magnetic field
Measurement reference	Indifferent or active reference	Need no reference
Number of channels	21~256 channels	About 300 channels
Spatial resolution	1.0~2.0 cm	0.2~0.5 cm
Temporal resolution	1.0 ms	1.0 ms
Preparation time for recording	Need some preparation time to attach electrodes	Shorter preparation time
Purchasing & maintenance price	Not expensive	Expensive
Shield room	No need	Required

can derive corresponding electric activities according to the electromagnetic relationships. However, perfect measurement is not possible. In the MEG measurements recording the activity of the living brain, constraints on measuring the signal limit the measurement range only to the upper surface of the skull. Magnetic fields on the lower side of the brain and the inner side of the skull cannot be measured. With partially recorded magnetic signals, electrical activities occurring within the brain cannot be derived accurately. This is also the same in the reverse direction. As EEG measurement is also not perfect, the MEG signal cannot be derived from the measured EEG. Even though electrical and magnetic activities of the brain are completely linked in their basics, MEG and EEG view the occurring activities only partially projected to different windows at different angles.

MEG has some advantages over EEG (Table 8.2). In contrast to EEG, the skull, scalp, and brain tissues do not affect the measurement of the magnetic field. The magnetic field measured over the head is almost not interrupted and not distorted by the intervening medium and it produces better spatial resolution in MEG. So, MEG is actively used in the inverse problem to localize the current sources from the measured signals. MEG does not need a reference electrode like EEG and examination is prepared in a shorter time by skipping the time attaching the multiple electrodes on the scalp as in EEG. However, MEG examination should be performed in a shielded room, and the purchasing and maintenance expenditure is much more expensive than EEG.

In conclusion, EEG and MEG are only partially independent and they complement each other. Additional examination to one of the MEG or EEG examinations brings extra information on the bioelectric activity of the brain. They have certain unique properties that make either of them more beneficial in certain applications.

7 Diagnosis and Applications

7.1 EEG in Clinical Diagnosis

Before the advent of functional magnetic resonance imaging (fMRI), EEG has been the unique diagnostic tool to evaluate the functional abnormalities and disorders of the brain. A routine EEG examination is performed in 20~40 min. During the examination, subjects are asked to rest quietly and open or close their eyes from time to time. Subjects are also asked to breathe in and out deeply for a few minutes. In most cases, a flashing light stimulation is presented to see if it causes any change in brain activity.

Among the many neurological disorders, epilepsy is the disorder most closely evaluated with EEG from the beginning. EEG is used as an essential tool in detecting any type of epilepsy, differentiating between epileptic and non-epileptic events, and classifying the different types of epilepsy. Before the surgery, ECoG is used to be recorded on the direct surface of the cortex to localize the epileptic region more accurately. In the diagnosis of nonepileptic seizures, video-EEG monitoring is used as the gold standard method. In video-EEG monitoring, EEG is recorded with closed-circuit video for a prolonged period. The recorded EEG and video are displayed simultaneously to easily evaluate the point-to-point correlations between EEG events and accompanying behavioral patterns. In addition to epilepsy, EEG is also used as a diagnostic tool in evaluating neurological disorders including brain tumors, brain damage due to head injury, brain dysfunction with a variety of causes, stroke, and inflammation of the brain.

EEG is the essential signal in sleep clinics to evaluate sleep-related disorders. Nocturnal polysomnography is performed during the whole night and records the multiple channels of biological signals including more than 2 channels of EEG. EEG does an essential role in classifying the sleep stages during sleep. Multiple sleep latency test (MSLT), which is also known as a daytime nap study, is another type of polysomnographic recording to check excessive daytime sleepiness. By measuring how quickly subjects fall asleep in a quiet environment during the day, MSLT is used to diagnose narcolepsy and idiopathic hypersomnia.

Continuous recording of EEG is used in the intensive care unit (ICU) setting. EEG can monitor the consciousness level and detect any changes in the brain states With continuous EEG monitoring, any problematic conditions in the brain can be treated in the earliest stage. EEG monitoring is useful in the detection of nonconvulsive status epilepticus, assessment of anesthetic states, and detection of secondary cerebral ischemia associated with subarachnoid hemorrhage.

EEG is mandatory in determining brain death. To determine brain death, EEG should record electrocerebral inactivity (ECI). The ECI can be confirmed when a 30-min good-quality EEG recording shows complete electrocerebral silence, which is defined as no cerebral activity greater than 2 μV with intense somatosensory or audiovisual stimuli (Szurhaj et al., 2015).

7.2 EEG and MEG as Research Tools

As EEG or MEG represents the brain processes and the responses to the environmental stimulations, they are actively used as evaluating tools of the brain responses in the disciplines of neurocognitive science, neurolinguistics, cognitive psychology, and psychophysiology. MEG is specialized to determine the location of electric activity within the brain, inversely from the multiple channels of recorded signals outside the head. By solving the inverse problems, the electrical activity of the brain is localized by a single dipole or distribution of multiple dipoles. Location, distribution, and the strength of dipoles are recognized as the activity of the brain to the corresponding events or stimulations. In neuromarketing research, analysis of EEG is used to understand the consumer's preferences, behaviors, and emotions, and forecast their decision-making processes. Brain-computer interface (BCI) is one of the most common applications of EEG. BCI uses real-time EEG data to control external devices and to help individuals having severe disabilities who are difficult to be supported by other assisting methods. Using the recently developed mobile EEG recording units, BCI is introduced actively in the entertainment where the brain responses are used directly in solving the suggested tasks. In biometric studies, EEG is used in identifying and authenticating people using physiological properties such as unique features like fingerprints. EEG devices have been applied in the research to make a comfortable environment, improve well-being and life quality, and boost the learning process. Neurofeedback data can be used for either clinical or non-clinical research (Soufineyestani et al., 2020).

References

Acharya, J. N., & Acharya, V. J. (2019). Overview of EEG montages and principles of localization. *Journal of Clinical Neurophysiology, 36*(5), 325–329. https://doi.org/10.1097/WNP.0000000000000538

Aminoff, M. J., Braunwald, E. F., Fauci, A. S., Kasper, D. L., Hauser, S. L., Longo, D. L., & Jameson, J. L. (2001). Electrophysiologic studies of the central and peripheral nervous systems. *Harrison's Principles of Internal Medicine, 2*, 2331–2336.

Andrillon, T., Nir, Y., & Staba, R. (2011). Sleep spindles in humans: Insights from intracranial EEG and unit recordings. *The Journal of Neuroscience, 31*(49), 17821–17834. https://doi.org/10.1523/JNEUROSCI.2604-11.2011

Boston, J. R. (1981). Spectra of auditory brainstem responses and spontaneous EEG. *IEEE Transactions on Biomedical Engineering, 28*(4), 334–341. https://doi.org/10.1109/TBME.1981.324801

Boto, E., Holmes, N., Leggett, J., et al. (2018). Moving magnetoencephalography towards real-world applications with a wearable system. *Nature, 555*, 657–661. https://doi.org/10.1038/nature26147

Brienza, M., & Mecarelli, O. (2019). Neurophysiological basis of EEG. In O. Mecarelli (Ed.), *Clinical electroencephalography*. Springer. https://doi.org/10.1007/978-3-030-04573-9_2

Chu, N. S. (1986). Median and tibial somatosensory evoked potentials. Changes in short- and long-latency components in patients with lesions of the thalamus and thalamo-cortical radiations. *Journal of the Neurological Sciences, 76*(2–3), 199–219. https://doi.org/10.1016/0022-510x(86)90169-3

Coleman, H., & Brookes, M. (2021). Quantum sensing the brain. *Physics World, 34*(2), 23–27.

Drung, D. (2016). Introduction to Nb-based SQUID sensors. *IEEE/CSC & ESAS Superconductivity News Forum.*

Freunberger, R., Klimiesch, W., Doppelmayr, M., & Holler, Y. (2007). Visual P2 component is related to theta phase-locking. *Neuroscience Letters, 426*, 181–186.

Iivanainen, J., Stenroos, M., & Parkkone, L. (2017). Measuring MEG closer to the brain: Performance of on-scalp sensor arrays. *NeuroImage, 147*, 542–553.

Kothari, R., Bokariya, P., Singh, S., & Singh, R. (2016). A comprehensive review on methodologies employed for visual evoked potentials. *Scientifica, 1*. https://doi.org/10.1155/2016/9852194

Light, G. A., & Braff, D. L. (2003). Sensory gating deficits in schizophrenia, can we parse the effects of medication, nicotine use, and changes in clinical status? *Clinical Neuroscience Research, 3*, 47–54.

Näätänen, R., Paavilainen, P., Rinne, T., & Alho, K. (2007). The mismatch negativity (MMN) in basic research of central auditory processing: A review. *Clinical Neurophysiology, 118*(12), 2544–2590. https://doi.org/10.1016/j.clinph.2007.04.026

Oostenvelda, R., & Praamstrac, P. (2001). The five percent electrode system for high-resolution EEG and ERP measurements. *Clinical Neurophysiology, 112*, 713–719.

Osterhout, L., & Holcomb, P. J. (1992). Event related brain potentials elicited by syntactic anomaly. *Journal of Memory and Language, 31*, 785–786.

Soufineyestani, M., Dowling, D., & Khan, A. (2020). Electroencephalography (EEG) technology: Applications and available devices. *Applied Sciences, 10*, 7453. https://doi.org/10.3390/app10217453

Stone, J. L., & Hughes, J. R. (2013). Early history of electroencephalography and establishment of the American Clinical

Neurophysiology Society. *Journal of Clinical Neurophysiology, 30*(1), 28–44. https://doi.org/10.1097/WNP.0b013e31827edb2d

Sur, S., & Sinha, V. K. (2009). Event-related potential: An overview. *Industrial Psychiatry Journal, 18*(1), 70–73. https://doi.org/10.4103/0972-6748.57865

Szurhaj, W., Lamblin, M. D., Kaminska, A., & Sediri, H. (2015). EEG guidelines in the diagnosis of brain death. *Neurophysiologie Clinique, 45*(1), 97–101. https://doi.org/10.1016/j.neucli.2014.11.005

Umbrichta, D., Kollera, R., Schmida, L., Skraboc, A., Grübeld, C., Huberb, T., & Stassenb, H. (2003). How specific are deficits in mismatch negativity generation to schizophrenia? *Biological Psychiatry, 53*(12), 1120–1131.

Wilkinson, R. T., & Morlock, H. C. (1967). Auditory evoked response and reaction time. *Electroencephalography and Clinical Neurophysiology, 23*(1), 50–56. https://doi.org/10.1016/0013-4694(67)90140-X

Electrical Signals from the Muscles and Nerves 9

In addition to the heart and the brain, all other organs within the body are innervated and controlled by the nerves. When the organs are functioning according to the signals from the brain through the nerves, a group of muscles is activated to perform the predetermined function. When the volume of muscles or nerves is large and their locations are close to the body surface, their electrical activities can be recorded from the body surface noninvasively. Figure 9.1 shows the bioelectrical signals being recorded from the body surface.

Table 9.1 and Fig. 9.2 show the amplitude and frequency range of bioelectrical potentials.

1 Electromyogram (EMG)

Electromyogram (EMG) is a record of the electrical activities generated from the muscles by the stimulation of motor neurons. It is associated with the spread of action potential along the muscle fiber membranes. The amplitude of the EMG is interpreted in relation to the force exerted by the muscles. By analyzing the EMG, contractile activity and any abnormality of muscles and nerves can be evaluated. EMG is measured with surface or needle electrodes.

1.1 Types of Muscles

Muscle is a contractile tissue functioning to produce a motion by which to interact with the environment. In the human body, muscles are functioning with various roles of pumping blood, supporting postures, lifting heavy weights, speaking, and making facial expressions. Most of the muscle contractions seem to be voluntarily occurring with intention, but there are many involuntary muscles contracting without consciousness. Muscle cells use the chemical energies of adenosine triphosphate (ATP), which is derived from the metabolism of food, to produce force and movement. A muscle can be interpreted as a machine that converts chemical energy into mechanical energy under the control of electrical signals. And the efficiency of the machine is about 30~40%, as that portion of inputted energy is converted into mechanical energy and the rest of 60~70% is consumed by the heat.

Muscles are classified into three types. *Skeletal muscle* is the voluntary muscle contracting consciously according to the intention. Skeletal muscles are attached via tendons to the bones to cause the movements and constitute the musculo-skeletal system with bones. Muscle tissues of skeletal muscles are striated and look striped due to the arrangement of the sarcomeres. In the human body, there are 650 skeletal muscles which make up 40% of body weight. Skeletal muscles are present in bilateral pairs for the symmetrical movement of the body between the right and left sides. They are also grouped with opposing pairs, like the biceps and triceps of the upper arm, working together to carry out an action. EMG mostly measures the activities of these skeletal muscles.

Smooth muscle is the involuntary muscle contracting slowly and automatically without consciousness. Smooth muscle is composed of narrow spindle-shaped cells with a single nucleus. It is non-striated as it has no sarcomeres appearing striation. Smooth muscles are found in various places of the internal organs. The walls of hollow organs, like the stomach, intestines, bladder, and uterus, are composed of smooth muscles. Smooth muscles also constitute the walls of passageways, like arteries, veins, lymph vessels, and tracts of the respiratory, urinary, and reproductive systems. Smooth muscles are also found in the eyes for controlling the eyeballs and shaping the lens. The piloerection muscle in the skin also belongs to the smooth muscle.

Cardiac muscle is specific to the heart and constitutes the main volume of the heart wall. Cardiac muscle possesses contractile units of sarcomeres like skeletal muscle and is distinguished by it from the smooth muscle. However, cardiac muscle differs from skeletal muscle as it exhibits rhythmic contractions without voluntary control and is

K. S. Park, *Humans and Electricity*, https://doi.org/10.1007/978-3-031-20784-6_9

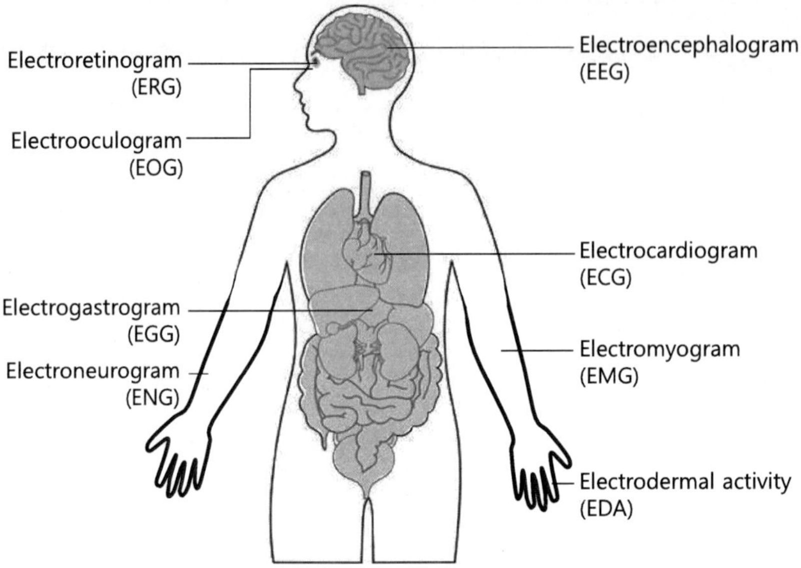

Fig. 9.1 Bioelectrical signals from the body surface
Electrocardiogram (ECG) is the record of the electrical activity occurring during the heart contraction
Electroencephalogram (EEG) is the record of the electrical activity of the brain at the scalp surface
Electromyogram (EMG) is the record of the electrical activity of the muscles during the contraction in response to a nerve's stimulation
Electroneurogram (ENG) is the record of the electrical activity of the neurons in the peripheral nervous system mostly to measure the conduction velocity of the neuron
Electrooculogram (EOG) is the record of the standing potential existing between the front and the back of the human eyeball
Electroretinogram (ERG) is the record of the electrical activity of the retina in response to a light stimulus
Electrogastrogram (EGG) is the record of the electrical activities of the stomach representing spontaneous muscular contraction
Electrodermal activity (EDA) is the record of the skin conductance reflecting the sweat secretion affected by the sympathetic activation

Table 9.1 Amplitude and frequency ranges of bioelectrical potentials

Bioelectrical signal	Amplitude	Frequency
ECG	0.5~4 mV	0.05~250 Hz
EEG	5~200 μV	0~150 Hz
EMG	0.1~5.0 mV	0~10 kHz
ENG	0.01~3.0 mV	100~1 kHz
EOG	0.1~2.0 mV	0~100 Hz
ERG	0.5 μV~1.0 mV	0~200 Hz
EGG	50~500 μV	0.015~0.3 Hz
EDA	0.1~40μS	0.045~0.25 Hz
Action potential	10 μV~100 mV	100~10 kHz

Varela et al. (2019)

distinguished by it from the skeletal muscle. The functioning of the cardiac muscles is measured and interpreted with ECG.

1.2 Structure of Skeletal Muscles

Typically, a skeletal muscle spans a joint and is connected to bones by tendons at both ends. One side of a joint remains relatively fixed and the other side moves by the muscle contraction. Skeletal muscles involved in the movements are varied considerably in their sizes, shapes, and fiber arrangements. The length of the smallest skeletal muscle, the stapedius muscle in the middle ear, is about 1 mm, and the longest skeletal muscle, the sartorius muscle which extends from the pelvis to the tibia across the hip joint, thigh and knee joint, is about 600 mm. While they are various in size and shape, skeletal muscles have the general microstructure shown in Fig. 9.3.

A skeletal muscle is composed of hundreds or even thousands of muscle fibers. The number of muscle fibers in a muscle is genetically determined and it does not increase with physical training. Muscle fibers are organized together into bundles called fascicles and wrapped in a connective tissue called perimysium. A muscle fiber is a single cylindrical muscle cell surrounded by connective tissue called the endomysium. The connective tissue covering muscle fiber

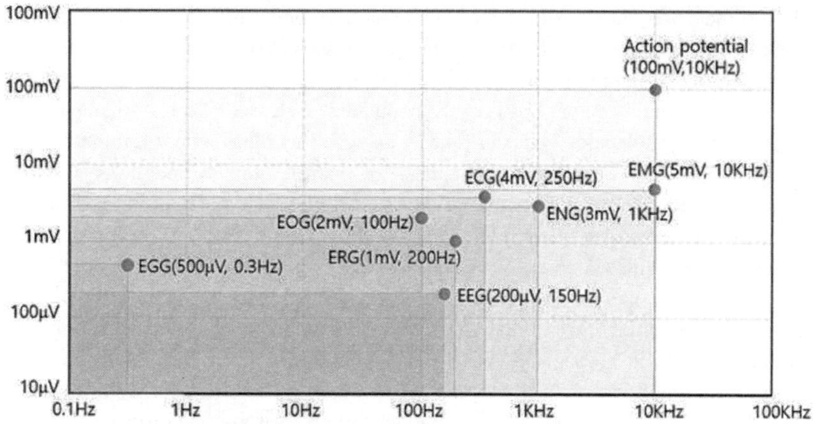

Fig. 9.2 Maximal range of amplitudes and frequencies of bioelectrical signals

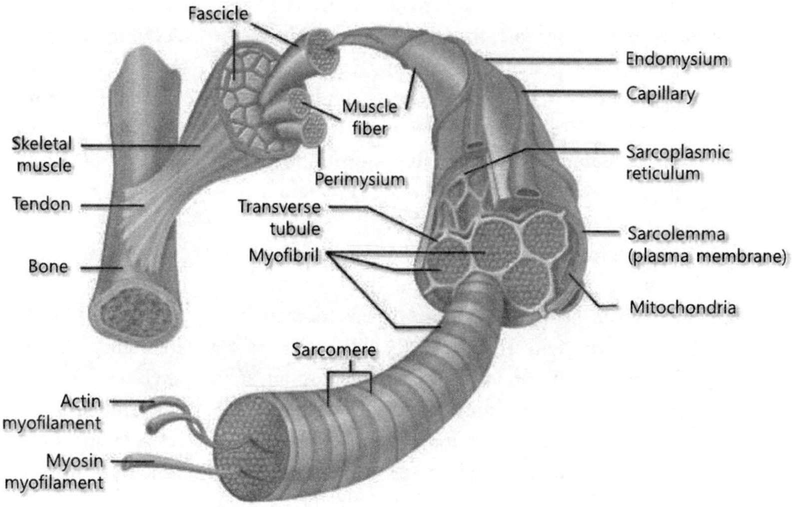

Fig. 9.3 Microstructure of a skeletal muscle

supports the muscle cell to withstand the forces of contraction. The coverings also provide pathways for an abundant supply of blood vessels and nerves. The plasma membrane of muscle fibers is called the sarcolemma and the cytoplasm inside the sarcolemma is referred to as sarcoplasm. Sarcolemma encloses all cellular organelles including the mitochondria and sarcoplasmic reticulum. The sarcoplasmic reticulum stores, releases, and retrieves calcium ions, Ca^{2+}, which does an important role in muscle contraction. A muscle fiber is mostly filled with hundreds to thousands of myofibrils which are organized along the length of the cell. The strength of the muscle is directly proportional to the number of myofibrils within a muscle fiber. Physical training increases the number of myofibrils within a muscle fiber and results in increased strength, mass, and bulk of skeletal muscle. Each myofibril contains sarcomeres, the smallest functional unit of a skeletal muscle fiber, connected in series. The sarcomere is a highly organized arrangement of structural and contractile proteins. Each sarcomere is composed of two main protein filaments, actin and myosin, which are the active elements responsible for muscular contraction. Shortening of an individual sarcomere occurs when actin filaments slide past the myosin filaments. Shortening of the sarcomeres leads to the contraction of myofibrils, muscle fibers, and ultimately to the contraction of a whole muscle.

1.3 Generation of Muscle Contraction

Muscle fibers are electrically excitable like neurons. That is, muscle fibers can generate action potentials and the action of muscle fibers is determined by the potential across the membrane of the muscle fiber, sarcolemma. Muscle contraction begins with the arrival of action potential from the central nervous system via a motor neuron. A motor neuron is connected to multiple numbers of muscle fibers through synaptic junctions called the neuromuscular junction. When an action potential arrives at the presynaptic side of the neuromuscular junction, neurotransmitter acetylcholine is released to the synaptic cleft and diffuses across the synaptic cleft to the postsynaptic side, the sarcolemma membrane of muscle fiber (Fig. 9.4a). The motor endplate is the muscle fiber side of the neuromuscular junction and the site where the acetylcholine receptors are densely populated. Released acetylcholine molecules are anchored on acetylcholine receptors and they open the ion channels for positive ions to increase the membrane potential of the muscle fiber. This potential is called *end plate potential.* As the membrane potential across the sarcolemma increases, voltage-gated ion channels start to open and action potential is generated if the membrane potential crosses the threshold. The generated action potential is propagated along the membrane to the end of a muscle fiber.

T(transverse)-tubules are transverse extensions of sarcolemma deep into the center of muscle fiber. The membrane of T-tubules also contains the ion channels and permits a rapid transmission of the action potential into the inner part of the muscle fiber, the sarcoplasmic reticulum (Fig. 9.4b). The sarcoplasmic reticulum is an endoplasmic reticulum organelle and contains high concentrations of calcium ions. As sarcoplasmic reticulum structure is surrounding the myofibrils, it can directly release and retrieve calcium ions at sites of actin and myosin filaments.

Calcium ions released into the myofibril cause the contraction by forming a bridge between the thin actin filament and the thick myosin filament (Fig. 9.4c). Thin filament also includes tropomyosin and troponin in addition to actin molecules. Tropomyosin shields the binding sites on actin molecules and prevents cross-bridge formation with myosin filament. The binding of calcium ions to troponin deforms tropomyosin and exposes the binding sites of the actin filament. By the anchoring of the myosin head into the exposed binding site of the actin filament, a cross-bridge is formed between the two filaments and the myosin filament pulls the actin filament by changing the head angle. The pulling of actin filament by myosin filament shortens the distance between the filaments by approximately 10 nm, and multiple repetitions of this action are required to increase the shortening distance to an anatomical limit.

Cycling of cross-bridge is the repeating action of binding and detaching of myosin head to the binding site of the actin filament. To increase the shortening distance, the bound head of myosin should be detached from the current binding site to form a new cross bridge to the next site. To detach the myosin head from a binding site, ATP is required. ATP causes the myosin head to detach from the actin filament. The energy released by ATP returns the angle of the myosin head into a preparing position for further movement. In the absence of ATP, the myosin head does not detach from actin. The rigor mortis observed soon after someone dies is the result of ATP depletion. Without the supply of ATP to detach myosin heads from the actin-binding sites, the cross-bridges formed to stay in place causing rigidity in the skeletal muscles.

1.4 Motor Unit

A motor unit is the smallest unit of the force to produce a movement. It is composed of a motor neuron and multiple skeletal muscle fibers (Fig. 9.5). When an action potential is transmitted through a motor neuron, it is connected to many muscle fibers by the synaptic junctions called neuromuscular junctions. Muscle fibers innervated to the same motor neuron contract together. To exert a force on a single muscle, a group

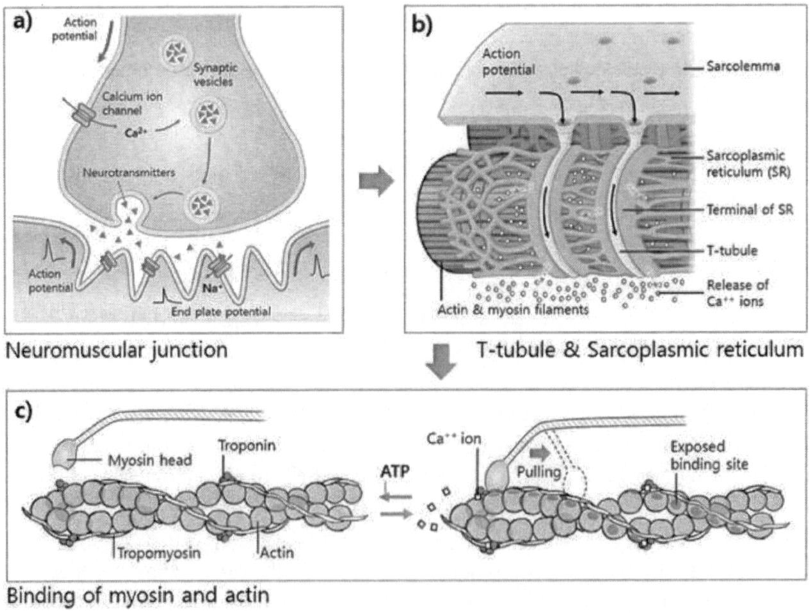

Fig. 9.4 Contraction of muscle fiber. (**a**) Action potential is transmitted from a motor neuron to multiple muscle fibers through neuromuscular junctions. (**b**) Action potential in the muscle fiber propagated along the sarcolemma, T-tubules, and arrives at the sarcoplasmic reticulum and causes the release of calcium ions into the sarcoplasm. (**c**) Released calcium ions act on troponin and make the shielded binding sites of actin filament to be exposed for the binding with the myosin head

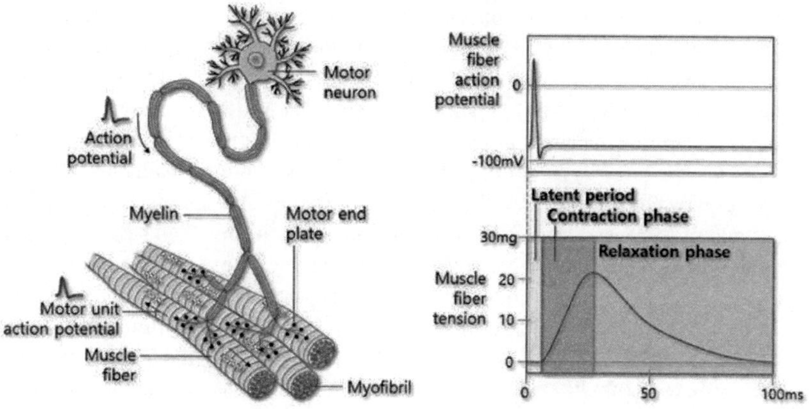

Fig. 9.5 Motor unit and phases of a muscle twitch

of motor units is recruited together to coordinate the targeted force. Muscle fibers innervated to a single motor neuron are distributed over a wide area within a muscle to spread the exerted force evenly in the muscle and to minimize the loss of muscle function when a single motor neuron is damaged. The number of muscle fibers for each motor neuron varies from muscle to muscle from 10 to 1000. Generally, motor neurons in the large muscle masses have innervated with many muscle fibers and the motor neurons in the smaller muscles are innervated with fewer muscle fibers. When a muscle possesses more motor units, it can control the muscle force more finely.

When action potentials are generated at neuromuscular junctions triggered by a motor neuron, they propagate along all the muscle fiber membranes toward the tendon regions. Motor unit action potential (MUAP) is the integrated activity of these action potentials in a motor unit originating from the same motor neuron and represents the contraction of a motor unit (Fig. 9.5). A single contraction of the motor unit is called a *muscle twitch*. It occurs in response to the action potential transmitted via a motor neuron and appears with temporal delay. A temporal course of a muscle twitch is composed of a latent period, contraction phase, and relaxation phase. The latent period is a short delay of less than 5 ms from the time when the action potential reaches the motor endplate to the time tension is started to be observed in the muscle. The contraction phase is the period that the tension is developed and last about 10~100 ms and the relaxation phase is the period that the muscle is returning to its normal length and last around about 10~100 ms.

Motor units are classified into three types by the muscle fibers constituting the motor units. *Slow motor units* innervate to small "red" muscle fibers which contract slowly and produce relatively small forces. And they are activated in a relatively long period of time, about 60–120 milliseconds. With plentiful mitochondria and extensive blood capillary supply, slow motor units are resistant to fatigue and do an important role in requiring sustained muscular contraction, like maintaining a posture. Slow motor units are recruited firstly when a muscle is activated, and they continue to be recruited while the muscle remains active. *Fast fatigable motor units* innervate to larger pale muscle fibers that produce stronger force in a relatively short period of time, about 10–50 ms. However, they are easily fatigued due to the sparsity of mitochondria. They are recruited after slow motor units to provide short bursts of higher muscle tension if they are required. They do an important role in requiring brief exertion of a large force, like running or jumping. *Fast fatigue – resistant motor units* innervate to the muscle fibers of intermediate size and has the properties lying between the other two types of the motor unit. They are substantially resistant to fatigue and produce stronger force than the slow motor units. But they are not fast as the fast-fatigable motor units. They are the motor unit type that loses size and force-producing capability most rapidly with the decreased physical activities, especially in the elderly.

Each individual muscle contains all three types of motor units. This composition of motor unit types provides each muscle to respond from a low-level force to a high-level force smoothly with a certain degree of fatigue resistance. The relative portions of three types of motor units determine the overall capability of the muscle, degree of strength and speed in force production, and endurance to the fatigue. The higher portion of slow motor units makes the muscle have better endurance to the fatigue and the higher portion of fast fatigable motor units makes the muscle have better power and speed capability.

Motor unit potential (MUP) is a summated response of all motor unit action potentials belonging to the same motor unit. It represents the discharge of a single motor neuron magnified by the muscle fibers it activates. As muscle fibers belonging to the same motor unit are relatively homogeneous in diameter and their motor endplates are relatively focalized close to the motor neuron, motor unit action potentials are generated in similar amplitude and in a relatively synchronous manner (Nandedkar, 2014). MUP can be characterized by the duration, amplitude, configuration, and recruitment by a voluntary contraction. When MUP is recorded with needle electrodes, MUPs are less than 2 mV in amplitude and have a duration of 5~10 ms with 3~4 phases. Alteration in MUP shape and recruitment pattern provide basic diagnostic information for the underlying disease processes. Neurogenic processes are characterized by the increased duration and amplitude of MUP and by a reduced recruitment pattern. In contrast, myopathic processes are characterized by the decreased amplitude, shorter duration, and polyphasic morphology of MUPs with early recruitment.

1.4.1 Motor Unit Recruitment

Motor unit recruitment describes the procedure that the additional motors unit are recruited as the targeted contractile force is increased (Fig. 9.6). In an individual muscle, there are three types of motor units of different properties. Each motor neuron has innervated to many muscle fibers within a muscle and corresponding muscle fibers are distributed broadly within a muscle. These muscle fibers are intermingled with the muscle fibers of the other motor units. As the muscle fibers connected to the same motor neuron are activated simultaneously, this broad distribution of muscle fibers provides the spread of induced tension by a motor unit widely within a muscle. A force that a motor unit can produce is dependent on the size and the number of muscle fibers. The activation of a single motor unit produces only a weak muscle contraction. The activation of additional motor units will increase the strength of produced muscle contraction. The higher the number of recruited motor units activated, the stronger force the muscle produces. When motor units are recruited, generally recruitment starts with smaller and slow motor units and toward large and fast motor units as the required contractile force is increased.

There are two district and parallel methods in recruiting the motor units to increase the muscle force produced. They are *spatial recruitment* and *temporal recruitment*. Spatial recruitment is to increase the produced force by increasing the number of recruited motor units at a given time (Fig. 9.7). As the number of spatially recruited motor units is increased, produced forces by the recruited muscle fibers are spatially summated to produce increased contractile muscle force. Temporal recruitment is to increase the produced force by increasing the rate of motor unit recruitment (Fig. 9.8). Consecutive activation of a motor unit before the end of a previous twitch produces additional muscle force to the contracting force produced by the previous twitch. As the rate of motor unit activation is increased, produced force is gradually increasing, and sequentially produced twitches are fused temporarily. The shape of produced force is changing from the quantized waveform into a smooth waveform of sustained contraction, which is called tetanic contraction of the muscle. Temporal recruitment increases the produced forces by the temporal summation in the same number of

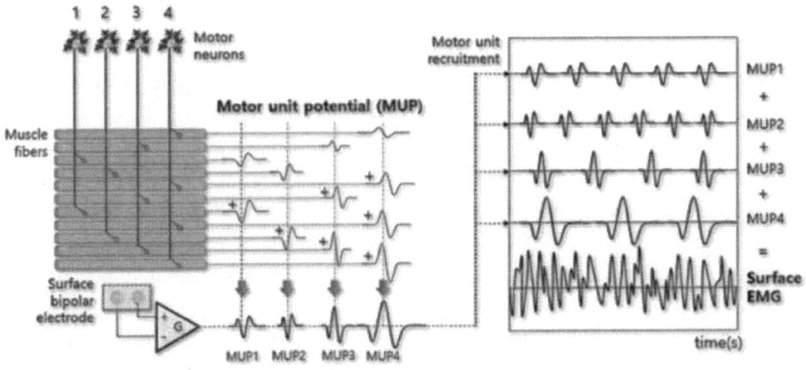

Fig. 9.6 Motor unit potential (MUP) and surface EMG

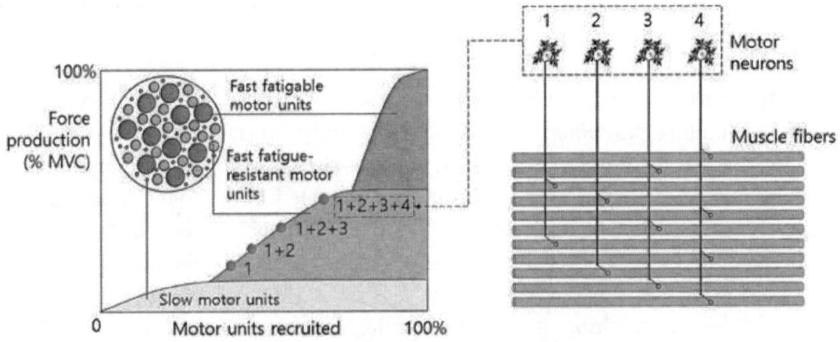

Fig. 9.7 Spatial recruitment of motor units. Smaller motor units in the size and the number of muscle fibers are recruited first. *MVC* Maximum voluntary contraction

Fig. 9.8 Temporal recruitment of motor units

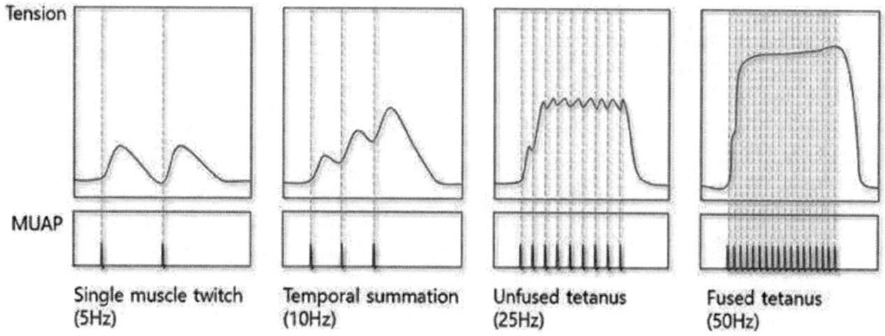

Fig. 9.9 Isometric and isotonic muscle contractions

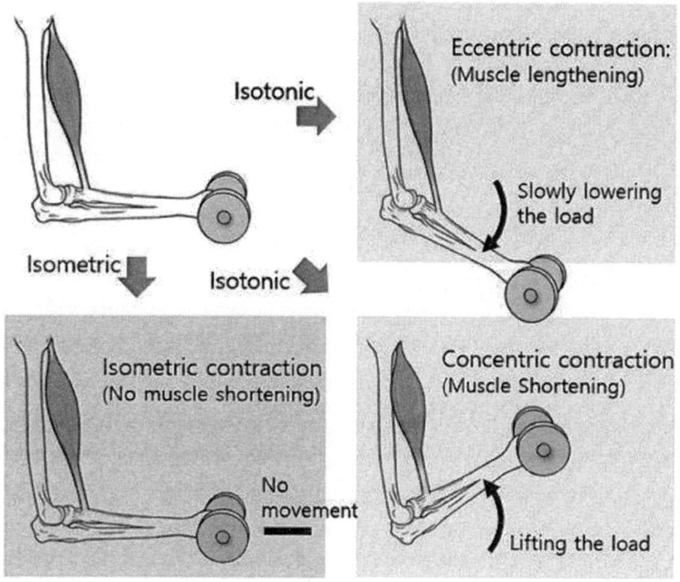

motor neurons. Generally, the primary recruitment at lower levels of muscle contraction is spatial recruitment. After the first motor unit increases the rate to a recruitment frequency of 10~11 Hz, the second motor neuron is recruited. Spatial recruitment takes precedence over temporal recruitment until nearly all motor units are recruited. At this level and beyond, all motor units are driven to their secondary firing range greater than 50 Hz.

1.5 Types of Muscle Contraction

Muscle contractions are defined by the changes in the muscle length during the contraction (Fig. 9.9). In *isometric contractions*, force is generated in the muscle but the muscle length is not changed. Muscles that are maintaining body posture are mainly functioning in isometric contraction. Isometric contractions are also common in muscles responsible for gripping or squeezing with hands and arms. Externally, we experience isometric contractions when we are exerting a force on non-moving objects, like pushing a wall or trying to lift a heavy object while they are not lifted. During the isometric contraction, the allowable maximum force of a muscle can be generated. In *isotonic contractions*, muscle contraction causes the shortening of the muscles and the force is exerted by the changes in muscle length. Isotonic contraction can be concentric or eccentric. A concentric contraction is an isotonic contraction in which the muscles are shortened while the generated force overcomes the external resistance. For example, when we lift a heavy object, a concentric contraction of the biceps causes the arm to bend at the elbow towards the shoulder. An eccentric contraction is an isotonic contraction in which the muscles are elongated while the muscle is still generating force against the external load. In the eccentric contraction, the external load is greater than the force generated, and movement is occurred in opposite direction to the force exerted. For example, eccentric contraction occurs when we carefully lower the heavy objects.

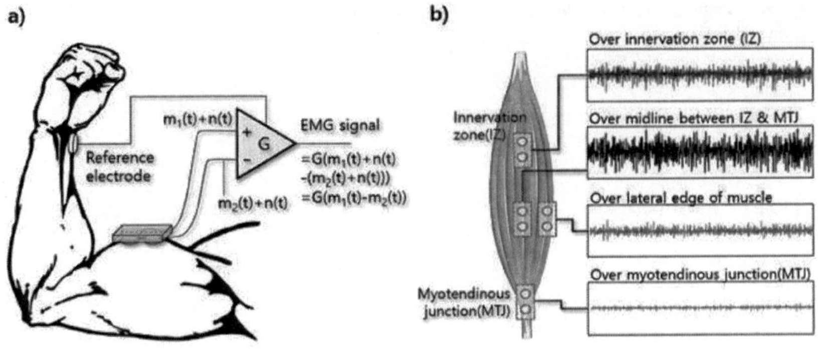

Fig. 9.10 Recording of surface EMG. (**a**) Surface EMG recorded with the bipolar electrode and differential amplifiers which minimizes the effect of artifacts from the body and the environment. (**b**) Amplitude of recoded surface EMG is dependent on the location of electrodes relative to the innervation zone and the myotendinous junction

1.6 Recording of EMG

There are two types of EMG recording. *Surface EMG* is the recording of muscle activity from the surface of the muscle with bipolar electrodes. *Intramuscular EMG* is a recording of muscle activity with needle-type electrodes inserted into the muscle.

Surface EMG records muscle activities indirectly, with a pair of surface electrodes placed on the skin surface immediately above the muscle tissue (Fig. 9.10). A bipolar electrode with an electrode distance of about 15 mm is commonly used for surface EMG recording. When this electrode is connected to a differential amplifier, artifacts from the body and environment commonly interfering with the measurement are canceled. The reference electrode is positioned to the body point which is not affected by the muscle contraction. The location of the surface electrode over the muscle affects the recorded amplitude of the EMG. Maximal amplitude is measured with an electrode on the midline between the innervation zone (IZ) and the myotendinous junction. Recorded surface EMG is the integration of activated motor unit action potentials (MUAPs) weighted by the distance from the recording electrode. Surface EMG does not impede the movement of a muscle. Due to its non-invasiveness and convenience, surface EMG is popularly preferred for muscular function evaluation in various domains including sports medicine. Surface EMG provides information regarding the electrical activation of muscles in relation to the intensity of contraction force, muscle fatigue, and the recruitment of motor units. The sensitivity of the surface EMG is dependent on the depth of the muscle and it restricts surface EMG recording mostly to the superficial muscles.

Intramuscular EMG is recorded invasively with needle-type or wire-type electrodes inserted into the muscle. Needle electrodes usually have one or two electrode surfaces within a needle shaft insulating each other. The monopolar needle electrode is typically a conductive metallic shaft insulated except the tip part and used with the surface electrode for reference. The concentric needle electrode is the common bipolar-type needle electrode. It has a fine wire of active electrode in the central part of the needle insulated from the circumferential shaft acting as a reference electrode. Concentric needle electrode records signals originated from a volume of the radius of about 1 mm containing some 100 muscle fibers. As the hundreds of muscle fibers belonging to the same motor unit are widely distributed throughout the muscle, only 4~6 fibers from a single motor unit belong to this volume. And, the activity of multiple motor units is recorded simultaneously from a concentric needle electrode (Mills, 2005). Recorded EMG with concentric needle electrodes tends to be smaller in amplitude than monopolar recording, but it is more reliable as it is minimally affected by the electrical artifacts from the body and the environment. To record EMG in multiple spots, the needle is retracted by a few millimeters each time while a muscle is in a relaxed state. In many cases, recorded EMG is also converted into sound to help the subject and examiner know when the muscle fibers are activated during the recording period. Needle EMG is highly selective to a small number of motor units and it can identify the activity of each motor unit. As movement is not allowable during the needle EMG recording, it is suitable for the recording of isometric contraction. Needle EMG is mainly recorded in the hospital for clinical diagnosis by a licensed electromyographer.

Fine wire EMG is recorded using flexible fine wire as an electrode which is injected via a hypodermic guide needle. It allows the recording of the deeply located muscles and the flexibility of the fine wire permits the movement of muscle. It is suitable for kinesiological studies which mostly assess the timing of muscle activities.

EMG is assessed with the voluntary contraction of the target muscle (Fig. 9.11). The amplitude of recorded EMG is variant upon many factors, like the location of the electrode and the difference in muscle masses between subjects. Normalized amplitude rather than absolute amplitude is used to compare the response of muscle activity within a

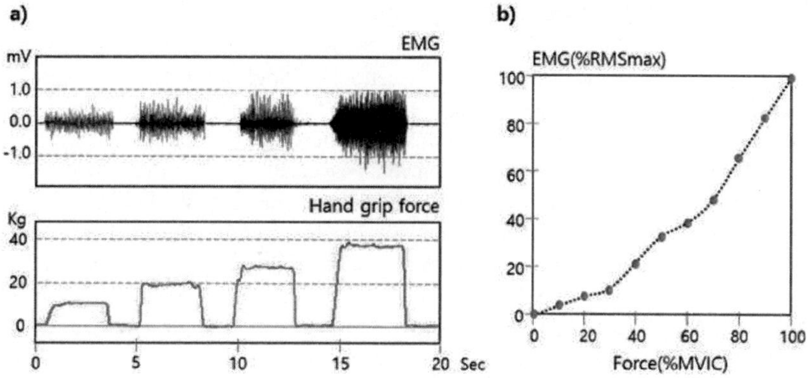

Fig. 9.11 Relationship between recorded EMG and applied force. (**a**) As the handgrip force increases, the amplitude of recorded EMG increases. (**b**) Graphic relationship between recorded EMG and applied force. MVIC: Maximum voluntary isometric contraction, RMSmax: Maximum value of root mean square amplitude

subject of a target muscle. Maximum voluntary isometric contraction (MVIC, MVC) is commonly used as the reference for the normalization of the measured EMG in a given muscle.

1.7 Application of EMG

Clinically, needle EMG evaluation is the standard method for assessing the neurophysiological characteristics of neuromuscular diseases. It is used to distinguish myogenic problems from neurogenic muscle problems. *Fibrillation* is the spontaneous production of action potential in individual muscle fibers. Its amplitude is 10~100 mV, duration is 1~2 ms and it occurs around 10 Hz in frequency. It occurs when the muscle fibers lose their connection to the motor neurons and sarcolemma becomes sensitive due to the broad dispersion of acetylcholine receptors. So, fibrillation is myogenic and not related to neuronal activation. Fibrillations under the skin are not visible and are detectable only by the needle EMG. *Fasciculation* is a visible spontaneous contraction of all or part of muscle fibers belonging to the same motor unit. Its amplitude is 50~500 mV, duration is 2~4 ms and it occurs around 2~20 Hz in frequency. Fasciculation is neurogenic problem. EMG is used to detect abnormalities such as chronic denervation or fasciculations in clinically normal muscle. EMG supports differentiating neuromuscular pathologies by determining the distribution of neurogenic abnormalities. In motor neuron diseases, EMG is an obligatory examination to secure diagnosis by confirming the widespread denervation and fasciculation.

Muscle atonia is the loss of muscle tone resulting in the paralysis of skeletal muscles. It does not occur during wakefulness in normal subjects. However, it does occur during sleep in the REM stage. The REM stage is characteristically differentiated by the totally diminished muscle tone in EMG recording with the presence of rapid eye movement (REM). During the REM stage, normal subjects dream, and their brain activity shows patterns like a waking state with rapid eye movements. However, their body becomes immobilized and it is considered to avoid subjects from performing the actions that occur during the dreaming. Subjects with REM sleep behavior disorder (RBD) do not experience muscle atonia during REM sleep. With alive muscle tone, their REM stage is characterized by significant muscular activity ranging from limb twitching to violent movements that can injure the co-sleeper. Cataplexy is the abnormal muscle atonia appearing during wakefulness. Sudden and transient episode of muscle weakness but with conscious awareness is triggered by emotions or terror. Cataplexy is typically accompanied by narcolepsy patients.

Kinesiology studies the mechanics of human movement to help recover patients' mobility and improve their wellness through exercise. To monitor and evaluate human movement objectively, EMG is used as the primary and unique tool for the recording electrical activity of skeletal muscles. Surface EMG is actively used in studying kinesiological tasks with its noninvasiveness and convenience. Kinesiological electromyography is an established subfield of modern locomotion biomechanics. In many sports training, surface EMG is actively used to reduce the incidence of soft tissue injury and improve the performance of the player.

2 Electroneurogram (ENG)

Electroneurogram (ENG) can be defined as a record of the electrical activity of neurons in the central nervous system or in the peripheral nervous system. As the term EEG is used for the record of the neuronal activity in the central nervous system, the term ENG is used mostly for the record of peripheral neuronal activity. ENG is actively used in nerve conduction studies to check the electrical activity of the peripheral nerve and muscle in conjunction with EMG.

2.1 Nerve Conduction Study

Nerve conduction study (NCS) is an electrophysiological examination commonly used to measure how fast an action potential travels along with the motor or sensory nerve of the human body. By evaluating the nerve function, especially the nerve conduction velocity (NCV), NCS provides diagnostic information helpful in the identification and characterization of disorders involving nerve roots, peripheral nerves, muscle, and neuromuscular junction. NCS is frequently accompanied by a needle EMG examination.

Generally, peripheral nerves contain many afferent and efferent nerve fibers of different diameters and degrees of myelination. NCS examination is to record the fastest part of these nerve fibers and to investigate the focal or extended abnormalities along the length of the nerve. Peripheral nerves are stimulated over the skin with surface stimulators placed close to a nerve. Stimulation applies electrical pulses mildly to the skin over a peripheral nerve and depolarizes the axonal fibers. And, the surface recording electrodes distal to the stimulating electrodes record a nerve action potential propagated over the nerve and a compound muscle action potential arising from the activated motor units. Recorded ENG gives information about the physiologic data of the fastest fibers by the time taken to conduct an impulse. NCS is performed on the sensory nerve, motor nerve, or mixed nerve, and findings of NCS often show specific patterns of abnormalities indicating neuromuscular disorders (Ahn, 2018).

Motor nerve conduction study is performed by electrical stimulation of a motor nerve and by recording the compound muscle action potential (CMAP) using the surface electrodes located over a muscle innervated (Fig. 9.12). The CMAP is the summation of nerve action potential and activated motor unit potentials and has an amplitude of around 10 mV. The latency is measured from the recorded CMAP waveform by the elapsed time from stimulus onset to the rising edge of CMAP.

To record CMAP, the active electrode is placed closer to the stimulation site over the muscle belly and the reference electrode is placed farther from the stimulation site at the electrically inactive site. A ground electrode is also placed between the stimulating and recording sites to provide a zero-voltage reference. Stimulating electrodes are placed over the targeted nerve with the cathode electrode closer to the recording site. Motor nerve conduction studies are performed routinely on the median nerve and ulnar nerve in the arm, and on the tibial nerve and the peroneal nerve in the leg. Stimulations are performed at two sufficiently separated sites to derive recognizable latency between the recorded CMAP, for example, one at the elbow and another at the wrist. The stimulating current is slowly increased from a baseline to the level that causes no more amplitude increase in recorded CMAP. At this supramaximal stimulation, CMAP is recorded reproducibly and the latency between the stimulation onset and rising edge of CMAP can be measured accurately.

Normally, stimulation of the motor nerve at the wrist and the elbow shows two CMAPs of similar shape and amplitude. However, the latency between the stimulation and the CMAP is increased in the elbow stimulation compared to the wrist stimulation due to its much longer distance traveled from the stimulating site. The difference in latency between the recordings represents the conduction time of the fastest nerve fibers between the two stimulation sites. All other factors including neuromuscular junction and muscle activation commonly affect both stimulations and are canceled by latency subtraction. The nerve conduction velocity of the fastest nerve fiber is calculated by dividing the distance between stimulating sites by the latency difference.

$$\text{Nerve conduction velocity} = \frac{\text{Distance between stimulating sites}}{\text{Proximal latency} - \text{Distal latency}} \quad (9.1)$$

The duration of the CMAP represents the temporal dispersion. If there is a demyelinating disease or a severe axonal loss, duration increases by the increased portion of fibers with lowered conduction velocity.

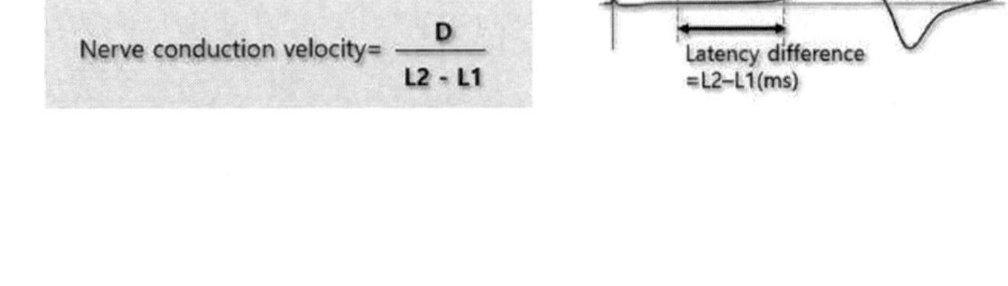

Fig. 9.12 Measurement of motor nerve conduction velocity from the recorded compound muscle action potentials (CMAPs) by the stimulations at distal and proximal sites

A sensory nerve conduction study is performed by electrically stimulating sensory fibers and recording the nerve action potential at a distal point along the same nerve (Fig. 9.13). Generally, the procedure for the recording of sensory nerve conduction study is the same as the motor nerve conduction study. The difference is the amplitude of the recorded potential. Since the recorded potential originates from a smaller volume of neuronal fibers instead of muscle fibers, recorded sensory nerve action potential (SNAP) is much smaller in amplitude than CMAP around 10 mV.

SNAP can be recorded in two directions. SNAP refers to the recording along the normal direction as the sensory electrical signal travels from the distal to the proximal. In the orthodromic SNAP, the sensory nerve is stimulated at a distal site like a finger and the SNAP is recorded at a proximal site like a wrist or elbow. In the antidromic SNAP, SNAP is recorded in reverse to normal direction that sensory electrical signal travels. The sensory nerve is stimulated at a proximal site like a wrist or elbow and the SNAP is recorded at a distal site like a finger. The amplitude of antidromic SNAP is slightly larger than orthodromic SNAP as the nerves at the recording site are more superficial. The sensory latency of the SNAP is measured and the measured latency directly correlates with the sensory conduction velocity. As the time delay in the neuromuscular junction and for muscle activation is not included in the onset latency of measured SANP, nerve conduction velocity can be estimated directly by dividing the distance between the stimulating and recording site by the onset latency. Estimated conduction velocity represents the functioning of only the fastest conducting sensory fiber which has the largest diameter and smallest stimulating threshold. Carpal tunnel syndrome, which occurs often in heavy users of a computer mouse or keyboard combined with awkward working postures of the hands and wrist, is one example of a disease examined with a sensory nerve conduction study on the median nerve.

3 Electrooculogram (EOG)

Electrooculogram (EOG) is the record of the standing potential between the cornea, the front side of the eyeball, and the retina, the backside of the human eyeball. As EOG is measured in positive potential on the front side of the cornea, the eyeball is modeled as a dipole directing forward from the retina to the cornea. With the movement of the eyes, this dipole is rotating along the movement direction and the variation of the generated electric field can be measured by the electrodes attached in the vicinity of the eyes. Due to this characteristic, EOG is mainly used to evaluate eye position and movements. Currently, EOG is the most practical and widely applied technique for eye movement recording in various fields including clinical routine and sight tracking studies.

3.1 Generation of EOG

When EOG was first discovered in 1920 with the electrodes placed in the skin near the eyes, and recorded potentials were varying synchronously to the movement of the eyes, it was initially believed that EOG is the record of the muscular action potentials participating in the movement of the eyes. However, it is now generally agreed that EOG potential is generated by the permanent potential set-up appearing as 0.4~1.0 mV between the cornea and retina. This potential is an indirect recording of the transepithelial potential (TEP), the potential difference between the upper surface and lower surface, of the retinal pigment epithelium (RPE). RPE is the pigmented layer of the retina just outside the neurosensory retina (Fig. 9.14). RPE nourishes retinal visual cells by firmly attaching to the underlying choroid. The developed standing potential changes as the retina adapt to the ambient illumination. The voltage becomes smaller with the dark adaptation

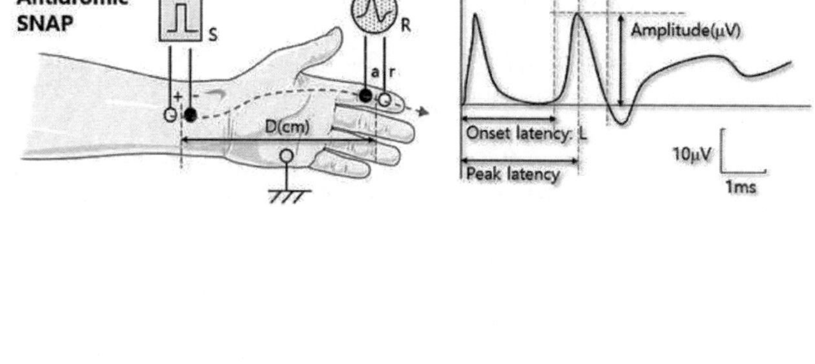

Fig. 9.13 Measurement of sensory nerve conduction velocity from the recorded sensory nerve action potential (SNAP)

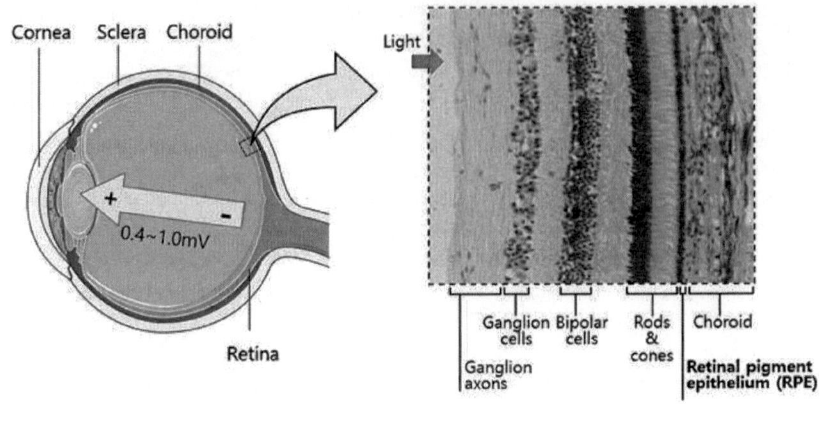

Fig. 9.14 Generation of EOG potential from the retinal pigment epithelium (RPE)

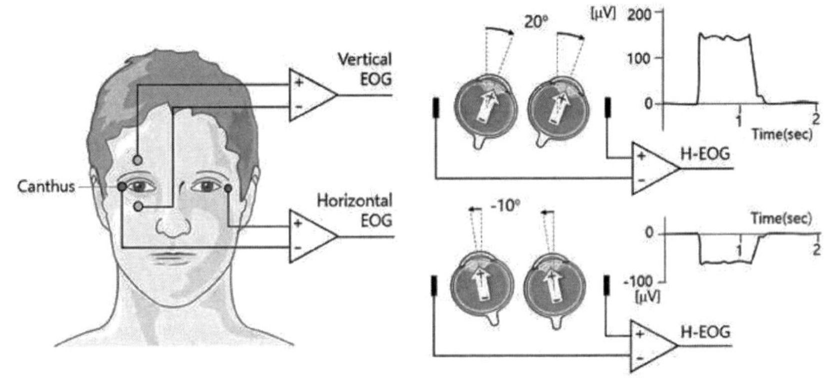

Fig. 9.15 Recording of EOG

and reaches its lowest potential after 8~12 min adaptation. With the lights on, the potential rises and reaches a peak in about 10 min. The ratio of peak value to lowest value is near 2:1. If this ratio is lower than 1.7, RPE is considered not functioning normally (Constable, 2017).

3.2 Recording of EOG

Typically, EOG is recorded with two pairs of silver-silver chloride electrodes. One pair of electrodes are placed to record the horizontal movement of the eyes and another pair of electrodes is placed to record the vertical movement of the eyes. Electrodes for the horizontal EOG are placed at each canthus, outer corner, of both eyes and record the right-left movement of the eyes. For the recording of the up-down movement of the eyes, one electrode is placed above the eye and the other electrode is placed below the eye. When the eyes move in the direction of the electrode, the potential recorded at the electrode becomes positive, since the front pole close to the surface has a positive charge. The potential difference between the pair of electrodes directly depends on eye position and is measured by means of a differential amplifier. The recorded potential is proportional to the sine value of the rotated angle from the primary eye position. It shows an approximately linear relationship for angles up to 30° with a sensitivity of about 5~20 mV/degree. EOG can record the eye movement with the eyes closed and during free head movements. EOG permits accurate recording of a large range of horizontal movements up to 40° with a resolution of about 1~2° (Heide, 1999).

Vertical EOG is easily contaminated by the eyelid closure or eye blinks. These artifacts make it difficult to determine the rotated angle in the vertical direction. Torsional eye movement of the eyes around the line of sight cannot be recorded by EOG, as the rotational eye movement does not induce the potential difference along the horizontal or vertical axes (Fig. 9.15).

3.3 Eye Movements

3.3.1 Saccadic Movement

EOG is used to record the saccadic eye movements, which are generated to move the eye rapidly to a specific target of interest in a visual space. In contrast to the smooth pursuit movement of the eyes, saccadic eye movement is a quick jerk motion simultaneously occurring in both eyes. Saccadic

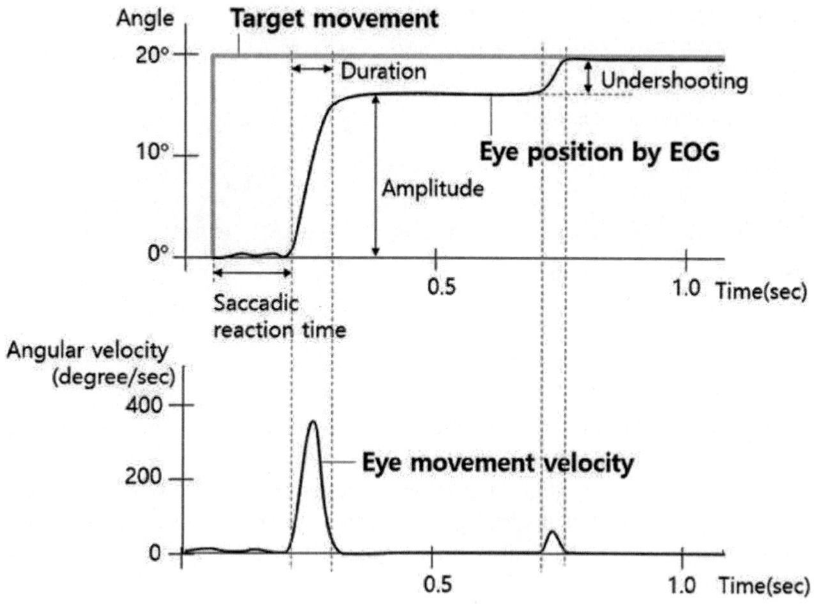

Fig. 9.16 Saccadic eye movement response following a target movement

movement is a voluntary action, but the speed of movement is not controlled. It last only a fraction of a second and the rotational speed is in the range of 20~700°/s.

During our daily activities, we constantly generate saccadic eye movements during reading the books, viewing the scenes, and scanning the surroundings. When we are seeing an object, we unconsciously try to focus the point of interest on the fovea, the small high-resolution area in the retina. Since the fovea covers only a small visual angle of 1~2°, we need to move the fixation points over the scene to analyze all the interesting points by focusing them on the fovea. Saccadic movement is the quick eye movement connecting these fixation points of the eyes. Saccadic movement is measured with EOG by recording the movement of eye position to a given target stimulus (Fig. 9.16). Saccadic reaction time is the latency between the target stimulus onset to the rising edge of the saccadic response. It takes less than 200 ms in normal subjects. Duration is the time to finish a jerk of the eye movement and it takes about 80 ms for angular movement of 20°. The velocity of the eye movement is calculated by differentiating the eye position response, and maximal movement velocity is reached in the middle of the movement duration. For example, the maximum velocity of eye movement reaches about 400°/s during an angular movement of 20°. For a large angle of movement, saccadic movement usually undershoots the target in a first jerk and requires an additional smaller saccade to reach the target angle.

The amplitude of saccadic movement can be reached up to 90° if the head is fixed. But in ordinary situations, the range of saccadic movement is far smaller as we move the head for a large shift of gaze angle greater than 20°. For a gaze shift of a larger angle, usually, the eyes produce saccadic movement first and the head movement follows slowly. The vestibulo-ocular reflex causes the eyes to roll back in the head while keeping the gaze on the target.

Slow saccadic reaction time might be originated from pathological conditions like imbalance in the vestibular system, and abnormality in the brainstem which normally keeps the eyes in position. Recording of eye movement is also used to evaluate the psychiatric disorders, like attention deficit hyperactivity disorder (ADHD), which is characterized by the increased saccadic reaction time.

3.3.2 Nystagmus

Nystagmus is an involuntary movement of the eyes, rhythmically side-to-side, up and down, or in a circular pattern with a variety of conditions. Generally, nystagmus is a pathological symptom and often results in reduced vision and reduced depth perception. It can be accompanied by disturbed balance and coordination. Most commonly, nystagmus is related to dizziness or vertigo.

We can focus the eye consistently on the interesting point of the scene even though the head is moving around. This is possible by the appropriate sensing mechanism of the vestibular system located in the inner ear. The semicircular canals in the vestibule of the ear sense the angular acceleration caused by the head movement and send signals to the nuclei controlling the eye movement in the brainstem. This information is integrated with the visual information and information from the other parts of the body and produces the vestibulo-ocular reflex (VOR). From this part of the brainstem, electrical signals are sent to the extraocular muscles controlling the eye position (Fig. 9.17a). For normal gaze, appropriate signals are sent to the muscles to move the

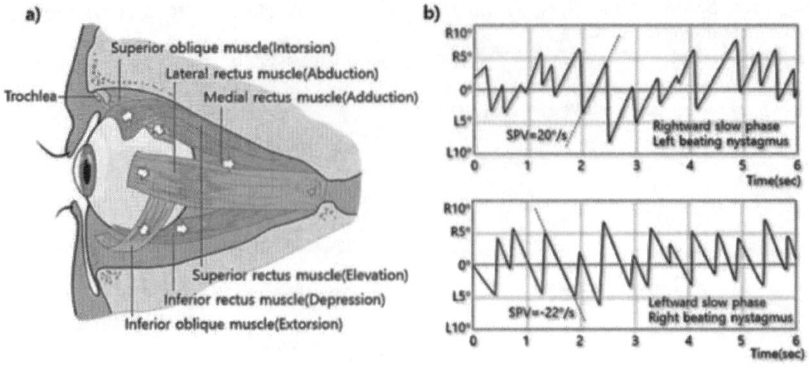

Fig. 9.17 Structure of extraocular muscles and ENG. (**a**) Six extraocular muscles controlling eye movements, horizontally by lateral and medial rectus muscles, vertically by superior and inferior rectus muscles, and circumferentially by superior and inferior oblique muscles. (**b**) Example of ENG recorded in a static position. *SRV* Slow phase velocity

eye position in a direction compensating the head motion and fixing the gaze stationary at the point of interest. However, if semicircular canals were stimulated, by the disease or intentional examination, while the head is stationary, inappropriate signals are sent to the brainstem and to the extraocular muscles of the eyes. Extraocular muscles move the eyes in one direction repeatedly, with a sudden return as the eyes reach the limit of the movement. This repeated movement is nystagmus.

Electronystagmogram (ENG) is a special type of EOG to evaluate eye movement regarding nystagmus and provides an objective assessment of the oculomotor and vestibular systems (Fig. 9.17b). ENG is recorded with some provocative maneuvers to demonstrate the normality and to investigate the origin of abnormality, whether the abnormality is central or peripheral, right or left.

Clinical ENG is performed in several procedural tests. The *oculomotor evaluation* is a subset of ENG assessing the eye movement function in the absence of vestibular stimulation. For oculomotor evaluation, ENG is recorded while the subject is presented with various types of visual stimuli. The *positioning and positional testing* evaluate the dizziness associated with positions of the head. Positional nystagmus occurs and persists while a subject maintains a new static head position. Positioning nystagmus is provoked by the act of moving the head from one position to the next and eventually decays away when the new position is maintained. This test is to localize the site of dizziness origin among the three semicircular canals. The *caloric test* is to discover the degree to which the vestibular system is responsive and how symmetric the responses are, between the left and right sides. As the electrical signals from the semicircular canals are dependent on the temperature, in the caloric test, ENG is recorded with temperature-changing stimulation to each ear. Usually, a constant flow of water or air that is 7 °C below (30 °C) and 7 °C above body temperature (44 °C) is irrigated for a fixed duration of 30~40 s.

While the ENG is traditionally used for nystagmus evaluation based on its convenience, in clinics, video nystagmography (VNG) is replacing the ENG with its additional capability of evaluating the torsional movement and recording the eye image.

3.3.3 Rapid Eye Movement During Sleep

Rapid eye movement (REM) is the rapid saccadic movements of the eyes occurring during a specific stage of sleep. This sleep stage is called REM sleep as the REM characterizes this sleep stage. During REM sleep, brain activity is almost like the brain in a waking state. The differences are shown in EMG by muscle atonia or greatly reduced muscle tone in addition to REMs in the EOG record. This sleep stage is differentiated physiologically from the rest of the Non-REM stage. During this REM sleep stage, the sleepers tend to dream and they report vivid dreams awakened from the REM sleep. This sleep is "paradoxical" as the EEG resembles that of waking, but behaviorally the subjects remain asleep and unresponsive.

EOG is used in two channels of polysomnography recording to examine the occurrence of REM (Fig. 9.18). To minimize the number of electrodes attached to the subject during the polysomnographic recording, two Ag-AgCl electrodes are attached obliquely, one on the upper left side of the left eye and the other on the lower right side of the right eye. With this electrode setup, both horizontal and vertical eye movements can be recorded. REMs are rather easily recognized as they always appear in antiphase. When the eyes move toward one electrode, the channel potential of that electrode increases, but the channel potential of the other electrode decreases. And, the potentials of two channels are moving in the reverse direction if they were originated by the eye movements.

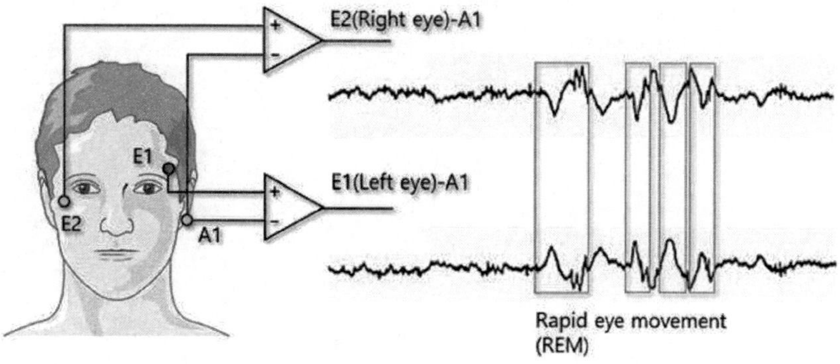

Fig. 9.18 EOG for the recording of rapid eye movement (REM) during sleep

4 Electroretinogram (ERG)

The electroretinogram (ERG) is the record of the electrical activity of the retina in response to a light stimulus. The ERG is recorded from currents generated directly by the retinal neurons and the glial cells. It is recorded externally at the corneal surface of the eye with a special type of electrode. ERG is an objective measure of retinal function that can be measured noninvasively in various physiological and pathological conditions. Stimulation to evoke ERG is applied in a diffuse flashlight or in visual patterns. The ERG is an important examination tool in ophthalmology to diagnose various disorders related to the retina.

4.1 History of ERG

Human ERG was first recorded by James Dewar in 1877. In 1908, Einthoven and Jolly separated the waveform of ERG into three components of a-wave, b-wave, and c-wave. However, the popular use of ERG in clinics began in 1941 after the introduction of electrodes in contact-lens type by Lorin Riggs. Ragnar Granit (1900~1991), a Finnish-born Swedish physiologist, conducted many observations and demonstrated the physiology underlying different ERG sources. His work was basically measuring the electrical response of the retina using the ERG. He measured the fusion frequency and found that the contrast in background illumination like an area of illumination plays an important role. He developed micro-electrodes for retinal recording. He also measured the response to lights of colors and found properties related to the rods and cones in several animals, including humans. With his contribution "for the discoveries concerning the primary physiological and chemical visual processes in the eyes," he received the Noble Prize in 1967 for Physiology or Medicine.

4.2 ERG Wave Components

The electrical potential of the ERG arises from the different layers of the retina (Fig. 9.19). The generated potential is conducted through the volume conductor constituting the eyes and surrounding tissue and recorded at the cornea surface. When a full-field ERG is recorded with the flash stimulation by the LEDs or strobe lamp, the recorded potential is the integrated activity of the whole retina area.

Early receptor potential (ERP) is the biphasic component recorded in the earliest part of the ERG less than 1 ms latency from the stimulus onset prior to a-wave. ERP reflects the earliest chemical responses to light in the photopigment molecules of the photoreceptor outer segments. The peak R1 of a positive phase has a latency of 120μsec, and the peak R2 of a negative phase has a latency of 800μsec (Yonemura, 1967). As an ERP is not produced unless the light is spread over a wide area of the retina with a brief intense flash and has the difficulty in recording due to its short latency, ERP is not widely used clinically.

The two components most often measured clinically are the a- and b-waves. The $a-wave$ is the first large negative component, followed by the b-wave. Compared to the ERP, these components are collectively called late receptor potential (LRP). The a-wave is generated from both rods and cones. With an appropriate stimulation set-up, components from the rods and cones can be separated. ERG with a dim blue flash to the dark-adapted eyes records a component by the rods and ERG with a bright red flash to the light-adapted eyes records a component by the cone cells. The $b-wave$ reflects the activities of the amacrine, horizontal, ON-type bipolar cells following the photoreceptors and Müller cells, which are glial cells having no synaptic connection to the retinal cells. So, the a-wave reflects the general physiological integrity of the photoreceptors in the outer retina and the b-wave reflects the integrity of the inner layers of the retina except for the ganglion cell layer. Two other

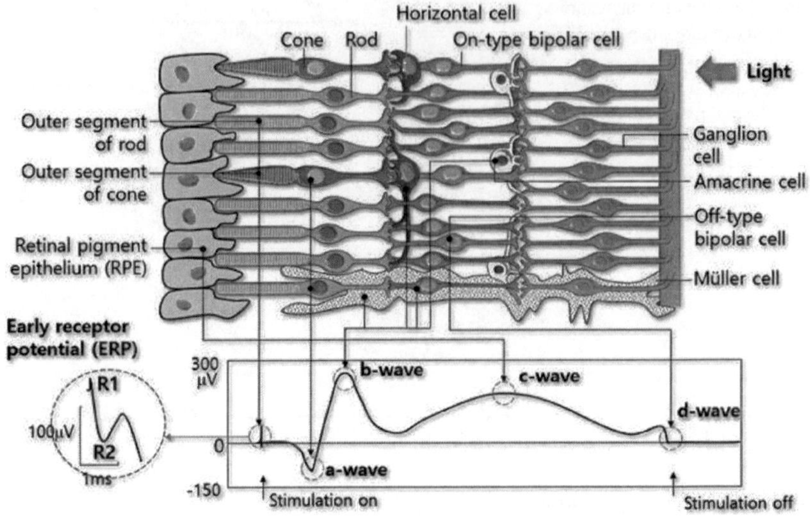

Fig. 9.19 ERG wave components and the related retinal cells

components are the c-wave and the d-wave. The c-wave is originating from the retinal pigment epithelium (RPE) through its interaction with the rods. The d-wave appears with the termination of light and indicates the activity of the OFF-type bipolar cells. The timing and the amplitude of each component are dependent on the type and the strength of the stimulation.

4.3 Full-field ERG

The full-field ERG (ffERG) is a response of the whole retina area summed over several retinal cell layers and the glial cells surrounding the retina. The ffERG is effective in the diagnosis of diseases with widespread retinal dysfunction. By the same token, the ffERG is not so effective in detecting the small localized retinal dysfunctions. The waveform components of ffERG are dependent on the strength of the stimulus flash and the state of adaptation (Fig. 9.20). For scotopic measurements to evaluate the retinal pathway beginning from the rods, ERG is recorded on the dark-adapted eyes. For photopic measurement to evaluate the retinal pathway beginning from the cones, ERG is recorded on the light-adapted eyes. For general evaluation of retinal function, specified ffERG subsets are commonly used under dark- and light-adapted conditions.

To record the function of the rod pathway of the retina, ffERG is recorded under a dark-adapted condition. When a weak and diffuse full-field flash of light is stimulated on the dark-adapted eye, ERG records a slow cornea-positive component. This waveform is the b-wave primarily generated by ON-type bipolar cells. The response is quantified by the peak amplitude and the peak latency from the stimulation onset.

Another subset of ffERG under a dark-adapted condition is performed with a stronger flash of stimulating light. With stronger stimulation, ERG records both of a negative-going a-wave and a positive-going b-wave. This response might be considered as the integrated response of rods and cones, as there are contributions from the cones as well as rods in the generation of a-wave. However, the contribution by the rods surpasses the contribution by the cones, as rods are much more broadly populated over the whole human retina. Each of the human eyes contains about 60~120 million rod cells and 3.2~6 million cone cells. The b-wave is generated by both the ON- and OFF-type bipolar cells. A series of wavelets, termed oscillatory potentials (OPs), can be seen on the ascending slope of the b-wave. Oscillatory potentials are thought to be generated primarily from the amacrine cells layer. The response to the stronger light stimulation is quantified by the peak amplitude and the peak latency of the a-wave, the peak latency of the b-wave, and the peak-to-peak amplitude from the negative peak of the a-wave to the positive peak of the b-wave. Markedly reduced amplitude of the b-wave reflects the dysfunction related to some retinal diseases including melanoma-associated retinopathy.

To record the response by the cone pathway with suppressing the response by the rods, ffERG is recorded under a light-adapted condition. The ffERG with a strong flash under a light-adapted condition normally records a negative a-wave and a positive b-wave, much like a response to a strong flash stimulation under a dark-adapted condition. With this photopic condition, the a-wave is generated by cone photoreceptors. The b-wave is generated by a combination of ON- and OFF-type bipolar cells. The response is quantified by the peak amplitudes and peak latencies of the a-wave and the b-wave.

Another subset of ffERG is performed with flickering light stimulation around 30 Hz. This examination is to assess the cone-pathway separately, as the rods generally cannot follow the rapid light stimulations. With periodic flickering light stimulation, recorded ffERG also shows a rhythmic pattern of the same fundamental frequency around 30 Hz, and the response is quantized by the peak-to-peak amplitude and latency of the peaks from each flash stimulation time. Responses under a light-adapted condition have generally smaller amplitudes than the responses recorded under a dark-adapted condition, as they are caused by the cones having a lower population than rods.

4.4 Multifocal ERG

The multifocal ERG (mfERG) is the record of the electrical activity of many focal sections of the retina to the light stimulation. In contrast to the ffERG, the mfERG provides activities of each smaller localized section simultaneously to assess the spatial integrity or any dysfunction of the retina (Fig. 9.21). The mfERG is recorded usually under the light-adapted condition to assess the cone-pathway. The central part of the retina including the macula is divided into 37, 61, or 103 smaller hexagonal sections, and individual responses represent the retinal function of these smaller sections. The mfERG provides clinically important information by localizing abnormalities and allows the separate assessment of the central and peripheral function of the retina.

During the examination, approximately half of the sections are illuminated at each stimulation, and each section follows a distinct pseudorandom "m-sequence" pattern of alternating bright and dark states. The electrode used for mfERG measures a single cornea potential as in the ffERG recording and does not provide any spatial information by itself. That is, the recorded response is not truly a response of an individual retinal section but rather a combined response of the multiple sections stimulated simultaneously. Response of individual section is calculated mathematically by correlating the recorded responses with the stimulating pattern and minimizing the effect from the other sections.

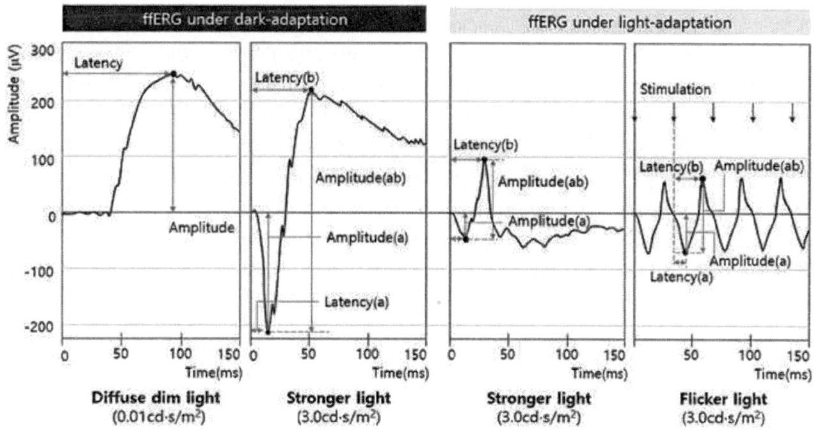

Fig. 9.20 The normal ffERG responses under dark- and light-adapted conditions with different stimulation strengths and patterns

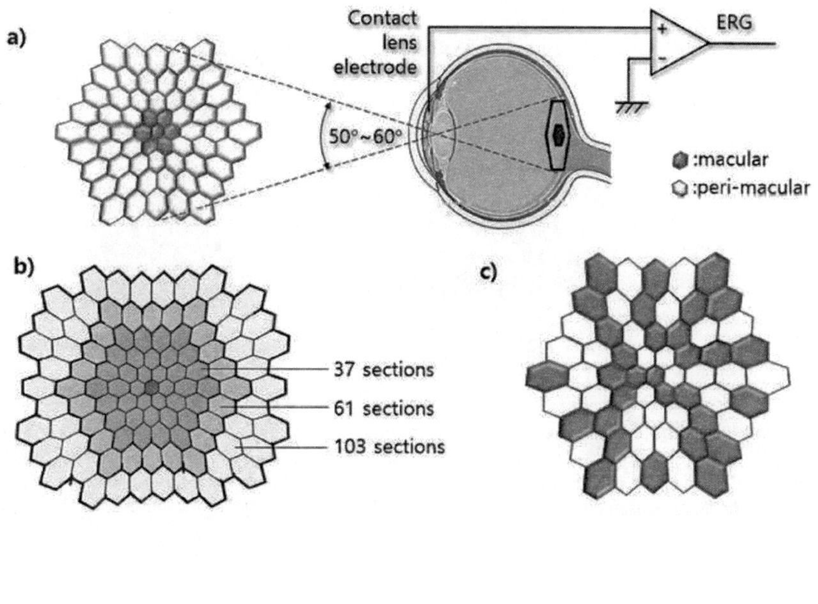

Fig. 9.21 The recording of multifocal electroretinogram (mfERG). (**a**) mfERG is recorded from the cornea with a single electrode under the visual pattern stimulation hexagonally divided into many smaller sections. (**b**) Clinically using visual patterns with a different number of regional sections. (**c**) Example of one visual pattern stimulating almost half of the divided regional sections

The recorded individual focal ERG waveform is similar in shape to the ffERG waveform recorded under a light-adapted condition, and the summation of focal potentials from all sections shows a similar waveform as of ffERG. However, the difference is that the mfERG is recorded with the same overall luminance with the difference only in regional contrast while the ffERG is totally the response to the change in the luminance. So, the wave components are named differently. The initial negative deflection resembling the a-wave is called N1, the following positive deflection resembling the b-wave is called P1, and a subsequent negative deflection is called N2 sequentially.

As the potential corresponding to each smaller section is calculated by the algorithm, there is a wealth of information can be provided. The recorded results are displayed in several summarized ways (Fig. 9.22). Potentials corresponding to individual hexagonal sections can be displayed in a single map to show all the recorded potential in a single picture. The feature, for example, the peak amplitude of the P1 component, can be displayed in a 3D map, by augmenting the points with the interpolation. This 3D plot shows the largest amplitude at the center representing the fovea and the amplitudes decline isotropically as the visual angle increases toward the periphery of the retinal area. To summarize the recorded potentials into a smaller number of potential waveforms, responses can be averaged over the regions divided into four quadrants or rings with different eccentricities. The first ring represents the response of the fovea, the second ring just next to the center hexagon represents the activity of the macular region, and the other rings represent the activities of the retinal area with the increase of visual angle.

4.5 ERG Electrodes

There are several types of ERG electrodes commonly used to record potentials of the cornea surface (Fig. 9.23). These electrodes have unique shapes and structures compared to the electrodes used for recording other bioelectrical potentials. Throughout history, inventors have created a variety of electrode designs. An ideal electrode does not exist for ERG recording and each type of electrode is a compromise with different considerations for the factors of recording. ERG electrodes can be categorized into three types by their location of placement, contact lens, conjunctival, and skin.

Contact lens electrode constantly contacts with the cornea by encircling the apex and records the largest amplitude compared to other types of electrodes. *Burian – Allen (BA) electrode* is a reusable contact lens electrode and most commonly used for ERG recording. It consists of an annular ring of stainless steel surrounding a contact-lens core. It incorporates a lid speculum, which helps to minimize eye blinks/closure. BA electrodes are available in eight different sizes to fit the eye sizes from adult to premature infant. *Jet electrode* is a disposable plastic contact lens with a gold-plated peripheral circumference. Potential is sensed by the thin gold ring of a few angstrom thicknesses on the concave plastic lens surface. Four protruding "bumps" on the anterior surface are intended to discourage the subject from closing their eyelids. It is smaller and slightly more comfortable than the BA electrode, but less secure on the eye and can be more easily blinked out by the subject. Recording ERG on the apex of the cornea using a contact-lens type electrode is the least comfortable for the subject and generally topical anesthesia is required.

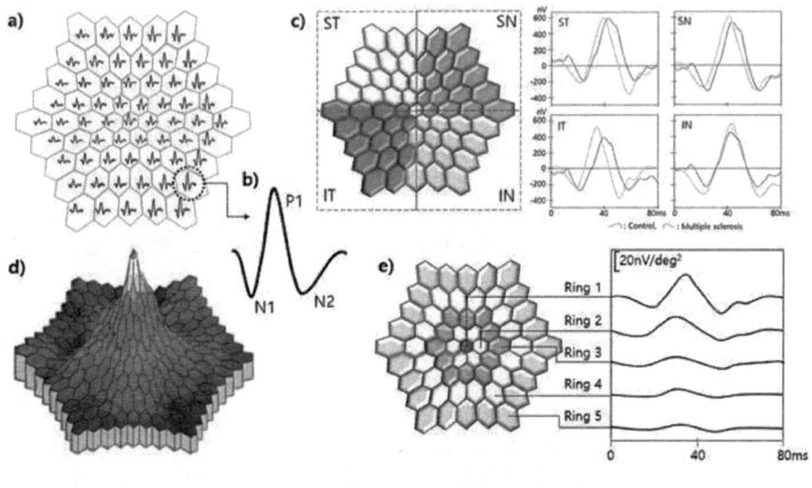

Fig. 9.22 Display of mfERG records. (**a**) The display of all calculated mfERG responses at corresponding hexagonal sections in a single picture. (**b**) Wave components of recorded mfERG potential. (**c**) Averaged mfERG responses over each quadrant. The display shows the difference between multiple sclerosis (MS) patient and a normal subject (Boquete, 2019). (**d**) 3D plot of recorded mfERG responses showing the overall distribution of amplitude at a given time. (**e**) Averaged mfERG responses over the rings having different eccentricities

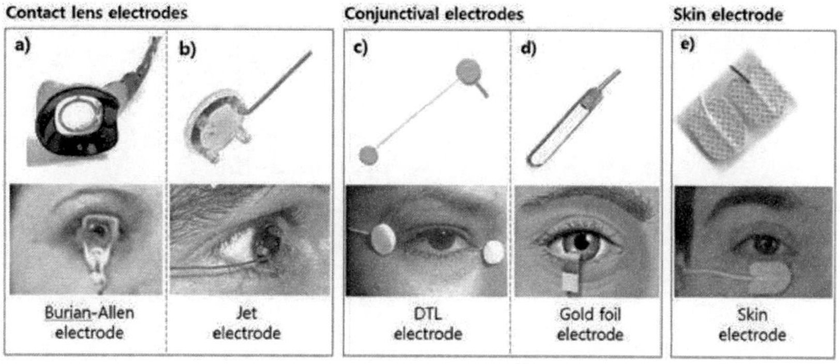

Fig. 9.23 ERG electrodes. (**a**) Burian-Allen (BA) electrode, (**b**) Jet electrode, (**c**) Dawson-Trick-Litzkow (DTL) electrode, (**d**) Gold-foil electrode, (**e**) Skin electrode

Conjunctival electrodes places the sensing part of the electrode in the conjunctiva instead of covering the cornea. These electrodes are designed to get a more patient-friendlier alternative to the contact lens electrode. They are considered to be optimized for the trade-off between the performance and comfort of the subject. Topical anesthesia is generally not required. As their small size allows subjects to blink freely, they can be worn comfortably for longer periods than contact lens electrodes. *Dawson – Trick – Litzkow (DTL) electrode* senses corneal potential via a fine flexible thread coated with conductive metals. Recently, DTL electrodes of multi-thread design improve the integrity of the recording. DTL threads are very flexible, easily fit corneal surfaces, and give little rise to irritation on the cornea. Used DTL electrodes are discarded after each examination, and there is no worry about the infection accompanying reusing the electrodes. *Gold foil electrode* is a very thin film electrode in which a thin gold layer is coated over the polyester film. It is placed over the lower lid and makes minimal touch with the corneal-scleral junction on the midline. This electrode is inexpensive and simple to fabricate. As the sight of the eye is not compromised, it can be used in circumstances that require prolonged testing of retinal function or with diverse visual stimulation types. Conjunctival electrodes record the same ERG waveform with approximately 80% amplitude compared to the contact lens electrodes.

Skin electrode is an alternative to the corneal electrodes. It is placed on the skin near the lower eyelid. ERGs recorded using a skin electrode are small and quite variable in amplitude due to the varying resistances of the eye-to-skin current pathway. While recorded ERG amplitudes tend to be small and noisy, skin electrodes are better tolerated in pediatric subjects. Also, skin electrodes collect greater background noises as the signal travels through the skin before reaching the electrode. With skin electrodes, the recorded amplitude of ERG is roughly 25~35% of the same ERG recorded with contact lens electrodes. In many cases, an increased number of light stimulations is required to get a recognizable waveform.

Most of the ERG electrodes are monopolar and recorded referred to another electrode site. A reference electrode is placed on the location indifferent to the ERG response, including the temple and ear lobe. A ground electrode serves as the zero reference for the polarity of the recorded potential and mostly it is placed at the center of the forehead.

5 Electrogastrogram (EGG)

Electrogastrogram (EGG) is the record of gastric myoelectrical activity from the abdominal surface. EGG provides objective evidence on stomach motor dysfunctions related to pathological conditions like gastroparesis and functional dyspepsia.

5.1 Gastric Electrical Activity

The gastric electrical activity represents a spontaneous rhythmic function of the stomach controlling the timing and the rate of the contractile activity including gastric emptying. This spontaneous gastric rhythmic activity is originating from the gastric pacemaker cells, which are located in the longitudinal muscle in the greater curvature region of the fundus (Fig. 9.24). By generating the highest rate of spontaneous rhythm, this pacemaker region dominates overall gastric activity. From the pacemaker cells, action potentials are spreading through gap junctions toward the antrum and the pylorus. In normal subjects, the spontaneous gastric rhythm of the entire stomach is very slow and is approximately 3 cycles per minute (cpm). This rhythm has a period of around 20 s, and it has a period longer than the heart rhythm of around 1 s and respiratory rhythm of around 5 s which superimposes EGG in the recording. As in the heart, impairment of pacemaker activity can induce physical or functional uncoupling of pacing activity. Uncoupling of pacemakers affects the efficiency of gastric motility and it alters the rhythm and the rate of recorded gastric electrical activity (Riezo et al., 2013).

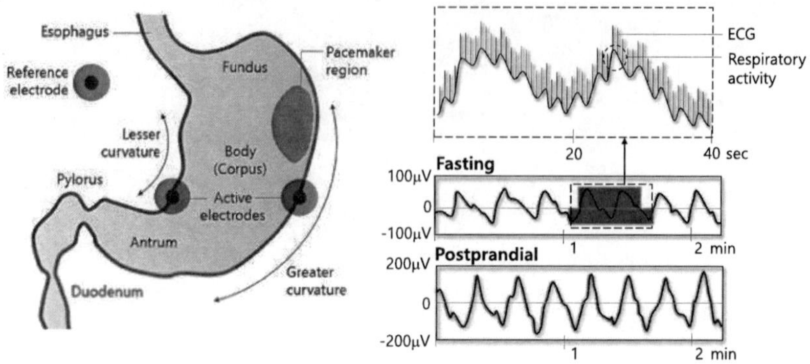

Fig. 9.24 Electrogastrogram (EGG)

5.2 Recording of EGG

EGG is recorded with the adhesive Ag/AgCl electrodes used for ECG recording. Both unipolar and bipolar techniques are employed to record the EGG. Unipolar recording displays the potential variation across the region and bipolar recording gives a better signal-to-noise ratio with the decreased effect of movement artifacts, respiration, ECG, and other background noises. While the bipolar recording is preferred for EGG recording, the number and position of electrodes are variant upon the investigators' preferences. Some electrode configurations use cutaneous reference points as landmarks for stomach shape and some other configurations are based on the actual position and shape of the stomach evaluated by medical images like X-rays or ultrasound.

EGG recording consists of two periods, a fasting period and a postprandial period. Approximately 30 min recording of the fasting period is followed by a 60 min recording of the postprandial period to acquire reliable and predictable results. As motion artifacts still remain a problem in EGG recording, examinations should be taken with minimizing the motion artifacts and the contaminated portion of the record should be preprocessed before or deleted from the analysis. Recorded EGG shows the amplitude in the range of 50~500 μV.

EGG is evaluated by the dominant frequency. Mean dominant frequency and power of dominant frequency are calculated from the spectrum of the recorded EGG. In normal subjects, the mean dominant frequency is approximately 3 cpm ranging from 2.0 to 4.0 cpm and it takes about 70% of the whole recording time. Bradygastria is the slowed gastric rhythm showing the peaks in frequencies ranging from 1.0 to 2.0 cpm in absence of a peak in the normal range. Tachygastria is the fastened gastric rhythm showing the peaks in frequencies ranging from 4.0 to 10.0 cpm in absence of a peak in the normal range. EGG power is evaluated as the ratio of postprandial to fasting EGG power values. As meal ingestion increases the contractility of the stomach, a power ratio lower than 1 is believed correlated to an impaired gastric emptying time.

6 Electrodermal Activity (EDA)

Electrodermal activity (EDA) is the record of electric conductance of the skin. EDA is also called many different names, electrodermal response (EDR), Galvanic skin response (GSR), Galvanic skin potential (GSP), skin conductance response (SCR), skin conductance level (SCL), sympathetic skin response (SSR), and psychogalvanic reflex (PGR). EDA is recorded by making an unrecognizable low level of constant current flow through the skin and measuring the voltage developed by this current. As the conductance, or the resistance, is varied under constant current flow, the voltage measured directly reflects the variation of skin conductance. Basically, skin conductance varies with the activity of sweat glands in the skin. Sweating is controlled by the sympathetic nervous system and skin conductance is an indication of sympathetic activation by the neuropsychological arousal. Skin conductance measured with EDA is represented with the unit of micro-Siemens (μS). Siemen is the unit of conductance and reciprocal of resistance. 1 Siemen (S) of conductance is equivalent to 1 Ohm (Ω) of resistance and 1 μS of conductance is equivalent to 1 MΩ of resistance. Typically measured values of skin conductance are in the range of 0.1–40 μS.

6.1 Sweat Gland

Sweat glands are coiled tubular structures distributed over the skin. Humans have two distinct types of sweat glands, apocrine and eccrine sweat glands. *Apocrine sweat glands* secrete oily substances containing lipids, proteins, and steroids into hair follicles. Their secretion has an odor determining the body odor of each individual. They are found in the skin containing hairs, the armpits, the groin, and the area around the nipples of the breast. Rather than responding to temperature, apocrine glands often respond to emotional stimuli including anxiety and fear. *Eccrine sweat glands* are widely distributed over the body, most densely at palms and soles. Humans have 2,000,000 to 5,000,000 eccrine sweat glands, with 150~340/cm^2 of average distribution and 600~700/cm^2 in palms and soles of the feet. They have a coiled, tubular structure extending from the epidermis to the lower dermis and open by duct directly to the skin surface. They secrete water and some electrolytes through the surface of the skin and control the body temperature by cooling the skin when the water evaporates. As the eccrine sweat glands are innervated by neurons, sweating is controlled by the central nervous system. Thermosensitive neurons in the brain integrate the internal body temperature and external skin temperature and control sweat glands to respond to decrease the core body temperature. By seeing that the skin of monkeys and apes remains dry even in a hot environment, profuse thermal sweating in humans seems to be a new function of highly advanced species. Among the eccrine sweat glands, some eccrine glands, those on the hairy surfaces, respond to thermal stimulation for thermoregulation, and some other eccrine glands, those on the palms and soles, respond to psychological stimuli and keep friction surfaces moist. By recording the electrodermal activity, we mostly measure the secreting variation of eccrine glands.

6.2 Nervous Control of EDA

The EDA reflects the innervated activity of the sympathetic nervous system on eccrine sweat glands. But, there is no antagonistic innervation of the parasympathetic nerves and the secretion from the eccrine sweat gland is indifferent to the parasympathetic activation. And, the effector synapse is cholinergic and mediated by the neurotransmitter acetylcholine. This is different from other sympathetic effector synapses mediated by the noradrenergic neurotransmission and it makes EDA independent from the circulating levels of adrenaline or noradrenaline (Critchley, 2013).

Within the brain, psychological influences on EDA are thought to emerge from activity within reticular formation centers of the brainstem. However, the contribution of higher brain regions implicated in emotion, attention, and cognition on EDA responses has been confirmed by the studies of patients with discrete brain lesions and more recently by functional imaging techniques. The sympathetic activation of sweat glands in the skin closely reflects the changes in mental states. EDA indices change by the cognitive arousal level and emotional states under the constant temperature at rest. They are expressed as the sustained shifts in tonic level, or in phasic responses during the transient evolving course of a few seconds. EDA responses enable the quantitative understanding of the mechanism between the state of body arousal and cognition or motivational behavior.

6.3 Recording of EDA

For EDA examinations, any preparatory procedures which might modify the skin conductance artificially should be avoided, because the conductivity of the skin at the electrode site is the main variable. There are no established standard points for electrode placement. Electrodes are placed at two points between which conductance is of interest. In many studies, the following sited are preferred with their higher sensitivity to skin conductance variation (Fig. 9.25).

1. Electrode pair 1: Between the tips of the index and middle finger. Electrodes are placed on volar surface of distal phalanges of fingers.
2. Electrode pair 2: Between the middle parts of the index and middle finger. Electrodes are placed on volar surfaces of the middle phalanges of fingers.
3. Electrode pair 3: Within a palm. Electrodes are placed on the thenar and hypothenar eminences of a palm.

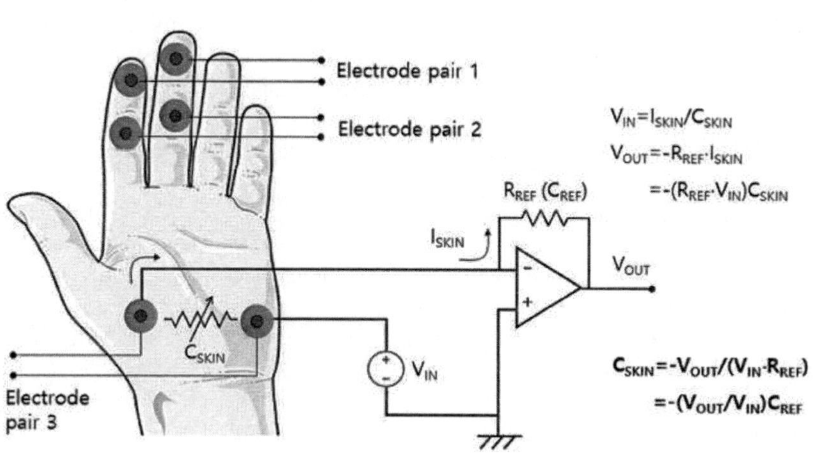

Fig. 9.25 Recording of electrodermal activity

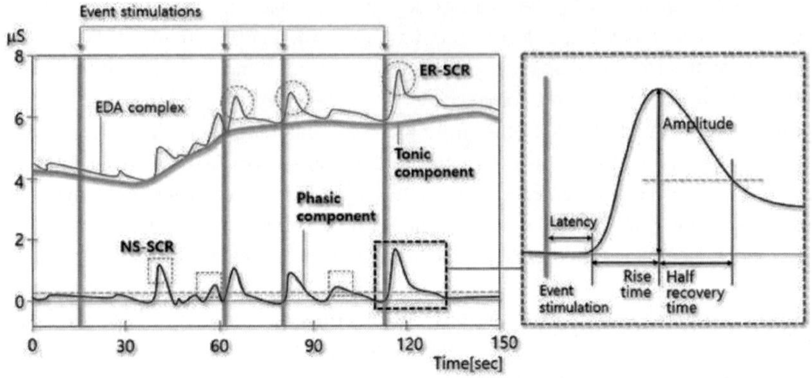

Fig. 9.26 Tonic and phasic components of EDA

6.4 Tonic and Phasic Components of EDA

There are two main components in the complex referred to as EDA (Fig. 9.26). One component is the general tonic EDA and it relates to the slowly acting components and background characteristics of skin conductance. The other component is the phasic component which occurs as faster rising and falling components. Both components are important and rely on different neural mechanisms.

Tonic component is the level of conductance activity when there are no explicit stimuli from the external environment. It varies only slowly with the temporal resolution of tens of seconds or minutes by autonomic regulation involving the psychological state of the subject or by the hydration or dryness of skin along the time. The measure of this slowly changing tonic level is specially referred to as skin conductance level (SCL). This tonic component is constantly changing within an individual subject and can differ markedly between subjects. The absolute value of the tonic component is not much informative and normalization is required to derive any comparison. Simply averaging across the recorded signal is not adequate as a measure of skin conductance level, because it likely overestimates the tonic component by including the multiple phasic components. The amplitude of the tonic component is in the range of 1~40 μS and its average value is in the range of 2~16μS.

Phasic component is typically associated with a short-term event and occurs in the presence of discrete external stimuli. The stimuli can be a sensory type or psychological type like cognitive processes prior to the event such as anticipation and decision-making. Phasic response changes abruptly showing peaks in the recorded signal. These phasic peaks, discriminately to tonic level, are referred to as skin conductance response (SCR). There are two types of skin conductance responses. *Non − specific skin conductance response* (NS-SCR) is a phasic component occurring without an external event stimulation. NS-SCR is typically defined as a spontaneous increase of skin conductance over the thresholding level of 0.01~0.05μS (Braithwaite et al., 2015). They occur 1~3 times/min on average, but 10 times/min sometimes during a period of quietness. *Event − related skin conductance response* (ER-SCR) is the phasic component generated as a response to a specific event like a visual stimulus or an unexpected question. ER-SCRs are the most common measure used in studies using EDA to relate changes in emotional arousal to specific stimuli. Discriminating ER-SCR from naturally occurring NS-SCR is critical for the accurate evaluation of the stimulation. As there is the latency from the event stimulation to the onset of response, ER-SCR can be discriminated by the occurring time window. Normally, the latency from the stimulus onset to the first significant deflection in the response is between 1 and 3 s. Deflections in the response earlier or later than this latency period can be regarded as NS-SCR instead of ER-SCR. The amplitude of the phasic component is typically in a range of around 2~3 μS, but, rarely in experiments with highly threatening stimuli, maximum amplitude response can increase to around 8 μS.

6.5 Application of EDA

EDA is a particularly useful examination since its responses are easy to get. As EDA is an unconscious response and very difficult to suppress intentionally, recorded responses are reliable as physiologic responses of the subject. However, careful interpretation is required as EDA is sensitive to many types of stimuli. As EDA changes typically in response to a complex or patterned autonomic reaction, EDA can be used as a circumstantial index of a physiological response.

The tonic components of the EDA of subjects are used to track individual differences in the general level of arousal and the integrity of diurnal rhythms. Phasic components are easily elicited by emotional stimulation or a change in attention. Thus, EDA is widely applied as an index of physiological and emotional arousal in the studies of stress responsivity, emotional films, and mental and physical effort.

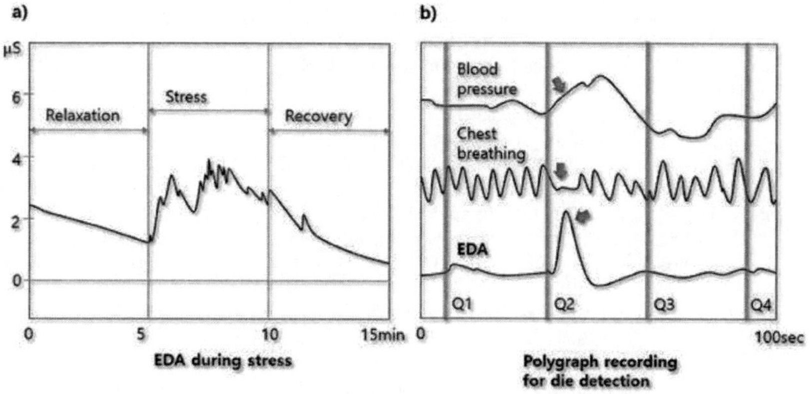

Fig. 9.27 EDA response during the stress test and lie detection. (**a**) Tonic and phasic EDA components are increasing during stress activity and return to pre-stress levels during the recovery period. (**b**) Polygraph recording for lie detection. Increased EDA response to the question 2 strongly suggests that the subject telling a lie and the blood pressure and chest breathing response also support the EDA response

EDA is reflecting stressful conditions nicely. When a human is in danger or injured, various physiological changes occur, which are termed stress reactions. In stress reaction, the autonomic nervous system is deeply involved and it is well represented by the activation of sweat glands which are exclusively innervated by the sympathetic nervous system. Consequently, EDA response varies with sympathetic activation and the stress level of the subject (Fig. 9.27a). During the stress test with a prior relaxation period, the skin conductance decreases during the relaxation task, increases during the stressful activity, and returns toward the pre-stress level during the recovery period. The response occurs in tonic components and phasic components during the stress period.

EDA is used also as a major signal in a polygraph for lie detection. It is based on the premise that telling a lie causes specific physiological responses manifested by the autonomic nervous system. These physiological responses are beyond the voluntary control of the subject. When a polygraph is used for lie detection, the EDA response represents the sympathetic branch of the autonomic nervous system. EDA generally shows phasic increases in response to the proposed questions. Respiration, blood pressure, and body movement are used together and support the response of EDA for lie detection (Fig. 9.27b).

During sleep, EDA shows increased tonic and phasic components compared to the daytime response. Their occurring pattern is dependent on the sleep stages. In the deep sleep stage, the tonic component is at the highest level and phasic components are occurring most frequently. The amplitude of the tonic level increase is somewhat reduced in light sleep and REM sleep. The occurrence of the phasic component is minimal during REM sleep. EDA shows generally increased activity during the first half of the sleep and a much-reduced pattern during the latter half of the sleep. It smoothly returns to day time pattern with wakening. Opposite to daytime, EDA is more sensitively measured on wrists than on palms during sleep.

EDA is also used in the clinical domain. Upon the physiological and psychological disorders, measured EDA shows increased or decreased responses. Increased tonic skin conductance levels and increased number of phasic components are reported in anxiety patients and in patients with chronic stress. Decreased tonic and phasic responsivity are reported in subjects with depression, in subjects having a higher occurrence of suicidal attempts, and in psychopathy.

EDA is also used as a therapeutic tool in biofeedback treatment. Biofeedback is a behavioral treatment approach where a subject learns to gain volitional control over an involuntary bodily process, like emotional arousal. In contrast to other biofeedback approaches like neurofeedback using an EEG, EDA biofeedback is easy to implement. Typically, EDA biofeedback is used to train subjects to reduce their sympathetic arousal to induce psychological and physiological relaxation states that alleviate stress-related tension. EDA biofeedback has successfully been applied in the treatment of tension headaches and irritable bowel syndrome. By training subjects to increase sympathetic arousal in the skin, EDA biofeedback reduced the seizure frequency in drug-resistant epilepsy patients (Nagai, 2004).

EDA has been used in forecasting the seizure occurrence in combination with other physiological signals, like body temperature, photoplethysmogram, and body movement, and it reduces the false alarm rate in the wristband-type alarming unit. Seizure forecasting provides patients with timely warnings to prepare for their activities and helps clinicians deliver more objective, personalized treatments (Meisel, 2020).

References

Ahn, S. W., Yoon, B. N., et al. (2018). Nerve conduction studies: basic principal and clinical usefulness. *Annals of Clinical Neurophysiology, 20*(2), 71–78. https://doi.org/10.14253/acn.2018.20.2.71

Boquete, L., Lopez-Guillen, E., Vilades, E., Miguel-Jimenez, J. M., Pablo, L. E., De Santiago, L., et al. (2019). Diagnostic ability of multifocal electroretinogram in early multiple sclerosis using a new signal analysis method. *PLoS One, 14*(11), e0224500. https://doi.org/10.1371/journal.pone.0224500

Braithwaite, J. J., Watson, D. G., Jones, R., & Rowe, M. (2015). A guide for analyzing electrodermal activity (EDA) & skin conductance responses (SCRs) for psychological experiments. *Psychophysiology, 49*, 1017–1034.

Constable, P. A., Bach, M., Frishman, L. J., et al. (2017). ISCEV standard for clinical electro-oculography. *Documenta Ophthalmologica, 134*, 1–9. https://doi.org/10.1007/s10633-017-9573-2

Critchley, H., & Nagai, Y. (2013). Electrodermal activity (EDA). In M. D. Gellman & J. R. Turner (Eds.), *Encyclopedia of Behavioral Medicine*. Springer. https://doi.org/10.1007/978-1-4419-1005-9_13

Heide, W., Koenig, E., Trillenberg, P., KoÈmpf, D., & Zee, D. S. (1999). *Electrooculography: technical standards and applications* (pp. 223–229). Recommendations for the Practice of Clinical Neurophysiology.

Meisel, C., El Atrache, R., Jackson, M., Schubach, S., Ufongene, C., & Loddenkemper, T. (2020). Machine learning from wristband sensor data for wearable, noninvasive seizure forecasting. *Epilepsia, 61*(120), 2653–2666. https://doi.org/10.1111/epi.16719

Mills, K. R. (2005). The basics of electromyography. *Journal of Neurology, Neurosurgery, and Psychiatry, 76*, ii32–ii35. https://doi.org/10.1136/jnnp.2005.069211

Nagai, Y., Goldstein, L. H., Fenwick, P. B. C., & Trimble, M. R. (2004). Clinical efficacy of biofeedback treatment on reducing seizures in adult epilepsy, A preliminary randomized controlled study. *Epilepsy & Behaviour, 5*, 216–223.

Nandedkar, S. D., & Barkhaus, P. E. (2014). Quantitative EMG analysis. In *Neuromuscular disorders in clinical practice* (2nd ed). Springer.

Riezzo, G., Russo, F., & Indrio, F. (2013). Electrogastrography in adults and children: The strength, pitfalls, and clinical significance of the cutaneous recording of the gastric electrical activity. *BioMed Research International, 2013*, Article ID 282757. https://doi.org/10.1155/2013/282757

Varela, D., Gudino-Penaloza, F., & Rodelas, D. (2019). Characterized bioelectric signals by means of neural networks and wavelets to remotely control a human-machine interface. *Sensors, 19*, 1923. https://doi.org/10.3390/s19081923

Yonemura, D., & Kawasaki, K. (1967). The early receptor potential in the human electroretinogram. *The Japanese Journal of Physiology, 17*, 235–244.

Brain-Computer Interface

10

Brain-computer interface (BCI) is the technology interfacing subjects directly by their brain signals to the computer or external environment. The term "brain-machine interface" (BMI) is also used almost in the same concept. A BCI system measures central nervous system (CNS) activities and converts them into artificial output. When the converted output can transmit the intent of subjects, they can respond to the stimuli or questions and interact with the environment in some possible ways. It bypasses the ordinary communication channels of the human, like speaking, moving body parts, and facial expressions, which are accomplished through the peripheral muscles (Fig. 10.1). As it bypasses the ordinary peripheral nervous system in transmitting the brain activity, it can replace or supplement the blocked communication channels of patients who cannot express their intents normally through peripheral pathways due to different types of disease. BCI is the method that the human can communicate with the external world electrically.

When the BCI is interpreted as the interfacing technology between the brain and the computer, it can be understood as the bidirectional interface between the brain and the computer, including the interface from the computer to the brain as well as from the brain to the computer. Some are using the term BCI including the technology for both directions, and some are using another term computer-brain interface (CBI) to focus on the reverse direction of the interface from the computer to the brain. Controlling brain activity from the external side by artificial electrical or magnetic signals is known as the neural prosthesis of the brain. If the BCI is interpreted as a bidirectional technology, the neural prosthesis can be part of it. However, currently, it is studied in a different domain mainly for medical purposes to recover the impaired function of the brain. As most of the BCI studies are on the communication from the brain to the external world, this chapter describes the BCI by concentrating on the technologies interpreting the brain activity signals to derive artificial outputs toward the external world. When a subject constitutes a closed loop using the BCI system, a feedback loop from the environment to the brain is accomplished through the ordinary sensory pathway of a subject. Electrical signals toward the control of brain functions will be described in the other chapters of this book.

1 Overview of BCI

1.1 History of BCI

The history of BCI starts from the discovery of EEG by Hans Berger in 1924. Even though he recognized the possibility of using EEG for deciphering the brain commands, EEG has been mostly used for detecting the abnormal states of the brain for a long time. However, for the normal brain, EEG is a good and unique source representing the activities of the brain electrically and provides a possibility of deciphering the brain activity if the connected deciphering system is powerful enough.

In 1965, before the term BCI is coined, Alvin Lucier, an American composer for experimental music, performed the piece "Music for solo performer." He recorded the EEG with two electrodes in the frontal region of the head and processed the recorded signals with analog hardware of filters, amplifiers, and mixing boards. He tried to compose innovative music by connecting the processed output to speakers and several percussion instruments. For example, a certain strength of alpha waves triggered the sequence of percussion rhythm.

In 1969, Eberhard Fetz, an American neuroscientist at the University of Washington, found in his animal experiment with monkeys that neuronal activity of the brain can be reinforced with appropriate feedback and used to activate external control. He confirmed that the operant can quickly condition the brain activity to expect the reward. He showed that monkeys could increase the activity of newly isolated cells by 50–500% above rates before reinforcement after several training sessions (Fetz, 1969). Since then, BCI

K. S. Park, *Humans and Electricity*, https://doi.org/10.1007/978-3-031-20784-6_10

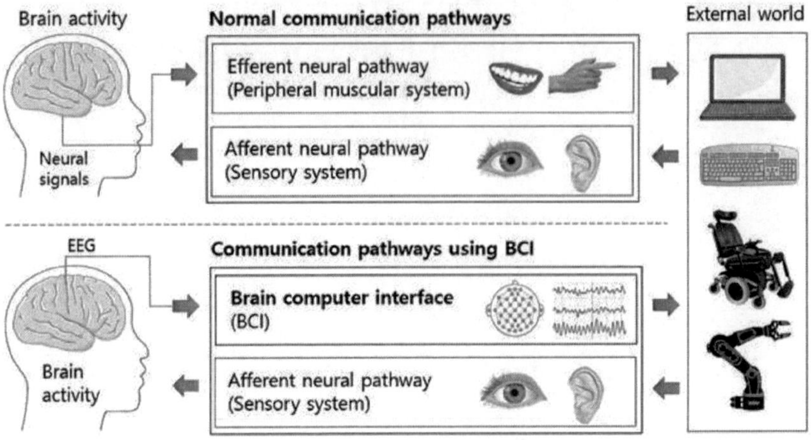

Fig. 10.1 Communication pathways using BCI in comparison to normal communication pathways

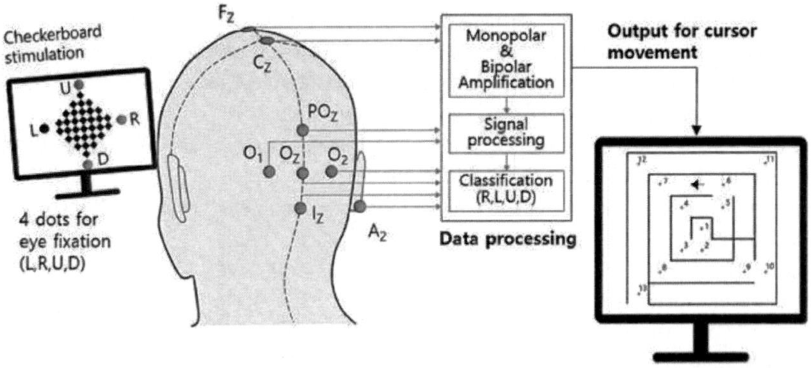

Fig. 10.2 First BCI demonstration using VEP to move the cursor in a maze. (Vidal, 1977)

technology has advanced and verified the performance in animal studies with the electrodes implanted in the brain. In 2008, Andrew Schwartz, an American neurobiologist at the University of Pittsburgh, strikingly demonstrated that monkeys could feed themselves by controlling the robotic arm only by their thought (Velliste et al., 2008). By controlling the neural activities, which were picked up by implanted electrodes in the motor cortex, monkeys could reach and grab a food positioned at an arbitrary point in 3D space within an arm's reach.

In 1973, Jacques Vidal, a professor at UCLA in the USA, first used the term "brain-computer interface" in his first BCI paper published in a peer-reviewed journal. He stated the "BCI challenge" of controlling the external objects using EEG signals. In his experiment with visual evoked potential (VEP), a subject fixed the eyes to one of the 4 points directing the movement of a cursor, and VEPs were recorded with 5 electrodes over the occipital area (Fig. 10.2). As the fixation point is imaged to the fovea in the retina, a checkerboard pattern is supposed to be imaged into different parafovea areas in the retina for each fixation point. This difference in stimulated retinal regions causes the difference in recorded VEPs among the channels and the output moving a cursor was made by classifying the processed VEPs.

Since then, various types of BCIs are suggested based on different brain signals and different responding characteristics of the signals. Noninvasive BCI using EEG was most actively studied with its convenience and possible easier deployment. During the 1980s and 1990s, BCIs using slow cortical potential (SCP), event-related potential (P300), sensorimotor rhythm (SMR), motor imagery (MI) response, and steady-state visual evoked potential (SSVEP) have emerged sequentially (Table 10.1). Even though these BCIs are commonly using EEGs, they have different characteristics and performances in their applications. Since 2000, different recording modalities of brain function other than EEG have been tried as new noninvasive types of BCIs. Functional magnetic resonance imaging (fMRI), functional near-infrared spectroscopy (fNIRS), and magnetoencephalography (MEG) have been used to build a BCI system. Even though they provide unique signals reflecting the brain activity, the bulky sizes of the recording unit limit the studies within the laboratory for research purposes and greatly prevent the extension of their practical applications.

In 1999, Niels Birbaumer, a Czechoslovakian psychologist at Padova University in Italy, developed a BCI speller as a new means of communication for completely paralyzed patients. The subjects were required to produce either

Table 10.1 First appearance of various BCI types

Year	Inventor	Type of BCI
1980	Elbert	Slow cortical potential (SCP)-based BCI
1988	Farwell & Donchin	P300 speller
1991	Wolpaw	Sensorimotor rhythm (SMR)-based BCI
1993	Pfurtscheller	Motor-imagery (MI)-based BCI
1995	McMillan	Steady-state visual evoked potential (SSVEP)-based BCI
2003	Weiskopf	Functional magnetic resonance imaging (fMRI)-based BCI
2007	Sitaram	Functional near-infrared spectroscopy (fNIRS)-based BCI
2009	Bai	Magnetoencephalogram (MEG)-based BCI

negative or positive greater potentials than a specific criterion amplitude and the system recognized the thought of the subjects using the slow cortical potential (SCP) of EEG. Even though the spelling rate was about 2 characters/min, the developed BCI system has operated sufficiently accurately as the first BCI spelling device (Birbaumer et al., 1999).

In 2006, John Donoghue, an American neuroscientist at Brown University, and his colleagues published an article reporting a clinical trial of BrainGate. BrainGate is the first BCI system based on implanted microelectrode array in humans and is designed to help patients who have lost control of their limbs, or other bodily functions, such as patients with amyotrophic lateral sclerosis (ALS) or spinal cord injury (SCI). They demonstrated the BrainGate system with tetraplegia and showed that patients can control a cursor on a computer screen, open emails, and operate devices such as a television in real-time just by thinking (Hochberg et al., 2006). With the advancement of technology, the BrainGate research team was reformed including multiple universities and medical centers in 2008. They progressed the performance of the intracortical BCI using the implanted microelectrode array. In 2012, they demonstrated that long-standing tetraplegia can control the robotic arm by performing 3-dimensional reaching and grasping movements. By decoding the recorded signals from 96-channel microelectrodes in the motor cortex, the implanted patient can manipulate a robotic arm to drink coffee from a bottle without explicit training (Hochberg et al., 2012). In 2017, they combined the intracortical BCI with functional electrical stimulation (FES) technology and restored the function of a paralyzed arm of a tetraplegia patient to reach and grasp a target object. Decoded signals of brain activity have been conveyed to the 36 implanted electrodes to stimulate the muscles in the paralyzed arm of the patient instead of the robotic arm. Using his paralyzed arm, the patient volitionally performed self-paced reaches to drink a mug of coffee and fed himself (Ajiboye et al., 2017). In 2021, they attempted decoding the imagery handwriting movement from neural activity in the motor cortex and translated it to text in real time. Using a recurrent neural network (RNN) decoding approach in a patient whose hand was paralyzed from spinal cord injury, they successfully translated the brain activity of imagery writing into texts with a typing speed of 90 characters/min (Willett et al., 2021).

In 2017, Masaki Nakanishi and his colleagues achieved the highest information transfer rates (ITRs) up to 325.3 bits/min which is equivalent to the transfer rate of 62.5 characters/min with noninvasive EEG. Their BCI system is based on SSVEP signals with a new joint frequency-phase modulation method to tag 40 characters with 0.5 s long flickering signals. To implement a high-speed speller, they applied the synchronous modulation and demodulation paradigm and tuned the target identification algorithm for each user. Their study demonstrated that BCIs can provide a truly naturalistic high-speed communication channel using noninvasively recorded brain activities (Nakanishi et al., 2018).

In 2019, Edward Chang, an American neurosurgeon at UCSF, and his colleagues demonstrated the BCI system that produces speeches from the recorded human cortical activities while a subject is speaking and reads mimed sentences silently. They recorded ECoG with subdural electrode patches and interpreted the recorded activity with kinematic and acoustic models. They decoded the recorded signals based on the deep learning technologies and produced synthesized speeches that can be recognized and comparable to original ones (Anumanchipalli et al., 2019). They applied the developed system to a real patient with anarthria (Moses et al., 2021), the loss of the ability to articulate speech, and spastic quadriparesis who had been unable to speak for over 15 years, and they decoded the sentences at the rate of 15.2 words per minute.

Studies on the BCI have diversified and progressively increased in number during the last 20 years. Recently, studies are very active with the advent of supporting technologies, and especially deep learning technology is actively and effectively used to understand brain activity by measured signals. There are many types of BCI and they can be summarized by grouping them according to some criteria.

1.2 Need of BCI Systems for the Locked-In Syndrome

Why do we need BCIs? While we have normal communication pathways via the nervous system, we do not need an additional system of BCI for ordinary communication. However, for persons who have lost some part of the ordinary communication pathway, BCI can be an effective alternative communication pathway. So, firstly, BCI is needed for the

Table 10.2 Remained body functions according to the severity in disorders of consciousness

Disorders	Self-awareness	Motor function	Respiration	Pain & suffering	Sleep-wake cycle	EEG activity
Dementia	Yes (lost in late stage)	Variable	Yes	Yes	Yes	Nonspecific slowing
Locke-in syndrome	Yes	No (except eye movement)	Yes, variably depressed	Yes	Yes	Normal, minimally abnormal
Coma	No	No	Variably depressed	No	Yes	Polymorphic delta & theta
Vegetative state	No	No	Yes	No	Yes	Polymorphic delta, theta & slow alpha
Brain death	No	No	No (only spinal reflex)	No	No	Electrocerebral silence

Multi-society task force on PVS (1994)

rehabilitation of the disabled who lost their ordinary commutation functions.

There are different types and different degrees of disabilities humans can have as a result of accident, disease, or congenitally. When some body functions are impaired or disturbed, those can be replaced partly by the remained body functions with the help of assisting devices. If subjects have their legs broken, they can use the remained function of the hands to control the wheelchairs and replace some portion of the lost function of the legs. If they lost the function of their hands, they can use the function of the mouth. Currently, voice recognition is already popular within our daily living environment and efficient enough to transfer our command or thought accurately to the external world. If the disabilities include speech, what can be remained? Then, the disabled person cannot move their legs, or hands and cannot speak either. They have only intact cognition and brain function with some controllable function of eye gazing. This clinical condition of *locked in syndrome* (LIS) commonly becomes the most immediate candidate for BCI applications.

Locked-in syndrome is a neurological disorder with complete paralysis of all voluntary muscles except the control of the movements of the eyes. Subjects with locked-in syndrome are conscious and awake, but voluntary motor paralysis prevents them from communicating by body movements or speeches. Their cognitive function is usually unaffected and the communication with the external world is only possible through eye movements or eye blinking.

Locked-in syndrome (LIS) is caused by the damage to the pons, which is a part of the brainstem and contains nerve fibers relaying information from the motor cortex to peripheral muscles. Blockade of relay station makes peripheral muscles unconnected and paralyzed. LIS can be caused by a variety of clinical disorders. They can be acute events like ischemic or hemorrhage infarction of the brainstem or traumatic brain injury resulting from the mechanical damage to the brainstem structures. LIS can be also developed from slowly progressing or chronic diseases like brainstem tumors or motor neuron diseases. Amyotrophic lateral sclerosis (ALS) is the most notable etiology of slowly progressing LIS.

While LIS patients have paralyzed voluntary motor functions, they have almost intact sensory systems. They can see and hear like a normal person. They have a normal sleep and wake cycle. They are differentiated by self-awareness from other severe disorders of consciousness like coma, vegetative state, and brain death (Table 10.2).

Locked-in syndrome (LIS) has three different subcategories in severity (Fig. 10.3). *Classic LIS* patients have quadriplegia and anarthria with preserved consciousness and eye movement. Patients of this subcategory are most efficiently benefited by the application of BCI. *Incomplete LIS* patients have the same dysfunctions as classic LIS patients but with some other type of voluntary movements other than eye movement like finger twitch. Patients of this group also can be benefited from the BCI, but other assisted communicative alternatives also can be applied according to the level of remaining voluntary movements. *Total LIS* patients are totally immobile with the immobility of eye movements and are unable to communicate. However, they are fully conscious. Applications of BCI for patients of this group are very limited in types and the performances of most trials are at the level of not so satisfactory.

There are some communication options used for individuals with motor impairments that result in limited voluntary movement as in LIS patients. They can communicate with the help of an assistant or a partner with the low-tech augmentative and alternative communication (AAC) tools. The currently available AAC tools for LIS patients are EyeLink, E-tran, and partner-assisted scanning (PAS) (Fig. 10.4).

An EyeLink is a transparent board with letters or symbols on it. It is positioned between a patient and a communication patterner. After a patient determines the gestures for YES and NO response with the possible motor functions like eyes blinking or eye wink, a letter is selected by the gaze of a patient. The communication partner moves the board until

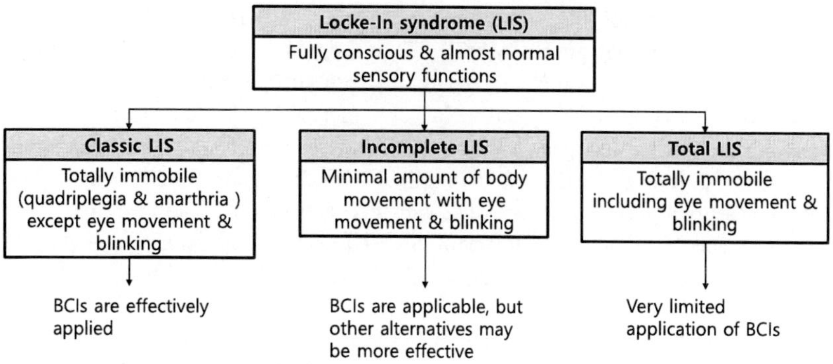

Fig. 10.3 Subcategories of locked-in syndrome (LIS)

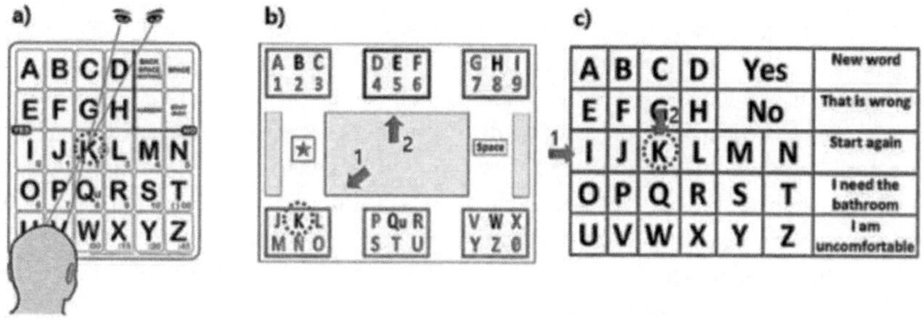

Fig. 10.4 Low-tech augmentative and alternative communication (AAC) options using accessible eye gaze. (**a**) Eye-link: a letter is selected by positioning a letter between the straight line connecting the eyes of a patient and a partner. (**b**) E-tran: a letter is selected by two eye gaze movements. The first eye gaze points toward the section a letter belongs to, and the second eye gaze points toward the section having the color of the selected letter. (**c**) Partner-assisted scanning: a letter is selected by scanning lines vertically and scanning horizontally within a selected line

their eyes meet the eyes of the patient and both people are looking at the same letter on the transparent board. It takes 6~10 s for selecting one letter, which is equivalent to 6~10 letters/min, upon the level of experience.

E-tran, which stands for "Eye Transfer," is another AAC tool for motor & speech impaired patients. E-tran is a board with the alphabet and numbers displayed in eight sections. It is positioned between a patient and a communication partner and they can see through the open space in the center of the board. The patient uses the E-Tran board by pointing with their eyes to indicate a letter and the communication partner follows the patient's eye gaze to identify the letter. To increase the rate of communication, each section is composed of a section color and group of letters and letters in a section written in all different colors. One letter is selected by two movements of eye gaze. The first gaze selects the section where the letter is grouped, and the second gaze selects the section having the color of the selected letter written. It takes 8~15 s for selecting one letter, which is equivalent to 4~8 letters/min, upon the level of experience.

Partner-assisted scanning (PAS) method also uses a transparent board with the letters written on it. After a patient determines the YES and NO response gestures, letters are scanned from the upper left corner. When the partner asks if the letter is in the row, the patient responds with predetermined gestures of YES or NO. If the patient responds with YES for the row, the partner asks for each of the letters in the row sequentially and identifies the designated letter by the second YES response of the patient. It takes 10~18 s to select one letter, which is equivalent to 3~6 letters/min, upon the level of experience.

Even though these low-tech eye gaze AAC options are low in transferring the information at most 10 letters/min and need a partner or an assistant, they are preferred in some clinical patients instead of high-tech devices. Unlike high-tech devices, these low-tech AAC options are simple, inexpensive, and quickly attainable, and they have no malfunctions and no liability issues for facilities. A high-tech system using BCI should provide a higher information transfer rate replacing the advantages accompanying the low-tech options.

1.3 Types of BCI Systems

BCI is largely grouped by the types of signals used. Brain activities are recorded in two ways. One is the group of *electromagnetic signal* which accompanies the neural activity of the brain. Electrical activities of the brain are measured in different electrical signals, local field potential (LFP), electrocorticogram (ECoG), electroencephalogram (EEG), and magnetoencephalogram (MEG). These signals are recorded at different locations of the brain with different levels of invasiveness. LFP and ECoG are measured by opening the skull via surgery to place the electrodes within the skull. So, they are classified as invasive BCI. EEG and MEG measure brain electrical activities with the electrodes or sensors placed externally over the skull. So, BCI using these signals are classified as noninvasive BCI. EEG measures conducted currents electrically with multiple electrodes at the scalp surface which are generated during the brain functions and transmitted through the volume conductor of the brain tissues. EEG is the most actively used signal in BCI applications due to its noninvasiveness and relative compactness of the measuring system. MEG measures the magnetic field with many magnetic sensors located over the scalp surface. Magnetic fields are accompanying electric fields generated during the neural activities of the brain and they are almost transparently transmitted out of the skull. The other group of signals is *metabolic signal*. As brain function is an energy-consuming activity, neuronal cells producing electrical signals need energy and oxygen supply according to their activities. Positron emission tomography (PET), functional magnetic resonance imaging (fMRI), and functional near-infrared spectroscopy (fNIRS) generate metabolic images and signals reflecting the metabolic activity of the regional neural tissues. All these metabolic signals are noninvasive. PET is the medical imaging system producing metabolic images using positron-producing radioactive isotopes. PET is mostly used to evaluate the regional glucose consumption of brain tissues which are used as the energy source of neuronal activities. The fMRI measures the small changes in oxygen-rich blood flow accompanying the neuronal activities of the regional brain tissue. fNIRS measures oxygenation of regional cerebral tissue occurring in response to neural activity using near-infrared light.

BCI can be classified according to the type of operating mode. *Asynchronous BCI* is the system that can be started at any time asynchronously when the subject or patient wants to use the BCI system (Fig. 10.18b). The asynchronous system is always in preparation to receive a command from the subject to start or turn on the system. After starting the system, the BCI system can be operated in synchronous mode. As there is no time period determined for output generation, generally asynchronous BCI is more complex and performance is lower than synchronous BCI. In contrast, *synchronous BCI* expects the response of the subject within the predetermined time duration according to the cue imposed by the BCI system (Fig. 10.18a). The output of the system is determined by the signal collected during the predetermined time period. As the BCI output is always determined by selecting the most probable one from the output list, the performance of the synchronous system is usually higher than asynchronous BCI. Most of the BCI systems are belonging to the synchronous system. From the user's point of view, a synchronous system is less friendly than an asynchronous system as the subject should follow the time-dependent plan of the system (Rashid et al., 2020).

BCI can be classified by the protocols generating the output commands. In *goal selecting* protocol, the user simply selects a goal of the destination, such as a door of the room. Then the BCI application program decodes the intent of the user and moves the cursor or wheelchair to the destination. In this application, the BCI application program should find an optimal path to the destination and takes the larger portion of the burden for navigation. So, the intelligent capability is required for the application program. In *process selecting* protocol, the user controls all the processes to reach the goal of the destination. At each step of control, a user moves a cursor or wheelchair to right or left, up(forward) or down(backward). In this output paradigm, the application program does not recognize a final goal up to the moment of final selection, simply decodes the brain activities and executes the decoded control at each step. In this application, the BCI application program takes the smaller portion of the burden for navigation and the user takes the larger portion of the burden for navigation.

2 Invasive BCIs

Invasiveness of the BCI is determined by the way of acquisition of electrical signals from the brain. In contrast to the noninvasive signal of EEG, local field potential (LFP) and electrocorticogram (ECoG) cannot be recorded without the invasive process of surgery. LFPs record intracortical electrical activities with the electrodes penetrated into the cortical tissues and ECoGs are recorded intracranially with the surface electrodes laid directly over the brain cortical surfaces. Since LFPs and ECoGs represent the brain activities of a much smaller volume of brain cortical tissues with broader spectral bandwidth up to 500 Hz, they provide a lot of information with a higher spatial resolution and more accurate BCI controls are possible when they are decoded efficiently. Brain activities recorded within the brain are less affected by other biological artifacts like eye or muscle movements. Also, invasive BCIs do not use any predetermined paradigm of stimulators to evoke brain activities. They are based on the brain activities generated by the natural thoughts or trials for the movements.

2.1 BCI Using Intracortical Electrodes

The most important part of BCI using LFPs is the electrodes to record the intracortical brain activities chronically. To

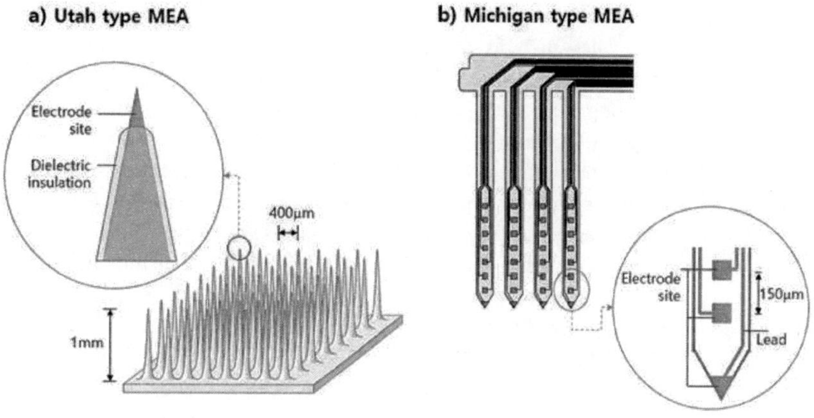

Fig. 10.5 Microelectrode array (MEA) for intracortical recording. (**a**) Utah-type microelectrode array. (**b**) Michigan-type microelectrode array

record the signal beneath the cortical surface, electrodes are designed in the form of small needles to penetrate the cortical tissues. Usually, a microelectrode array (MEA) is used to record neural activities in close proximity to the target neuronal population mostly around the motor-related areas. MEA is composed of contacting sites that directly contact brain tissues and transduce electrical signals, lead which is not in contact with tissue and guides contacting sites deep into tissues, and substrate which structurally integrates the multiple electrodes into a single array. Different technologies have been applied to develop efficient MEA.

Microwire-type electrodes are composed of thin metal wires, like tungsten, stainless steel, platinum, and platinum-iridium alloy, with insulation except for the tips of the diameter of 10~100 μm. Microwire electrodes are packed in a bundle to record the electrical signals in close proximity. Microwire electrodes can be easily fabricated, permits a high degree of flexibility in assembly, cost-efficient and convenient. However, they have serious problems in the shape disparities between the fabricated electrodes and difficulty in compromising positional accuracy due to buckling during implantation.

With the advancement of microelectromechanical systems (MEMS) technology, electrodes have become more integrated into silicon-based neural electrodes. MEMS technology reduced the size and increased the uniformity of electrodes, compared to microwire electrodes. As a result, signals are recorded with improved reliability in multiple places over the brain cortex. The Utah and Michigan electrodes are currently the most widely used microelectrodes (Fig. 10.5a). Utah MEA is a silicon-based microelectrode array, in the form of a square grid in a 10 × 10 pattern, and each of 100 needles is vertically positioned on the substrate. Each needle is 1 mm in length with a diameter of 80 μm at the base tapering to a tip and spaced at intervals of less than 400 μm. The tips of electrodes are coated with metals like platinum to enhance the electrical properties. Additionally, Utah electrodes were implanted using the pneumatic inserter system and could withstand long periods of implantation as a result of their biocompatibility and structure (House et al., 2006). In contrast to Utah MEA, Michigan MEA contains multiple contacting sites in a single needle-type shank (Fig. 10.5b). The probe length ranges from a few millimeters to centimeters and can be connected to the IC circuit based on the fabrication similarity. Based on their structures, Michigan MEA records the LPFs along with the depth while the Utah MEA records LFPs within the same depth.

One of the problems of silicon-based MEA is the mismatch of mechanical rigidity between soft brain tissues and the silicon-based electrode shanks limiting conformal contact. To avoid this mechanical mismatch, and increase the sensitivity of the electrodes, new technologies, conductive polymers, carbon materials, and nanowires, are actively applied to the development of MEA. To increase the sensitivity with low noise, these technologies can be efficiently combined. For example, conductive polymers of low modulus are useful as substrate materials in realizing conformal contact to the brain interface. Nanoscale carbon materials, like graphene and CNTs, are useful as conductive materials with their low impedance (Wu et al., 2021).

BCIs using LFP with intracranial electrodes have been applied in many experimental trials with animals. In 2008, Andrew Schwartz and his colleagues successfully demonstrated that monkeys could feed themselves by controlling the robotic arm only by their thought with the implanted intracortical electrodes. Monkeys have moved the robotic arm just like their own arms by the thought and could reach and grab a food positioned at an arbitrary point in 3D space within an arm's reach.

BrainGate is the intracortical BCI successfully applied to the patients, who have lost control of their limbs, using the Utah-type microelectrode array (MEA). Through their continuous development, BrainGate successfully demonstrated that patients can control cursors on the computer screen, open emails, operate devices such as a television in real-time, control the robotic arm by performing three-dimensional reaching and grasping movement, and drink coffee from a bottle without explicit training (Ajiboye et al., 2017). Accurate cursor control performance still has been obtained 1000 days after implantation. Instead of moving the robotic arm, BrainGate also demonstrated moving the paralyzed arm of a tetraplegia patient to reach and grasp a target object by electrically stimulating the paralyzed muscles of the patient with decoded brain signals. Recently, BrainGate successfully translated imagery handwringing thought into text with a typing speed of 90 characters/min with a 94.1% raw online accuracy (Willett et al., 2021).

While intracortical BCI has the merits of accurate control with a natural way of operation, it also has some limitations and issues to be solved in further advancement. Since the electrodes have to penetrate the live cortical tissues, methods should be prepared that can minimize the initial injury of neural cells and micro blood vessels. As electrodes are foreign bodies to the brain cells, there is a biocompatibility issue to be solved so that the implanted electrodes do not induce any deleterious biological responses or functional losses. Biofouling is the adsorption of molecules to the electrode sensing site and preventing proper contact of the electrode with the environmental tissues. It is the main cause of decreasing systemic performance over time. A mechanical mismatch between rigid electrodes and soft brain tissue resulting in performance degradation is a still remained issue to be improved. In addition to these biological issues, there are engineering issues minimizing the dimension of implanted electrodes and supporting circuits, minimizing power consumption, and increasing the rate of information transfer from the brain to the outside.

Neuralink, a neurotechnology company founded by Elon Musk in 2016, aims to provide BCI wirelessly using only their brain signals via implanted electrodes. The 32 electrodes are fabricated into a thread, 20 mm long and 4~6 μm thick, from biocompatible polyimides, and multiple threads, maximally 96 threads of 3072 electrodes, are to be plugged into brain areas responsible for coordinating movement. Each thread is implanted with a specially designed surgical robot avoiding surface vasculature and targeting specific brain zones. They are connected to the 23 mm-wide hermetically packaged electric unit which is placed neatly in the skull and transmits the recorded signal outward wirelessly. While the early experiments with small and large animals show promising results, no clinical trials have been conducted to date. Future human trials will be critical for the exploration and validation to be a successful neuro-prosthetic unit. Outside of the medical community, Neuralink hopes to achieve symbiosis with artificial intelligence by developing collective man-computer mutualism.

2.2 BCI Using ECoG

ECoG is recorded intracranially with the electrode array laid directly over the cerebral cortex surface (Fig. 10.6). The subdural electrodes are embedded in a thin plastic pad so that it is easily placed above the cortex. ECoG provides signals of lower temporal and spatial resolution compared to the LFP measured with intracortical electrodes. However, it provides higher temporal and spatial resolution as well as higher amplitudes than noninvasively recorded EEG. During the active movement, ECoG shows two characteristic changes. One is the sensory motor rhythm similarly to EEG in the mu band (8~13 Hz) and beta band (18~30 Hz). This rhythmic change occurs as the decrease in power in a narrow frequency band over a large area of the sensorimotor cortex. Another change is a broad power increase over the gamma band (40~500 Hz) compared to the resting state, which is not detectable in EEG. It is highly specific to motion and cortical location, in contrast to the spatially spread low band power. It

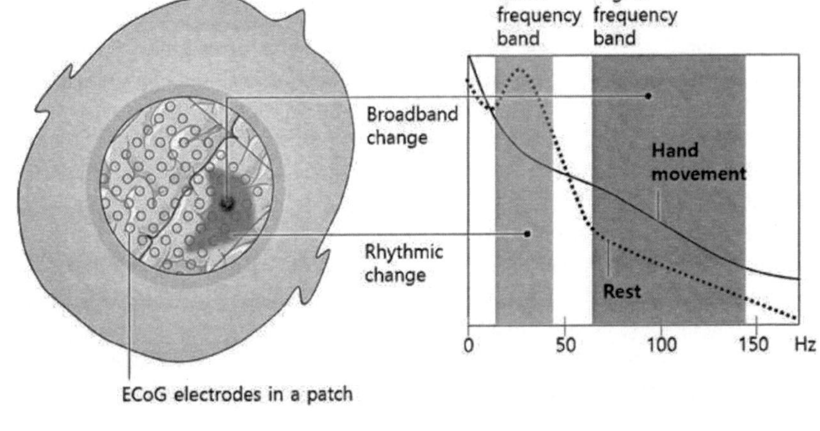

Fig. 10.6 ECoG electrode and the change of electrocorticogram (ECoG) during movement

represents the increased cortical activation of the localized region. ECoG is also less vulnerable to artifacts such as eyes and muscle movements like LFP. While the recording of ECoG is invasive with the surgery of craniotomy, it is less invasive compared to the intracortical recording as the electrodes are not penetrating and damaging the cortical tissues. So, ECoG is positioned intermediately between EEG and LFP in invasiveness as well as in signal quality.

Imagery motor movement induces ECoG-measured cortical activity change in primary motor areas, and it can successfully be used to control a cursor on a computer screen. This ECoG-based BCI enabled an ALS patient to spell independently on a tablet computer. By attempting to move the right hand, the subject can select a group of characters and an individual character. It took 33 s to spell 1 character with word prediction. This ECoG-based BCI has been stable in signal recording and translation algorithms while the patient was using the system for more than 43 months (Vansteensel et al., 2016). Wireless Implantable Multichannel Acquisition system for Generic Interface with Neurons (WIMAGINE) is a newly developed 64-channel ECoG recorder. With bilateral WIMAGINE recorders permanently placed over the upper-limb sensorimotor cortices, the upper arms of the exoskeleton have been controlled by imagery movements. The patient has successfully driven an exoskeleton to make 2-handed, multi-joint movements of the upper limbs and to simulate walking for a total distance of 145 m (Benabid et al., 2019). Phonemes, distinct units of speech, can be decoded by measuring ECoG while a subject's speech intention is recited. ECoGs in the area associated with movement of speech articulators in the vocal tract, like the larynx, tongue, lips, and jaw, provide rich information for generating speeches. Based on these earlier results, a speech synthesizer has been successfully implemented that translates the neural activity measured by ECoG into speeches (Anumanchipalli et al., 2019). This speech synthesis BCI has used two models for translation. The first model translates brain activities into articulatory movements, and the second model translates articulatory movements into synthesized speeches. Parameters of the two models have been estimated with collected and experimental data under the deep learning strategy. Listeners were able to understand the resulting synthesized speech, even when it was based on cortical activity generated by mimed speech (Miller et al., 2020).

3 Noninvasive BCIs Using EEG

Due to the inherent invasiveness of invasive BCI using LFP or ECoG, applications of invasive BCI are greatly limited. In contrast, BCI using EEG can be implemented without the invasiveness and it can extend the application more conveniently to the patients as well as normal subjects. There are many types of noninvasive BCI based on the EEG.

3.1 Slow Cortical Potential (SCP)-Based BCI

Slow cortical potential (SCP) is a brain electrical activity of slow voltage shifts lasting 300 ms to several seconds in recorded EEG (Fig. 10.7). SCP belongs to the lowest frequency range less than 1 Hz and is mostly analyzed in the time domain. SCP has been actively used in the early stage of BCI history. SCP is recorded over the sensorimotor cortex with relatively slow changes and typically associated with movement or movement imagery. As the SCP is time-locked and phase-locked to the specific sensorimotor event, SCP is regarded as an event-related potential. In relation to the event, it can occur before, during, or after the specific event (Birbaumer et al., 1990). SCP is negatively associated with the change in brain activity. It typically decreases toward the negative direction preceding the actual movement, imagery movement, or cognitive tasks like arithmetic calculation. So, SCP is considered to represent the cortical activation preparing the action, and the movement-related potential generally follows after the SCP.

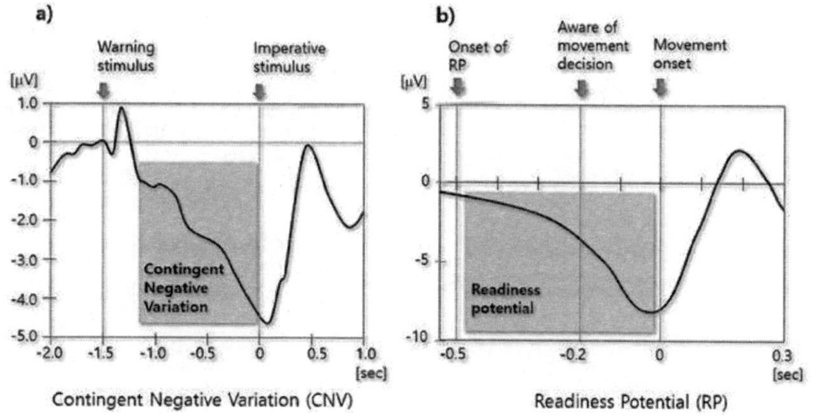

Fig. 10.7 Slow cortical potentials. (**a**) Contingent negative variation (CNV), (**b**) Readiness potential (RP)

Contingent negative variation (CNV) is a slow negative cortical potential appearing between a warning and an imperative stimulus in a reaction time task. It begins after 200~500 ms after the warning stimulus notifying that the imperative stimulus will occur some seconds later. CNV is recorded broadly over the regions involved in the motion and cognitive tasks and it is affected by the motivational and task-specific factors. While CNV reflects an index of cortical arousal during orienting and attention, its functional neuroanatomical basis is poorly understood.

Readiness potential (RP), which is also called pre-motor potential or Bereitschaftspotential, is another type of negative SCP. The RP is triggered by a cued or a self-initiated voluntary muscular movement. RP is a record of activity in the motor cortex and supplementary motor area of the brain preparing the voluntary muscle movement. It is recognized by negative potential about 500 ms before the time point of a movement, but its onset is 1.5~2.0 s before the movement. It is seemingly too early for the subject to be aware of the conscious decision to move. Based on the experiment that the conscious decision occurred 500~800 ms after the start of the RP, some researchers claim that free will does not exist. Its amplitude and topology are variant upon the movement types and the muscles involved as well as psychological variables.

SCP can be modulated by many cognitive activities like moving or performing arithmetics. SCP shifts have been applied in BCI systems by moving a cursor and selecting the target presented on a computer screen (Fig. 10.8). Subjects usually require prior training of repeated sessions to generate SCP effectively. Subjects can be trained to generate voluntary SCP changes using a thought-translation device, which presents visual and auditory feedback so that they can learn to shift the SCP. Typically, subjects have trained to move a cursor vertically by the amplitude of the SCP shifts. Subjects also can train to generate SCP by themselves without external feedback. In this self-regulated training, the subject's physical and psychological states, motivation, social context, and subject-trainer relationship drastically affect SCP modulation training.

In a typical SCP-based BCI, the subject communicates through a series of trials. In each step of the trial, a binary option is offered. Each active trial period is separated by baseline periods. During the baseline period, the patient rests and returns potential in resting level. During the active trial period for the selection of one option from the offered two options, the patient generates a SCP for one option or remains relaxed for the other option. In a spelling application, candidate letters are divided into two groups, and the patient selects one group that contains a targeted letter by SCP. Binary selections are performed sequentially subdividing the letters in a selected group until the patient finally selects one letter.

Typical accuracies achieved with the SCP-based BCI system are acceptable and vary between 70% and 80%. However, SCP-based BCI is slow as the patients take about 4~10 s for one binary selection.

The first BCI-based speller was developed based on SCP by Birbaumer and his colleagues in 1999 (Birbaumer et al., 1999). They applied the developed SCP-based speller to two severe ALS patients. They have learned to generate SCPs and to select a letter by controlling the language supportive program. They spelled 2 characters/min on average. One patient wrote a first full message with BCI reflecting that he was mentally active and very grateful for his BCI communication system. He is known to have continued using this BCI system up to his death in 2007.

In the early stage of BCI, SCP-based BCI was actively studied and developed. However, several fundamental issues have greatly limited further advancement and practical application of SCP-based BCI. The most significant problem is its low speed of information transfer. The information transfer

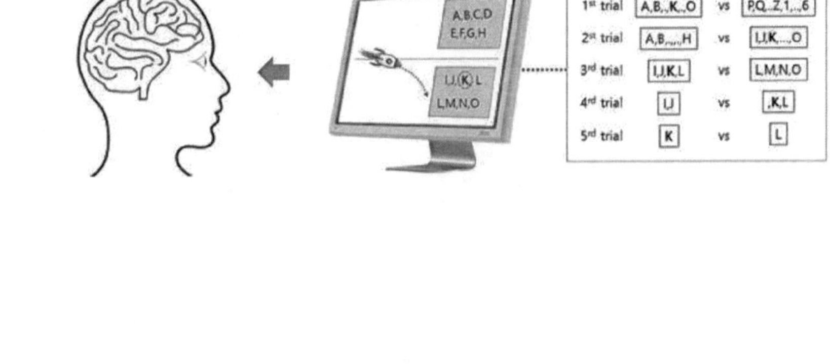

Fig. 10.8 BCI speller based on slow cortical potential (SCP). When subjects want to select a letter on the downside, they try to move a rocket downside by generating an SCP as they have trained

rate of 2 characters/min is lower than other noninvasive BCI systems based on different types of EEG signals and even lower than low-tech AAC options having 10 characters/min maximally. SCP-based BCI only allows binary selection or a selection in a one-dimensional direction and does not allow multidimensional control, which naturally results in a low information transfer rate. The accuracy of SCP-based BCI is limited to 70~80% and prone to error. Lastly, SCP-based BCI requires extensive training of the patient over weeks and months.

3.2 P300-Based BCI

Among the EEG records, event-related potential (ERP) is the recorded EEG in response to the sensory, motor, or psychological events. ERP usually appears synchronized to the applied event stimulus with determined time delay and its amplitude is superimposed on the spontaneous EEG. ERPs show different delays and amplitudes according to the physical type and the content of the applied event stimulus. Each ERP is characterized by the specific time delay and the location where it is dominantly recorded within the cortical region. Among the various ERPs, P300 and steady-state visual evoked potential (SSVEP) are the most actively used in BCI applications by producing higher performances.

P300 wave is the most representative ERP. It is recorded with a maximal positive peak at approximately 300 ms after a visual or auditory event stimulation is applied to a subject. P300 is generated in response to the event, which has special meaning, interest, or task to a subject in examination. Even though the stimulation is similar in a physical quantity, if the stimulation does not have any contents that can be regarded as an event, P300 is not generated. P300 is recorded with the highest amplitude in the central and parietal regions of the cortex. As the small amplitude of the P300 wave is superimposed on spontaneous EEG, it is difficult to recognize the occurrence of P300 in the record of single stimulation, and usually averaging is required with multiple event stimulations.

P300 is generated under "the oddball paradigm." In the oddball paradigm, a subject is stimulated by a series of stimulation composed of two types of events. One type of event stimulation is targeted or desired event stimulation which occurs more infrequently than the other type of event stimulation. For this type of stimulation, a subject is anticipated to produce a P300 response. The other type of stimulation is undesired or uninteresting event stimulation which occurs much more frequently than the former type of event stimulation. For this type of stimulation, a subject is anticipated not to produce a P300 response. For the acquisition of proper P300 responses, the subject should pay attention to the stimulations applied. Attention can be increased by counting the infrequently occurring target events.

Visual P300-Based BCI

As the P300 is the response to the infrequent event stimulation, event stimulation can be applied in different types of physical stimulations, visual, auditory, or somatosensory stimulation. The first P300-based speller, developed by Farwell and Donchin in 1988, used a visual stimulation for the "the oddball paradigm." In their study, a unique computer screen was suggested which serves as a keyboard or controller for the prosthetic device of the subject (Fig. 10.9). A computer screen is presented in the form of a 6 × 6 matrix containing 36 characters including 26 alphabets and several other symbols and commands. While the subject focuses attention successively on the character he wants to communicate and silently counts the number of times the desired character flashes, the speller system detects the selected character online and in real-time by P300. Each time, event stimulation is applied by flashing a single line of 6 characters vertically or horizontally. Each horizontal and vertical line is flashed in random order but with equal probability. P300 is

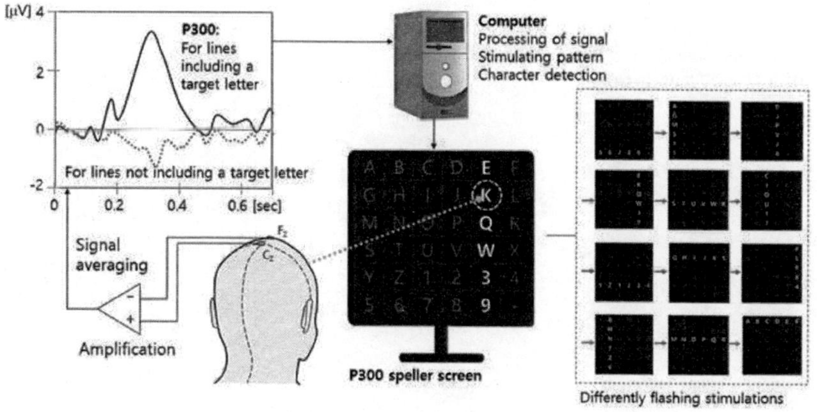

Fig. 10.9 BCI speller system based on P300

produced when the vertical or horizontal line containing the character wants to spell is stimulated. The character in the intersection of the vertical and horizontal lines producing the P300 wave is detected as the character that a subject wants to communicate by the processing in the computer. In this first P300-based speller, they transferred the information of the subject at a rate of 12.0 bits/min or 2.3 characters/min (Farwell & Donchin, 1988).

Auditory P300-Based BCI

P300 can be generated by auditory events. Hill and his colleagues introduced P300 BCI based on auditory event stimulations in 2005 (Hill et al., 2005). To select a symbol or a character based on auditory stimulations, 3 × 3 auditory matrix speller was suggested. The row of matrix is selected by focusing attention to the pitch of the sound, high, medium, or low. And, the column of matrix can be selected by focusing attention to the direction of the sound, from left, right or both sides of the speaker. The resulting 9 different control signals are used to spell the texts with predictive module and produced the ITR of 3.4 bits/min (Höhne et al., 2011). Auditory speller, named CharStreamer, was proposed in 2014 (Höhne & Tangermann, 2014). The auditory stimuli of CharStreamer comprise 30 spoken sounds and they are composed of 26 alphabet characters and 4 command items (Fig. 10.10). This sound stimuli set is split into three groups, group L, R, and B. Letters belonging to the group L is spoken to the left ear, R to right ear, and B to both ears. The letters in one group are read out by the same voice and from the same side of the ear. Letters in three groups are read in parallel without overlapping in the onset timing of the letter. For example, between the sound stimulation onsets of letters from group L, there are two sound stimulation onsets one from group R and one from group B. That is, sound stimulation onsets are interleaving in time. As the number of letters in each group are different by 9, 10, and 11, instantaneous sounding pattern is not repeated before 990 letters are stimulated. Subjects can use this speller simply by attending the letters that they want to spell in one step procedure. It is user-friendly as it is stimulated by the sound of the letter itself without using any artificial substitute. Best performing subject correctly spelled characters with 75% of accuracy and showed the information transfer rate of 4.3 bits/min (Höhne & Tangermann, 2014). Even though the BCI system based on auditory P300 can be applied to the patients with their vision impaired, lower information transfer rate greatly limits the practical application of the method.

Somatosensory P300-Based BCI

P300 BCI based on somatosensory stimulation was also developed. In this BCI system, multiple numbers of small motors are located circumferentially around the waist to generate the vibration at different locations. The subject is requested to attend to the vibration generated at the desired location and ignore all the other vibrations to elicit a proper P300 signal. Their results showed the classification accuracy well above the chance level and the information transfer rate around 2.0 bits/min (Brouwer & van Erp, 2010). In another BCI based on somatosensory P300, stimulations are applied to the fingertips of the subjects with short mechanical taps (Fig. 10.11). Similar to the 6 × 6 matrix used for visual P300 speller, 36 characters are divided into 6 groups, and each group is allocated to the different left or right fingers. Subjects can spell the letter by attending to the stimulations and silently counting the number of stimulations on the finger that the letter is allocated to. This BCI system produced an information transfer rate (ITR) of 7.8 bits/min on average and the best performance showed an ITR of 27 bits/min (van der Waal et al., 2012). As tactile stimuli have the advantage of not using the visual or auditory system, it is especially suitable for patients whose vision or eye gazes are impaired and it can be applied unnoticeably to other people. But, it requires an additional vibration unit put on the body and the information transfer rate is relatively lower than BCI based on the visual P300.

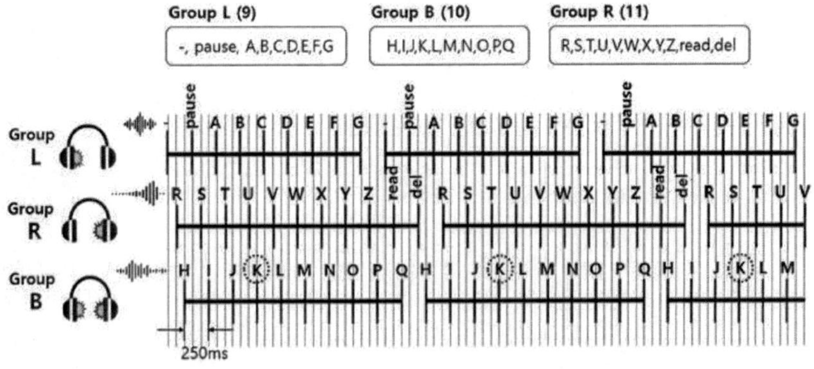

Fig. 10.10 Sound stimulation pattern in auditory P300-based BCI, CharStreamer. (Höhne & Tangermann, 2014)

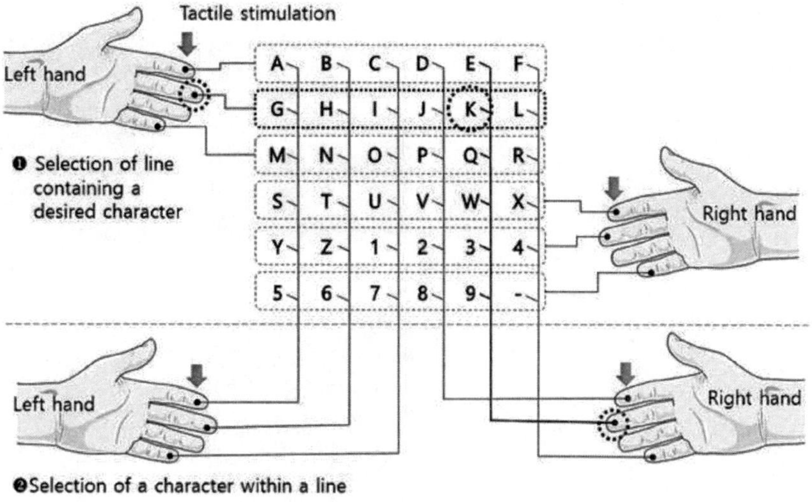

Fig. 10.11 Fingertip stimulating pattern in somatosensory P300-based tactile speller. (van der Waal et al., 2012)

Approaches to Improve P300 BCI Performance

P300 BCIs have several appealing features. They are relatively fast, effective for most users, and straightforward. Especially, they do not require intensive user training as the P300 component arises from endogenous attention-based brain function. However, P300 responses change their amplitudes and temporal patterns in response to several factors, like motivation, level of attention, fatigue level, mental state, learning, and other nonstationarities emanating from the brain. As each subject has a unique EEG and P300 patterns, the BCI system requires calibration for each individual to obtain notified performance. Several perceptual characteristics of the human, like attentional blink, temporal blindness caused by repetitive visual stimulation, and habituation are potential sources of lowering the detection rate of P300.

P300 BCI is based on eliciting large differences between target and nontarget responses and provides higher performance than the SCP-based BCI. Among P300 BCIs, a visual paradigm displaying stimulation patterns on a computer screen is most preferred and actively studied. The P300 BCI with row and column paradigm (RCP) originally introduced by Farwell and Donchin was the most common paradigm for a long time. But, it has some issues to be solved for the improved acceptance of P300 BCI. There have been various attempts to solve the raised issues beyond the first RCP.

In RCP, there is a chance that the same character flashes twice in succession, also called a double target item flash. As the successive stimulation does not generate a P300 component properly, there is a need to eliminate or minimize the chance of consecutive flashing of the same letters. Errors occurring in the RCP are typically in locations adjacent to the target letter in the same row or in the same column. This adjacency problem occurs because each time a target letter flashes, a P300 is produced for every item in the same row or column. If some other paradigm can reduce this type of error, BCI performance can be increased. A large number of flashing stimulations with intense light causes eye fatigue in subjects and sometimes makes eyes temporally blind. If other stimulations instead of light flashing can induce P300 similarly, it may reduce eye fatigue and increase the acceptance of P300 BCI. The RCP requires eye movement to focus the attention on the selected letter. If the patient has a limitation in eye gaze movement, it causes great degradation of BCI performance. If the P300 paradigm can be used without eye gaze movement, it can be applied also to patients whose eye gaze is impaired. The maximum information transfer rate of RCP is 12 bits/min. It takes around 26 s for a selection of a character on average. Even though it may not be slow a lot for a person with no other means of communication, it is very slow compared to the conventional typing speed of healthy people and speed improvement is persistently required. Upon these problematic issues and the requirement, many studies have been done advancing the P300 BCI in various types of approaches. The followings are some noticeable approaches that have advanced P300 BCI at some points (Rezeika et al., 2018).

1. Instead of flashing individual symbols which is uncomfortable to the subjects, *graphical effects* like translations, rotations, zoom in/out, pattern rotation, and sharpening types are used in the row and column paradigm (RCP) (Liu et al., 2010). With this type of stimulation, P300s are generated in greater amplitude and the ITR increased to 35.8bits/min with rotational stimulation (Fig. 10.12a).
2. By dividing a 6 × 6 matrix into submatrices of size 3 × 3 or 2 × 3, and stimulating randomly only one symbol

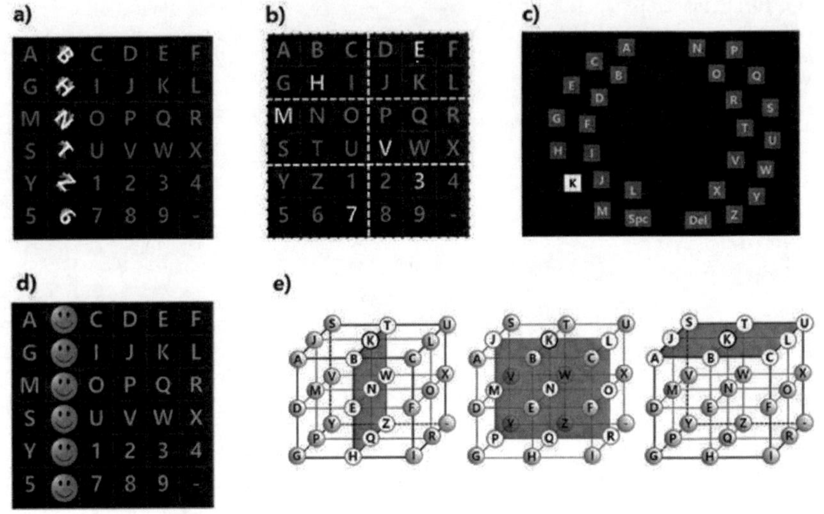

Fig. 10.12 The various paradigms of P300 BCI. (**a**) Graphic effect paradigm of rotating the characters instead of flashing the selected line, (**b**) 2 × 3 submatrix paradigm, (**c**) Lateral single character (LSC) paradigm, (**d**) Famous or familiar face paradigm, (**e**) 3D virtual matrix speller paradigm

within a *submatrix*, errors caused by stimulating the neighboring symbol was reduced (Fig. 10.12b). This paradigm results in an increased maximal ITR of 26.8 bits/min (Shi et al., 2012). The checkerboard paradigm (CBP) is also used in a similar concept. The 8 × 9 matrix is composed of a checkerboard pattern and the characters and symbols are divided into two groups according to the color of the checkerboard cell. Six characters or symbols are selected randomly from a group of the same color and flashed. As nonadjacent elements are pseudo-randomly flashed by alternating the group selection, this method avoided the adjacency-distraction problem and the double flash problems. CBP produced the ITR of 23 bits/min which are higher than RCP and confirmed the increased comfortability of the patients than RCP (Townsend et al., 2010).

3. Instead of a 6 × 6 matrix format, *various format* of spellers, like 3 × 3 or 7 × 7 matrix, have been suggested including the optimized letter arrangement. P300 BCI using the 7 × 12 matrix speller which is similar to a laptop keyboard produced the ITR of 27.1 bits/min (Jin et al., 2012).
4. The *lateral single character* (LSC) speller was proposed (Fig. 10.12c). LSC speller divided the characters into two groups and positioned the characters laterally so that a subject could focus a character on the one side of the screen at a time. Instead of a line or group of characters, a single character was flashed alternately and pseudo-randomly from each side. This layout is expected to be more visually attractive and comfortable and reduced the effect of the crowdedness problem as well as the adjacency-disturbance error, and achieved the ITR of 26.1 bits/min.
5. When each character of a 6 × 6 matrix speller was superimposed by a semitransparent picture of a *familiar or famous* face, P300 BCI showed faster character detection and comparably high accuracy since the familiar faces induced a higher ERP response (Fig. 10.12d). The P300 BCI using familiar faces showed a very high information transfer rate reaching approximately 80 bits/min (Kaufmann & Kubler, 2014).
6. Instead of a 2D matrix speller, *3D virtual matrix* was suggested (Fig. 10.12e). The characters were displayed in 3D matrix blocks and each plane of the block directing 3 different orthogonal directions flashed at a time. The speed of the 3D matrix interface was proven to be higher than the classic 2D matrix interface since it needed a smaller number of flashes per character. The 3D speller is more attractive and has the potential to be a more user-friendly interface (Noorzadeh et al., 2014) (Fig. 10.12).
7. P300 BCIs are also implemented in other $non - English$ *languages*. Each of these spellers requires innovative interfaces reflecting the unique characteristics of a specific language. The Chinese language written in logographic scripts is also implemented with P300 BCI. By transcribing a Chinese Pinyin into initial consonant and vowel components according to its Mandarin pronunciation, each sinogram is inputted by selecting the initial consonant and then the vowel components and subsequently selecting the sinogram itself. This P300 BCI achieved a mean information transfer rate of 39.2 bits/min, which takes 43.9 s on average to spell one Chinese character (Yu et al., 2016).
8. A *module predicting* and suggesting probable words have been added to make the BCI matrix speller faster and more efficient in writing sentences (Fig. 10.13a). The suggested words are flashed within the matrix in the same way as other characters or displayed in a separate smaller matrix to be selected in the second selection phase. By adding the word predicting module, the ITR

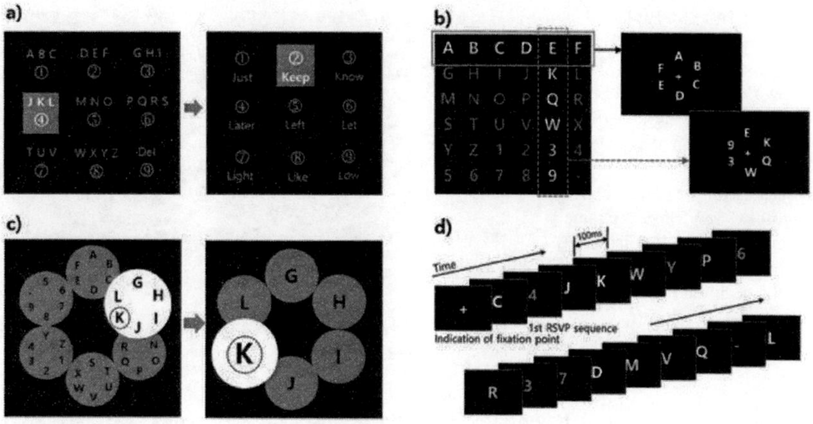

Fig. 10.13 Various paradigms of P300 BCI. (**a**) 3 × 3 matrix speller with word prediction module, (**b**) Geometric speller, (**c**) Hex-O-speller, (**d**) BCI speller with rapid serial visual presentation (RSVP) paradigm

of the P300 speller increased up to 26.1 bits/min (Akram et al., 2014).

9. Matrix speller based on 6 different *colors*, the Chroma Speller, was suggested. After selecting the color that a target character is grouped, the subject is required just to focus on the selected color while the color is sequentially flickering. It aimed to achieve a gaze-independent speller with a minimum workload, as the user is only required to focus on the selected color and not on the individual letter. This BCI system achieved an online spelling speed of 1.4 characters/min. This speller is suitable for patients in the advanced stage as they have limited eye gaze control (Acqualagna & Blankertz, 2013).
10. To minimize the eye movement, *GeoSpell*, speller displaying the selected characters geometrically in the center of the screen, is suggested (Fig. 10.13b). Instead of flashing a selected row or column line of the matrix, characters on the selected line are displayed sequentially on the central part of the blank screen. This paradigm is suitable for covert BCI, in which attention is deployed on the displayed characters while the eye gaze is fixed on the center of the screen and does not move. However, the system did not achieve great performance to compete with typical BCI spellers (Aloise et al., 2012).
11. *Hex − O − Speller* was suggested as a gaze-independent BCI speller (Fig. 10.13c). In this speller, characters are divided into 6 groups and displayed in one of the circles positioned in 6 different directions at the same distance from the center of the screen. Each group of circles is randomly intensified by enlarging the size of the circle and the characters within it. If the circle is selected, characters in the circle move into the central part of the screen and fill the 6 circles. In the second phase, each circle with the character in it is randomly enlarged while the subject is attending and focuses on one of the circles. This speller achieved higher performance than matrix-type spellers in over-attention conditions, which are dependent on eye movements, and in covert attention conditions, in which attention rather than gaze is deployed in the visual periphery (Schmidt et al., 2012). Hex-O speller produced the ITR of 2.5 characters/min on average.
12. Rapid serial visual presentation (RSVP) was developed as an efficient gaze-independent P300 speller (Fig. 10.13d). In this paradigm, individual characters appear in the center of the screen sequentially at a rate of 10 characters/s in a randomized order. The target letter evokes a P300 signal when it appears, while the subject looks at the screen and counts the number of times the target letter appears on the screen. The subjects showed better performance when the characters were presented in a different set of colors than in a monochrome color. Although the accuracy of the RSVP speller outperforms both of RCP speller and the Hex-O speller, the ITR of 1.43 characters/min is lower than expected. Long waiting time between the target character presentations might have caused the user to feel bored or lose focus. However, the system worked as a gaze-independent speller (Acqualagna et al., 2013) (Fig. 10.13).

3.3 SSVEP-Based BCI

The steady-state evoked potential (SSEP) is another type of event-related potential (ERP) actively used in BCI. SSEP is recorded by superimposing on the spontaneous EEG when the subject is exposed to steady stimulation of periodic sensory stimulations. The main characteristic of SSEP is that the same frequency and harmonics of the stimulating frequency appear as peaks in the recorded signals. For example, if the stimulation is applied as a steady 11 Hz fluctuation of sensory pattern, SSEP is recorded as a major peak at the identical frequency of 11 Hz stimulation with minor peaks at the harmonics of 11 Hz, like 22 Hz and 33 Hz. As there are different types of evoked potentials like visual evoked potential (VEP), auditory evoked potential (AEP), and

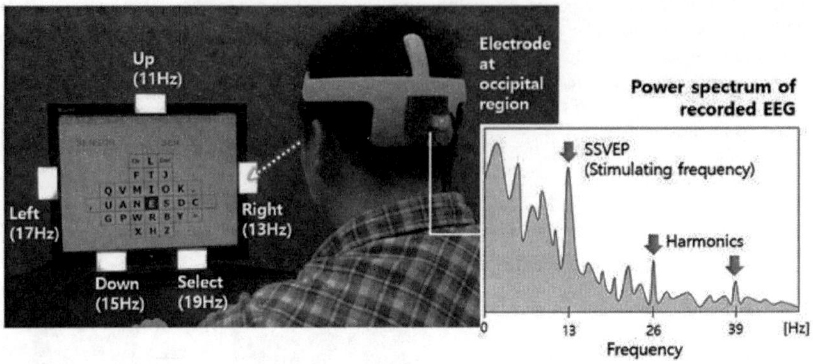

Fig. 10.14 SSVEP-based Bremen speller. Lightboxes of five control commands, Up, Down, Right, Left, and Select, are flashing with different stimulating frequencies. The user selects one of the control commands by focusing on the corresponding lightbox. SSVEP of the same frequency with stimulation appears as peaks in recorded EEG at the occipital region

somatosensory evoked potential (SEP), SSEP also can be divided, based on the physical type of sensory stimulation, into steady-state visual evoked potential (SSVEP), steady-state auditory evoked potential (SSAEP), and steady-state somatosensory evoked potential (SSSEP).

SSVEP-Based BCI

Steady-state visual evoked potential (SSVEP) is elicited by the steady stimulation of flashing light or visual pattern in specific frequencies (Fig. 10.14). Stimulation frequencies in the band of 3.5~75 Hz possibly elicit SSVEP, however the SSVEP response is easy to detect when the stimulating frequency is in the band of spontaneous EEG in 4~30 Hz. The subject using SSVEP-based BCI is just required to focus the eyes on the target generating a visual pattern of a specific frequency. Different commands can be induced by the different frequencies of visual stimulation within this frequency band. SSVEP is typically observed and recorded on occipital sites with electrodes at Oz. SSVEP is preferred in BCI applications due to its excellent signal-to-noise ratio and relative robustness to the artifacts.

SSAEP-Based BCI

Steady-state auditory evoked potential (SSAEP) can be elicited by steady stimulation of sound. As common audible sound in the range of 200~2KHz is beyond the spectral range of EEG, sound stimulation is applied in modulated form instead of continuous pure tone. Usually, sound stimulation is modulated with a rate 20~100 Hz in click sound, tone pulse, or amplitude-modulated tones. The resulting brain response can be localized at the primary auditory cortex and recorded with electrodes at temporal positions of T3~T6. Although SSAEP is promising for mitigating the problem of eye fatigue that often occurs during the SSVEP-based BCIs, it suffers from low performance. To improve the performance, training to control the subject's attention levels is required while simultaneously detecting the auditory stimulation frequency (Punsawad & Wongsawat, 2017).

SSSEP-Based BCI

Steady-state somatosensory evoked potential (SSSEP) can be elicited by the application of steady vibrotactile stimulation on pre-determined positions of the body surface. Vibrotactile stimulators at different positions generate stimulations of different frequencies. While the subject focuses attention on specific local stimulation, the vibration frequency of stimulation is reflected as SSSEP in recorded EEG. SSSEP can reduce the fatigue induced by visual attention and overcome the low reliability of motor activity. However, the performance of SSSEP is not high enough due to the difficulty in discriminating attention in a tactile stimulation task (Ahn et al., 2015).

Among SSEPs, SSVEP is most actively used for BCI applications. SSVEP has some advantages. It does not require any training and calibration for each subject. As the SSVEP is generated while the subject is focusing on a target stimulation continuously, it does not require a specific number of stimulating trials as in P300-based BCIs. Target can be detected in SSVEP as long as the detected SSVEP is strong and stable enough. Generally, SSVEP-based BCI provides a higher information transfer rate (ITR) than P300-based BCI.

Various Types of VEP for SSVEP-Based BCI

In a standard visual evoked potential (VEP) recording, VEP is elicited temporally synchronized to the light stimulation with a specific delay. Based on the specific delay in the recorded VEP, the time of stimulation application can be estimated from the recorded VEP reversely. In time-modulated VEP (t-VEP) BCI, each target is stimulated with a different time sequence strictly nonoverlapping (Fig. 10.15a). From the timing of peak amplitude in the recorded VEP, the time of stimulating light application and the targeted stimulation at that time can be identified. However, accurate target identification in the t-VEP BCI requires averaging with multiple stimulation trials. Also, as recorded VEP should not be overlapped to identify the peak time, stimulation interval cannot help becoming longer than

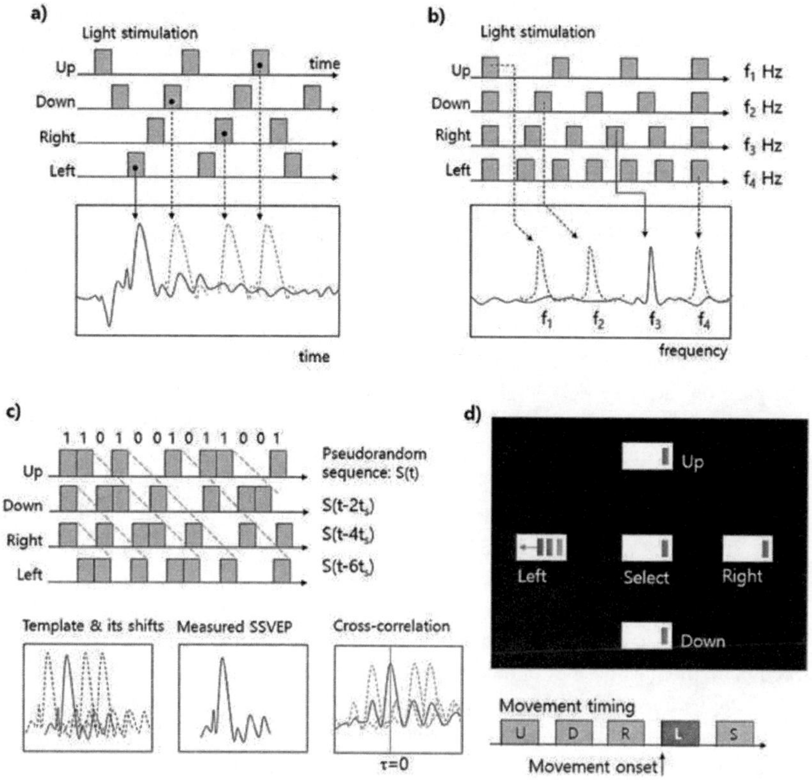

Fig. 10.15 Various types of VEP are used for SSVEP-based BCI. (**a**) Time-modulated visual evoked potential (t-VEP), (**b**) Frequency-modulated visual evoked potential (f-VEP), (**c**) Code-modulated visual evoked potential (c-VEP), (**d**) Motion onset visual evoked potential (m-VEP)

250 ms and usually stimulation rate is lower than 4 Hz. Due to these limitations, t-VEP BCIs have a lower information transfer rate (ITR) of around 30 bits/min (Bin et al., 2009).

The frequency-modulated visual evoked potential (f-VEP)-based BCI is the typical type of SSVEP BCI (Fig. 10.15b). In f-VEP-based BCI, targets of individual commands or characters are differentiated by different flickering frequencies. Power spectral analysis is most widely used for target identification in the f-VEP-based BCI since the f-VEP appears as the peaks at the frequencies of stimulating light and its harmonics.

Another type of SSVEP BCI is the one based on the code-modulated visual evoked potential (c-VEP) (Fig. 10.15c). Instead of using continuously flickering light of a single frequency, light stimulation is applied according to the predetermined pseudorandom sequence. This pseudorandom sequence is generated by the shift register not randomly, but its behavioral characteristics are similar to random sequences. Autocorrelation of the pseudorandom sequence is close to impulse function as in the case of white noise sequence and the cross-correlations with the shifted versions of the pseudorandom sequence are close to 0. For each target, this pseudorandom code is applied with different shifts to generate a pseudorandom temporal stimulating pattern. After deriving the template VEP response by averaging the response of one target stimulation, this template is used for the identification of the target. The recorded response is cross-correlated with the shifted versions of the template and the target is identified by the one showing the maximal correlation to the corresponding shift of the template. SSVEP BCI based on c-VEP requires precise timing between stimuli presentation and data acquisition for higher accuracy and ITR.

Instead of the light flashes, the motion of the target can elicit a VEP. Among visual motion-related VEPs (Fig. 10.15d), motion-onset VEP (m-VEP) shows the largest amplitudes and the lowest inter- and intra-subject variability compared to other motion-related VEPs, like motion-offset VEP, motion-reversal VEP, and steady-state motion-related VEP. Motion is generated within multiple target boxes on the screen by moving a small vertical line toward the left side at different timing without temporal overlapping. N2 and P2 components of VEP are used as features representing the attended target that the subject has selected by gazing. The computer determines the attended target by comparing the timing of prominent N2/P2 components with the movement onset timing of each target (Guo et al., 2008).

Bremen speller is the first type of SSVEP-based BCI. In Bremen speller, a virtual keyboard is presented on the screen with 32 characters displayed in a diamond-shaped form (Fig. 10.14). Each character is selected by the movement of the cursor which is located at the center of the keyboard at the start. The subject can move the cursor in four directions, up, down, left, and right, for each step and access a character wants to spell. The direction of the cursor is selected by the SSVEP. There are five simulating lightboxes, which are all stimulating in different frequencies. 4 boxes are for each direction and one is for the selection of the character. When the subject focuses the gaze on one of the stimulating lightboxes, the SSVEP BCI system identifies the gazing lightbox by the peak frequency in the recorded SSVEP signal. The cursor moves according to the selected directions in several trials and the character is finally selected by focusing the gaze on the "SELECT" lightbox. To minimize the average moving distance, characters are arranged according to the statistical occurrence rate positioning the most frequently using character "E" at the center of the virtual keyboard. SSVEP BCIs generally provide higher ITR than P300 BCIs. In the first study of using SSVEP for BCI application, Bremen speller produced the ITR of 25.67 bits/min in unfavorable experiment conditions of the high level of noise and surrounding distractions. With the integration of the word prediction module, the ITR of the Bremen speller increased to 32.71 bits/min. After improving the processing of the recorded signal by the adaptive time segment length and the novel type of visual feedback, the ITR of the Bremen speller increased up to 61.70 bits/min and the best performing subject reached a peak ITR of 124 bits/min in an online experiment (Volosyak, 2011).

Instead of a speller that selects the character by moving the cursor sequentially, the character can be selected by multi-phase block selection. In a multi-phase speller, characters are grouped into several groups and the individual character is spelled by the sequential selection of groups that the targeted character is allocated. According to the number of groups that can be displayed and differentiated by the different frequencies, the individual character can be selected in 2 phases or 3 phases. When the speller of 27 characters was implemented in 3 phases with three groups of character blocks, the BCI system produced the maximal ITR of 27.36 bits/min. In another study on a multi-phase SSVEP speller, a 2-phase speller was designed with 16 target stimulations. 42 characters are dived into 3 groups of 14 characters (Fig. 10.16a). Each screen is composed of 16 stimulating blocks, 14 blocks for the selection of an individual character and 2 blocks for screen switching to other groups of characters. The stimulating frequencies of these 16 blocks start from 8 Hz to 15.5 Hz with an interval of 0.5 Hz. Each stimulating block was implemented on the screen having a frame rate of 60 Hz by frame-based encoding. Each frame of a block is turned on or off so that the average stimulating frequency is converged to the predetermined stimulating rate on average. Two-phase spellers provide higher ITR than three-phase spellers since each character is selected with a smaller number of trials. The highest ITR of the two-phase speller produced the mean ITR of 61.64 bits/min (Cao et al., 2011).

If the character can be selected in a single step, the spelling speed is anticipated to be increased significantly. One-phase speller is the SSVEP-based BCI system that selects a character in a single-step trial (Fig. 10.16c). To implement many different frequencies in a single conventional monitor with a fixed frame rate of 60 Hz, a sampled sinusoidal stimulation method was used. Visual stimulator of different stimulating frequencies is implemented by manipulating the intensity of the screen according to the sampled signal strength of the corresponding frequency (Fig. 10.16d). With an advanced signal processing method considering the task-specific and subject-specific responses, the ITR of the one-phase spellers can be increased up to 325.3 bits/min in a cue-guided task using a 40-characters speller and 198.7 bits/min in free spelling task (Nakanishi et al., 2018). One-phase speller was also implemented with code-modulated visual evoked potential (c-VEP) with 32 target stimulations in a 4 × 8 matrix. The ITR achieved with c-VEP is 144.0 bits/min, which is slightly lower than SSVEP spellers based on f-VEPs (Spüler et al., 2012).

SSVEP BCI system can be implemented similarly to P300 BCI based on a 6 × 6 row and column paradigm (Yin et al., 2015) (Fig. 10.16b). In this application, stimulating frequencies are the same for characters in the same row and in the same column as they are intensified at the same time in the P300 BCI. And, the characters are stimulating in rows and in columns with different frequencies. Only 6 stimulating frequencies are used alternatively stimulating 6 rows or 6 columns. This two-step stimulation occurs without stopping between the steps and the stimulating frequencies of individual characters dynamically alternate between row and column frequencies. This RCP-based SSVEP BCI achieved an ITR of 41.1 bits/min which is higher than P300 BCI based on RCP (Fig. 10.16).

3.4 Motor Imagery-Based BCI

Event-Related Synchronization (ERS) and Event-Related Desynchronization (ERD)

The sensorimotor rhythm (SMR) is the brain activity recorded over the motor cortex by the contribution of some somatosensory areas in relation to the movement or imagination of movement. Interestingly, the brain activity of the cortical area in relation to the movement is decreasing during the movement, and its amplitude is in increased level during the resting state or after the movement. An increase in SMR amplitude is caused by the synchronous activity of cortical neurons and this increased brain activity is referred to as event-related synchronization (ERS) and a decrease of SMR

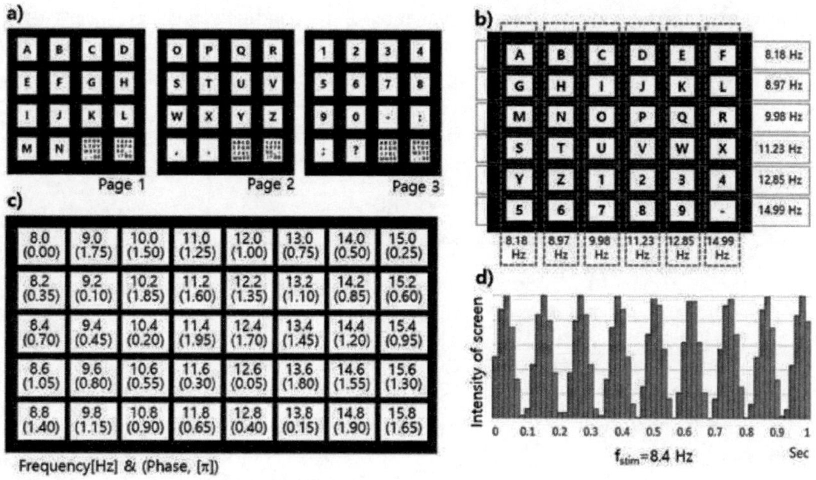

Fig. 10.16 Different paradigms of SSVEP-based BCI. (**a**) Multi-phase speller. Each page is composed of 16 stimulating blocks of different frequencies, 14 blocks for characters, and 2 for switching to other pages. (**b**) SSVEP BCI based on row and column paradigm. 6 rows or 6 columns are stimulated simultaneously with different frequencies. Stimulating frequencies can be shared for rows and columns by alternating in time and the stimulating frequency of each block changes dynamically. (**c**) One phase SSVEP speller with 40 target blocks. Each stimulating block is stimulated with different frequencies and with different starting phases. The gazed block is identified by the frequency and the relative phase of recorded SSVEP. (**d**) Intensity of the screen block generating a visual stimulation of 8.4 Hz. Stimulating block of many different frequencies under a fixed frame rate of the screen is implemented by manipulating the intensity of the screen according to the sampled signal strength of the corresponding frequency

caused by losing the synchrony among the cortical neurons is referred to as event-related desynchronization (ERD) (Fig. 10.17). The locations of ERD and ERS vary depending on which limb is moving and on which side of the body the specific movement is taking place. ERD and ERS are recorded most prominently in the mu band of 8~13 Hz and often accompanied by a changed rhythm in the beta band of 18~30 Hz and gamma band of 30~200 Hz. EEG recording is limited to mu, beta, and low-frequency range of the gamma band, and higher frequency activities are only recordable in implanted electrode recording, ECoG, or MEG. SMR recorded in EEG is weak in strength compared to ERP and VEP, but the movement of the left arm, right arm, and legs can be distinguished.

Event-related desynchronization (ERD) is the blocking of mu band rhythms associated with sensory processing or motor behavior and reflects a decrease of oscillatory activity related to an internally or externally paced event. The rhythm of the mu band is overlapped in frequency with alpha rhythm, but it is found over the motor cortex unlike the alpha wave recorded over the visual cortex during the rest with eyes closed. Lower frequency mu band ERD of 8~10 Hz appears nonspecifically and widely over the entire sensorimotor cortex during almost any kind of motor behavior. It reflects general motor preparation and attentional process. Higher frequency mu band ERD of 10~13 Hz is topographically restricted to a specific region functionally related to the task (Pfurtscheller et al., 2000). ERD also occurs in the beta rhythm. In addition to ERD, beta rhythm also appears as brief ERS just after the movement which is called beta rebound.

When mu ERD occurs in a specific region related to the movement, it is usually accompanied by the simultaneous ERS in neighboring cortical areas. It is known as focal-ERD/surround-ERS. For example, during the hand movement, the hand-related cortex shows ERD, and the foot-related cortex shows ERS. Whereas during the foot movement, the hand-related cortex shows ERS, and the foot-related cortex shows ERD. This mechanism reflects the brain accentuating the attention to a specific region related to the task by inhibiting other cortical areas that are not directly involved in the specific behavior.

ERD and ERS are similarly recorded in imagery movements as well as in actual movements. Mu ERD is also recorded when subjects with amputated limbs imagine moving their missing limbs. The strong similarity between actual movement and imagery movement reflects that imagery movement is realized via the same brain structure in movement preparation and performing. SMR generated by the imagery movement allows the implementation of BCI within a subject-paced paradigm rather than a response to an external cue like visual stimulators used in P300 BCIs or SSVEP BCIs. This characteristic of the user-initiated paradigm is appropriate for the asynchronous BCI.

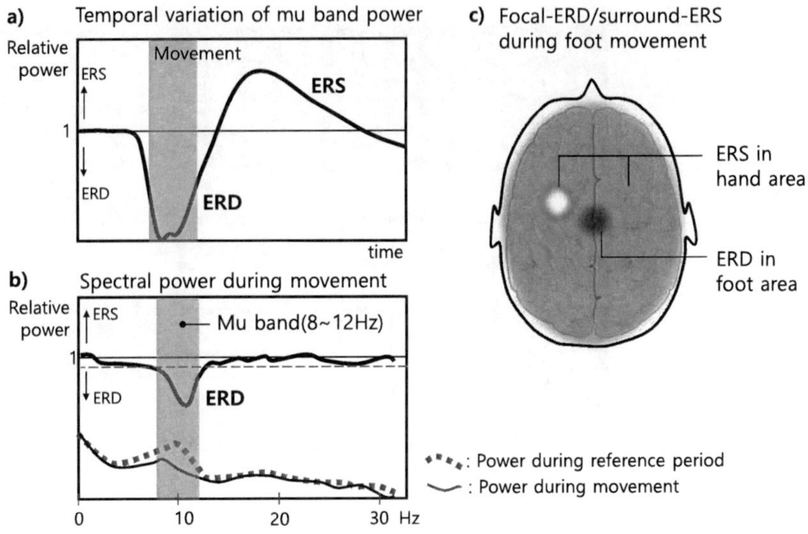

Fig. 10.17 Sensorimotor rhythm (SMR). (**a**) Event-related desynchronization (ERD) and event-related synchronization (ERS) of mu band occurring in relation to the movement onset and offset. (**b**) Relative spectral power during the movement and occurrence of ERD in the mu band. (**c**) Focal-ERD and surround-ERS SMR during the movement of the foot

In summary, ERD and ERS are temporally synchronous activities to the movement. ERD and ERS are also specific to frequencies and prominently noticed only in specific frequency bands. ERD and ERS are localized to the specific region in the motor cortex corresponding to the moving part of the body. So, ERD and ERS can be quantified in time-frequency and spatial domains combinedly to differentiate the imagined movements in BCI applications.

BCI Based on Sensory Motor Rhythms

BCI based on SMR has several merits compared to the event-related potential (ERP)-based BCI. First of all, it can avoid uncomfortable stimulation and doesn't need to synchronize the behaviors to external cues. It can be used at any time when the subject wants to use it while the system is prepared for the response. As gaze movement is not required for the operation, it is more appropriate for the patient in the later stages of the disease.

BCI based on motor imagery (MI) moves the cursor by different imagery motions. Two controls can be derived from the two imagery movements. One is the imagery movement of the right hand and the other is the movement of the foot. In Hex-O speller, 30 characters are divided into 6 groups having 5 characters and displayed around the center of the screen. By imagining the right-hand movement, the subject can rotate the pointer 60° each time and reach the hexagon containing the character wants to spell. Then, the subject selects the hexagon by imagining the foot movement. If the group is selected, characters within the group expand and replace the screen display of 6 groups. In the second phase, the character is selected in the same way, rotating by imagining the right-hand movement and selection by the imagining foot movement. Hex-O speller based on motor imagery SMR achieved the ITR of 7.6 characters/min which is higher than the ITR of Hex-O speller based on P300, 2.75 characters/min.

The number of controls to select the boxes can be increased to four different sides. The boxes on the left and right sides can be selected by imagining the movement of the left and right arms, respectively. The box on the upper side can be activated by imagining the movement of both arms, and the box on the lower side by imagining the movement of both legs. In contrast to the arm or hand movement, leg or foot movement of the left and the right side is difficult to differentiate in the recorded SMR since the related cortical area of the right and left leg and foot are located closely at the central part of the motor cortex. Movement of the tongue also produces significant and discriminative changes in the recorded SMR and can be used as another control option (Schlögl & Bischof, 2005). These imagery movements can be applied in the appropriate BCI paradigm as 4 or 5 control commands. ITR of 3 characters/min has been achieved in a study using 4 imagery movements (D'Albis et al., 2012).

BCI can be asynchronous, in a self-paced protocol, without external cues for the selection of command. In asynchronous BCI, the user determines the timing of the command. Asynchronous BCIs are more natural and convenient for the users, but they are much more challenging to implement. Motor imagery (MI)-based SMR can be used to control the external command buttons asynchronously. MI-based BCI has applied asynchronously in video games, spelling applications, and navigating the virtual environment. MI-based BCI is applied in controlling the drone asynchronously in 3D virtual space (Choi et al., 2021) (Fig. 10.18c).

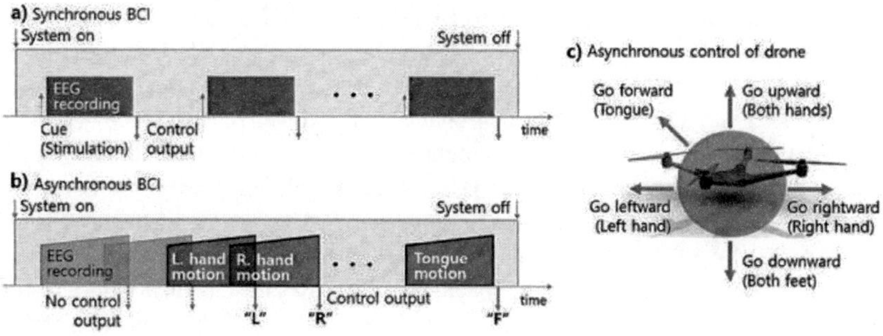

Fig. 10.18 Types of BCI operating mode. (**a**) Synchronous BCI operation mode. (**b**) Asynchronous BCI operation mode for the control of drone. (**c**) Imagery motion allocation for the directional control of drone in 3D space

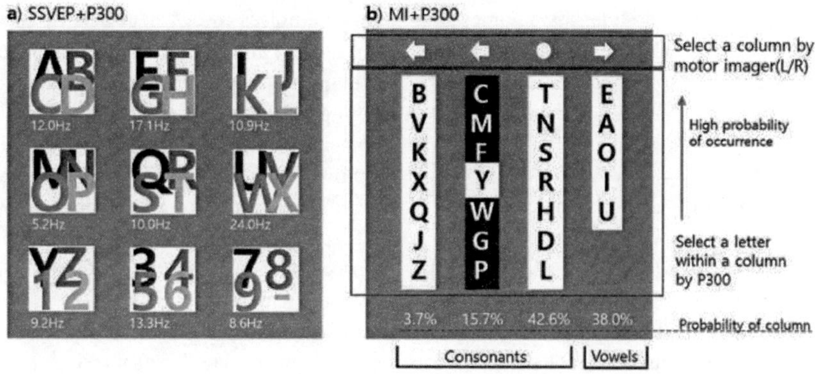

Fig. 10.19 Examples of Hybrid BCI. (**a**) Hybrid BCI combining SSVEP and P300 paradigm. 9 blocks are flashing in different frequencies and induce SSVEP. Each of the four characters within a block is flickering pseudo-randomly to induce P300 based on the "odd-ball" paradigm. (**b**) Hybrid BCI combining motor imagery (MI) and P300 paradigm. A column of character group is selected by moving the cursor left or right with the MI paradigm and the character within a column is selected by the P300 paradigm. Characters are grouped and displayed upon the occurrence frequency

To move the drone in a 3D space, five commands, go forward, turn left, turn right, go upward and go downward, are allocated to 5 imagery movements out of possible 6 imagery movements, tongue, left hand, right hand, both hands, both feet, and both hands with both feet. The subjects have been trained to produce each type of imagery movement and allocated one imagery movement for each directional control based on the subject's preference. To analyze the occurrence of SMR, EEGs are recorded with 33 electrodes with most of them near the sensorimotor cortex. Signals in the frequency band of 8~36 Hz have been used for classification after filtering with 7 bandpass filters each having 4 Hz bandwidth. For asynchronous control, the last 3 s of EEG is used to perform motor imagery classification in real-time with a 1 s stride. Classification is performed once every second in order to asynchronously control the simulated drone, and the drone moves only when the two most recent classification results indicate the same command. In simulated navigation in 3D space, subjects navigated the suggested path with an average elapsed time of 715 s which is comparable to an average elapsed time of 411 s controlled with a keyboard.

3.5 Hybrid BCI

To increase the performance or convenience of the user, BCI can be implemented in a hybrid type combining different paradigms based on their advantages. SSVEP paradigm can be combined with the P300 paradigm (Fig. 10.19a). When 36 characters are divided into 9 groups with 4 characters for each group, a group can be selected by the SSVEP paradigm and the character in a group can be selected by the P300 paradigm. Each of the 9 groups is flashing with different frequencies to elicit an SSVEP and each character in a group is pseudo-randomly flickering to produce a P300 component based on the "Odd-ball paradigm" while a subject attends the gaze on a character in the group. The hybrid system showed higher performance compared to individual SSVEP or P300 BCI systems. The individual unmixed SSVEP and P300 BCI system produced the ITR of 13.0 bits/min and 19.9 bits/min respectively. By combining the two interfacing paradigms, ITR can be increased to 31.8 bits/min (Chang et al., 2016).

MI paradigm can be combined with SSVEP or P300 paradigms (Fig. 10.19b). MI paradigm can simply be used to turn on and turn off the synchronous BCI system which is based on SSVEP or P300 paradigms. MI paradigm also can be combined to spell a character with the P300 paradigm. 26 characters are divided into 21 consonants and 5 vowels. 21 consonants are subdivided into 3 columns of 7 characters based on the occurrence frequency. Users select a column from 4 columns of characters, 1 for vowels and 3 for consonants, by moving a cursor right or left by the MI paradigm starting from the most probable column. After the column is selected, a character within a column is selected by the P300 paradigm. Characters within a group are also displayed based on their occurring frequency to minimize the selecting time. When a word module is combined by suggesting probable words after selecting three characters, an overall ITR of 5.5 characters/min has been achieved (Roula et al., 2012).

4 Data Processing in BCI

In BCI systems, the performance of the system is determined largely by the data processing on the recorded signals. By the methods applied in sequential stages of processing for the recorded EEG data, the accuracy of final classification is variant. BCI is one of the areas where various types of advanced signal processing techniques are proactively applied to increase performance. Generally, collected raw EEG data result in final classifications through a series of sequential processing of preprocessing, feature extraction, and selection and classification (Fig. 10.20).

4.1 Preprocessing of Raw Data

Electrical signals measured for BCI application have only low SNR and are contaminated by different sources of noise in many cases. Some noise sources are external or environmental, like power lines noise and noises from the electronic equipment. Some of the noise sources are physiological and originate from the body, like eye blinking, EOG, ECG, EMG, and respiration. They make it difficult to derive accurate classification results as a final result.

Pre-processing is a procedure to eliminate the unwanted noise components embedded in collected EEG data. Good preprocessing of raw data leads to an increased signal quality, which in turn results in better feature separability and classification performance. Different approaches have been used in the preprocessing step to reduce the effects of noise. Simple low, high, and bandpass filters are the primary attempts to attenuate noises in the measured EEG. But, when the spectrum of EEG and the noise are overlapped, various alternative techniques are used. They include adaptive filtering, Wiener filtering, Bayes filtering, wavelet transform, surface Laplacian transforms, regression, blind source separation, empirical mode decomposition, nonlinear mode decomposition, and canonical correlation analysis (CCA). For real-time BCI applications, automatic and computationally efficient preprocessing methods are required. The CCA algorithm is one of the widely used methods in real-time applications with a little computational time. The fewer number of measurement channels is another factor that should be considered for computational cost and the practical usage as a tradeoff with the performance. Recently, machine learning-based approaches have gained due attention, particularly for their ability to identify artifacts (Rashid et al., 2020).

4.2 Feature Extraction and Selection

After the noises and artifacts are maximally removed in preprocessing step, the most salient signal characteristics are extracted from the EEG data through different feature extraction techniques. Features can be extracted in time, frequency, time-frequency, and spatial domain techniques.

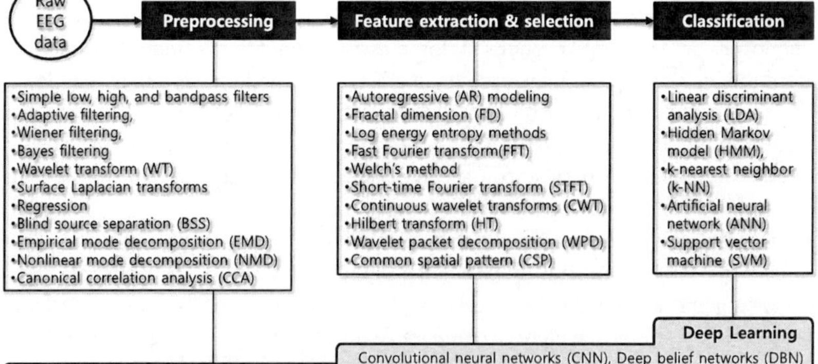

Fig. 10.20 Signal processing methods for BCI

Autoregressive (AR) modeling is a typical time-domain feature extraction approach. Fractal dimension and log energy entropy method are also used to extract features in the time domain. Frequency features are derived from the spectrum of the signal. The spectrum of signals is calculated by fast Fourier transform (FFT) or Welch's method. When the selection of features in the time domain or in the frequency domain independently is inefficient, time-frequency domain methods, which are both in the time domain and frequency domain, are applied. Short-time Fourier transform (STFT), continuous wavelet transforms (CWT), Hilbert transforms (HT), and wavelet packet decomposition (WPD) are widely used time-frequency methods for feature selection. Another feature selection technique in the spatial domain is the common spatial pattern (CSP), which is extensively utilized in EEG-based BCI. By using CSP, EEGs of different sources, like right-hand movement and left-hand movement, can be computationally separated into additive subcomponents having maximum differences in variance in the transformed domain. Generally, the time-domain features tend to provide superior performance compared to the frequency and spatial domain features.

4.3 Classification

The purpose of the BCI system is to identify the intention of the user by the recorded signals. This goal is finally achieved by accurately classifying the recorded signals based on the selected features. To make a command for the control of an external device, generated brain signal should be classified appropriately into one of the classes corresponding to the command. Numerous classification algorithms have been applied to different types of BCIs based on their own characteristics. As conventional machine learning methods, linear discriminant analysis (LDA), hidden Markov model (HMM), k-nearest neighbor (k-NN), artificial neural network (ANN), and support vector machine (SVM) are the most widely used classification methods for BCI applications.

Linear discriminative analysis (LDA) linearly combines selected features describing the characteristics of the corresponding signal. The LDA tries to separate signals into two or more classes by the linearly combined results. Hidden Markov model (HMM) is a Bayesian classifier that produces nonlinear decision boundaries. HMM is appropriate for the classification of time series data and provides the likelihood of the signal with given feature values observed in each class. The k-nearest neighbor (kNN) algorithm depends on the principle that the feature values of the signal belonging to the same class will be close and clustered in the feature space while those belonging to different classes will form a separated cluster in the feature space. A new signal is allocated to a class whose feature distances of k nearest samples are minimal. Artificial neural networks (ANN) classify the signal with selected features as input of the trained network. ANN is trained with formerly classified data with just input data and classification results. Multilayer perceptron (MLP) is the primitive form of ANN and numerous advanced variants have been applied in BCI classifications. The support vector machine (SVM) is a classifier that builds a set of nonlinear hyperplanes for separating the feature space into several classes. Classification performance is enhanced by adopting the appropriate kernel and optimally separating the classes in transformed space.

4.4 Deep Learning Approaches

As deep learning (DL) technology has been introduced in many processing-intensive domains, DL technology has also been introduced in the signal processing for BCI applications and produced performances that surpass the results with conventional methods. DL methods uncovered relevant information from the signals that were not possible to acquire via conventional approaches. In contrast to traditional sequential processing steps, DL technology can be applied in one step for joint tasks of feature selection and classification, and even tasks including preprocessing. A number of prevalent DL architectural models including convolutional neural networks (CNN), deep belief networks (DBN), restricted Boltzmann machines (RBM), long short-term memory (LSTM), gated recurrent units (GRU), stacked autoencoder (SAE), generative adversarial networks (GAN), and Riemannian geometry classifier (RGC) have selectively applied for various applications depending on the characteristics of BCI paradigms. As the performance of the BCI has increased significantly by adopting the DL technology and the role of signal processing is important for the implementation of an efficient and successful BCI system, it is expected that DL technology will be used as a major data processing method and produce advanced results consistently.

5 Application of BCI

5.1 BCI Systems for the Patients

The primary application of the BCI systems is to provide an alternative communication channel to the external world for the patients who are not possible with other methods like locked-in syndrome (LIS) patients. For those, BCIs have developed in several types of applications.

The first attempt of BCI was to control a cursor on a computer screen vertically using slow cortical potential. Since then, several cursor control BCI systems are developed mainly based on the MI paradigm with the additional use of P300 or SSVEP. They could control a cursor vertically, horizontally, and in a diagonal direction by executing both sorts of activities simultaneously. BCI speller is the most efficient BCI type to transmit a large amount of information. When a subject cannot make speeches, writing sentences is the most efficient way of communication with the outer world. BCI speller can replace the low-tech eye gaze-based augmentative and alternative communication (AAC) options. BCI speller is the most commonly studied application and many different types of BCI speller have been developed based on diverse paradigms. Patients can select a character or control command, spell a word and write sentences with brain activities according to the designed paradigm. Currently best performing noninvasive and invasive BCI spellers can spell characters with a speed of 62.5 and 90 characters/min respectively which are comparable to the speed of normal subjects. Instead of using a computer screen, the interface unit can be applied directly to the daily use devices in the environment. By using flickering buttons and causing SSVEP or P300 responses, the subject can operate home electronics, including a TV, DVD player, and lights with commands turn-on, turn off, play, pause, scroll, volume-up & down and stop. They can be similarly used to move a bed up and down and to control the power of the massaging cushion.

BCI has applied for wheelchair control to upgrade the life quality and the autonomy of patients. Various paradigms using EEG, MI, P300, SSVEP, and the hybrid of MI and SSVEP, have been applied to control the wheelchair. From the simplest three commands of right, left and forward, the number of command types can be increased to 18 including turn-on, turn-off, stop, acceleration, deceleration, uniform velocity, and additional destined locations. To supplement the loss of body motion, BCI can be applied to control the robotic arm or paralyzed arm of their own. With the brain activities recorded with intracortical electrodes, the BrainGate system successfully controlled the robotic arm and the paralyzed arm of the subject by stimulating the muscle appropriately. This concept of BCI can be extended to move the patient by controlling the arms and legs of the exoskeleton.

5.2 BCI Systems for the General Population

Even though the BCIs are most beneficial for the patients like LIS, they also can be usefully applied to the general populations who are maintaining their ordinary communication pathways intact. For the general population, BCIs provide an alternative and secondary communication pathway in addition to the intact ordinary communication pathway and can be used supportively or discriminately depending on the applications for the enhancement of human performances.

BCIs provide a real-time response to brain activities which is not possible with other conventional analytic methods of human response based on questionnaires and video recordings. The real-time response of the brain can be used in immediately reacting to the environment by optimizing and increasing the performances of the users.

BCI can be used in the real-time evaluation of attention levels. As human attentional level is often variable, real-time monitoring of attention level is particularly important in safety-related applications. Car accidents, which are caused by visual distraction and a decrease in vigilance, are the leading cause of death worldwide. During the monitoring of EEG activities, the changes in the attention level can be recognized at an earlier time and can be used as an alert or informative index controlling human behaviors. In addition to assessing the attention level of the driver, BCI can be used to brake a car at an earlier time in an emergency situation. Driver's recognition of an emergency and the intention to perform an emergency braking can be estimated with high accuracy from the driver's EEG (Hernández et al., 2018). Because activities are less variant than neuromuscular responses under unfavorable cognitive states such as stress, workload, and fatigue, this system can be used more effectively in these human cognitive states.

BCI for virtual reality and games is the application attracting the greatest interest from the healthy general populations. BCI-based video games are implemented with a wearable EEG recording headset and a specified paradigm of SSVEP, P300, or mental imagery (Kaplan et al., 2013). In the game of the virtual world, the player can move a character from one field to another on a game board and control the settings of the virtual environment. Also, the players can use motor imagery and enjoy the game with others by competing or collaborating. Games designed with SSVEP or P300 paradigm are preferred by the players to MI paradigm because these paradigms do not require training of the players.

References

Acqualagna, L., & Blankertz, B. (2013). Gaze-independent BCI-spelling using rapid serial visual presentation (RSVP). *Clinical Neurophysiology, 124*, 901–908. https://doi.org/10.1016/j.clinph.2012.12.050

Acqualagna, L., Treder, M. S., & Blankertz, B. (2013). Chroma Speller: Isotropic visual stimuli for truly gaze-independent spelling. In *Proceedings of the 6th international IEEE/EMBS conference on neural engineering (NER)* (pp. 1041–1044).

Ahn, S., Kim, K., & Jun, S. C. (2015). Steady-state somatosensory evoked potential for brain-computer interface-present and future. *Frontiers in Human Neuroscience, 9*, 716. https://doi.org/10.3389/fnhum.2015.00716

Ajiboye, A. B., Willett, F. R., Young, D. R., Murphy, B. A., Miller, J. P., et al. (2017). Restoration of reaching and grasping movements through brain-controlled muscle stimulation in a person with tetraplegia: A proof-of-concept demonstration. *Lancet, 389*(10081), 1821–1830. https://doi.org/10.1016/S0140-6736(17)30601-3

Akram, F., Han, H. S., & Kim, T. S. (2014). A P300-based brain computer interface system for words typing. *Computers in Biology and Medicine, 45*, 118–125. https://doi.org/10.1016/j.compbiomed.2013.12.001

Aloise, F., Arico, P., Schettini, F., Riccio, A., Salinari, S., Mattia, D., Babiloni, F., & Cincotti, F. (2012). A covert attention P300-based brain-computer interface: Geospell. *Ergonomics, 55*, 538–551. https://doi.org/10.1080/00140139.2012.661084

Anumanchipalli, G. K., Chartier, J., & Chang, E. F. (2019). Speech synthesis from neural decoding of spoken sentences. *Nature, 568*, 493–498. https://doi.org/10.1038/s41586-019-1119-1

Benabid, A. L., Costecalde, T., Eliseyev, A., et al. (2019). An exoskeleton controlled by an epidural wireless brain-machine interface in a tetraplegic patient: A proof-of-concept demonstration. *Lancet Neurology, 18*(12), 1112–1122.

Bin, G., Gao, X., Wang, Y., Hong, B., & Gao, S. (2009). VEP-based brain-computer interfaces: Time, frequency, and code modulations. *IEEE Computational Intelligence Magazine, 4*(4), 22–26. https://doi.org/10.1109/MCI.2009.934562

Birbaumer, N., Elbert, T., Canavan, A. G., & Rockstroh, B. (1990). Slow potentials of the cerebral cortex and behavior. *Physiological Reviews, 70*, 1–41. https://doi.org/10.1152/physrev.1990.70.1.1

Birbaumer, N., Ghanayim, N., Hinterberger, T., et al. (1999). A spelling device for the paralyzed. *Nature, 398*, 297–298. https://doi.org/10.1038/18581

Brouwer, A. M., & van Erp, J. B. (2010). A tactile P300 brain-computer interface. *Frontiers in Neuroscience*. https://doi.org/10.3389/fnins.2010.00019

Cao, T., Wang, X., Wang, B., Wong, C. M., Wan, F., Mak, P. U., Mak, P. I., & Vai, M. I. (2011). A high rate online SSVEP based brain-computer interface speller. In *Proceedings of the 5th international IEEE/EMBS conference on neural engineering* (pp. 465–468).

Chang, M. H., Lee, J. S., Heo, J., & Park, K. S. (2016). Eliciting dual-frequency SSVEP using a hybrid SSVEP-P300 BCI. *Journal of Neuroscience Methods, 258*, 104–113. https://doi.org/10.1016/j.jneumeth.2015.11.001

Choi, J. W., Kim, B. H., & Jo, S. (2021). Asynchronous motor imagery brain-computer interface for simulated drone control. In *9th international winter conference on brain-computer interface (BCI)* (Vol. 2021, pp. 1–5). https://doi.org/10.1109/BCI51272.2021.9385309.

D'Albis, T., Blatt, R., Tedesco, R., Sbattella, L., & Matteucci, M. (2012). A predictive speller controlled by a brain-computer interface based on motor imagery. *ACM Transactions on Computer-Human Interaction, 19*, 1–25. https://doi.org/10.1145/2362364.2362368

Farwell, L. A., & Donchin, E. (1988). Talking off the top of your head: Toward a mental prosthesis utilizing event-related brain potentials. *Electroencephalography and Clinical Neurophysiology, 70*(6), 510–523.

Fetz, E. (1969). Operant conditioning of cortical unit activity. *Science, 163*(3870), 955–958. https://doi.org/10.1126/science.163.3870.955

Guo, F., Hong, B., Gao, X., & Gao, S. (2008). A brain-computer interface using motion-onset visual evoked potential. *Journal of Neural Engineering, 5*(4), 477–485. https://doi.org/10.1088/1741-2560/5/4/011

Hernández, L. G., Mozos, O. M., Ferrández, J. M., & Antelis, J. M. (2018). EEG based detection of braking intention under different car driving conditions. *Frontiers in Neuroinformatics, 29*. https://doi.org/10.3389/fninf.2018.000.29

Hill, N. J., Lal, T. N., Bierig, K., Birbaumer, N., & Schölkopf, B. (2005). An auditory paradigm for brain-computer interfaces. *Advances in Neural Information Processing Systems*, 569–576.

Hochberg, L., Serruya, M., Friehs, G., et al. (2006). Neuronal ensemble control of prosthetic devices by a human with tetraplegia. *Nature, 442*, 164–171. https://doi.org/10.1038/nature04970

Hochberg, L., Bacher, D., Jarosiewicz, B., et al. (2012). Reach and grasp by people with tetraplegia using a neurally controlled robotic arm. *Nature, 485*, 372–375. https://doi.org/10.1038/nature11076

Höhne, J., & Tangermann, M. (2014). Towards user-friendly spelling with an auditory brain-computer interface: The charstreamer paradigm. *PLoS One, 9*, e98322. https://doi.org/10.1371/journal.pone.0098322

Höhne, J., Schreuder, M., Blankertz, B., & Tangermann, M. (2011). A novel 9-class auditory ERP paradigm driving a predictive text entry system. *Frontiers in Neuroscience, 5*, 99.

House, P. A., MacDonald, J. D., Tresco, P. A., & Normann, R. A. (2006). Acute microelectrode array implantation into human neocortex: Preliminary technique and histological considerations. *Neurosurgical Focus, 20*(5), E4.

Jin, J., Sellers, E. W., & Wang, X. (2012). Targeting an efficient target-to-target interval for P300 speller brain-computer interfaces. *Medical & Biological Engineering & Computing, 50*, 289–296. https://doi.org/10.1007/s11517-012-0868-x

Kaplan, A. Y., Shishkin, S. L., Ganin, I. P., Basyul, I. A., & Zhigalov, A. Y. (2013). Adapting the P300-based brain–computer interface for gaming: A review. *IEEE Transactions on Computational Intelligence in AI and Games, 5*, 141–149. https://doi.org/10.1109/TCIAIG.2012.2237517

Kaufmann, T., & Kubler, A. (2014). Beyond maximum speed-a novel two-stimulus paradigm for brain-computer interfaces based on event-related potentials (P300-BCI). *Journal of Neural Engineering, 11*, 056004. https://doi.org/10.1088/1741-2560/11/5/056004

Liu, Y., Zhou, Z., & Hu, D. (2010). Comparison of stimulus types in visual P300 speller of brain-computer interfaces. In *Proceedings of the 9th IEEE international conference on cognitive informatics* (pp. 273–279).

Miller, K. J., Hermes, D., & Staff, N. P. (2020). The current state of electrocorticography-based brain–computer interfaces. *Neurosurgical Focus, 49*(1), E2.

Moses, D. A., Metzger, S. L., Liu, J. R., Anumanchipalli, G. K., Makin, J. G., Sun, P. F., Chartier, J., Dougherty, M. E., Liu, P. M., Abrams, G. M., Tu-Chan, A., Ganguly, K., & Chang, E. F. (2021). Neuroprosthesis for decoding speech in a paralyzed person with anarthria. *New England Journal of Medicine, 385*(3), 217–227. https://doi.org/10.1056/NEJMoa2027540

Multi-society task force on persistent vegetative state. (1994). Medical aspects of the persistent vegetative state (1). *The New England Journal of Medicine, 330*(21), 1499–1508. https://doi.org/10.1056/NEJM199405263302107

Nakanishi, M., Wang, Y., Chen, X., Wang, Y. T., Gao, X., & Jung, T. P. (2018). Enhancing detection of SSVEPs for a high-speed brain speller using task-related component analysis. *IEEE Transactions on Biomedical Engineering, 65*, 104–112. https://doi.org/10.1109/TBME.2017.2694818

Noorzadeh, S., Rivet, B., & Jutten, C. (2014). Beyond 2D for brain-computer interfaces: Two 3D extensions of the P300-Speller. In *Proceedings of the IEEE international conference on acoustics, speech and signal processing (ICASSP)* (pp. 5899–5903).

Pfurtscheller, G., Guger, C., Müller, G., Krausz, G., & Neuper, C. (2000). Brain oscillations control hand orthosis in a tetraplegic. *Neuroscience Letters, 292*(3), 211–214. https://doi.org/10.1016/S0304-3940(00)01471-3

Punsawad, Y., & Wongsawat, Y. (2017). Multi-command SSAEP-based BCI system with training sessions for SSVEP during an eye fatigue state. *IEEJ Transactions on Electrical and Electronic Engineering, 12*, S72–S78. https://doi.org/10.1002/tee.22441

Rashid, M., Sulaiman, N., Majeed, A., Musa, R., Nasir, A., Bari, B., & Khatun, S. (2020). Current status, challenges, and possible solutions of EEG-based brain-computer interface: A comprehensive review. *Frontiers in Neurorobotics, 14*. https://doi.org/10.3389/fnbot.2020.00025

Rezeika, A., Benda, M., Stawicki, P., Gembler, F., Saboor, A., & Volosyak, I. (2018). Brain-computer interface spellers: A review. *Brain Sciences, 8*(4), 57. https://doi.org/10.3390/brainsci8040057

Roula, M. A., Kulon, J., & Mamatjan Y. (2012). Brain-computer interface speller using hybrid P300 and motor imagery signals. In *Proceedings of the 4th IEEE RAS & EMBS international conference on biomedical robotics and biomechatronics (BioRob)* (pp. 224–227).

Schlögl, A. L. F., & Bischof, H. P. G. (2005). Characterization of four-class motor imagery EEG data for the BCI-competition. *Journal of Neural Engineering, 2*, L14–L22.

Schmidt, N. M., Blankertz, B., & Treder, M. S. (2012). Online detection of error-related potentials boosts the performance of mental typewriters. *BMC Neuroscience, 13*, 19. https://doi.org/10.1186/1471-2202-13-19

Shi, J. H., Shen, J. Z., Ji, Y., & Du, F. L. (2012). A submatrix-based P300 brain-computer interface stimulus presentation paradigm. *Journal of Zhejiang University-SCIENCE C, 13*, 452–459.

Spüler, M., Rosenstiel, W., & Bogdan, M. (2012). Online adaptation of a c-VEP brain-computer interface (BCI) based on error-related potentials and unsupervised learning. *PLoS One, 7*, e51077. https://doi.org/10.1371/journal.pone.0051077

Townsend, G., LaPallo, B. K., Boulay, C. B., Krusienski, D. J., Frye, G. E., Hauser, C. K., Schwartz, N. E., Vaughan, T. M., Wolpaw, J. R., & Sellers, E. W. (2010). A novel P300-based brain-computer interface stimulus presentation paradigm: Moving beyond rows and columns. *Clinical Neurophysiology, 121*(7), 1109–1120.

van der Waal, M., Severens, M., Geuze, J., & Desain, P. (2012). Introducing the tactile speller: An ERP-based brain–computer interface for communication. *Journal of Neural Engineering, 9*(4). https://doi.org/10.1088/1741-2560/9/4/045002

Vansteensel, M. J., Pels, E. G. M., Bleichner, M. G., et al. (2016). Fully implanted brain–computer interface in a locked-in patient with ALS. *The New England Journal of Medicine, 375*(21), 2060–2066.

Velliste, M., Perel, S., Spalding, M., et al. (2008). Cortical control of a prosthetic arm for self-feeding. *Nature, 453*, 1098–1101. https://doi.org/10.1038/nature06996

Vidal, J. (1977), Real-time detection of brain events in EEG, *Proc. IEEE 65*, 633–664.

Volosyak, I. (2011). SSVEP-based Bremen-BCI interface: Boosting information transfer rates. *Journal of Neural Engineering, 8*(3). https://doi.org/10.1088/1741-2560/8/3/036020

Willett, F. R., Avansino, D. T., Hochberg, L. R., et al. (2021). High-performance brain-to-text communication via handwriting. *Nature, 593*, 249–254. https://doi.org/10.1038/s41586-021-03506-2

Wu, N., Wan, S., Su, S., Huang, H., Dou, G., & Sun, L. (2021). Electrode materials for brain–machine interface: A review. *InfoMat*. https://doi.org/10.1002/inf2.12234

Yin, E., Zhou, Z., Jiang, J., Yu, Y., & Hu, D. (2015). A dynamically optimized SSVEP brain-computer interface (BCI) speller. *IEEE Transactions on Biomedical Engineering, 62*, 1447–1456. https://doi.org/10.1109/TBME.2014.2320948

Yu, Y., Zhou, Z., Yin, E., Jiang, J., Liu, Y., & Hu, D. (2016). A P300-based brain–computer interface for Chinese character input. *International Journal of Human Computer Interaction, 32*, 878–884. https://doi.org/10.1080/10447318.2016.1203529

11 Authentication with Bioelectrical Signals

Authentication is the process of recognizing a user's identity to confirm whether someone is, in fact, who it says. Its mechanism associates an incoming request with a set of authorized identifying items. The authentication system checks the requested trial by comparing the incoming item with the stored authorized item of the user. By rejecting unmatched trials, authentication technology provides secure access control for the connected system. Various items are used for authentication. Bioelectrical signals, like electrocardiogram (ECG) and electroencephalogram (EEG), also can be used as valuable items to identify and authenticate a person. Based on their unique characteristics, biosignals are anticipated to be more effectively integrated into the secure management of healthcare and medical data as well as in other applications of authentication.

1 Types of Authentication

Identification occurs when a user notifies an identity, like a username, and authentication occurs when a user proves the identity with an authentication item, like a password. For example, a user is authenticated by providing both the correct username and the correct password. After proving the identity by authentication, the following procedures of the system are then granted to the user. While the identification is done commonly by notifying a user by name, username, e-mail address, or phone number, authentication can be performed in various ways by the types of used items. They can be grouped into different types based on their common characteristics, knowledge-based authentication, possession-based authentication, biometric authentication, and multifactor authentication (Fig. 11.1).

1.1 Knowledge-based Authentication

Knowledge-based authentication is the technique of authenticating a person by "something you know," like a password and pin number. Some knowledge is required to keep confidentially in mind. While this type of authentication is the most commonly used, it also can be defeated most easily. When a password is used, it should be made strong enough not easily be attacked.

Attacks are being tried in different types via password guessing, user observation, or impersonation. Guessing the password is a time-consuming attack, as the intruder should try all combinations of characters constituting the used language. Increasing the length of a password and the number of possible characters is suggested as a solution to this type of attack. Stealing a password is another type of attack. By shoulder surfing, video recording, or keyloggers, intruders can steal a password when it is entered or sent. Keyloggers are often software programs responsible for sending the user's keystrokes to the attacker. Intruders can steal a password by eavesdropping. An attacker can impersonate both parties to each other and takes all roles in the active transaction. In a phishing attack, an attacker steals a password by masquerading as a website that requests the user's authentication information (Sahu et al., 2014).

The strength of a password can be increased by increasing the number of characters and mixing the different types of characters, like upper case, lower case, number, and special characters. To minimize the possibility of password cracking, many systems tend to require longer passwords of more than 12 characters. As the passwords become longer, they become hard to remember. To reduce the burden of remembering, a longer password used to be built with some meaningful order, which might provide the clue for the cracking. If a long password is built without any meaningful order, it is difficult to remember and used to be written down somewhere, which also weakens the security. We cannot but have many passwords, as many organizations and systems require passwords not used in other systems and even the same system requests the update of its password periodically.

K. S. Park, *Humans and Electricity*, https://doi.org/10.1007/978-3-031-20784-6_11

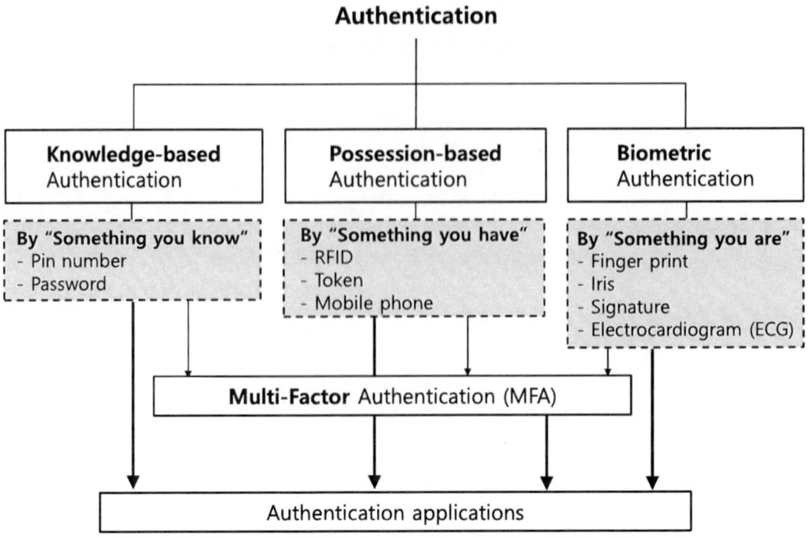

Fig. 11.1 Types of authentication

1.2 Possession-based Authentication

Possession-based authentication is the technique of authenticating a person by "something you have." There are many types of objects used to authenticate a person. Driver's license, passport, mobile phone, smart card, RFID, and hand-held token are some examples. Radiofrequency identification (RFID) tags become a popular item to secure or check the identity in the world of the "Internet of Things" (IoT). RFID can be attached to access cards, badges, contactless credit cards, and e-passports for authentication purposes. A token is a widely used device as an authentication item, especially in the web-based environment. A token, also called a one-time password (OTP) card, is a small hand-held device with LEDs displaying a number which is a one-time password synchronized with an authentication server. The number shown on the token changes regularly, such as every 60 s, and the authentication server always knows the currently shown number on the token display. Users can use a generated number as a one-time password to the affiliated site within a given time limit. Token provides merit of increased security, but it should be carried and there is a possibility that the carried token might be lost or stolen. Also, multiple tokens are required when each site does want to use an independent token not shared with other sites. Recently mobile phones are used efficiently as a possession-based authentication device since it is carried during the whole day. They are already widely used in authenticating ongoing bank transactions and providing mobile payment services.

1.3 Biometric Authentication

Biometric authentication is the technique of authenticating a person by "something you are." Biometric authentication does not require anything to remember and carry. It uses traits or behavioral characteristics to prove the claimed identity. It uses the body parts of the users as an authentication item. The fingerprint is the most widely used biometric authentication method. There are many more methods using biometric items from the body, like iris image, retinal image, hand geometry, handwritten signature, voice, and more. Authentication using biological signals also belongs to biometric authentication as it uses the signals originating from the human body without remembering or carrying anything.

While biometric authentication has some merits including the convenience, it is relatively susceptible to errors. Many factors can influence the accuracy of authentication. Fingerprints can be contaminated by the dirt, voices can be combined with environmental noises, and scanned images are strongly affected by the illuminations. Errors are appearing in two types. A false rejection error occurs when a system falsely recognizes a legitimate user as an intruder and rejects a user. And, false acceptance error occurs when a system falsely recognizes and accepts an intruder as a legitimate user. It is difficult to decrease the two types of error rates, false rejection rate (FRR) and false accept rate (FAR), simultaneously. When the sensitivity is increased to reduce the FRR, FAR tends to increase with the increased sensitivity. Also, when the specificity is increased to decrease the FAR, FRR tends to increase with the increased specificity.

The optimal point is determined by minimizing the combined errors, like equal error rate (EER), where the error curves of FAR and FRR cross over by having the same value. Among different biometric authentication methods, accuracies are compared by this EER, and the authentication method of a lower EER value is considered a more reliable authentication method.

1.4 Multifactor Authentication

Multifactor authentication is a hybrid type of authentication that combines more than two types of authentication items. Compared to multifactor authentication, the aforementioned three types of authentication are called single-factor authentication. To increase the accuracy, combining types of authentication comes from different types of single-factor authentications. For example, using a password and a token can be an effective multifactor authentication, because the two factors have different characteristics of "something you know" and "something you have." While using multi factors for authentication increases the performance, it also can increase the inconvenience, difficulty, and cost of the device with the additional factors.

2 Required Properties as a Biometric Authentication Item

Which biometric item is optimal as an authentication method? It is difficult to define simply in a single measure as there are various required properties that biometric authentication should satisfy for practical acceptance. The following are the commonly required properties of the authentication method as a valuable item (Fig. 11.2).

- Universality: The biometric item should be universal. Each person should have a biometric item and it should be found commonly in each individual. If the biometric item is not found in some individuals, it is not appropriate as an authentication item.
- Uniqueness: The biometric item should be unique. Any two individuals should be sufficiently different in terms of the item characteristics. If it is not distinct enough, it is difficult to separate one person from another, and not appropriate as an authentication item.
- Permanence: The biometric item should be permanent. That is, the biometric item should be sufficiently invariant for the short time period as well as over the aging. If the measured biometric feature varies along the time or depends on the body state of the user, it is not appropriate as an authentication item.
- Collectability: The biometric item should be quantized and easily collectible. If the biometric measurement is invasive, technically difficult, expensive, or time-consuming, it is not easily collectible and not appropriate as an authentication item.
- Performance: Authentication using the biometric item should provide sufficient performance for the applications. Performance includes the authentication accuracy, speed of the authentication, and robustness. Performance should be verified in adverse environments where the biometric measurements can be contaminated by unwanted sources.
- Acceptability: Biometric measurement should be acceptable so that people are willing to accept the use of a particular biometric item as an authentication tool. If it has some privacy issue or makes a person hesitate to use it for some reason, it is not appropriate as an authentication item.
- Circumvention: Authentication using the biometric feature should not be fooled by others and be able to

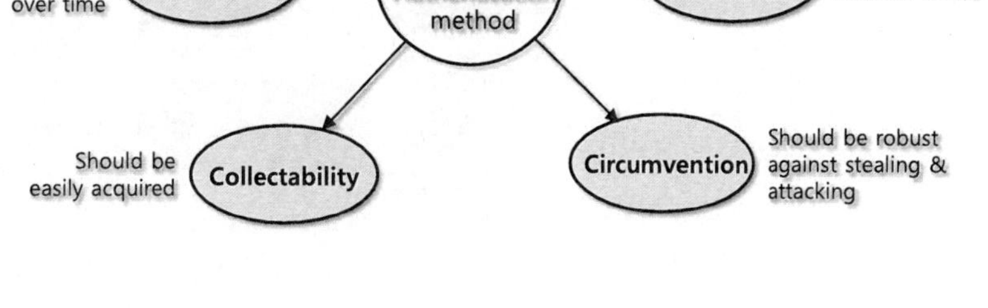

Fig. 11.2 Required properties of effective biometric authentication item

circumvent fraudulent methods. If someone can easily steal or get the biometric item of the user without notice of the user, it can make the authentication system insecure and is not appropriate as an authentication item.

3 Biometric Authentication

The word biometric is a combination of two words, "bio" and "metric." Bio represents the biology including the human, and metric represents the measurement. Biometrics are any measurements of human features and it can quantify the differences of individuals as well as represent the personal characteristics of the individual.

In biometric authentication, unique biological features or characteristics measured from the body part of individuals are used as items for the authentication. Various characteristics can be used as authentication items if they sufficiently contain the required properties as an authentication item. Depending on the characteristics of the biometric items, they also can be subdivided into three categories, anatomical items, behavioral items, and physiological items (Fig. 11.3).

Anatomical biometric items are body parts that constitute the body anatomically. They are maintaining their shapes and do not change upon the body states or functions. Among the many body parts, only a small number of body parts are satisfying the uniqueness and the collectability that are required as authentication items. If the body part is not unique, that is, if it is not distinct among the person, it is not an appropriate biometric item for authentication. Even though it has the property of uniqueness, it is not appropriate if it is difficult to measure or collect. For example, bone structure can have the uniqueness of differentiating the people by the X-ray image, but it is not appropriate as it is difficult to collect. As the anatomical features generally do not change over time, they also have the property of permanence. Currently used and studied anatomical biometric items for authentication include fingerprint, iris scan, retinal scan, hand geometry, vessel geometry of the hand, ear shape, and the face.

Behavioral biometric items are the characteristics that are generated while the people are behaving during their ordinary living activity. Behavioral patterns can be unique as they are developed based on the anatomical and physiological characteristics of each individual. If they can be collected easily, those behavioral patterns can be used as an appropriate authentication item. The signatures of the individuals are different from the other people. The voice of each individual is different and can be easily differentiated by the listener. Walking patterns of the individuals are different among the people and we can recognize the coming person by their walking patterns before we see their faces.

Physiological biometric items are the biometric items that represent the physiological characteristics of the body organs of the individual. In contrast to the anatomical biometric items, physiological items vary on the function of the body organ and the physical and mental states of the individual. Authentication using physiological biometric items is authentication using "something you make." Biological signals belong to physiological items. The electrocardiogram (ECG) represents the functional activity of the heart. ECG is generally used for a diagnostic purpose in the medical and healthcare domain to differentiate different types of abnormalities from the normal. However, within the normal individuals, ECGs show different patterns that can discriminate the individuals and can be effectively used as an authentication item. As recorded ECG reflects the anatomical and functional structure of the heart and the volume conductor of an individual body, it can be sufficiently distinct among people. The electroencephalogram (EEG) is the record of

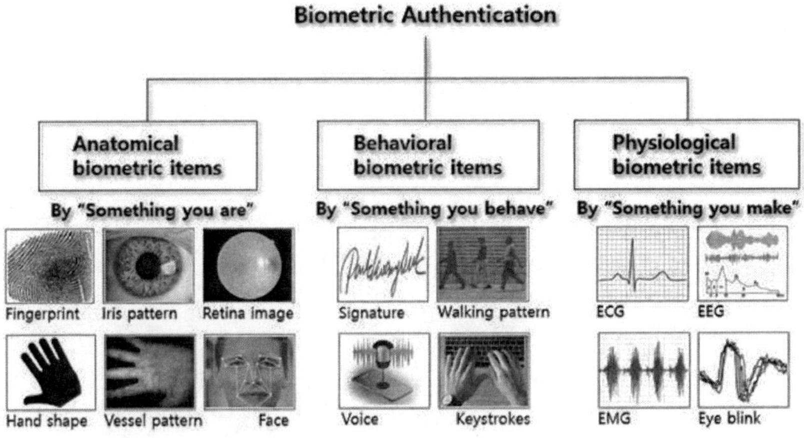

Fig. 11.3 Subcategories of biometric authentication items based on their characteristics

brain activity. EEG is widely used for diagnostic purposes in hospitals and also actively used in the studies for the brain-computer interface. As EEG is widely varying on the functioning of the brain, it might seem to be inappropriate for authentication at first sight. However, the generated brain activity has its own permanence specific to mental states. For example, during the resting state, each individual generates a resting EEG spectrum that can be sufficiently differentiated from the others. EEG patterns generated when a subject sees the image of his/her mother are unique to the subject. Any other person cannot generate the same EEG pattern by seeing the image of someone other's mother. Like this, EEG has a unique aspect that other authentication items do not have. Depending on the stimulating task and the mental thought, a subject can generate a unique EEG rhythm and it can be used as a "pass-thought" like a password.

The authentications based on the biological signals of physiological characteristics have very special properties which are not possible with other types of biometric items having anatomical and behavioral characteristics.

- Anatomical biometric items like fingerprints or iris may be forged or imitated. For instance, an intruder can simply make a fake fingerprint or contact lens by just getting a picture of one's finger or iris pattern. However, bioelectrical signals are not exposed to the intruder and are not possible to obtain without the notice of each individual. It makes authentications using biological signals very strong against spoofing attacks.
- Biological signals are dependent on the activation of the autonomic nervous system. Ordinary EEG or ECG cannot be recorded under stressful, fearful, or forged situations. As a result, an authentication system using biological signals cannot be passed under force and coercion situations, while in the other authentication systems like fingerprint or iris, the authentication system can be passed by the forced contact of the genuine individuals against their own will (Bidgoly et al., 2020).
- In the other authentication systems, the liveness of the claimed subject is not guaranteed. Intruders can use a dead body or artifact to pass the authentication system. However, as a dead body cannot make biological signals, recordings of biological signals like ECG or EEG guarantees that the authentication is acting on the live individual.

The above characteristics of biological signals as authentication items are very unique and cannot be found in other types of authentication items like knowledge-based, possession-based, or other anatomical or behavioral biometric items. While biological signals can provide sufficient performances as a single authentication item, it is also expected that they can increase the robustness of the authentication system against the spoofing attack in a multi-factor authentication system combined with other types of authentication items.

3.1 Biometric Authentication Using Anatomical Items

Anatomical biometric items are the most widely used ones in the applications of biometric authentication. These features have enough inter-user differences and individual stability and meet most of the required properties as a biometric authentication item. They are practically being used as efficient tools in many applications including PCs, smart phones, and banking systems.

3.1.1 Fingerprint

Fingerprint is the oldest and most commonly accepted biometric authentication item in both academia and industry. The fingerprint is the pattern of friction ridges, which is the raised portion of the dermis on the fingers consisting of one or more connected ridge units. Fingerprint pattern contains the global feature and local features (Fig. 11.4). The global feature of the fingerprint is the overall flow of the ridges

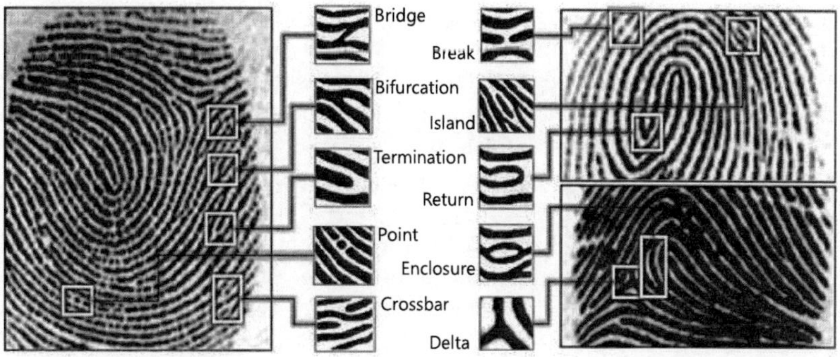

Fig. 11.4 Local features of fingerprint used for authentication

which can be seen with the naked eyes. Local features, which are also called minutia, are the unique characteristics of tiny ridges formed into loops, arches, and whorls. Local features are characterized prominently by the pattern of ridge endings, bifurcations, divergences, and enclosures. It is known that there are about 60–70 local feature points on each finger. While the global feature can be shared among the people, local features are so diverse and no pair of fingerprints having the same local features have been found. The distinction of local features appears even in identical twins, and also between fingers of the same person.

A fingerprint image is obtained simply by scanning the printed image of a finger with ink or directly scanning the finger surface optically. As authentication with a fingerprint is considered to have medium universality, high uniqueness, permanence, and acceptance, many sensing devices are developed and applied in various applications. Also, the sensor units do not introduce much cost to the authentic application based on the fingerprint. With its high overall usability, authentication based on fingerprint also has been embedded into most smartphones (Rui & Yan, 2019).

The performance of the fingerprint-based authentication varies depending on the number of fingerprints from an individual used for the matching. National Institute of Standards and Technology (NIST), a U.S. agency that works in the area of standardization and technology, evaluated the performance of the 34 commercially available modern fingerprint scanners. When the FAR is fixed to 0.01%, the best performance of accuracy is 98.6% (FRR = 1.4%) on one finger test, 99.6% (FRR = 0.4%) on the two-finger test, and 99.9% (FRR = 0.1%) for the test with four or more fingers.

Among the errors, false rejection error is the type that frequent users encounter. This error occurs when an authentication system refuses to let the user pass through, even though the user is legitimate with the enrolled fingerprint. These errors are caused mostly by the suboptimal fingerprint collection process producing degraded image quality. Errors can be originated from dirty, wet, or dry fingers. Or it also can be originated from the dirty, scratched, or broken sensor surface. A false acceptance error occurs when a fingerprint-based authentication system fallaciously accepts an unregistered fingerprint scan and lets the user pass through. False acceptance is more concerning than false rejection, as it can let unauthorized users access sensitive and confidential resources. It occurs because of erroneous matching between the fingerprint of an unauthorized user and the fingerprint of the enrolled users.

However, authentication based on a fingerprint item still has some limitations. The biggest limitation of the fingerprint is that it is not safe from stealing. As we remain fingerprints unintentionally as traces on the surfaces of daily contacting devices, it is almost exposed to the outer world. While these traces of fingerprints are used to find the criminals in many crimes, it means that fingerprint is easily exposed to whom want to obtain them. It is impossible to live with gloves and not to remain any trace of fingerprints for a whole day. An exposed fingerprint can be used in spoofing.

The biggest challenge to fingerprint-based authentication is spoofing, in which a fake replica of the fingerprint is presented in order to deceive the system. This replica used to have the exact biometric pattern to deceive the systems. A fake fingerprint can be easily copied into a sheet of silicone or latex material and overlaid on the finger surface without noticing of the examiner.

Fingerprint cannot be changed. While it is the required property for an authentication item, it can be a critical weak point for wider use in multiple applications. Unlike the authentication with passwords, a fingerprint cannot be changed or updated when it is exposed or stolen.

3.1.2 Iris

Iris is the intricate annular structure of tissue and muscle fibers surrounding the pupil in the eye. Iris consists of the epithelium layer and the stromal layer. The epithelium layer contains dense pigmentation cells and the stromal layer lies above the epithelium layer and contains blood vessels, pigment cells, and iris muscles. The color of the eyes is the color of the iris determined mainly by the density of melanin pigmentation cells in the stromal layer. When the iris is used as a biometric authentication item, the visual pattern of the scanned iris is analyzed. This complex pattern of the iris is universal, unique, permanent, stable, and can be seen accurately at less than a meter distance. As the iris pattern has a high degree of freedom, authentication using the iris can provide an increased accuracy compared to the authentication based on the fingerprint. As fingerprints are different over 10 fingers for each individual, the irises of both eyes are not the same and provide an increased amount of entropy as authentication data supporting the uniqueness of each individual. Human has all different iris patterns. Even identical twins and blind person have their own different and unique iris patterns. And, it is not possible to enroll with the left eye and authenticate with the right eye (Weaver, 2006). Iris authentication is robust to the variation of scanning conditions, like the iris change due to pupil dilation and the use of eyeglasses or contact lenses.

The striated trabecular meshwork of the iris creates the predominant texture under visible light or near-infrared (NIR) illumination of wavelengths in the range 700~900 nm. Iris scan with NIR provides deeper and somewhat more slowly modulated stromal features and reveals rich and complex features even in darkly pigmented irises.

The most common iris patterns are pigmented rings, crypts, and furrows. Pigmented rings are wide, colored bands around the pupil that are of a different shade compared

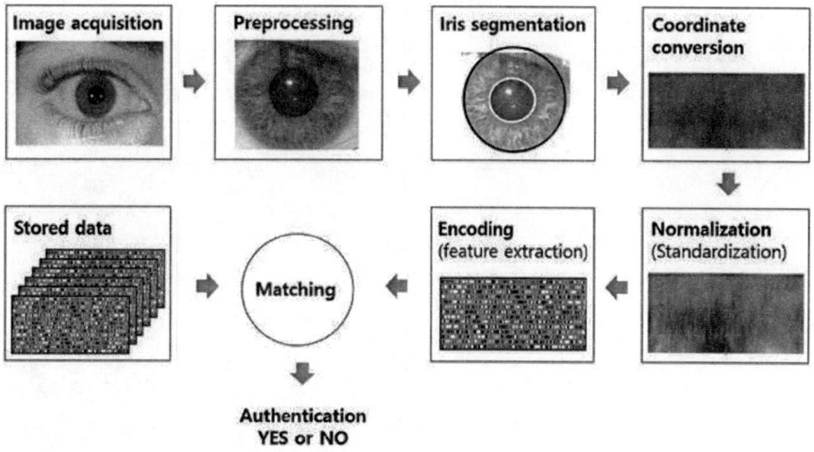

Fig. 11.5 The procedure of authentication using iris

to the rest of the iris. Crypts are tiny diamond-shaped holes in the iris that can range in size. Furrows are a series of pale lines that curve around the iris, and they're produced as the pupil opens and closes in response to the light. For authentication, the iris image is scanned with a resolution higher than 70 pixels per iris radius (Fig. 11.5). The Iris part of the image is segmented from the whole image by Hough transform based on the circular characteristic of the inner and outer border to the pupil and sclera region in the eye image. Eyelids and eyelashes can block or shade the iris and make it difficult to get a full shape of the iris and many times, they result in erroneous authentications. After the iris part is segmented, it is calibrated and normalized to produce consistent results under the illumination or pupil size change. A normalized scanned image is converted from the polar format into the rectangular format and encoded into two-dimensional code representing the scanned iris. Various algorithms can be applied in the encoding procedure. Traditionally, scanned iris images have been encoded with a Gabor filter into a 20×240 binary image. The authentication is performed by calculating the distance between the two encoded images. In the experiment for the accuracy of the iris authentication, the best performance has resulted in FAR of 0.005% and FRR of 0.238% (Benaliouche & Touahria, 2014).

The greatest limitation of authentication based on the iris is that it can be easily fooled by a high-quality image of the iris, face, and fake iris lenses. Also, the image quality and illumination condition can affect the quality of the scanned iris image and degrade the performance of the authentication system. Iris patterns can be deformed temporarily by alcohol consumption. And, the sensing units tend to be more expensive than the sensing units for the fingerprint as they require a high-resolution camera with intelligent focusing.

3.1.3 Face

Authentication using the face is the most natural method and similar to the ordinary recognition of a person. Authentication using the faces is accomplished based on digitized image and image processing technologies. The facial metric method uses the relative geometry of facial components, like eyes, nose, ears, and mouth. After the image is rescaled to the standardized form, facial features are extracted by the position, distance, and angles of facial components from the face image. Estimated facial features are encoded into data for authentication. Another method is to decompose the face image using the eigenfaces. Eigenfaces are the eigenvectors representing the face over the high-dimensional vector space of face images. Even though eigenfaces are derived statistically from the collection of face data, each eigenface contains a characteristic feature of the face, like brightness, symmetry, face line, hairline, and eyes. Every face can be composed of a weighted sum of eigenfaces and the face image can be matched by the distance between the major weights. Authentication based on faces contains a privacy issue and the image quality affects the authentication performance significantly. There is also a chance that the authentication can be faked by the picture of the face instead of the real face. Authentication based on face has reported an EER of 2.5% (Hassan et al., 2014).

3.1.4 Hand Geometry

Hand geometry is used as an authentication item based on the fact that every person possesses their own unique hand shape and the geometry does not change over time after a certain age in adults. Typically, hand geometry is analyzed with the contour of the hand by measuring the length, and width of each finger, palm width, and surface area of the hand from a black-and-white bitmap image of the hand. In this case, only 2D features are used for authentication. By measuring the hand in 3D space with an additional sensor, vertical measurements can be integrated into the authentication and increase the performance (Bhattacharyya et al., 2009). The accurate recognition rate was 95% for the identification of 10 people based on the artificial neural networks

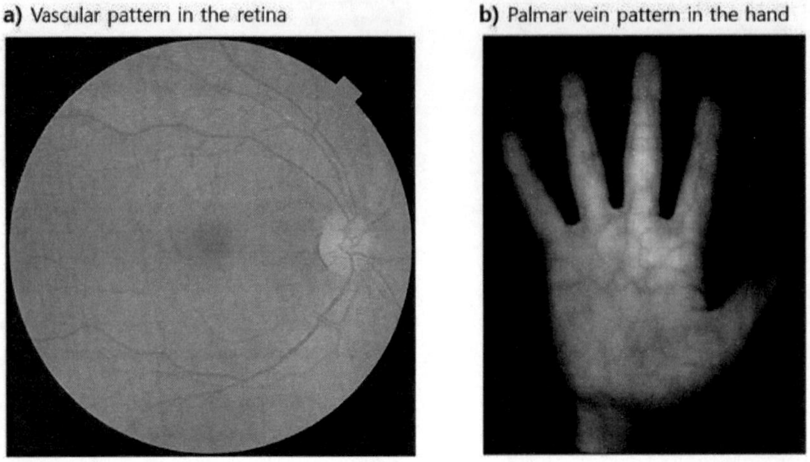

Fig. 11.6 Anatomical biometric items used for biometric authentication. (**a**) Vascular pattern in the scanned retinal image, (**b**) Veins structure in the palmar side of the hand

(Mohammed et al., 2021). Hand geometry is considered very difficult to fake when it requires creating an entire 3D physical hand volume. The greatest limitation is on the permanence of the hand geometry. As the hands are exposed externally, they are prone to the harshness of the external environment. Also, the geometry of the hand can be influenced by weight loss and weight gain, injuries, and even by the diseases like rheumatism.

3.1.5 Retinal Scan

Retinal scan is also used for authentication. Scanned retinal images show the pattern of blood vessels in the retina, at the back of the eye (Fig. 11.6a). The complex structure of the capillary supplying the blood in the retina produces a unique visual pattern of blood vessels in the retinal image. The pattern of blood vessels in the retina varies among the subjects and between the eyes. It is only partially determined by the genes and thus even identical twins do not have a similar pattern, as iris patterns are different between the twin. A retina scan requires a special recording unit to image the retina with coherent infrared illumination. Users are required to stare at a specific point for approximately 10~15 s with the bare eye. The process of authentication includes pre-processing, feature extraction, and feature matching. Features derived from the distributed pattern of blood vessels provide higher performance than features derived from other than blood vessels. Authentication unit with retinal scan becomes bulky and expensive compared to the unit for fingerprint. However, the retinal scan has a great advantage against faking, as the retina is not exposed externally and it is not possible to forge a human retina. The retinal scan provides very high accuracy with the EER of 0.06% for the identification of 159 subjects (Fatima et al., 2013).

3.1.6 Hand Blood Vessels

As the blood vessels in the retina provide uniqueness among the person, *blood vessel in the hand* also provides uniqueness and can be used as an authentication item (Fig. 11.6b). The vein structure of the hands is different even in identical twins and between left and right hands in the same person. For the light in the infrared spectral window of 700~900 nm, which can penetrate deep into the tissue, deoxyhemoglobin in the vein vessel has a higher absorption rate compared to the tissues surrounding and the oxyhemoglobin in the arterial vessels. So, hand veins beneath the skin absorb a high amount of infrared light and thus produce a darker shadow in the images by the infrared camera. Hand vessel images can be obtained in reflective mode as well as in transmission mode. Obtained hand vessel images are processed through the preprocessing, feature extraction, encoding, and matching procedures for the authentication results. Authentication with hand veins produced the EER of 0.44% in 20 subjects with 207 hand images (Savitha & Ramegowda, 2017). Veins in the fingers also can be used for authentication as well as veins in the palm of the hand. Authentication using the hand vessels is strong against faking as hand veins are hard to duplicate as they are internal to the human body. However, the sensing units for authentication become bulky and expensive.

3.2 Biometric Authentication Using Behavioral Items

Behavioral biometric authentication recognizes a person by the unique patterns exhibited when they behave during ordinary daily activities, like speaking, writing, walking, and typing. There are many behavioral biometric items that are unique to each person and can differentiate the person. As a voice contains a unique speaking pattern and characteristics, we can recognize a person by their voice. As a signature contains a unique writing pattern and is considered difficult to forge, it is widely accepted socially as an identifying and confirming item of a person. Gait is unique to each person by the body shape and walking pattern, and we can recognize a person from a distance before we see their faces. Keystrokes contain the characteristic typing pattern of each person and can be used to differentiate the person. These behavioral patterns can be obtained via ordinary using devices, like microphones, cameras, tablets, keyboards, mouses of PC, and smartphones, and the extra sensing device is not required. Behavioral biometric items can be used conveniently even without the notice of the subjects.

3.2.1 Signature

Handwritten signature authentication is a most demanding and leveraging behavioral biometric item. A signature is any handwritten specimen in someone's own writing that defines the individual's first name, last name, surname, nickname, or combination of them used for identification. Signature-based authentication can be categorized into two classes, offline and online signature authentication. The offline signature authentication system analyzes the scanned signatures from written pages obtained by the optical detector or camera. The offline system tries to authenticate only with the 2D static signature images after they are written down. It only contains the features like slope, bounding box, length, height, baseline, width, and size of strokes. Order of strokes and other dynamic characteristics cannot be identified with this offline signature. Authentication with an offline signature is relatively difficult due to the absence of dynamic parameters for the detail of the signature and performance is lower than authentication using the online signatures. In online signature authentication, a signature is obtained as serial data using electronic devices like an electronic pen, tablet, smartphone, and magnetic pad. The online signature can capture the dynamic features of signature detail, like the order of strokes, the direction of a stroke, positions of pen tips, speed, and writing pressure in addition to the static features used for offline signature authentication. Temporal information and mechanical pressure information during writing the signature increase the dimension of collected signature data higher than the two-dimension of static signature. Therefore, authentication based on the online signature produces higher performance than that based on the offline signature. Even though the imposters can see and copy the written static signature, they cannot get any information on how the signature is written down and cannot make a forgery of a signature without dynamic details. The authentication using the dynamic signatures obtained with smartphones achieved the best EER of 0.525% which is at a similar level to the signature authentication based on ergonomic stylus-based pen tablets (Krish et al., 2013).

3.2.2 Voice

Voice can be used as a behavioral biometric authentication item enabling users to access services by their voices. It can also be performed on-site manner or in an online manner in remote locations via the phone line. Voice properties required for authentication are often obtained using text-dependent mode prompting the user to speak a predetermined phrase for the current transaction. They also can be obtained text independently using any spoken phrase or other speech content. When the voice is obtained in digital format, characteristic features are derived reflecting the duration, pitch, intensity, and dynamics of each vocal element. Mel-frequency cepstral is the method widely used in voice authentication to transform input voice into the domain where the features can be efficiently derived. Mel-frequency spectrum is calculated by mapping the transformed voice spectrum into the Mel-frequency, which is the rescaled frequency based on the perceptual pitches judged by listeners. Mel-frequency cepstrum is calculated by inverse transforming the logarithm of the Mel-frequency spectrum. Mel-frequency cepstral coefficients can be used as features for the following procedure for authentication of a person. With the recent popularity of machine learning and deep learning, authentication based on the voice is becoming more powerful. One advantage of voice authentication is that it does not require any special and expensive hardware for authentication. While a high sampling rate is not required for authentication, environmental noises cause a significant error and decrease the performance of the authentication. So, it is not applicable in all environments. As the voice can be easily recorded and replayed, liveness detection is required to verify that the voice is from a live speaker and not from a recording. The performance of the authentication with voice is not high enough like the authentications using other items. In various studies on voice authentication, EERs in the range of 0.01~9.2% have been achieved (Abozaid et al., 2019).

3.2.3 Gait

Gait is a relatively new behavioral biometric item used for authentication. Gait refers to the locomotion achieved through the movement of limbs and it is recorded by a sequence of received images. Due to the different properties of an individual's muscular-skeletal structure, gait patterns are fairly unique among individuals and very difficult to mimic. Gaits cannot be shared and forgotten. Gait authentication utilizes the shape of the body and the gait pattern of each individual. For gait authentication, the subject's image is taken using a video camera from a distance, and gait patterns are extracted from the sequence of images and processed. Characteristic features of gaits include the pace, length of steps, angles of different parts of arms and legs, and static parameters of body shape. In robbery cases, where the thieves mask their faces and use gloves to avoid leaving fingerprints, they cannot be identified by their faces and fingerprints. Nevertheless, gait recognition might be exploited for the identification of thieves based on the recorded videos in the security system. Instead of a video camera recording the gait, a set of sensors installed on the floor can be used for analyzing the gait pattern. When subjects walk over the floor, gait pattern is measured by the heel strike, length of each step, pace, and so on. This type of sensor can be installed in the front of the building entry to collect the gait information for authentication and security purpose. Wearable sensors like the accelerometer in a smartphone can be used to record the body movements accompanying the gait and can be used for authentication. Individual gait pattern is recorded and characterized in 3-D acceleration signals. These gait sensors can be utilized in smartphones and in portable electronic gadgets for support, protection, and authentication purposes. Compared to other biometric items requiring contact with the sensing device at least within some distance, the gait of a person can be obtained without any contact in a longer distance without asking for any cooperation from the users. As the subjects cannot conceal the gait from the observers, authentication based on gait is acceptable as a part of secret surveillance systems of smart environments in security paths and airports. It also makes the authentication with a gait very difficult, because there are various factors affecting the recording of the gait, like distance to the person, the relative direction of walking, occlusion by the clothing and other people, and different lighting conditions. While the video camera-based gait authentication achieved the minimal EER of 3.6% (Vera-Rodriguez et al., 2013), gait authentication using the wearable sensors achieved the minimal EER of 5.31% for normal walking.

3.2.4 Keystrokes

Keystroke dynamics is another behavioral biometric item that can be used for authentication. Keystroke patterns are unique to individuals depending on their proficiency in typing and many behavioral habits for typing. Some trained typists are typing at high speed with all 10 fingers and many amateur users only use two fingers. For the same typed text, each individual can use different keys, like the left shift or right shift key, has a different typing pace with some pause between the typed words or characters, and has unique repetitive errors in typing. The seek times to get and depress a key and the hold time a key is held down are very characteristic to each individual regardless of their typing speed. Primarily, the statistical approach is applied for keystroke pattern analysis. Statistics of typing behavior of users at a specific time is used for the predetermined phrase like passwords. General statistical measures, like mean, median, standard deviation, and statistical t-test, are applied for the derived keystroke features, like pressing time, dwell time, and total typing time for the password. Recently, machine learning including deep learning technology is extensively used for keystroke analysis for authentication. While the authentication using the keystroke does not require any authentication device, it requires the user to actively interact with the system's keyboard or keypad for authentication and authentication is only possible when the user types on the keyboard. One limitation of authentication using the keystroke is that keystroke dynamics vary substantially during a day and between different days, and maybe affected by a number of external factors. The EER using the keystroke authentication was achieved at 2.67% (Lu et al., 2020).

4 Authentication Using ECG

Various items are used for authentication. Within the biometric authentication category, several anatomical biometric items like fingerprint, iris, and face are widely accepted in practical authentication applications. One of the limitations of anatomical and behavioral biometric features is that they are relatively weak to the attack using the fake authentication items. As these anatomical and behavioral items are exposed externally, they can be easily obtained by intruders when they are intended. Faking or spoofing technologies by intrudes is also progressing. While the recent advancement of artificial intelligence and deep learning technology has enhanced the performance of authentication, they also affect adversely by making authentication systems more vulnerable to spoofing attacks. One method of increasing the robustness against spoofing is to add liveness detection or continuous biometric authentication functions. Liveness detection is the ability to detect if a proposed biometric item is a real one from a live person present at the point of capture or a fake one by a spoofing artifact or lifeless object. Continuous biometric authentication is continuously checking the identity of the subject by using noninvasive sensors or unconstrained devices that can collect users' biometric data smoothly.

In this point of view, biometric authentication using biological signals is an attractive option that can provide liveness detection and continuous authentication functions. In contrast to the anatomical biometric items, like fingerprint and iris pattern, and behavioral biometric items, like signature and voice, biological signals, like ECG and EEG, are physiological biometric items. They are produced by the physiological body functions. Authentication using the biological signals can be regarded as a type of authentication "By something you make." Biological signals are not exposed externally and only can be measured in live subjects.

Among the various biological signals, electrocardiogram (ECG) is most promising as an authentication item based on its deterministic waveform and high signal quality.

4.1 Characteristics of ECG as an Authentication Item

Are ECG signals unique for authentication? That is the most important factor as an authentication item. Yes, ECG is sufficiently unique among the subjects. Since we are accustomed to seeing the standard ECG pattern of normal subjects, we tend to think that all the normal subjects have the same undistinguishable ECG waveforms. However, that is not correct. As normal faces are distinguished from the wounded faces, normal ECG waveforms are distinguished from the abnormal ECG waveforms. Also, as the faces of normal subjects are different from the others and they can be identified by their faces, ECGs of the normal subjects are different from the others and they can be identified by their ECGs. Variability of ECG waveform is caused by several factors, like the position of the heart within the chest, size, and geometry of the heart, and the conductive properties of the body volume from the electrical source to the body surface contacting the recording electrodes. These factors are unique to each subject and provide uniqueness in the recorded ECG. While the ECG is recorded in 12 leads in standard ECG recording for clinical purposes, single-channel ECG is considered to provide sufficient information to discriminate the subjects for an authentication purpose. Figure 11.7 shows the ECG waveforms collected from 50 normal subjects. Each subject shows a unique waveform pattern that can be differentiated from others in various details.

ECG is universal. There is no live subject having no ECG. This is very contrast to the fingerprint or iris pattern which are not collectible from the live subjects in some special cases. ECG is collectible as it easily can be recorded by contacting the fingers of each hand to the electrodes of the sensing device. ECG is acceptable. It does not have a privacy issue like face and people have no special reason to hesitate to record ECG for authentication. With the wide deployment of health monitoring devices, like the ECG recoding function in warble watches, ECG is recorded for many other purposes to monitor health levels. Already prevail use of ECG makes it more acceptable as an authentication item. With reliable performance, the subject of recorded ECG can be identified by the recorded ECG itself without entering demographic information for the recorded ECG.

Is ECG permanent? It is difficult to say 'Yes' since heart rate is variable upon the exercise and emotional level of the subject and the ECG waveform slightly varies as the heart rate varies. If we resister the ECG waveform in a resting state and try authentication just after hard exercise, the rate of rejecting authentication seems to be increased. However, if ECG is collected for registration and used only in a normal resting state, it is sufficiently permanent for authentication. Even in increased heart rate, the waveform of the QRS complex of ECG does not vary and this characteristic can be used to increase the robustness upon the change of heart rate. The variable nature of ECG is totally different characteristic not found in other anatomical biometric items. It can be used as an advanced function protecting forced authentication as the subject can be authenticated only in the resting and calm emotional state and not authenticated under a disrupted emotional state forced by others. This type of authentication is not possible with other types of authentication items.

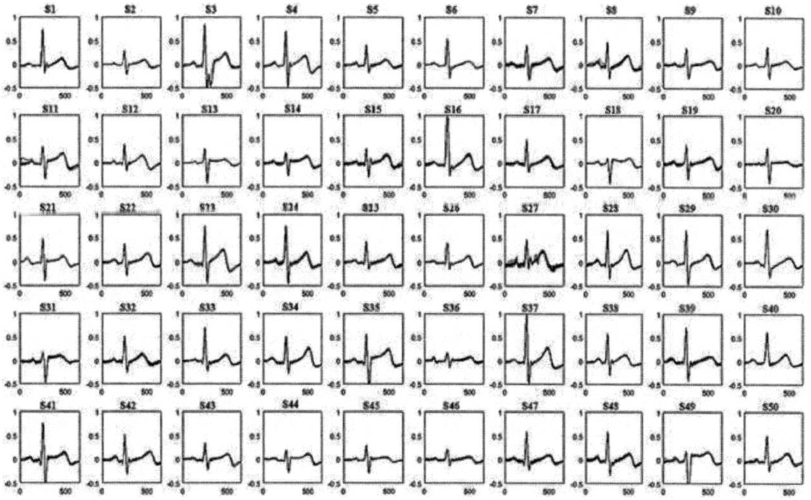

Fig. 11.7 ECG waveforms collected from 50 normal subjects. 10 ECG waveforms are overlapped in synchronization to the R-peak for each subject

Table 11.1 ECG properties as an authentication item

	Property	Level	Comments
1	Uniqueness	Medium	Unique among the normal person. But the simplicity of waveform can limit the number of enrolled subjects.
2	Universality	Very high	Some person can have no fingerprint and no iris. But, no one has no ECG.
3	Collectability	High	Easily collected by two fingers of each hand. Wearable devices also can be used.
4	Acceptability	High	No reason to hesitate to record ECG for authentication. ECG is also used widely for health monitoring.
5	Permanence	Medium	ECG waveform varies with heart rate. But, QRS complex is consistent. Variation in normal range is a unique feature that other authentication items do not have.
6	Circumvention	Very high	Not exposed externally. Produced only while the subject is alive. Difficult to steal and fake by artifact.
7	Performance	Medium	Sufficiently high performance for single-factor authentication. Increase the performance and the robustness for multifactor authentication.

ECG shows the strongest advantage related to the circumvention property. ECG is difficult to steal or eavesdrop on as it is not exposed externally. Even though the intruder got the ECG waveform of the other subject, it is difficult to fake the authentication system with an artifact as the ECG is recorded from the subject directly. The performance of authentication using ECG is not higher than other biometric authentication systems using fingerprint or iris patterns. This is caused by the smaller entropy in a simpler waveform of ECG compared to the fingerprint and iris pattern. As a single-factor authentication, it produces an EER of 1.48% under a time-varying situation (Kim et al., 2019). As ECG is an effective item that can provide liveness detection and continuous authentication, ECG can be more effectively used to increase performance and prevent spoofing in multifactor authentication in combination with other authentication items. ECG can be collected together with fingerprints at the same time from the fingers by combining a fingerprint scanner with ECG electrodes. Table 11.1 summarized the properties of ECG as an authentication item.

4.2 ECG Leads for Authentication

When the ECG is recorded from the person, it should be clearly defined which lead of ECG is used for the authentication. In the medical domain, the ECG of standard 12 leads collected from the 9 electrodes is used for diagnostic purposes. As the ECG waveform of each lead is different from the others, if multiple leads are used, performance would increase. However, as there is a redundancy among the 12 leads of ECG waveforms, increasing the number of leads does not provide a proportional amount of additional information. Also using the precordial leads is not convenient for authentication as it limits the collectability and the acceptability. In this respect, using the ECG measured from the limb leads is appropriate for the authentication. Among the limb leads, Lead I, Lead II, and Lead III, lead II shows the maximum amplitude and highest SNR and it is used as a representative single ECG waveform of the cardiac activity. However, lead II is somewhat inconvenient to record as it is the potential between the right hand and left foot. In contrast, Lead I is recorded between the right hand and left hand and is much easier to collect in a wide range of applications. It can be recorded just by contacting a finger of each hand to the electrodes of the measuring device. If ECG is used with the fingerprint for multifactor authentication, lead I is the most suitable as ECG can be collected simultaneously when the fingerprint is collected from the fingers. Also, lead I ECG can be collected using a wearable watch by contacting the right hand to the exposed surface of the watch, while the one electrode is already contacting the surface of the left hand. So, the ECG of lead I is most preferable for authentication.

4.3 Structure of Authentication System Using ECG

Authentication with ECG follows the general framework for authentication (Fig. 11.8). The system should enroll the ECG of the legitimate persons in the authentication database. Signal collected from a subject is preprocessed to refine a waveform. Various filters are applied in this stage for increasing SNR, artifact rejection, normalization, or segmentation. After the collected ECG waveform is prepared for processing, feature extraction is applied. Features are selected based on the variance of a feature distributing the person along the feature axis. Features having high variance are preferable for authentication purposes. From the preprocessed ECG waveform, features are extracted and encoded into a vector form representing a person in a database. When a user applies an ECG for authentication, the collected signal from a claimed user follows the same procedure of preprocessing and feature extraction. Matching is a process that compares the features extracted from the current authentication trial and the features of the person stored in the database. Authentication is decided by the closeness of the features. The closeness between features is quantified by some measures, like the

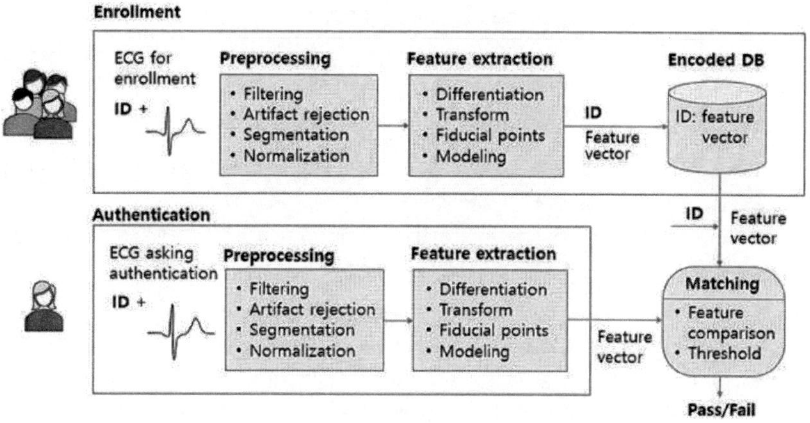

Fig. 11.8 Structure of authentication system using ECG

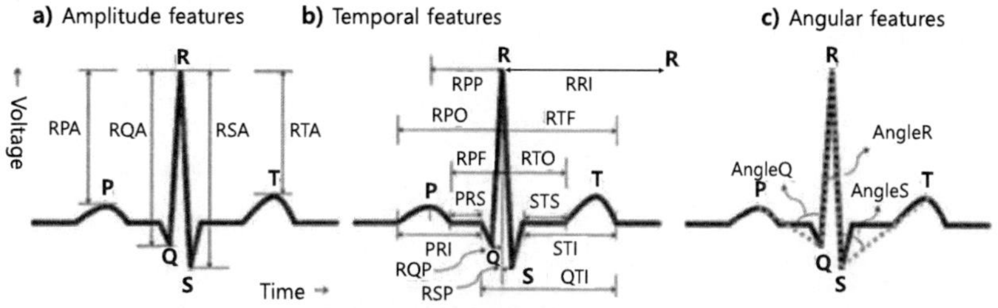

Feature	Definition
RPA	P-R amplitude
RQA	Q-R amplitude
RSA	S-R amplitude
RTA	T-R amplitude

Feature	Definition	Feature	Definition
RRI	R-R interval	RPO	R-P_{onset} interval
PRI	P-R interval	RPP	R-P_{peak} interval
PRS	P-R segment	RPF	R-P_{off} interval
STI	S-T interval	RQP	R-Q_{peak} interval
STS	S-T segment	RSP	R-S_{peak} interval
QTI	Q-T interval	RTO	R-T_{onset} interval
		RTF	R-T_{offset} interval

Feature	Definition
AngleQ	P-Q-R angle
AngleR	Q-R-S angle
AngleS	R-S-T angle

Fig. 11.9 Deterministic features from the fiducial points in ECG waveform

distance in the feature space or the matching score, and the decision is made by the threshold. If the distance between the two feature vectors is shorter than the threshold distance or the quantified matching score is higher than the threshold, the authentication of the submitted trial is issued. Threshold determines the performance of the system. A lower threshold, for example, a lower matching score, increases the sensitivity by minimizing the false rejection rate (FRR) for the legitimate persons. However, it increases the false acceptance rate (FAR) for the illegitimate intruders. On the contrary, a higher threshold, for example, a higher matching score, increases the specificity by minimizing the false acceptance rate (FAR) for the illegitimate intruders. However, it reversely increases the false rejection rate (FRR) for legitimate users and blocks them to pass the authentication system. The threshold of the authentication system should be carefully determined considering several factors, like the number of enrolled persons, the chance of attack from intruders, and risks and inconveniences caused by the false decisions, to optimize system function (Uwaechia & Ramli, 2021).

4.4 ECG Features for Authentication

4.4.1 Features from the Fiducial Points of ECG Waveform

As the ECG waveforms are different among the persons, features for authentication can be extracted directly from the ECG waveform. These features represent the local characteristics of the waveform and the corresponding function of the heart. There are 5 wave components of P, Q, R, S, and T in one period of ECG waveform. Features are extracted based on the relative position of the wave components in amplitude and time. They can include the amplitude difference between wave components, temporal duration of wave components, time interval between wave components, and the angle or slope between the wave components. Figure 11.9 shows some example features extracted from the fiducial points of the ECG waveform. These features are intuitive to understand and distinctive enough as there are differences in amplitudes and temporal durations in waveforms from each individual. Feature extraction based on the fiducial points

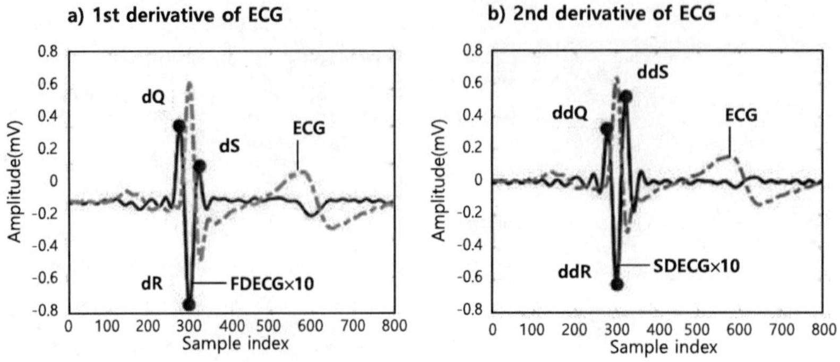

Fig. 11.10 Fiducial points from the derivatives of ECG. (**a**) 1st derivative of ECG (FDECG) and fiducial points. dQ, dR, and dS represent the peaks in 1st derivate waveform and are named after the nearest wave component in raw ECG. (**b**) 2nd derivative of ECG (SDECG) and fiducial points. ddQ, ddR, and ddS represent the peaks in 2nd derivate waveform and are named after the nearest wave component in raw ECG

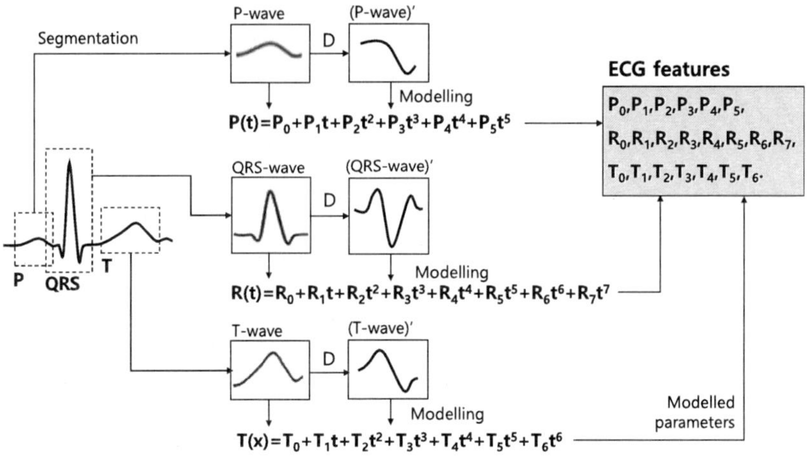

Fig. 11.11 Features derived by modeling the ECG wave components. *D* differentiation

relies on the accurate detection of wave components. If the waveform is corrupted by the noise, fiducial points are changed and the extracted features become sensitive to the noise. When one fiducial point is not detected from the waveform by unclear waveshape or absence of a wave component, related features cannot be extracted from the waveform and can result in a significant error.

Fiducial points can be derived from other than the raw ECG waveform. They can be derived from the derivative waveforms of the raw waveform. Figure 11.10 shows the 1st and 2nd derivates of the ECG waveform and notifies the peak points in the derivatives as fiducial points. dQ, dR, and dS represent the maximal and minimal slope of the raw ECG waveform, and ddQ, ddR, and ddS represent the maximal and minimal change of slope of the raw ECG waveform. They are named after the wave component near the corresponding peak. As these features contain information on the change of slopes, they provide additional information on the waveform characteristics. When the features from the derivative waveform are included, the performance of the authentication increases (Kim & Chun, 2019). During the preprocessing, segmentation is performed, which is selecting the portion of the waveform to be used in the feature extraction step. This segmentation is performed usually in reference to the R-peak, which is the most dominant peak and easy to detect. However, in certain cases, R-peak can be ambiguous in the raw waveform, and it can be smaller than the peak amplitude of the P-wave or T-wave. However, fiducial point dR, the minimum peak in FDECG, is always clear and much better to employ as a reference position for segmentation.

4.4.2 Features by Modeling ECG Waveform

Instead of directly measuring the features after detecting the fiducial points from the ECG waveform, features can be derived by modeling each ECG wave component into polynomial equations. When the wave components are modeled and approximated to polynomial equations, wave components are encoded into model parameters and these model parameters can be used as features differentiating the individuals. Figure 11.11 shows the procedure of feature derivation based on the polynomial modeling. Each wave component of P, QRS, and T is segmented from the raw ECG signals. Then segmented parts of wave components are modeled by a certain degree of polynomial equations. The depth of the model is determined by the complexity of the wave components to be modeled. For example, QRS component is modeled up to the 7th order of polynomials while the P-wave is modeled up to the 5th order of polynomials based on their complexity. Model parameters can be estimated either from the raw ECG signal or using the derivative of the wave components. Estimated model parameters combinedly represent the ECG waveform to be analyzed for authentication. Then, derived model parameters are used as authentication features and the closeness of the ECG waveforms is evaluated by estimating the distance between the polynomial parameters of two ECG waveforms. Authentication using features of polynomial parameters is faster than using fiducial points, easy to implement, and shows a degree of robustness in dealing with abnormal beats that may occupy some normal ECG signals (Sufi et al., 2010).

4.4.3 Features from Dimension Reduction of ECG Waveform

Low dimensional output produced by dimensional reduction methods for the high dimension of ECG data can be used for features discriminating the ECG waveforms. Linear discriminant analysis (LDA) and principal component analysis (PCA) are the simplest and most widely used methods for dimensional reduction.

LDA is a linear statistical method used for dimensionality reduction as well as for classification with data of known input and output pairs. In LDA, output, y, is calculated by a linear combination of input data, $\boldsymbol{x}$.

$$y = wx \tag{11.1}$$

Matrix **w** transforms input data of dimension N into the output data of dimension M, which is much smaller than N for the dimensional reduction. Linear transform matrix, **w** of size N × M, is derived from the paired data during the training to maximize the variance of output between the ECG data from different persons while minimizing the variance of output between the ECG data from the same person. LDA finds the most significant M components from the N data samples. If the M samples of output represent input ECG data of N samples sufficiently accurate, M sample values can be used as authentication features. After the matrix **w** is determined, output **y** is calculated simply by matrix multiplication.

PCA is another method used for dimensional reduction for ECG data. In contrast to the LDA method, PCA is an unsupervised learning method not using the paired data set. PCA also linearly transforms the high-dimensional input data into lower-dimensional output data just like the LDA method. However, the linear transform matrix is derived differently by solving an eigenvalue problem. PCA finds a first variable so that the transformed single output has the least reconstruction error. The second variable is added so that the transformed two outputs have the least reconstruction error while maintaining the orthogonality to the first variable. The third variable is added so that the transformed three outputs have the least reconstruction error while maintaining the orthogonality to the two former variables. Output variables can be calculated similarly up to N variables.

Practically, the transforming matrix is calculated from the covariance matrix of the unlabeled ECG dataset.

$$S_{\mathrm{COV}} = \frac{1}{K}\sum_{i=1}^{K}(x_i - \bar{x})(x_i - \bar{x})^T \tag{11.2}$$

where x_i is i-th ECG data of N sample points, $\bar{x}$ is the averaged ECG of K ECG data, K is the number of ECG data, and $\mathbf{S}_{COV}$ is the covariance matrix. A transformation matrix **w** is calculated by using eigenvalue analysis for the covariance matrix $\mathbf{S}_{\mathbf{COV}}$.

$$\lambda_i w_i = S_{\mathrm{COV}} w_i \tag{11.3}$$

where λ_i is i-th eigenvalue of the covariance matrix $\mathbf{S}_{\mathbf{COV}}$,$\mathbf{w}_i$ is the eigenvector corresponding to eigenvalue λ_i. M eigenvectors corresponding to the highest eigenvalues are actually used for dimensional reduction to produce output by the linear transformation.

$$y = w^T(x - \bar{x}) \tag{11.4}$$

Matrix $\boldsymbol{w}^T$ has the size of N × M and transforms input ECG of N samples into the output of M PCA components. While the input ECG is reconstructed with M PCA components with a small enough reconstruction error, M PCA components can be used as authentication features. Performance of authentication increases as the number of principal components used for authentication features increases. Figure 11.12 shows the distribution of ECG waveforms from 10 subjects represented using 3 PCA components as features. With only three features, ECG waveforms from the same individual are clustered together and ECG waveforms from different persons are separated in distance.

Instead of using the linear dimensional reduction method, the nonlinear dimensional method can be used for authentication for increased performance. Dimensional reduction using linear methods may fail to find the data structure of nonlinear patterns in input ECG waveforms. Nonlinear dimensional reduction is performed by kernel-based techniques. The nonlinear kernel function transforms input data into data in another space where the input ECG

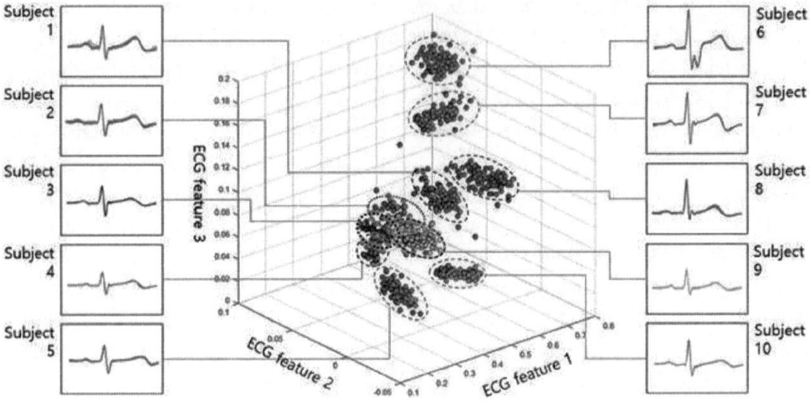

Fig. 11.12 Distribution of ECG feature values derived using principal component analysis (PCA) for 10 subjects

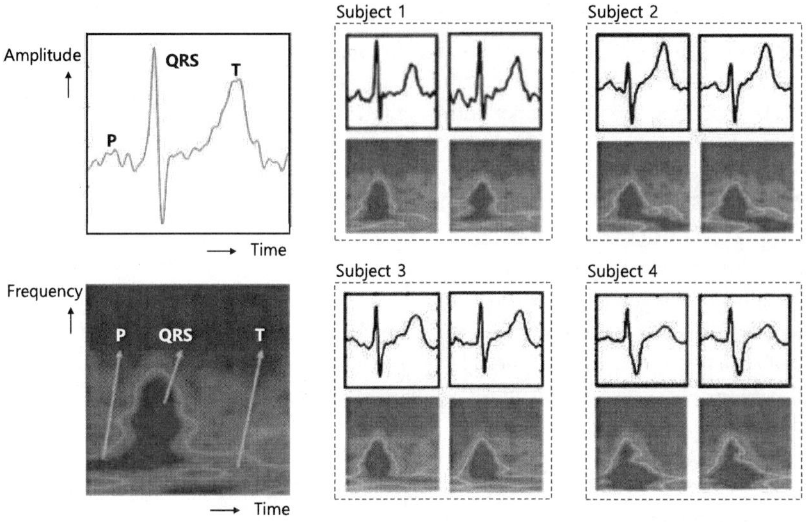

Fig. 11.13 Transformed 2D ECG waveform by continuous wavelet transforms (CWT) using Morlet wavelet. CWT results are different among four subjects while they are analogous within a subject

waveform is more clearly separated. Then the dimensional reduction is applied similarly to the linear dimensional reduction methods in the transformed data domain. Kernel-based methods can be used for LDA and PCA as kernel linear discriminant analysis (KLDA) and kernel principal component analysis (KPCA). Polynomial kernel and Gaussian radial basis function kernel are most commonly used.

4.4.4 Features from the Transform of ECG Waveform

ECG features for authentication can be derived from the frequency domain based on the frequency characteristics of the ECG waveform. After the ECG waveform is transformed into a frequency domain using a fast Fourier transform (FFT) or periodogram, the magnitude and phase of a certain frequency and power of a certain frequency band can sufficiently differentiate the ECG waveform from other individuals.

As ECG characteristics are very distinct on the temporal scale, combining temporal information with frequency representation is expected to be more effective for authentication. Spectrogram using the short-time Fourier transform (STFT) is one method for the time-frequency representation and the magnitude and phase values of the spectrogram can be used as features (Chun et al., 2016). A 1-D ECG waveform is transformed into a 2-D image representing time on the horizontal axis and frequency on the vertical axis. Wavelet transform also can represent the ECG waveform in time and frequency combinedly. Wavelet transform linearly transforms and decomposes an ECG waveform into components that appear at different time scales using a special type of wavelet. Wavelet does the role of sinusoidal wave in Fourier transform and it produces more focalized transform results when the using wavelet resembles components of the input signal. Figure 11.13 shows ECG waveforms transformed by the continuous wavelet transform with the

Morlet wavelet. ECG wave components are distinctly appearing in time scales with different frequency characteristics. ECG waveforms from different subjects are clearly differentiated in a visual pattern. Features can be derived from this 2-D time-frequency image or the image can be directly used as an input to the classifier for authentication, like the one based on the deep neural network.

4.5 Deep Learning for ECG Authentication

Instead of selecting features in a handcrafted manner by applying various analytic methods to ECG signals, non-handcrafted features started to be used for ECG authentication with the advent of deep learning technology. The deep neural network (DNN) can be used by taking the raw ECG waveform as an input to the network without the feature selection step. Then, the deep neural network can do the whole procedures of authentication including preprocessing, feature selection, and the decision for authenticating the input ECG signal.

Deep neural network structure can derive various features in multi-resolution as internal coefficients of layer filters, which are not easily found by the handcrafted methods, and these features are anticipated to increase the performance and robustness against the perturbation (Ingale et al., 2020). ECG signal can be applied directly to the 1-D deep neural networks designed for ECG authentication or can be converted to a 2-D image to use pre-developed 2-D convolutional neural networks (CNNs), which have proven their performance in many applications for image recognition and classification.

DNN can include the matching procedure. When the two ECGs are applied as inputs to the DNN, DNN calculates the matching score of two ECG waveforms by calculating the distance between the features derived in the former layers of the network, and provides the authentication result as an output of the network.

4.6 Matching the Features

Derived features from the ECG waveform are matched with the enrolled features in the database. For authentication purposes, the derived features are compared to the features of the claimed identity, and the authentication is issued when the closeness or matching score of two feature sets is higher than the threshold level. If the features are close enough, a claimed user is recognized as a legitimate user, and authentication is issued. If the features are not close enough, the user is regarded as an illegitimate intruder and the authentication is rejected. For identification purposes, submitted features are compared to all feature sets enrolled in the database to find the person having the most similar values of the feature set.

The closeness of features can be implemented in several methods. Simply the *correlation* of two waveforms can be used to evaluate their closeness. However, the correlation method does not care about meaningful ECG wave components and simply calculates the numeric closeness of the entire waveform. A small temporal shift by the minute change of heart rate can induce a relatively larger degradation in correlation and would decrease the performance. *Dynamic time warping* (DTW) method can be applied to attenuate the error by allowing the temporal warping in ECG waveforms. DTW can produce an increased matching score when two ECG waveforms are not of the same heart rate. DTW method finds the optimal alignment between two ECG waveforms, such that the sum of the differences between aligned points is minimal.

When the features are derived in the feature selection stage, the closeness or matching score of the two feature sets is calculated by the distance. Euclidean distance is the simplest form of distance. When the ECG waveform is represented with N features, it is calculated as the distance of the two points in the N-dimensional feature space. When the features of the submitted ECG $\boldsymbol{F^q}$ and the features of the claimed identity enrolled in the database is $\boldsymbol{F^e}$, the Euclidean distance is calculated as the root mean square distance of two points representing each ECG waveform.

$$\text{Euclidean distance}(\boldsymbol{F}^q, \boldsymbol{F}^e) = \sqrt{\sum_{i=1}^{N} (\boldsymbol{F}_i^q - \boldsymbol{F}_i^e)^2} \quad (11.5)$$

When the distance between two feature sets is less than the threshold distance, authentication is issued or the trial for authentication is rejected. There are many variants of distance measures to reduce unexpected errors and increase the allowable range of the distance calculation. For example, as features might have different units and scales, they need to be normalized. Hamming distance is also used when many features are used for comparison. Hamming distance measures the distance of feature sets by counting the number of features having a difference lower than certain values.

The threshold to classify the acceptance or rejection of a submitted trial is optimally determined by the analytic classification method. Various classifiers have been used in ECG-based authentication including linear discriminant analysis (LDA), support vector machine (SVM), k-nearest neighborhood (kNN), multilayer perceptron (MLP), and random forest (RF). Classifiers are trained to find the optimal threshold under training data set composed of legitimate users and intruders under the given cost function.

Features can be connected directly to the classifier as input data. Many studies have applied deep learning methods for authentication instead of deciding the authentication by the threshold.

The performance of the authentication is dependent on the threshold level. Performance of the authentication system is described by several measures including accuracy, false rejection rate (FRR), false alarm rate (FAR), and equal error rate (EER), which are defined as follows:

$$\text{Accuracy} = \frac{\text{Number of truely authneticated attemps}}{\text{Number of total attemps}} \tag{11.6}$$

$$\text{FRR} = \frac{\text{Number of falsely rejected attemps}}{\text{Number of attemps by the legitimate}} \tag{11.7}$$

$$\text{FAR} = \frac{\text{Number of falsely accepted attemps}}{\text{Number of attemps by intruders}} \tag{11.8}$$

Accuracy describes accepted attempts and correctly rejected attempts combinedly by counting the number and calculating the ratio. It describes two types of performance in a single measure. However, it is strongly dependent on the relative number of legitimate attempts and illegitimate attempts by intruders.

As the threshold of matching score for accepting the authentication is increased, FRR is increasing, and FAR is decreasing. Reversely, as the threshold of the matching score is lowered, FRR is decreasing, and FAR is increasing. FAR and FRR are strongly dependent on the threshold level, and FAR and FRR cannot describe the performance of the system in a simple way. EER is the error rate at the threshold level where two errors, FRR and FAR, become equal. EER can represent the performance of an authentication system in a single measure and is most widely used when comparing the performance among the authentication systems.

4.7 Performance of Authentication Based on ECG

There have been active studies using ECG for authentication. Table 11.2 summarized the results of the studies showing the high performance based on the equal error rate (EER). Maximum performances are around 1.48% of EER. Many different methods of feature selection and matching have been used to increase the performance of ECG authentication. In addition to the authentication algorithms, the performance of the authentication is dependent on various factors. The number of subjects in the experimental database affects the performance. A small number of subjects generally shows increased performance, and it can be lowered with a higher number of subjects. The authentication system should be verified with enough subjects. Abnormal subjects included in the database affect the performance. As they have district ECG patterns, the inclusion of subjects having heart diseases can increase the performance and the results cannot be properly extended to the performance of the healthy individuals. The type of ECG recording also affects the performance. To practically use ECG authentication, it should be recorded conveniently from the wrists or fingers. The quality of these ECGs is not high as the ECG recorded in the chest leads for medical diagnostic uses and they can be much more easily contaminated with the noises. The number of beats used for authentication significantly affects the performance. Performance increases with the increase in the number of beats used. Authentications using 5 or 10 beats of ECG show increased performance than using a single-beat ECG. However, as it takes a longer time to record the ECG instead, optimal acquisition time should be determined by compromising the performance with acceptability and collectability. Considering that heart rate is also variant on the respiration phases of inspiration and expiration, recording of a heart rhythm of 5 to 10 beats seems to be appropriate.

Table 11.2 Performance of authentication based on ECG

Reference	Performance (EER)	Number of subjects	Feature selection	Matching	ECG recording
Choi et al. (2016)	4.61%	175	Fiducial	Support vector machine	Fingers
Li et al. (2020)	4.26%	90	GNMF	Residuals	Lead I
Kim et al. (2019)	3.80%	89	CS	GLRT	Lead I
Labati et al. (2019)	3.37%	92	CNN	Euclidean distance	Holter ECG
Hammad et al. (2019)	3.20%	100	CNN	QG-MSVM	Chest lead II
Wahabi et al. (2014)	2.57%	1020	AC/LDA	Euclidean distance	Fingers
Wang et al. (2020)	2.17%	46	MSDF, CMF	Euclidean distance	Fingers
Ranjan (2019)	2.00%	800	CNN	Cosine distance	Wrist-finger
Chu et al. (2019)	2.00%	90	MS-CNN	Feature vector similarity	Lead I
Kim et al. (2019)	1.48%	105	SWT, IFS, PCA, LDA	Euclidean distance	Fingers

AC Autocorrelation, *CS* Compressive Sensing, *GLRT* Generalized Likelihood Ratio Test, *GNMF* Graph regularized Non-negative Matrix Factorization, *IFS* Infinite Feature Selection, *LDA* Linear discriminant Analysis, *MS-CNN* Multi-Scale Convolutional Neural Network, *PCA* Principal Component Analysis, *QG-MSVM* Q-Gaussian Multi-class Support Vector Machine, *SWT* Stationary Wavelet Transform

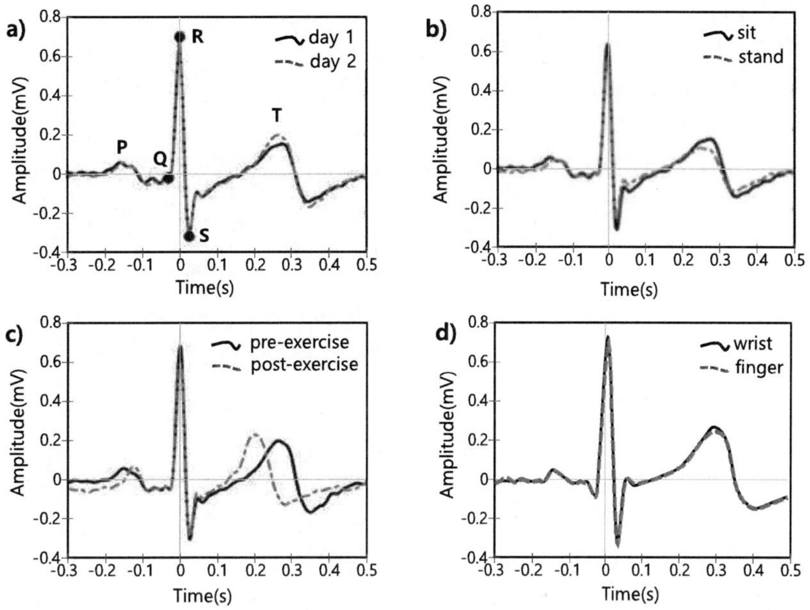

Fig. 11.14 Variation of ECG waveform on various factors. (**a**) On different days, (**b**) On different postures, (**c**) Before and after exercise, (**d**) On different recording sites

There are innate variable characteristics in ECG as a representation of the biological system. ECGs recorded in different sessions or on different days can show a minor discrepancy in recorded waveforms. As the discrepancy tends to increase when the temporal gap is increasing, regular or repetitive template enrollment can provide the consistent performance of the ECG authentication. ECGs recorded from the same individual with the same lead can show differences depending on the posture during the ECG recording. Standing from a supine position to an upright position causes a clear shift in the P-wave and QRS complex. Minor morphological changes occur in an ECG when subjects alter their position from sitting to standing. Most significantly, heart rate affects the morphology by degrading the temporal and frequency contents of the ECG waveform. Various factors, including exercise, emotional variation, stress, pregnancy, and caffeine intake, can alter the heart rate of each individual. Among these factors, exercise usually results in the most abrupt changes. Depending on the grade of exercise, the heart rate can be increased to more than twice the normal resting rate. The heart-rate increase causes the T-wave to approach the QRS complex, as shown in Fig. 11.14c, but the shape of the QRS complex is relatively consistent. The performance of the authentication is degraded with the addition of variable factors. When ECGs recorded in the same session with a sitting position are used for the enrollment and the test, an EER of 0.03% has resulted. However, the EER is increased to 1.48% when the ECGs recorded on another day are used for the test, to 1.74% when the ECGs recorded in a different posture of standing are used for the test, and to 5.61% when the ECGs recorded after exercise is used for the test (Kim et al., 2019).

5 Authentication Using EEG

In addition to ECG, EEG is another type of bioelectrical signal that can be used for authentication purpose. As EEG is a type of bioelectrical signal like ECG, it has many common features with ECG as an authentication item. EEG also has its own uniqueness as ECG has.

5.1 Properties of EEG as an Authentication Item

EEG is universal like ECG. All subjects have their own EEG responses that can be recorded. EEG is unique among the subjects like ECG. Recorded EEG patterns are different among the subjects under the same situation, like resting state or visual stimulation. Figure 11.15 shows the EEG spectrum recorded from 50 subjects when they are in the resting state with their eyes closed. They have a similar structure of decreasing power with increased frequency and peak of the alpha rhythm. Overlaid 30 epochs of each subject follow the same spectral pattern and they are differentiated from the others.

However, EEG has some different properties compared to the ECG. The greatest difference point is that EEG can be changed by thought or stimulations, while the ECG cannot be changed intentionally by any means. The brain is a highly active organ that receives and processes signals from the body parts and the environment, generates responses accordingly, and recalls the stored information when needed. Activations of brain regions reflect the behavioral and

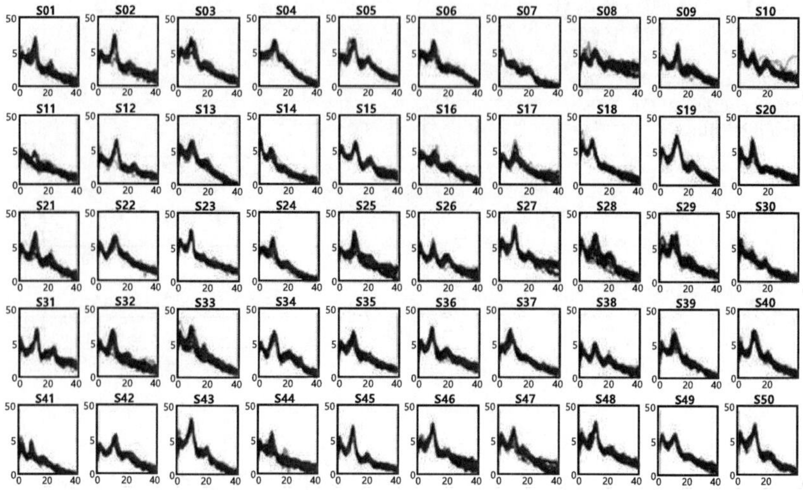

Fig. 11.15 Power spectrum of 50 subjects in the resting state with their eyes closed. For each subject, a power spectrum of 30 epochs of 10 s has been overlaid

physiological characteristics of each subject in processing the information. There are differences in calculating power, stored information, conduction of signals, understanding of received information, reflections of emotion, and so on among the subjects. Therefore, corresponding records in EEG are different among the subjects and have the potential as an authentication item identifying each subject. Within a subject, EEG changes dynamically reflecting the brain states in tasks, like responding to the external stimulations, preparing actions, and doing mental thoughts. That is, recorded EEG can be changed with external stimulations or by specific mental thoughts. As we change a "password" when it is exposed or no more secure, we can change a "pass-thought" to elicit an EEG response when it is required. This changeability of recorded EEG is the advantageous difference of EEG as an authentication item.

In collectability and acceptability, EEG has relative disadvantages compared to the ECG. EEG can be collected with the electrodes attached to the scalp surface. Standard Ag/AgCl electrodes are attached to the scalp using the paste, and it makes a subject inconvenient and hesitant to record EEG. Portable EEG devices with a single or lower number of channels using dry electrodes are regarded as alternatives to increase the collectability and the acceptability of the EEG as an authentication item. However, the quality of recorded EEG with dry electrodes can be deteriorated and is easily affected by noises. These properties of collectability and acceptability limit the use of EEG in applications where a high level of confidence is required.

Is EEG permanent? It is difficult to say 'Yes' like ECG. But under the same brain state, it generally produces similar signals in amplitude and frequencies. Even though EEG changes dynamically upon the brain states, the EEG of each subject responds with a consistent pattern under the given brain state, like resting, responding to the stimulation, and during the mental thought. In this regard, EEG has a sufficient permanence property like ECG as an authentication item.

EEG has a very high-level circumvention property like ECG. EEG is not exposed externally and cannot be easily stolen. Even though recorded EEG is exposed to an intruder, it cannot be copied into the recorded EEG of another subject. In addition to the robustness to spoofing attacks, recording the ordinary EEG response is not possible under pressure and coercion states.

Performance of authentication using EEG varies dependent on the type of EEG signals and the procedures for authentication. When only a short time record of spontaneous EEG during the resting is used for authentication, performance is lower than the performance using the ECG. But, when the series of stimulations and mental thoughts are combined, performance can be increased to a very high level enough as a single task authentication. Considering the complicated procedure and time required for authentication with high performance, EEG authentication is most appropriate where the highest confidence is required.

5.2 Tasks for EEG Acquisition

Contrary to the ECG, EEG should be recorded under a predetermined task for authentication. EEG can be recorded in the resting state without any special task, with a mental task, or with an external sensory or event stimulation. The selected task is included as a part of the authentication procedure and affects the performance of the authentication. While the resting and the mental task can be performed without additional equipment other than an EEG recorder, tasks responding to sensory or event stimulations require additional devices to provide appropriate stimulation. EEGs recorded with tasks provide a higher signal-to-noise ratio (SNR) than EEGs in the resting state. EEG should be recorded above the corresponding brain region related to the applied task. For instance, event-related potential (ERP) responding to the visual event stimulation is mostly observed in the occipital lobe of the visual cortex and central cortex.

Resting states are the easiest and most popular task in acquiring EEG signals. In the resting state, subjects are asked to be completely relaxed and sit on a comfortable chair in a quiet environment. EEG can be recorded with "Eyes-closed" and "Eyes-open" conditions. When the EEG is recorded with the "Eyes-closed" condition, the dominant peak in the alpha frequency band is recorded and it increases the uniqueness of individuals in their EEG. Recording from the central region and parietal region is most effective for "Eyes-closed" and "Eyes-open" conditions respectively. As the resting states are easily affected by environmental noise and artifacts, the environment should be quiet. While the recording EEG in the resting state is simple without using an additional device, the subject should have no mental activity or preoccupation not to deflect the recorded EEG.

When the EEG is recorded with a mental task, the individual is asked to imagine some physical body movements or do mental activities. Imaginary movement of hand or foot provides the unique EEG rhythm as it is actively used in brain-computer interface (BCI). Various mental activities are possible to produce a specific EEG pattern, which include silently singing a favorite song, counting numbers in mind, mental multiplication, visualizing the rotation of a geometric figure, composing a letter to a friend, generating words beginning with the randomly selected letter, imagining a person, imagining a gesture or taste of food, and concentrating on a desired thought. These mental tasks can be used as a "Pass-thought" like a password to pass the authentication system. A mental task requires some simple equipment to generate a clue for the individual to initialize the task.

To record an EEG responding to a task with stimulation, external stimulation can be applied visually or acoustically. Visual stimulations are most common and can be applied in various types. The task can be reading words in strange forms like unconnected text, acronyms, illegal strings, and pseudowords. Pictures invoking special feelings or emotions, like images of a favorite food or celebrity, can be presented. Visual stimulation can be applied in the sequence of pictures and a task of remembering the picture contents or counting the special items in the picture can be added. Presenting pictures can be selected in the registration stage like a password for each subject to invoke a subject-specific EEG pattern.

Sensory stimulation protocols used for the brain-computer interface can be used to record EEG involved in stimulation tasks. Event-related potential (ERP) is recorded under the oddball paradigm to develop subject-specific responses to rare and interesting events. Rapid serial visual presentation (RSVP) is another protocol to elicit a P300 ERP appearing after an infrequent stimulus. Steady-state visual evoked potentials (SSVEP) are also used to derive EEG under task-related stimulation. A periodic visual stimulus, in the frequency range of 4–60 Hz, induces enhanced periodic oscillations in the same frequency and in its harmonics over visual areas. The rate of increase in the power spectrum and composition of harmonics also varies in different individuals. SSVEPs are useful with their simplicity, excellent SNR, and robustness against artifacts. Visual stimulation requires additional stimulating devices for authentication. However, it provides consistent EEG waveform and deterministic temporal waveform patterns which are not possible in EEG from the mental tasks or from the resting state.

Acoustic stimulations can be used similarly to invoke ERP. Subjects are asked to listen to a piece of music or a special tone. Different genres of music and sound tones played for individuals induce different emotions and interests. To induce a more prominent "Pass-thought" by the acoustic stimulation, preferred sounds or favorite songs can be registered for each individual.

5.3 Structure of Authentication System Using EEG

The structure of the authentication system using EEG also follows the general structure of authentication systems like the authentication system using ECG (Fig. 11.16). During the enrollment phase, each individual enrolls their EEG records under the specific task. Enrolled EEGs follow the procedure of preprocessing and feature selection. The values of selected features of each individual are stored in the database. In the authentication phase for the login of the subject, EEG is recorded under the same task, and recorded EEG follows the same procedure of preprocessing and feature selection as in the enrollment phase. Derived feature values are compared with the stored feature values of the subject in the database. Matching the feature sets for authentication is performed by calculating the closeness of two feature value sets using their correlation or some distance measures in the feature space. When the feature values are close enough over the threshold, the authentication is issued. Otherwise, the trial for authentication is rejected.

The unique part of the authentication system using the EEG is the task to produce a prominent EEG response. EEG can be recorded in a resting state without any tasks or under the same task of mental thought or sensory stimulation. To record the prominent EEG and increase the performance of the authentication, specific tasks using an image, sound, or sensory stimulation can be pre-enrolled for each individual.

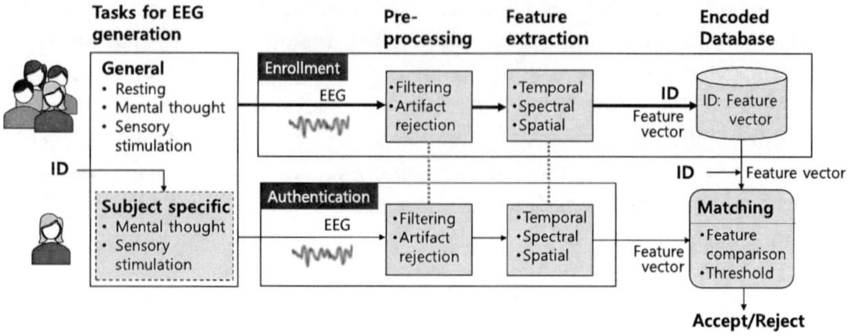

Fig. 11.16 Structure of authentication using EEG

Invoked EEG in response to the pre-enrolled task can be used as a "Pass-thought." Then, this "Pass-thought" can be changed by changing the enrolled task as we do with the passwords. This is a special characteristic that is not possible with other biometric items used for authentication.

5.4 Features Selection for Authentication Using EEG

Methods used for selecting features from the recorded EEG for authentication are quite similar to the methods used for BCI. BCI using EEG and authentication using EEG are the same in using the same signal and the difference is only the results of classification. While EEG is used for the classification of subjects' intention in BCI, EEG is used for the authentication that the recorded EEG is from the genuine subject or from an intruder. Features from the EEG can be selected in temporal, spectral, and spatial domains.

In the temporal domain, fiducial points can be detected in event-related potentials having deterministic wave shapes. Amplitudes and time durations of each wave component can be selected as features in relation to the event stimulation. The entire waveform of EEG, like evoked potentials to the sensory stimulation, can be used as a template, and the recorded EEG is directly compared without using intermediated features. Statistical features, like mean, variance, and higher moments can be derived in the temporal domain as features for EEGs not synchronized to the stimulation. The statistical nature of EEG can be modeled using the autoregressive model, which represents the current sample values as a combination of previous sample values.

$$x_n = -\sum_{i=1}^{K} a_i x_{n-i} + w_n \tag{11.9}$$

where x_n is n-th sample value of EEG, a_i is i-th coefficient of the AR mode, and w_n is n-th sample value of residual noise. AR coefficients are used as features representing the characteristic of recorded EEG.

EEG features are most commonly derived from the power spectrum. Power spectrum can be derived from the recorded EEG in several methods including fast Fourier transform (FFT) and periodogram. Absolute and relative power in each spectral band of alpha (8–13 Hz), beta (13–30 Hz), gamma (higher than 30 Hz), delta (lower than 4 Hz), and theta (4–8 Hz) derived from the spectrum are the most common and effective features of EEG as in other applications of EEG. To derive features based on the time-varying nonstationary characteristic of EEG, features are extracted in the time-frequency domain using the transform like a short-time Fourier transform (STFT) and wavelet transform (WT). These transforms convert 1-D time-domain EEG into a 2-D spectral image in the time-frequency domain. This representation is very suitable as an input for a deep neural network showing high performance in many image classification applications.

In contrast to the ECG signal, EEG can be recorded in multiple channels with each channel representing the activities of a different region of the brain. The spatial distribution of electrical potentials and relationships among the channels can be selected as spatial features. Spatial coherence is one method to develop features representing spatial characteristics. In many cases, temporal, spectral, and spatial features are combined and fused together to improve the performance. Features can be fused directly or via the feature reduction methods like the principal component analysis (PCA) and canonical correlation analysis (CCA).

5.5 Matching the Features for Authentication Using EEG

Like the feature selection procedure, similar classification methods used for BCI can be applied for matching the EEG features derived for authentication.

Most simply, the closeness of the two feature vectors can be quantified by a closeness score like the cross-correlation or the distance between two vectors in the feature vector domain. If the evaluated closeness is higher than the predefined threshold the authentication is issued or the trial is rejected. By setting the threshold to determine the authentication, error characteristics are changing. If the threshold is increased to a higher closeness level, the false rejection rate (FRR) is increasing while the false alarm rate (FAR) is decreasing. And, if the threshold is decreased to a lower closeness level, the false alarm rate is increasing (FAR) while the false rejection rate (FRR) is decreasing. Distance between two feature vectors can be evaluated with Euclidean distance, cosine distance, Manhattan distance, Mahalanobis distance, and dynamic time wrapping.

The matching of the feature vectors can be determined by using the classifier methods. Various classification methods can be applied to the derived features. A support vector machine (SVM) is the classifier most effectively used for EEG authentication. Linear discriminant analysis (LDA), Hidden Markov model (HMM), artificial neural network (ANN), k-nearest neighbors (kNN), and random forest (RF) are the other classification methods that are used in the EEG authentication.

With the rapid development of deep learning technology, it is also actively applied in EEG authentication. While the deep learning technology can be used just for classification using the derived feature values, it can directly be applied to raw EEG data as it has the capability to extract features within the network structure during the training phase. A convolutional neural network (CNN) is the most commonly accepted structure for authentication using the EEG and it provides increased performance compared to the previous results using other analytic types of classifiers. In most cases, the input to the network is applied in 2D or 3D format, representing time, frequency, and spatial for each dimension. While CNN provides high performance with the static images, it does not provide the results reflecting the temporal relations among the series of time samples. Recursive neural networks (RNNs) models, like long short-term memory (LSTM) and gated recurrent unit (GRU), are used to extract time attributes in recorded EEG sequences.

5.6 Performance of Authentication Based on EEG

Various factors affect the performance of the authentication using the EEG (Table 11.3). The number of channels affects the performance results. As it increases, performance is getting better. However, too many electrodes to collect the EEG prevent the deployment of authentication using EEG. A Portable Emotive EEG recording unit is widely used with its convenience in collecting 14 channels of EEG. Recording time also affects the performance and the collectability. In most cases, recording time is longer than one minute and it is also dependent on the type of tasks involved to invoke EEG. The number of subjects who participated in the experiments also affects the resulting performance. With a small number of subjects, very high performance can result. However, it is difficult to generalize the results to an increased number of subjects. Tasks to invoke EEG recording are also important. Generally, subject-specific tasks invoke prominent EEGs specific to each subject and result in increased performance. Various combinations of features and classification methods for matching the features can be used to derive a high level of performance. Deep learning technology introduced recently combines feature selection and matching procedures in a single deep network and produces generally increased performance with the same data set.

Table 11.3 Performance of authentication using EEG

References	EER	No. of subjects	Task	Features	Matching	No. of channels
Miyake et al. (2020)	8.10%	20	Invisible visual stimuli	Wavelet transform	Ensemble of 30 NN	4
Sun et al. (2019)	5.59%	109	Motor imagery	Raw EEG	CNN	4
Riera et al. (2008)	3.40%	51	Eyes-closed resting	AR, PSD, MI, Coherence	ANN	2
Sun et al. (2019)	3.25%	109	Motor imagery	Raw EEG	LSTM	32
Chen et al. (2020)	2.94%	157	RSVP	Raw EEG	GSLT-CNN	28
Yamashita et al. (2018)	2.00%	31	Image stimulation	Cross correlation	SVM	14
Thomas and Vinod (2018)	1.96%	109	Resting	Gamma band PSD	Cross correlation	19
Sun et al. (2019)	0.41%	109	Motor imagery	Raw EEG	1C-LSTM	64
Gopal and Shukla (2021)	0.14%	26	Video, type	PSD, FD, DFA	ANN	14
Mukai and Nikanishi (2020)	0.00%	10	Ultrasound stimulation	LS, SE, MLE, PE, FD	SVM	14

1C-LSTM 1-D Convolutional Long Short-term Memory, *ANN* Artificial Neural Network, *CNN* Convolutional Neural Network, *DFA* Detrended Fluctuation Analysis, *FD* Fractal Dimension, *GSLT-CNN* Convolutional Neural Network with Global Spatial and Local Temporal filter, *LS* Logarithmic Spectrum, *MLE* Maximum Lyapunov Exponent, *PE* Permutation Entropy, *RSVP* Rapid Serial Visual Response, *SE* Sample Entropy, *SVM* Support Vector Machine

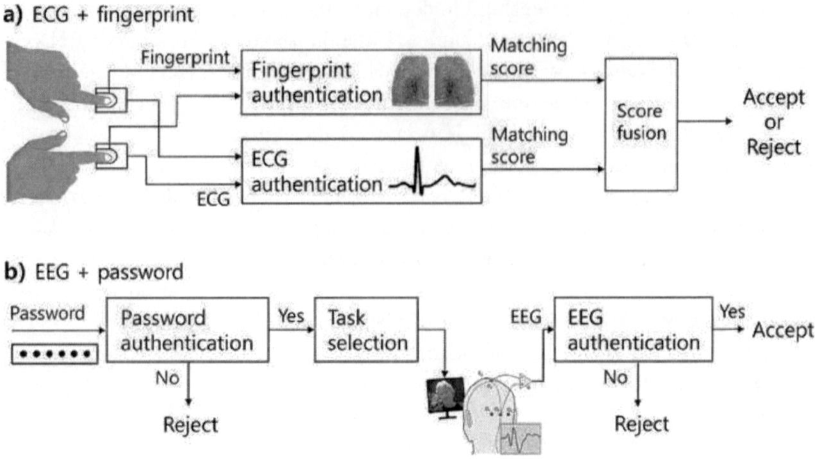

Fig. 11.17 Multifactor authentication using bioelectrical signals. (**a**) Multifactor authentication using ECG and fingerprint by score fusion. (**b**) Multifactor authentication using EEG and password by serial fusion

6 Multi-factor Authentication Using Biological Signals

Biological signals have unique characteristics as authentication items. They are made only from the live subjects and they are almost impossible to forge or imitate. These characteristics are not found in other types of authentication items based on knowledge, possession, and anatomy of the body. When the biosignals are combined as an additional factor, it can greatly enforce the circumvention property of authentication which is not easily evaluated in performance.

ECG can be combined with the fingerprint as an additional authentication factor (Fig. 11.17a). ECG complements the vulnerability of fingerprints, leaving traces behind. With fingerprint only, an intruder can get the traces left behind and can use them to pass the authentication system. However, an intruder cannot easily get an ECG record without prior information about the subject. ECG makes the authentication possible while only the subject is alive and it provides a new dimension of security to the authentication based on fingerprint biometrics. Adding the ECG to the fingerprint also increased the performance by decreasing EER from 1.18% of the fingerprint only to 0.46% (Arteaga-Falconi et al., 2018).

ECG can be combined with other biometric types of authentication items like passwords or smartphone. Password is the authentication item based on the knowledge factor and smartphone is based on the possession factor. ECG nicely complements the disadvantage of a password or smartphone as an authentication item. When the ECG recording device is wirelessly connected to the authentication system, two factors can be fused serially. When a subject possessing a smartphone wants to access an application requiring authentication, the user can receive a dynamic password from the server via the smartphone and input the ECG via the ECG recorder wirelessly connected to the smartphone via Bluetooth (Neves et al., 2016).

EEG can be combined with a password (Fig. 11.17b). When two authentication factors, a physiological biometric factor of EEG and a knowledge factor of password, are combined, the authentication system becomes much safer from the attack of the intruders. Password itself is used as one factor for authentication. If the users are failed to submit a correct password, they are easily rejected as in a one-factor authentication system. When the password is correct, it can be used as a code to allocate a specific task for each subject. For example, a password selects the image presentation task to invoke an EEG and the image of the subject's mother is used as a stimulating image. Even though the password is used by the intruders, they cannot produce the same EEG pattern as a genuine subject by the same presented image. By adding the EEG as an additional factor, the security of the authentication system is increased by preventing the cases of password cracking.

References

Abozaid, A., Haggag, A., Kasban, H., & Eltokhy, M. (2019). Multimodal biometric scheme for human authentication technique based on voice and face recognition fusion. *Multimedia Tools and Applications, 78*, 16345–16361. https://doi.org/10.1007/s11042-018-7012-3

Arteaga-Falconi, J. S., Osman, H. A., & Saddik, A. E. (2018). ECG and fingerprint bimodal authentication. *Sustainable Cities and Society, 40*, 274–283. https://doi.org/10.1016/j.scs.2017.12.023

Benaliouche, H., & Touahria, M. (2014). Comparative study of multimodal biometric recognition by fusion of iris and fingerprint. *The Scientific World Journal, 2014*, 829369. https://doi.org/10.1155/2014/829369

Bhattacharyya, D., Ranjan, R., Alisherov, F., & Choi, M. (2009). Biometric authentication: A review. *International J. u- and e- Service, Science and Technology, 2*(3), 13–28.

Bidgoly, A. J., Bidgoly, H. J., & Arezoumand, Z. (2020). A survey on methods and challenges in EEG based authentication. *Computers & Security, 93*, 101788. https://doi.org/10.1016/j.cose.2020.101788

Chen, J. X., Mao, Z. J., & Yao, W. X. (2020). EEG-based biometric identification with convolutional neural network. *Multimedia Tools and Applications, 79*, 10655–10175. https://doi.org/10.1007/s11042-019-7258-4

Choi, H. S., Lee, B., & Yoon, S. (2016). Biometric authentication using noisy electrocardiograms acquired by mobile sensors. *IEEE Access, 4*, 1266–1273.

Chu, Y., Shen, H., & Huang, K. (2019). ECG authentication method based on parallel multi-scale one-dimensional residual network with center and margin loss. *IEEE Access, 7*, 51598–51607.

Chun, S. Y., Kang, J. H., Kim, H., Lee, C. H., Oakley, I., & Kim, S. P. (2016). ECG based user authentication for wearable devices using short time Fourier transform. *International Conference on Telecommunications and Signal Processing, 2016*, 656–659. https://doi.org/10.1109/TSP.2016.7760964

Fatima, J., Syed, A. M., & Akra, A. M. (2013). A secure personal identification system based on human retina. *IEEE Symposium on Industrial Electronics and Applications (ISIEA), 2013*, 90–95.

Gopal, S. R. K., & Shukla, D. (2021). Concealable biometric-based continuous user authentication system: An EEG induced deep learning model. *IEEE International Joint Conference on Biometrics (IJCB), 2021*, 1–8. https://doi.org/10.1109/IJCB52358.2021.9484345

Hammad, M., Liu, Y., & Wang, K. (2019). Multimodal biometric authentication systems using convolution neural network based on different level fusion of ECG and fingerprint. *IEEE Access, 7*, 26527–26542.

Hassan, N., Ramli, D. A., & Suandi, S. A. (2014). Fusion of face and fingerprint for robust personal verification system. *International J. of Machine Learning and Computing, 4*(4), 372–376.

Ingale, M., Cordeiro, R., Thentu, S., Park, Y., & Karimian, N. (2020). ECG biometric authentication: A comparative analysis. *IEEE Access, 8*, 117853–117166. https://doi.org/10.1109/ACCESS.2020.3004464

Kim, J. H., Sung, D. S., Koh, M. J., Kim, J., & Park, K. S. (2019). Electrocardiogram authentication method robust to dynamic morphological conditions. *IET Biometrics, 8*, 401–410.

Kim, H., & Chun, S. Y. (2019). Cancelable ECG biometrics using compressive sensing-generalized likelihood ratio test. *IEEE Access, 7*, 9232–9242.

Krish, R. P., Fierrez, J., Galbally, J., & Martinez-Diaz, M. (2013). *Dynamic signature verification on smart phones* (pp. 213–222). International Workshops of PAAMS.

Labati, R. D., Muñoz, E., Piuri, V., Sassi, R., & Scotti, F. (2019). Deep-ECG: Convolutional neural networks for ECG biometric recognition. *Pattern Recognition Letters, 126*, 78–85.

Li, R., Yang, G., Wang, K., Huang, Y., Yuan, F., & Yin, Y. (2020, January). Robust ECG biometrics using GNMF and sparse representation. *Pattern Recognition Letters, 129*, 70–76.

Lu, X., Zhang, S., Hui, P., & Lio, P. (2020). Continuous authentication by free-text keystroke based on CNN and RNN. *Computers & Security, 96*, 101861. https://doi.org/10.1016/j.cose.2020.101861

Miyake, T., Kinjo, N., & Nakanishi, I. (2020). Wavelet transform and machine learning-based biometric authentication using EEG evoked by invisible visual stimuli. *IEEE Region 10 Conference, 2020*, 573–578. https://doi.org/10.1109/TENCON50793.2020.9293772

Mohammed, H. H., Baker, S. A., & Nori, A. S. (2021). Biometric identity authentication system using hand geometry measurements. *Journal of Physics: Conference Series, 1804*, 012144. https://doi.org/10.1088/1742-6596/1804/1/012144

Mukai, K., & Nakanishi, I. (2020). Introduction of fractal dimension feature and reduction of calculation amount in person authentication using evoked EEG by ultrasound. *IEEE Region 10 Conference, 2020*, 567–572. https://doi.org/10.1109/TENCON50793.2020.9293921

Neves, P., Nunes, L., & Lourenço, A. (2016). Multi-factor authentication for improved efficiency in ECG based login. *International Conference on Physiological Computing Systems, 2016*, 67–74. https://doi.org/10.5220/0005936500670074

Ranjan, A. (2019). Permanence of ECG biometric: Experiments using convolutional neural networks. *Proc. Int. Conf. Biometrics (ICB), 2019*, 1–6.

Riera, A., Soria-Frisch, A., Caparrini, M., Grau, C., & Ruffini, G. (2008). Unobtrusive biometric system based on electroencephalogram analysis. *EURASIP Journal on Advances in Signal Processing*, 143728. https://doi.org/10.1155/2008/143728

Rui, Z., & Yan, Z. (2019). A survey on biometric authentication: Toward secure and privacy-preserving identification. *IEEE Access, 7*, 5994–6009. https://doi.org/10.1109/ACCESS.2018.2889996

Sahu, S. K., Dala, A., & Jena, S. K. (2014). Varying password based scheme for user authentication. *Advanced Computing, Networking and Informatics, 2*, 361–368. https://doi.org/10.1007/978-3-319-07350-7_40

Savitha, A. P., & Ramegowda. (2017). A comparative study of palm vein feature extraction and classification. *Materials Today: Proceedings, 4*(11), 11882–11887.

Sufi, F., Khalil, I., & Habib, I. (2010). Polynomial distance measurement for ECG based biometric authentication. *Security and Communication Networks, 3*, 303–309. https://doi.org/10.1002/sec.76

Sun, Y., Lo, F. P. W., & Lo, B. (2019). EEG-based user identification system using 1D-convolutional long short-term memory neural networks. *Expert Systems with Applications, 125*, 259–267. https://doi.org/10.1016/j.eswa.2019.01.080

Thomas, K. P., & Vinod, A. P. (2018). EEG-based biometric authentication using gamma band power during rest state. *Circuits Systems Signal Processing, 37*, 277–289. https://doi.org/10.1007/s00034-017-0551-4

Uwaechia, A. N., & Ramli, D. A. (2021). A comprehensive survey on ECG signals as new biometric modality for human authentication: Recent advances and future challenges. *IEEE Access, 9*, 97760–97802. https://doi.org/10.1109/ACCESS.2021.3095248

Vera-Rodriguez, R., Gabriel-Sanz, S., Fierrez, J., Tome, P., & Ortega-Garcia, J. (2013). Analysis of gait recognition on constrained scenarios with limited data information. *International Workshops of PAAMS*, 231–239.

Wahabi, S., Pouryayevali, S., Hari, S., & Hatzinakos, D. (2014). On evaluating ECG biometric systems: Session-dependence and body posture. *IEEE Ransactions on Information Forensics Security, 9*(11), 2002–2013.

Wang, K., Yang, G., Huang, Y., & Yin, Y. (2020). Multi-scale differential feature for ECG biometrics with collective matrix factorization. *Pattern Recognition, 102*, 107211. https://doi.org/10.1016/j.patcog.2020.107211

Weaver, A. C. (2006). Biometric authentication. *Computer, 39*(2), 96–97. https://doi.org/10.1109/MC.2006.47

Yamashita, M., Nakazawa, M., & Nishikawa, Y. (2018). The proposal and it's evaluation of biometric authentication method by EEG analysis using image stimulation. *11th International Conference on Mobile Computing and Ubiquitous Network (ICMU), 2018*, 1–4. https://doi.org/10.23919/ICMU.2018.8653605

Part III

Electricity Toward Human Body

Electrical Heart Stimulations

12

The heart is the main organ that pumps and supplies blood to all the body parts through the blood vessels of the circulatory system. The pumped blood carries oxygen and nutrients to the body through the arteries and carries carbon dioxide and metabolic waste from the body through the veins. To supply blood up to the end parts of the body, blood should be pumped with enough pressure and volume. And, the amount of blood needs to be carefully adjusted by the demand of the body parts upon the physical exercises and emotional variations.

Pumping the blood adequately upon the demand of the body is performed by the muscular function of the heart. This muscular function of the heart is precisely controlled by the nerve fibers innervated into the muscle layers of the heart and the electrical conduction between myocardial cells. The electrical signal originating from the sinoatrial (SA) node is propagated to every corner of the cardiac muscles with programmed delay and stimulates the muscles to work cooperatively for efficient pumping of the blood. While the electrical conduction of the heart functions normally, the rate and strength of the blood pumping are controlled by the heart rhythm which is reflecting the demand of the body via the autonomic nervous system.

However, in many heart disorders, the heart does not work properly. There are many reasons that make the function of the heart deviate from the normal ones. Abnormal function of the heart can be caused by damaged cardiac muscles, insufficient blood supply through the coronary arteries, heart valve diseases, and abnormal electrical control of the heart muscles. Each type of heart disorder is treated with an appropriate method of using the medicine, surgical procedures of valve substitution, or coronary artery catheterization.

Stimulation of the heart is another option for the treatment of the heart when the electrical control of the heart is disturbed. As the heart is electrically controlled, there are ways that electrical intervention recovers or replaces the normal electrical control of the heart. There are several specific disorders for which the electrical stimulation of the heart is the only or the optimal treatment solution. These methods are also very effective to treat the disorder and became the essential methods of treating the applicable heart disorders.

Electrical stimulation of the heart can be applied in two different concepts. One approach is to replace some parts of the normal conduction system of electrical signals. This is the approach taken by artificial cardiac pacemakers. An artificial cardiac pacemaker generates electrical signals and stimulates the cardiac muscles or nerves instead to recover the normal sinus rhythm. The other approach is to recover the disrupted electrical controls of the heart back to the normal state. This is the approach taken by the defibrillator. A defibrillator supplies a large amount of stimulating current to depolarize most of the muscle cells with the anticipation that the normal rhythm returns with the termination of the stimulating current.

1 Artificial Cardiac Pacemaker

Artificial cardiac pacemakers are electronic devices used primarily to replace the part of cardiac rhythms in patients whose intrinsic heart rate is inadequate to sustain a normal life. There has been astonishing progress in the technology for artificial cardiac pacemakers during the last 60 years and it became the pioneering achievement in recovering the body function through electrical stimulations as well as in the biomedical engineering industry. An artificial cardiac pacemaker is the first active implantable device in the human body and the invention of the pacemaker gave a significant impact on the quality of life and longevity. Physicians regard the use of an artificial cardiac pacemaker as a preferable treatment option, and over one million cardiac pacemakers are implanted every year worldwide (Bhatia & El-Chami, 2018) (Fig. 12.1).

When the heart rate of subjects is too slow, they may need a pacemaker. That is an artificial cardiac pacemaker mainly used to treat symptomatic bradycardia. Normally hearts are

K. S. Park, *Humans and Electricity*, https://doi.org/10.1007/978-3-031-20784-6_12

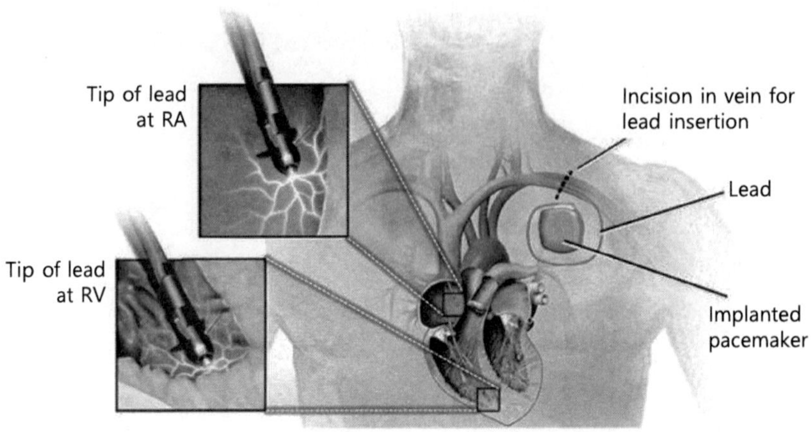

Fig. 12.1 Artificial cardiac pacemaker. *RA* right atrium, *RV* right ventricle. (Blaus & Blausen.com staff, 2014, modified for B/W conversion)

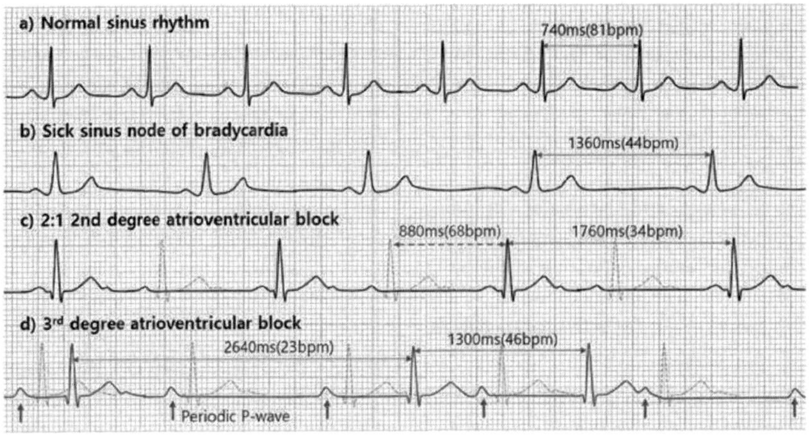

Fig. 12.2 Indications for pacemaker implantation in reference to the normal sinus rhythm. (**a**) Normal sinus rhythm of 60–100 beats/min. (**b**) Bradyarrhythmia by the sick sinus node with a heart rate of 44 beats/min, (**c**) 2:1 2nd degree atrioventricular block. By alternating skip of ventricular contraction, RR interval is increased from 880 to 1760 ms, (**d**) 3rd degree of atrioventricular block. QRS-waves are generated irregularly and completely dissociated from the periodic P-wave. Dashed waveforms represent the normal QRS waveform if the conduction system functioned normally. Bpm: beats per minute

beating at the rate of 60~100 beats per minute in a resting state. Bradycardia is the heart rhythm beating abnormally slower than 60 beats/min. Symptomatic bradycardia can be life-threatening as the slow heart cannot supply enough oxygen-rich blood throughout the body and it is responsible for symptoms such as syncope or near syncope, dizziness, confusion, fatigue, exercise intolerance, and heart failure.

Figure 12.2 shows the ECG recordings for the indications for pacemaker implantation in reference to the normal sinus rhythm. Symptomatic bradycardia can be caused by two types, sinus node dysfunction, and heart block. The sick sinus node does not function properly and fires too slowly lower than 60 beats/min. The sick sinus node is common in patients in their 70s or 80s and is considered a result of sinus node aging. Also, bradycardia can occur with atrioventricular node dysfunction when there is a block in the conduction of electrical signals from the atria to the ventricles. In this type of bradycardia, only part of the heartbeats are properly conducted from the atria to the ventricles and the remaining part of the heartbeats are delayed or blocked disturbing the timely contraction of the ventricles. Second- or third-degree heart block is associated with symptomatic bradycardia.

1.1 Brief History of Artificial Cardiac Pacing

The possibility of recovering the heart beating by electrical stimulation was first confirmed in 1899 by John MacWilliam (1857–1937), a physiologist at the University of Aberdeen in England. He reported the experimental results that the electrical impulses of 60–70/min caused the ventricular contraction with the same rate in the asystolic human heart.

Following the supporting experiments and studies on the use of electrical stimulations on reviving the heart rhythms,

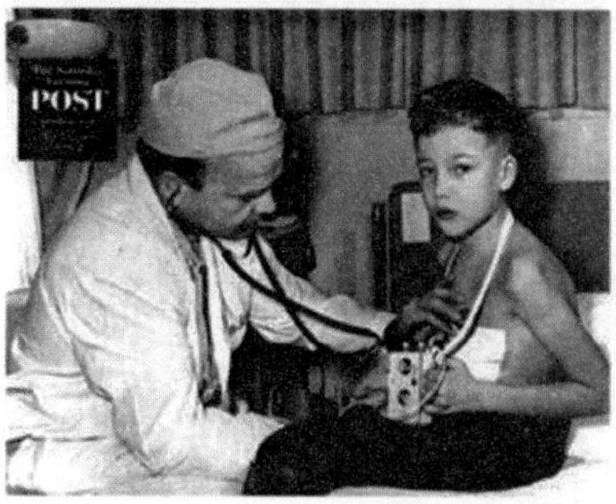

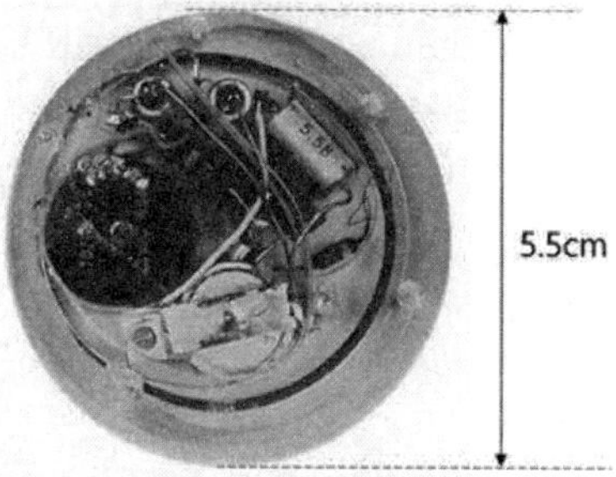

Fig. 12.3 The first artificial pacemakers. (**a**) Surgeon Lillehei and a child being paced with the first wearable artificial pacemaker. (**b**) First implanted artificial pacemaker. (Aquilina, 2006)

the first instrument of cardiac stimulation was developed by Albert Hyman (1893–1972), an American cardiologist, in 1932. When he became interested in reviving the "stopped heart" by means of "intracardial therapy," initially he started this therapy with the intracardiac injection of stimulant drugs like epinephrine. However, he soon realized that it was not really the drug that restarted the heart but the needle that sets up an action current of injury as it punctured the cardiac wall. Based on his experience, he developed the first artificial pacemaker instrument. This instrument delivered electrical impulses directed into the patient's right atrium through a bipolar needle electrode introduced via an intercostal space. The instrument was powered by the spring-wound hand-cranked motor. Hyman noted that his method was successful in most of the deaths caused by the shock. However, his instrument was adversely accepted by society and the community. They criticized it as an infernal machine interfering with the will of God and he could not find the manufacturer for the instrument. He named the instrument an "artificial pacemaker" and this name is used up to this day.

Paul Zoll (1911–1999), an American cardiologist, is given credit for ushering in the modern era of clinical cardiac pacing. In 1952, he first developed an external electrical pacemaker for clinical use and successfully performed cardiac pacing for a duration of 20 min. It was comprised of an electrocardiograph to monitor the cardiac rhythm and an electric pulse generator to pace the heart. It delivered periodic electric impulses of 2 ms pulse width and 50–150 V in alternating current (AC) pulse amplitude. The electric current was delivered transcutaneously with a pair of 3 cm^2 metal electrodes strapped to the patient's chest directly over the heart. The instrument was the desktop size and only portable on a wheeled cart. However, the electrodes irritated the skin, and the patients used to complain of the pain from the repeated electric pulses. The instrument unit was still bulky and heavy and it only can move from one AC wall socket to another AC wall socket as it is powered by a household AC line (Aquilina, 2006).

The first external wearable pacemaker was developed in 1958 by Earl Bakken (1924–2018), an American engineer at Medtronic, and Walton Lillehei (1918–1999), an American surgeon. Their cooperation started from the tragedy that happened in 1957 with a municipal power failure lasting 3 hours. It caused a death of a baby connected to a pacemaker powered by the AC line. The hospital had an emergency power system only in surgical suites and recovery areas but not in its patient rooms. Upon the need for the pacemaker to work freely from the AC power line, Lillehei asked for the cooperation of Bakken. They initially tried to convert battery DC power into AC power to supply AC-powered pacemakers. However, they soon recognize that power conversion is highly inefficient and a 10 V of DC is sufficient to stimulate the heart. By using the new technology of transistors, the developed pacemaker can be miniaturized into a small aluminum box of size 10 × 10 × 4 cm. It weighs about 300 g and can be worn by a patient with a necklace (Fig. 12.3a). It had controls to adjust the pace of heart rate and output voltage and was connected to electrode leads. Electrode leads passed through the skin of the patient and terminated in electrodes attached to the surface of the myocardium of the heart.

The first implantation of a pacemaker was accomplished in the same year 1958 in Sweden. This was also possible with the advent of the transistor which replaced the vacuum tubes and made the electric circuits small enough to be implanted within the body. Ake Senning (1915–2000), a Swedish surgeon, and Rune Elmqvist (1906–1996), a Swedish engineer, were able to incorporate the transistor technology into a pacemaker and built a successive story of pacemaker implantation. Electrodes are attached to the myocardium of the heart by thoracotomy and the stimulating unit was implanted under the skin in the epigastrium. The first patient was Arne Larsson (1915–2001), 43 years old man of the same age as surgeon Senning. At the time of pacemaker implantation, the viral infection had disrupted the normal electrical conduction of cardiac rhythm and Larsson's heart could no longer contract synchronously. His heart beats as slowly as 28 beats/min causing decreased blood flow to the brain and producing potentially fatal fainting spells, known as Stokes-Adams attacks. They developed an implantable pacemaker of small size, 5.5 cm in diameter and 1.6 cm in thickness, which is the size of a shoe polish can (Fig. 12.3b). The first implanted device functioned only 3 hours and they had to replace it with a second device which lasted 2 days. Since then, Larsson replaced his pacemakers 26 times with 11 different models and lived up to age 86 longer than surgeon Senning and died of melanoma which is totally unrelated to cardiac problems. The earliest model of the 26 pacemakers used a nickel-cadmium rechargeable battery and once a month pacemaker should be recharged by induction for an entire night.

The first pacemaker using a lithium-iodine battery has implanted in 1972. Nickel-cadmium rechargeable battery has soon been replaced by mercury-zinc batteries. This type of battery has been used extensively up to the mid-70s. However, mercury-zinc batteries can last the pacemakers only up to 3 years and tend to fail very quickly. And, even though they are coated with epoxy resin, they are invaded by the body fluids eventually. Based on the associated disadvantages of the mercury-zinc batteries, lithium-iodine batteries were developed for the pacemaker use by Wilson Greatbatch (1919–2011). They have a longer life up to 10 years and have more predictable end-stage characteristics. And, they can be sealed in hermetic packages using titanium as an encasing metal so that no waste product can invade the device and have smaller weights. Using lithium-iodide batteries became the standard for pacemaker designs since then (Ward et al., 2013).

Based on the many complications related to pacing leads, the leadless pacemaker was introduced. Even though it was conceptualized in the 1970s and experimented with animals, it was first clinically implanted in humans in 2014 (Reddy et al., 2014) and the first model was FDA approved in 2016. They successfully implanted 32 leadless pacemakers out of 33 trials. In addition to the great advantage of leadless operation, it can be delivered via a femoral vein transcatheter approach without chest incision. Advances in battery technologies, communication capability, and catheter-based delivery systems made leadless pacemakers a reality (Bhatia & El-Chami, 2018).

1.2 Stimulations for Cardiac Pacing

1.2.1 Stimulations for Temporary Cardiac Pacing

When the heart is in an asystole state or too slow to supply blood up to the farthest part of the body, stimulation methods for cardiac pacing are considered a preferable and urgent option to recover or replace the heart rhythm. Stimulations for cardiac pacing can be applied temporarily or permanently. Temporary stimulations are applied up to the time when the normal rhythm is recovered or permanent stimulation is available. There are various stimulating methods for cardiac pacing (Fig. 12.4).

Percussive pacing is the nonelectrical stimulating method for cardiac pacing. It applies mechanical energy by transthoracic force to induce a ventricular beat and a following myocardial contraction. Force is applied with a clenched fist usually on the left lower edge of the sternum over the right ventricle, by striking from a distance of 20–30 cm. Percussion must be done to raise the ventricular pressure by 10–15 mmHg to induce electrical activity. The mechanism of the percussive pacing is not clearly known. However, it is a method instantly available and can be easily performed as an emergency cardiac pacing in hemodynamically unstable bradycardias. It is only temporarily used before the electrical stimulation methods for cardiac pacing are available. If no cardiac stimulation is achieved in pulseless subjects, cardiac pulmonary resuscitation (CPR) must be started immediately.

Transcutaneous pacing is the method inducing the cardiac beats by externally applied pulse currents. Pacing currents are applied through the pacing pads placed in the anterior and posterior positions of the chest. After selecting the pacing rate, the pacing current is slowly increased until the electrical

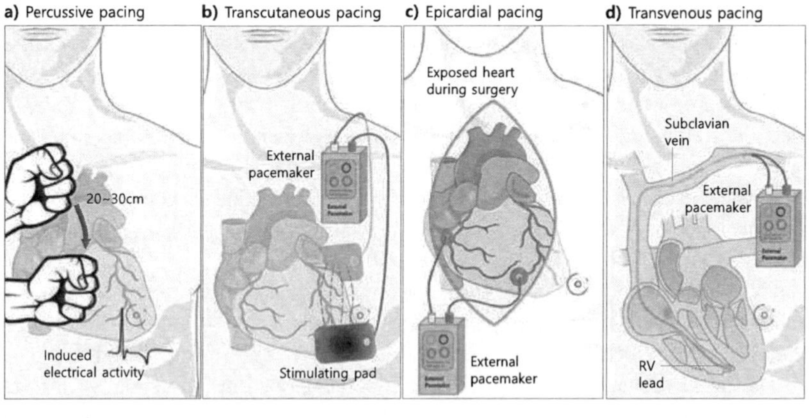

Fig. 12.4 Stimulations for temporary cardiac pacing. (**a**) Percussive pacing, (**b**) transcutaneous pacing, (**c**) epicardial pacing, (**d**) transvenous pacing

QRS beats are captured with corresponding hemodynamic pulses. It is a noninvasive method that can be applied without surgery or an incision. This method is commonly applied when bradycardia is indicated with signs and symptoms of shock that requires emergency treatment. This is the cardiac pacing method adopted in the earlier stage of the artificial cardiac pacemakers by Zoll. However, with the relatively large amount of current, skin burn can result at the sites of pad contact when it is used for an extended period of time. It is a temporary and emergency procedure as a bridge until transvenous pacing or other stable therapies are prepared.

Epicardial pacing is the stimulating method to induce cardiac pacing from the outer side of the myocardial wall. It is often used during open-chest surgery, when there occurs sudden complete heart block (CHB), to maintain satisfactory cardiac output until a transvenous electrode has been inserted. For small infants, it is preferred over endocardial stimulation via a vein, which partially blocks the blood flow through the small lumen of the infants. And, the use of epicardial leads have the advantage that it avoids the risks associated with the use of endocardial leads. However, implantation of epicardial leads is invasive because a thoracotomy is required and sometimes these leads also can be problematic.

Transvenous cardiac pacing is the endocardial pacing method, stimulating the heart from the inside of the myocardial wall. It can be used as a temporary option to treat symptomatic bradycardia especially when it is not responding to transcutaneous pacing or to drug therapy using atropine or epinephrine. Transvenous pacing is achieved by inserting a pacing lead through a vein, usually a subclavian vein, into the right atrium, right ventricle, or both and by connecting the leads to an external pacemaker outside the body. Pacing lead is advanced through the vein under fluoroscopic and ECG monitoring. Placement of the pacing tip of the lead is confirmed by the X-ray after the insertion procedure. Transvenous pacing of the heart is not popular as other temporary pacing methods since it involves a similar level of bleeding risk as a permanent placing of the implanted pacemaker.

1.2.2 Stimulations for Permanent Cardiac Pacing

Permanent cardiac pacing is achieved with the implanted artificial cardiac pacemaker. For permanent pacing, leads for stimulating the myocardial wall are inserted through a vein like a lead in the transvenous cardiac pacing. In addition, the pacemaker unit is totally implanted inside the skin usually under the clavicle. Electrode lead is inserted through an incision point of a suitable vein and passes through a vein under the guidance of fluoroscopy. It is advanced until it is placed on the wall of the right atrium or through the tricuspid valve to the wall of the right ventricle. After the placement of the electrode at the tip of the lead is confirmed, the other end of the electrode lead is connected to the pulse generator unit to be implanted. Permanent pacing can be applied in different types depending on the stimulating locations and the number of stimulating channels (Fig. 12.5).

In *single chamber pacing*, only one pacing lead is placed into one chamber of the heart. Depending on the subject's symptom and required pacing type, one lead is placed either in the upper chamber of the right atrium (RA) or in the lower chamber of the right ventricle (RV). Early pacemakers were designed to sense and pace in a single chamber. Single-chamber pacing on RV can prevent ventricular bradyarrhythmia or asystole of any etiology. Single-chamber pacing on RA can be used in patients with isolated sinus node dysfunction and intact atrioventricular conduction.

Dual chamber pacing is used when the separate pacing of two chambers is required. After two electrode leads are inserted through a vein, one is placed in the right atrium (RA) and the other one is placed in the right ventricle (RV). This type of pacing helps the upper and lower chambers work together, contracting and relaxing in a proper synchronized rhythm. It allows efficient blood flow from the atria to the ventricles by contracting heart chambers at proper timing. By combining the sensing and pacing capabilities in both the atria and the ventricles, dual-chamber pacing provides physiologic pacing closer to the intrinsic pumping of the heart. Synchronized pushing of blood from atria to ventricles is considered to prevent the loss of 30% cardiac output.

Biventricular pacing is an advanced pacing method by stimulating the right and left ventricle separately. In biventricular pacing, three electrodes are used. Two electrodes are inserted in the same fashion as in the dual-chamber pacing via the subclavian vein, vena cava, and into RA and RV. The third electrode lead is inserted to stimulate the left ventricle separately and positioned over the epicardial wall of the left ventricle through the coronary sinus vein. The coronary sinus is a collection of veins joined together to form a large vessel. It collects deoxygenated blood from the heart muscles directly to the right atrium. The third electrode is inserted via the subclavian vein, vena cava to RA like the other two electrode leads. However, it advances reversely through the coronary sinus vein from the RA up to the point of the left ventricle is located beneath. Biventricular pacing is an excellent option in heart failure subjects with a left bundle branch block or right bundle branch block, which causes a cardiac desynchrony by the delay between two ventricles. Biventricular pacing is also known as cardiac resynchronization therapy (CRT).

In His bundle pacing (*HBP*), His-Purkinje fiber network is directly stimulated instead of myocardial muscles. Conventional pacing leads are usually anchored on the myocardial

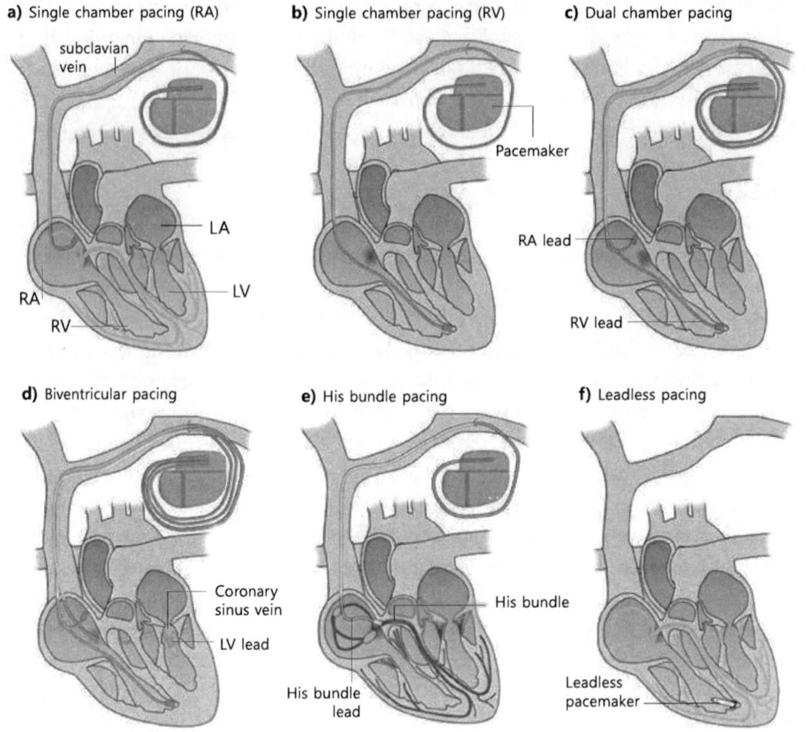

Fig. 12.5 Pacing methods for the permanent artificial pacemaker. (**a**) Single-chamber pacing on the right atrium (RA), (**b**) single-chamber pacing on the right ventricle (RV), (**c**) dual-chamber pacing on the right atrium (RA) and right ventricle (RV), (**d**) biventricular pacing on right atrium (RA), right ventricle (RV), and left ventricle (LV), (**e**) his-bundle pacing, (**f**) leadless pacing

wall and stimulate myocardial muscles directly bypassing the nervous conduction through the fiber networks. It can cause a negative effect on the function of the heart and increase the risk of heart failure and atrial fibrillation, and life expectancy can be shortened with the weakening of the heart muscles. As stimulating the nerve fiber network is closer to the natural way of conducting the stimulating pulses, it can cause a synchronized and effective ventricular contraction and avoid the long-term effects of heart muscle burdens.

Leadless pacing can be delivered with the leadless pacemaker. The leadless pacemaker is a small pacemaker that includes a pulse generator and battery and directly contacts the stimulating site of the endocardial wall of the chamber, usually the right ventricle (RV), without using electrode leads. It has a volume of less than 1 cc with a diameter of less than 6.7 mm and a length of less than 42 mm, and its weight is about 2 g. As the lead failure of the pacemaker is one of the major reasons for preventing the long-term use of an implanted pacemaker, avoiding the lead problem ultimately offers a great theoretical advantage. The leadless pacemaker can be put in place using a steerable, long, and thin tube called a catheter. It is fed into the femoral vein via a small incision in the groin under the guidance of fluoroscopy. After confirming the placement and functioning of the leadless pacemaker within the chamber, the catheter is withdrawn. The implantation procedure takes about 30 min to complete.

1.3 Artificial Pacemaker Device

Implantable artificial pacemakers are composed of mainly three parts, pulse generator, leads, and battery (Fig. 12.6).

1.3.1 Pulse Generator of Artificial Pacemaker

The pulse generator is the electrical part of the artificial pacemaker. It generates electrical output pulses to stimulate the myocardium according to the programmed basic stimulating rate. Usually, the pulse generator includes the sensing circuit, logic circuit, pulse generation circuit, and communication circuit.

The sensing circuit is the first part that receives the recorded signals from the electrodes integrated into the inserted leads. The sensing circuit amplifies the recorded signals for the detection of the occurrence of the intrinsic cardiac activity of the chambers.

Recorded and amplified signals are sent to logic circuits. Within this logic circuit, the pacemaker determines whether the recorded signal in the sensing circuit is physiological that can be used to trigger the following pulse or inhibit the pulse output not to interrupt an ongoing intrinsic cardiac event. Logic circuit should discriminate the recorded signals whether it is from the environmental interference or noise from composed components like leads. According to the results, the logic circuit can force the pacemaker to work in asynchronous pacing to avoid periods of asystole in

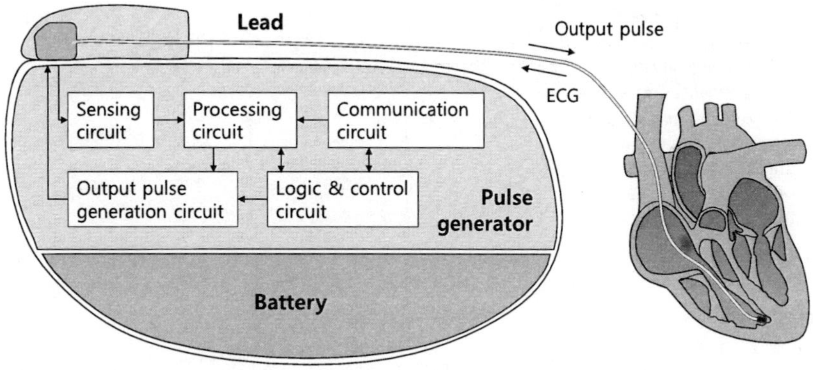

Fig. 12.6 Basic components of implantable pacemaker

pacemaker-dependent patients (DeForge, 2019). It is also connected with the communication circuits to set the programmable parameters for pace making, like basic heart rate and stimulating voltage. If the pulse generation is determined, pulse generating command is sent to the pulse generation circuit.

The sensing circuit is to record the intrinsic cardiac activity and prepares responses of the artificial pacemaker appropriately. To confirm the inherent depolarizations of each chamber of the heart, separate sensing of the P-wave and QRS wave is required. Sensing T-wave is not required as it reflects the repolarization of ventricles. To sense P-wave and QRS-wave accurately, each sensing electrode records the current generated within the time and amplitude window. Current generated only near the electrode is recorded and the current generated in the far-field does not contribute to sensed current. Current generated in inappropriate times like repolarization current is ignored. Atrial sensing is off from the beginning of the QRS-wave until the completion of the QRS complex. This refractory period, called the postventricular refractory period (PVARP), ignores the atrial sensing caused by the backward transmission of current generated in ventricular depolarization. Also, currents induced by external electrical sources, like mobile phones and computers, must be ignored. Sensitivity in the pacemaker is optimized to sense the intrinsic activity reliably by adjusting the temporal and amplitude threshold. When the sensing threshold is set to a higher value to ignore irrelevant activities, there is a chance that it misses the proper intrinsic activity and thus causes inappropriate cardiac pace making. The sensitivity setting is in the range of 0.2–10 mV for atrial pacing and in the range of 2–20 mV for ventricular pacing.

The pulse generation circuit generates an electrical pulse according to the output from the logic circuit with predetermined voltage and time duration under the basic stimulating rate. The basic rate is the lowest heart rate programmed in the pacemaker. If the sensed intrinsic heart rhythm is too slow below this basic rate, an electrical stimulation pulse is triggered from the generator circuit. Usually, the basic rate is set to 60 beats/min, which is the minimal rate of normal heart rhythm, meaning that the pacemaker waits 1000 ms before it delivers a pulse. If the inherent rhythm is sensed during this period, the programmed pulse is ignored and no stimulation is applied. When a pacemaker functions in triggering mode, which generates a ventricular stimulating pulse triggered by the sensed intrinsic atrial activity, the maximum rate is set so that the pacemaker does not stimulate the heart at a too high rate. Figure 12.7 shows the waveform of stimulating output pulse and recorded ECG while stimulating the ventricle triggered by sensed intrinsic P-wave.

The output of the pulse generator is determined by the threshold to stimulate the myocardium to evoke muscular contraction. The intensity of the electrical stimulation is described by the voltage amplitude and the pulse duration. And this shape of the electrical pulse output is determined to minimize the consumption of the battery while ensuring the depolarization of the stimulated cells. Typically, the amplitude of stimulating pulses is less than 2.8 V and the duration of the pulse is less than 0.5 ms. When the electrical pulses are applied through the leads, the flowing current and the consumed energy are determined by the resistance of the lead by the relationship of Ohm's law. Typically, the resistance of the lead tips is in the range of 400~1200 Ω, and the smaller current flows as the resistance of the tips increases. For example, consumed energy by a single-pulse output can be evaluated as follows.

$$\text{Pulse amplitude}: V_p = 2.5\ \text{V} \tag{12.1}$$

$$\text{Pulse duration}: t_p = 0.5\ \text{ms} \tag{12.2}$$

$$\text{Lead resistance}: R_l = 500\ \Omega \tag{12.3}$$

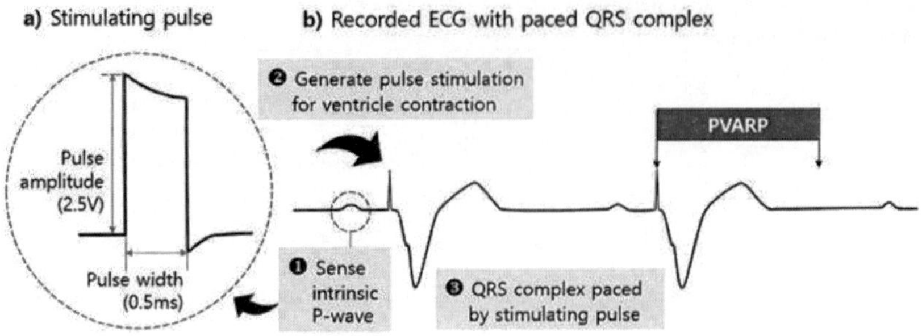

Fig. 12.7 Output pulse and recorded ECG while stimulating the ventricle triggered by sensed intrinsic P-wave. (**a**) Waveshape of stimulating output pulse, (**b**) recorded ECG showing stimulating spike and paced QRS complex. PVARP: postventricular atrial refractory period

$$\text{Stimulating current: } I_s = V_p/R_l = 5 \text{ mA} \quad (12.4)$$

$$\begin{aligned}\text{Energy delivered: } E_p &= V_p \cdot I_s \cdot t_p \\ &= 2.5(\text{V}) \cdot 5(\text{mA}) \cdot 0.5(\text{ms}) \quad (12.5) \\ &= 6.25 \times 10^{-6} \text{ J}\end{aligned}$$

A single pulse of the amplitude of 2.5 V and duration of 0.5 ms delivers a stimulating current of 5 mA over the lead to the myocardium and consumes 6.25 μJ (Fig. 12.7).

The communication circuit provides access to the implanted pacemaker from the outside of the body. Parameters of the pacemakers can be set by the external operator for the basic stimulating rate, voltage for stimulating pulses, and thresholds of sensing the intrinsic activities. Also, the data of the recorded statistics and the battery information can be transmitted from the pacemaker to the outside.

1.3.2 Leads of Artificial Pacemaker

Pacemaker leads are thin wires leading the output pulses from the pulse generator to the myocardium where the stimulation is required. The length of endocardial leads is usually in the range of 54~64 cm and the diameter is in the range of 1.67~2.67 mm.

Pacemakers can pace the chamber with a unipolar lead or bipolar lead (Fig. 12.8). In unipolar pacing, electrons flow from the pacemaker through the lead to the cathode of the lead and contact the myocardium of the heart. And, the electrons are collected by the metal case of the implanted pacemaker located in the pocket beneath the skin. The metal case of the pacemaker itself works as another electrode, anode, in unipolar mode pacing. Even though unipolar leads have a smaller diameter and are easier to place, unipolar pacing can stimulate other than myocardial tissues around the implanted pocket and has a large chance of interference due to the large distance between the two electrodes. Due to the simplicity, unipolar pacing has been used in the earlier stage of artificial pacemakers.

Leads for bipolar pacing have two wires for each cathode and anode electrode. The cathode electrode at the tip of the lead contacts the myocardium and the anode electrode is located 1 cm below the tip in a ring shape surrounding the lead. The stream of electrons flows through the lead wire of the cathode electrode and enters the myocardium. Then, a stream of electrons flows through myocardial tissue and blood, and it is collected via the anode electrode of the lead. In both unipolar and bipolar pacing, the electrode contacting the myocardium is the cathode, where the electrons flow out from the pacemaker to the myocardium. As practical moving particles are electrons, contacting the myocardial tissue by cathode has higher efficiency than contacting by the anode. Even though the diameter of the lead is increased and the device becomes complex in bipolar leads, they have the advantage of not stimulating other tissues, like pectoral muscles around the pocket, and are not susceptible to external EMI interferences. Also, when the leads are used including the sensing capability, bipolar lead has a clear advantage to differentiate intrinsic physiological cardiac rhythm, which is localized within heart chambers, from other extraneous electrical events occurred outside of the heart.

Lead wires are the weakest part of the pacemaker against the continuous motion of the heart. At a heart rate of 80 beats/min, lead wires are flexed more than 100,000 times a day. If the wires are not flexible enough, they cannot be used for a long duration over 10 years. To increase the flexibility, bipolar leads are made of concentrically placed conducting metal coils embedded in insulating material (Fig. 12.9a). And conducting coils are made of a metal alloy having good tensile strength in addition to good electrical conductivity. Coil's structure provides reversible flexibility in all directions to withstand a long period of bending and extending caused by the heart motion. Usually, the internal coil is used for the cathode pole and the outer coil is used for the anode pole. This coaxial structure provides redundancy for emergency pacing. If the outer coil is no more usable for pacing by its fracture or by the breach of external insulation, the pacing can be continued in unipolar mode with the still intact internal coil. Modern pacemakers have the functionality to detect the problems in the leads and the pacing mode can be switched upon the detection of the problem until the lead is replaced.

To ensure accurate electrical stimulation, lead wires should be affixed to the myocardium stably. Fixation can be achieved in a passive manner or in an active manner. Leads for passive fixation have tines at their ends (Fig. 12.9b).

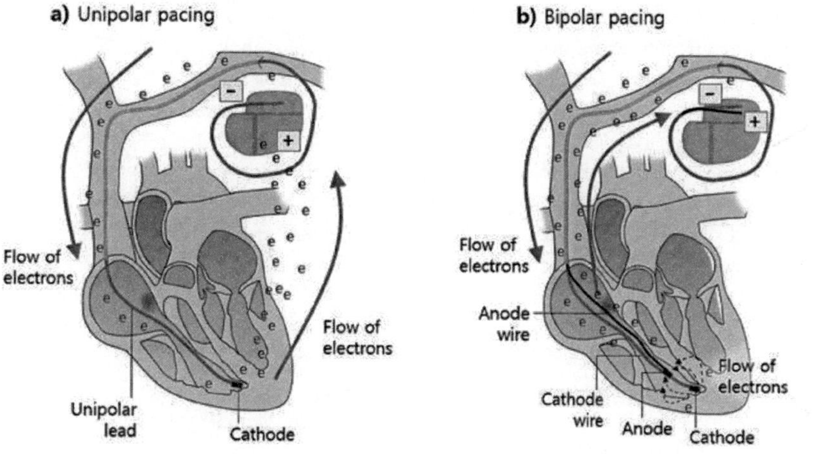

Fig. 12.8 Types of pacing leads. (**a**) Pacing with a unipolar lead, (**b**) pacing with a bipolar lead

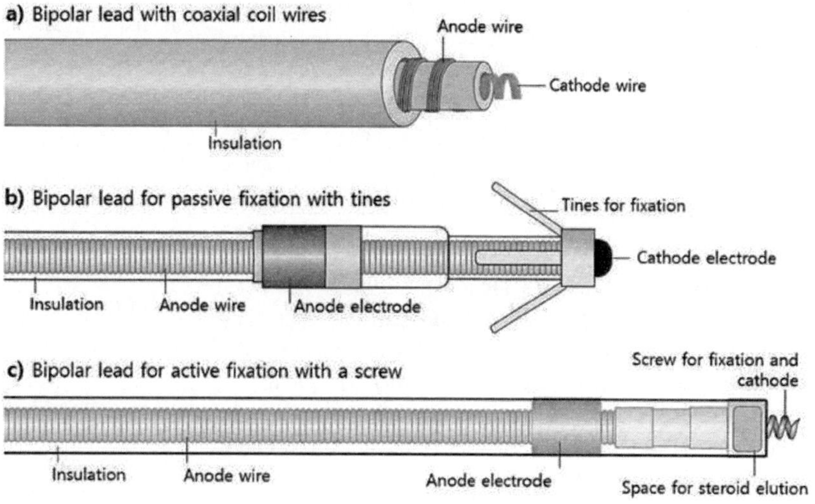

Fig. 12.9 Bipolar leads. (**a**) Concentrically placed conducting metal coil wires embedded in insulating material, (**b**) a bipolar lead for passive fixation with tines, (**c**) a bipolar lead for active fixation with a screw at the tip

These tines get entangled in the RV trabeculae, which are the rounded or irregular muscular columns projecting from the inner surface of the right ventricle of the heart. Therefore, passive fixation is not possible in any location other than the apical position of the chamber. Leads for active fixation have a screw in the tip instead of tines (Fig. 12.9c). The screw is inserted into the myocardial tissue by the manipulation from the other end of the lead and fixation in other places than the apical position is possible. The screw itself can be used as a stimulating cathode electrode.

As the time passes after fixation, the resistance of the leads used to be increased and it may problematically increase the consumption of the battery and shorten the lifetime of the implanted pacemaker. This problem occurs by the fibrosis or inflammation around the lead tip-muscle tissue interface with scar tissue formation. As a solution to the problem, a minute amount of steroid is slowly eluted at the lead tip to prevent excessive scarring. This can help maintain lead resistance in an acceptable range and extend the battery life. So, modern leads have a small space at the lead tips for steroid eluting.

1.3.3 Battery of Artificial Pacemaker

Battery built-in within the pacemaker unit supports power for the pulse generator. Pulse generator includes electrical components responsible for generating stimulating pulses, sensing and amplifying the intrinsic heart rhythms, processing, and control of the pulse generation, testing the components, wireless communication to outside, and data storage for programming and reporting. Usually, a cardiac pacemaker consumes half of its battery power for stimulating pulse generation and the other half for remaining sensing and housekeeping tasks.

The battery is the essential component determining the lifetime of the pacemaker. If the failure of the device does not occur, the replacement of the pacemaker is determined by the depletion of the built-in battery within the pacemaker. As batteries of the pacemakers are hard-wired at the time of manufacture before the device is hermetically sealed, they cannot be replaced. There have been unique challenges for developers and manufacturers to increase the longevity of the pacemaker to avoid frequent replacements. Technological advances in leads and electrodes have reduced energy requirements by two orders of magnitude. And, technological advances in microelectronics sharply reduced internal current drain concurrently decreasing the size and increasing the functionality, reliability, and longevity (Mallela et al., 2004). As the battery occupies a major portion of the pacemaker unit in terms of weight, volume, and size, technology for the battery also has progressed by increasing the capacity within a restricted volume as well as removing the problems that occurred in long-term uses.

A rechargeable nickel-cadmium battery was used at the beginning of pacemaker implants. The voltage of the battery was 1.25 V and the capacity was 190 mAh. The first battery used in 1958 lasted only 3 hours, and the successor lasted only 6 weeks. The major problems were the very short lifetime and the safety in recharging the battery. As nonrechargeable batteries give a longer lifetime compared to rechargeable batteries, they soon replaced the rechargeable battery.

Mercury-zinc batteries were introduced and widely used as nonrechargeable batteries around the 1960s. However, mercury-zinc batteries released hydrogen gas and they could not be hermetically sealed. Fluid leakage into the pacemaker used to cause electrical shorting and premature failure. And, abrupt decay of the voltage in the terminal stage of battery life makes it difficult to predict the remaining lifetime of the battery. This battery has improved the lifetime of a battery up to 2 years.

Lithium-iodine battery was introduced in 1972. It greatly extended the pacemaker battery lifetime, more than 10 years, and replaced the mercury-zinc battery totally. It produces no gas and hence it can be hermetically sealed. Also, slow decay of voltage at the terminal stage makes remaining life of the battery predictable in earlier time sufficient to prepare a replacement. Lithium batteries have a long shelf life, typically equivalent to a 10% loss of capacity over 5 years. The technical advantages of lithium-iodine battery in terms of its longevity, no gas generation, adaptable shapes and sizes, corrosion resistance, reduced weight, and excellent current drain characteristics are suitable to cardiac pacemakers and the currently adopted as a power source in most cardiac pacemaker models with a capacity of 0.6~3.0 Ah.

The longevity of the battery can roughly be calculated as follows.

$$\text{Battery capacity: } C_b = 1.0 \text{ Ah} \tag{12.6}$$

$$\begin{aligned}\text{Average pulse current: } I_p &= I_s \times (t_p/1 \text{ s}) \\ &= 5 \text{ mA}(0.5 \text{ ms}/1.0 \text{ s}) \\ &= 2.5\ \mu\text{A}\end{aligned} \tag{12.7}$$

$$\text{Total current drain: } I_t = I_p \times 2 = 5.0\ \mu\text{A} \tag{12.8}$$

$$\text{Longevity in hours: } L_h = C_b/I_t = 1.0 \text{ Ah}/5.0\ \mu\text{A} = 2 \times 10^5 \text{ h} \tag{12.9}$$

$$\begin{aligned}\text{Longevity in years: } L_y &= L_h/(24 \times 365) \\ &= 2 \times 10^5/(24 \times 365) = 22.8 \text{ years}\end{aligned} \tag{12.10}$$

In this example case, a battery of 1 Ah is sufficient to support the pacemaker for up to 22 years. During the above calculation, the stimulating current is 5 mA, the duty cycle of the pulse is 0.5% and the total current consumed is assumed two times the current consumed by the stimulating pulses. If any of these values are multiplied by two times, the expected life of the battery decreases abruptly from 22 years to 11 years. As the life of the battery and pacemaker is critically determined by the consumption of the power per pulse stimulation, the amplitude and width of the stimulating pulse should be minimized while it guarantees the reliable stimulation of the heart chambers.

1.4 Modes of Cardiac Pacing

When the heart chambers are paced with pacemakers, pacing can be applied in various modes. As most modern pacemakers have an additional function of sensing the intrinsic rhythm of each chamber, the stimulating pulse is only required when the inherent rhythm is absent during the predetermined period. If it does sense intrinsic activity, it will hold off electrical stimulations. This sensing and stimulating procedure, which is called demand pacing, continues on a beat-by-beat basis. As sensing and stimulation are performed within each chamber, they can cooperate across the chambers in dual-chamber pacing. When an intrinsic contraction of the right atrium (RA) is sensed, the pulse generator waits for the occurrence of the following contraction of the right ventricle (RV) and delivers an electrical pulse stimulation only if it is not sensed up to the end of waiting interval. In general, pacing modes are designed to be overridden by the heart's intrinsic rate whenever it gets back to a nonpathologic normal sinus rhythm and to reinitiate artificial electrical pacing when the pathological event happens again. Depending on the possible situations, there can be various types of pacing modes.

Table 12.1 Standard NBG coding system for pacemaker pacing mode

Position I		Position II		Position III		Position IV		Position V	
Chambers paced		Chambers sensed		Response to sensing		Rate modulation		Multisite pacing	
A	Atrium	A	Atrium	T	Triggered	R	Rate adaptive	A	Atrium
V	Ventricle	V	Ventricle	I	Inhibited			V	Ventricle
D	Dual(A + V)	D	Dual(A + V)	D	Dual(T + I)			D	Dual(A + V)
O	None	O	None	O	None	O	None	O	None

1.4.1 NBG Coding System

In order to standardize and facilitate the use and understanding of pacemakers, NBG (NASPE/BPEG Groups) coding system, has been devised (Table 12.1). NASPE is an acronym for the North American Society for Pacing and Electrophysiology and BPEG is an acronym for British Pacing and Electrophysiology Group. The NBG code has five positions that denote pacemaker functions. The code is generic and does not describe specific functional characteristics for each device. When a code includes only three or four codes, it is assumed that the positions not mentioned are absent.

- *Position I*: The first position represents the chamber paced. "A" represents the atrium, "V" represents the ventricle, and "D" represents the dual chambers of the atrium and ventricle. "O" represents the nonpacing mode, and the implanted pacemaker does not pace the heart at all and the heart beats with its intrinsic rhythm. This mode is used for temporarily turning off the pacemaker.
- *Position II*: The second position represents the location where the pacemaker senses intrinsic rhythms. The codes are the same as those for the first position. "A" represents the atrium, "V" represents the ventricle, and "D" represents the dual chambers of the atrium and ventricle. And, "O" represents the absence of the sensing. When the pacemaker is paced with "O" in the second position, it will pace asynchronously at a specified rate, ignoring any intrinsic rhythm.
- *Position III*: The third position represents how the pacemaker responds to sensed intrinsic rhythms. "I" represents that a sensed rhythm inhibits the output pulse and makes the pacemaker wait until the next cycle. "T" represents that an output pulse is triggered in response to a sensed rhythm. "D" represents that there are dual modes of response. This type of pacing is restricted to the dual-chamber systems. A rhythm sensed in the atrium inhibits the atrial stimulation and triggers ventricular stimulation. Delay is programmed between the sensed atrial rhythm and the triggered ventricular stimulation to approximate the normal PR interval. If the ventricular lead senses a native ventricular rhythm during the programmed delay, it inhibits the ventricular stimulation. "O" represents no response to sensed input and it is most commonly used when there is no sensed chamber in the second position with code "O."
- *Position IV*: The fourth position represents the rate modulation, also referred to as rate-responsive or rate-adaptive pacing. "R" in the fourth position represents that the pacemaker modulates its pacing rate adaptive to meet the physiological needs of the subject like physical activity. It also means that the pacemaker has an additional sensor to monitor the activity of the body by movement, temperature, or respiration. "R" is the only code commonly used in the fourth position. "O" represents that rate modulation is either unavailable or disabled. When the code in the fourth position is "O," it is usually omitted from the fourth position. That is "DDD" is the same as "DDDO."
- *Position V*: The fifth position is rarely utilized, but it represents the location or presence of multisite pacing. Multisite stimulation is defined as multiple stimulation sites in any single chamber, or a combination of these. "O" represents no multisite pacing. "A" represents multisite pacing in the atrium or atria. "V" represents multisite pacing in the ventricle or ventricles. "D" represents dual multisite pacing in both the atrium and ventricle. The most common multisite pacing is the biventricular pacing for the management of heart failure, which is called cardiac resynchronization therapy (CRT).

1.4.2 Asynchronous Pacing

Pacemakers may be programmed simply to pace at a fixed rate, without attempting to sense or trigger intrinsic cardiac rhythm (Fig. 12.10). These modes are referred to as asynchronous pacing. Pacing modes "AOO," "VOO," and "DOO" belong to the asynchronous pacing. In these modes, the atrium, ventricle, or both chambers are periodically paced, but the pacemaker does not have sensing capability and it stimulates each chamber ignoring the occurrence of intrinsic cardiac rhythms. It is the most basic pacing mode and was the only pacing mode in the first pacemakers. These modes are rarely used long-term. They are typically used when the subject is undergoing a surgical procedure using electrocautery. Electrical artifacts during the electrocautery can be sensed by the pacemaker as intrinsic cardiac rhythms, which would inhibit pacemaker output if the pacemaker is paced in sensed modes. Extended inhibition of stimulations can make the subject have profound bradycardia or even asystole. Also, there is a possibility that a pacing pulse could be delivered during the vulnerable period of the T

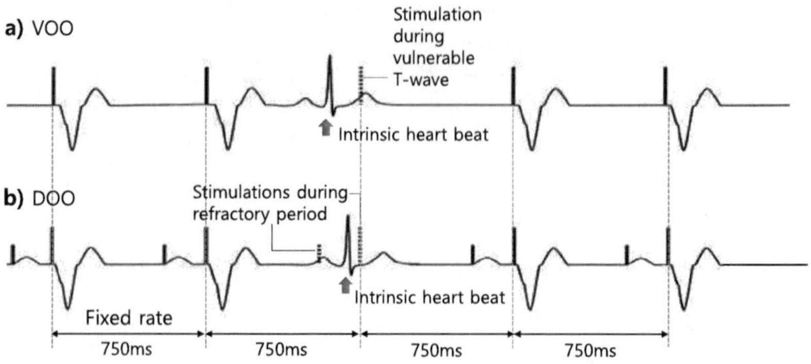

Fig. 12.10 Asynchronous pacing modes. (**a**) Asynchronous stimulation mode for the ventricle, (**b**) asynchronous stimulation mode for dual chambers of the atrium and ventricle

wave, which might induce a lethal ventricular tachyarrhythmia.

1.4.3 Single-Chamber Pacing

As the intrinsic beats of the heart are much more effective than the artificially paced beats, it is preferable to inhibit the stimulation when the intrinsic rhythms are occurring spontaneously. Artificial stimulation is applied only if there is a no intrinsic beat, by sensing the intrinsic rhythms. "AAI" and "VVI" are pacing modes of this principle. "VVI" mode delivers ventricular pacing at the programmed pacing interval, but inhibits pacing and resets the waiting interval when an intrinsic QRS complex is sensed (Fig. 12.11b). "VVI" mode is the basic pacing mode of the ventricle on demand. As pacing "VVI" mode does not provide synchronized contraction to the atrium, some subjects paced with this mode appeal to the symptoms of pacemaker syndrome. This mode is suitable when no synchronization with the atrial beat is required, as in atrial fibrillation. "VVIR" mode is a ventricular pacing mode with the rate-adaptive response. The pacing rate is adjusted by the activity of the subject detected by an activity monitoring sensor.

Similarly, "AAI" and "AAIR" mode delivers atrial pacing when intrinsic atrial rhythms are sensed during the programmed interval (Fig. 12.11a). In sensing the atrial event, sensitivity should be set to detect a smaller amplitude of atrial event and there should be a refractory period not to sense ventricle events as atrial events. "AAI" and "AAIR" are the modes of choice when atrioventricular conduction is intact but the intrinsic pacemaker is unreliable as in sinus node disease (SND) or sick sinus syndrome.

When the "AAT" or "VVT" mode is programmed, the sensing of an intrinsic chamber rhythm triggers immediate pacing of the same chamber (Fig. 12.11c). These modes correspond to the opposite of the "AAI" or "VVI" pacing modes. If the sensed intrinsic rhythms are the normal ones, delivered pulses are within the refractory period of the intrinsic event and do not produce additional events and only drain the currents from the battery. While the pacemaker senses normal intrinsic rhythms accurately, these modes are of no use. These modes are used when there is frequent sensing of myopotentials or electromagnetic interference. On these events, "AAT" or "VVT" modes can deliver pacing pulses instead of inhibiting the pacing when each artifact is sensed outside the refractory period. Subjects can avoid a possible extended pause of pacemaker pacing. These pacing modes used to be selected in the unipolar mode, but they are rarely used these days.

1.4.4 Dual-Chamber Pacing

Pacing modes "DDD" and "DDDR" are most commonly used as they cover all the options (Fig. 12.12a). They are standard modes in dual-chamber pacing. They increase the performance of artificial pacing by atrioventricular synchrony even though the pacemakers require separate atrial and ventricular leads and careful programming of their functions. Intrinsic atrial rhythm sensed during the programmed V-A interval inhibits atrial stimulation and intrinsic ventricular rhythm sensed during the programmed A-V interval inhibits ventricular stimulation. A ventricular pulse is triggered by the sensed intrinsic or paced atrial rhythm after programmed atrioventricular (AV) delay if the ventricular rhythm is not sensed during this waiting period. AV delay can be shortened in higher pacing rates or can be lengthened to favor intrinsic atrioventricular conduction.

In "DDI" and "DDIR" pacing modes, the atrium and ventricle are paced with pulses at a programmed rate as in "AAI" and "VVI" pacing modes (Fig. 12.12b). And, atrial or ventricular pulses are inhibited respectively when the intrinsic atrial or ventricular rhythm is sensed during the waiting interval. Sensed intrinsic rhythm is not used for triggering the pulses. However, dual chambers are paced synchronously with programmed AV delay. In the case of atrial arrhythmia, these modes do not increase the ventricular pacing rate as sensed atrial events do not trigger an AV delay and ventricular pacing. So, "DDI" or "DDIR" modes are suitable in subjects with frequent atrial arrhythmias like tachycardia-bradycardia syndrome.

"VDD" pacing mode delivers AV synchronized pacing for the ventricle (Fig. 12.12c). Sensing occurs in the atrium and

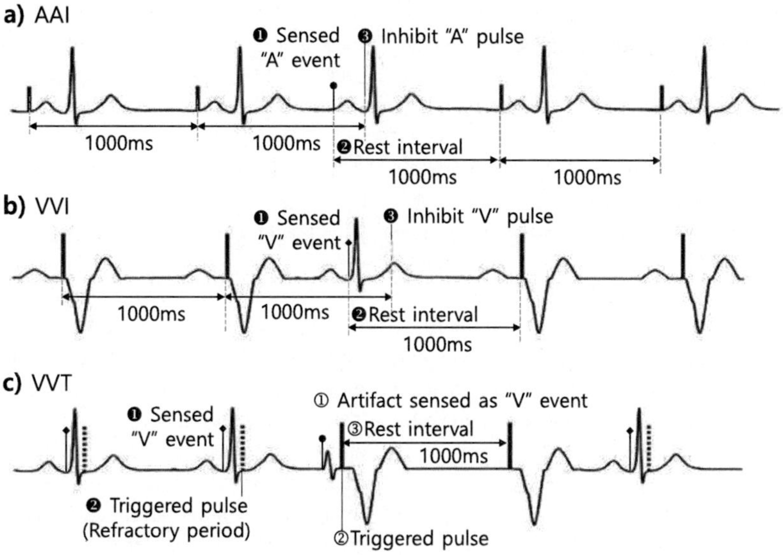

Fig. 12.11 Single-chamber pacing modes. (**a**) "AAI" pacing mode. The atrium is periodically paced with the interval of 1000 ms while an atrial event is not sensed. If an intrinsic atrial event is sensed, the waiting interval is reset and the programmed pulse is inhibited. (**b**) "VVI" pacing mode. The ventricle is periodically paced with the interval of 1000 ms while a ventricle event is not sensed. If an intrinsic ventricle event is sensed, the waiting interval is reset and the programmed pulse is inhibited. (**c**) "VVT" pacing mode. The ventricle is periodically paced with the interval of 1000 ms while a ventricle event is not sensed. If an intrinsic ventricle event is sensed, a pulse is triggered and the waiting interval is reset. When an artifact is sensed as a ventricular event, it triggers a pacing pulse and can induce a ventricular contraction

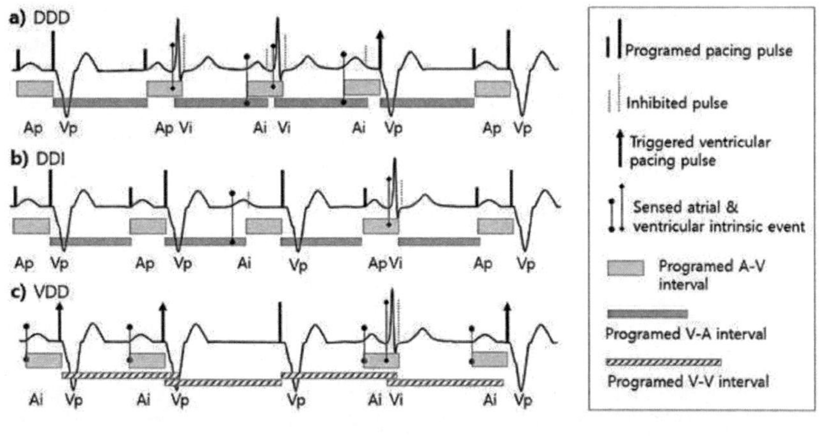

Fig. 12.12 Dual-chamber pacing modes. (**a**) DDD mode, both of atrium and ventricle are paced with a programmed pacing rate if there is no sensed intrinsic rhythm. If the intrinsic atrial rhythm is sensed, programmed atrial pacing is inhibited and ventricular pacing is triggered. If the intrinsic ventricular rhythms are sensed, programmed ventricular pacing is inhibited. (**b**) DDI mode. The atrium and ventricle are paced with programmed rate and AV delay. Intrinsic rhythm inhibits the pacing pulse of the same chamber. (**c**) VDD mode. If the intrinsic atrial rhythm is sensed, the ventricular pulse is triggered. If the intrinsic ventricular rhythm is sensed, the ventricular pulse is inhibited. Otherwise, the ventricle is paced with a programmed rate. *Ai* intrinsic atrial contraction, *Ap* paced atrial contraction, *Vi* intrinsic ventricular contraction, *Vp* paced ventricular contraction

in the ventricle but inhibited and triggered pacing occurs only in the ventricle. If the intrinsic atrial rhythm is not sensed, the pacemaker paces the ventricle as in "VVI" mode. Thus, "VDD" is most appropriate for patients with complete AV block but with normal sinus node function.

1.4.5 Rate Adaptive Pacing

Rate adaptive pacing is to pace the heart adaptive to the patient's activity instead of a fixed programmed rate. This is to help a patient respond to the physiological stress even though the patient's intrinsic sinus node normally cannot follow. For patients having an intact sinus node, dual-chamber pacemakers allow the adaptive increase of the beating rate of the heart. However, many patients have sick sinus syndrome or chronic atrial fibrillation that cannot modulate the sinus rhythm in response to physical exercise, mental stress, and emotional change. From the early stage of rate-adaptive pacing, the pacing modes of "AAIR," "VVIR," and "DDDR" have provided hemodynamic advantages over corresponding fixed-rate pacing, "AAI," "VVI," and "DDD."

Rate adaptive pacing is performed based on the sensed information from the additional sensor to evaluate the body activity. A variety of signals, like mechanical movement, minute ventilation, body temperature, and change in right ventricular impedance, are used to determine when a patient increased the activity.

1.5 Magnetic Resonance Imaging Compatibility

The lifestyle of subjects after pacemaker implantation is usually not modified to any great degree. However, activities under intense magnetic fields are the one that belongs to the unwise. Activities involving intense electromagnetic fields should be avoided. However, magnetic resonance imaging (MRI) can be an inevitable diagnostic option. When a pacemaker is exposed to a strong magnetic field, a variety of issues are arising, like heating at the lead-tissue interface, force & torque on devices, alteration of programming, damage to the pacemaker circuitry, inhibition of pacing output, rapid atrial/ventricular pacing, and induction of ventricular fibrillation.

With the technological progress regarding this issue, pacemaker implantation is no longer an absolute contraindication to magnetic resonance imaging. Pacemaker units are typically labeled as magnetic resonance imaging (MRI) "Compatible," "Conditional," or "Non-MRI conditional." Pacemakers labeled as "Compatible" are safe for whole-body scans without regard to patient positioning. Pacemakers labeled as "Conditional" are safe for scans with proper programming and generally may not be safe for scanning the implant region. Imaging protocol should be planned to ensure that absorbed energy does not breach the thresholds for each sequence. The pacemaker may need reprogramming on the pacing mode and parameters before the MRI scan and it should be returned to its standard settings at the completion of the scan. Typically, the recommendations are applied for MRI scans of 1.5 T or less. "Non-MRI conditional" pacemakers would be contraindicated for MRI scanning. However, several studies have shown that patients with these devices can safely have an MRI scan over the nonthoracic region at 1.5 T (DeForge, 2019). No complications to either patients or devices occurred during the 564 MRI scans with pacemakers. There were no significant differences in common clinical parameters between pre- and post-MRI scans (Biederman et al., 2020).

1.6 Leadless Artificial Pacemaker

The leadless pacemaker is a self-contained generator and electrode system. Figure 12.13 shows the structure and specifications of a currently available leadless pacemaker. The small main unit of the pacemaker is implanted directly in the right ventricle without using the leads. It is another significant advancement in pacemaker technology. The leadless pacemakers have diameters less than 6.7 mm, and lengths less than 42 mm. Their weight is about 2 g and their volume is less than 1 cc. They are almost 90% smaller than a transvenous pacemaker in weight and volume. It can be implanted via a femoral vein transcatheter approach. Implantation of a leadless pacemaker does not require a chest incision or subcutaneous pocket for the placement of the pulse generator unit. Leadless pacemakers are fixed to the right ventricle myocardium via self-expanding tines or via an active screw-in helix. The leadless pacemaker also has cosmetic appeal because there is no chest incision or visible pacemaker pocket (Ashley & Kristen, 2019).

Among the complications and failures of implanted pacemakers, 12% of them are related to the transvenous leads and subcutaneous pockets. In short-term, these complications include pneumothorax, cardiac perforation, lead dislodgement, and pocket infection or hematoma. In the long-term, pocket infection, tricuspid regurgitation, venous obstruction, lead fractures, and insulation failure.

The primary advantage of a leadless pacemaker is escaping from the complications associated with transvenous pacemakers and their leads. Leadless pacemakers provide only single-chamber ventricular pacing. So mostly leadless pacemaker is paced with VVI mode. The ventricle is paced, sensed, and the pulse generator inhibits pacing output in response to a sensed intrinsic ventricular event. This type of pacing is primarily indicated in patients with atrial fibrillation with bradycardia or bradycardia-tachycardia syndrome. By recognizing the body demand with the temperature sensor or 3D accelerometer, the ventricle can be paced in rate adaptive mode. By recognizing the atrial contraction with the 3D accelerometer, the ventricle can be paced in VDI mode. It gives ventricular pacing, while sensing both chambers of the atrium and ventricle, and inhibits ventricular pacing if an intrinsic wave is sensed.

The two leadless pacemaker models are available for clinical trial at the time of writing. Their battery life is around 10 years, comparable to that of a transvenous pacemaker. The future of leadless device technology is promising by

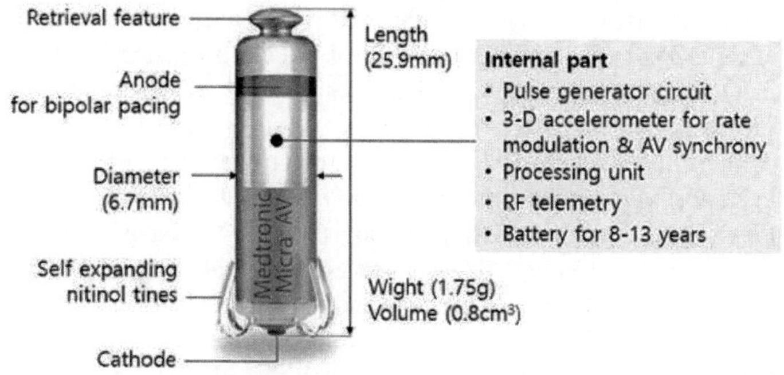

Fig. 12.13 Structure and specifications of a currently available leadless pacemaker

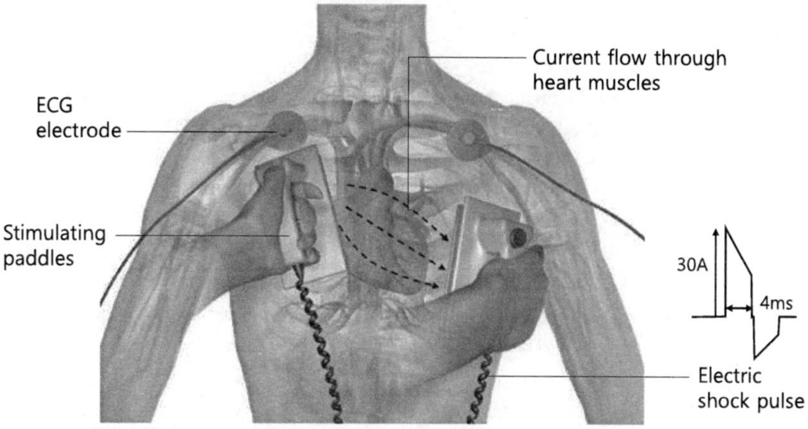

Fig. 12.14 Defibrillation with an external defibrillator. A short pulse of electrical energy is applied through the paddles to depolarize the heart muscles synchronously with anticipation of rerunning the normal sinus rhythm. (https://commons.wikimedia.org/wiki/File:External_Defibrillator.png)

eliminating the weakest point of implanted pacemakers and might eventually lead to expanded pacing capabilities.

2 Cardiac Defibrillator

Defibrillation is the administration of electric shocks to the heart in order to regain normal heart rhythm in subjects who are experiencing cardiac arrest or whose heart function is endangered because of severe arrhythmia. Figure 12.14 shows the defibrillation with an external defibrillator. Electrical shocks are delivered by defibrillators. In contrast to pacemakers which pace the heart permanently, defibrillators deliver electric shocks in a very brief period of time. Even though the mechanism of returning to the normal sinus rhythm is not clearly understood, the defibrillation procedure is considered to depolarize most of the heart muscles at the same time and increase the possibility to end ongoing dysrhythmia and regain the normal sinus rhythm. The analysis shows that 1700 lives were saved each year only in the USA by the use of automated external defibrillators (AED). About 3.2 million AEDs currently deployed throughout the USA are only 10% of the required number of 30 million AEDs. With the increased availability of AEDs and increased use by bystanders, the number of subjects saved by the bystander AED use are estimated at 3,459 per year in the USA.

2.1 Brief History of Defibrillator

From the age when electricity first became popular with electrostatic generators and storage, electricity has been applied to humans for many purposes. Electricity has been applied with the hope to resuscitate the dead persons. The first use of electric shock as a resuscitative experimental method was reported in 1776. A girl fell from a window and was in a state of apparent death revived by the applied electricity to various parts of the body and finally through the chest (Acierno, 1994). And, the electricity was considered the unique tool to stimulate the heart in an instant and should be used where others have failed especially for the recovery of people apparently drowned. In 1872, surgeon Thomas Green described six successful resuscitations out of seven attempts following cardiac arrest from chloroform using a galvanic battery installed in his operating room (Green, 1873).

On the contrary, with the increased use of electricity in the 1870s after practical generators have developed, there reported many accidental deaths, particularly among power company employees, caused by the electricity contact. In the experiments with dogs, it was confirmed that electrical stimulation can cause ventricular fibrillation leading to death. Electrical stimulation is also used as an execution method.

The effect of electric shocks that can revive lives and lead to deaths was clearly demonstrated in 1899 by Jean-Louis Prévost and Frédéric Batelli, physiologists at the University of Geneva, Switzerland. They observed that weak shocks cause fatal ventricular fibrillation and that large charges restore sinus rhythm in some animals from ventricular fibrillation. At the time, ventricular fibrillation was thought to occur only in animals. With the advent of ECG in the 1920s, many arrhythmias including ventricular fibrillation had been clearly described in humans. With ECG, various studies revealed that hearts are very susceptible to ventricular fibrillation when the electric shocks are applied during a refractory period (Ball & Featherstone, 2019).

William Kouwenhoven, an electrical engineer at Johns Hopkins University USA, invented the first defibrillator in 1930, which was operated in AC mode. He definitively established that electrically induced ventricular fibrillation could be reversed with an appropriate counter-shock. His defibrillator was first applied to a human subject in 1947 by Claude Beck, at the University Hospitals of Cleveland, USA. A 14-year-old patient, whose heart was stopped during surgery, regained life by the second shock applied from the AC defibrillator. It consisted of a transformer to separate the patient from the 110 V AC wall supply, a variable resistor to limit the current to a heart-safe value, and two metal tablespoons with wooden handles to deliver the shocks to the exposed heart. Even though it saved lives, the AC system persisted for only several years. Patients tended to appeal to unpleasant side effects of the large steady currents, including broken ribs and damage to the heart muscles.

The first external defibrillation was conducted by Paul Zoll, a cardiologist at Harvard Medical School, USA. In 1956, he successively returned the sinus rhythm from ventricular fibrillation in four patients. He used a bulky AC external defibrillator, which can be moved on wheels with a large transformer converting 120 into 720 V.

In 1962, Bernard Lown and colleagues first reported the use of DC current from a synchronized capacitor. Generated short period of electrical pulses from the capacitor can be delivered in synchronization with the phase of recorded ECG. And, the equipment has evolved toward a small portable defibrillator having a standard feature of modern defibrillators.

The portable defibrillator was invented in the 1960s by Frank Pantridge, an Irish cardiologist. His first design was powered by a car battery and weighed 70 kgs. But, in his second design, weight can be significantly reduced to around 3 kg. This invention was the predecessor of today's modern automated external defibrillator (AED). Gradual improvements in defibrillators have led to the wide availability of AEDs. These devices can analyze the heart rhythm by themselves, diagnose the shockable rhythms, and charge for treatment. This means that no clinical skill is required in their use, allowing lay people to respond to emergencies effectively. Frank Pantridge also first made it available for all emergency responders such as police officers, firefighters, emergency medical technicians (EMT), and paramedics.

The implantable cardioverter-defibrillator (ICD), implantable defibrillator, is the first-line treatment and preventive therapy for patients at risk for sudden cardiac death due to ventricular fibrillation and ventricular tachycardia (Mirowski et al., 1980). Triggered by the death of his friend due to the sudden cardiac death, Michel Mirowski built the concept of the implantable defibrillator in 1969 and continued its development. ICD was firstly implanted in humans by Levi Watkins, an American heart surgeon at Johns Hopkins Hospital, USA, in 1980.

2.2 Indications for Defibrillation

Subjects who fell in sudden cardiac arrest (SCA) are the most probable candidates for defibrillation. Figure 12. 15 shows different types of cardiac arrest. During the SCA, a subject cannot pump the blood sufficiently from the heart due to the abnormal heart control rhythm called ventricular fibrillation (VF) or ventricular tachycardia (VT). Unsynchronized or too fast contraction of ventricle muscles cannot pump blood appropriately and it produces almost zero cardiac output. This sudden loss of pumping function of the heart blocks the oxygen supply to the rest of the body tissues, which results in the death of tissues, organs, and whole life.

Hypoxia is a state in which oxygen is not sufficiently available at the tissue level to maintain adequate functioning. Hypoxia is mostly caused by inadequate oxygen delivery to the tissues. Hypoxia can be resulted from the low oxygen content in the normally supplied blood or from the low blood supply itself as in SCA. Low oxygen contents below the normal level in the arterial blood, which is called hypoxemia, can be caused by respiratory or circulatory dysfunctions. Compared to the hypoxia caused by hypoxemia, hypoxia caused by cardiac arrest is much more serious and life-threatening. Cardiac arrest can lead to complete loss of oxygen supply, which is called anoxia.

SCA is a different event from a heart attack. Heart attack is caused mostly by the problem in a coronary artery supplying the blood to the heart itself without the intervention of disorganized electrical activity and the subject usually remains awake and able to talk. However, SCA is caused by the problem of the electrical system controlling the heart contraction and there is rarely any warning before the SCA. When

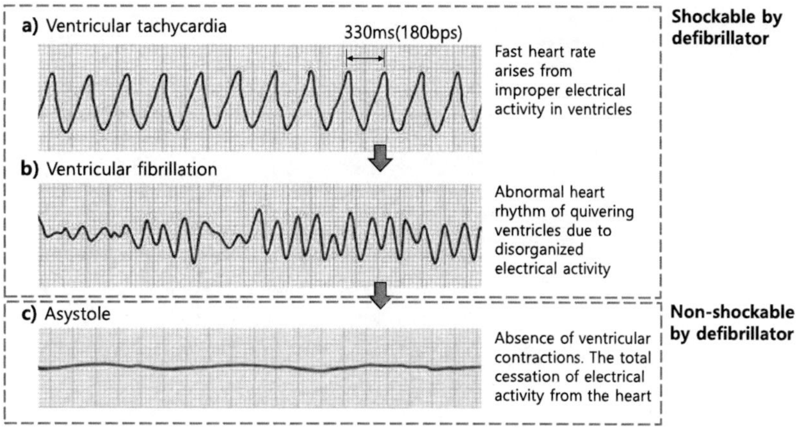

Fig. 12.15 ECG recordings for each type of cardiac arrest. (**a**) Ventricular tachycardia (VT) which is shockable with a defibrillator, (**b**) ventricular fibrillation (VF) which is shockable with a defibrillator, (**c**) asystole which is not shockable with a defibrillator

cardiac arrest occurs, the subject becomes unconscious quickly within 20 s.

Under the anoxia condition, body tissues and organs cannot withstand a long time. The brain is the most rapidly damaged organ by anoxia. Without oxygen and glucose supplied by the blood, neurons in the brain cannot deliver electrical signals needed to maintain breathing and other body functions. So, when cardiac arrest occurs, cardiopulmonary resuscitation (CPR) must be started rapidly within 2 min. Three minutes of anoxia can lead to brain injury that worsens progressively. In 9 min, severe and permanent brain damage can result. After 10 min, the chance of survival also decreases to a lower level. Even though a subject is resuscitated, 80% of them will sustain some level of brain damage like a coma. Simply, the longer the brain is deprived of oxygen, the worse the damage will be.

More than 400,000 people suffer from SCA each year in the USA only. Overall survival rates are low as 22% in in-hospital cases and 10% in out-of-hospital cases, respectively (Ramiro & Kumar, 2015). SCA is a leading cause of death with a rate higher than breast cancer, lung cancer, and HIV/AIDS combined. When SCA occurs, the heart needs immediate treatment using a defibrillator to restart the electrical system.

Ventricular fibrillation (VF) is the first indication of defibrillation. It is the most critical abnormal heart rhythm. During the VF, heart muscles twitch randomly instead of synchronously under a coordinated electrical conduction system and the heart just quivers without pumping blood normally. It is caused by the disrupted electrical conduction system. VF is clearly diagnosed by electrocardiogram (ECG) with no discernable pattern and just shows irregular unformed QRS complexes without any clear P wave. Disruption of the cardiac electrical system leading to VF can be caused by various heart diseases including ischemic heart disease, cardiomyopathy, and myocarditis. VF also can be caused by electrolyte imbalance, overdoses of cardiotoxic drugs, drowning, and trauma (Mogayzel et al., 1995). These diseases can increase the probability of individual heart muscle fiber initiating an impulse spontaneously. The extended period of the hypoxic condition of heart muscle fibers by ischemia can make them hyperirritable and act as a local pacemaker cell. Also, tissues around damaged cells easily become hypoxic and hyperirritable. Disturbance of sodium or potassium ion concentration in and out of the muscle fibers may drive them to easily generate impulses with a practically lowered threshold and they can act as problematic pacemaker cells.

Ventricular tachycardia (VT) is the second indication of defibrillation. VT is the fast heart rhythm higher than the normal rate in ventricles. VT is different from the temporary tachycardia of normal sinus rhythm induced by the exercise. VT originates abnormally from the local region of the ventricle. Rate may increase highly to 250 beats/min. While this rhythm returns to a normal rate within a short period, it does not result in a problematic symptom. However, prolonged or repeated periods of VT are dangerous and may result in ventricular fibrillation and sudden cardiac arrest. VT is diagnosed by an electrocardiogram (ECG) showing a rate of higher than 120 beats/min and widened QRS complexes at least three in a row. It is classified into nonsustained and sustained VT depending on the lasting time threshold of 30 s (Baldzizhar et al., 2016). A locally developed reentry circuit is the major cause of ventricular tachycardia.

The reentry circuit is the abnormally developed pulse generating circuit in the local region of the myocardium. It is developed when the conducted pulse is routed to the myocardial cells just after their refractory period and makes them contract again. For the development of a reentry circuit, there should be an anatomical closed loop of myocardial cells. It can be developed around the damaged heart muscles from previous heart diseases like myocardial infarction. As damaged tissue cannot conduct electrical pulses, there can be developed a conductive ring around this damaged area.

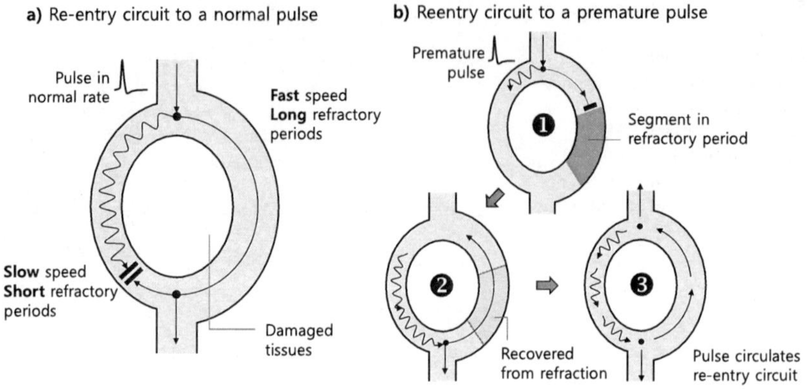

Fig. 12.16 Development of ventricular tachycardia by reentry circuit. (**a**) Reentry circuit to a normal pulse entered after all cells are recovered from refraction. (**b**) Development of ventricular tachycardia by a premature local pulse entered reentry circuit

Figure 12.16 explains one example of a reentry circuit. When an anatomical ring circuit is developed, each branch may have different characteristics. The left branch may have a slow conduction speed along the branch with a short refractory period and the right branch, in contrast, may have a fast conduction speed along the branch with a long refractory period. When an electrical pulse is arrived through the upper path at the branching point after the cells in both branches finished the refractory period as in the case of a normal rate of sinus rhythms, pulses can be conducted via both branches smoothly. Due to the difference in conduction speed, a pulse conducted through the right branch arrives at the joining point earlier and branches into two directions of a lower path and left branch reversely. Pulse conducted through the left branch meets this pulse in the middle of the left branch and they terminate conduction as pulses cannot conduct over the meeting point where the cells are just fallen into the refractory period. That is, even though the anatomical ring circuit is developed, a pulse is conducted normally through the branches from the upper path to the lower path while a pulse arrives at a normal rate after all cells have passed their refractory periods.

However, a problem can occur when an abnormal premature beat generated from a local region is conducted at an earlier time before both branches are recovered from their refractory period. Arrived pulse can be conducted through the left branch which has recovered from its short refractory time. However, a pulse cannot conduct through the right branch where some part of the branch is still in its long refractory period. Conduction through the right branches stops there. Meanwhile, the pulse conducted slowly through the left branch arrives at the joining point and branches into two directions of the lower path and right branch reversely. There is a high chance that the cells in the right branch recovered from their refractory period when a pulse arrives since the pulse is conducted through a slowly conducting left branch. Then the cells in the right branch can be fired and the pulse is conducted through the right branch reversely. When a pulse arrives at the branching point, it can reenter the left branch and the pulse starts to circulate the loop and produces a series of pulses with a short interval. This reentry mechanism produces ventricular tachycardia when the circuit is developed in the ventricular area.

Asystole is another type of cardiac arrest which is not effective to the electrical shocks. It is a total absence of ventricular contractions and the most serious form of cardiac arrest. Subjects with asystole are usually irreversible. Asystole is recorded as flatline in ECG recording representing the total cessation of electrical activity in myocardial muscles and no blood outflow from the heart. While the defibrillator is required for the subject in ventricular fibrillation before they progress into the asystole state, the use of defibrillation is not recommended in the asystole state. Asystole is treated by cardiopulmonary resuscitation (CPR) combined with an intravenous vasopressor such as epinephrine (Kempton et al., 2019).

2.3 Generation of Electrical Shock Pulse

A defibrillator generates an electrical pulse which is enough to depolarize most of the myocardial cells in the heart. It is a pulse of high energy delivered through the electrodes with a large surface area. The strength of the applied pulse is described in energy with the unit of Joules (J), which is a product of applied voltage (V), current (I), and time (T).

$$\text{Energy of a pulse (J)} = \text{Voltage}(\text{V}) \times \text{Current}(\text{I}) \times \text{Time}(\text{T}) \quad (12.11)$$

For example, a pulse of 200 J can be delivered with 1000 V and 20 A during 10 ms. Extremely high voltage and current are needed as a large amount of energy should be delivered in a very short period of time.

Electrical pulses for defibrillation are delivered by discharging the stored charges in the capacitor. Since the transthoracic resistance of the body is very low in the range of 50~100 Ω, stored charges are discharged very quickly. Pulses can be delivered in monophasic mode or in biphasic mode.

The monophasic defibrillator is the simplest type and uses a single capacitor for storing the electrical energy needed for stimulation. A capacitor is charged with high voltage and the current flows only in one direction during the discharging up to the depletion of stored charges. As total energy is discharged for a single shock, the amount of energy applied is determined by the stored energy in the capacitor during the charging period. When only a small amount of energy is wanted, the capacitor is charged up with only a required small amount of energy. Usually, the monophasic defibrillator can deliver electrical shocks in the energy range 2~360 J. The energy stored in a capacitor is calculated by half of the capacitance multiplied by the voltage squared.

$$\text{Energy stored in a capacitor} = \left(\frac{1}{2}\right) \times \text{Capacitance}(\text{C}) \times \text{Voltage}^2\left(\text{V}^2\right). \quad (12.12)$$

For the fixed value of capacitance in a defibrillator, the stored energy is determined by the charging voltage. In other words, if the required energy of electric shock is determined, the charging voltage can be calculated by this equation. To apply shocking pulses of 200 and 360 J from a defibrillator having a capacitance of 100 μF, required charging voltages are calculated respectively as follows.

Charging voltage for a 200J shock

$$= \sqrt{2 \times 200(\text{J})/\left(100 \times 10^{-6}(\text{F})\right)} = 2000\ \text{V} \quad (12.13)$$

Charging voltage for a 360J shock

$$= \sqrt{2 \times 360(\text{J})/\left(100 \times 10^{-6}(\text{F})\right)} = 2683\ \text{V} \quad (12.14)$$

Due to the low resistance, the peak amplitude of discharging current can reach 40~50 A, but it decays rapidly and totally vanishes within 10 ms as shown in Fig. 12.17a.

Instead of discharging the whole stored charge in the capacitor, the discharging pulse can be truncated by electrical switches, like high-voltage thyristors. Then discharge curve can get quick rising and falling edges with more condensed energy during a shorter period of time. In this case, applied energy is calculated in the control unit by monitoring the actual voltage as well as the real current through the subject. The Control unit integrates the applied energy and turns off the switches at the time when the total energy applied is equal to the set energy level.

When the discharging pulse can be truncated, the discharging pulse can be delivered in biphasic mode by adding a negative pulse following the positive pulse. A negative pulse can be applied using a secondary capacitor or by changing the connection routes from a capacitor to electrodes as in Fig. 12.17b. Biphasic defibrillation shows increased clinical efficiency and less amount of output energy is required. The success rate of first shock of biphasic defibrillation is about 90%, which is higher compared to the 60% of monophasic defibrillators. As defibrillation can be achieved with a reduced amount of energy in biphasic mode, the maximal setting of the biphasic defibrillators is

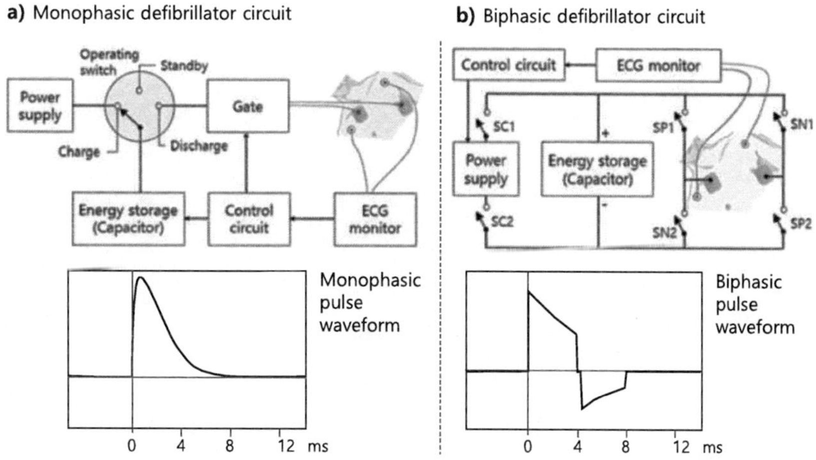

Fig. 12.17 Defibrillator circuits for stimulating pulse generation and the shape of generated waveform. (**a**) Monophasic stimulating pulse generation circuit and generated monophasic pulse waveform. (**b**) Biphasic stimulating pulse generation circuit and generated biphasic pulse waveform. SC1 & SC2: switches for charging the capacitor, SP1 & SP2: switches for stimulation in a positive phase, SN1 & SN2: switches for stimulation in a negative phase

usually 200 J instead of 360 J of monophasic defibrillators. A reduced level of applied energy also can reduce the complications like cardiac injuries and burns caused by a strong current. Defibrillator units get smaller, lighter, faster, and cheaper. As it is known that the biphasic defibrillation is more effective, most of the current defibrillators are designed in biphasic mode.

2.4 Types of Defibrillators

2.4.1 Manual Defibrillator

Manual defibrillator is the standard type of defibrillator which provides functions for diverse applications. These defibrillators require experience and training to effectively handle them. Hence, they are commonly used in hospitals and in ambulances where professionals are present. In conjunction with a recorded ECG, a provider first diagnoses the cardiac rhythm of a subject and then manually determines the energy level of the electrical shock. Electric shock is applied through external paddles to the subject's chest for external defibrillation.

Manal defibrillation can perform cardioversion, which is defibrillation synchronized to the ongoing cardiac rhythms. Cardioversion is to restore normal heart rhythm in subjects with a certain type of abnormal heartbeats other than ventricular fibrillation, including supraventricular tachycardia, atrial fibrillation, atrial flutter, and ventricular tachycardia. Based on the recorded ECG, the defibrillator detects the optimal time for the application of electric shock, which is 20~30 ms after the large R-peak of the QRS complex. By applying the controlled shock at the optimal time, defibrillation becomes more efficient with less amount of energy. Also, the vulnerable repolarization period of T-wave, during which electric shock can cause ventricular fibrillation reversely, can be avoided. While these defibrillators have connectors for separate ECG electrodes, ECG can be recorded through stimulating paddles by using them as electrodes and sharing the times.

Manual defibrillators also include the pacing mode. This makes the defibrillator work as an artificial pacemaker that paces the heart transcutaneously. This is performed as temporary pacing of the heart for subjects, in situations like awaiting cardiac surgery or undergoing cardiac catheterization.

The manual defibrillator can be used for internal defibrillation during open-chest surgery. Electrodes consist of metal plates and contact directly with the exposed heart surface. For internal defibrillation, a reduced amount of energy, usually one-tenth of external fibrillation, is applied and the maximum energy is limited to 50 J to prevent injury to the heart muscles.

2.4.2 Implantable Cardioverter-Defibrillator

Implantable cardioverter – defibrillator (ICD) is also called an automatic internal cardiac defibrillator (AICD). ICD is a battery-powered device implanted under the skin and thin transvenous lead wires connect the ICD to chambers of the heart (Fig. 12.18). It is similar to the implantable pacemaker in appearance; however, it also works for defibrillation and cardioversion. It continuously monitors the heart rhythm of a subject and detects any abnormal heart rhythms, including ventricular fibrillation, ventricular tachycardia, supraventricular tachycardia, and atrial fibrillation. When an abnormal heart rhythm is detected, ICD automatically determines the required amount of energy and provides an electric shock from the internal side of the myocardium to regain the normal sinus rhythm. To deliver an electric shock by the stored energy in a capacitor, the capacitor within the ICD should be charged to a high voltage level equivalent to the required energy level. As the ICDs are operated with a battery of lower voltages, the capacitor is charged using a special DC to DC

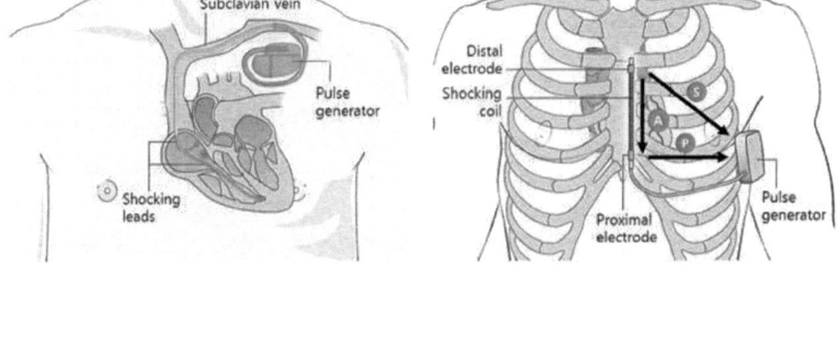

Fig. 12.18 Types of an implantable cardioverter-defibrillator (ICD). (**a**) Transvenous implantable cardioverter-defibrillator, (**b**) subcutaneous implantable cardioverter-defibrillator (S-ICD). *P* primary vector, *S* secondary vector, *A* alternative vector for cardiac rhythm recording

converter and step-up transformer, for example from 3.2 to 800 V. Then, a capacitor charged to a high voltage can deliver a single shock. ICDs usually can store energy up to 40 J for a single shock and are operated in biphasic mode. The battery in the ICDs can provide more than a hundred shocks. ICD is implanted in subjects who have a high potential of cardiac arrest to prevent it. Some ICDs may have a pacemaker function additionally and can function in dual mode, as a pacemaker for too slow heart rate and as a defibrillator for any shockable abnormal beats.

2.4.3 Subcutaneous Implantable Cardioverter-Defibrillator

Subcutaneous implantable cardioverter-defibrillator (S-ICD) is another type of ICD. Instead of ICD which connects the device via transvenous leads from inside of the chamber, S-ICD connects the device to the heart from the outside of the heart. S-ICD does not require leads inside of the heart and leaves the vasculature untouched. It can be implanted via a simpler procedure of small incisions based on anatomical landmarks without X-ray fluoroscopic guiding. S-ICD consists of a pulse generator and a lead which contains a shocking coil and two sensing electrodes. The tip of a lead is positioned parallel to the left of the sternal midline, and the pulse generator is positioned in the lower left part of the chest (Bardy et al., 2010). Stimulating current flows from the shocking coil between the sensing electrodes to the can of the pulse generator through the volume of the heart muscles. Electrical activities are recorded in three channels by the potential difference among distal electrode, proximal electrode, and the can of pulse generator (Sideris et al., 2017). Based on the unique feature of entirely extracardiac implantation by avoiding the transvenous access, S-ICD has obtained many advantages. It does not cause any risk of vascular injury during the implantation, risk of systemic infection in the venous system, and risk of extracting leads for replacement. S-ICD is equally effective with transvenous ICD for terminating induced ventricular fibrillation with the delivery of 65–80 J shocks (Bardy et al., 2010). S-ICD is the alternative in many cases to transvenous ICD. Major candidates for S-ICD include the subjects with an ICD indication not requiring pacing for bradycardia, with complex anatomy and venous access problems, and at high risk for infections (Kaya et al., 2019).

2.4.4 Automated External Defibrillator

The automated external defibrillator (AED) is the defibrillator especially designed for the out of hospital use by lay persons. It is a simplified and automated version of a manual defibrillator so that it can be easily carried in a portable box and used without any experience and training. It includes the ECG recording part and processing part which analyzes the recorded rhythm and determines the suitability of defibrillation shock. AEDs are widely deployed in public places to be easily accessed and used in emergent situations. AED first instructs the user to attach the pad electrodes to the upper right and lower left of the patient's chest. Once the electrodes are attached, AED automatically records the ECG of the patient and analyzes the recorded cardiac rhythm. If the AED determines that the patient has a shockable rhythm of ventricular fibrillation (VF) or ventricular tachycardia (VT), it announces the decision and starts to charge the capacitor in preparation for the delivery of a shock. When the AED finishes charging the capacitor, it instructs the user to press the button to deliver a shock. Everyone should avoid touching the patient during the ECG recording and delivering a shock. Even though pulses are delivered automatically right after finishing the charging in some units, generally human intervention is required in delivering the shock to avoid the possible accidental injury to attending persons. After delivering a shock, AED records and analyzes cardiac rhythm again and instructs a cardiopulmonary resuscitation (CPR) to be performed during the period to prepare for an additional shock if it is necessary.

AEDs do a critical role in increasing the survival rate from cardiac arrest occurring outside of the hospital environment. While the average survival rate is only 10%, it increases to 30% if bystanders use cardiopulmonary resuscitation (CPR). However, if bystanders use AED with CPR, the average survival rate increases up to 50%. As the minutes to a first defibrillation shock increase, the chance of survival decreases 7~10% per every minute from 90% at 1 min. The chance of survival becomes 50% after 5 min and it is lowered to less than 10% if 10 min have passed without delivering a defibrillation shock. The goal of deploying AED in public places is to make an AED available within 90 s. Having an AED at home provides a life-saving treatment at the time of cardiac arrest much sooner before emergency medical service (EMS) arrives. Considering the 9 min of average EMS arrival time, which may be increased by various factors, the chance of survival decreases to less than 20%. Having an AED at home increases the chance of survival from a cardiac arrest by more than 50% (Fig. 12.19).

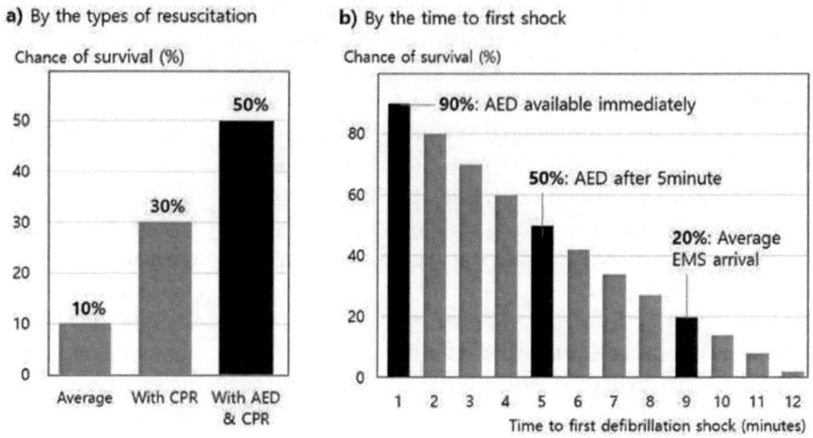

Fig. 12.19 Chance of survival from a cardiac arrest occurred outside of the hospital. (**a**) Chance of survival depending on the type of applied resuscitation methods, (**b**) chance of survival depending on the time to a first defibrillation shock

References

Acierno, L. (1994). *The history of cardiology*. Parthenon Publishing Group.

Aquilina, O. (2006). A brief history of cardiac pacing. *Images in Paediatric Cardiology, 8*(2), 17–81.

Ashley, G., & Kristen, G. (2019). The leadless pacemaker: An innovative design to enhance pacemaking capabilities. *Journal of the American Academy of Physician Assistants, 32*(6), 48–50. https://doi.org/10.1097/01.JAA.0000554750.85170.d4

Baldzizhar, A., Manuylova, E., Marchenko, R., Kryvalap, Y., & Carey, M. G. (2016). Ventricular tachycardias: Characteristics and management. *Critical Care Nursing Clinics of North America, 28*(3), 317–329. https://doi.org/10.1016/j.cnc.2016.04.004

Ball, C. M., & Featherstone, P. J. (2019). Early history of defibrillation. *Anesthesia and Intensive Care, 47*(2), 112–115. https://doi.org/10.1177/0310057X19838914

Bardy, G. H., Smith, W. M., Hood, M. A., et al. (2010). An entirely subcutaneous implantable cardioverter–defibrillator. *The New England Journal of Medicine, 363*, 36–44. https://doi.org/10.1056/NEJMoa0909545

Bhatia, N., & El-Chami, M. (2018). Leadless pacemakers: A contemporary review. *Journal of Geriatric Cardiology, 15*(4), 249–253. https://doi.org/10.11909/j.issn.1671-5411.2018.04.002

Biederman, R. W., Gevenosky, L., Rayarao, G., Williams, R., Lombardi, R., Shah, M., Farah, V., Dahiya, G., Wetzel, A., & Doyle, M. (2020). MRI effects on pacemaker/ICD within the magnetic field; an intra-observe variability study. *Circulation, 142*, A14497. https://doi.org/10.1161/circ.142.suppl_3.14497

Blaus, B., & Blausen.com staff. (2014). Medical gallery of Blausen Medical 2014. *WikiJournal of Medicine, 1*(2). https://doi.org/10.15347/wjm/2014.010. ISSN 2002-4436.

DeForge, W. F. (2019). Cardiac pacemakers: A basic review of the history and current technology. *Journal of Veterinary Cardiology, 22*, 40–50. https://doi.org/10.1016/j.jvc.2019.01.001

Green, T. (1873). On death from chloroform: Its prevention by galvanism. *British Medical Journal, 1*(595), 551–553.

Kaya, E., Rassaf, T., & Wakili, R. (2019). Subcutaneous ICD: Current standards and future perspective. *International Journal of Cardiology: Heart & Vasculature, 24*, 100409. https://doi.org/10.1016/j.ijcha.2019.100409

Kempton, H., Vlok, R., Thang, C., Melhuish, T., & White, L. (2019). Standard dose epinephrine versus placebo in out of hospital cardiac arrest: A systematic review and meta-analysis. *The American Journal of Emergency Medicine, 37*(3), 511–517. https://doi.org/10.1016/j.ajem.2018.12.055

Mallela, V. S., Ilankumaran, V., & Rao, N. S. (2004). Trends in cardiac pacemaker batteries. *Indian Pacing and Electrophysiology Journal, 4*(4), 201–212.

Mirowski, M., Reid, P. R., Mower, M. M., Watkins, L., Gott, V. L., Schauble, J. F., Langer, A., Heilman, M. S., Kolenik, S. A., Fischell, R. E., & Weisfeldt, M. L. (1980). Termination of malignant ventricular arrhythmias with an implanted automatic defibrillator in human beings. *The New England Journal of Medicine, 303*(6), 322–324. https://doi.org/10.1056/nejm198008073030607

Mogayzel, C., Quan, L., Graves, J. R., Tiedeman, D., Fahrenbruch, C., & Herndon, P. (1995). Out-of-hospital ventricular fibrillation in children and adolescents: Causes and outcomes. *Annals of Emergency Medicine, 25*(4), 484–491. https://doi.org/10.1016/s0196-0644(95)70263-6

Ramiro, J. I., & Kumar, A. (2015). Updates on management of anoxic brain injury after cardiac arrest. *Missouri Medicine, 112*(2), 136–141.

Reddy, V. Y., Knops, R. E., Sperzel, J., Miller, M. A., et al. (2014). Permanent leadless cardiac pacing: Results of the leadless trial. *Circulation, 129*, 1466–1471. https://doi.org/10.1161/CIRCULATIONAHA.113.006987

Sideris, S., Archontakis, S., Gatzoulis, K. A., Anastasakis, A., et al. (2017). The subcutaneous ICD as an alternative to the conventional ICD system: Initial experience in Greece and a review of the literature. *Hellenic Journal of Cardiology, 58*(1), 4–16. https://doi.org/10.1016/j.hjc.2017.01.010

Ward, C., Henderson, S., & Metcalf, N. H. (2013). A short history on pacemakers. *International Journal of Cardiology, 169*(4), 244–248. https://doi.org/10.1016/j.ijcard.2013.08.093

13 Electrical Brain Stimulations

Can electricity act on the brain directly? As we know that the cells in the brain work basically on electrical signals, we can easily expect that the brain might modulate its functions and respond in some ways under electrical exposure. Definitely, the brain is affected by any type of electricity applied directly to the brain. However, the range of effects is various depending on the strength and the type of applied electrical energy. It can be of some effect or it can be of no discernable effect. Effects might be positive and can be negative. The concern is that application of electrical energy in the appropriate type can manipulate the specific cells or a region of the brain and induce the effect anticipated.

The brain is known to have about 86 billion neurons. We do not know the roles and the responses of each neuronal cell of the whole brain at this moment. And, it seems difficult to know even after some tens of years later. While we only know the function of each brain cell with limited knowledge, manipulating the brain with electricity at the cell level is not possible. At most, we can consider the access of the brain at the regional level. The brain is composed of many constituting components and countless studies have revealed the function of each brain component and subregions of each component. Parallelly, there have been endeavors to manipulate the brain function in a positive direction by focalizing the applied energy at the possible smallest functional region. Recent imaging technologies including magnetic resonance imaging (MRI) have greatly contributed to improving the functional resolution of the brain stimulation.

There are many types of brain stimulation, which are delivered electrically or magnetically. The mechanisms of these electrical stimulation methods are not clearly known. Studies are actively being done on the mechanism, the efficacy of the current stimulating and new stimulating methods. Among those, some are clinically accepted and used routinely in the current clinical settings. Deep brain stimulation (DBS) is the most representative type of brain electrical stimulation. It is an invasive neurosurgical procedure that implants electrodes in the brain and stimulates electrically to treat movement disorders associated with Parkinson's disease (PD), essential tremor, dystonia, and other neurological conditions. Transcranial magnetic stimulation (TMS) is another type of brain stimulation that is performed in a concentrated magnetic field. The magnetic field stimulates the regional cells of the brain to improve the symptom of depression. Electroconvulsive therapy (ECT) is a medical treatment performed with brief electric shocks while the subject is under anesthesia. It is most commonly used in subjects with severe depression or bipolar disorder who are not responding to other treatments. Transcranial direct-current stimulation (tDCS) is a noninvasive and painless brain stimulation treatment. It uses a constant low intensity of DC current to stimulate specific parts of the brain. It may be a valuable tool for the treatment of neuropsychiatric conditions such as depression, anxiety, Parkinson's disease, and chronic pain. There are also many trials to improve the cognitive function of the brain with noninvasive electrical and magnetic stimulation.

1 Brief History of Brain Electrical Stimulation

Electrical stimulation of the brain has a long history. It starts almost with the history of electricity itself. The first electricity known to humans was electrical energy originated from electric fishes like an electric torpedo and electric ray. They did not understand the mechanism of stimulation at all and that it is electricity. The first application of electricity using an electric torpedo and electric ray to remedy the headache is reported as dating back to 46 AD (Sironi, 2011). Then, electric shocks with the electric fishes were used extensively for the treatment of seizures, depression, and pain as a unique treatment method until the eighteenth century.

Up to the beginning of the nineteenth century, practically nothing was known about the brain except the gross anatomical structure and the existence of gray and white

K. S. Park, *Humans and Electricity*, https://doi.org/10.1007/978-3-031-20784-6_13

matter. When there was no idea about the constitution and function of each brain region, electrical stimulation then could open the doors to the experimental investigation. Giovanni Aldini (1762–1834), an Italian physicist in Bologna, used one of the earliest forms of transcranial direct-current stimulation (tDCS). With newly developed battery piles by Volta, he did many experiments of electrical stimulation on the brain, especially on the exposed human cerebral cortex of recently decapitated prisoners. He demonstrated how electrical stimulation applied on the surface of the head evokes responses like blinking and opening the eyes, facial grimaces, tongue, eye, and limb movements of the dead person. In 1804, Aldini reported the results of cortical stimulation and this finding led him to conclude that electricity could have therapeutic effects in the treatment of many neuropsychiatric disorders. Aldini's experimentations and hypotheses activated the following studies during the nineteenth and twentieth centuries on electrical stimulations to the brain for neurophysiologic investigation and for therapeutic purposes.

Electroconvulsive therapy (ECT) was first used for humans in 1938 by Ugo Cerletti (1877–1965), an Italian neurologist. At those times, schizophrenia and epilepsy were considered antagonistic disorders. So, convulsive therapy, which triggers a seizure intentionally, was used to treat the subjects with schizophrenia as a method to relieve the hallucination and confusion and return subjects to a normal state of mind. Even though a drug, Metrazol, was effective in inducing a seizure and alleviating the symptoms of subjects with mental illness, it was a deeply unpleasant and some uncontrolled seizures were difficult to avoid. Cerletti used a series of electrical shocks as a controllable alternative. He found that subjects had significantly improved after 10~20 electroconvulsive treatments. Retrograde amnesia, which occurs as a side effect of the ECT treatment, made the subjects have no adverse memory of the endured treatment. Since ECT was cheaper, less frightening, and more convenient, ECT soon replaced the drug within a few years later and it was being widely applied in mental hospitals all over the world since then.

José Delgado (1915–2011), a Spanish neurophysiologist at Yale University USA, is a pioneer in brain electrical stimulation. He is famous for the demonstrational experiment with a bull. In an experiment in 1963, he stopped a charging bull in its track of attack by sending a radio signal to a device implanted into a bull's brain. He has developed the implanted device, which he called "stimoceiver," as small as a quarter coin to be fully implanted in the brain of a bull and it is connected with the battery pack strapped to the head or worn around the neck by transdermal wires. In the experiments with other animals, like dogs, cats, monkeys, chimpanzees, gibbons, and humans, he could evoke smiles, snarls, bliss, terror, hunger, lust, and other responses by pushing a button (Horgan, 2005). Over the next two decades, he implanted electrodes in 25 subjects with schizophrenia and epilepsy. He achieved his best results in treating subjects with chronic pain, including a man injured in an automobile accident. The "stimoceiver" relieved both subject's pain and the depression caused by the accident so that the subject could return to work (Horgan, 2005). He also developed a device that could deliver electromagnetic pulses noninvasively to specific regions of the brain. Through the experiments with animals and human volunteers, he discovered that it could induce drowsiness and alertness, and treat tremors of Parkinson's disease successfully.

In 1985, Anthony Barker (1950-), an English electrical engineer at the University of Sheffield, developed the first transcranial magnetic stimulation (TMS) devices that are able to stimulate the brain in an awake adult focally and noninvasively. He first used time-varying magnetic fields to induce a current flow in tissue to depolarize a patch of neurons. An earlier method to induce neuronal depolarization was to use direct transcranial electrical stimulation (TES) with electrodes placed on the scalp. However, it has several problems and the high intensity of electrical stimulation is often painful. As a magnetic field, in contrast, pass through the scalp and skull unimpeded and undistorted, it could give much more precise localizations. It also can be applied painlessly. He demonstrated his invention by producing twitches in a specific area of the hand of a human volunteer. He applied transcranial magnetic stimulation (TMS) to the motor cortex in the opposite hemisphere that controls the movement of the hand. This demonstration revealed that TMS was capable of stimulating precise areas of the brain without the pain of electrical stimulation. Moreover, all the procedures of TMS had performed while subjects are in an awake and alert state. With technological progress after him, bursts of electrical current can be sent continuously by TMS to stimulate neural activity in the parts of the brain associated with mood control. Many following studies supported the effectiveness of TMS to treat depression and several mental health conditions including obsessive-compulsive disorder (OCD) and anxiety. TMS gained US Food and Drug Administration (FDA) approval in 2008.

In 1987, Alim Benabid (1942-), a French-Algerian neurosurgeon, discovered that an invasive method of deep brain electrical stimulation could improve the symptom of tremors in Parkinson's disease. Parkinson's disease is commonly treated with drugs like levodopa to improve muscle control, balance, and walking. However, the needed dosage of a drug tends to increase progressively over time along the course of the neuropathological conditions, and long-term use tended to induce motional abnormalities like tremor, rigidity, or slowness (Lee, 2001). Surgical treatment, ablating

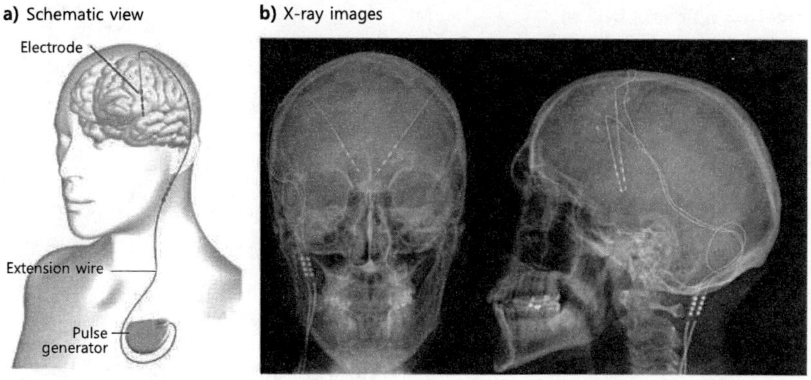

Fig. 13.1 Deep brain stimulation (DBS). (**a**) Schematic view of DBS with an electrode, extension wire, and a pulse generator, (**b**) frontal and lateral X-ray images of the head with DBS device implantation. (https://commons.wikimedia.org/wiki/)

the corresponding region, was considered an alternative option for the treatment of Parkinson's disease. Ablating is to inhibit involuntary abnormal motion by irreversibly destroying the target cells associated with severe Parkinson's disease not controlled by optimal drug dosage. Efficacy of electrical stimulation for the treatment of Parkinson's tremors was found accidentally during the surgical procedure to find the target location for ablation. To find the correct and accurate target location for ablation, an electrode used to be positioned around the target area and cells were stimulated with electrical current in the frequency range of 20–50 Hz while observing the movement of a subject. As electrical stimulation itself to examine the target area sometimes has stopped the quivering of the movement of a subject, Benabid reasoned that electrical stimulation might be a more proper surgical solution than ablation without destroying tissue. With the following experiments, he tried to find the optimal stimulating frequency and the optimal location. He later indicated that the subthalamic nucleus (STN) might be a more effective stimulation target than the thalamus through a series of animal experiments. As technology advances, stimulations could be applied continuously for a long time. In the 1990s, DBS has widely accepted as a treatment option for subjects with movement disorders like Parkinson's disease, essential tremor, and dystonia. DBS gained US Food and Drug Administration (FDA) approval in 2002 to treat Parkinson's disease.

Transcranial direct-current stimulation (tDCS) regained interest among neuroscientists at the beginning of the twenty-first century as an effective treatment with fewer side effects than antidepressants and other drugs. The device is developed mostly for use at home, by delivering a low-energy electrical current to increase brain activity and reduce symptoms of depression.

These days, mental health professionals often prescribe medication and some form of psychotherapy alongside brain stimulation techniques to maximize the chances of subjects being effective.

2 Deep Brain Stimulation

Deep brain stimulation (DBS) is the most successful electrical stimulation procedure for the brain. It is a neurosurgical procedure that uses implanted electrodes and electrical stimulation to treat neurological or psychiatric diseases including Parkinson's disease (PD). Electrodes to stimulate the specific target tissue of the brain are inserted deep into the target area and are connected via electrical wires to the pulse generator placed under the chest (Fig. 13.1). Sometimes DBS device is called a neurostimulator or brain pacemaker. DBS directly changes brain activity with electrical stimulation in a controlled manner. It is estimated that more than 160,000 subjects have received DBS, and the number of new subjects treated worldwide are growing by more than 12,000 per year (Lee et al., 2019).

2.1 Mechanism of Deep Brain Stimulation

In spite of extensive basic science and clinical studies, the exact mechanism of DBS is still not entirely understood. However, symptoms responding to DBS on different timescales suggest that DBS is functioning with several different mechanisms that have their intrinsic time courses (Fig. 13.2). Symptoms responding rapidly are mediated by reversible DBS mechanisms such as the immediate electrical neuromodulation of pathological network activity. Symptoms responding more slowly are at least partly mediated by longer-term mechanisms like neuroplasticity and ultimately anatomical reorganization (Agnesi et al., 2013).

Electrical neuromodulation is the most basic and traditional understanding of the DBS mechanism. The basic hypothesis is that applied electrical stimulation modulates the responses of the nearby cells from the abnormal state toward the intrinsic normal physiological state (Fig.13.3). In

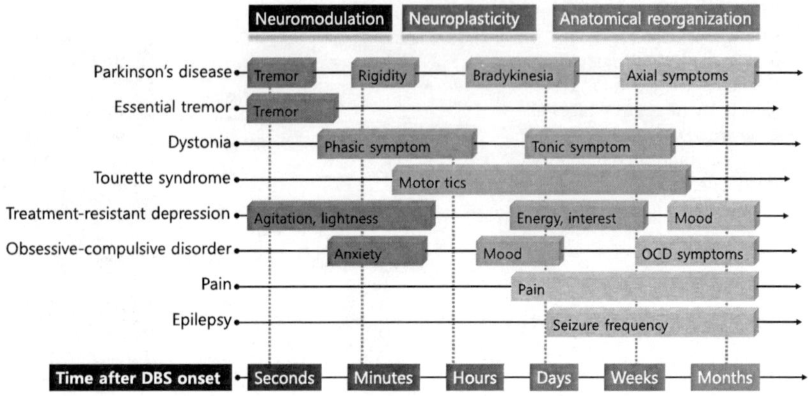

Fig. 13.2 Effects of DBS on various diseases along the time

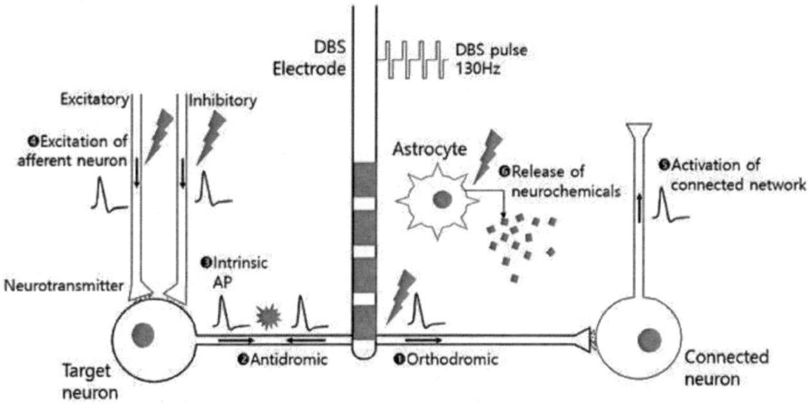

Fig. 13.3 DBS mechanism of immediate electrical neuromodulation

the basic level of mechanism, the electric field developed by the applied current stimulates the neural cell and results in the opening and closing of voltage-gated ion channels. DBS is usually applied to the target cells via the cathode, negative electrode, and it drives the potential of the neighboring area toward the negative direction. Strong negative potential out of the nerve cells depolarizes the neuronal membrane and the action potential can be generated in the stimulated site.

Developed electric potential stimulates predominantly axons and dendrites near the electrode, rather than the soma, which have substantially higher stimulation thresholds. When an action potential is generated in the axon, it can propagate both antidromic, toward the soma, as well as orthodromic directions, away from the soma. Antidromically propagating action potential may collide with the intrinsic action potential which is generated from the soma and propagating in the orthodromic direction. This collision of action potentials blocks further propagation of intrinsic action potential, which might belong to an abnormal pattern caused by the disease. On the contrary, action potential generated by DBS in the orthodromic direction propagates along with the axonal fiber up to the end of the axon terminal. It is transferred to the connected neurons via synapses replacing the intrinsic action potential and the following neural circuit can be modulated by the DBS.

An additional mechanism of DBS is the action on the afferent axons coming to the neurons in the target area and connected through the synapses. If the developed electrical potential acts on the axonal terminals of the afferent presynaptic neurons, it can increase the release of neurotransmitters which may be excitatory or inhibitory, and can modulate the postsynaptic neuron in either direction. The delayed time course of neuronal inactivation, 100 ms after stimulation onset, supports this hypothesis that inhibition occurs through the release of the inhibitory neurotransmitter through the synapses. This mechanism can be functional in inhibitory direction and excitatory direction depending on the type of related synapses to stimulation.

From the early stage of DBS, applied high-frequency electrical stimulations showed similar effects to an ablative surgery performed on the same regions. So, DBS was thought to function by inhibiting neurons with a reversible lesion near the stimulating electrode just like an ablative surgery (Herrington et al., 2016). Intraoperative recordings in human subjects showed decreased neuronal spike rates during stimulation. The inhibitory function can be resulted predominantly from local cells near the electrode or by the activation of connected neuronal networks.

For example, in Parkinson's disease, bradykinesia is related to the increase of rhythmic activity of 13~30 Hz within the basal ganglia. This rhythmic activity in the beta band is thought to signal "Don't move" to the other area of the brain. In normal subjects, this beta rhythm reduces before any voluntary movement allowing the subsequent movement. However, in PD subjects, the beta activity is not reduced and continues synchronized to the movement. So, this rhythm is considered abnormal and is thought important in the production of bradykinesia (Silverdale, 2017). During the DBS, reduction of this beta rhythm in basal ganglia has been confirmed and it supports the inhibitory mechanism of DBS. As in this example, DBS seems to work by dissociating inputs and outputs in the stimulated target cells and disrupting abnormal information flow through the connected neural network.

In addition to the local electrical effects of the developed electric potential, DBS also induces some neurochemical changes including the release of neurotransmitters locally and throughout the stimulated network. DBS induces glutamate and adenosine triphosphate (ATP) release from astrocytes, which are non-neuronal cells. Released chemicals include neurotransmitters, like dopamine, noradrenaline, serotonin, and gamma-aminobutyric acid (GABA), and they inevitably affect the behavior of the neural functions of the nearby network in the following period. Preclinical and clinical studies show that neurochemical effects of DBS appear in short-term effects over seconds or minutes.

DBS effects that emerge over minutes to days likely result from synaptic plasticity-related changes in the stimulated neural network and such network changes occur over similar timescales in natural behaviors such as learning. Many studies demonstrated that continuous DBS results in network-level changes in the activity.

Rather than a single unifying mechanism, DBS likely acts via several, nonexclusive mechanisms including local and network-wide electrical and neurochemical effects of stimulation, modulation of oscillatory activity, synaptic plasticity, and, potentially, long-term anatomical reorganization. A better understanding of the mechanisms underlying DBS will permit the progress of supporting technologies and fine-tuning of surgical methods to maximize benefits and reduce adverse effects.

2.2 Indications for Deep Brain Stimulation

DBS is one of the fastest-growing treatment in neurosurgery and indications for the application of DBS have extended (Table 13.1). Currently, DBS is an FDA-approved treatment for Parkinson's disease (PD), essential tremor (ET), dystonia, obsessive-compulsive disorder (OCD), and medically refractory epilepsy.

Parkinson's Disease

Parkinson's disease (PD) is a major target disease for DBS treatment. PD is a long-term degenerative disorder of the brain that mainly affects the motor system leading to shaking, stiffness, and difficulty in walking, balance, and coordination. PD symptoms usually begin slowly and get worse over time. As the disease progresses, subjects may have difficulty in walking and talking and nonmotor symptoms become more common. They may have mental and behavioral changes, sleep problems, depression, memory difficulties, and fatigue. PD typically occurs in subjects over the age of 60. And about 1% of subjects over age 60 are affected by PD. Men are 50% more affected than women. Globally, the number of subjects with PD is estimated at 9.4 million in the year 2020 (Maserejian et al., 2021). The average life expectancy after diagnosis of PD is 7~15 years (Sveinbjornsdottir, 2016).

PD occurs when nerve cells in the substantia nigra of the basal ganglia, an area of the brain that controls movement, become impaired and/or die. Normally, these nerve cells, produce an important brain chemical known as dopamine. When the neurons die or become impaired, they lead a brain to a dopamine deficit, which causes the movement problems of PD. Even though some cases of PD appear to be related to genetic origin, most PD occurs randomly and does not seem to be related to their families. PD is believed to result from a combination of genetic factors and environmental factors such as exposure to toxins. However, the mechanism of PD, how the cells producing the dopamine die, is not known clearly.

PD has four main symptoms. They are tremors in hands, arms, legs, jaw, or head, stiffness of the limbs and trunk, slowness of movement, impaired balance, and coordination, sometimes leading to falls. Other symptoms may include depression and other emotional changes, difficulty in swallowing, chewing, and speaking; urinary problems or constipation, skin problems, and sleep disruptions. Symptoms of PD and the rate of progression differ among subjects. As there is no medical test accurately detecting PD, it is difficult to diagnose in the earliest stage. PD is usually diagnosed based on a subject's medical history and a neurological examination for motor symptoms of chief complaint.

PD belongs to a disease that cannot be cured. PD is only treated to relieve the symptoms so that subjects can return to their ordinary life maximally. Before the advent of medicine, levodopa, stereotactic functional surgery, which ablates the symptom-corresponding thalamic region was the major treatment. However, since levodopa, also called L-dopa, was introduced in 1968, it became the main therapy for

Table 13.1 Indications for DBS implantation

Type of disorder	Disorder	Symptoms	DBS location
Movement disorders	**Parkinson's disease** (PD)	Tremors, stiffness of limbs & trunk, slowness of movement, impaired balance & coordination	Subthalamic nucleus (STN), internal region of the globus pallidus (GPI), ventral intermediate nucleus (VIM)
	Essential tremor	Tremors obviously in hands, difficulty in using hands, quivering voice, uncontrollable head-nodding	Ventral intermediate nucleus (VIM)
	Dystonia	Involuntary movements like muscle contractions & spasms, repetitive movements, unusual & awkward postures	Internal region of the globus pallidus (GPI)
	Tourette's syndrome (TS)	Tics (sudden, brief, intermittent movements or sounds) like eye blinking, facial grimacing, shoulder shrugging, and throat clearing	Thalamic centromedian nucleus (CMT), internal region of the globus pallidus (GPI), external region of the globus pallidus (GPE), anterior limb of the internal capsule, nucleus accumbens (NAcc).
Psychiatric disorders	**Treatment-resistant depression** (TRD)	Depression ineffective to antidepressants, increasingly severe & longer episodes of depression, high anxiety or anxiety disorder	Subgenual anterior cingulate cortex (sACC), ventral capsule/ventral striatum (VC/VS), nucleus accumbens (NAcc), lateral habenula (LH)
	Obsessive-compulsive disorder (OCD)	Pattern of unwanted thoughts & fears (obsessions) leading to repetitive behaviors (compulsions) which interfere daily activities & cause significant distress	Anterior limb of the internal capsule (ALIC), ventral capsule/ventral striatum (VC/VS)
Neurological disorders	**Epilepsy**	Recurrent epileptic seizures, uncontrolled drug-resistant seizures	Anterior thalamic nucleus (ATN), thalamic centromedian nucleus (CMT)
	Pain	Chronic, neuropathic pain occurring without an obvious pain-inducing event or factor, pain refractory to medication	Ventral posterolateral nucleus (VPL), ventral posteromedial nucleus (VPM), periventricular gray matter (PVG), periaqueductal gray matter (PAG), internal capsule (IC), anterior cingulate cortex (ACC)

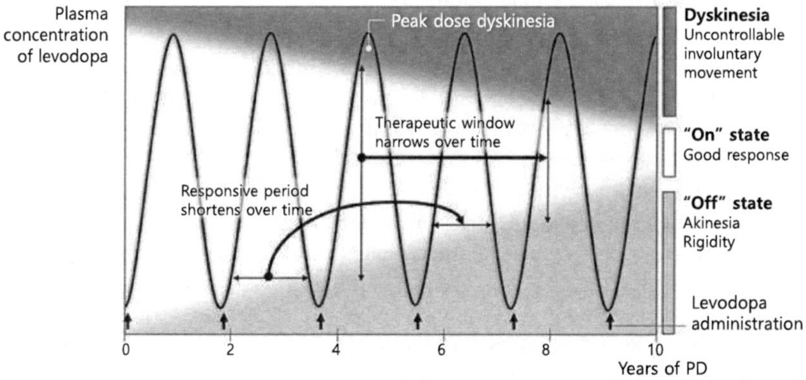

Fig. 13.4 The response pattern of levodopa with the progression of PD over time. Levodopa provides stable symptomatic control during the early years. However, it causes motor fluctuations and dyskinesias in the progressed PD state. The therapeutic window narrows and the responsive window shortens over time

PD. Levodopa is absorbed in the blood from the small intestine and travels through the blood to the brain, where it is converted into dopamine replenishing the deficit production. However, as the disease progresses, it becomes less effective with increased involuntary muscle movements and has some side effects, like nausea, vomiting, low blood pressure, and restlessness.

Levodopa works very well and provides stable symptomatic control in the early years of medication for the PD subjects. Levodopa acts like dopamine. However, it creates an environment where there's too much, or too little, dopamine in the body, with increased probability over time (Fig. 13.4). This can cause uncontrollable movements. The initial complication is commonly wearing-off, which occurs in the environment of too little concentration of levodopa. PD symptoms reemerge or worsen before the next scheduled levodopa administration as the effect of levodopa diminishes. The time when their medication is working is referred to as the "On" state and the time when their medication is not working is referred to as the "Off" state.

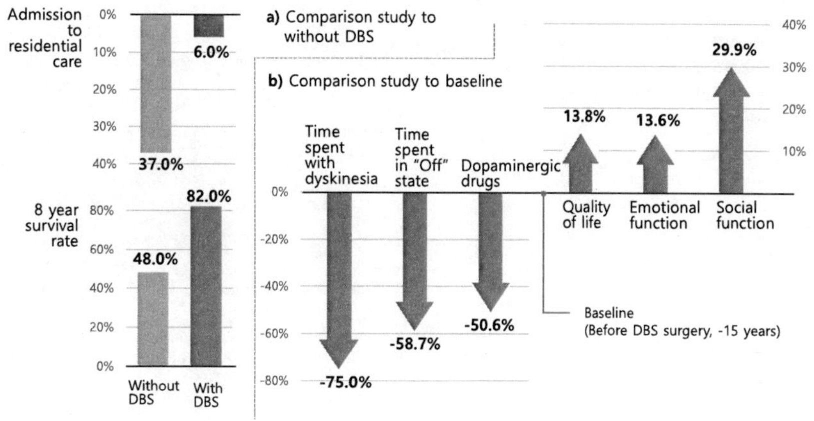

Fig. 13.5 Effects of DBS treatment in PD subjects. (**a**) Comparison results to subjects who denied the DBS surgery, (**b**) comparison of symptoms at 15 years after DBS surgery to the baseline of before DBS surgery

Dyskinesia is the major side effect occurring in the environment of too much concentration of levodopa with long-term use of levodopa. Dyskinesia is uncontrolled, involuntary movement that can include twitches, jerks, twisting, or writhing movements. Dyskinesia can affect various parts of the body such as the arms, legs, and torso. With the progress of PD and increased dose of levodopa, the therapeutic window of the "On" state narrows, and the responsive period of levodopa is shortened with increased wear "Off" time. This dyskinesia, medication refractory tremor, and unpredictable "Off" time appearing with long-term use of levodopa are the major indications for DBS in PD subjects.

There are three target areas in the deep brain that are stimulated for PD subjects (Fig. 13.6). They are the subthalamic nucleus (STN), the globus pallidus internus (GPI), and the ventral intermediate nucleus (VIM) of the thalamus. Treatment with DBS improves PD symptoms in PD subjects quite similarly to the levodopa medication. It can increase ease of movement and reduce the freezing of gait, disabling tremor, dyskinesia, severe cramps, and pains. DBS also can reduce the amount of medication dosage. Some symptoms like tremor, slowness, and stiffness, may improve immediately after the device is switched on, while some symptoms take more time to respond. DBS works more effectively than medications alone in controlling Parkinson's symptoms. The effects of DBS are confirmed in many facets (Fig. 13.5). DBS can extend lifespan. The survival rate after 8 years of PD subjects with DBS surgery was 82% and it is higher than 48% of PD subjects who denied the DBS surgery and received only medication treatment. The rate of admission to residential care was significantly reduced in DBS PD subjects to 6% from 37% of PD subjects who only cared with medications (Ngoga et al., 2014). Long-term outcomes of 15 years compared to the baseline, DBS PD subjects show a 75% reduction in time spent with dyskinesia, 58.7% reduction in time spent in the "Off" state, and 50.6% reduction in dopaminergic drugs. And, they also show a 13.8% increase in quality of life, 13.6% in emotional function, and 29.9% increase in social function domains (Bove et al., 2021). In effect, DBS is providing better and longer lives to PD patients than the treatment using medication only.

Essential Tremor

In addition to PD, *essential tremor* (*ET*) and dystonia are disorders that belong to movement disorders like PD and are actively benefited by DBS. Essential tremor (ET) is one of the most common movement disorders causing involuntary and rhythmic shaking. It can affect almost any part of the body, but the tremors occur most often in the hands, mostly when the subjects do simple tasks, like drinking from a glass or tying shoelaces. ET can cause significant functional disability in some subjects. Although medications can relieve the subject's tremors, approximately half of them are medication-resistant. For this group of ET subjects with marked disability due to the tremor, DBS with a target at ventral intermediate (VIM) is a viable treatment option (Fig. 13.6). Multiple studies of DBS treatment on ET subjects have shown its efficacy in the treatment of ET hand tremors, often with secondary improvement in voice and head tremors. Long-term studies have reported that a majority of patients continue to experience improvement in tremors (Lyons & Pahwa, 2004). Up to 90% of subjects with ET have at least moderate symptom relief after DBS treatment.

Dystonia

Dystonia is another movement disorder that is characterized by patterned directional and sustained muscle contractions and can cause twisting, repetitive movements, and abnormal postures. Usually, medical treatments are unsatisfactory and limited by side effects and some patients become resistant to medical treatment later (Hu, 2014). Bilateral globus pallidus internus (GPI) DBS has demonstrated convincing clinical

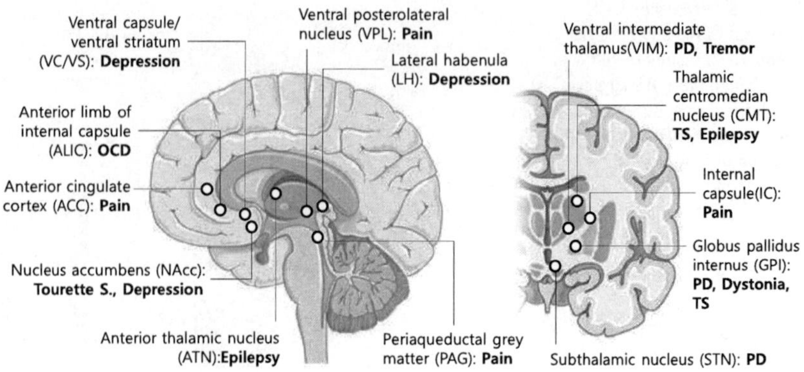

Fig. 13.6 DBS target regions for each disorder. *OCD* obsessive-compulsive disorder, *PD* Parkinson's disease, *TS* Tourette's syndrome

benefit in a large number of patients with primary generalized/segmental dystonia (Fig. 13.6). DBS treatment for PD, essential tremor, and dystonia have proved by FDA in the year 2002, 1997, and 2003 respectively.

Tourette's Syndrome

Tourette's syndrome (TS) is a disorder that involves repetitive movements or unwanted sounds, which are called tics, that cannot be easily controlled. Subjects might repeatedly blink their eyes, shrug shoulders, and blurt out unusual sounds or offensive words. Mostly TS onsets in childhood and the symptoms fade away. If symptoms are persisted toward adulthood, pharmacological therapies are required in association with cognitive behavioral therapies. Rarely, when subjects experience severe and disabling symptoms refractory to conventional treatments, DBS can be considered an effective option for symptomatic control. A multiplicity of targets including the thalamic centromedian nucleus (CMT) and the internal region of the globus pallidus (GPI), reflects the fact that there is no consensus on which target is the most effective yet. DBS is partially effective in reducing the motor and nonmotor symptoms of TS with around 40% improvement on tic severity scales (Casagrande et al., 2019).

Psychiatric Disorders

The safety and efficacy of DBS treatment in movement disorders have increased the interest in the potential application of DBS for psychiatric disorders. *Obsessive – compulsive disorder* (*OCD*) is a common, disabling psychiatric disease characterized by persistent, intrusive thoughts, and ritualistic, repetitive behaviors. OCD subjects spend nearly 6 hours per day occupied with intrusive obsessions and performing compulsions or rituals. Stimulation of the internal capsule (ALIC) and ventral capsule/ventral striatum (VC/VS) received FDA approval in 2009 for patients with severe, treatment-refractory OCD (Karas et al., 2019). DBS has shown promising outcomes for patients with *treatment – resistant depression* (*TRD*), for whom antidepressant or psychological counseling does not ease depression symptoms. DBS treatment is statistically beneficial for TRD by inducing a robust antidepressant effect that is sustained over a long period of time. However, further basic scientific research is required to find optimal targets for TRD subjects to increase the efficacy and reduce the side effects via the DBS treatment.

Pain

DBS is also considered a viable option for subjects with chronic or neuropathic pain. About 5% of the population is estimated to suffer from chronic pain despite pharmacotherapy. Neuropathic pain is pain caused by a lesion or disease of the somatosensory system or occurring without an obvious pain-inducing event or factor. DBS for chronic pain is most successful for pain after amputation, cranial and facial pain, and poststroke pain (Farrell et al., 2018). When invasive options are considered for the treatment of chronic pain, currently DBS is considered next to spinal cord stimulation (SCS), which is another successful and common strategy for treating chronic pain, first reported half a century ago.

Epilepsy

Epilepsy is a neurological disorder in which brain activity becomes abnormal by causing seizures or periods of unusual behavior, sensations, and sometimes loss of awareness. First, they are treated with antiepileptic drugs (AEDs). AEDs can control seizures in most patients with epilepsy. However, at least 30% of adults with epilepsy, which is called drug-resistant epilepsy (DRE), do not achieve seizure control with AEDs. Surgery to remove or disconnect the epileptogenic zone is not always an appropriate option. These patients may be appropriate candidates for DBS treatment (Zangiabadi et al., 2019). DBS tends to have better results in patients with generalized epilepsy, although it has been used with success in some patients with focal onset seizures. While the optimal target region is variable for each epileptic

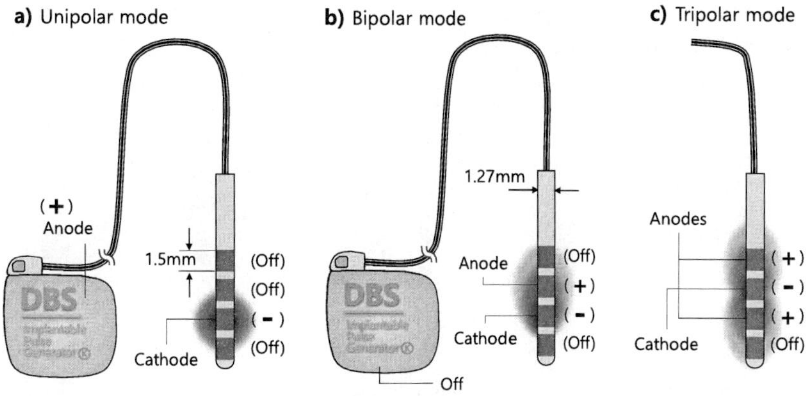

Fig. 13.7 Stimulating modes in DBS. (**a**) Unipolar stimulation mode with a cathode from a lead electrode and anode at a metal case of IPG, (**b**) bipolar stimulation mode with a cathode and an anode from lead electrodes, (**c**) tripolar stimulation mode with one cathode and two anodes from lead electrodes

syndrome, the anterior thalamic nucleus (ATN) is most recommended over the other targets.

DBS is currently under active investigation for a wide range of other treatment-resistant conditions, including Alzheimer's disease, schizophrenia, addiction, dementia, vegetative state, anorexia, restless legs syndrome, camptocormia, obesity, tinnitus, arterial hypertension, sleep disorders, and stroke recovery (Lyons, 2011). Figure 13.6 shows the DBS target regions for each disorder.

2.3 Components of Deep Brain Stimulation Device

A DBS unit is composed of the implanted parts and the external units that control the implantable pulse generator. Implantable components of the DBS device include an implantable pulse generator (IPG), stimulating DBS leads, and extension wires that connect the DBS leads to the IPG. Leads having four stimulating contacts are most commonly used for DBS and they are stereotactically implanted in the target region deep in the brain. Leads are typically placed bilaterally, although unilateral stimulations are dictated upon the clinical conditions. Implanted leads are connected via subcutaneous extension wires to an implantable pulse generator (IPG). IPG is the titanium unit containing all the electronics, sensing, processing, pulse generation and control circuits, and battery for the circuits. It is placed in a subcutaneous pocket on the chest wall underneath the collarbone as in the case of cardiac pacemaker implantation. External units interact wirelessly with the IPG to program the parameters and manipulate the functions. Clinicians use a handheld external unit for communication with the IPG to adjust the parameters of stimulation and for tuning stimulation to maximize symptom relief and minimize side effects.

Stimulating leads for DBS most commonly used contain four platinum/iridium electrodes encased in polymer-based insulation. The diameter of the leads is 1.27 mm and each of the four electrodes has a length of 1.5 mm with 0.5 mm or 1.5 mm gaps between them providing a total span of 7.5 mm or 10.5 mm. Using the four electrodes, electrodes can be paired to provide optimal electrical stimulation to provide maximal symptom suppression with minimal stimulation-induced adverse effects. Stimulation can be applied in unipolar mode or in bipolar mode (Fig. 13.7). In unipolar mode stimulation, a cathode, a negative electrode, is selected from the lead electrodes and the countering anode, a positive electrode, is designated to the metal case of the IPG. In bipolar mode stimulation, cathode and anode electrodes are selected from the lead electrodes and the metal case of the IPG is not connected. Unipolar stimulation mode provides a spherical stimulation area around the cathode, whereas bipolar stimulation mode provides a more elongated but constricted stimulating area between the cathode and anode. Bipolar stimulation mode provides more focused stimulation which can limit the spread of the electric current and reduce the side effects of stimulating structures adjacent to the intended stimulation target. Among the electrodes, active electrodes, alone or in combination, are selected to have a negative polarity, that is, to serve as the cathode. Polarity and the position of stimulating electrodes can be programmed by the clinician using the external unit via wireless communication.

Technological advances and innovations extended the flexibility of DBS with various types of leads (Fig.13.8). Leads of four electrical contacts have been used as a standard lead. By extending the gap between electrodes from 0.5 mm to 1.5 mm, the possible range of stimulation can be increased from 7.5 mm to 10.5 mm. One stimulating electrode can be selected as a cathode for unipolar stimulation, and two consecutive electrodes also can be selected as cathode electrodes for unipolar stimulation to extend the stimulating volume. Stimulation can be applied in alternating mode by stimulating two interleaved electrodes at temporally different times. The number of electrical contacts have increased to eight

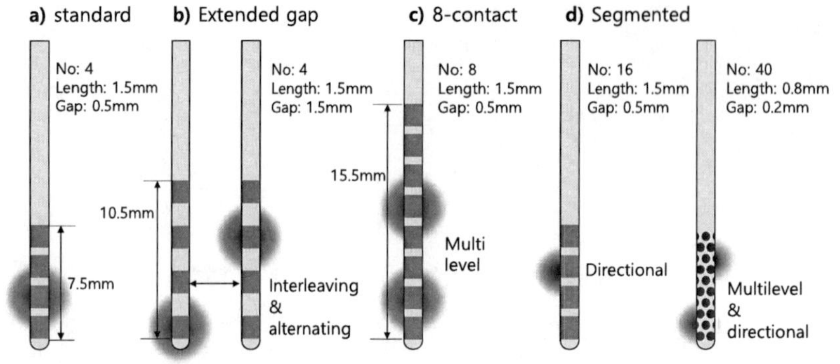

Fig. 13.8 Different types of advanced DBS leads. (**a**) Standard DBS lead with 4 contact and 0.5 mm gaps, (**b**) lead with an extended gap of 1.5 mm showing spatially interleaved and temporally alternating stimulating pattern, (**c**) lead with eight electrical contacts showing multilevel stimulation at the same time, (**d**) segmented leads with 16 and 40 electrical contacts showing directional stimulating patterns

doubling the range of stimulation to 15.5 mm by a single lead. With this lead, multilevel stimulation, which stimulates more than one electrode at the same time, can be selected to excite multiple locations simultaneously.

A recent innovation is the development of the segmented lead which divided a single electrical contact surface into three or four segments along the circumference. The number of electrical contacts have increased to 16 and 40 as shown in Fig. 13.8d. The greatest advantage of this segmented lead is to allow directional stimulation. Instead of stimulating the whole volume of neuronal cells along the ring electrode, only a small portion of the neighboring volume is selected by the direction of stimulated electrical contact. By selecting cathode and anode electrodes from electrical contacting segments in the same depth, the flow direction of the stimulating current can be 90° rotated to a horizontal direction along the circumference of the lead from the vertical direction along with the lead. In fact, current flow can be steered in an arbitrary direction based on the needs of individual patients by selecting the cathode and anode electrodes from the contacting segments in appropriate depths and directions. By selecting and programming the optimal directional segment of electrical contact, the volume of tissue activated (VTA) can be precisely decided with increasing the selectivity maximally and decreasing the stimulation-induced adverse effect minimally.

External units can be classified by the users. An external unit for clinicians is a handheld or portable device to communicate wirelessly with IPG using radiofrequency telemetry. It includes diverse programming functions that clinicians easily use to set the electrode configurations, adjust the various stimulation parameters, and activate various features of the DBS system. On the contrary, an external unit for patients and their caregivers is a smaller portable kind of remote control to interact with the IPG. Patients can use the external unit to turn their implanted device on and off, adjust stimulation parameter levels within a range specified by the clinician, select a setting among different stimulation settings, and check the battery status.

2.4 Stimulating Pulse Patterns

Even though the shape of the lead and the number of electrical contacts have been determined, the stimulation effect can be differentiated greatly by the characteristics of applied pulse shapes and their pattern. Characteristics of stimulating pulses can be recognized by their amplitudes, frequencies, widths, shapes, and applying patterns.

The amplitude of the stimulating pulse determines the volume of stimulation. Increased amplitude will cause further spread of the stimulating current and affect neural tissues at further distances from the electrode. Usually, the amplitude of stimulation is set to limit the range of current diffusion from the electrode which is less than 3 mm. Stimulating pulses can be controlled by the voltage or by the current. As the impedance around the stimulating electrodes can be varied over time according to the environmental states like glial encapsulation or protein adsorption on the electrode, controlling the stimulating pulse amplitude by the current is considered to provide more stable stimulations. When the stimulation is applied in a current-controlled manner, the maximum amplitude of the pulse is usually maintained at lower than 4 mA.

DBS pulses are applied usually in high-frequency stimulation (HFS) mode and a pulse rate of 130 Hz has been widely used from the beginning of DBS. HFS may induce better responses generally and especially for tremors. Low-frequency stimulation (LFS), below than 100 Hz, may be superior to HFS in akinesia, gait, and freezing of gait. Using LFS is also associated with a decrease in power consumption and may help extend battery longevity. Typically, lower pulse rates below 50 Hz are not efficacious for most of the applications to reduce symptoms accompanying the disorders. On the contrary, stimulating the brain target with this low rate may actually cause stimulation-induced worsening of tremors and other symptoms accompanying movement disorders (Zauber et al., 2015). Current DBS devices provide programmable stimulating pulse rates in the range 2~250 Hz.

The width of the stimulating pulse is related to the portion of neurons in the volume of tissue activated (VTA). The wider pulse width increases the probability to excite an increased number of different neuronal cells within the same volume. Usually, the pulse width is set in the range of 60~90 μs and incremented to optimize symptom control. However, considering the stimulation-induced side effects and the battery drainage, a shorter width pulse is strategically employed. Current DBS devices provide programmable stimulating pulse width in the range of 20~450 μs.

The waveform and temporal pattern of stimulation pulses can influence the number of stimulated neuronal cells within the volume of tissue activated. Different shapes of waveforms of stimulating pulse have been designed and used (Fig. 13.9a). Pulses are usually delivered in biphasic mode to balance the applied charges between phases. Conventional biphasic pulse has the first short and high amplitude negative phase which is followed by a longer and lower-amplitude positive discharge phase. This waveform is also called pseudo-monophasic since it looks similar to a negative monophasic pulse overall. Symmetric biphasic pulse shape has produced greater suppression of symptoms in PD and ET subjects, even though they increase the battery drainage. Generally, at any given stimulation intensity, symmetric biphasic pulses are likely to activate a larger number of neuronal cells than asymmetric pulses. Both negative and positive phases are considered to contribute to net activation (Krauss et al., 2021). Studies show that nonrectangular pulse shapes, like a triangular waveform shape, may be more energy-efficient than rectangular pulses, but that their effects are dependent on pulse width and electrode geometry (Ballestero et al., 2015).

The polarity and temporal gap between positive and negative phases also can be determined to improve the activation of neurons. The threshold to induce a reduction in PD tremor and the threshold producing the adverse side effects are higher with anodic stimulation than with cathodic stimulation. However, at the amplitude just below the threshold producing an adverse effect, anodic stimulation produces a greater suppression of symptoms than cathodic stimulation (Kirsch et al., 2018). Introducing an interphase gap produces an increase in stimulus performance when compared to a simple biphasic pulse with equal amplitude and charge per phase (Carlyon et al., 2005).

Stimulation pulses can be delivered regularly with a fixed stimulation rate and nonregularly with random interpulse intervals (Fig. 13.9b). Increasing pieces of evidence are suggested that the temporal pattern of stimulation can influence the clinical performance of DBS (Grill, 2018). As pathological excessive synchronization is a hallmark of several neurological disorders impairing brain function, stimulation techniques have been computationally designed to counteract this abnormal synchronization of neural activity. Coordinated and distributed burst stimulating pattern as in Fig. 13.9b, which delivers brief high-frequency pulse trains through different stimulation contacts, might cause acute desynchronizing effects during stimulation preventing the development of pathological synchronization. The effect of desynchronization was shown to accumulate and persist after the cessation of coordinated reset stimulation (Popovych & Tass, 2014).

Adaptive Closed-Loop DBS

As the cardiac pacemakers progressed with rate-adaptive stimulation mode, DBS, brain cardiac pacemaker, is also progressing with adaptive stimulating mode. In the adaptive

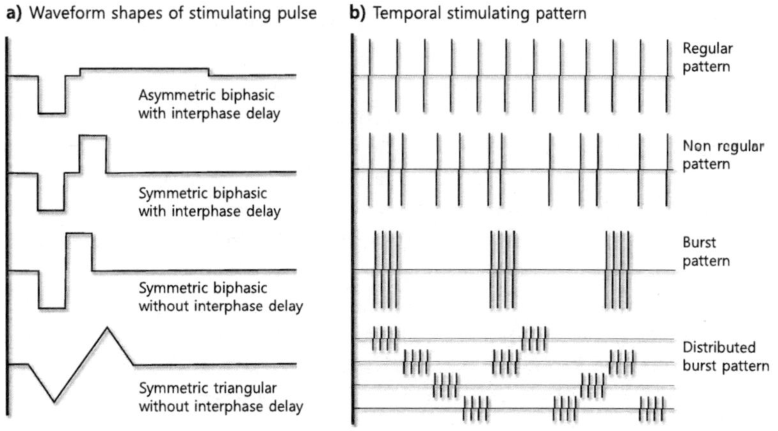

Fig. 13.9 Waveform shapes and temporal patterns of stimulating pulses. (**a**) Four different shapes of stimulating pulse waveforms, (**b**) different temporal patterns of stimulating pulses

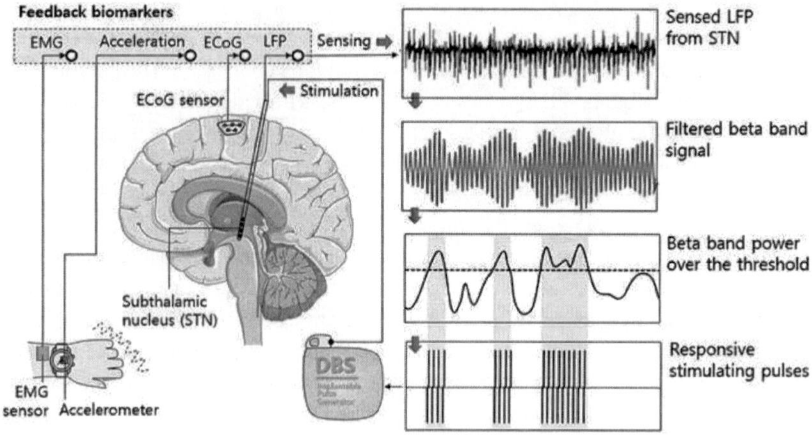

Fig. 13.10 Adaptive closed-loop DBS. Local field potential (LFP) is sensed via the electrical contact of the implanted lead into the subthalamic nucleus (STN) and stimulation pulses are generated when the power of the filtered signal in the beta band crosses the threshold

DBS, the systemic signal flow loop is closed with the addition of the sensing part (Fig. 13.10). Closed-loop DBS can optimize stimulations proactively, reactively, and adaptively to the brain states and accompanying symptoms. If the sensed signals have a causal relationship with the symptoms of the disorder, they can be used to prevent symptoms by enabling DBS earlier than reactive symptom management. Currently, closed DBSs have been studied actively in PD, essential tremors, and epilepsy.

Local field potentials (LFP) sensed from contacts of implanted lead electrodes can be used to differentiate the various states of disorders. As beta activity in the range of 13~30 Hz is characteristically recorded in the implanted target region of the subthalamic nucleus (STN) and internal region of the globus pallidus (GPI) accompanying the occurrence of slow movement and muscle stiffness, increased power in the beta band in the recorded signal can be interpreted as the increased probability of movement and muscle stiffness. In a closed-loop DBS, DBS is delivered when the amplitude of beta band power crosses over the threshold.

Gamma activity recorded at the cortical level can be applied as a biomarker for evaluating dyskinesia. Electrocorticographic (ECoG) recorded over motor cortex with surface electrode shows gamma activities clearly with large amplitude and more easily distinguished from stimulation artifacts differently from LFP recordings. When a patient is in a dyskinesia state equivalent to the hypersensitive state on levodopa, prominent gamma activities in the range of 30~70 Hz are picked up in ECoG sensors. According to the power level of gamma activities, DBS can be stopped or reduced to ameliorate the dyskinetic symptom. This adaptive mechanism can reduce the power requirements by up to 50% while effectively reducing the on-state dyskinesia.

For epileptic patients, an abnormal electrocorticographic activity that precedes a possible clinical seizure can be decoded from recorded electrical signals via the stimulating electrode or cortical surface electrode. Then, DBS provides short-term stimulation in response to the detected abnormal electrocorticographic activity. This adaptive DBS for epilepsy is reported to reduce the frequency of partial-onset seizures and is well-tolerated and acceptably safe (Morrell, 2011). To detect the onset of tremor or tremor-triggering voluntary movement, EMG sensors or accelerometers recorded in symptom-affected parts like hands can be used in replacement of LFP or ECoG. Three channels of gyroscopic data collected from a smartwatch worn on the affected limb can be processed and fed into a control algorithm to adjust the stimulation according to the magnitude of the tremor band power (4–8 Hz). This tremor-modulated approach reduces the average tremor band amplitude by about 59% when compared to open-loop DBS, but it used only 16% of baseline stimulation (Herron et al., 2017). Adaptive feedback control is also likely to expand to include multiple signals related to the neural and physical state, supported by more sophisticated processing and control algorithms.

2.5 Battery for DBS

The battery does a key role to determine the size of an implantable IPG unit and its operating service life. Compared to the cardiac pacemaker, DBS takes more power consumption since it stimulates with a higher frequency of around 100 Hz compared to around 1 Hz of the cardiac pacemaker. Consumed power is determined similarly to a cardiac pacemaker.

When the pulses are delivered in voltage-controlled mode, total electrical energy delivered (TEED) is calculated

$$\text{TEED} = \left(\text{voltage}(\text{V})^2 \times \text{frequency}(\text{Hz}) \times \text{pulse\_width}(\text{s})\right)/Z_{\text{load}} \tag{13.1}$$

$$Z_{\text{load}} = Z_{\text{e1}} + R_{\text{s}} + Z_{\text{e2}} \tag{13.2}$$

where Z_{load} is the impedance composed of impedances of two stimulating electrodes Z_{e1}, Z_{e2} and impedance of stimulated tissue volume R_s.

When the pulses are delivered in current-controlled mode, total electrical energy delivered (TEED) is calculated

$$\text{TEED} = \text{current}(\text{I})^2 \times Z_{\text{load}} \times \text{frequency}(\text{Hz}) \times \text{pulse\_width}(\text{s}) \tag{13.3}$$

For a battery of capacity 5.0 Ah, the expected life of the battery can be calculated as follows, when the stimulation is delivered with 130 Hz and 60 μs pulses.

$$\text{Battery capacity}: C_b = 5.0Ah \tag{13.4}$$

$$\text{Average current}: I_a = I_s \times \text{frequency} \times (t_p) = 4mA \times 130Hz \times 60\mu s = 31.2\mu A \tag{13.5}$$

$$\text{Total current drain}: I_t = I_s \times 2 = 62.4\mu A \tag{13.6}$$

$$\text{Longevity in hours}: L_h = C_b/I_t = 5.0Ah/62.4\mu A = 80{,}128h \tag{13.7}$$

$$\text{Longevity in years}: L_y = L_h/(24 \times 365) = 80{,}128/(24 \times 365) = 9.14 \text{ years} \tag{13.8}$$

In this example case, a battery of 5.0 Ah, which is an almost equal capacity of a smartphone battery, is sufficient to power the DBS for approximately 9.14 years. During the above calculation, the peak stimulating current is assumed as 4 mA, stimulation is delivered with 130 Hz and 60 μs pulses, and the total current consumed is assumed two times the current consumed by the stimulating pulses. Most of the current DBS units are programmable in the stimulating frequency in the range of 2~250 Hz and pulse width in the range of 20~450 μs. If the same battery of 5.0 Ah is operated in maximal strength of 250 Hz and 450 μs, it can supply power for only 0.63 years. A smaller battery will reduce overall device size, but with shorter service life, and a bigger battery will allow longer service life, but with the cost of larger implanted devices. On average, nonrechargeable batteries have supplied DBS units for about 5 years.

Since 2008, rechargeable batteries have started to be used for DBS. It has the advantage to reduce the size and the weight of the DBS unit to less than 40 g compared to the 60 g that includes the nonrechargeable battery. Also, it can be used long time of 10~25 years which is 2~5 times longer. It makes subjects not be exposed to the risk and inconvenience of the replacement surgeries every 5 years. However, as it takes a smaller capacity of the battery, 250 mAh instead of 5000 mAh for example, it requires regular recharging, 15~30 min daily or 3~4 h every 1 or 2 weeks. There are also chances of difficulty caused by the charger and it may not be appropriate for some patients, like patients with a conscious issue or patient with no caregivers.

2.6 Technologies for Target Placement

During the surgery of DBS, accurate placement of stimulating lead is the primary task to get the optimal surgical outcomes. The volume of the target is very small of less than 1 cm^3. For example, the subthalamic nucleus (STN), which is the major target region for PD subjects, has a volume of 5.9 mm, 3.7 mm, and 5.0 mm in three directions, respectively. To locate the stimulating leads within this small volume accurately, various techniques are combinedly used during each surgical procedure by experienced neurosurgeons (Fig. 13.11). Identifying and localization of anatomic targets for DBS initially relied on the experience and knowledge gained from surgical ablative procedures (Abosch et at., 2013).

First, to quantify the position in the 3D space, a stereotactic frame is placed around the head. Then, any position in the head can be represented in reference to this frame. Multiple slices of images are taken wearing this stereotactic frame with medical imaging systems like computed tomography (CT) or magnetic resonance imaging (MRI). Before beginning the surgery, stimulating target and trajectory to reach the target is accurately planned based on the acquired subject's images, anatomical atlases, electrophysiological databases, computer model, and all available information. The surgery begins by making an incision and burr hole in the skull with a stereotactic frame on. Prior to placing the stimulating lead on the guided target position by the plan, the target is once more confirmed by the microelectrode recording. Target is confirmed by the coherent electrical activities to the symptoms like a tremor. After the target position is confirmed, stimulating lead is paced in the target position, and its position is examined by fluoroscopy or X-ray during the procedure. After lead placement is finished, implanted stimulated lead is confirmed postoperatively by CT or MRI. When the right placement is confirmed, surgery placing the IPG in subclavicular region or abdomen is followed. When the available technologies are combinedly used, the target region can be localized within an average error of 1.7 mm (Guo et al., 2006).

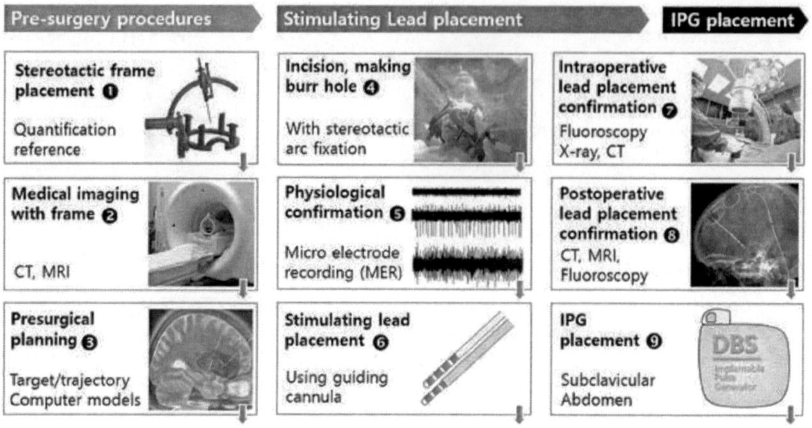

Fig. 13.11 Surgical procedure of DBS implantation using various target localization techniques. (Zhang et al., 2020)

2.7 Future Directions and Concerns

Implanting the stimulating electrodes in the brain is a great advance in treating related disorders. And, advanced DBS surgical procedures and applicable indications are progressively being sought. Accompanying progress in technologies will improve the DBS system toward more productive, convenient, and safe directions. As in the leadless cardiac pacemaker, leadless DBSs are expected in near future with progress in battery and integrated electronic technologies. Then, without using an extension wire, IPG might be placed within the skull directly connected to the stimulating leads. Advanced software technology based on artificial intelligence can improve the performance by optimizing the volume of tissue activated with a large number of directional electrodes and intelligent pulse train programming. It can maximize the therapeutic effect of DBS without accompanying adverse side effects,

Currently, DBS has been used wholly for medical purposes, to restore brain functions of the decreased or distorted brain to a level of normal brain. Is it not possible to enhance brain function above the level of normal subjects with DBS? As the brain is recovered from the lowered level to the normal level, there is a strong chance that the brain level is enhanced higher than the normal level. Intelligence, memory, and muscular responses might be increased with stimulation of an appropriate target with an appropriate pulse train. Humans can be benefited from DBS in several senses.

DBS is controlled wirelessly by an external program unit for setting the parameters and stimulating modes. It means that the brain is connected with the external world electrically and wirelessly. In the world of the internet of things (IoT), where all the items can be connected electrically and wirelessly, the brain also can be a part of it. Activities of the brain can be interpreted externally by electrical signals, as in the brain-computer interface. Now, modulation of brain functions by DBS accomplishes an electrical closed loop of the brain with the external world without using ordinary sensory and motor systems. Humans can join the electrically connected world as electrical beings just like a computer.

The connection of DBS to the outer world also produces safety and privacy problems as in the connected computers. Simply draining the battery or switching off the DBS could cause a symptom rebound or tissue damage. The attack might be targeted to specific indications, leading to motor impairment, increased pain, altered impulse control, or unpleasant emotions (Krauss et al., 2021). As hacking on digitally connected identity, brain jacking might be possible which modulates the subject's cognitive, emotional, and decision-making processes to move against the subject's own will. We need to prepare in many respects for the unseen future of electrically connected humans.

3 Electroconvulsive Therapy

Electroconvulsive therapy (ECT) is a treatment that sends an electric current through the patient's brain to cause a seizure, a brief surge of electrical activity within the brain (Fig. 13.12). The treatment aims to relieve the symptoms of refractory mental disorders. While DBS is similar to a cardiac pacemaker in stimulating the targeted cells in the organ periodically, ECT is similar to a cardiac defibrillator in stimulating a large volume of cells for a short period to reset the organ electrically. It is estimated that between 80,000 and 100,000 patients receive ECT annually only in the USA (Kellner et al., 2020).

3.1 Indications for Electroconvulsive Therapy

Since the first human trial of ECT in 1938 by Ugo Cerletti, its technique has changed dramatically during the last 80 years making it the refined procedure of today. ECT still remains an important treatment in psychiatric medicine for subjects with

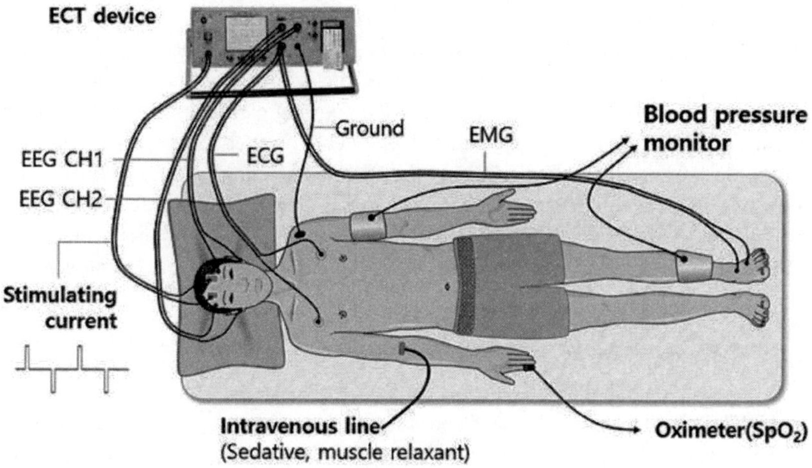

Fig. 13.12 Electroconvulsive therapy (ECT) with patient monitoring during the treatment

severe mood and psychotic disorders. Even though newer brain stimulation techniques and novel pharmacological agents have emerged, ECT maintains better efficacy in subjects with the following indications.

Severe and treatment-resistant *depression* is the main indication for ECT, in most European and North American countries. Depression is a mood disorder that causes a persistent feeling of sadness and loss of interest that interferes subject's daily activity. Among the depression patients, patients who have severe and debilitating symptoms that are not responsive to multiple medications and cognitive behavioral therapy (CBT) are indicated for ECT. The urgency of the clinical situation, often marked by intense suicidal preoccupation and drive, commonly compels the recommendation for ECT (Kellner et al., 2020). Considering the long period suffering by the sequential trials of antidepressant medications and possible loss of life by suicide, ECT after only two failed antidepressant medication trials would be regarded as cost-effective (Ross et al., 2018). Among treatments for severely depressed pregnant women, ECT is one of the least harmful to the fetus (Pompili et al., 2014). The response rate to ECT in depression patients is 80~90% when it is used as first-line treatment and 50~60% in medication-resistant depression (Khalid et al., 2008).

Schizophrenia is the most common indication for ECT worldwide, more accepted in Asian countries. Schizophrenia is a chronic mental disorder with symptoms including delusions, hallucinations, disorganized thinking or speech, movement behavior, and trouble concentrating. Schizophrenia also has negative symptoms, which refer to an absence or lack of normal mental function. Negative symptoms of schizophrenia include diminished facial and emotional expression, a decrease in verbal output, decreased social relationships, reduced ability to feel or express pleasure, and lack of motivation. ECT is effective to treat exacerbations of psychotic illness and mainly positive symptoms. However, it is less effective on negative symptoms.

ECT is used to treat refractory *mania*, which is a mental illness manifested by mental and physical hyperactivity, disorganization of behavior, and elevation of mood. Since most manic episodes can be successfully treated with medications, ECT is typically reserved for patients, who are not successful with medications, for adequate treatment of the symptoms. Delirious mania, a particularly severe and potentially lethal variant, also referred to as catatonic excitement, may be considered an urgent indication for ECT (Luchini et al., 2015).

Catatonia is a psychomotor disorder characterized by abnormal movements, immobility, abnormal behaviors, and withdrawal. It is one of the clinical indications that typically responds very well to ECT. ECT may be used as a first-line treatment, or more commonly, after medications have been shown to be inadequately helpful (Fink et al., 2016).

The most common adverse side effects are confusion and transient memory loss immediately following treatment. Subjects may not know where they are or why they are there. And, subjects may have trouble remembering events before the treatment weeks or months ago. However, these memory problems usually improve within a couple of months after treatment ends. Compared to the ECT using steady current flow, ECT based on pulse trains seemed to incur less memory loss.

3.2 Mechanism of ECT

The mechanism of ECT is not clearly known. Simply, ECT can be thought of as a form of electrical lobotomy. Or, ECT might have derived the effects by wiping out the memories that make patients sad or have unrealistic thoughts. There is a consensus that the main mechanism of electroconvulsive therapy (ECT) is the induction of a generalized clonic seizure. The beneficial effect of reducing the symptoms in patients with severe depression and schizophrenia is considered that it has been caused by disrupting the ongoing electrical nervous system with the induced seizure. However, there are various theories explaining the effects resulting from the ECT. Neurotransmitter theory supposes that ECT works like an antidepressant medication by restoring disturbed transmitter functions either by elevating transmitter concentrations or by activating receptors. The anticonvulsant theory is based on the observation that seizure threshold often rises and seizure duration shortens as treatment sessions elapse over a course of ECT. It leads to the hypothesis that the inhibitory brain process associated with the progressive rise in seizure threshold is also responsible for depression relief. And, ECT-induced seizures train the brain to resist seizures and relieve depression. In neuroendocrine theory, induced seizure causes the hypothalamus to release hormones into the blood and cerebrospinal fluid regulating the dysfunction in many patients with a severe mood disorder. In brain damage theory, induced seizure causes an illusion of euphoria that the problems are gone, which is frequently observed as an illusional breakthrough after brain injury or physical trauma. The psychological theory explains that depressed patients are satisfied by escaping from their guilty feeling, which they often feel, with the punishment-like ECT treatment.

As ECT is effective to treat very different symptoms, like depressive, manic, delusional, hallucinatory, and catatonic, treatment effects may be caused by distinct pathophysiologic mechanisms. As each symptom concerns particular neuronal circuits in different levels of brain regions as well as neurotransmission and intracellular chemistry, several mechanisms can be involved for each effect from the molecular level to the circuit level.

Eliciting a generalized seizure seems crucial for ECT efficacy. The larger the neuronal networks get stimulated, the better the clinical response. The primary site of action seems to be the excitable cortex, preferably the frontal lobe, although it seems essential to stimulate subcortical structures such as the hypothalamus and brainstem (Michael, 2009).

Overall, it is not the induced seizure that treats the symptoms, but the brain that restores innate function from a series of ECT seizures. Biological systems have mechanisms to rebuild balanced functions by homeostasis. Neuronal tissues in the brain are considered to restore their balanced function after repeatedly induced seizures by ECT.

3.3 ECT Procedure

Usually, ECT treatment is performed 2~3 times per week and a total of 6~12 times until the patient no longer suffers from symptoms. A single ECT session usually lasts about an hour, which includes 15~20 min in the treatment room as well as 20~30 min in the recovery room.

While ECT was performed without anesthesia formerly, current ECT is performed under full general anesthesia with a muscle relaxant. During the procedure, the patient's blood pressure, heart rate, muscular activation, and oxygen saturation are carefully monitored (Fig. 13.12). A stimulation of brief electrical pulses is delivered. Applied electrical stimulation produces a mild seizure which can effectively reduce the symptoms of disorders. As applied current also stimulates the motor cortex, muscle contractions are accompanied which last less than 1 min.

The placement of the electrodes delivering stimulations has a significant impact on both the efficacy and side effects of ECT treatment. There are three different treatment placements for current stimulation, right unilateral (RU), bitemporal (BT), and bifrontal (BF) (Fig. 13.13). In most cases, treatment begins with a right unilateral (RU) placement to minimize the chance of memory loss. In right unilateral placement, the electrical current is concentrated away from the language areas of the left hemisphere and it thus lessens memory impairment caused by

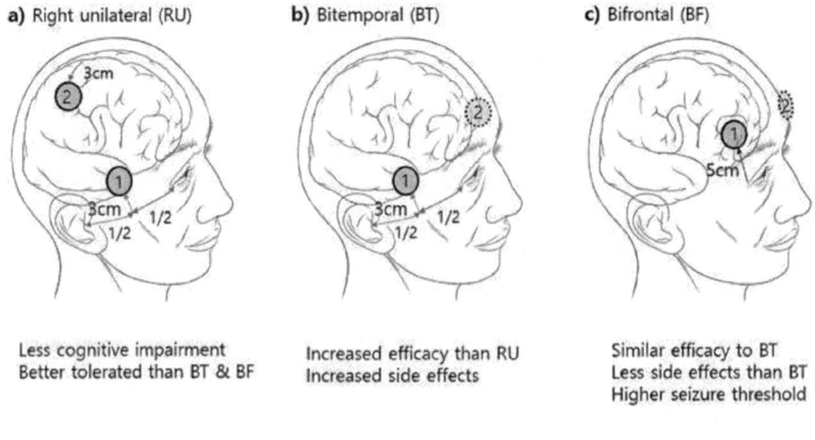

Fig. 13.13 Three different ECT electrode placement modes. (**a**) Right unilateral (RU), (**b**) Bitemporal (BT), (**c**) Bifrontal (BF). (https://commons.wikimedia.org/wiki/)

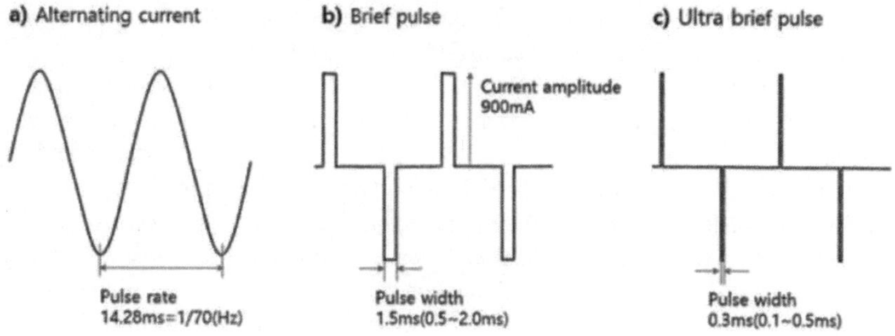

Fig. 13.14 Types of stimulating pulse waveforms. (**a**) Alternating current, (**b**) Brief pulse, (**c**) Ultra brief pulse

Table 13.2 Electrical parameters of stimulating pulse for ECT

Parameter	Range	Typical value	Comment
Voltage	70~450 V	120 V	Impedance: 120 Ω
Current	500~900 mA	900 mA	
Pulse width	0.25~2.0 ms	1.0 ms	Ultra-brief, brief pulse
Pulse frequency	10~120 Hz	70 Hz	
Treatment duration	0.03~8 s	6 s	
Charge applied	7.5μC~1728mC	756mC	Impedance: 120 Ω
Impedance	120~350 Ω	120 Ω	

ECT treatment. This placement is better tolerated than bitemporal or bifrontal placements in most patients. If the symptoms are not improved with several trials of unilateral stimulation, stimulation placement can be changed to bilateral or bifrontal placements. Response of bitemporal stimulation placements is faster and efficacious but with an increased chance of cognitive side effects like amnesia. Bifrontal placement is used as an alternative to bitemporal placement. It gives similar efficacy to bitemporal placement with fewer side effects but with a higher seizure threshold.

Electric stimulations are applied in brief pulses rather than in alternating current used in earlier times of ECT (Fig. 13.14). Brief pulses, whose pulse width is less than 2 ms, and ultra-brief pulses, whose pulse width is less than 0.5 ms, can produce a seizure with less energy than conventional sine-wave stimulation, and it concurrently reduces the memory loss associated with ECT. Stimulation dose, amount of charge delivered by pulse train is determined by the product of pulse amplitude, pulse width, pulse frequency, and duration of stimulation. The amount of stimulating charge is the strongest predictor of treatment outcome. The larger amount of applied energy tends to increase the efficacy of treatment and reduce cognitive degradations. The threshold for seizure occurrence is established in the first session using a low charge. And, electrical stimulations of subsequent sessions are delivered in reference to the determined threshold. The guideline recommends charges of 5~8 times larger than the seizure threshold for right unilateral ECT and 1.5~2.5 times larger for bitemporal ECT (Surya & Mori, 2015).

3.4 Electric Properties of ECT

Typically, ECT stimulation is applied by a series of pulses having an amplitude of 120 V and a current of 900 mA. As the stimulations are applied in brief pulses having small widths, the total stimulation dose is determined also by the pulse width and pulse frequency. Typically, the pulse width of 1.0 mA and frequency of 70 Hz are used.

Electrical stimulation is applied by the constant current amplitude. As electrode-scalp contact can be variable among patients and upon the electrode contacting pressure, applying stimulation with constant current is safer than constant voltage-driven stimulation. The applied voltage can be varied by the contacting impedance of the patient being treated and it ranges between 120 and 350 Ω during the stimulation. Stimulation dose can be increased by any of one among four variable parameters, pulse width, pulse frequency, pulse amplitude, and duration of treatment. Generally, the amplitude of pulse current is fixed to a maximal value to stimulate the maximum volume of brain tissue.

$$\begin{aligned}\text{Stimulation dose} &= \text{Amount of applied charge (mC)} \\ &= \text{pulse\_amplitude}(\text{A}) \times \text{pulse\_width}(\text{ms}) \\ &\quad \times 2 \times \text{frequency}(\text{Hz}) \times \text{duration}(\text{s})\end{aligned} \tag{13.9}$$

Table 13.2 shows the stimulating parameters and their variable ranges to get the appropriate stimulation dose. The applied amount of charge can be minimal of 7.5 μC with the lowest values of each parameter, and maximal value of 1728 mC with the highest values of each parameter. To find the threshold inducing a seizure, these parameter values are combinedly increased up to the point where the specified strength of seizure is induced.

Determining the seizure threshold is important as it is variable by 50 folds according to various factors including gender, electrode placement, and ECT experience. During the titration session, the threshold of stimulation dose is determined by 4~5 repeated stimulation sessions. Specified seizure appears generally and is recognized by EMG and EEG simultaneously with the durations 20~60 s in EMG and 30~120 s in EEG. This seizure threshold tends to increase with the experience of the ECT treatment and needs to be determined newly in the following treatments. After the seizure threshold is determined, stimulations are applied at a suprathreshold level in the following session with different ratios according to the electrode placements.

4 Transcranial Direct-Current Stimulation

Transcranial direct-current stimulation (tDCS), is a noninvasive, nonconvulsive, nonfocal, and painless brain stimulation treatment using direct electrical current. Constant and low-intensity direct current is passed through two electrodes placed over the head to modulate the neuronal activity of specific parts of the brain (Fig. 13.15). Treatment is performed while the subjects are in a conscious state with minimal discomfort. There are two types of stimulation with tDCS, anodal stimulation and cathodal stimulation. Anodal stimulation acts to excite or enhance neuronal activity while cathodal stimulation inhibits or reduces neuronal activity.

4.1 Indications for tDCS

Transcranial direct-current stimulation (tDCS) is a noninvasive neurostimulation technique that may help treat various neurological and psychiatric disorders, even when they are refractory to medications. Specifically, it has a nonsignificant level of low risks and typical side effects are transient like minor skin irritation, itching, tingling, and redness of the skin. Additionally, tDCS is easy to use with a portable device, painless, and has the advantage of low cost. These advantages made the clinicians consider tDCS as a new treatment option, particularly for conditions having limited options. The official regulatory status of tDCS is in development in many countries with early EU clearances for depression and pain, and many centers worldwide use tDCS as an investigational therapy (Fregni et al., 2021).

Since the tDCS was originally developed to help patients with brain injuries or neuropsychiatric conditions such as major depressive disorder, continued research and clinical trials revealed supporting evidence of tDCS for various indications. While they came to a consensus on the efficacy of tDCS for depression, evidence is not unanimously supporting the efficacy in many other indications. Recently, evidence-based guidelines summarized the recommendations of tDCS for various disorders and their conditions. They recommended tDCS for depression with "Level A" of definitely effective. They recommended with "Level B" of probably effective for some conditions of disorders, like pain, Parkinson's disease, stroke, epilepsy, schizophrenia, and addiction. Obsessive-compulsive disorder is recommended with "Level C" of possibly effective. Table 13.3 shows the summary of recommendations for various indications (Fregni et al., 2021).

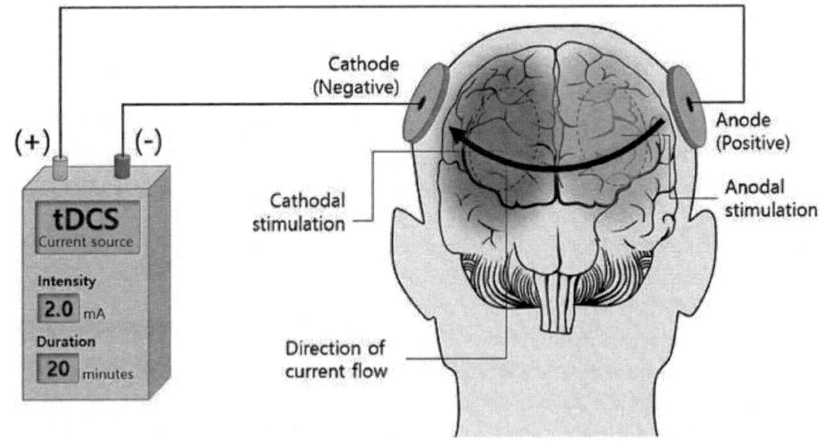

Fig. 13.15 Transcranial direct-current stimulation (tDCS)

Table 13.3 Recommendations on tDCS efficacy according to clinical indications

Disorder (condition)	Recommendation	Type (location)
MDD (Depression)	A: Definitely effective	Anodal (Left DLFPC)
Pain (Neuropathic pain, fibromyalgia, migraine, postoperative acute pain)	**B**: Probably effective	Anodal (M1)
Schizophrenia (Auditory hallucinations)	**B**: Probably effective	Anodal (Left prefrontal) + Cathodal (Left temporoparietal)
Addiction (Alcohol)	**B**: Probably effective	Anodal (Right DLFPC) + Cathodal (Left DLFPC)
Epilepsy (Seizure frequency)	**B**: Probably effective	Cathodal
Subacute stroke (Motor function)	**B**: Probably effective	Anodal (Ipsilateral M1)
PD (Cognitive function)	**B**: Probably effective	Anodal (DLFPC)
PD (Motor function)	C: Possibly effective	Anodal (M1, premotor, SMA)
Chronic stroke (Motor function)	**B**: Probably effective	Anodal (Ipsilateral M1)
Chronic stroke (Aphasia)	C: Possibly effective	Anodal (Broca's area)
Obsessive-compulsive disorder	C: Possibly effective	Anodal (Pre-SMA)

DLFPC dorsolateral prefrontal cortex, *M1* primary motor & lateral premotor cortex, *MDD* major depressive disorder, *OCD* obsessive-compulsive disorder, *PD* Parkinson's disease, *SMA* supplementary motor area (Fregni et al., 2021)

In healthy populations, tDCS has been shown to temporarily enhance behavior, accelerate learning, and boost task performance. Anodal stimulation has been shown to enhance facial expression recognition or inhibit aggressive responses (Thair et al., 2017). And cathodal stimulation has been shown to foster implicit motor learning by suppressing working memory activity (Zhu et al., 2015). However, evidence about whether tDCS is useful for cognitive enhancement in healthy people is mixed.

4.2 Mechanism of tDCS

As in other electrical stimulation methods, tDCS stimulates and activates brain cells by delivering electrical signals. However, due to the low intensity used, tDCS does not stimulate neuronal cells directly to trigger action potentials. A small amount of current only modulates them more or less likely to trigger action potentials. This modulation of cortical excitability can act effectively to facilitate rehabilitation, treat some neuropsychiatric symptoms, and enhance cognitive functions. tDCS can modulate the neurons in two ways. When anodal stimulation of positive potential is delivered, it increases the excitability of neurons by raising the membrane potential in the direction of depolarization and makes neurons more easily trigger action potentials. On the contrary, when cathodal stimulation of negative potential is delivered, it decreases the excitability of neurons by hyperpolarizing the membrane potential and makes neurons less likely trigger action potentials.

In addition to the neuromodulation effect during the stimulating period, the effects of tDCS can last even after the stimulation is ended. tDCS can induce the neuroplastic effect of long-lasting alterations of cortical excitability and activity, which share features with long-term potentiation and depression. These neuroplastic processes are important foundations for enhancing the various cognitive functions like learning and memory formation that are pathologically altered by neurological and psychiatric diseases. The duration of the long-lasting effects generally increases with the length and intensity of tDCS (Utz et al., 2010).

4.3 Procedures for tDCS

tDCS device is simple compared to other electrical stimulation units and its price ranges 15~500$. tDCS devices also can be assembled at home with a DIY kit. While it is generally safe when used by the experts, there should be caution in using tDCS by nonexperts, for example, at home.

tDCS can be applied in unilateral setup or in bilateral setup (Fig. 13.16). In a unilateral setup, the active electrode is placed over the target region of the brain, and the reference electrode is commonly placed opposite the active electrode. The target region stimulated using the active electrode depends on the hypothesis and the task. If the hypothesis concerns aggression, the prefrontal cortex might be the target region, and if the task concerns motor outputs, motor cortices can become the target regions. For tDCS stimulation, target regions are recruited for the task at hand for upregulating or downregulating. In unipolar electrode placements, the reference electrode is placed at the extracephalic place, for example, on the upper arm of the contralateral side. Extracephalic reference helps to exclude the effect of the reference electrode on cortical modulation, and the effect of tDCS is derived solely by the current focalized around the active electrode. In order to modulate cortical activity, the current density should be above a minimum threshold of 0.017 mA/cm^2 (Nitsche & Paulus, 2000). For example, a stimulating current of 1.5 mA is applied with an active electrode of 35 cm^2 and a reference electrode of 100 cm^2, the current density becomes 0.043 mA/cm^2 in the active electrode and

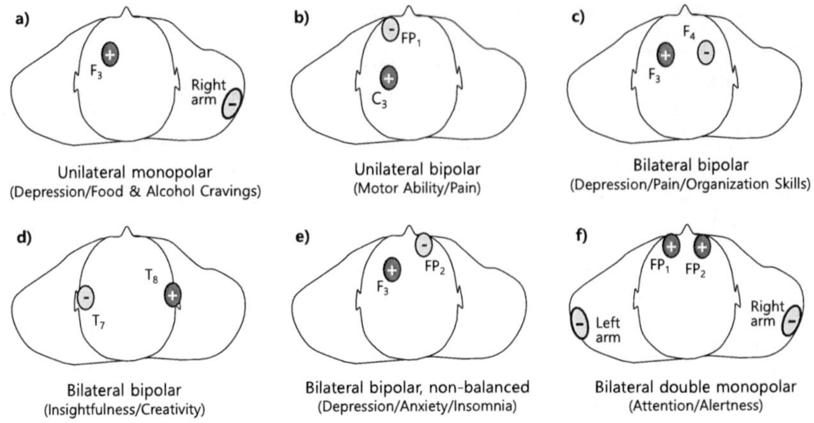

Fig. 13.16 Examples of tDCS electrode placement for various conditions. (**a**) Unilateral monopolar, (**b**) Unilateral bipolar, (**c**) Bilateral bipolar, (**d**) Bilateral bipolar, (**e**) Non-balanced bilateral bipolar, (**f**) Bilateral double monopolar

0.015 mA/cm^2 in the reference electrode. So, only the region around the active electrode gets stimulated while the region around the reference electrode remains under an effective stimulating level. In a bihemispheric setup, two cortical regions are parallelly stimulated. The cortical region beneath the anodal electrode gets upregulated and the region beneath the cathodal electrode gets downregulated.

Even though the applied current is for the stimulation of brain cortices within the skull, a large portion of the current does not enter the skull. Current may be dissipated along the surface of the scalp of lower impedance as a shunting current and only decreased concentration of current reaches the cortical region. If the distance between electrodes is less than 5 cm, the current would be highly susceptible to a shunting effect. And, if the distance between the electrodes is too small, the current entering the brain flows mostly through the cerebrospinal fluid (CSF), which is more conductive than cortical tissues, instead of stimulating the cortical region. For efficient supply of nominal current, good contact between the electrode and the scalp surface is critical. Conductive paste or sponge soaked with saline is used to establish good contact and distribute current more effectively.

tDCS is applied with the current intensity of 1~2 mA for the duration ranges between 5 and 30 min. Duration of stimulation affects the returning time of the excited cortical tissues to baseline level after the end of stimulation. For example, 9 min of tDCS accompanies after-effect time of up to 30 min, and 13-min tDCS increases after-effects time to 90 min. Understanding the after-effect is important to design tasks appropriately and for the safety of subjects. After-effects can appear in reversed modulation. After anodal stimulation, an inhibitory effect can result rather than an excitatory effect. Similarly, cathodal stimulation can result in excitatory changes. tDCS can be used simultaneously with drug treatments to speed recovery and improve motor and cognitive performance.

5 Transcranial Magnetic Stimulation

Transcranial magnetic stimulation (TMS) is a noninvasive method of brain stimulation that relies on electromagnetic induction at a specific area of the brain using an insulated coil placed over the scalp (Fig.13.17). Repetitive transcranial magnetic stimulation (rTMS) refers to the application of recurring TMS pulses, instead of a single or a pair of pulses, to a brain region mostly for therapeutic purposes. The coil generates a focused magnetic field by the changing electric current flowing through the coil. The generated magnetic field passes easily and painlessly through the skull into the brain. Then penetrated magnetic field into the focalized cortical region of the brain reversely induces a current which stimulates the cortical tissues of the region. TMS was FDA-approved in 2008 for treating depression, in 2018 for obsessive-compulsive disorder (OCD) and now is widely used at clinics and hospitals.

5.1 TMS as a Diagnostic Tool

TMS can be used clinically as a diagnostic tool to measure the activity and function of specific brain circuits in humans, most commonly with single or paired magnetic pulses. TMS specifically provides methods to examine the integrity of the fast-conducting corticospinal pathways and excitability of the motor cortex. A wide range of diseases associated with motor dysfunction, like multiple sclerosis (MS), amyotrophic lateral sclerosis (ALS), stroke, movement disorders, disorders affecting the spinal cord, and facial and other cranial nerves, can be aided in their diagnoses by the patterns of findings with TMS.

Determining the cortical motor threshold (CMT) is the first step in the clinical use of TMS for the target muscles. Determined CMT reflects the integrated excitability at the

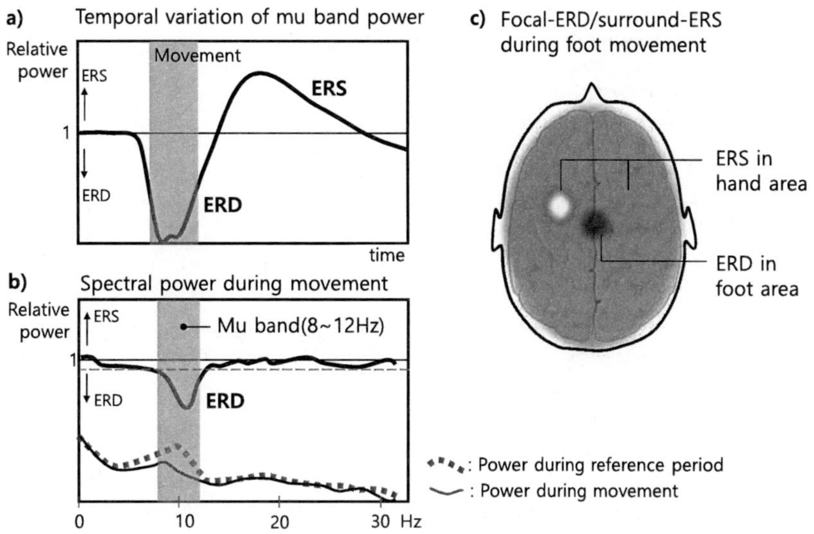

Fig. 13.17 Transcranial magnetic stimulation (TMS)

spinal level as well as the corticomotor projection. The easiest method is to measure the TMS threshold intensity that evokes motor evoked potential (MEP) while the target muscle is completely relaxed. CMT is defined as the lowest stimulus intensity that is required to induce a MEP in five out of ten trials with an amplitude higher than 50 μV peak-to-peak. This threshold is commonly referred to as the resting motor threshold (rMT) and is used as a reference intensity in the treatments with rTMS.

Single-trial response of the MEP is used as a diagnostic parameter. The largest amplitude and the shortest latency from a single-pulse TMS are measured from the recorded MEP. These parameters of MEP are similar to the parameters derived from compound muscle action potentials (CMAP) in response to peripheral nerve electrical stimulation. However, the amplitude of MEP is smaller than the corresponding CMAP due to substantial phase cancelation caused by desynchronized muscle activation. This desynchronization is thought to be caused by the randomness in muscle excitation by motor neurons, by different timing of motor nerve excitation with TMS, and by different conduction times along the axons. Generally, 5~6 consecutive MEP recordings are sufficient to assess corticomotor conduction of the target muscle.

If the TMS is applied while the target muscle is tonically contracted, it disrupts the ongoing voluntary EMG activity for some duration after the recording of MEP. This duration is called the cortical silent period (CSP). CSP increases with TMS intensity following the shape of the sigmoid. And, it lasts several hundred milliseconds at high stimulus intensity in hand muscles. CSP largely originates in the primary motor cortex. Spinal mechanisms contribute to the early part of EMG inhibition, while the latter part is caused by a suppression of corticospinal output at the cortical level (Inghilleri et al., 1993). CSP is a measure of motor cortical inhibition, mediated by GABA receptors, but influenced by a large variety of other physiological factors (Groppa et al., 2012).

Central motor conduction time (CMCT) is the conduction time of neural impulses transmitting through the central nervous system on their way to the target muscles. CMCT is measured by the time difference between the motor evoked potential latency elicited by motor cortical TMS and the peripheral motor latency (Fig. 13.18). CMCT varies on the various neurological conditions related to disorders. CMCT increases when the conduction of impulses is slowed down through the corticospinal fibers. As prolonged CMCT is recorded in certain clinical conditions, it is of diagnostic value for various disorders such as demyelinating diseases including multiple sclerosis (MS), amyotrophic lateral sclerosis (ALS), structural lesions in the corticospinal tract including stroke and compressive myelopathy, and neurodegenerative disorders including multiple system atrophy (Udupa & Chen, 2013) (Fig. 13.18).

5.2 Indications for Therapeutic TMS Treatment

TMS has shown therapeutic potential in a wide variety of disease states in neurology and psychiatry. Currently, FDA approved several TMS devices for the treatment of major depressive disorder (MDD), obsessive-compulsive disorder (OCD), migraine with aura, and smoking addiction with a specific setup for each symptom.

TMS therapy is majorly indicated for the treatment of major depressive disorder (MDD) in adult patients who have failed to receive satisfactory improvement from prior antidepressant medication. TMS using high-frequency

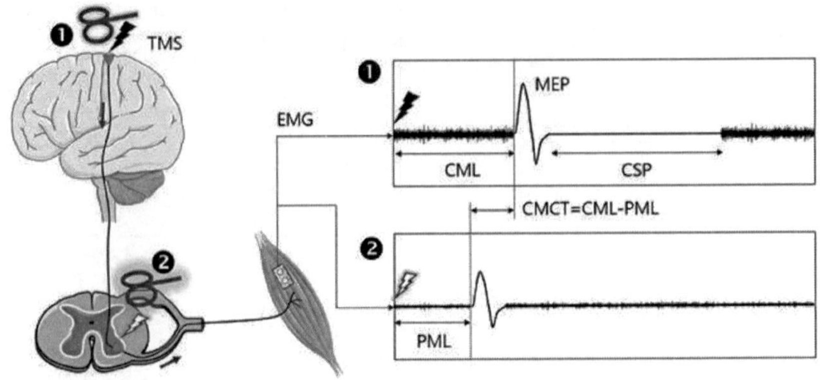

Fig. 13.18 Measurement of CMCT with TMS. *CMCT* central motor conduction time, *CML* central motor latency, *CSP* cortical silent period, *EMG* electromyogram, *MEP* motor evoked potential, *PML* peripheral motor latency, *TMS* transcranial magnetic stimulation

electromagnetic induction targeting the dorsolateral prefrontal cortex (DLPFC), which is the part of the brain dysregulated in patients with major depression, effectively treats the behavioral dysregulation. Patients are treated with five daily treatment sessions per week over 4~6 weeks, resulting in a total of 20~30 sessions during the treatment. A literature review reported that the response rates to TMS range between 50% and 55%, and the remission rates range between 30% and 35% in patients with major depression (Reddy & Vijay, 2017).

TMS also effectively alleviates the symptoms of obsessive-compulsive disorder (OCD), particularly in patients who are not responsive to medications or cognitive behavioral therapy (CBT). The TMS treatment for OCD uses stimulating coils which can penetrate the magnetic field into deeper brain structures, the anterior cingulate cortex, and the medial prefrontal cortex associated with OCD. TMS is well-tolerated, and it does not cause any adverse or long-lasting side effects. Since it does not require a significant recovery period, 18-min treatment in an outpatient clinic can easily be integrated into the patient's day-to-day schedule. Studies show that TMS is responsive with more than a one-in-three ratio for treated OCD patients and restores their quality of life.

TMS can relieve the pain caused by migraine headaches preceded by an aura, a visual, sensory, or motor disturbance immediately preceding the onset of a migraine attack. At least 30% of patients with migraines experience an aura. Migraine headaches are characterized by an intense pulsing or throbbing pain in one area of the head accompanied by nausea and/or vomiting and sensitivity to light and sound. Without treatment, migraine can last 4~72 hours. The study showed that 38% and 34% of patients were pain-free at 2 hours and 24 hours after the treatment respectively, compared to 17% and 10% in the control group.

TMS therapy has proven its efficacy in treating smoking addiction. By stimulating deeper regions of the brain with a specially designed stimulating coil, TMS treatment can significantly improve the continuous quit rate, reducing craving, and the average number of cigarettes smoked per week. In a multicenter randomized controlled trial of 262 subjects, who have a history of smoking over 26 years on average and multiple failed attempts to quit, the continuous quit rate was increased from 12% to 28% and the average number of cigarettes smoked per day were decreased from 47 to 32 after 6 weeks of treatment (Zangen et al., 2021). TMS application as an aid in short-term smoking cessation in adults was FDA approved in 2020 with specially designed coils for deeper region stimulation.

In addition to current FDA approved TMS treatments, TMS can be an effective treatment for other disorders like addictions, schizophrenia, post-traumatic stress disorder (PTSD), Parkinson's disease, multiple sclerosis, chronic pain, poststroke rehabilitation, bipolar depression, autism, Alzheimer's disease, dementia, eating disorders, postpartum depression, and tinnitus. Research and clinical trials are actively being done, especially based on the effectiveness based on the noninvasiveness and low risk of side effects.

TMS treatment is generally well-tolerated by patients and associated with few side effects. The most common side-effect is a mild headache which generally diminishes over the course of the treatment. Patients may experience mild painful scalp sensations, facial twitching, tingling, or sleepiness. The risk of a seizure is exceedingly low, less than 0.1%. Compared to the antidepressant medications having side effects, like gastrointestinal upset, dry mouth, sexual dysfunction, weight gain, and sedation, TMS has a great advantage of lowered risk of side effects. However, the TMS coil produces an acoustic noise that may exceed 140 dB and the use of earplugs is highly recommended to avoid hearing damage.

5.3 Mechanism of TMS

TMS uses induced current by the generated magnetic field, which is basically described by Faraday's law of induction. TMS device generates a magnetic field by flowing the high current in the excitation coil. In contrast to other electrical stimulating systems, TMS applies stimulating energy in the form of magnetic flux without direct physical contact with the subjects via electrodes. While the intensity of the magnetic field decreased exponentially along with the depth, it is generally transparent to the scalp and skull and smoothly penetrates into the cortical depth almost unimpeded. TMS does not induce a high electrical current in the scalp and subcutaneous tissue, and it explains why TMS does not induce pains and therefore is usually well-tolerated by the patients. Depth of penetration is dependent on the intensity of generated magnetic field and the geometry of the exciting coils.

The exciting coil is positioned parallel to the head surface, and the magnetic field is generated in the perpendicular direction to the head surface. Current induced in the cortical tissue is perpendicular to a magnetic field, which is parallel to the head surface and the coil plane. Finally, TMS induces a current in the tangential direction of the brain. Even though the magnetic flux is used during the transmission of stimulating energy, TMS stimulates the cortical neurons with an induced electrical current through the same mechanism as other electrical stimulating systems.

When the intensity of the induced current is sufficiently strong to depolarize the membrane potential in the focalized region of the cortex, TMS can stimulate the cortical neurons and trigger action potentials. Triggered action potentials in cortical neurons propagate across synapses to other neurons, resulting in a spread of neuronal activation to connected cortical and subcortical areas. For diagnostic TMS, triggered action potentials travel along the corticospinal tract and peripheral motor nerve resulting in motor evoked potential (MEP) in the target muscle.

Repetitive TMS (rTMS) is the type of TMS used for therapeutic applications. Generally, high-frequency (HF) stimulation over 5 Hz is used to induce facilitatory or excitatory effects and low-frequency (LF) stimulation lower than 1 Hz to induce suppressive or inhibitory effects in the brain. The widely accepted mechanism for the long-lasting after-effect of rTMS treatment is that it can alter synaptic plasticity, mainly the long-term potentiation/depression (LTP/LTD) of excitatory synaptic transmission (Ma et al., 2014). In addition to synaptic plasticity changes in excitatory transmission, modulations of inhibitory interneuron activity and membrane potentials co-occur with excitatory alterations. The balance of these effects underlies the ultimate effect of rTMS (Muller et al., 2014).

5.4 TMS System

TMS system is composed of an exciting magnetic coil and a generator that supplies the electric current to the coil. TMS generator circuit charges a capacitor into a high voltage of 400~3,000 V and discharges the stored charges into the exciting coil by making a temporarily high current of 4000~20,000 A flow through the coil. By the rapid change of current, the strong magnetic field of 1.0~2.5 Tesla, which is in the range of current magnetic resonance imaging (MRI) use, is generated during a very short period of time of less than 1 ms.

To induce a current in the targeted cortical region exclusively, precise TMS is required which focuses magnetic flux in the target area while minimizing the stimulation in neighboring regions. The magnetic field generated by TMS coils decreases rapidly and diffuses broadly as it penetrates the brain. A simple circular coil tends to stimulate a relatively large area in the brain cortex (Fig.13.19). The coil configuration to increase the focality and depth of penetration while minimizing the consumed energy is required. The Figure-of-eight coil is the most widely used coil configuration for focalization of stimulating target region in TMS. The Figure-of-eight coil is composed of two circular wire loops composing a figure of 8 in geometry. Current flows in the reverse direction for each loop, one in a clockwise and the other in a counterclockwise direction. This configuration generates a magnetic field in different directions between two current loops and induced currents in the cortical tissue are in the reversed rotational direction. Current density becomes

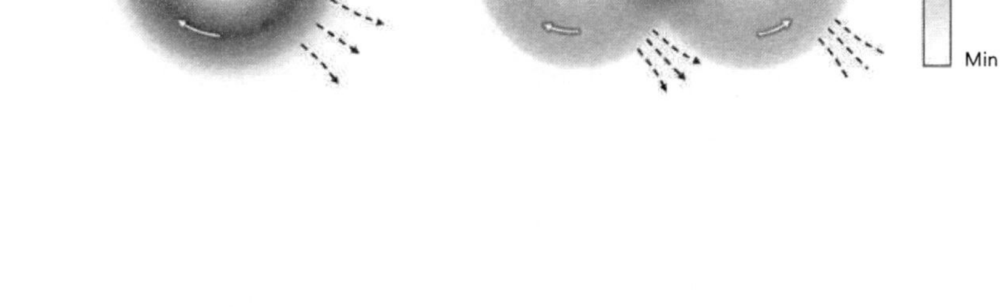

Fig. 13.19 Distribution of induced electrical intensity of TMS coil. (**a**) Circular coil, (**b**) figure-of-eight coil

maximum where the two current loops meet making the stimulated region more focalized to a targeted area.

Recently, various TMS coil configurations are designed based on computationally optimizing the generated magnetic field distribution. Even though the structure of the coils becomes complex, they can shape the induced current and increase the penetration depth. These types of coils increase the penetration depth over 1.8 cm from the cortical surface compared to the depth of 1.1 cm in the Figure-of-eight coil, and they make TMS more applicable to an extended range of disorders related to the deeper areas of the brain. FDA has approved TMS for the treatments of OCD and smoking addiction with coils that can excite deeper areas.

5.5 Stimulating Pulse Protocols for rTMS

In contrast to a single- or paired-pulse TMS, rTMS is to induce long-term after-effects, facilitatory or suppressive, that last after the sessions from tens of minutes to days and weeks. These after-effects using rTMS are different depending upon the stimulating pulse protocols. Frequency of stimulating pulses, intervals between pulse trains, the intensity of stimulation, total duration of a stimulating session, and location of the stimulating area are the factors determining the after-effects. Low-frequency stimulation less than 1 Hz has inhibitory effects and high-frequency stimulation over 5 Hz has excitatory effects. Usually, TMS devices provide the stimulating frequency in the range 1~50 Hz.

To induce suppressive after-effects, pulses are applied in two protocols. The most common suppressive protocol is the continuous application of 1 Hz rTMS for a period of 15~30 min (Fig. 13.20a). It results in a suppression of the targeted cortical region for a period lasting up to several minutes following the stimulation. Another suppressive protocol is continuous theta-burst stimulation (cTBS) (Fig. 13.20b). Theta burst is 3 pulses of 50 Hz which simulated the brain's natural theta rhythm occurring in the hippocampus. cTBS applies theta bursts at 5 Hz for the 20 s or 40 s. This protocol induces suppression in motor cortical excitability for a period of approximately 30~40 min following less than a minute of stimulation.

Two protocols are used to induce facilitatory after-effects. High-frequency protocols are most commonly applied for the excitatory effect. However, the pulses are applied in an intermittent fashion. Exact protocols may differ among clinicians. One example is to apply a 10 Hz pulse in suprathreshold intensity, 110~120% of resting motor threshold (rMT) (Fig. 13.20c). One stimulation period is composed of a stimulation period of 2~5 s and an intervening resting period of 25~55 s. rTMS with this protocol to the left dorsal lateral prefrontal cortex (DLPFC) is effective as available antidepressant medications (Schutter, 2009). Intermittent theta-burst stimulation (iTBS) is another protocol for faciliatory after-effect (Fig. 13.20d). It is effective for a number of neurological conditions including depression, multiple sclerosis, and schizophrenia. iTBS protocol involves the theta-burst of 3 pulses of 50 Hz applied at 5 Hz. However, pulses are applied intermittently with a 10 s period, which is composed of 2 s stimulation and 8 s interval. Application of iTBS for 190 s can induce an after-effect of approximately 30~40 min.

6 Comparison of Brain Stimulation Methods

Table 13.4 summarizes and compares the brain stimulation methods described in this chapter.

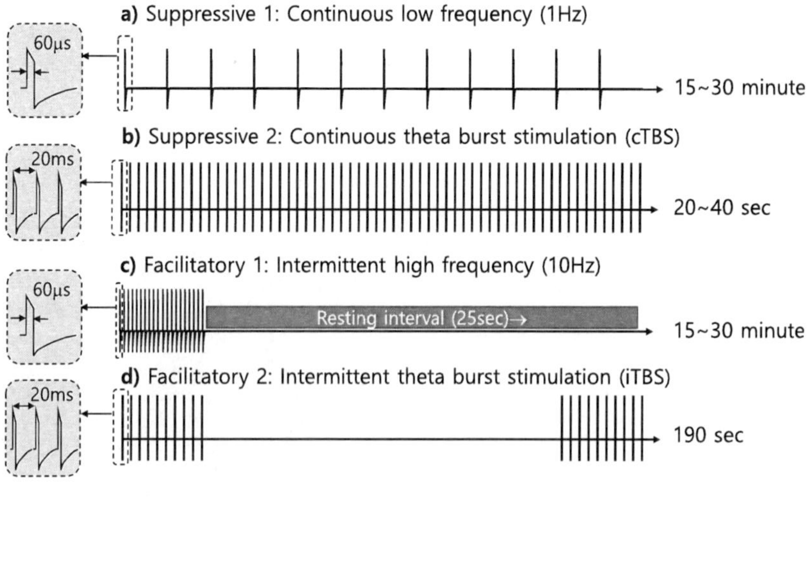

Fig. 13.20 Suppressive and facilitatory rTMS pulse protocols. (**a**) Continuous low-frequency stimulation of 1 Hz for 15~30 min, (**b**) continuous theta-burst stimulation (cBTS) for 20 or 40 s, (**c**) intermittent high-frequency stimulation of 10 Hz for 15~30 min, (**d**) intermittent theta-burst stimulation (iTBS) for 190 s

Table 13.4 Comparison of brain stimulation methods

	DBS	ECT	tDCS	TMS
Invasiveness	Invasive	Invasive	Noninvasive	Noninvasive
Stimulated area	Focalized volume	Whole brain	Selected region	Selected region
Energy type	Electrical	Electrical	Electrical	Magnetic
Current type	Pulses	Pulses	DC	Pulses
Stimulating frequency	130 Hz	70 Hz	0 Hz	1~50 Hz
Stimulation amplitude	4 mA	900 mA	1.5 mA	1.0~2.5Testa
Session period	Continuous	15~20 min	5~30 min	15~30 min
Treatment duration	Permanent	3~4 weeks	10 days	4~6 weeks
Major indication	PD, movement disorder	Depression, schizophrenia	Depression, pain	Depression, OCD
First FDA approval	1997 for essential tremor	1976, grandfathered approval	CE mark in 2015 for depression	2008 for depression

References

Abosch, A., Timmermann, L., Bartley, S., Rietkerk, H. G., Whiting, D., Connolly, P. J., Lanctin, D., & Hariz, M. I. (2013). An international survey of deep brain stimulation procedural steps. *Stereotactic and Functional Neurosurgery, 91*(1), 1–11. https://doi.org/10.1159/000343207

Agnesi, F., Johnson, M. D., & Vitek, J. L. (2013). Deep brain stimulation: How does it work? *Handbook of Clinical Neurology, 116*, 39–54.

Ballestero, J., et al. (2015). Reducing current spread by use of a novel pulse shape for electrical stimulation of the auditory nerve. *Trends in Hearing, 19*, 1–12.

Bove, F., Mulas, D., Cavallieri, F., Castrioto, A., et al. (2021). Long-term outcomes (15 years) after subthalamic nucleus deep brain stimulation in patients with Parkinson disease. *Neurology, 97*(3), e254–e262. https://doi.org/10.1212/WNL.0000000000012246

Carlyon, R. P., et al. (2005). Effect of inter-phase gap on the sensitivity of cochlear implant users to electrical stimulation. *Hearing Research, 205*, 210–224.

Casagrande, S., Cury, R. G., Alho, E., & Fonoff, E. T. (2019). Deep brain stimulation in Tourette's syndrome: Evidence to date. *Neuropsychiatric Disease and Treatment, 15*, 1061–1075. https://doi.org/10.2147/NDT.S139368

Farrell, S. M., Green, A., & Aziz, T. (2018). The current state of deep brain stimulation for chronic pain and its context in other forms of neuromodulation. *Brain Sciences, 20(8)*, 158. https://doi.org/10.3390/brainsci8080158

Fink, M., Fricchione, G., Rummans, T., & Shorter, E. (2016). Catatonia is a systemic medical syndrome. *Acta Psychiatrica Scandinavica, 133*, 250–251.

Fregni, F., El-Hagrassy, M. M., Pacheco-Barrios, K., et al. (2021). Evidence-based guidelines and secondary meta-analysis for the use of transcranial direct-current stimulation in neurological and psychiatric disorders. *International Journal of Neuropsychopharmacology, 24*(4), 256–313. https://doi.org/10.1093/ijnp/pyaa051

Grill, W. M. (2018). Temporal pattern of electrical stimulation is a new dimension of therapeutic innovation. *Current Opinion in Biomedical Engineering, 8*, 1–6.

Groppa, S., Oliviero, A., Eisen, A., et al. (2012). A practical guide to diagnostic transcranial magnetic stimulation: Report of an IFCN committee. *Clinical Neurophysiology, 123*(5), 858–882. https://doi.org/10.1016/j.clinph.2012.01.010

Guo, T., Finnis, K. W., Deoni, S. C. L., Parrent, A. G., & Peters, T. M. (2006). Comparison of different targeting methods for subthalamic nucleus deep brain stimulation. *Medical Image Computing and Computer Assisted Intervention, 2006*, 768–775.

Herrington, T. M., Cheng, J. J., & Eskandar, E. N. (2016). Mechanisms of deep brain stimulation. *Journal of Neurophysiology, 115*(1), 19–38. https://doi.org/10.1152/jn.00281.2015

Herron, J. A., Thompson, M. C., Brown, T., Chizeck, H. J., Ojemann, J. G., & Ko, A. L. (2017). Chronic electrocorticography for sensing movement intention and closed-loop deep brain stimulation with wearable sensors in an essential tremor patient. *Journal of Neurosurgery, 127*, 580–587.

Horgan, J. (2005). The forgotten era of brain chips. *Scientific American, 293*, 66–73.

Hu, W., & Stead, M. (2014). Deep brain stimulation for dystonia. *Translational Neurodegeneration, 21*(3), 2. https://doi.org/10.1186/2047-9158-3-2

Inghilleri, M., Berardelli, A., Cruccu, G., & Manfredi, M. (1993). Silent period evoked by transcranial stimulation of the human cortex and cervicomedullary junction. *The Journal of Physiology, 466*, 521–534.

Karas, P. J., Lee, S., Jimenez-Shahed, J., Goodman, W. K., Viswanathan, A., & Sheth, S. A. (2019). Deep brain stimulation for obsessive-compulsive disorder: Evolution of surgical stimulation target parallels changing model of dysfunctional brain circuits. *Frontiers in Neuroscience, 12*, 998. https://doi.org/10.3389/fnins.2018.00998

Kellner, C. H., Obbels, J., & Sienaert, P. (2020). When to consider electroconvulsive therapy (ECT). *Acta Psychiatrica Scandinavica, 41*(4), 304–315. https://doi.org/10.1111/acps.13134

Khalid, N., Atkins, M., Tredget, J., et al. (2008). The effectiveness of electroconvulsive therapy in treatment-resistant repression. *The Journal of ECT, 24*(2), 141–145. https://doi.org/10.1097/YCT.0b013e318157ac58

Kirsch, A. D., Hassin-Baer, S., Matthies, C., Volkmann, J., & Steigerwald, F. (2018). Anodic versus cathodic neurostimulation of the subthalamic nucleus: A randomized-controlled study of acute clinical effects. *Parkinsonism & Related Disorders, 55*, 61–67.

Krauss, J. K., Lipsman, N., Aziz, T., et al. (2021). Technology of deep brain stimulation: Current status and future directions. *Nature Reviews. Neurology, 17*, 75–87. https://doi.org/10.1038/s41582-020-00426-z

Lee, C. S. (2001). Levodopa-induced dyskinesia: Mechanisms and management. *British Columbia Medical Journal, 43*(4), 206–209.

Lee, D. J., Lozano, C. S., Dallapiazza, R. F., & Lozano, A. M. (2019). Current and future directions of deep brain stimulation for neurological and psychiatric disorders. *Journal of Neurosurgery, 131*, 333–342.

Luchini, F., Medda, P., Mariani, M. G., Mauri, M., Toni, C., & Perugi, G. (2015). Electroconvulsive therapy in catatonic patients: Efficacy and predictors of response. *World Journal of Psychiatry, 5*, 182–192.

Lyons, M. K. (2011). Deep brain stimulation: Current and future clinical applications. *Mayo Clinic Proceedings, 86*(7), 662–672. https://doi.org/10.4065/mcp.2011.0045

Lyons, K. E., & Pahwa, R. (2004). Deep brain stimulation and essential tremor. *Journal of Clinical Neurophysiology, 21*(1), 2–5. https://doi.org/10.1097/00004691-200401000-00002

Ma, J., Zhang, Z., Kang, L., et al. (2014). Repetitive transcranial magnetic stimulation (rTMS) influences spatial cognition and modulates hippocampal structural synaptic plasticity in aging mice. *Experimental Gerontology, 58*, 256–268. https://doi.org/10.1016/j.exger.2014.08.011

Maserejian, N., Vinikoor-Imler, L., & Dilley, A. (2021). Estimation of the 2020 global population of Parkinson's disease (PD). *Movement Disorders, 35*(suppl-1), S1–S599.

Michael, N. (2009). Hypothesized mechanisms and sites of action of electroconvulsive therapy. In C. Swartz (Ed.), *Electroconvulsive and neuromodulation therapies* (pp. 75–79). Cambridge University Press. https://doi.org/10.1017/CBO97805115763393.005

Morrell, M. J. (2011). RNS system in epilepsy study group. Responsive cortical stimulation for the treatment of medically intractable partial epilepsy. *Neurology, 77*, 1295–1304.

Muller, P. A., Dhamne, S. C., Vahabzadeh-Hagh, A. M., et al. (2014). Suppression of motor cortical excitability in anesthetized rats by low-frequency repetitive transcranial magnetic stimulation. *PLoS One, 9*(3), e91065. https://doi.org/10.1371/journal.pone.0091065

Ngoga, D., Mitchell, R., Kausar, J., et al. (2014). Deep brain stimulation improves survival in severe Parkinson's disease. *Journal of Neurology, Neurosurgery & Psychiatry, 85*, 17–22.

Nitsche, M. A., & Paulus, W. (2000). Excitability changes induced in the human motor cortex by weak transcranial direct-current stimulation. *The Journal of Physiology, 527*(3), 633–639. https://doi.org/10.1111/j.1469-7793.2000.t01-1-00633.x

Pompili, M., et al. (2014). Electroconvulsive treatment during pregnancy: A systematic review. *Expert Review of Neurotherapeutics, 14*(12), 1377–1390. https://doi.org/10.1586/14737175.2014.972373

Popovych, O. V., & Tass, P. A. (2014). Control of abnormal synchronization in neurological disorders. *Frontiers in Neurology, 5*, 268.

Reddy, M. S., & Vijay, M. S. (2017). Repetitive transcranial magnetic stimulation for depression: State of the art. *Indian Journal of Psychological Medicine, 39*, 1–3.

Ross, E. L., Zivin, K., & Maixner, D. F. (2018). Cost-effectiveness of electroconvulsive therapy vs pharmacotherapy/psychotherapy for treatment-resistant depression in the United States. *JAMA Psychiatry, 75*, 713.

Schutter, D. J. (2009). Antidepressant efficacy of high-frequency transcranial magnetic stimulation over the left dorsolateral prefrontal cortex in double-blind sham-controlled designs: A meta-analysis. *Psychological Medicine, 39*(1), 65–75.

Silverdale, M. (2017). Deep brain stimulation for movement disorders other than Parkinson's disease. In *Electroceutical* (pp. 75–105). Springer.

Sironi, V. A. (2011). Origin and evolution of deep brain stimulation. *Frontiers in Integrative Neuroscience, 18*(5), 42. https://doi.org/10.3389/fnint.2011.00042

Surya, S., & Mori, N. (2015). Recent literature on electroconvulsive therapy techniques for the treatment of depression. *The American Journal of Psychiatry Residents' Journal, 10*, 3–5.

Sveinbjornsdottir, S. (2016). The clinical symptoms of Parkinson's disease. *Journal of Neurochemistry, 139*(Suppl 1), 318–324. https://doi.org/10.1111/jnc.13691

Thair, H., Holloway, A. L., Newpor, R., & Smith, D. (2017). Transcranial direct-current stimulation (tDCS): A beginner's guide for design and implementation. *Frontiers in Neuroscience, 11*, 641. https://doi.org/10.3389/fnins.2017.00641

Udupa, K., & Chen, R. (2013). Central motor conduction time. *Handbook of Clinical Neurology, 116*, 375–386. https://doi.org/10.1016/B978-0-444-53497-2.00031-0

Utz, K. S., Dimova, V., Oppenländer, K., & Kerkhoff, G. (2010). Electrified minds: Transcranial direct-current stimulation (tDCS) and Galvanic vestibular stimulation (GVS) as methods of non-invasive brain stimulation in neuropsychology—A review of current data and future implications. *Neuropsychologia, 48*(10), 2789–2810. https://doi.org/10.1016/j.neuropsychologia.2010.06.002

Zangen, A., Moshe, H., et al. (2021). Repetitive transcranial magnetic stimulation for smoking cessation: A pivotal multicenter double-blind randomized controlled trial. *World Psychiatry, 20*(3), 397–404. https://doi.org/10.1002/wps.20905. PMID: 34505368; PMCID: PMC8429333.

Zangiabadi, N., Ladino, L. D., Sina, F., Orozco-Hernández, J. P., Carter, A., & Téllez-Zenteno, J. F. (2019). Deep brain stimulation and drug-resistant epilepsy: A review of the literature. *Frontiers in Neurology, 6*(10), 601. https://doi.org/10.3389/fneur.2019.00601

Zauber, S. E., Smith, P. A., & Metman, L. V. (2015). Fundamentals of deep brain stimulation programming. In *Deep brain stimulation management*. Cambridge University Press.

Zhang, C., Ramirez-Zamora, A., et al. (2020). An international survey of deep brain stimulation utilization in Asia and Oceania: The DBS think tank east. *Frontiers in Human Neuroscience, 14*, article:162. https://doi.org/10.3389/fnhum.2020.00162

Zhu, F. F., Yeung, A. Y., Poolton, J. M., Lee, T. M., Leung, G. K., & Masters, R. S. (2015). Cathodal transcranial direct-current stimulation over left dorsolateral prefrontal cortex area promotes implicit motor learning in a golf putting task. *Brain Stimulation, 8*, 784–786. https://doi.org/10.1016/j.brs.2015.02.005

Check for updates

14 Sensing by Electricity

Can the deaf hear a sound by some means? Can the blind see some lights? Is electricity able to do a critical role for these persons? These are the oldest questions that many scientists, clinicians, and biomedical engineers have investigated throughout history. Now, these are partly solved and this chapter describes the current status of technologies on this issue, sensing by electricity for the sensory impaired.

When I first entered the discipline of biomedical engineering, many of us were motivated by the bionic persons who were popularized in the 1970s by the US TV series, "The six-million-dollar man" and the following "The bionic woman." Bionic persons, Steve Austin and Jaime Sommers have enforced some of their abilities to a superhuman level using the cutting-edge technologies during the recovery from the accidents. Among their superhuman ability, Steve had restored with a superhuman level bioelectronic eye which can see the objects at longer distances with a zooming function and Jaime had revitalized with a superhuman level bioelectronic ear which can hear incredibly minute sounds. In those times, these bioelectronic sensory organs were totally fictional stories. However, these days, these bioelectronic ears and eyes are no more fictional items. Even though their functions are not advanced to a superhuman level, they are actively used in the medical domain to restore the sensory functions of disabled subjects maximally approaching a normal human level. The cochlear implant is the bioelectronic ear which senses the sound and translates it into electrical signals to be transmitted along the auditory neural pathway and the retinal prosthesis is the bioelectric eye which senses the image and translates it into electrical signals to be transmitted along the visual pathway to the brain.

1 Brief History of Sensing by Electricity

1.1 History of Artificial Hearing

Experiments showing that hearing can be elicited by electrical stimulation go back more than 200 years. Around 1790, Alessandro Volta first did the experiment stimulating the ears with a battery pile he developed. By connecting the battery piles of about 50 V to his own ears by metallic rods, he heard "boom within the brain," a noisy sound similar to the sound of boiling soup. In 1865, Duchenne de Boulogne (1806~1875), a French neurologist, reported a buzzing and ringing sensation, achieved by using alternating current. In the 1930s, Ernest Glen Wever (1902~1991), an American psychologist at Princeton University, first recorded electrical potentials in the cochlea synchronized to sound stimulus waveform. His work provided the vision that replicating those potentials could restore hearing.

In 1957, André Djourno (1904~1996), a French neurophysiologist, and Charles Eyriès (1908~1996), a French surgeon, had an interest in the direct stimulation of the auditory nervous system. In the re-operating on a patient with recurring ear problems, they placed an electrode on the stump of the patient's auditory nerve and return electrode at the temporalis muscle. After implantation of electrodes, the patient responded to different sound intensities and small closed sets of spoken words. But they did not extend the cochlear implantations further in patients after their failure in the second implantation and just reported their findings in the French medical journal.

William House (1923~2012), an American otologist who is considered "the Father of Neurotology," has been

K. S. Park, *Humans and Electricity*, https://doi.org/10.1007/978-3-031-20784-6_14

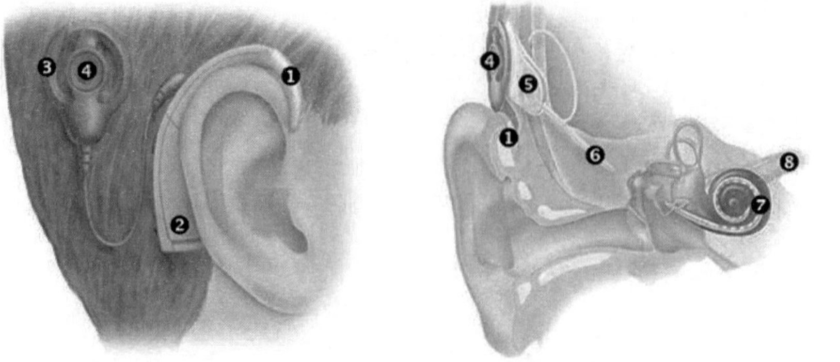

Fig. 14.1 Cochlear implant with information flow along with its external and internal components. (https://commons.wikimedia.org/wiki/)

introduced to the cochlear implant by his patient who has read the story of the cochlear implant by André Djourno and Charles Eyriès picked up in a newspaper. In 1961, his team placed implants of a single-channel gold electrode in a few patients and the patients could successfully identify words in small closed sets. Even though there was not any treatment for sensorineural deafness in those early days, most of the experts in hearing science were not optimistic about cochlear implants. They felt that success was unobtainable. However, House actively progressed the clinical implantations with the belief that proving the efficacy in the implanted patients is most important and it successfully resulted in the current acceptance of cochlear implants.

Georg von Békésy (1899~1972), a Hungarian biophysicist, has clarified how the sounds of different frequencies are converted into electrical impulses of the auditory nerves in a series of studies from 1940 to the 1960s. By visualizing the minute movement of the basilar membrane within the cochlea in the inner ear, he verified that different sound wave frequencies are spatially dispersed and excite different auditory nerve fibers that lead from the cochlea to the brain. His finding became the base for the multichannel stimulation of cochlear implants to increase the interpretability of sound. In 1961, he received the Nobel prize in physiology or medicine for his discoveries of the physical mechanism of stimulation within the cochlea.

Based on the theory of the interpretation of sound in the basilar membrane and in the auditory nerves, increasing the number of stimulating channels has been pursued in the following years from the mid-1960s through the 1970s. The number of channels increased to 2, 3, 4, 5, 6, 12, and 20 progressively. In 1978, Graeme Clark (1935–), an Australian otolaryngologist at the University of Melbourne, and his team successfully implanted a multichannel prototype with an array of 20 electrodes, which they developed through the years in the 1970s. Technological progress in related fields supported the development of safe and wearable forms of cochlear implants. Hermetic sealing technology, which was adapted in cardiac pacemakers, solved a major problem of body fluid leaking into the implanted electronics and extended the period of implantation not limited to a few months. Signals and power for the electronics started to be transferred with an induction system with a radiofrequency (RF) coil and attaching magnets (Fig. 14.1). The cochlear implant system became totally implant and wearable during this period in the 1970s. During the 1980s, technologies transformed into industrial companies in the USA and Australia and the commercialization of cochlear implants began.

Single-channel cochlear implant for adults got FDA's first approval in 1985, and the multiple-channel cochlear implant was FDA approved for adults in 1987. Cochlear implants extended the application to younger generations quickly with following FDA approval. The multiple-channel implant was approved for children older than 2 years in 1990. The initial age for implantation went to 18 months in 1998 and 12 months in 2000. As of December 2019, approximately 736,900 registered devices have been implanted worldwide.

1.2 History of Artificial Vision

The idea of using electricity to make a vision is quite old, more than 250 years, as the idea of using electricity to develop a hearing. However, the retinal prostheses of current use are almost recently progressed during the last 20 years. In 1800, Alessandro Volta also did a similar experiment on electrical stimulation on the eyes and optic nerves with his battery pile and noted that it could induce the sensation of light. However, 45 years earlier in 1755, before the battery was developed by Volta, Charles Le Roy (1726~1779), a French chemist and physician, tried to cure his patient of blindness with crude transcranial electrical stimulation. After winding a wire around the patient's head, he sent a current pulse through the wire from the Leiden jar. The blind patient could not be cured, but he perceived the vivid flash of light, called phosphenes, during the stimulation. This experiment of electrical stimulation on humans was done 26 years earlier than the famous experiment of electrical stimulation on a frog's leg done by Luigi Galvani in 1781.

During the eighteenth and nineteenth centuries, there was a great advance in understanding cerebral anatomy. The connection of the occipital cortex from the eyes via optic nerves, lateral thalamus, and optic radiation was revealed with the role of the occipital cortex for vision. With this understanding, phosphenes have developed by electrically stimulating the occipital cortex. In 1929, Carl Foester (1825~1902), a German ophthalmologist, confirmed that direct electrical stimulation of the occipital cortex can elicit spots of light in a completely blind man. In 1931, Fedor Krause (1857~1937), a German neurosurgeon, confirmed that the phosphenes are elicited in the same pattern in the blind as well as in the normal and showed that the phosphenes elicited by the stimulation of a fixed point on the cortex were localized to a corresponding point in visual space. In 1968, Giles Brindley (1926~), a British physiologist, did the first experiment implanting a prosthesis containing 80 intracranial electrodes in a silicon cap for a 52-year-old woman blinded from glaucoma and retinal detachment. Stimulating electrodes are connected wirelessly to an extracranial radio frequency transceiver. She reported the appearance of phosphenes when the electrodes were activated and the brightness of the phosphenes was dependent on the frequency and duration of stimulation. When the two positions in the cortex apart 2~4 mm were simultaneously stimulated, generated phosphenes were spatially distinguishable. In the late 1990s, William Dobelle (1941~2004), an American biomedical researcher, did clinical trials of cortical implants over ten patients. However, these early cortical implants were not fully successful with poor resolution, significant surgical challenges, and accompanying medical or psychological complications.

Visual perception also can be generated by electrically stimulating other locations than the visual cortex along the visual pathway. Since 1998, electrical prostheses interfacing the visual system at various locations along the visual pathway, retina, optic nerve, and geniculate nucleus have been investigated. Optimal implantation is largely dependent on the degree of the subject's visual pathway liveness. For patients with retinitis pigmentosa (RP) and age-related macular degeneration (AMD), retinal prostheses are considered optimal, as their inner retina and the rest of the visual pathway remain relatively intact and functional except for the outer retinal degeneration. However, it is not indicated for diseases such as glaucoma in which the optic nerve is completely damaged, or for the blindness caused by the damage to the higher visual centers in the brain (Yue et al., 2018).

Among the possible stimulating locations along the visual pathway, recent developments in visual prostheses are focused on the retinal prosthesis. Retina has some advantages compared to other locations. Retina prosthesis is surgically easy to access and stimulating images can simply be generated using the well-reserved retinotopic mapping. And, remaining visual circuitry can be used to make the stimulated images follow the normal visual processing. In 1999, Mark Humayun, a Pakistani-American ophthalmologist, demonstrated the generation of phosphenes by delivering electric currents through the multielectrode array to retinal cells in blind subjects. The subject reported the movement of phosphene locations as different electrodes are stimulated. His experiment revealed the retinotopic correlation between the stimulation pattern in the electrode array and generated pattern of phosphenes in the visual field. Retinal prosthesis with a two-dimensional electrode array gained attention as an optimal location to elicit visual perception representing the real images with minimal distortion. Since then, the field of retinal prosthesis has rapidly evolved, spawning commercially available retinal prosthetic systems. Currently available retinal prosthetic systems have successfully restored useful artificial vision in blind subjects. While the restored vision is lower in quality compared to the normal vision, it extends the range of vision-related activities including guided movement and object recognition. Retinal prostheses got FDA approval in 2013. Currently, the number of patients who have received the implantation of a retinal prosthetic device are approaching 350.

2 Human Hearing System

2.1 Sound That Humans Can Hear

Sound is a longitudinally propagating vibrational acoustic wave through the transmission mediums like air, water, and metals. It is longitudinal as the particles of the medium vibrate along the direction of propagating direction, which is in contrast to the light of a transversal wave. Sound is not generated and transmitted without a medium, also different from light. Physiologically, the sound is the perception of such waves in the brain through the processing of the auditory system.

Physically, sound can be represented as a waveform having frequency and amplitude components. From a wide range of sound frequencies, humans can only hear sound in the frequency range of 20~20 KHz. Ultrasound is the sound in higher frequency over 20 KHz that humans cannot hear and infrasound is the sound in lower frequency less 20 Hz that humans cannot hear. The highest audible frequency of 20 KHz is 1000-folds from the lowest frequency of 20 Hz.

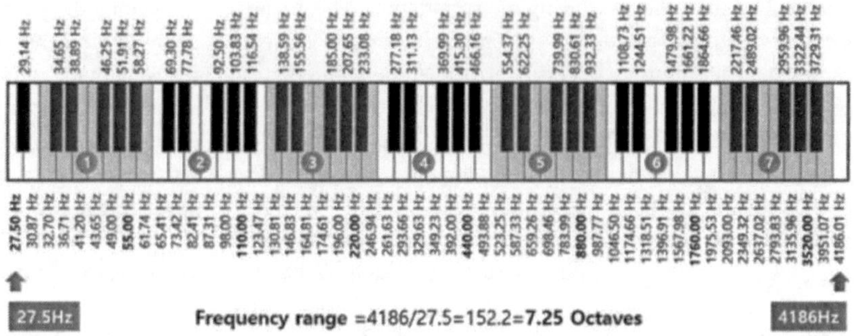

Fig. 14.2 Audible frequency range of the piano

The range of audible sound covers almost 10 octaves as $2^{10} = 1024 \cong 1000$. The lowest key of the piano is 27.5 Hz and the highest key is 4186 Hz. And, piano keys cover almost 7.25 octaves by $\log_2(4186/27.5) = \log_2(152.2) = 7.25$ (Fig. 14.2).

Hearing acuity rapidly drops at both ends, and it is very variable among people. Even though some people can hear the highest and lowest sounds, most people cannot hear the highest or lowest sound clearly. Hearing acuity declines gradually with age. Presbycusis, age-related hearing loss, is the gradual loss of hearing in both ears. It occurs commonly with aging. One in three adults over age 65 has hearing loss. Due to its gradual change, many people age without noticing the hearing change in the earlier stage. Most often, it affects high-frequency sound earlier while lower frequency sound is usually not affected. Even though the audible range of sound is declined to some level, it may not affect the normal activities. The telephone line only transmits sounds in the frequency range of 200~2 KHz. However, we easily recognize the person over the line by the transmitted voice. The frequency of sound is an important parameter in the recognition of sound in the physiological auditory system.

Another parameter of sound is the amplitude of sound waves which is recognized as the loudness by the human auditory system. The perceived loudness of sound is increasing with the amplitude of the sound wave, which is the magnitude of pressure applied by the medium particles. Pressure is expressed with the unit Pascal, abbreviated as Pa, which is defined as N/m^2, Newton per square meter.

The human ear can perceive a very wide range of sound pressure. The weakest sound a normal human ear can hear is a pressure variation of 20 μPa. This level of sound pressure is called the weakest audible sound (WAS) and it is used as the reference to represent sound in decibels. On the other hand, humans can endure the sound having the loudness of a flying jet airplane, which is approximating the pressure of 200 Pa, without damaging the auditory system. Beyond the sound pressure level of 2000 Pa, humans feel the applied sound as a physical pressure instead of sound and it causes damage to the auditory system.

Expressing the sound pressure level (SPL) in terms of Pa is quite inconvenient since it required too many digits or exponents from 0.000,02 Pa to 2000 Pa. A simpler way is to express the loudness using the logarithm with 10 as the base, which is called decibel (dB). Also, the decibel scale gives a much better approximation to the human perception of relative loudness than the Pascal scale. This is because the human ear responds to the loudness of sound in logarithmic change, as it responds to the frequency of sound. That is, the human hearing system recognizes the change of sound similarly when the frequency or amplitude of sound has doubled. The sound pressure level in decibel (dB) is defined as

$$\text{Sound pressure level (SPL)} = 20 \times \log_{10} \frac{\text{Sound pressure[Pa]}}{\text{WAS[Pa]}} \tag{14.1}$$

$$\text{WAS} = \text{Weakest audible sound} = 2 \times 10^{-5}\,\text{Pa} \tag{14.2}$$

WAS is the weakest audible sound level that a young human with undamaged hearing can detect at 1000 Hz. The sound pressure level represented in the decibel is a unitless variable. It just represents the loudness of sound by how many folds it is larger than the reference sound of WAS.

Sound intensity, also known as acoustic intensity, describes the power carried by sound waves per unit area and the unit is Watt per square meter, W/m^2. Sound intensity is related to sound pressure by the relation

$$\text{Sound intensity (I)} = \text{Soundpressure(P)} \cdot \text{Particle\_velocity (v)} \tag{14.3}$$

$$\text{Sound pressure (P)} = \text{Particle\_velocity (v)} \cdot \text{Acoustic\_impedance } (z_0) \tag{14.4}$$

Combining two equations, sound intensity can be calculated by the square of sound pressure divided by the acoustic impedance.

Sound intensity (I)

$$= \text{Soundpressure } (P)^2/\text{Acoustic\_impedance } (z_0) \quad (14.5)$$

For the weakest audible sound of 2×10^{-5}Pa, the weakest sound intensity (WSI) can be calculated with an acoustic impedance of 400 Pa·s/m in air.

$$\begin{aligned}&\text{Weakest sound intensity (WSI)}\\&= \left(\frac{2\ 10^{-5}\text{N}}{\text{m}^2}\right)^2/(400\ (\text{N}\cdot\text{s})/\text{m}^3)\\&= 10^{-12}\text{N}/(\text{m}\cdot\text{s}) = 10^{-12}\text{W}/\text{m}^2\end{aligned} \quad (14.6)$$

Sound intensity level also can be represented in logarithm unit similarly to sound pressure level.

$$\text{Sound intensity level (SIL)} = 10 \times \log_{10}\frac{\text{Sound intensity}[\text{W}/\text{m}^2]}{\text{WSI}[\text{W}/\text{m}^2]} \quad (14.7)$$

For any sound pressure level, the sound intensity level is calculated with the same value. That is, 70 dB of SPL is equivalent to 70 dB of SIL and 100 dB of SPL is equivalent to 100 dB of SIL. So, the sound pressure and intensity are used interchangeably when it is described in decibels. Table 14.1 shows the examples of sounds of different intensities.

Most of the sounds we hear are not pure tones of a single frequency. They are complex sounds, which are made up of many frequency components having different amplitudes. If the sound is composed of three sinusoidal components of different frequencies as in Fig. 14.3a, sound signal x(t) can be represented as the sum of three components,

$$\begin{aligned}x(t) &= x_1(t) + x_2(t) + x_3(t)\\&= A_1\cos(2\pi f_1 t + \varphi_1) + A_2\cos(2\pi f_2 t + \varphi_2)\\&\quad + A_3\cos(2\pi f_3 t + \varphi_3)\end{aligned} \quad (14.8)$$

When the sound signal is described in the time domain, the recorded waveform reflected the combined change of sound amplitude along the time. However, as this sound is a mixture of three components, we can equivalently represent the sound signal exactly just by knowing the amplitude of components A_1, A_2, A_3, and the phase of components φ_1, φ_2, φ_3, for the frequencies f_1, f_2, f_3, while the cosine function is assumed as the basis function. That is, if we know the A(f), amplitude spectrum, and φ(f), phase spectrum, of signal x(t), we know equivalently all the variations of x(t). A(f) and φ(f) can be calculated by Fourier transform mathematically and they widely used in sound signal processing.

$$X(f) = X_R(f) + jX_I(f) = \int_{-\infty}^{\infty} x(t)e^{-2\pi ft}dt \quad (14.9)$$

$$A(f) = \sqrt{X_R^2(f) + X_I^2(f)} \quad (14.10)$$

$$\varphi(f) = \tan^{-1}(X_I(f)/X_R(f)) \quad (14.11)$$

Figure 14.3b shows the example sound of the violin in the time domain and in the frequency domain after being transformed. In a signal segment, while the violin string is bowed, the sound shows a complex waveform with multiple harmonics in the frequency domain. In contrast, in a signal

Table 14.1 Sound examples of different intensities

Sound pressure, (Pa = N/m^2)	Sound intensity (W/m^2 = N/m·s)	SPL, SIL (dB)	Sound example	Classification
0.000,02	0.000,000,000,001	0	Hearing threshold	Faint
0.000,063	0.000,000,000,01	10	Rustle of leaves	
0.000,2	0.000,000,000,1	20	Background in TV studio	
0.000,63	0.000,000,001	30	Quiet bedroom at night	Soft
0.002	0.000,000,01	40	Quiet library	
0.006,3	0.000,000,1	50	Average home	Moderate
0.02	0.000,001	60	Normal conversation	
0.063	0.000,01	70	Noisy office, busy traffic	Loud
0.2	0.000,1	80	Loud radio	
0.63	0.001	90	Inside a heavy truck	Very loud
2.0	0.01	100	Noisy factory	
6.3	0.1	110	Chain saw, 1 m	
20	1.0	120	Threshold of discomfort	Uncomfortable
63	10	130	Threshold of pain	Painful
200	100	140	Jet airplane flight, 50 m	
630	1000	150	Jet airplane take-off, 25 m	
2000	10,000	160	Bursting of ear drum	Dangerous

SPL sound pressure level, *SIL* sound intensity level

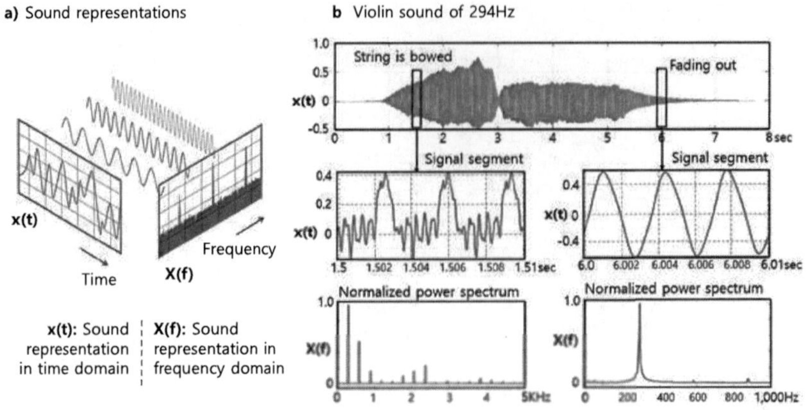

Fig. 14.3 Representation of sounds. (**a**) Sound is a mixture of many different frequencies having different amplitudes. It can be equally represented in the time domain as well as in the frequency domain. (**b**) Example violin sound of 294 Hz. During the string is bowed, the sound is composed of fundamental and many harmonics, while the sound is fading away almost with a pure tone

segment, while the sound is fading out, the sound waveform is almost in sinusoidal form and it can be confirmed in its transformed frequency components.

Transforming the sound from the time domain to the frequency domain and representing the sound by its components' amplitudes and phases is what the human auditory system does to sense and interpret the sound. The auditory system of humans transmits the sound from the ear canal to the inner ear and produces maximal fluctuation of the basilar membrane at different locations according to the frequency of input sound.

2.2 Sound Transmission in the Auditory System

The human auditory system collects sound, conducts sound, and converts it to a series of electrical pulses for transmission to the brain along the auditory nerves. The auditory system is largely composed of three parts of the outer ear, middle ear, and inner ear (Fig. 14.4).

The outer ear is composed of the auricle, the external auditory canal, and the tympanic membrane. The auricle is the only visible part of the human auditory system and it collects and funnels sound to the external auditory canal. The convolutional structure of the auricle has evolved so that it collects and transmits higher frequencies effectively. This structure provides selective boosting of sound pressure by 30~100 folds around 3 KHz and this amplification makes humans more sensitive to the sound around 3 KHz, where the most human sounds are distributed. However, it also makes acoustical injury more easily occur around this frequency range.

The external auditory canal is a passageway that guides sound from the outside of the head to the tympanic membrane. It is about 2.5 cm in length and 0.7 cm in diameter. Its outer part consists of cartilage, while the wall of its inner part is made of the temporal bone. Tiny hairs in the outer cartilaginous portion are directed outward and produce earwax, which helps to protect the ear canal by trapping dust and debris and to discourage insects from entering the ear.

The tympanic membrane, also called the eardrum, separates the outer ear from the middle ear. The tympanic membrane blocks the auditory canal at the end and looks like a flattened cone with its tip pointed inward and attached to tiny bones of the ossicle. It receives sound vibrations from the auditory ear canal and transmits them to the auditory ossicles, which are the tiny bones in the middle ear cavity. The mechanical resistance of the tympanic membrane is similar to that of the auditory canal, so it vibrates at approximately the same magnitude as the air molecules. The magnitude of vibration is incredibly small. For the weakest audible sound, the magnitude of vibration of the tympanic membrane is 0.1 Å, $0.1{\times}10^{-10}$ m, which is even much smaller than the diameter of a water molecule of 3 Å. On the contrary, for the maximal intensity of 140 dB sound before bursting the tympanic membrane, the magnitude of vibration increases 10^7-folds and becomes 0.1 mm. To maintain the sensitivity of the tympanic membrane, there should be no pressure difference between both sides of the tympanic membrane, between the external auditory canal and the tympanic cavity of the middle ear. While the external auditory canal is open to atmospheric pressure, the closed tympanic cavity can deviate from atmospheric pressure sometimes as we experience during the departure or arrival period in the airplane. Then, the Eustachian tube opens and connects the tympanic cavity to the nasal cavity of atmospheric pressure. It clears the pressure difference across the tympanic membrane and returns the sensitivity of the tympanic membrane to the optimal level. By the tympanic membrane, the displacement of air particles is converted to the displacement of tiny bones in the middle ear.

The middle ear is the cavity from the internal part of the eardrum to the external surface of the oval window of the cochlea. The middle ear includes ossicles in the tympanic cavity, which transfer the vibrations of the eardrum into the

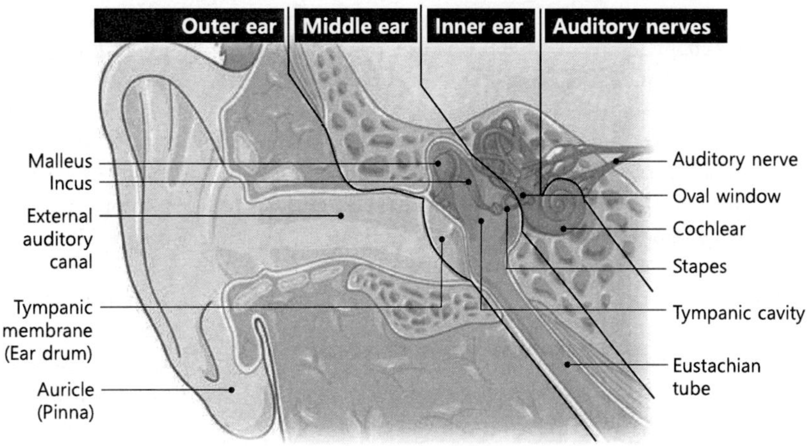

Fig. 14.4 Anatomy of the auditory conduction system

vibration of fluid within the cochlea. Ossicle is a tiny set of bones composed of the smallest bones within the body, malleus, incus, and stapes, which have the shapes of hammer, anvil, and stirrup respectively. These three bones form a kind of bridge. Malleus is connected to the tympanic membrane; the stapes is connected to the oval window of the cochlea and the incus connects the malleus and stapes. These three bones combinedly work like a lever and increase the force by a factor of 1.3 when it is applied to the oval window. As the effective vibratory area of the tympanic membrane is about 14 times larger than that of the oval window, the pressure concentrated on the oval window increases by $1.3 \times 14 = 18.2$ times. When the sound waves in the air collide with fluid directly, most of the energy is reflected off the surface of the interface. In this point of view, the middle ear provides a kind of impedance matching between the sound traveling in the air and the acoustic waves traveling in fluids within the cochlea. The middle ear contains a Eustachian tube that connects the tympanic cavity to the nasal cavity by opening and returning the pressure of the tympanic cavity to the atmospheric level. For the sound of high intensity, the incoming sound is attenuated by the feedback through the attached muscles to the ossicle. This ossicular attenuation starts from the sound intensity of 100 dB and attenuates the intensity of 5 dB for every 10 dB of input sound intensity higher than 100 dB.

The inner ear is the bony structure of the labyrinth located in the deepest part of the ear. It contains sensing organs for equilibrium as well as hearing. The bony labyrinth is composed of three sections, the semicircular canals, the vestibule, and the cochlea. Semicircular canals detect rotational acceleration for the balance of the body. The vestibule includes the saccule and utricle and detects linear acceleration in three axes. The cochlea is the part in the inner ear that works for hearing.

The cochlea is a part of a bony labyrinth spiral-shaped with two and a half turns. It plays the important role in sensing and processing the sound and transducing the mechanical signal of sound into the electrical signal of the action potential train. The cochlea is innervated with auditory nerves and the transduced action potentials are transmitted via these nerve fibers to the brain.

The cochlea is connected with the middle ear via the oval window, which is the only inlet of the cochlea not covered by the bone and covered by the membrane with its counter outlet, the round window. Vibrational pressure of sound is conducted through the middle ear and concentrated on the oval window via connected stapes. The cochlea is filled with fluids and two membranes divide the internal space longitudinally into three parts along with the spiral turns. Vibrational pressure applied to the oval window by the stapes induces a vibrational movement of fluids inside the cochlea and the internal membranes fluctuate according to the movements of the fluids. As the fluid in the cochlea is incompressible, the round window, which is also covered by a membrane, works as a countering outlet allowing easy fluid fluctuation within the cochlea. If the round window is calcified, it leads to a relatively common hearing deficiency. Hair cells located on the membranes sense the membrane fluctuations and generate action potentials synchronized to membrane fluctuation. Transduction of electrical action potentials in the hair cells by the fluctuation of the basilar membrane is described in the "auditory receptor" section of Chap. 6.

2.3 Sound Processing in the Cochlea

The most unique feature of intrinsic sound processing in the cochlea is that it does a kind of Fourier transform and works as a spectrum analyzer. As well in a cochlear implant, it is very important to generate effective multiple channels of electrical pulses representing the frequency characteristics to restore hearing capability for the sensorineural hearing loss.

The cochlea has two and a half spiral turns. Its whole length is about 35 mm when unfolded linearly and its diameter is about 2.0 mm at the base. The spiral shape is considered to yield a packing of compact structure. Its internal space is divided into three spaces, the scala vestibuli, scala media, and scala tympani as shown in Fig. 14.5. The vestibular membrane divides scala media from scala vestibuli and the basilar membrane divides scala media from scala tympani. Within the space of scala media, the organ of Corti, which includes hearing sensors of hair cells, is located on the top of the basilar membrane. When the basilar membrane fluctuates, there occurs an accompanying movement in the organ of Corti and hair cells. And, the transduction of electrical signals is derived from this movement. Hair cells do not recognize the frequency of the fluctuation. They only generate action potentials simply synchronized to the basilar membrane movement. This characteristic is the same for all the hair cells along the whole length of the cochlea. As the difference in sound is represented by the different patterns of basilar membrane fluctuation, the basilar membrane does a key role in discriminating the difference in sounds.

The basilar membrane works as a mechanical filter. Depending on the fluctuation frequency of fluids, it acts differently. For the sound of high frequency, maximal fluctuation of the basilar membrane occurs at the basal part of the entrance and for the sound of lower frequency, maximal fluctuation of the basilar membrane occurs at the tip part, at the top of the cochlea farthest from the entrance. That is, each location of the basilar membrane has finely tuned to a different resonance frequency. This characteristic is caused by the mechanical geometry of the basilar membrane including regional mass and stiffness. The basilar membrane is not of the same width along the cochlea length. Its width is smallest at the base, 0.1 mm, and it becomes wider as it spirals up to the top. It is widened to 0.5 mm at the top. So, the narrow and stiff basal part of the basilar membrane responds maximally to high-frequency sounds. And, the wide and soft apical part of the basilar membrane responds maximally to low frequencies. It is just like a piano as in Fig. 14.6. If the sound has a higher frequency, it generates maximal fluctuation at the base, at the right part of the cochlea in Fig. 14.6. As the hair cells in this location are innervated with auditory nerves, which are connected to the brain area recognizing high-frequency components, any movement in this location induces the recognition of higher frequency in the sound. For lower frequency sound, maximal fluctuation and action

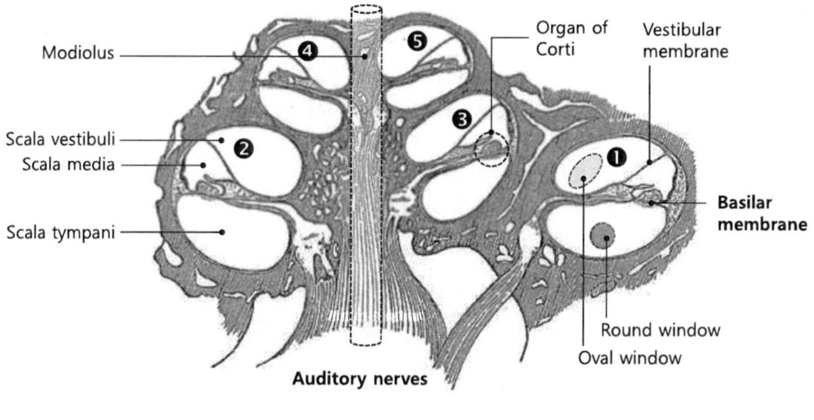

Fig. 14.5 Cross section of two and half turns of the cochlear spiral

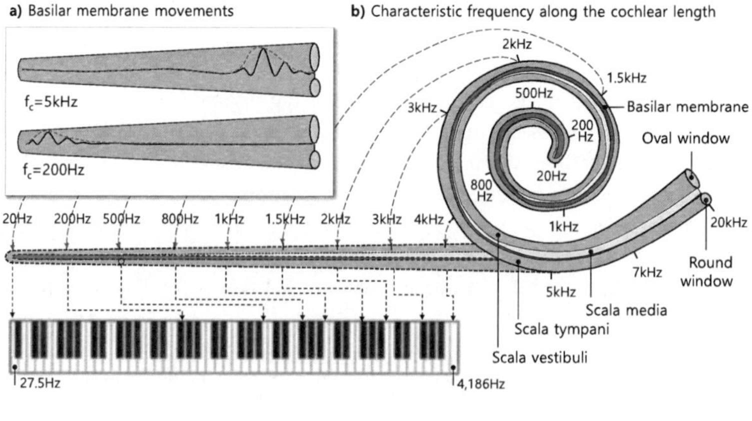

Fig. 14.6 Basilar membrane fluctuation and the characteristic frequency distribution along the longitudinal axis of the cochlea

potentials are generated in the left part of Fig. 14.6. Their auditory nerves are connected to the brain area recognizing low-frequency components. So, the major role of sound discrimination is performed by the basilar membrane. Characteristic frequencies along the basal membrane are tonotopically mapped to the brain through the bundle of auditory nerves. The basilar membrane fluctuates its basal part for high-frequency sound and makes it be recognized as a high-frequency component and vice versa for the low-frequency sound.

2.4 Hearing Loss and Indication for Cochlear Implant

Hearing loss is a diminished acuity to sounds that would otherwise be heard normally. Hearing loss may be partial or total, be temporary or permanent, present at birth or acquired, and occurred in one or both ears. Hearing loss affects language learning ability in children and makes difficulties in social interaction and work in adults.

Hearing loss is diagnosed by measuring the hearing threshold at each frequency and comparing that with those of normal subjects. For example, if a subject detects a 500 Hz sound only when the sound intensity becomes 50 dB, while normal subjects hear a 500 Hz sound from a sound intensity of 10 dB, this subject has 40 dB of hearing loss at 500 Hz. Audiogram is the graphic record of hearing thresholds for at different frequencies, 125, 250, 500, 1000, 2000, 3000, 4000, and 8000 Hz. The degree of hearing loss is classified by the amount of hearing loss into a slight (15~25 dB), mild (25~40 dB), moderate (40~55 dB), moderate-severe (55~70 dB), severe (70~90 dB), and profound (90 dB~). Deafness is a degree of hearing loss such that a person is unable to understand speech, even in the presence of amplification. Hearing loss may be caused by various factors, including aging, genetics, exposure to intense noises, chronic ear infections, birth complications, trauma to the ear, and certain medications or toxins. Hearing loss can be classified into three types, conductive hearing loss, sensorineural hearing loss, and mixed hearing loss.

Conductive hearing loss is caused by a disturbance or blockage of sound conduction along with the auditory conduction system. It can occur when problems occurred in the external ear or in the middle ear disturb the normal conduction of sound to the inner ear. Ear infections, colds, and allergies can cause problems in the middle ear with the tympanic membrane, ossicles, muscles, or ligaments. Physical blockage in the external auditory canal may decrease the vibrational amplitude of the incoming sound. In most cases, conductive hearing losses are treatable by treating the underlying infection or removing the obstruction.

Sensorineural hearing loss is a more common type of hearing loss and is caused by damage in the inner ear or auditory nerve, in the sensorineural part of the auditory system. In contrast to conductive hearing loss, sensorineural hearing loss is permanent and, in most cases, it is not treatable with medicine or surgery. It is caused mostly by aging and exposure to loud noises, and less commonly by a viral infection, injury, genetics, and medication. Most cases of sensorineural hearing loss progress with a gradual deterioration of hearing thresholds over years to decades. Damages can occur in the hair cells, the sensory part of the cochlea, or in the auditory nerves in the neural part of the auditory nervous system.

In the cochlea, there are two types of hair cells, inner hair cells and outer hair cells. Inner hair cells are the major sensory cells and detect mechanical fluctuation caused by the sounds and convert it to electrical pulses of the action potential. If the inner hair cells are damaged, sounds cannot be sensed. As the inner hair cells are populated along the whole length of the cochlea, cell damages can occur locally or over the whole length. If damage is overall, a subject cannot hear at all and if the damage is local, the subject cannot hear the sound having the frequency responding maximally to the dead region. When the inner hair cells are damaged, hearing aids are of no use. As a hearing aid is a kind of frequency-selective acoustic amplifier, it cannot induce electrical signals from the damaged inner hair cells even with amplified fluctuation of the basilar membrane. In this case, a cochlear implant can be the optimal treatment solution. If the connected auditory nerves are intact, a cochlear implant can provide equivalent electrical signals to auditory nerves. However, if auditory nerves are also damaged, a cochlear implant cannot be applied and there is no other option to restore sound perception.

Outer hair cells are different from inner hair cells. They work in an efferent manner than in an afferent manner. They receive signals through auditory nerves and work like tiny muscles. Sensed movement of the basilar membrane by the inner cells of the region is feedbacked to outer hair cells. Their cilia are connected to the tectorial membrane, which is covering the organ of Corti. Through these connections, outer hair cells modulate the movement of the tectorial membrane according to the sensed movement of the basilar membrane to increase the sensitivity of the sound. That is, outer hair cells do the role of positive feedback increasing the hearing sensitivity so that even the smallest sound can be sensed by inner hair cells. If the outer hair cells are damaged, hearing sensitivity is decreased and subjects become difficult to hear smaller sounds with hearing. For this type of sensorineural hearing loss, hearing aids can be of help. If the small sounds are acoustically amplified, inner hair cells can sense the amplified basilar membrane movement without the support of outer hair cells. These outer hair cells are more easily damaged than inner hair cells.

For the subject with sensorineural hearing loss, hearing aid and a cochlear implant can be the options to restore normal hearing capability. Selection between two devices is carefully decided by the otolaryngologists with examinations. It depends on the type of damage and the degree of hearing loss. Various examinations are used to identify the type of damage and degree of hearing loss. For adults, audiometry and an open-set speech recognition test are used. To evaluate the hearing loss of prelingual children, physiological confirmation with auditory brainstem response and otoacoustic emission is used in addition to initial diagnosis with behavioral audiometry. Current guidelines for hearing aid prescription include at least 30 dB hearing loss between 500 and 4000 Hz and less than 80% speech understanding score at 65 dB. In contrast, indications for cochlear implantation in adults are bilateral postlingual deafness, or bilateral sensorineural hearing loss greater than 70 dB between 500 and 4000 Hz with less than 50% speech understanding score at 65 dB (Szyfter et al., 2019).

3 Cochlear Implant System

A cochlear implant is a surgically implanted electrically stimulating device that provides the subject with restored hearing from moderate to profound sensorineural hearing loss. Instead of providing amplified sound to decrease the perceptible hearing threshold as in hearing aids, a cochlear implant bypasses the acoustic conduction pathway of the outer ear and middle ear and stimulates the auditory nerves at the cochlea directly with generated electrical pulses. With restored sound perception and the following session of auditory training, subjects can greatly improve their hearing-related abilities including speech understanding and discriminating environmental sounds.

3.1 Functional Architecture of Cochlear Implant

The cochlear implant is largely composed of two parts, an external unit and implanted internal unit (Fig. 14.7). Input to the cochlear implant is a sound, which is received by the microphone in the external unit and the final output is the electrical pulses stimulating auditory nerves through the array of electrodes inserted into the cochlea.

The external unit is composed of a microphone, speech processor, data and power transceiver, and battery. The microphone receives the sound and is located just over the auricle to receive the sound with the same intensity as the normal ear. Received sound is converted to a digital signal and transferred to a speech processor. Instead of sound being received and converted into an electrical signal by a microphone, recent external units can receive electrical signals directly by wire or wireless connection to smartphones or sound generating units. The sound processor does the main role of sound processing, which is performed by a basilar membrane in the normal ear. Two purposes of the speech processor are to extract the information from the sound and convert it to an appropriate form of the signal that can be used for stimulation. Through the series of processing, the sound processor extracts features reflecting the distribution of spectral components in the sound and converts features into a stream of bits that can be transmitted to the internal unit through the RF link. During the processing, the sound processor reflects the subject-specific parameters, which can be set and modified by the fitting program (Zeng et al., 2008). The dynamic range of input sound is increased up to 75 dB to match the range of intensity variation of usual speech and environmental sounds. The frequency range has broadened into the range of 100~8000 Hz to perceive lower components and to improve language understanding based on the pitches in the sound. Various speech coding strategies are used to optimize the perception of processed sound including the continuous interleaved sampling (CIS) method.

The cochlear implant is in contrast to cardiac pacemakers and deep brain stimulators, which are totally implanted. Separation of the stimulating device into two parts of the external and internal unit can minimize the volume of the implanted unit. However, it requires an efficient method that can connect the two units. Radiofrequency (RF) transmission link allows the efficient connection between two parts of the cochlear implant. It eliminates the risk of contamination and infection that might accompany the percutaneous wire connection between two parts and the repetitive surgeries to replace the battery of implanted unit. As the roles of

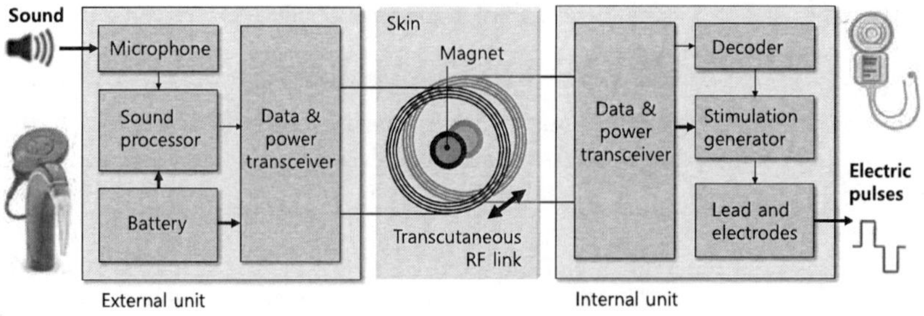

Fig. 14.7 Functional architecture of the cochlear implant

connecting wires and the battery are replaced by the RF transmission link, it should transmit power and data to the internal units efficiently. The RF link uses a pair of inductively coupled coils to transmit both power and data. Several factors should be considered for the design of RF coils. External and internal coils are accurately positioned over the skin using the magnets in their centers. To optimize the power transmission over the coil, the use of a single-resonant frequency is optimal. However, the data in the pulse waveform include a wide range of frequencies. For power transmission, a larger coil size is required for efficiency, but it is restricted by the size of the head. The central frequency of RF carrier is different within the 5~49 MHz range and data transfer rates are in the range of 0.5~1.0 Mbits/sec. RF transmission link has about 40% transmission efficiency and delivers 20~40 mW power over the skin thickness of 4~10 mm (Zeng et al., 2008).

The battery in the external unit is free from replacement. Both disposable and rechargeable batteries can be used to fuel the cochlear implant. Due to the small volume of the external unit, currently these batteries support the device for about 16-60 hours.

The implanted internal unit is hermetically sealed and composed of a data and power transceiver, decoder, stimulation generator, and an array of electrodes at the tip of the connecting lead. As the internal unit includes no battery, the data and power transceiver part first derives power from the transmitted RF signal to fuel the whole circuits in the internal unit. And, it also separates the signal part which can be used for stimulating each electrode in the array. The decoder recovers digital stimulation signal information from the RF signal with error and safety checks to ensure proper decoding. The stimulation generator converts decoded digital stimulating information into analog-type current pulses via D/A conversion. Electrical circuits in the internal unit are implemented with ASIC technology to minimize consumed power. An array of electrodes is the contacting place where the cochlear implant meets the neural tissues in the cochlea. Stimulating electrical current flows from the electrode to the surrounding tissues including auditory nerves. With the progress of technology, the number of electrodes have increased from a single channel to multiple active channels of 22 electrodes. Usually, the array of electrodes is inserted through the scala tympani, under the basilar membrane. To access the auditory nerves responsible for lower frequency in the range of 200~1200 Hz, the electrode array is required to advance about one and a half turns, into the depth of about 26 mm by the active length. Increasing the coupling of electrodes to the auditory nerves is required to increase the stimulating efficiency and reduce the consumed power. So, the electrode array is advanced along the inner wall close to the modiolus, the axis of the spiral of the cochlea where the auditory nerves are densely populated. During the insertion of the electrode array, intracochlear damage might occur where the insertion force exceeds the resisting strength of the tissue. Damages result in decreased coupling efficiency and inconsistent channel-to-channel performance (Rebscher et al., 1999).

3.2 Sound Processor of Cochlear Implant

The sound processor does a key role in processing for the input sound to generate interpretable electrical stimulations. As the cochlear implant bypasses the ordinary acoustic conduction and sensing system from the outer ear to the inner hair cells in the cochlea, it is required to replace the function of bypassed parts of the auditory system by mimicking their function maximally. Various sound processing strategies have been developed to maximize the interpretability of restored hearing.

When the sound processor was implemented in analog circuits for a single-channel stimulation, simply amplitude of the sound is compressed, modulated with a 16 kHz sinusoidal carrier, and applied as the effective electrical stimulations (Zeng, 2004). Current cochlear implants adopt a digital sound processor for the processing of sound to generate the electrical pulse signals appropriate for the array of multiple electrodes. The basic concept of processing is to take advantage of the tonotopic organization in the cochlea, namely, the basal part of the cochlea responds to high frequencies, and the apical part of the cochlea responds to low frequencies selectively. Based on this concept, most sound processors include a bank of bandpass filters that separate the input sound signal into a number of different signals representing different frequency bands. However, following signal processing to extract features to generate the stimulation pulses are significantly different among the various strategies. Among the various strategies, the continuous interleaved sampling (CIS) method is most widely adopted in current cochlear implants.

In the continuous interleaved sampling (CIS) method, the sound processor generates stimulating signals equal to the number of electrodes. For the array of 22 electrodes, the CIS method generates 22 electrical signals which are used to generate stimulating current pulses in the generator circuit of the internal unit. Figure 14.8 shows the sound signal processing example with the CIS method for the stimulation of an array of four electrodes. In the first step, the sound received from the microphone in the external unit is divided into the number of frequency bands using the bandpass filter bank. The output of each bandpass filter contains sound components that only belong to the passband of each filter. This filter bank separates the input sound signal into individual signal components to stimulate each electrode based on the tonotopic organization of the cochlea. The channel output

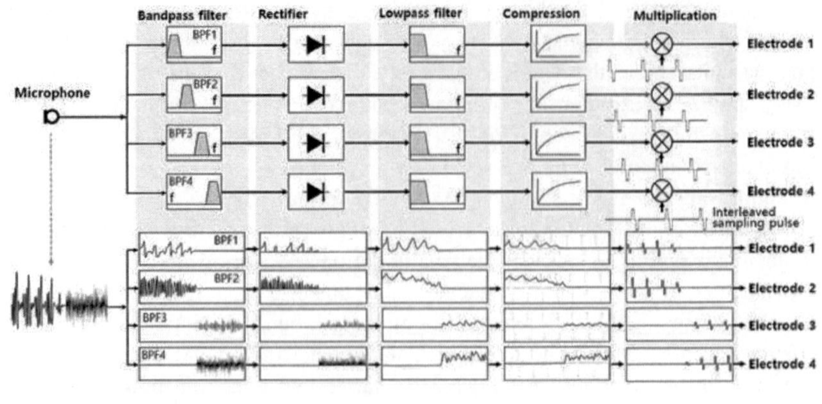

Fig. 14.8 Continuous interleaved sampling (CIS) method for the generation of stimulating signal for the array of electrodes

signal of the lowest passband is used to stimulate the most apical electrode and the channel output of the highest passband is used to stimulate the most basal electrode. The following procedure is to generate an optimal stimulating signal pattern that is effective to restore sound.

The output of each bandpass filtered signal is rectified and lowpass filtered with a cutoff frequency of 160~320 Hz. This is to detect the envelope of the bandpass filtered signal. This is equivalent to continuously detecting the powers of each band. That is, after detecting the powers of each band, a stimulating signal is generated in proportion to the continuous variation of power of each band. For these envelope signals, the amplitude compression procedure, usually logarithmically, is performed. This is to mimic the responding characteristic of the human auditory system to the amplitude of the sound. Instead of perceiving intensity linearly, perceived sound intensity increases with the logarithmic scale, as the sound pressure level (SPL) is described in decibel (dB). It also limits the maximal stimulating level to match the narrow dynamic range in electric stimulation and not to cause any discomfort by the high intensity of input sound. As the cochlear implant bypasses the middle ear, it can be considered partly replacing the ossicular attenuation in the middle ear.

After the amplitude of the signal is compressed, this signal is used to modulate the interleaved sampling pulses. This can be done by simply multiplying two signals. Then the amplitudes of interleaving sampling pulses, which are higher than 800 Hz for each channel, are attenuated according to the compressed envelope signal of each frequency band. The key point of the CIS method is using interleaved sampling pulses rather than simultaneous pulses among the different channels. In interleaved sampling pulses, there is only one active pulse at a time among the whole number of channels. Sampling pulses of all channels are temporally interleaved. This is to stimulate the auditory nerves only at a single location at a time. If the electrical stimulations are applied simultaneously through multiple electrodes, generated electric fields can be overlapped and the extracted features tend to smear with the features of neighboring channels. Previous methods using the simultaneous sampling pulses were not effective by causing confusion in understanding the sound. This also can be considered more closely mimicking the action of the basilar membrane, which only generates a single peak along the whole length of the cochlea. Finally, stimulating signals for each channel corresponding to the electrode of a special position along the cochlea are produced in a form of pulse series, whose amplitude is modulated by the continuous power of the frequency band and whose temporal occurrences are not overlapping each other. These generated stimulating signals are transmitted to the internal unit via RF data and power transmission link, where these signals are decoded and used to generate stimulating current pulses for each electrode.

To optimize the sound processing for each subject, the speech processor is fitted during the training period. Basically, fitting is performed by setting the threshold level and comfort level so that the perceived sound is loud enough for each subject to be aware of the sound, and not too loud as to cause discomfort.

3.3 Electrodes of Cochlear Implant

Electrical signals from the internal unit of the cochlear implant are transferred to the auditory nerves through the array of electrodes. The internal unit has two leads. One lead guides a wire bundle for an intracochlear array of electrodes and another lead is for an extracochlear electrode. Intracochlear electrodes are used as active electrodes located close to the auditory nerves and provide a focused high intensity current. The number of electrodes inserted into the cochlea are different among the models with a maximal number of 22 electrodes. Figure 14.9a shows an example of implanted internal unit with its intracochlear electrodes. While the whole length of the cochlea is 35 mm over two and a half turns of spiral rotation, the intracochlear electrodes only can be inserted to the depth of about one and a half turns

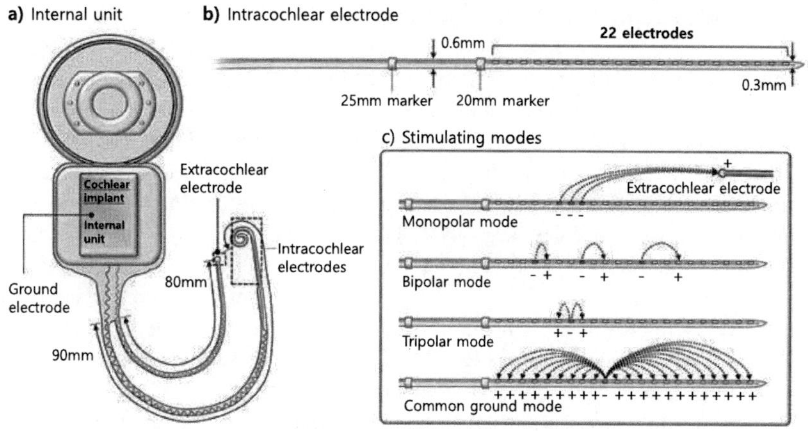

Fig. 14.9 Internal unit and electrodes of the cochlear implant. (**a**) Example of an internal unit with its intracochlear, extracochlear, and ground electrodes, (**b**) an example of uncoiled intracochlear electrodes with 22 contacting sites, (**c**) different modes of cochlear implant stimulation

due to the difficulty of inserting the electrode along the curvature without cell damage. In Fig. 14.9b, 22 electrodes are aligned throughout the length of 19 mm with an average distance between electrodes less than 1 mm. The diameter of inserted lead is very small, 0.6 mm on the basal side and 0.3 mm on the apical side, to be inserted via the round window in the space of scala tympani. The extracochlear electrode is not inserted into the cochlea and is used as a returning electrode in pairing with active intracochlear electrodes. The external case of the internal unit can be used as an extracochlear electrode collecting currents. Due to the distance between the cochlea and the location of the internal unit case, paths of current become longer and there increases a chance that other nerves and muscles can be stimulated. Another type of extracochlear electrode has a long wire to position the ball-shaped extracochlear electrode near the cochlea to minimize the path of current and not disturb other nerves and muscles around.

Electrodes can be stimulated in different modes (Fig. 14.9c). In monopolar stimulation mode, current flows from one active intracochlear electrode to the extracochlear electrode, which is placed either as a ball electrode close to the cochlea or a plate electrode on the case of the internal unit. In monopolar stimulation mode, the impedance of the electrode decreases as the surface area of returning electrode increases. Monopolar stimulation is the most commonly used stimulation mode in current sound processing strategies and provides the largest current spread (Litovsky et al., 2017). In bipolar stimulation mode, current flows from an active electrode to a returning electrode within the cochlea. Bipolar stimulation can minimize the stimulating volume with a smaller current spread and increase the selectivity of the stimulation. The stimulating volume of tissue can be managed by increasing the distance between active and return electrodes. In tripolar stimulating mode, current flows from one active electrode parallelly to two neighboring returning electrodes. In common ground stimulation mode, current flows from one active electrode to all other intracochlear electrodes. This stimulating mode provides the lowest electrode impedance.

3.4 Pulse Parameters for Loudness Control

The loudness of the sound is determined by the strength of the stimulating pulse. The parameter most closely related to the loudness is the charge. When the stimulating frequency is fixed, the charge of applied stimulation can be varied by the amplitude of the current pulse and the phase duration of the pulse (Fig. 14.10). So, the loudness can be controlled by manipulating the amplitude of the current pulse with fixed phase duration or by manipulating the phase duration of the pulse with fixed current amplitude. Generally manipulating amplitude produces a greater change in loudness compared with the change due to manipulation of the phase duration. Loudness can grow differently across electrodes. Current devices provide an amplitude of current stimulation up to 1.75 mA and phase duration in the range of 9.6~400 μs.

The interphase gap is the time between subsequent phases in a biphasic pulse. Biphasic pulse has two phases, the anodic and cathodic phase, providing opposing charges. The balance of charge between the two phases is to prevent irreversible corrosion of electrodes and the potential deposit of metal oxides at the electrode-tissue interface. Loudness also increases as the interphase gap increases (Prado-Guitierrez et al., 2006). Increasing the stimulation rate can also induce an increase in loudness as more charge is provided per unit time. Change in loudness by increasing stimulation rate is relatively effective at low rates (Hong & Rubinstein, 2003). When the sound is generated with multichannel stimulation, loudness summation occurs. The degree of loudness summation is dependent both on the level of stimulation within the subject's dynamic range and the stimulation mode.

3.5 Effectiveness of Cochlear Implant

Subjects who use cochlear implants show relevant improvement in the quantitative perception of sound and in their quality of life. Benefited subjects broadly range from young children to the elderly over 70 years. When the hearing threshold of subjects older than 70 years was evaluated with audiometry by pure tone average (PTA) measured at 500, 1000, and 2000 Hz, PTA significantly decreased from 109 dB to 37 dB after implantation of a cochlear implant. This level of improvement is not correlated with age and is similar in postlingual deaf adults less than 70 years. Speech perception is also greatly improved with a cochlear implant. The comparison of speech perception between preoperative testing and postoperative testing with and without lipreading is shown in Fig. 14.11a (Sanchez-Cuadrado et al., 2013). After the surgery of the cochlear implant, the speech perception score increased from 40% to 88% with lipreading, and from 34% to 73% without lipreading postoperatively.

Children implanted by 12 months of age demonstrate better language development compared with children who received their cochlear implant between 13 and 24 months. This implies that presenting raw speech signals of multiple electrodes without preprocessing might result in better performance for very young prelingual children. When the brain is trained with millions of words over a lifetime, it can be far better at extracting the most important features of the sound.

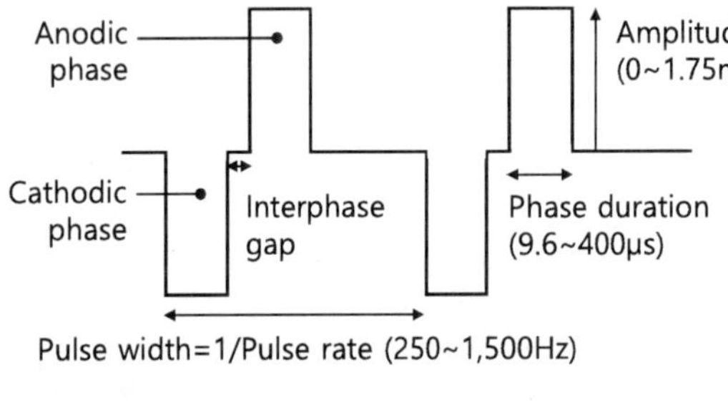

Fig. 14.10 Electric stimulating pulse shape and its parameters controlling the loudness

The earliest implantation even within the first year enhances the probability that children with severe-to-profound hearing impairment decrease the language learning gap and commence elementary school with age-appropriate quality of life (Clark, 2015) (Fig. 14.11b).

Adult subjects are often benefited immediately and continually improve their speech perception for about 3 months after the initial tuning sessions. Subjects report that they can perceive different types of sounds, like footsteps, slamming of doors, sounds of engines, ringing of the telephone, barking of dogs, whistling of the tea kettle, the rustling of leaves, the sound of a light switch being switched on and off, and so on. Many subjects can make telephone calls and understand familiar voices over the telephone. And, they also watch TV more easily, especially when they see the speaker's face. Some can enjoy music with the sound of certain instruments and certain voices. While the restored hearing ranges broadly after implantation, in many successful cases subjects can understand speech with near-normal ability.

3.6 Electroacoustic Stimulation

Electric acoustic stimulation (EAS), also known as hybrid stimulation, is the synergistic combination of two technologies, cochlear implant technology and a hearing aid. EAS is indicated for subjects with high-frequency hearing loss, who have intact low-frequency hearing, but with profound high-frequency hearing loss. Audiometric criteria are pure tone thresholds better than 60 dB at 125, 250, and 500 Hz, and worse than 75 dB at frequencies over 1000 Hz (Fig. 14.12). These subjects are usually able to detect vowels, but few or no consonants, miss out on hearing high-pitched sounds, and have difficulty in word understanding and hearing in a noisy environment. Subjects with EAS implantation demonstrate pronounced benefits in speech perception, listening under the background noise, and music appreciation over subjects with implants relying on electrical stimulation alone (Irving et al., 2014).

Fig. 14.11 Effectiveness of cochlear implant. (**a**) Postoperative speech perception score increases significantly with and without lipreading test (Sanchez-Cuadrado et al., 2013). (**b**) Language age follows chronological age more closely as early as the cochlear implant is implanted. (Clark, 2015)

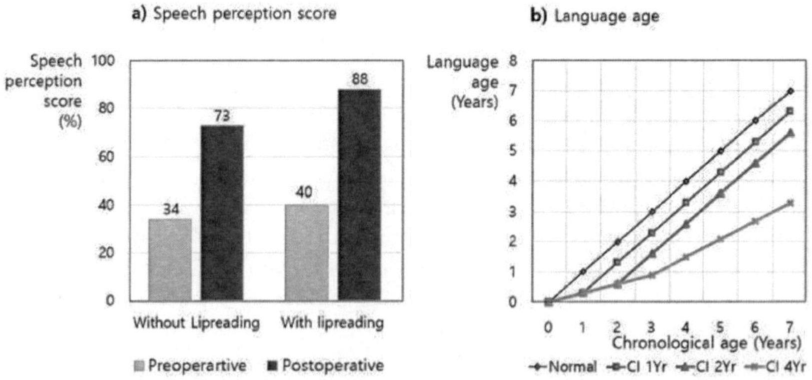

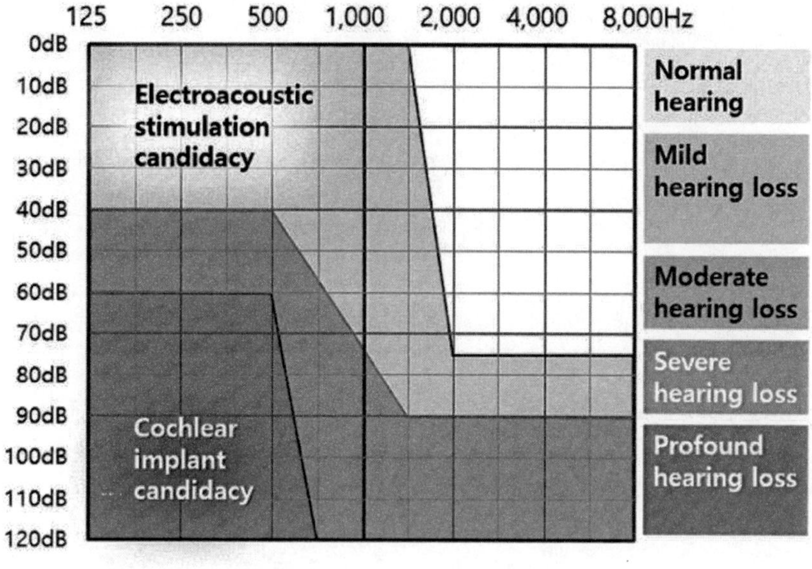

Fig. 14.12 Candidacy for cochlear implant and electroacoustic stimulation

For the middle and high frequencies, EAS uses the cochlear implant electrode array to electrically stimulate the auditory nerves within the basal turn of the cochlea. And, EAS uses the hearing aid to convey the low frequencies with acoustic amplification at the apical turn of the same cochlea. It is considered that the interaction of the electric and acoustic signals plays a synergistic role in the auditory pathway enhancing the sound perceptibility. Keeping the residual sensorineural cells intact during the implantation surgery is critical for successful EAS implantation.

EAS uses a single microphone, but the sound processor has two parts for different processing, one for hearing aid and another for the cochlear implant. These two parts of the sound processor perform separate processing optimized for hearing aid and cochlear implant, respectively. Amplified acoustic signals are forwarded to the auditory pathway with an ear mold as in conventional hearing aid. As in a cochlear implant, assistive listening allows direct connection of electrical signals from mobile phones, TVs, or external microphones rather than being picked up by a microphone improves the sound quality by bypassing the environmental noises.

4 Human Visual System

The human visual system is a nervous system from both eyes to the occipital cortex of the brain including all the neuronal cells along the visual pathway (Fig. 14.13). Lights from the objects enter the eyes through the lens and are sensed by the photoreceptor cells in the retina. After sensed signals are processed through multiple layers of the retina, generated action potentials are transmitted via optic nerves. Optic nerves are splitting into halves toward both hemispheres and half of the nerve fibers cross over at the optic chiasm and proceed to the opposite hemisphere. Optic nerves from both sides of the eyes are combined at the lateral geniculate nucleus (LGN) in the thalamus with other sensory signals and projected to the primary visual cortex via optic radiations. The optic radiations are axons conveying the action potential to the primary visual cortex. The primary visual cortex starts processing the perceived image for interpretation and recognition and sends the processed signals into higher layers for extended visual processing.

Information through the visual system takes 90% of information transmitted to the brain. About 40% of nerve fibers are linked to the retina and the brain process information in visual form 60,000 times faster than the same information in text form. Compared to other animals, like dogs characterized by the sense of smell and bats characterized by the hearing sense, humans are mostly visual sensing being. As much human visual system takes an important role, the impairment of the visual system is devastating. Electricity can help the blind to see with the method of the visual prosthesis. Visual prosthesis using electrical stimulation is possible at several locations along the visual pathway. Figure 14.13 shows the locations of electrical stimulations where the studies are currently going on for visual prostheses. Among these, currently, visual prosthesis at the retina is the most successful.

4.1 Light that Humans Sense

The light that humans can see is only a small segment of the electromagnetic spectrum, typically in the range of 375~750

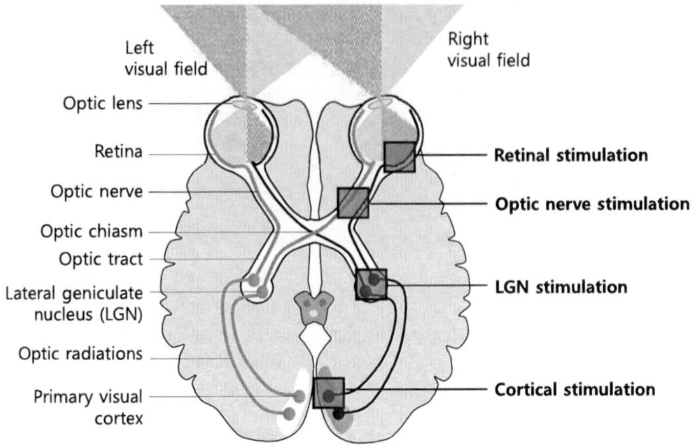

Fig. 14.13 Human visual pathway and the possible visual prosthetic locations

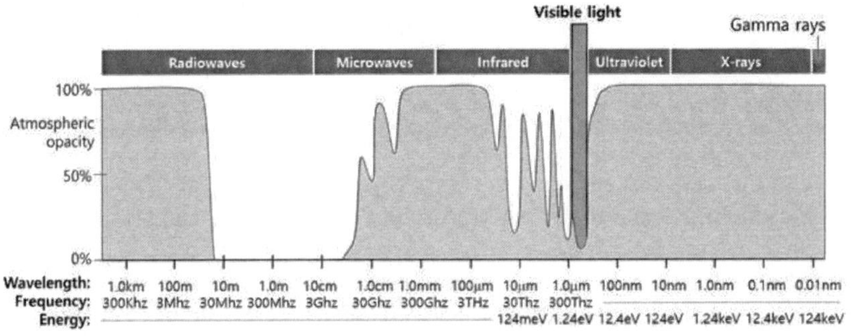

Fig. 14.14 Visible light in the electromagnetic wave spectrum

nanometers in wavelength and 400~800 Terahertz (TH) in frequency. As light has characteristics of a particle as well as a wave, the visible range can be represented with light photons having energy in the range of 1.65~3.3 electron Volt (eV). Electromagnetic waves in this segment of visual light are the band of the highest frequency among the electromagnetic waves that arrived at the surface of the earth with minimal attenuation through the atmosphere (Fig. 14.14). Electromagnetic waves of higher frequency than visible light, like ultraviolet, X-ray, and gamma-ray, are almost blocked in the upper atmospheric space. And, for some electromagnetic waves in lower frequency which also arrive at the surface of the earth, the human requires too big antennae to sense the electromagnetic waves.

The brightness or intensity of the light can mean different things (Fig. 14.15). The amount of light emanating from a light source is luminous flux and is described by the unit of lumen (lm). The intensity of light falling on a surface is illuminance and is described with the unit of lux. And, the intensity of light reflected off a surface is luminance and is measured with cd/m^2. Lumen is used to describe the total luminous flux over the whole 4π solid angle. One lumen is defined as 1.46 mW of radiant electromagnetic (EM) power measured at the frequency of 540 THz mono frequency light. Light of 540 THz is the green light having a wavelength of 555 nm, which is the most sensitive wavelength of human photopic vision. As photon energy of 555 nm is 2.23 eV, 1.46 mW is equivalent to the 4.08×10^{15} green light photons per second. If the light is of a different wavelength other than 550 nm, it requires higher radiant electromagnetic power than 1.46 mW for 1 lumen luminous flux, because the sensitivity at any other frequency is lower than the maximal sensitivity of 555 nm. Lumen is used to describe the brightness of light bulbs. For example, a LED bulb can have a brightness of 800 lumens with electric consumption of 8~12 W and it lights with an equivalent brightness to an incandescent light bulb of 60 electric Watts. Candela(cd) is the unit for the luminous intensity which means the amount of light that travels in certain directions from the source. Candela(cd) is equal to luminous flux divided by solid angle. 1 candela is the luminous intensity having 1 lumen per solid angle. As a full sphere has a solid angle of 4π steradians, a light source that uniformly radiates one candela in all directions has a total luminous flux of $1\ \text{cd} \times 4\pi\text{sr} = 4\pi\ \text{cd}\cdot\text{sr} \cong 12.57\ \text{lm}$.

As the illuminance is the intensity of light falling on a surface, for the same light source, illuminances are different between a closer place and a far place. One lux is defined as a number of lumens falling on one square meter of a surface. If a light amount of 1000 lumens from a light source totally falls on one square meter surface, it generates an illuminance of

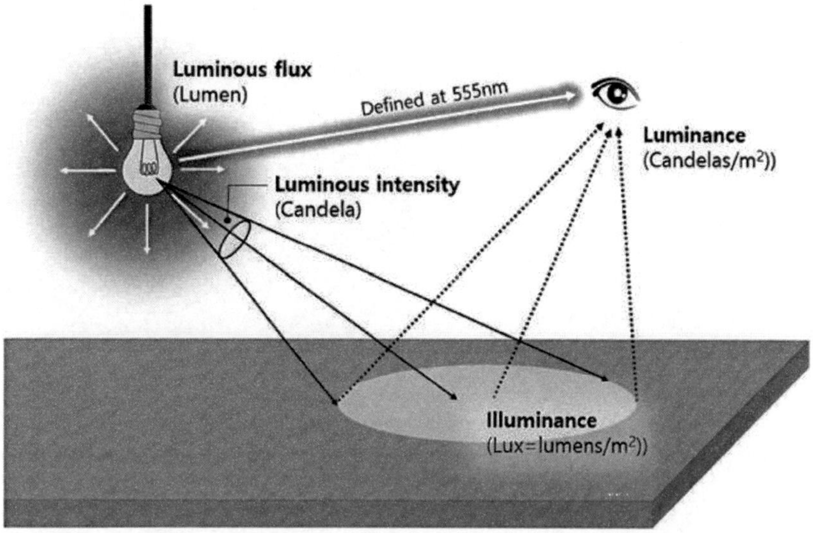

Fig. 14.15 Units for describing light intensities

1000 lux. However, the same source illuminates 10 square meters, it only provides an illuminance of 100 lux over the whole surface area. Illuminance can be used to determine the appropriate light shining onto the operating table for surgery. Luminance is the amount of the emitted or reflected light from a surface per unit area in a particular direction. Luminance is the brightness of light that humans sense. The unit for luminance is candela per square meter (cd/m^2). As the solid angle is unitless, 1 cd/m^2 can be converted to 1 lux (Table 14.2 shows the responses of human visual system on different levels of luminance.

Table 14.2 Human vision responding difference level of luminance

Luminance (Cd/m^2)	Vision type	Cones	Rods	Light example (cd/m^2)
1000,000,000				Sun at noon (1,600,000,000)
100,000,000				Retinal damage
10,000,000				Tungsten filament at 2700 °K
1000,000	Photopic	O	×	Upper limit of vision
100,000		O	×	Bright summer day
10,000		O	×	Full daylight
1000		O	×	Average cloudy sky (2000)
100		O	×	Very dark day
10	Mesopic	O	O	Sunrise or sunset (25)
1.0		O	O	Deep twilight
0.1		O	O	Full moon
0.01		O	O	Quarter moon
0.001	Scotopic	×	O	Night sky
0.000,1		×	O	Darkest sky (0.000,4)
0.000,01		×	O	Grass in starlight
0.000,001		×	O	Threshold of vision

4.2 Retina

Within the eyes, the retina is the outermost structure composed of layers of neuronal cells to sense the light. Despite its peripheral location, the retina belongs to the central nervous system and about 100 million neurons are densely populated within the thickness of 285~335 μm comprising complex processing neuronal circuitries and layers. It comprises five different cell types, photoreceptor cells, horizontal cells, bipolar cells, amacrine cells, and ganglion cells. These retinal cells combinedly process and convert the graded electrical activity from photoreceptors into action potentials of ganglion cells that travel to the brain via axons in the optic nerves. The pigmented layer of retinal pigment epithelium (RPE) supports and nourishes retinal neuronal cells and this pigmented layer is firmly attached to the underlying choroid.

The neuronal cell bodies and processes of these neurons are stacked in five alternating layers which are easily differentiated by hematoxylin and eosin staining (Fig. 14.16). Neuronal cell bodies are populated in nuclear layers and synaptic connections and dendrites are composing the plexiform layers. The outermost layer is the layer of photoreceptor outer segments. The outer nuclear layer is mostly populated with the bodies of photoreceptor cells of rods and cones. The inner nuclear layer contains three types of cells, horizontal cells, bipolar cells, and amacrine cells. The inner nuclear layer and outer nuclear layer are intervened by the outer plexiform layer, where synaptic connections among photoreceptor cells, horizontal cells, and bipolar cells are densely populated. Bodies of ganglion cells are located in the innermost layer of the ganglion cell layer. Ganglion cells are connected via synapses to amacrine cells and bipolar cells in the inner plexiform layer. And the axons of ganglion cells

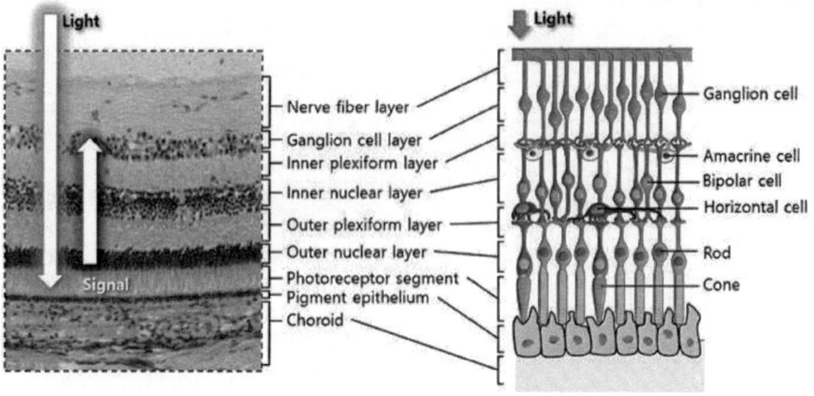

Fig. 14.16 Cell types and layers of the retina

exit the retina by the optic nerve and these nerve fibers compose the superficial layer of the nerve fiber layer (NFL).

At first glance, the spatial arrangement of retinal layers seems counterintuitive, as the photoreceptor cells are expected to be located at the innermost layer where the lights first arrive. However, contrary to a simple thought, photoreceptor cells are located in the outermost layer, in the deepest layer along the path of light. That is, the light should pass through nonlight sensitive layers, nerve fiber layer, ganglion cell layer, inner and outer plexiform layers, and inner and outer nuclear layers before it reaches the outer segments of photoreceptors. Is it efficient? Is this really the result of evolution? There is no agreed explanation for this reversed spatial arrangement of retinal layers. One explanation is that it is an inevitable selection through evolution. When there was a single layer of photoreceptor cells, adding a neuronal layer beneath the photoreceptor cell layer might be impossible or difficult. Adding just one layer on top of the photoreceptor layer is considered an easier selection while it does not greatly interrupt light transmission. Once the second layer was structured in the reverse direction, developing a third layer beneath the photoreceptor cell layer looks much more difficult and the third and following additional layers are considered inevitably mounted on top of the previous layers. Another explanation comes from the relationship of photoreceptor cells with the choroid and retinal pigment epithelium (RPE) layer neighboring the outer segment of photoreceptors. As photoreceptors require a large amount of oxygen and nutrient for signal transduction, positioning in the closest position to the choroid having a dense network of blood vessels beneath the retina is the optimal selection for overall functioning. And, RPE plays an essential role in removing the used receptor disks for replacement and supporting the regeneration of photopigment molecules. In conclusion, light penetrates several cell layers and is sensed by the photoreceptor cells. And, the sensed signals are processed, proceed backward in the direction to the ganglion cell layer, and leave the retina from ganglion cells via optic nerves toward the primary visual cortex.

There are two types of photoreceptors in the retina, rods and cones. The plasma membrane of the photoreceptor is divided into two compartments, the outer segment and the inner segment. The outer segment is light sensitive and filled with disk membranes and located close to the retinal pigment epithelial layer. The inner segment contains the soma, nucleus, and axon and has synaptic connections with bipolar cells and horizontal cells. Light absorption by the photopigment in the outer segment initiates a series of events resulting in the membrane potential change of the photoreceptor, and therefore reduces the amount of released inhibitory neurotransmitters in the synapses connecting the horizontal cells and bipolar cells. Phototransduction in photoreceptors is described in Chap. 6.

In the outer plexiform layer, terminals of photoreceptors synapse with bipolar cells and horizontal cells, whose bodies are located in the inner nuclear layer. Horizontal cells are primarily responsible for lateral interactions by their lateral interconnections between the photoreceptor cells and horizontal cells. These lateral interactions are largely responsible for local averaging to adapt luminance contrast over a wide range of light intensities. Bipolar cells are one of the major retinal interneurons and provide the shortest and most direct pathways from photoreceptors to ganglion cells, between the input and output of visual signals in the retina. Bipolar cells have synapses with rods or cones and also with horizontal cells. Bipolar cells then transmit the output to the ganglion cells directly or indirectly via amacrine cells. Transmitted signals from bipolar cells are still in graded potentials.

In the inner plexiform layer, short axonal processes of bipolar cells have synaptic connections with the dendrites of ganglion cells, whose bodies are located on the inner surface of the ganglion cell layer. Ganglion cells receive signals from bipolar cells and amacrine cells in graded potential and convert them into a series of action potentials. Generated action potentials are carried over the larger axons of ganglion cells which form the optic nerves to the lateral geniculate nucleus (LGN) in the thalamus. Amacrine cells also have synaptic connections with ganglion cells in the inner plexiform layer. Amacrine cells mostly serve to integrate, modulate, and interpose temporal variations in the visual signals. There are several subtypes in amacrine cells. For example, one type of amacrine cell plays a role in light adaptation transforming the persistent light responses into brief transient responses. Another type of amacrine cell plays a role in direction selectivity and detection of directional motion. The diversity of amacrine cells suggests the parallel processing in the retina by providing multiple pathways conveying different sorts of information to central targets.

Retinal pigment epithelium (RPE) is the pigmented cell layer just beneath the layer of the outer segment of photoreceptors and nourishes retinal cells. RPE is firmly attached to the underlying choroid and overlying retinal cell layers. This pigmented layer serves as a pass-through between the light-sensitive photoreceptors and the choroid, which is a layer of blood vessels below. RPE nourishes the fragile retinal nerve cells and supports their normal functions by transporting molecules, removing dead cells, secreting hormones, and modulating immune factors. Due to these roles, inflammation of RPE results in macular degeneration impairing visual acuity and requiring visual prosthesis.

The choroid is the thin vascular layer of thickness of 250~350 μm in the wall of the eye beneath the retina. The choroid is made up almost entirely of blood vessels supporting 85% of blood flow within the eyes. These blood vessels supply oxygen and nutrients to the outer part of the retina keeping the retina healthy and functioning. It also absorbs light and limits reflections within the eye that could harm vision. Red eyes when a photograph is taken with a flash are due to the rich blood supply of the choroid.

4.3 Visual Pathway

The visual pathway is a path that electrical action potentials travel from the retina to the primary visual cortex at the occipital part of the brain. As shown in Fig. 14.13, action potentials generated from the ganglion cells in the retina, pass through optic nerves, optic chiasm, lateral geniculate nucleus (LGN), optic radiations, and arrive at the primary visual cortex.

Optic nerves do the work of transmitting the sensed signals out of the eyes toward the brain. Axons of the ganglion cells which are distributed two-dimensionally over the whole retina are collected to the optic disc forming a bundle of the optic nerve. The optic disc, which has a diameter of 1.5 mm, is purely formed by the nerve fibers without any cell body including photoreceptors. It is also known as a blind spot, as the image focused on the optic disc cannot be sensed. The optic nerve then exits the eye through the small hole in the sclera and enters the bony optic canal to emerge intracranially on the underside of the front brain. The average length of optic nerves is 40 mm. The optic nerve is the second cranial nerve composed of more than one million nerve fibers, responsible for transmitting the special sensory information for vision.

Optic nerves coming from the left and right eyes then converge at the optic chiasm. At the optic chiasm, fibers from the nasal side of each retina cross over to the contralateral optic tract, while fibers from the temporal side of the retina remain in the optic tract on their sides respectively. From here, signals from the left visual fields of both eyes are transmitted and processed by the optic tract on the right side and vice versa. It allows the visual cortex of each hemisphere to get information from both eyes.

The vast majority of nerve fibers in optic tracts, about 90%, extend from the optic chiasm to the lateral geniculate nucleus (LGN) in the thalamus. Thalamus is the place where all the sensory nerves stop to connect secondary sensory neurons by synapses and enter the brain. There are two LGNs, one on the left hemisphere and another on the right hemisphere. At LGN, afferent sensory nerves of optic tracks from the eyes synapse with the second-order sensory neurons. Part of nerve fibers travels to various portions of the brain other than LGN to provide information on perceived vision. For example, some nerve fibers leave the optic tract without entering the LGN and enter the brainstem to provide information for the control of pupil size.

Optic radiations are the axons of secondary sensory neurons which are synapsed with afferent optic nerves at LGN. Their average length is 98 mm (Peltier et al., 2006). Information in action potentials is directly projected to the primary visual cortex via optic radiations.

The primary visual cortex located at the occipital lobe of the brain is the final destination of action potential generated in the retina. Axons constituting the optic radiation terminate in the primary visual cortex in a retinotopic manner, where axons carrying information from a specific part of the visual field terminate in a specific location in the primary visual cortex that corresponds to that location in the visual field. Optical radiation travels along at least three distinct pathways. These different types of neurons of different pathways are specialized for specific categories of stimuli. They are thought to be involved in the detection of movements, recognizing shapes, and detecting colors respectively. The primary visual cortex, often called V1, is a structure essential to conscious processing and is the very beginning place of visual processing for the recognition and interpretation of the perceived visual scenes. Within the layers in the primary visual cortex, elementary processing for the perception of depth, form, color, and motion of objects is performed and the processed information is conveyed to the next cortical area for the following visual processing. There are two main pathways by which information travels from V1 to the surrounding visual areas. One pathway is the "What" pathway to the inferior part of the temporal lobe which carries information involved with object form and recognition. Another pathway is the "Where" pathway to the posterior parietal lobe which carries information involved with perceiving motion and spatial relationships between objects in the visual field.

5 Visual Prostheses

The visual prosthesis is a device to restore lost visual function for the blind based on electrical stimulation. For about 36 million blind subjects globally, the visual prosthesis is a unique solution while there is no other medical treatment to restore lost visual function. As the visual information is transferred in action potentials of electrical characteristics from the retina, replacing the intrinsic signal with the electrical signal is possible in several locations along the visual pathway from the retina to the visual cortex. Depending on the cause of blindness and the remaining healthy part of the visual pathway, location and the type of possible visual prosthesis can be determined. Electrical stimulations for visual prosthesis have been tried in several places along the visual pathway as in Fig. 14.13, at the retina, optic nerve, lateral geniculate nucleus (LGN) in the thalamus, and visual cortex. Currently, trials of the visual prosthesis are focused on the retinal prosthesis while the prosthetic applications in other locations are remaining somewhat more experimental (Mirochnik & Pezaris, 2019).

5.1 Retinal Prosthesis

Currently, the retina is a major focus for visual prosthesis. Retina has some advantages for a prosthetic place over other places. Its extracranial placement and simple architectural organization allow relatively easy surgical access. An implant in a single eye can potentially cover the whole left and right visual fields simultaneously. Generation of an image pattern for stimulation is easy due to the minimized geometrical image distortion. Within the retinal structure, three locations are targeted as the places for electrical stimulation, the epiretinal surface of the retina, the subretinal place between the layer of photoreceptors and choroid, and the suprachoroidal place between the choroid and sclera or in the sclera pocket (Fig. 14.17). However, the retinal implant requires a significant amount of remaining vital retinal ganglion cells, and it limits the range of applicability. Subjects with blindness affected by the later part of the visual pathway, from glaucoma, and traumatic injuries, cannot be benefited from the retinal implant.

The epiretinal prosthesis is the method of positioning the stimulating electrode array by fixing it with tacks on the surface of the retina in the vitreous cavity. Image is captured with an external camera installed on eyeglasses, processed, and transmitted into the internally implanted stimulating unit. However, in contrast to the normal who scan the view with smooth eye and head movement, the implanted subjects should move the head to scan the view. As vital ganglion cells are required to generate appropriate electrical stimulation, blind subjects with retinitis pigmentosa (RP) and outer retinal degeneration are indicated for retinal implantation. As the electrode array is placed at a close distance to ganglion cells, the required current for stimulation is very low with minimal power consumption. The quality of acquired vision

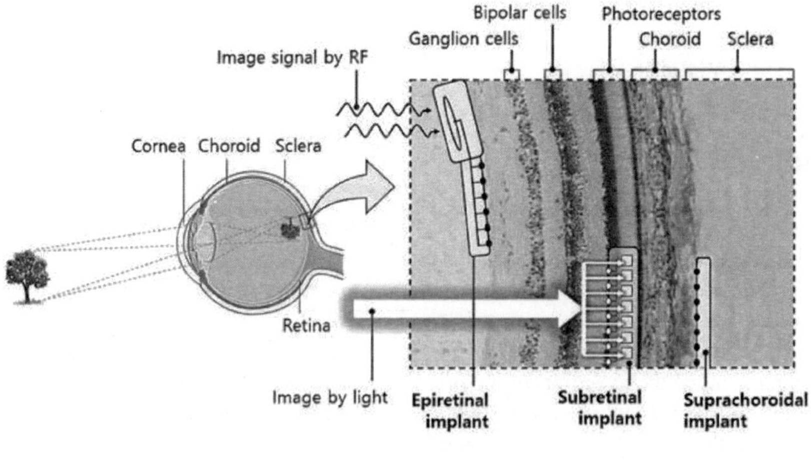

Fig. 14.17 Three different types of retinal implantations

is very low compared to normal vision and color is not manipulated. An array of 60 electrodes provides the visual acuity of 20/1260, which means that an implanted subject can see an object 20 ft away that an average normal subject can see at 1260 ft away. Due to the size of the electrode array, the covered visual field is restricted to a small portion of the whole field. However, with this visual acuity, the implanted subject can read, recognize words, and detect motion even though it is slow. Currently, the epiretinal implant is the unique approach that received both FDA and CE approval for general use.

The subretinal prosthesis is the approach to locate the stimulating electrode array on the outer retina at the position where the photoreceptors are populated. In the subretinal prosthesis, a stimulating array of electrodes is combined with a microphotodiode array (MPDA) doing the role of a camera. Image sensed by the MPDA can be used directly with electronic amplification and minimal processing to stimulate the multielectrode array (MEA). The implanted device is inserted into the position of the photoreceptor layer and does the same role as the photoreceptor, sensing the light and generating electrical signals. Subretinal implantation stimulates bipolar cells instead of ganglion cells and can use the remaining neural network of the retina beyond bipolar cells. Substantial intraretinal signal processing generates more naturalistic phosphenes. One great advantage is that it allows view scan with eyes and head combined just like normal subjects as the sensor is fixed on the back of the eyeball and it allows little training effort after implantation. Conceptually, the subretinal prosthesis is closer to the natural vision. While limited available space restricts the thickness of the implanted unit, it makes the unit be held in position without using any fixing tacks, which can damage the retina. The implanted device has electrodes up to 1600 as the output of photodiodes can be directly used (Gabel, 2017). Even though the spreading of stimulating current degrades the visual acuity, a subretinal implant provides a relatively higher visual acuity of 20/460 than other prosthetic approaches with 378-pixel MEA (Palanker et al., 2019). With this visual acuity, implanted subjects are able to read letters, perceive and localize light, detect motion, navigate familiar locations, and identify and grasp objects (Edwards et al., 2018). Another approach to a subretinal implant is to use an external camera and transmit the processed image via near-infrared (880 nm) image to the implanted sensing and stimulation combined unit. Using the external camera allows a full range of image processing like epineural implants. Like the epiretinal implant, the subretinal implant also cannot manipulate the color of phosphenes and is limited in the portion of the viewing field. Currently, the subretinal visual prosthesis has received governmental approval only in Europe.

Suprachoroidal retinal prosthesis is implanted over the choroid, between choroid and sclera. The selection of a suprachoroidal position is to minimize the retinal damage by escaping direct contact with the retinal cells. However, due to the distance to the retinal cells over choroid and retinal pigment epithelium (RPE), it requires a higher stimulating current. High-stimulating current increases the spread of stimulating current and decreases the visual contrast and resolution. Currently, suprachoroidal prostheses have proceeded with phase 1 clinical trials (Fujikado et al., 2016).

5.2 Indications for Retinal Implants

Blindness is a loss of vision that cannot be corrected with glasses or contact lenses. Usually, blindness means complete blindness that cannot see anything and does not see light. In North America and most of Europe, legal blindness is defined as the visual acuity worse than 20/200, even with glasses or contact lenses. Blindness can happen suddenly or gradually over a period of time. There are many causes leading to blindness including injury, lesions in the visual pathway, cataracts, glaucoma, diabetes, retinal detachment, infection, dietary deficiency, and heredity. Among the blind with these causes, subjects whose damages are restricted within the retina, like retinitis pigmentosa (RP) and age-related macular degeneration (AMD), are indicated for the retinal implant.

Retinitis pigmentosa (RP) is a rare inherited disorder that involves a breakdown and loss of photoreceptor cells in the outer layer of the retina. Retinal dysfunction is characterized by diffuse and progressive loss of rods, cones, and the retinal pigment epithelium (RPE). However, retinal cells in the inner layers, like bipolar cells and ganglion cells are largely preserved. Visual impairment usually manifests as night blindness, progressive visual field loss, and eventually progresses to blindness. At present, RP is the only indication for which retinal implant has been approved by FDA.

Age-related macular degeneration (AMD) is an acquired degeneration of the retina that most commonly leads older adults to vision loss by significant central visual impairment. It happens when aging causes damage to the macula, which is the central part of the retina with densely populated cone distribution. Losing central vision makes it harder to see faces, read, drive, or do close-up work like cooking or fixing things around the house. While the peripheral vision remains intact and useful, there is currently no cure or treatment for returning the vision already lost for most of its subtypes. For AMD subjects, retinal implant pursuits the coexistence of prosthetic vision with the residual vision to increase the overall visual perception. This is quite similar to the electro-acoustic stimulation which combines prosthetic high-frequency hearing with residual lower frequency hearing for overall hearing restoration. Currently, clinical trials are actively being performed with AMD subjects for their general use approvals.

5.3 Visual Prostheses at Other than Retina

The visual prosthesis can be implemented in locations other than the retina along the visual pathway. Table 14.3 summarizes and compares various visual prosthetic approaches. The optic nerve is an interesting potential target as it can be accessed with minimally invasive surgery before it enters the cranium. As a bundle of optic nerves cover the whole visual field, it can be another place for vision restoration with electrical stimulation at a single location. Even with a high level of retinal diseases or retinal detachment, visual prosthesis in optic nerves can provide the full view (Luo & da Cruz, 2014). Electrical stimulation can be applied with two types of electrodes, cuff electrodes and penetrating electrodes. Cuff electrodes with four equidistant 200 mm^2 contacts implanted around the optic nerve of a human subject can elicit phosphene clusters. Phosphenes can be elicited in different sizes and in various locations by controlling the stimulation parameters like pulse duration and pulse train frequency (Veraart et al., 2003). Optic nerve prosthesis is restricted to serial stimulation and difficult to adjust the phosphene luminosity (Farnum & Pelled, 2020). Penetrating electrode arrays can enhance fiber selectivity by inserting the electrode into the optic nerve. Optic nerve prosthesis demonstrated only in low resolution and low apparent brightness with high current stimulation. Studies on optic nerve prostheses have demonstrated the phosphenes in RP patients by localizing and identifying the objects, albeit with very long recognition times (Sakaguchi et al., 2009).

The lateral geniculate nucleus (LGN) is the place where the optic nerves terminate and visual information is relayed to the secondary nerve fibers to be transmitted to the visual cortex. With this configuration, LGN prosthesis has the prosthetic potential for blind subjects with either retinal or optic nerve pathologies. Since the successful applications of deep brain stimulation (DBS) on the thalamic area for movement disorders, LGN in the thalamus became an attractive site for the visual prosthesis. In contrast to the retina, spatial density over the receptive field of LGN is consistent regardless of the visual field eccentricity. The compact structure of LGN provides a wide visual field for prosthesis and a well-characterized receptive field with the separated visual subdivisions may support color artificial vision (Panetsos et al., 2011). As 60% of the LGN volume is devoted to processing the central 3° of the visual field, lower density MEAs can be used. It can reduce tissue damage due to mechanical insertion and electrical current delivery of higher density MEA (Farnum & Pelled, 2020). As LGN is behind the optic chiasm, for full visual field coverage, both of left and right visual field, two arrays of stimulating electrodes are required, one in each hemisphere. Studies on LGN prostheses have proved that electrical stimulation can produce

Table 14.3 Comparison of visual prosthetic approaches

Location	Epiretinal	Subretinal	Supra choroidal	Optic nerve	LGN	Visual cortex
Target	Ganglion cells	Bipolar cells	Ganglion cells	Optic nerves	LGN	Cortical neurons
Stimulating electrode	Surface	Surface	Surface	Cuff, penetrating	Penetrating	Surface, penetrating
Visual field	Entire visual field/eye				Half/hemisphere	
Advantage	Intact geospatial image mapping, ease of surgical accessibility, early positioning in visual pathway			No retina damage, accessible with minimal invasive surgery	Possible use of DBS technology, even field density, stable positioning.	Applicable to various pathologies, larger area
	Easy access,	Naturalistic percepts	No retinal damage, mechanical stability			
Dis-advantage	Retinal damage by fixing	Uneven rod & cone density	High-stimulating current	Difficult to control intensity	Half field stimulation	Complex encoding, craniotomy
Camera	External	Internal, external	External	External	External	External
View scan	By head	By eyes	By head	By head	By head	By head
Indication	Retinitis pigmentosa (RP), age-related macular degeneration (AMD).			Retinal diseases & detachment	Optic nerve disease	Except cortical injury
Approval	FDA (2013), CE(2011)	CE(2013)	In clinical trials	Human experiment	No human trial yet	In clinical trials
Visual acuity	20/1262	20/546, 20/460	20/8397		20/240 (by model)	
Pixels	60	1600, 37,	20, 49			

phosphenes and that such stimulation produces similar responses in the visual cortex as natural visual stimulation (Panetsos et al., 2011). However, LGN prosthesis is not applied to humans yet. A model of bilateral 400-electrode implants is estimated to provide visual acuity up to 20/240 (Kyada et al., 2017).

The visual cortex is the site where the first electrical stimulation has tried to restore vision and phosphenes have successfully generated in the earlier stage of the visual prosthesis. Since then, many researchers have focused their efforts based on the advantages of the visual prostheses in the cortex. The visual cortex has a large surface area and a stimulation strategy can be applied straightforwardly. Similar to the LGN, the spatial density of the receptive field has fairly consistent organization across the surface of the visual cortex. Central receptive fields can be easily accessed with their locations near the surface of the occipital lobe compared to the peripheral fields. Cortex prosthesis can be applied to all forms of diseases leading to blindness while the visual cortex is intact. Penetrating electrodes can be used as well as surface electrodes and it can allow closely spaced electrical stimulation with greatly lowered current levels (Schmidt et al., 1996). Early simulations for restoring functional vision via cortical prostheses estimate a total of at least 625 discrete phosphenes (Cha et al., 1992). Similarly, to all other visual prostheses except the subretinal implantation, the cortex prosthesis needs also head movement to scan the view as it cannot be benefited by the eye movement. The rapid advancement of microelectronic device fabrication and information processing has made cortical prostheses a viable prosthetic option. There are several ongoing clinical trials testing the effectiveness and the risk associated with the cortical implants (Farnum & Pelled, 2020).

5.4 Retinal Implant Systems

The retinal prosthetic system consists of an external unit and an internal unit similar to the cochlear implant (Fig. 14.18). The external unit is composed of three functional parts, camera, video processing unit (VPU), and transmitter. A small miniature video camera is usually mounted on a pair of glasses worn by the subject to capture the images at the position of the subject's eyes view and connected to the VPU, which is usually worn on a shoulder strap or a belt, by a cable. VPU converts the recorded images by the camera into real-time brightness maps of smaller pixels with appropriate image processing. Image processing like edge detection can be applied intelligently in VPU to improve visual acuity leading to better shape and object recognition. Processed images are encoded and connected by another cable to the transmitter part integrated into the glasses. Energy to drive internal units and the data including the encoded images are transmitted wirelessly by radiofrequency waves via coil antennae.

The internal unit is surgically implanted in and around the eye. It consists of three parts of a transceiver, electronic module, and microelectrodes array (MEA). The transceiver receives wirelessly transmitted energy and encoded data via matching coil antennae. The electronic module is secured in case to the outside of the eye using a standard scleral band and sutures. It contains electronic circuits to decode and extract the data from the received signal and generates electrical stimulating signals for each electrode. The microelectrode array is connected to the electronic module by a flexible parallel cable extended from the outside of the eye into the internal part of the eye, epiretinal or subretinal positions where the MEA is located. The electrodes stimulate the retinal cells with electrical stimulating pulses whose amplitude increases according to the brightness of the scene. Currently, FDA-approved epiretinal implant system uses MEA of 60 electrodes arranged in a 6 × 10 grid. Each microelectrode has a diameter of 200 μm, separated by 525 μm each other and each electrode covers an area equivalent to hundreds of photoreceptors. Total MEA size measures 5.5 mm in width and 6 mm in length and spans approximately 20° of a diagonal visual angle. The number of electrodes is planned to be increased to 240 in the next-generation implant and the eventual goal is to minimize the size of each electrode to the size of retinal ganglion cells to activate each cell individually (Dagnelie et al., 2017).

5.5 Spatial Resolution for Prosthetic Vision

Even though visual prosthesis makes the subjects sense the light and helps them in doing their ordinary activities, the current quality of the vision restored by visual prosthesis is much poor compared to the normal vision. The best visual acuity of 20/460 currently restored with a visual prosthetic system is still worse than the legal threshold of the blindness. And the wearers of a visual prosthesis device still more benefited by traditional vision aids for the blind, like guiding dog, the white cane, Braille reading, and voice-over programs on computers or mobile phones. To increase the efficacy and usability of the visual prosthesis, increasing the quality of restored vision is significantly required.

Quality of vision can be affected by several factors in the provided stream of images. These parameters include spatial resolution, brightness, color, the field of view, and frame rate.

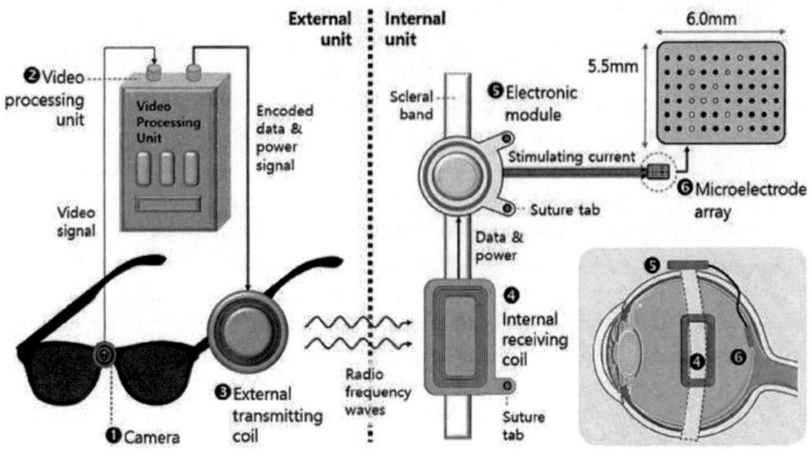

Fig. 14.18 External and internal units of retinal implant system

Table 14.4 Minimal spatial resolution required to reach useful vison for some daily living tasks

Task	Task description	Required phosphenes	Field of view	Spatial resolution
Mobility	Move along in well predictable environment	~200	33° × 23°	~0.25 pixels/deg^2
Mobility	Move along in simple but unpredictable environment	≥500	33° × 23°	≥0.7 pixels/deg^2
Object recognition	Point one figure out of 20 figures	~500	17° × 12°	~2.5 pixels/deg^2
Reading	Standard newspaper reading even with slower speed	~500	1.6° × 1.1°	~300 pixels/deg^2
Retinal implant	Currently approved for general use	60	20° × 12°	0.25 pixels/deg^2

Sommerhalder and Fornos (2017)

Among these parameters, spatial resolution is most critical to the quality of vision and is very difficult to increase technically. Even though the restored vision does not provide the normal quality of vision, minimal visual acuity is required in doing each activity during ordinary life. Table 14.4 shows the approximate minimal spatial resolution required to reach the useful vision for some daily living tasks. With the spatial resolution of 0.25 pixels/deg^2, subjects can move around in a well predictable environment like in their houses. However, to read a newspaper even with a much slower speed than normal subjects, a higher spatial resolution of about 300 pixels/deg^2 is required. Even though the required spatial resolutions are different greatly, this discrepancy can be manipulated by restricting the field of view. With 500 phosphenes, many living tasks can be performed under useful vision by manipulating the field of view appropriately. This manipulation of the field of view can be done intelligently in VPU by analyzing the contents of the recorded images by the camera and VPU can provide stimulating maps in the optimal combination of spatial resolution and field of view.

Another point that should be considered is that part of the visual information is lost in the electrode-nerve interface. Current retinal implant with 60-pixels MEA provides the spatial resolution of 0.25 pixels/deg^2 which can support mobility living tasks usefully. However, the experimental results with the recipients of the retinal implants are poor than the simulated results. Image supplied in the MEA by stimulating electrode does not effectively reach the brain as it is stimulated. This means that a considerable part of the visual information provided by the stimulating device does not reach the brain and further. Simply increasing the spatial resolution with the increased number of microelectrodes might not fundamentally improve the performance of the visual quality. New and better stimulation strategies have to be developed to optimize information transfer from the prosthesis to the upper visual centers and to limit information loss at this site (Sommerhalder & Fornos, 2017).

In addition to spatial resolution, the temporal properties of visual perception are also important. To reconstruct meaningful images in the brain, perceptions should last long enough as well as they should be sharp and well localized in the space. Generally, the current perception of vision does not last an equal period of time as the electrical stimulation is continued. As visual perception tends to adapt to temporal variations, continued electrical stimulations are used to produce simple transient perceptions. After analyzing the temporal characteristics, compensating temporal processing is required in VPU to provide perception close to normal vision.

References

Cha, K., Horch, K. W., & Normann, R. A. (1992). Mobility performance with a pixelized vision system. *Vision Research, 32*, 1367–1372. https://doi.org/10.1016/0042-6989(92)90229-c

Clark, G. M. (2015). The multi-channel cochlear implant: Multidisciplinary development of electrical stimulation of the cochlea and the resulting clinical benefit. *Hearing Research, 322*, 4–13. https://doi.org/10.1016/j.heares.2014.08.002

Dagnelie, G., Christopher, P., et al. (2017). Performance of real-world functional vision tasks by blind subjects improves after implantation with the Argus® II retinal prosthesis system. *Clinical & Experimental Ophthalmology, 45*(2), 152–159.

Edwards, T. L., Cottriall, C. L., Xue, K., et al. (2018). Assessment of the electronic retinal implant alpha AMS in restoring vision to blind patients with end-stage retinitis pigmentosa. *Ophthalmology, 125*(3), 432–443.

Farnum, A., & Pelled, G. (2020). New vision for visual prostheses. *Frontiers in Neuroscience, 14*, 36. https://doi.org/10.3389/fnins.2020.00036

Fujikado, T., Kamei, M., et al. (2016). Outcome of 49-channel suprachoroidal-transretinal stimulation prosthesis in patients with advanced retinitis pigmentosa. *Investigative Ophthalmology & Visual Science, 57*(14), 6147–6157.

Gabel, V. P. (2017). *Artificial vision: a clinical guide*. Springer.

Hong, R. S., & Rubinstein, J. T. (2003). High-rate conditioning pulse trains in cochlear implants: Dynamic range measures with sinusoidal stimuli. *Journal of the Acoustical Society of America, 114*(6 Pt 1), 3327–3342. https://doi.org/10.1121/1.1623785

Irving, S., Gillespie, L., Richardson, R., Rowe, D., Fallon, J. B., & Wise, A. K. (2014). Electroacoustic stimulation: Now and into the future. *BioMed Research International, 2014*, 350504. https://doi.org/10.1155/2014/350504

Kyada, M. J., Killian, N. J., & Pezaris, J. S. (2017). *Artificial vision* (Vol. 177-89). Springer.

Litovsky, R., Goupell, M. J., Kan, A., & Landsberger, D. (2017). Use of research interfaces for psychophysical studies with cochlear-implant. *Users. Trends in Hearing, 21*, 1–15. https://doi.org/10.1177/2331216517736464

Luo, Y. H. L., & da Cruz, L. (2014). A review and update on the current status of retinal prostheses (bionic eye). *British Medical Bulletin, 109*(1), 31–44.

Mirochnik, R. M., & Pezaris, J. S. (2019). Contemporary approaches to visual prostheses. *Military Medical Research, 6*, 19. https://doi.org/10.1186/s40779-019-0206-9

Palanker, D. V., Le Mer, Y., et al. (2019). Restoration of sight in geographic atrophy using a photovoltaic subretinal prosthesis. *Investigative Ophthalmology & Visual Science, 60*, 970.

Panetsos, F., Sanchez-Jimenez, A., et al. (2011). Consistent phosphenes generated by electrical micro-stimulation of the visual thalamus. An experimental approach for thalamic visual neuroprostheses. *Frontiers in Neuroscience, 5*, 84.

Peltier, J., Travers, N., Destrieux, C., & Velut, S. (2006). Optic radiations: A microsurgical anatomical study. *Journal of Neurosurgery, 105*, 294–300. https://doi.org/10.3171/jns.2006.105.2.294

Prado-Guitierrez, P., Fewster, L. M., Heasman, J. M., McKay, C. M., & Shepherd, R. K. (2006). Effect of interphase gap and pulse duration on electrically evoked potentials is correlated with auditory nerve survival. *Hearing and Research, 215*(1–2), 47–55. https://doi.org/10.1016/j.heares.2006.03.006

Rebscher, S. J., Heilmann, M., et al. (1999). Strategies to improve electrode positioning and safety in cochlear implants. *IEEE Transactions on Biomedical Engineering, 46*(3), 340–352.

Sakaguchi, H., Kamei, M., Fujikado, T., et al. (2009). Artificial vision by direct optic nerve electrode (AV-DONE) implantation in a blind patient with retinitis pigmentoa. *Journal of Artificial Organs, 12*, 206–209.

Sanchez-Cuadrado, I., Lassaletta, L., Perez-Mora, R. M., et al. (2013). Is there an age limit for cochlear implantation? *Annals of Otology, Rhinology & Laryngology, 122*(4), 222–228.

Schmidt, E. M., Bak, M. J., Hambrecht, F. T., Kufta, C. V., O'Rourke, D. K., & Vallabhanath, P. (1996). Feasibility of a visual prosthesis for the blind based on intracortical microstimulation of the visual cortex. *Brain, 119*(Pt 2), 507–522.

Sommerhalder, J., & Fornos, A. P. (2017). Prospects and limitations of spatial resolution. In *Artificial vision: A clinical guide*. Springer.

Szyfter, W., Karlik, M., Sekula, A., Harris, S., & Gawęcki, W. (2019). Current indications for cochlear implantation in adults and children. *Otolaryngologia Polska, 73*(3), 1–5. https://doi.org/10.5604/01.3001.0013.1000

Veraart, C., Wanet-Defalque, M. C., Gérard, B., Vanlierde, A., & Delbek, J. (2003). Pattern recognition with the optic nerve visual prosthesis. *Artificial Organs, 27*, 996–1004. https://doi.org/10.1046/j.1525-1594.2003.07305.x

Yue, L., Weiland, J. D., & Humayun, M. S. (2018). Retinal prostheses: A brief history. In *Retinal prosthesis, essentials in ophthalmology*, Springer. https://doi.org/10.1007/978-3-319-67260-1_1.

Zeng, F. G. (2004). Trends in cochlear implants. *Trends in Amplification, 8*(1), 1–34. https://doi.org/10.1177/108471380400800102

Zeng, F. G., Rebscher, S., Harrison, W., Sun, X., & Feng, H. (2008). Cochlear implants: System design, integration, and evaluation. *IEEE Reviews in Biomedical Engineering, 1*, 115–142. https://doi.org/10.1109/RBME.2008.2008250

15 Electrical Stimulation of Nerves and Muscles

In addition to electrical stimulation used to pace the heart and modulate brain functions, it can be applied in various ways to the nerves and muscles around the body. As the nervous system is functioning based on the electrical signal of the action potential, modulation of its functioning by electrical intervention is reasonably probable. Nerves and muscles are the cells constituting the nervous system and are the targets of electrical intervention. Nerves generate action potentials and transmit them along their axonal fibers. While the neuronal cell bodies are located deep in the brain or in the spinal cord, their fibers are extending to the target organs in the distal part of the body and are easier to access than the cell bodies. So, most of the electrical stimulation is performed on the nerve fibers. While nerve fibers are accessible during their connection to the periphery, usually thousands of fibers go parallelly in a nerve trunk. Instead of stimulating a single nerve fiber, currently, a branch of the nerve trunk which contains the target nerve fiber maximally is selected for stimulating target. As the other fibers within the same nerve trunk can be stimulated simultaneously, it may cause unintended side effects.

When stimulating nerves is not possible due to some damage, muscles can be directly stimulated by electrical pulses. Muscle contraction is also achieved through the excitation of intramuscular nerve fibers. As they are distributed over the whole muscle volume, generally muscle stimulation requires a higher level of electrical strength than nerve fiber stimulation.

As many of the electrical stimulation approaches on the nerves and muscles are invasive using implantable devices for permanent use, they are mostly targeted to treat and restore the impaired functions by some disorders or diseases. However, it can enhance the normal functions of the body above the normal level, and some methods are being used to improve the health level and enhance the body function like muscular power.

Various terminologies are used for the electrical intervention for body functions. Functional electrical stimulation (FES) describes the electrical manipulation of body functions. In a broad sense, it can include all the electrical stimulation approaches like cardiac pacing and deep brain stimulation. However, FES is also used narrowly to describe the electrical stimulation for the restoration of movement by stimulating skeletal muscles. Neuromodulation describes the aspect of electrical stimulation in the alteration of nerve activity through targeted delivery of stimulations. However, neuromodulation includes various stimulation methods like chemical and optical stimulations in addition to electrical stimulation. Electroceutical describes the electrical stimulation in the aspect of its efficacy similar to the pharmaceutical approach.

1 Functional Electrical Stimulation (FES)

Functional electrical stimulation (FES) is a treatment that uses mild electrical stimulations to improve paralyzed or weakened muscular function. To restore a movement of a muscle, electrical stimulation is used to stimulate the motor nerve connected to a muscle or directly muscle itself.

FES was first developed in the 1960s and initially tested in stroke patients. Even though FES has not been used regularly in clinical practice until much later, there have been active studies extending its application. FES has been used to treat the movement of the lower limb, upper limb, and ambulation in paralyzed subjects. And, it successfully made a patient with spinal cord injury raise from his wheelchair and walk in the 1990s. FES is currently being investigated to treat swallowing function, arm movement, and breathing function.

1.1 Electrical Activation of Muscles

Activation of muscles can be achieved by stimulating the motor nerve connected or by stimulating the muscle directly. However, the stimulating current required to directly elicit a

K. S. Park, *Humans and Electricity*, https://doi.org/10.1007/978-3-031-20784-6_15

muscle contraction in denervated muscle is much higher than the current required to elicit action potentials in the innervating motor nerves. So, FES applications for the restoration of motor functions generally use electrical stimulations to excite the nerves rather than muscles.

For the application of FES, intact motor nerves are necessary. Even though the upper motor neurons from the brain to the spinal cord were damaged, the lower motor neurons must be intact from the spinal cord to the neuromuscular junctions in the targeted muscles. Additionally, the neuromuscular junction and muscle tissues must be healthy for electrical stimulation to be effective. FES is not applicable in some diseases which accompany the damage in the lower motor neurons like polio, amyotrophic lateral sclerosis (ALS), and peripheral nerve injuries and diseases accompanying muscle damage like muscular dystrophies. FES has traditionally been applied when the subjects have intact lower motor neurons and healthy neuromuscular junction and muscles. Subjects with spinal cord injury (SCI), stroke, head injuries, cerebral palsy, and multiple sclerosis are belonging to this criteria and FES has most actively targeted subjects with SCI.

When the stimulating electrical current is applied, motor nerves are recruited parallelly and sequentially and the produced muscle tensions are summated spatially and temporally. Action potential generated by the stimulation has an effective duration of less than 5 ms. However, the muscle twitch of a motor unit triggered by an action potential lasts for the duration of 100 ms. When the stimulation is applied repetitively, trigged muscle twitches are also elicited repetitively. As the repetition rate increases, elicited twitches start to be overlapped and the fluctuation of generated tension by the muscle is decreasing (Fig. 15.1). In a higher repetition rate over 50 Hz, tetanus force is developed, where generated force is sustained almost smoothly without fluctuation. By stimulating motor nerves electrically with a high repetition rate over 50 Hz, targeted muscle can be continuously and smoothly contracted.

The strength of the generated force can be manipulated by the number of neurons recruited by the electrical stimulation. The more electrical charges are delivered, the larger number of neurons are recruited for the generation of force. In an intrinsic neuromuscular system, motor nerves having a smaller diameter and smaller number of connected muscle fibers are recruited first. However, when the nerves are stimulated artificially by the electrical current, the order of recruitment is reversed. As the motor fibers have a larger diameter and a large number of connected muscle fibers have a lower threshold to electrical stimulation, these thick and large motor units are recruited first. That is, while minute movements are controllable in an intrinsic neuromuscular system, it is difficult in an artificially controlled system as the large motor units are recruited first.

As the critical parameter controlling the strength is the charge, stimulation strength basically can be regulated by the current with its amplitude and pulse width. Current-based control can provide the expected amount of charge irrespective of the unknown tissue impedance and generally provide higher reproducibility in FES performance. However, in some cases using transcutaneous electrodes, there used to be high electrode-skin contacting impedance as the electrode dries or loses contact, and it can generate high electric density and skin burn. For this case, stimulation strength can be regulated by the applied voltage even though the motor response is more variable due to the impedance-dependent resulting currents.

1.2 FES for Foot Drop

Foot drop, sometimes called drop foot, is a problem in lifting the front part of the foot. Subjects with a foot drop might drag their front part of the foot on the ground when they walk. Foot drop can be caused by a problem in the peroneal nerve, muscle damage, or brain and spinal cord diseases. It is a

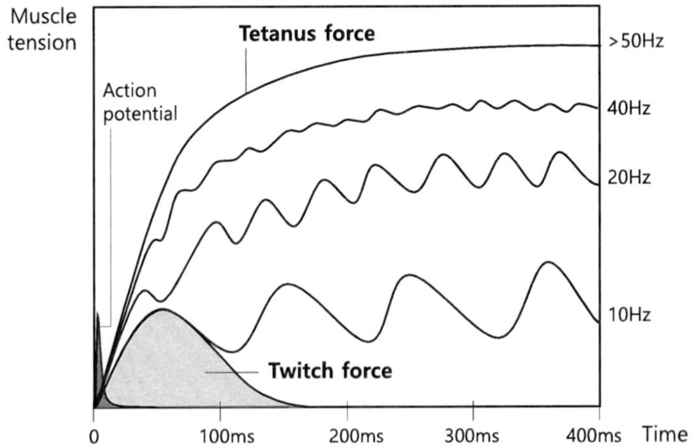

Fig. 15.1 Generated muscle tensions by the temporal summation when stimulating electrical pulses are applied with different repetition rates

Fig. 15.2 Functional electrical stimulation for a foot drop

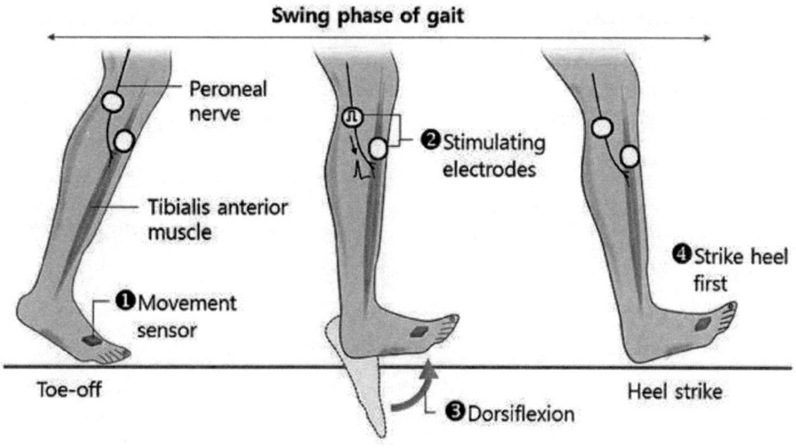

common problem in subjects who have hemiplegia as a result of a stroke. They have difficulty in performing ankle flexion and extension in the affected lower limb. As a consequence, the foot cannot take off the ground which makes walking difficult or impossible. FES can only be applied to subjects who still have intact peroneal nerves and muscles. Among the foot drops, foot drops caused by the diseases in the central nervous system, like stroke, SCI, multiple sclerosis (MS), and spinal cord injury (SCI), are belonging to this indication. Correction of foot drop using electrical stimulation is the first and the simplest application of successful FES (Fig. 15.2).

The front part of the foot is raised from the ground with the dorsiflexion of the ankle, which is resulted from the contraction of the tibialis anterior muscle in the lower leg. Tibialis anterior muscle is innervated by the peroneal nerve. To raise the front part of the foot from the ground, the tibialis anterior muscle should be contracted timely by the electrical stimulation of the peroneal nerve. So, the electrodes are appropriately positioned over the peroneal nerve. Electrodes can be positioned transcutaneously over the skin or can be implanted for permanent use. To apply electrical pulse effectively and make walking balanced and symmetrical, the stimulating time of electrical pulse should be accurately determined. This is done by the sensing part. With various types of sensors, including heel off switch, goniometer, accelerometer located in the shoe, or tilt sensor located in the knee, heel strike and lift of the foot can be detected and the appropriate time can be determined for the stimulation to cause the dorsiflexion of the ankle. There are several FDA- or CE-approved FES systems for foot drop. Some devices are implantable. To control the foot movement more accurately, 2 or 4 channels of stimulating electrodes are used (Marquez-Chin & Popovic, 2020).

1.3 FES for Upper Limb Movement

FES systems have also been developed to assist the upper limb function including reaching and grasping (Fig. 15.3). FES for reaching and grasping can be implemented using implanted electrodes or noninvasive transcutaneous electrodes. The objective of upper limb FES is to enable subjects with upper limb paralysis to use their hands in activities of daily living (ADL).

To activate hand grasping, usually, three channels of stimulations are used. The first channel stimulates the median nerve and the flexor muscles at the motor endplate to induce the flexion of fingers and pronation of the wrist. The second channel is used to stimulate extensor muscle via radial nerve and motor end plate for the extension of fingers and the wrist. The third channel is used to stimulate the muscle controlling the thumb for grasping. The start of the motion can be initiated by pushing the button of the control unit or using the intact functions of the patients. Grasping can be triggered by the movement of the wrist or shoulder of the opposite side of the paralyzed hand.

1.4 FES for Ambulation

FES can make paralyzed patients walk. With the help of FES, patients whose lower limbs are paralyzed and only move with wheelchairs can walk by themselves. This seems to be the ultimate goal of the FES. Simply, the FES system for ambulation can be implemented with four channels, two channels for the left leg and two channels for the right leg. For each leg, one pair of electrodes is placed on the peroneal nerve and the second pair of electrodes is placed over the quadriceps

Fig. 15.3 FES electrodes for hand grasping. (**a**) Electrode positions for the contraction of flexion and extension. (**b**) Example of hand grasping types

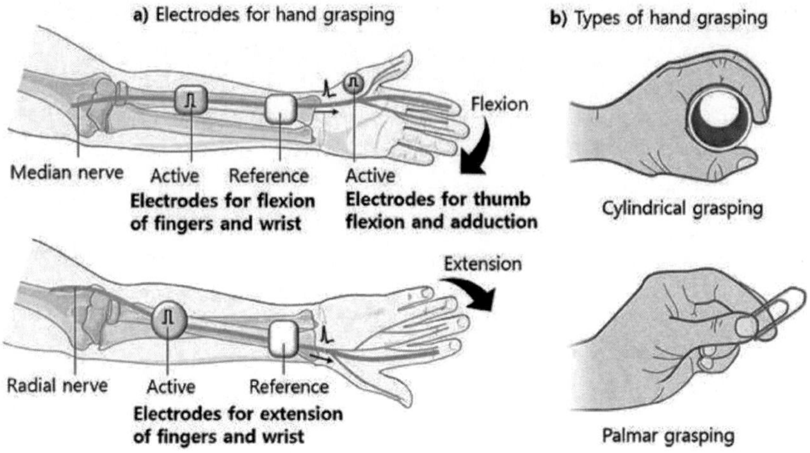

Fig. 15.4 Functional electrical stimulation for ambulation

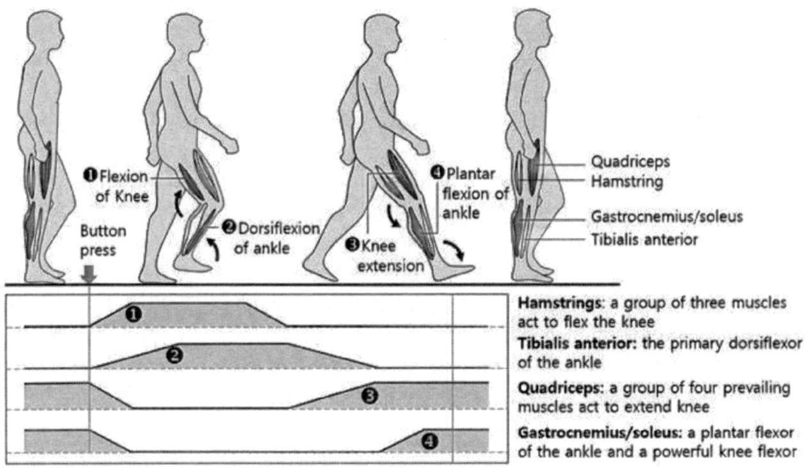

muscle. Stimulation on the quadriceps muscle, which is the large muscle at the front of the thigh composed of four distinct portions, acts to extend the leg. And, the stimulation of the peroneal nerve elicits the flexion withdrawal reflex which produces simultaneous ankle dorsiflexion and hip flexion. By timely alternating stimulations for flexion and extension of the leg, the swing phase of a gait can be facilitated with the corresponding upper body movement. The same sequence of stimulations is applied in a different phase to the leg on the opposite side to complete a stride cycle. Patients can trigger the stimulation of each leg using buttons placed on the handles of a rolling walker or crutches.

When four electrodes are used for each leg, gait movement is generated without using the flexion withdrawal reflex. Four electrode pairs are placed over the quadriceps, hamstrings, gastrocnemius/soleus, and tibialis anterior muscles of each leg. A sequence of stimulation pulses is initiated by pressing the button. Then, the stimulation is applied sequentially to hamstrings, tibialis anterior, quadriceps, and finally to gastrocnemius muscles with finely tuned timing and amplitude. Quadriceps and gastrocnemius stimulations are used during the late swing phase and during the stand phases of a gait. SCI patients can walk with a rolling walker, cane, or orthosis (Thrasher et al., 2006) (Fig. 15.4).

1.5 Functional Electrical Stimulation Therapy (FEST)

Functional electrical stimulation therapy (FEST) is a short-term treatment using FES, instead of extended use of FES, to assist patients in regaining their ability to move voluntarily. FEST is a kind of carry-over effect of the FES. This carry-over effect used to be reported from the early days of FES in the subjects treated with FES. FEST is used to support the patients when they try to relearn executing impaired functions through the rehabilitation of the central nervous system, instead of making them dependent on the lifelong use of prosthetic devices.

FEST is based on the concept of neuroplasticity. Neuroplasticity is the ability of the central nervous system to modify its synaptic connectivity to reorganize itself and incorporate new motor abilities during the acquisition, retention, and consolidation of motor skills. Neuroplasticity can be applicable in improving the restoration of motor functions during neurorehabilitation. Functional electrical stimulation therapy (FEST) takes advantage of neuroplasticity to restore the ability to perform voluntary movement after stroke and SCI (Marquez-Chin & Popovic, 2020).

In a FEST treatment, patients are asked to initiate functional movements of their affected limb. Soon after a few seconds, a therapist applies functional electrical stimulation purposed to move the attempted muscle. Following electrical stimulation facilitates the attempted task producing the movement and activates accompanying sensory feedback. In each session, this procedure is repeated multiple times and the session is performed multiple times over 40 to induce a restorative effect of the treatment. It is believed that the repeated simultaneous presence of a patient's intention to move and the sensory feedback resulting from the FES-assisted movement produces neuroplastic changes that ultimately lead to the restoration of voluntary motor function (Nagai et al., 2016). Application of FEST to foot drop patients due to the stroke and SCI results in sustained improvements in lower limb function with the increase in the generated torque and overall walking speed. FEST also can improve the upper limb function by increasing the grasping power and the range of movements.

Contralaterally Controlled FES

Contralaterally controlled FES is also a good example of FEST which improves hand dexterity in chronic hemiparesis patients (Fig. 15.5). In this treatment, EMG signal or bending sensed in the hand of the normal side is used to trigger the contraction of muscles in the opposite weaker side by FES. When a patient attempts to open both hands, neural activities are generated simultaneously in bilateral motor cortices and transmitted to hand muscles. Upon this attempt, the normal hand is open while the weaker hand is not open sufficiently. The device picks up the signal from the normal hand when the muscle is contracted for opening and the detected signal is used to trigger stimulation for the opposite hand. Then, the weaker hand can be open almost simultaneously by the electrical stimulation even though it cannot respond sufficiently through its intrinsic neural pathway. When both hands are open, receptors in the muscles and the skin in the hands sense the contraction and transmit generated signals to the brain via the afferent neural pathway. Generated motor signals and the responding afferent signals from both hands help to restore motor functions by developing neuroplasticity.

FES Bike

FES bike also works in a similar manner for the rehabilitation of complete or partial leg paralyzed patients due to the diseases like spinal cord injury, multiple sclerosis (MS), Parkinson's disease, and stroke. FES bike usually includes arm pedals as well as foot pedals. Even though patients cannot move foot pedals due to their paralysis, they can move arm pedals. When arm pedals are in motion, a chain connecting the arm and foot pedals ensures that the legs also have a pedaling movement. Sensors on the bike detect the angle and the speed of the wheel and forward the signals to the stimulator. The stimulator calculates which muscle should be activated and sends electrical pulses to the electrodes attached over each leg muscle. Electrical pulses stimulate the muscles in the same way as the intrinsic motor signals and a patient moves the foot pedals with the patient's own muscle force despite incomplete motor function. The number of electrodes can be increased to 12 to stimulate different muscles on both sides separately and accurately.

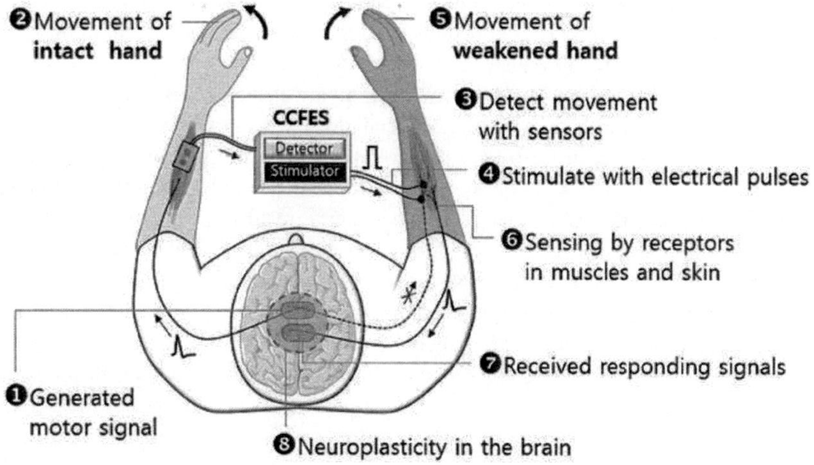

Fig. 15.5 Contralaterally controlled FES for the rehabilitation of paralyzed right hand

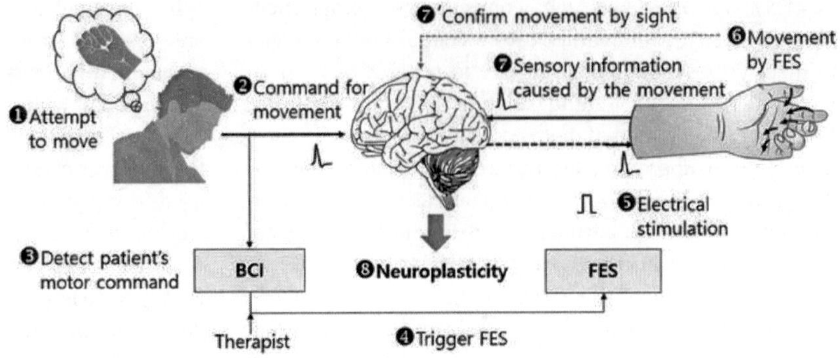

Fig. 15.6 Integration of BCI to increase the efficacy of neuroplasticity during the FEST

This FES exercise can build muscle bulk, and strength, improve fitness, and reduce muscle spasms producing circulation benefits. In addition, biofeedback signals from the receptors in the muscles and the skin are transmitted to the brain and activate neuroplasticity which encourages descending drive and causes faster recovery.

If FEST is used in combination with the brain-computer interface (BCI), an increased effect of neuroplasticity is expected (Fig. 15.6). The simultaneous presence of a motor command, which is resulting from a patient's attempt to move, and sensory information which is the proprioceptive and somatosensory resulting from the FES-produced movement, will promote changes in the nervous system that will ultimately lead to restoration of voluntary movement (Daly & Wolpaw, 2008). In this procedure, BCI systems can provide the opportunity to confirm and strengthen the intention of the patient to move, even though there is no overt movement due to the severity of the impairment.

1.6 Electrical Muscle Stimulation

Electrical muscle stimulation (EMS), also called neuromuscular electrical stimulation (NMES), is a method of delivering an electrical current to induce a muscle contraction. EMS is targeting to cause muscle contractions as in their intrinsic voluntary contractions by replacing motor action potentials with external electrical pulses. However, by applying electrical pulses many times in a row, EMS can do a role in strengthening and repairing tissue, particularly in muscles that have become shortened, weakened, or atrophied due to injury or disease. If a subject has a shoulder injury, resting, immobilizing, or favoring the shoulder may weaken the shoulder and connected muscles. Neuromuscular disorders also weaken muscles throughout the body progressively. For these cases, EMS can help to strengthen the muscles and to slow the weakening progress so that subjects improve their motor control.

EMS also promotes the secretion of myokine, which is the peptide produced and released by skeletal muscle cells in response to muscular contractions. Myokines are involved in metabolic changes associated with exercise and following training adaptation. They also participate in tissue regeneration and repair, maintenance of healthy bodily functioning, and immunomodulation. Repeated skeletal muscle contraction by EMS can induce the production and secretion of myokines, which have a wide range of positive functions in different tissues and organs.

Even in the artificial contraction of muscle, contraction of the muscle is also elicited by the action potentials in intramuscular nerve fibers which are evoked by the electrical current stimulation. As the intramuscular nerve fibers are populated densely in the belly of muscles, an electrode for muscle stimulation is positioned usually over the belly of the target muscle. When electric stimulation is applied transcutaneously, it generally requires considerably higher stimulation amplitude as intramuscular nerve fibers of the target muscle are not easily selectable and distributed through the muscle volume.

Typically, stimulating frequency is adjusted by the targeted contraction. A low frequency of 1~10 Hz generates only muscle twitches which cause less fatigue and is appropriate for the heavily atrophied muscles. Mid-frequency stimulation of 20~60 Hz generates tetanic muscle contraction sufficient to move joints and is appropriate for strengthening the muscle. With pulse frequency higher than 60 Hz, EMS is more effective on the sensory nerves rather than motor nerves, and the power of the muscle contraction tends to reduce. Pulse width is determined within the range of 50~500 μs. To induce a powerful contraction in a large muscle group, extended pulse width higher than 350 μs is used. Application of electrical stimulation, 20 Hz and 500 μs, can evoke a force equal to 70% of the peak force achieved during a maximal voluntary contraction (MVC) (Enoka et al., 2020). The amplitude of pulse current is determined in combination with pulse width up to the desired contraction within the tolerable level, which is usually in the range of 10~30 mA.

EMS is similar to transcutaneous electrical nerve stimulation (TENS) in its look and way of function. Both devices apply electrical current transcutaneously using the pad electrodes attached over the skin surface. In some units, they can be used for TENS as well as EMS after selecting the stimulation mode. However, TENS targets stimulating the nerves with the goal of pain relief and uses current pulses of small amplitude. TENS induces the pain relief effect by blocking the transmission of pain signals and by the extrasegmental analgesia including the release of endogenous opioids. So, TENS is used for treating both acute and chronic pains caused by various conditions. In contrast, EMS targets stimulating innervated nerve fibers in the muscle to cause a contraction of a muscle and uses a relatively higher stimulating current. EMS provides intense exercises by sending artificial signals to the target muscles and to the brain. EMS is used to enhance muscle strength and motor learning in normal subjects as well as in subjects with impaired muscles. EMS treatment is often used in sports medicine to strengthen and maintain muscles during prolonged periods of immobilization.

2 Electrical Stimulation for the Restoration of Body Function

In addition to stimulating skeletal muscles to restore movements, muscles working for other functions in various organs also can be stimulated electrically to restore their intrinsic functions.

2.1 Phrenic Nerve Stimulation

Phrenic nerve stimulation (PNS), also called diaphragm pacing (DP), is to restore breathing by periodically stimulating the diaphragm muscle and eliminating the dependency on the mechanical ventilator. Some subjects who have spinal cord injury lose their ability to breathe autonomously due to the blockage of nervous connection to the central nervous system or impaired respiratory centers in the brain. These subjects cannot but depend on mechanical ventilation through a tracheostomy at the base of the neck. Mechanical ventilators are associated with substantial morbidity, mortality, inconvenience, physical discomfort, fear of disconnection, difficulty with speech, impaired sense of smell, and disturbed mobility (DiMarco, 1999). Phrenic nerve stimulation can make these subjects breathe without mechanical ventilators.

Respiration is the self-controlled and unconscious movement of respiratory muscles, diaphragm, intercostal muscles, and scalene muscles. Among these, the diaphragm separating the chest and the abdominal cavity acts as the primary respiration muscle. In the resting state, diaphragm contraction accounts for 60~80% of the total amount of ventilation. As the diaphragm is innervated with phrenic nerve bilaterally, synchronous bilateral stimulation of phrenic nerves causes a contraction of each hemidiaphragm, which subsequently induces the expansion of thoracic volume, reduction of intrathoracic pressure, and inspiration of air. When the stimulation is stopped, phrenic nerves terminate the contraction diaphragm muscle. Relaxation of the diaphragm returns with increasing the intrathoracic pressure and expiration of air. The subject can control the repetition rate of breathing and the duration of inspiration by controlling the parameters of stimulating pulses.

Phrenic nerve stimulating systems consist of electrodes and receivers (Fig. 15.7). Implantable electrodes are placed bilaterally on phrenic nerves on each side of the cervical or thoracic region. Receivers are implanted in the accessible area over the anterior thorax or abdomen of both sides. Implantable receivers receive power and stimulating signals from external transmitters by radio frequency signal and convert them directly into stimulating electrical pulses and they do not include batteries. With a phrenic nerve

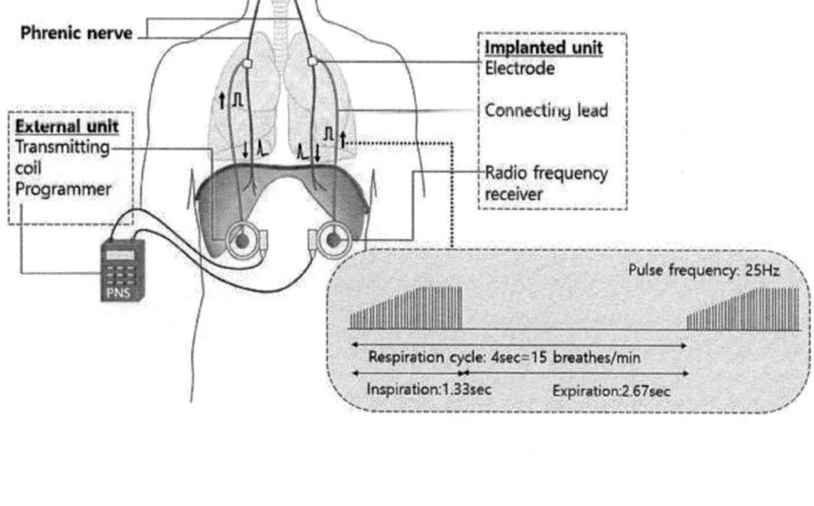

Fig. 15.7 Phrenic nerve stimulation for diaphragm pacing

stimulation system implanted, subjects are gradually reconditioned to improve their strength and endurance. Diaphragm pacing is first applied throughout the waking hours and it is extended during sleep hours. Subjects can achieve normal tidal volumes with full pacing time.

Stimulating pulses are applied only during the inspiration phase, and there is no electrical stimulation during the expiratory phase. To achieve enough inspiratory volume, multiple pulses should be delivered for each inspiratory phase. Tidal volume can be controlled by the pulse parameters. It increases monotonously with the increase of stimulating frequency and with the increase of inspiration period and converges to an asymptotic volume. As high frequency over 33 Hz and narrow pulse interval easily lead to rapid and obvious fatigue, it is safe and effective to adjust the stimulation pulse frequency to less than 25 Hz. Respiratory frequency of fewer than 15 breaths/min and inspiratory period with one-third of the respiratory cycle are preferred to make the diaphragm twitch slowly and fatigue-resistant (Gu et al., 2020).

2.2 Gastric Stimulation

Gastroparesis is a chronic symptomatic gastric motility disorder, defined as delayed gastric emptying of a solid meal in the absence of mechanical obstruction. Symptoms or signs of gastroparesis include nausea, vomiting, early satiety, fullness, anorexia, and/or weight loss (Cheng et al., 2021). Gastric electrical stimulation (GES) is the method of stimulating the smooth muscle of the stomach to treat the symptoms associated with gastropareses like chronic nausea and vomiting. It significantly reduces vomiting frequency in patients with gastroparesis of diabetic or idiopathic origin.

The main function of the stomach is to receive ingested food, to triturate successfully, and to empty nutrients into the small bowel. The contractile function of the stomach is performed under the neurological control of gastric rhythm originating from the interstitial cells of Cajal (ICC) located in the border of the fundus and the body of the stomach. This gastric rhythm has a rate of approximately 3 cycles/min and propagates circumferentially and distally down to the pylorus, the exit opening of the stomach. This rhythm is activated by the postprandial stimulation and also modulated by the vagus nerve and by the intestinal peptides, like motilin, ghrelin, and cholecystokinin. If there occurs a disruption in any of these processes, the timing and the strength of normal gastric contractility can be interrupted and it can result in gastroparesis. In diabetic patients, gastroparesis is thought to occur due to autonomic neuropathy, where the nerves controlling the involuntary bodily functions are damaged, and characterized by recurrent episodes of postprandial nausea, vomiting, epigastric pain, fullness, and early satiety (Reddymasu et al., 2010). While some minor cases can be treated with dietary change or some medications, severe cases are not treatable with these methods and invasive methods including gastrectomy are considered as remaining options. Gastric electrical stimulation can be considered an alternative treatment option less invasive than gastrectomy.

GES device consists of electrodes, leads, and a pulse generator (Fig. 15.8). A pair of electrodes is sutured to the muscular layer of the greater curvature of the stomach, approximately 10 cm proximal to the pylorus. A pulse generator is implanted subcutaneously in a pocket of the abdominal wall. Electrodes are connected to the implanted pulse generator through a lead. Electric pulses for gastric stimulation are delivered in two forms, gastric pacing and gastric stimulation. In gastric pacing, pulses are delivered in low frequency similar to the intrinsic gastric rhythm of 3 Hz but with higher strength of long pulse width of 300 ms enough to induce the contraction of gastric muscles. This is to restore intrinsic gastric contractile function by electrical stimulation. In gastric stimulation, pulses are delivered in higher than normal gastric rhythm but with lower current strength with

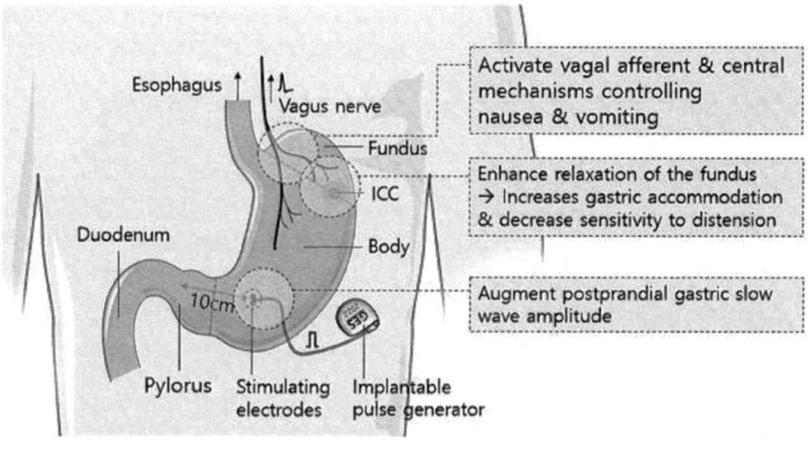

Fig. 15.8 Gastric electrical stimulation (GES) to reduce the severe vomiting and nausea accompanying gastroparesis. ICC: interstitial cells of Cajal

a shorter pulse width of less than 10 ms. This neuromodulation type of stimulation is targeted to stimulate the neural pathway, instead of directly stimulating efferently, to reduce the frequency of vomiting and nausea accompanying gastroparesis. It is considered that gastric stimulation enhances the relaxation of the fundus to increase gastric accommodation and decrease the sensitivity to distension. And, it also stimulates the vagus afferent nerves around the fundus area of the stomach resulting in the activation of the central mechanisms controlling nausea and vomiting. As the higher stimulating frequency has better efficacy, it is preferred over gastric pacing. For example, in high-frequency gastric stimulation, electrical pulses of 0.33 ms width and 5 mA amplitude can be delivered in 14 Hz frequency for the 100 ms of on-period and off for the following 5 s off-period. The actional mechanism of gastric stimulation is not completely understood. However, it is considered that the effects are mediated mostly by the activation of the afferent vagus nerve.

GES provides efficacy in gastric emptying and reducing upper gastrointestinal symptoms, even though it is not returned to the normal level. It also improves the quality of life, reduces the requirement for health care use, improves glycemic control in diabetic patients, reduces dependence on enteral or parenteral nutrition, and also improves nutritional status (Reddymasu et al., 2010).

2.3 Hypoglossal Nerve Stimulation

Hypoglossal nerve stimulation (HNS) is an electrical stimulation method for the treatment of obstructive sleep apnea (OSA). The hypoglossal nerve is the twelfth cranial nerve, and it starts at the base of the brain and travels down the neck, and branches out, ending at the base and underside of the tongue. The hypoglossal nerve controls the tongue and the muscles around the tongue and makes movement in the mouth for speaking and swallowing. Especially, the hypoglossal nerve controls muscles in the back of the throat. If they relax too much during sleep, the tongue can slide out of place and block the airway.

Obstructive sleep apnea (OSA) is a disorder occurring during sleep when the upper airway collapses and blocks the airflow to the lung. Apnea is defined as a complete breathing cessation for at least 10 s. Blockage of the airflow decreases the oxygen level in the bloodstream and reduces the oxygen supply to the brain. The brain senses a lack of oxygen and wakes the body up just long enough to take a breath, then falls back asleep. This cycle repeats more than 200 times in severe cases throughout the night and causes disruptive sleep. OSA can cause vehicle and workplace accidents due to daytime sleepiness. It also can cause worsening mood and memory, stroke, heart attack, and even death due to the insufficient oxygen supply during sleep. While continuous positive airway pressure (CPAP) is the primary treatment for the OSA, HNS is a beneficial option for some moderate and severe OSA patients, whose apnea-hypopnea indices (AHI)s are in the range of 15~65 events/h, with CPAP failure. HNS produces the mean AHI reduction of 18 events/h after 5 years (Costantino et al., 2020).

HNS devices consist of three implantable components, a respiratory sensor, a cuff electrode, and a pulse generator (Fig. 15.9). A respiration sensor is placed at the internal and external intercostal muscles in a lower intercostal space. The cuff electrode is carefully wrapped around the portion of the nerve that contains the fibers controlling protrusion. And, the pulse generator is placed over the pectoralis major muscle. During sleep, the respiratory sensor detects the initiation of inspiration and transmits the detected signal to the pulse generator. The pulse generator generates electrical pulses to stimulate the hypoglossal nerve. The stimulated hypoglossal nerve in turn controls tongue protrusion and upper airway dilation, ensuring the open airway for the inspiration.

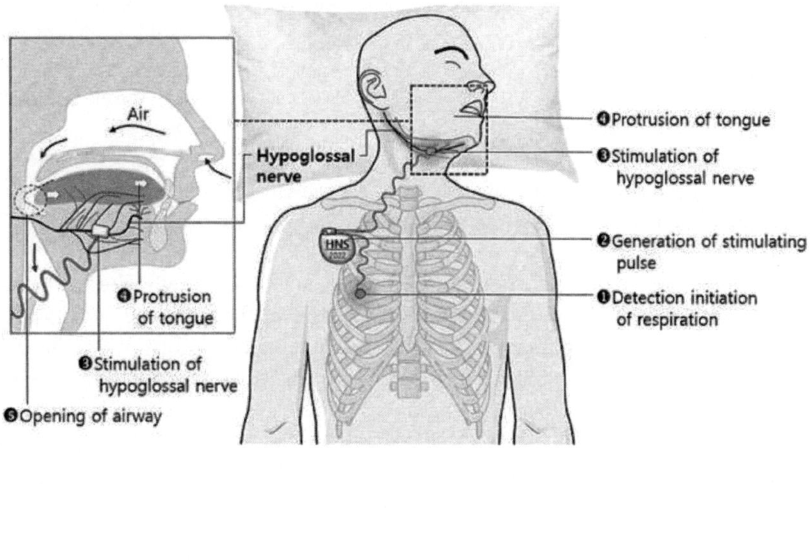

Fig. 15.9 Hypoglossal nerve stimulation to secure airway in an obstructive sleep apnea subject

Stimulating pulses are selected in the frequency range of 10~40 Hz with a pulse width range of 60~210 μs. One of the default settings is 33 Hz with 90 μs pulse width and the alternative is 40 Hz with 120 μs. As the amplitude of the stimulating current increases, three threshold levels are determined. The sensation threshold is the current amplitude level at which the patient first feels the stimulation. The functional threshold is the current amplitude level at which the patient's tongue motion is observed to protrude to the level of lower incisors. And, the sub-discomfort threshold is the current amplitude level at which the patient reports discomfort or declines further increases (Steffen et al., 2021). Stimulating amplitude is determined between the functional threshold and sub-discomfort threshold. As the alternative setting uses longer pulse width of 120 μs, it can induce the same function of tongue protrusion with lower current amplitude than the default setting of 90 μs pulse width. So, the alternative setting can be used for subjects who feel discomfort with the default setting.

3 Vagus Nerve Stimulation (VNS)

Vagus nerve stimulation (VNS) is a clinical therapy that treats diseases or symptoms by delivering electrical pulses to the vagus nerve. VNS is a unique method that uses peripheral intervention to treat the disorders occurring within the brain.

In the late nineteenth century, James Corning, an American neurologist, first introduced the idea of VNS. A series of experiments successfully stopped the induced seizures in dogs and promoted commercial VNS devices for the human subjects. In 1988, VNS for epileptic patients was first implanted clinically by Penry and colleagues (Penry & Dean, 1990). With several large clinical studies, VNS for epilepsy has received CE approval in 1994 and FDA approval in 1997, respectively. Currently, more than 80,000 epileptic patients have been treated with VNS. VNS for chronic treatment-resistant depression (TRD) is also FDA approved in 2005.

3.1 Vagus Nerve

The vagus nerve is the tenth cranial nerve (CNX) among the 12 pairs of cranial nerves that connect the brain to different parts of the body. It comprises a pair of left and right vagus nerves. The vagus nerve is the longest nerve of the autonomic nervous system connected to various organs including the heart, lungs, spleen, intestine, and other digestive tracts mainly for parasympathetic control. In addition to visceral connections, it also has branchial connections with muscles over the neck and face relating to social interactions like eye contact, voicing, speech, and recognizing and processing facial expressions. The vagus nerve has many branches. Actually, it is a family of neural pathways that originate from the brainstem and wander throughout the body to reach various organs in the body.

The vagus nerve includes afferent sensory fibers and efferent motor fibers. About 80% of the nerve fibers of the vagus nerve are sensory. Most of the sensory vagal fibers are small and unmyelinated, and they continually report the states of the visceral organs like the stomach, intestines, liver, pancreas, and spleen to the brain. Most of these visceral sensory fibers encode nonpain sensations, like hunger, satiety, and nausea. These fibers join with larger myelinated vagal afferents coming from baroreceptors and chemoreceptors in the heart at the cervical level. In addition to visceral sensory fibers, there are general sensory fibers that carry touch, pain, and temperature information from the ear and parts of the pharynx and larynx. Among these fibers, the auricular branch innervates the pinna and auditory canal, most superficially easy to be stimulated transcutaneously. The vagus nerve also includes 20% of nerve fibers as motor fibers and they mostly provide parasympathetic outflow to the organs. All these communications take place in an unconscious way out of a subject's thinking capacities. The brain is actually body-aware by the vagus nerve.

The vagus nerve plays an important role in balancing the body with parasympathetic controls against sympathetic activations. It does a key role in the subject's parasympathetic "rest and digest" nervous system. When a subject is activated in stressful, dangerous, and difficult times, it drives a subject toward a physiological state of relaxation. When the vagus nerve is not properly functioning, a subject is easy to be in excessive emotions and irrational states.

Vagal Tone

Vagal tone is a continuous activity of the vagus nerve representing the parasympathetic activation of the autonomic nervous system. This vagal tone is an internal biological process that is not controlled consciously. It is responsible for the regulation of many organs and body parts at rest and it results in various effects. It reduces heart rate and blood pressure. It increases vasodilation and promotes digestion.

Increasing the vagal tone activates the parasympathetic nervous system, and having a higher vagal tone means that the body can relax faster after stressful conditions. That is, a higher vagal tone is related to higher levels of psychophysical well-being, and a lower vagal tone is associated with states of chronic inflammation, negative mood, negative emotions, isolation, and heart attack. There is also a positive feedback loop between high vagal tone, positive emotions, and good physical health. In other words, the more subjects increase their vagal tones, the more their physical and mental health will improve, and vice versa (Kok et al., 2013). Accordingly, inappropriate or lowered vagal tones might lead the subjects to physical or psychological disorders more easily. It provides a clue that electrical stimulation of the vagus nerve can be effective when all other noninvasive treatments are ineffective for abnormally lowered vagal tones.

Vagal tone can be measured by monitoring the biological signals like heart rate, heart rate variability, breathing rate, and heart-lung coherence. For example, there is a slight increase in the heart rate during the inspiration phase of the lung and a slight deceleration during the expiration phase. The greater a subject has the difference in the heart rates between the respiration phases, the higher a subject is considered to have the vagal tone. Vagal tone is frequently used in assessing heart function, emotional regulation, and other processes altered by parasympathetic activation. Monitoring the vagal tone or parasympathetic activation via heart rate variability (HRV) is described in Chap. 7.

3.2 Vagal Nerve Stimulation for the Epilepsy

Epilepsy has a global prevalence of 0.5%~1.0%, and about one-third of epileptic patients are drug-resistant. VNS is currently guided as an adjunctive treatment for these patients of drug-resistant epilepsy, who are left with few therapeutic options and are not eligible for resective surgery. While the antiepileptic effect of VNS, interrupting the seizure and inducing an extended seizure inhibitory period, has been verified in multiple trials, its therapeutic mechanism is still incompletely understood. Studies demonstrated that the activation of afferent vagal fibers with electrical stimulation is influencing seizure-related circuitry directly and indirectly at subcortical and cortical levels (Krahl, 2012). VNS can increase the release of neurotransmitters and chemicals supporting the antiepileptic effect. The antiseizure effects of VNS may be also related to the altered activity in the limbic system, autonomic networks, and reticular activating system. Since hypersynchrony is a known feature of epilepsy, VNS might be effective in causing desynchronizing among the cortical cells. This mechanism is supported by the lowered functional connectivity among brain compartments during the VNS-on period compared to the VNS-off period in VNS-responding patients (Bodin et al., 2015).

The vagus nerve (VN) is a collection of axon fibers of afferent neurons and efferent neurons. It contains a large number of fibers and its number is dependent on the distance from the brain as a number of fibers branch out successively as it wanders around the body. Human VN contains 51,288 nerve fibers within a diameter of 2.017 mm at the cervical location and 28,714 nerve fibers within a diameter of 1.916 mm at the abdominal location on average (Stakenborg et al., 2020). These fibers are composed of myelinated fibers and unmyelinated fibers. In a cervical VN, where the stimulating cuff electrode is coiled for the epileptic patient, 54% of fibers are myelinated and 63% of myelinated fibers have small diameters.

As there are too many nerve fibers within a VN which are too small to be selected and each of their connections is not identified, electrical stimulation is delivered for the whole trunk of the vagus nerve. From the point of stimulation optimization, there are not enough selective parameters. Selectable parameters are restricted only to the location of electrical stimulation along with the VN, strength, and frequency of electrical stimulation and the side of the vagus nerve.

When electrical stimulation is delivered by an electrode encircling the VN, all the fibers within the VN trunk can be stimulated and they can induce effects not intended other than suppressing the seizure activity. To minimize the occurrence of unintended side effects, the selection of location should be determined not to include nerve fibers which might induce serious side effects. VNS electrodes are implanted on the cervical vagal trunk, approximately 8 cm above the clavicle. At this location, the auricular, laryngeal, and pharyngeal branches related to the voicing are not included in the VN trunk as well as the superior and inferior cervical cardiac branches related to the heart rhythm control. Electrical stimulation can be delivered to reduce the epileptic activity at this location with minimizing the expected side effects related to the speech generation and heart rhythm variation.

Another parameter that can be manipulated is the strength of the stimulation current. It is based on the different threshold characteristics depending on the types and the diameters of axon fibers. The threshold of unmyelinated fibers (C-fibers) for electrical stimulation is higher than myelinated fibers (A-, B-fibers). Also, thresholds tend to decrease as the diameter increases. While all the nerve fibers are activated with a strong current, only myelinated fibers selectively can be activated with the smaller strength of the electrical stimulating current. If the antiepileptic effect can be derived without the involvement of the unmyelinated C-fibers, VNS for epilepsy can be performed with smaller strength of electrical current without stimulating all of the unmyelinated fibers and reducing the side effects which can be caused by the activations of these fibers. Fortunately, VNS for epilepsy can be performed with a smaller electrical current without stimulating the unmyelinated fibers.

As the vagus nerve is composing a pair of left nerve trunk and right nerve trunk, stimulation can be applied on either side or both sides of the vagus nerve trunks. For most paired structures like lungs and bronchi, VN is connected only via ipsilateral nerve trunk while the midline and unpaired viscera are innervated bilaterally. Currently, VNS for seizure suppression is approved only for the left-sided vagus nerve. Right VN is innervated on the sinoatrial node and is closely associated with cardiac atrial function. Left VN relates to ventricular function, and its stimulation is thought less likely to have cardiac effects such as bradycardia. Experimental results on the dogs showed that right-sided VNS results in a greater degree of bradycardia as compared to left-sided VNS. However, electrical stimulation with left-sided VN does not produce any severe side effects and is equally effective as right-sided VNS (Krahl, 2012). Considering that electrical stimulations in a left-side VN are not optimal in many cases, right-sided or even bilateral VNS might be viable alternatives to increase the efficacy of VNS.

VNS reduces the incidence of complex partial seizures in the majority of patients. In 20~40% of patients, VNS results in a higher than 50% reduction in seizure frequency. Responsive rate increases as the period of stimulation increases. Also, the amount of antiepileptic drugs (AEDs) necessary to maintain satisfactory seizure control decreases with VNS (Tatum et al., 2001).

Vagal Nerve Stimulation Device

The VNS device consists of two components, a pulse generator and a lead (Fig. 15.10). The pulse generator is subcutaneously implanted under the left clavicle. Lead is connected to a pulse generator on one end and on another end, it has two spiral electrodes to be placed around the left vagus nerve. VNS can be implanted in a simple surgery of one and half hours in an outpatient setting.

VNS devices use a pair of helical cuff electrodes, anode and cathode electrodes. These electrodes cover the majority of the circumference of the vagus nerve and produce a largely uniform current density within the nerve. Current stimulation is applied in bipolar mode between anode and cathode electrodes. The cathode electrode is placed proximal to the head and the anode electrode is placed distal to the head. This electrode placement is to make an anode electrode block the transmission of generated action potential in an efferent direction and ensure stimulation of afferent fibers. The anchor tether also coils the vagus nerve and supports the fixation of electrodes. Electrodes are placed on the main vagus nerve below the junction where superior and inferior cervical cardiac vagus nerves are separated out.

The basic stimulation mode is intermittent 24 h stimulation. Pulses of 20~30 Hz having a pulse width of 250~500 μs are applied for a turn-on period of 30 s in a burst. And, a turn-off period of 5 min is followed. A turn-on period of 30 s and a turn-off period of 5 min are repeated consistently throughout the whole day. The amplitude of the current is increased gradually by 0.25 mA every month from 0.25 mA to the therapeutic levels of 1.25~2.0 mA. In addition to the basic stimulation mode, patients and caregivers can use the magnet to provide extra stimulation at the occurrence of an aura or seizure. By swiping the magnet over the pulse generator for 1~2 s, the extra routine of stimulation can be initiated. Recent models adopt a closed-loop stimulation mode. VNS device detects the seizure by increased heart rate accompanying the seizure. When the detected heart rate is above 100 beats/min and with at least a 55% or 35 beats/min above the baseline, the device recognizes it as a biomarker of seizure occurrence and initiates a routine of extra stimulation for improved seizure control (Pérez-Carbonell et al., 2020).

3.3 Vagal Nerve Stimulation for the Depression

Depression is a common and serious medical illness that negatively affects feeling, thinking, and acting. It is one of the major healthcare issues in modern times and is ranked as the single largest contributor of "Years lived with disability" worldwide. It is of high burden for the patients as the disease itself, impaired daily functioning, and social relationships cause significant adverse consequences for patients, their families, and society. As depression also increases the

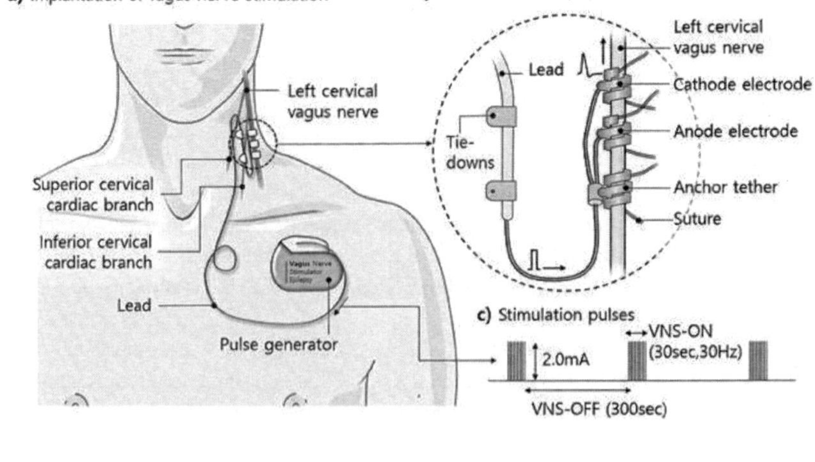

Fig. 15.10 Vagus nerve stimulation for epilepsy. (**a**) Implantation of vagus nerve stimulation device, (**b**) stimulating electrodes, (**c**) stimulating pulses

mortality risk significantly due to accompanying medical disorders and increased suicidality, treatment in the earliest stage is required.

Fortunately, there are three common modalities for the treatment of the major depressive disorder (MDD). Medication with antidepressant drugs is usually the first-line treatment. However, treatment-resistant depression (TRD) not responding to at least two different antidepressants occurs commonly in up to 30% of the treated MDD patient population. The second treatment modality is manual-based psychological therapy with the most persuasive evidence for cognitive behavioral therapy (CBT). The third possible treatment modality for MDD is to use neurostimulation strategies including electroconvulsive therapy (ECT) and repetitive transcranial magnetic stimulation (tTMS). This treatment modality is specially targeted the MDD subjects who are insufficiently responsive to medication and/or psychosocial interventions and require an immediate treatment response. VNS for depression is approved as an adjunctive treatment option for patients with treatment-resistant depression (TRD), who are experiencing a major depressive episode and not responding to four or more adequate antidepressant medications.

VNS was first approved for the treatment of epilepsy. The first efficacy of VNS for the subjects with depression was recognized during the clinical treatment of epileptic patients with VNS. In these patients, mood improvement was observed even in the absence of better control of seizures after VNS implantation. In the following clinical trials, significant mood improvement was found in the VNS group at three months, which appeared to be independent of any improvement in seizure control, suggesting that VNS was having a separate and distinct effect on depressive symptoms (Harden et al., 2000).

For the treatment of depression with VNS, the same VNS device used for the treatment of epilepsy is used. The placement of electrodes and stimulating parameters are almost similar to VNS for epilepsy. While the stimulating parameters can be varied depending on the patient's conditions, typically VNS is turned on for 30 s with 20 Hz pulses of 500 μs widths and turned off for the next 5 min. Pulse amplitude is gradually increased starting from 0.25 mA to a comfortable tolerance level between 0.25 and 3.5 mA. Subjects treated with VNS as an adjunctive treatment generally show better clinical outcomes compared with the subjects treated as usual. VNS group shows a significantly higher 5-year cumulative response rate of 67.6% compared with 40.9% and a significantly higher remission rate of 43.3% compared with 25.7% of the treatment-as-usual group. VNS also shows a significantly increased 5-year cumulative response for electroconvulsive therapy (ECT) nonresponders with 59.6% over 34.1%, as well as ECT responders with 71.3% over 56.9% compared to the treatment-as-usual group (Aaronson et al., 2017).

3.4 Transcutaneous Vagus Nerve Stimulation

Transcutaneous vagus nerve stimulation (tVNS) is the method of stimulating the vagus nerve electrically over the skin instead of surgically implanting the pulse generator and electrodes (Fig. 15.11). Even though classical VNS is performed in a minimally invasive procedure in the outpatient setting, it cannot entirely escape from the inherent risks of surgery related to the location of implantation. As the placement of electrodes around the vagus nerve requires a surgical intervention including the dissection of the vagus nerve from the carotid artery, several adverse side effects may occur during and after the implantation surgery. Even though their incident rates are not high, these potential side effects include bradyarrhythmia, peritracheal hematoma, vocal cord dysfunction, and dyspnea. The potential of adverse side

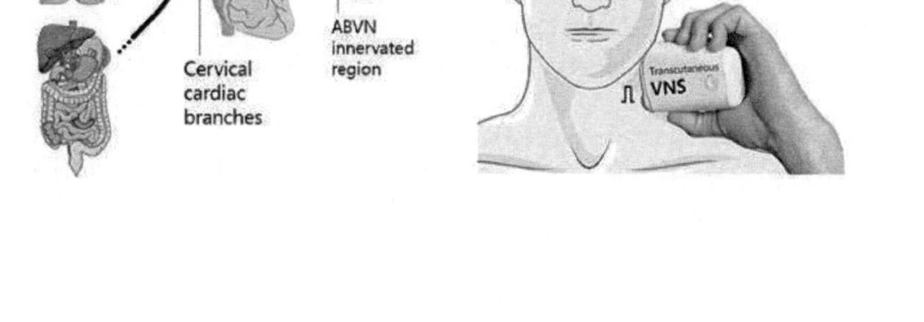

Fig. 15.11 Transcutaneous vagus nerve stimulation. (**a**) Auricular branch of the vagus nerve, (**b**) electrode positions for stimulation of auricular vagus nerve, (**c**) transcutaneous stimulation of the cervical vagus nerve

effects basically makes the candidate subjects hesitate and limits the extension of the application range. In addition to the risks accompanying the surgical intervention, the cost of the device and implantation procedure is prohibitively high to pay for VNS as an adjunctive treatment. As transcutaneous vagus nerve stimulation (tVNS) easily overcomes these disadvantages of implanting VNS, tVNS can be considered one of the first-line treatment options in many applications.

There are two locations where the vagus nerve can be accessed transcutaneously. On place is over the left neck where the left cervical vagus nerve passes underneath. External stimulating electrodes are placed over the place where the electrodes for implanting VNS are positioned. However, the location of the vagus nerve within the carotid sheath about 2 mm beneath the skin makes the transcutaneous stimulation use in increased current amplitude and difficult to selectively stimulate. Noninvasive tVNS on cervical VN has received the approvals in Europe for several types of headaches, and reactive airway diseases as well as an adjunctive treatment for epilepsy, depression, gastric motility disorder, and irritable bowel syndrome (Yuan & Silberstein, 2016).

Another place is within an auricle where the cutaneous afferent auricle branch of the vagus nerve (ABVN) is distributed. ABVN originates from the jugular ganglion and branches from the main vagus trunk earlier than cervical cardiac branches and it is most prominently distributed in the deepest part of the auricle, including concha cymba, concha cavum, and tragus. Electrodes for stimulating ABVN are positioned usually in these places. ABVN comprises general sensory fibers and is one of the few branches containing no motor fibers. Due to this totally afferent characteristic of ABVN, stimulation on ABVN is also performed on the right side as well as on the left side in contrast to the left-only stimulation in cervical VNS. Stimulation of these afferent fibers produces similar treatment effects to those by the cervical VNS. tVNS on ABVN was approved in Europe for the treatment of epilepsy and depression in 2010, and pain in 2012 (Yuan & Silberstein, 2016).

Transcutaneous cervical vagus nerve stimulation is FDA approved for the treatment of primary headache and medication overuse headache in adults. Stimulating modes are different from the implanting VNS. Generally, tVNS is applied in increased intensity with several sessions per day. Conductive gel is used to decrease the contacting resistance of the tVNS electrodes to the skin. The intensity of stimulation is controlled by the user under the range of 24 V and 60 mA and the stimulation duration typically lasts for 120 s per session. During the stimulation, pulses are applied at 25 Hz having a 1 ms pulse width. In many trials, subjects were treated with 6~12 sessions per day.

Transcutaneous stimulation of ABVN received the CE mark for the treatment of resistant epilepsy. Stimulation intensity is adjusted under 25 V by the user until a slight tingling or pulsating sensation is perceived at the stimulation site. Stimulation pulses are applied at 25 Hz having a 0.25 ms pulse width. Each session for the tVNS lasts at least an hour and 3~4 sessions are performed per day.

Stimulating parameters can be different between the locations of transcutaneous stimulation. Stimulating current is generally determined just below the pain threshold considering the individual sensitivity to the current. Due to the different stimulation tolerance among individuals, stimulation amplitudes can vary over a wide range from 0.5 mA to 12 mA (Yap et al., 2020). The frequency of stimulating pulses is determined typically in the range of 20~30 Hz, as a stimulating frequency higher than 50 Hz can cause major or irreversible damage to the vagus nerve during stimulation. Currently, optimal parameters for the tVNS stimulation are not determined yet, because the functional mechanisms inducing the efficacy of tVNS are not completely understood and tVNS is used for various ailments whose anticipated primary efficacy is different.

3.5 VNS and Electroceutical

The electroceutical is describing the devices that treat diseases and disorders with electrical stimulating pulses. This term is coined by their similarity to pharmaceutical in the way of therapeutic activation for the diseases. As pharmaceutical is dissolved in the bloodstream, conveyed to all parts of the body including the targeted locations along the vascular network, and do the intended pharmaceutical function there, applied electrical stimulating pulses are conveyed to various parts of the body including the targeted locations via neural fiber network and do the intended neural functions there. When the stimulating nerve is precisely selected, electroceutical can more specifically target the location related to the disease than pharmaceutical and minimize the unintended side effects.

In a broad sense, electroceutical includes cardiac pacemakers for the heart, deep brain stimulator for the brain, cochlear implant for the ears, and all the devices using the electrical stimulations for their treatment goals. However, VNS takes the main portion of electroceutical because it extends the range of electrical stimulation treatment to almost all organs innervated with the vagus nerves. Vagus nerve stimulation is attractive for electroceutical therapeutic approaches because the vagus nerve and its branches are typically located favorably for surgical interventions compared to the organs, like the brain and heart, requiring invasive surgical approaches.

In addition to the therapeutic effects on epileptic and depressive patients, the effects of VNS have been confirmed in relation to other disorders (Fig. 15.12). Cervical VNS has

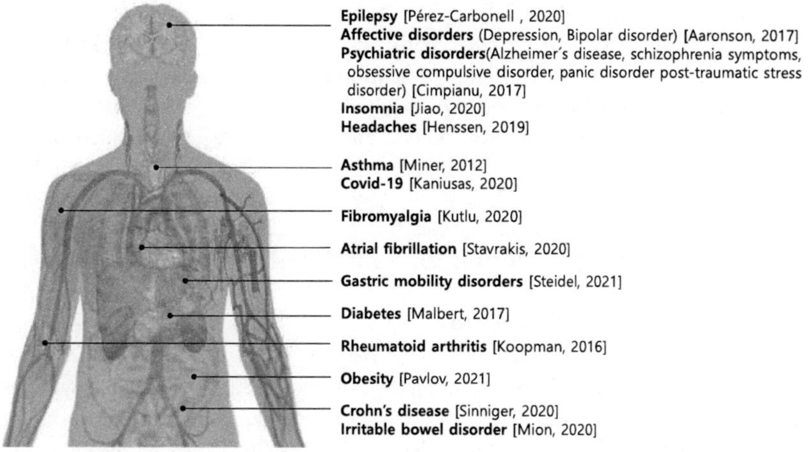

Fig. 15.12 Disorders expecting to be treated with electroceuticals of vagus nerve stimulations

induced anti-inflammatory effects, which are measured by a reduction in systemic inflammatory cytokines, which are signaling molecules that mediate and regulate immunity and inflammation. VNS has proved its effectiveness in the treatment of early-stage rheumatoid arthritis and Crohn's disease using a similar VNS stimulation setup for epileptic patients (Sinniger & Pellissier, 2020). Rheumatoid arthritis is a long-term autoimmune disorder primarily affecting joints with warm, swollen, and painful joints and Crohn's disease is another immune disease that involves inflammation of the intestines. The effectiveness of VNS treatment on these inflammatory diseases suggests the broad potential of VNS for many diseases with inflammatory components including cardiac disease, obesity, diabetes (Malbert 2017), and cancer (Horn et al., 2019). Unlike most electroceutical effects, the anti-inflammatory effect of the VNS extends for many hours after a brief stimulation. For example, in humans, 30 s of VNS produces and maintains a decreased level of inflammatory cytokines up to 4 h after stimulation (Koopman et al., 2016).

Stimulation effects also can be expected by stimulating the right side of the vagus nerve instead of the left vagus nerve and by targeting the efferent fibers instead of afferent fibers. Right cervical VNS is being studied for the treatment of heart failure. As there is an imbalance of parasympathetic and sympathetic activations in heart failure subjects, VNS can reestablish the diminished parasympathetic activation. In this trial, electrodes are designed and positioned to stimulate mostly the efferent fibers innervated in the atrium and ventricles of the heart. Anode electrode is positioned proximal to the head to block the afferent flow of stimulating pulses. Even though the efficacy of this trial for reducing heart failure is on its way to verification, it has extended the scope of VNS to right-side stimulation and efferent fiber stimulation.

VNS and the pharmaceutical have one common characteristic. They are applied at remote locations from the site of symptoms. Drugs are administered in various routes by swallowing, injection, and inhalation. Then, they flow following the transmitting network of the body and reach the target sites. This is the characteristic differentiated from the surgical treatments which directly act on the sites of the symptoms. VNS is also administered at remote locations away from the sites of symptoms like the brain or heart. Generated pulses by stimulations are transferred along with the neural network and reach the target site. This characteristic is also differentiated from other electrical stimulation methods like a cardiac pacemaker and a deep brain stimulator in which the stimulating electrodes are positioned directly on the sites of symptoms. In this point of view, electroceuticals can be restricted to the methods having the characteristic of stimulation in remote sites.

Applications of VNS are being extended. With the introduction of transcutaneous VNS, VNS is more easily tried to relieve the ailments bypassing the risk of implanting surgery. Currently, VNS is approved for many disorders as first-line treatment or adjunct treatments and is actively being studied for various disorders toward clinical approval starting from the classical application of VNS to epilepsy and depression.

VNS is expected to be used for the treatment of various diseases including primary headaches including migraine, medication overuse headaches (Henssen et al., 2019), reactive airway disease including asthma (Miner et al., 2012), gastric motility disorders (Steidel et al., 2021) including gastroparesis, irritable bowel syndrome (Mion et al., 2020), bipolar disease, Alzheimer's disease (Cimpianu et al., 2017), obesity (Pavlov 2021), impaired glucose tolerance, fibromyalgia (Kutlu et al., 2020), traumatic brain injury, stroke, cerebellar tremor, involuntary movement disorders, insomnia (Jiao et al., 2020), rheumatoid arthritis, osteoarthritis, atrial fibrillation (Stavrakis et al., 2020), and Covid-19 (Kaniusas et al., 2020). With this diversity of treatable diseases, a single treatment of VNS might be used to treat several disorders and symptoms, like heart disease, depression, insomnia, painful joints, and Covid-19, simultaneously.

4 Electrical Stimulation for Pain Relief

4.1 Spinal Cord Stimulation

Spinal cord stimulation (SCS) is the method of stimulating the spinal cord with surgically placed electrodes and a pulse generator (Fig. 15.13). A stream of electrical pulses flows through the thin wire from the pulse generator to the electrodes positioned over the nerve fibers in the spinal cord. SCS traditionally used to relieve various kinds of pains. However, recently its application is extending to various indications including neurological and cardiac disorders (Harmsen et al., 2021).

Relieving pain was the first practical application of electricity. It has a long history from ancient times when the people were first shocked by electric fishes. Electricity derived from the alive electric fishes has been used to treat the subjects having pains of gout, arthritis, and headaches. With the development of artificial electricity and understanding of electricity, the use of electricity for pain relieving has parallelly progressed. While the electrotherapy for pain was nearly eliminated in the middle of the twentieth century due to the insufficient evidence for therapy, its revival was encouraged by the successful application of cardiac pacemakers in the 1950s. The gate control theory of the pain proposed by Melzack and Wall in 1965 triggered the development of conventional Spinal Cord Stimulation (SCS). In 1967, Shealy first surgically implanted the electrodes and noted that paresthesia elicited by electrical stimulation inhibited deep pain due to metastatic lung cancer. Since then, technology sophisticatedly developed to the current epidural spinal cord stimulation (SCS) devices which can be implanted with minimized surgical intervention.

SCS is primarily indicated to treat the chronic pain arising from failed back surgery syndrome (FBSS), which is the persistent pain following the low back surgery, and the pain arising from angina or peripheral vascular disease. It is also actively used to treat pains of various causes including complex regional pain syndrome (CRPS), postherpetic neuralgia, postamputation pain syndromes, visceral pain, ischemic pain and to relieve pain in patients with spinal injury (North, 2008). In many of these cases, pains are devastating and intractable with medications including opioids. Treatments with SCS are very effective in most indications. Not only SCS can reduce the level of pain, but it also decreases or eliminates the use of pain medication. So, SCS has been successfully used over the last half-century as an adjunctive treatment of refractory pain syndromes not amenable to conservative therapy (Caylor et al., 2019).

4.2 Mechanism of Pain Relieving

Spinal cord stimulation is basically based on the management of pain by the gate control theory. The gate control theory of pain is proposed in 1965 by Ronald Melzack and Patrick Wall and states how pain is perceived in the brain. It states that in each dorsal horn of the spinal cord segment, there exists a neuronal connection acting as a "gate" for the passage of pain from the periphery to the brain (Fig. 15.14). The control of this "gate" depends upon the relative activation of large-diameter myelinated (Aβ) fibers and small-diameter unmyelinated (Aδ and C) fibers. Pain signals generated from the nociceptors are transmitted via small-diameter (Aδ and C) fibers at a slow speed toward the spinal cord. If there is no inhibitory input in the connection to the secondary neuron, the "gate" is not closed and pain signals are transmitted via the secondary neuron to the thalamus, and pain is perceived

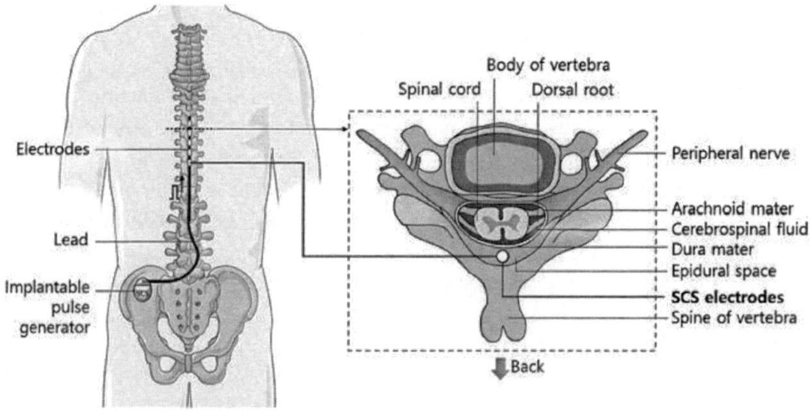

Fig. 15.13 Percutaneous spinal cord stimulation with electrodes implanted in epidural space

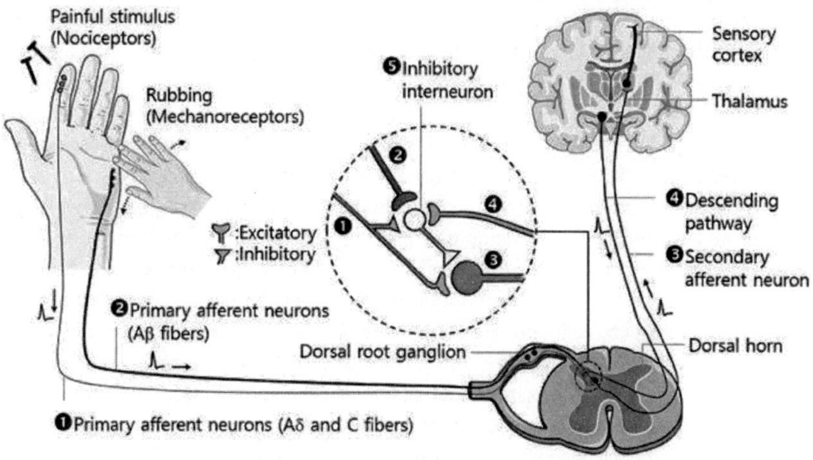

Fig. 15.14 Gate control theory of pain. If a large-diameter Aβ fiber (❷) is simultaneously activated by rubbing the injured region during the pain is transmitting through the small-diameter Aδ or C- fibers (❶), it excites inhibitory interneuron (❺) to close the gate for the transmission of pain signal to secondary afferent neuron (❸)

in the brain. However, if there is a simultaneous activation of large-diameter (Aβ) fiber transmitting the nonpainful signals with fast speed toward the spinal cord, like sensations from mechanoreceptors, it additionally activates inhibitory interneuron at the dorsal horn of the spinal cord on its own connection to the secondary neuron. The activated inhibitory interneuron closes the gate by releasing the transmitters to prevent the activation of the secondary neuron from transmitting the pain signal. Actually, the inhibitory interneuron also receives signals from the small-diameter fibers by inhibitory connection as well as from the large-diameter fibers by excitatory connection. Inhibitory interneuron generates the output based on the dominance of received inputs. When the pain is strong and the received input via an inhibitory connection from the small-diameter fiber is dominant, it does not generate the output signal to close the gate. However, when the received input from the large-diameter fiber is stronger than the input via an inhibitory connection from pain-transmitting fiber, the inhibitory interneuron generates an output signal to close the gate. Gate theory of pain explains why rubbing an injured area helps to reduce the pain. This inhibitory interneuron also receives input coming from the brain via descending pathways. That is, pain transmission can be gated by the brain functions, too. While optimistic feelings, concentration, and physical activity tend to close the gate, emotional stress and tension, boredom and lack of activity tend to open the gate for pain transmission. It also provides a physiological explanation for the observed effect of psychology on pain perception. This theory is the basic rationale to reduce pain by stimulating the neighboring neurons with electrical signals in spinal cord stimulation (SCS) and transcutaneous electrical nerve stimulation (TENS).

In addition to the gate theory of pain, separate mechanisms have been proposed for the therapeutic effects of SCS in the treatment of neuropathic and ischemic pain. Neuropathic pains are originating from damage to the nerves. As large fibers that supposedly close the pain gate are extremely sensitive to neuropathic pain, gate control theory does not explain all the therapeutic efficacies of SCS. Because 30 min of SCS can cause sustained pain relief lasting 1~3 h, SCS also likely alters the concentration of a multitude of neurotransmitters and neuromodulators in the dorsal horn of the spinal cord that interact in a complicated manner. Ischemic pain is originating when circulatory demands of tissues exceed blood supply. It is described as sharp, aching, heavy, and tiring pain. Relieving ischemic pain by SCS is partly explained by the vasodilatation effects of SCS (North, 2008). SCS increases blood flow by reducing vascular resistance by suppressing the sympathetic tone of vasoconstriction and increasing the concentration of vasodilator mediators.

4.3 Finding the Location for Electrical Stimulation

Nociceptors sensing the pain are distributed all over the body to sense noxious inputs from external and internal environments. Acute sharp pains are sensed and transmitted by the primary afferent of Aδ fibers, which are large and myelinated fibers. Slow-onset, dull, lingering, and achy pain are sensed and transmitted by the primary afferent of C-fibers, which are small and unmyelinated. Both of these primary afferent fibers have their cell bodies in the dorsal root of the ganglion and have synapses to the secondary afferent neuron in the dorsal horn of the spinal cord, where the gating for the pain transmission also occurs. The dorsal columns of the spine are also predominantly populated with projections of large-diameter pressure and touch sensing fibers that enter the spinal cord. To stimulate the nerves to gate the pain transmission, the dorsal side of the spine is the optimal place. So, the electrodes stimulating the spinal cord are placed over the dorsal side to stimulate the nerves entering the dorsal horn of the spine. At any vertebrate level, the dorsal column carries information from nearly all body parts below that level. The practical issue is how to find the optimal stimulating level and position on the dorsal surface of the spine which minimizes the unintended side effects of electrical stimulation.

In finding the optimal stimulating locations, a favorable side effect, paresthesia, is used. Paresthesia is an abnormal sensation of tingling or buzzing with variable secondary sensations expressed with thumping, pulsing, throbbing, sharpness, and coolness (Moffitt et al., 2009). When the afferent nerves of large diameter, which might transmit the sensation of pressure or touch, are stimulated, the generated pulses can propagate in three directions. The first direction is to the inhibitory interneuron in the dorsal horn of the spinal cord which can induce the intended effect of gate closing for pain blocking. The second direction is toward the periphery in reverse to the ordinary propagation direction of action potentials. And the third direction is to the brain along the ordinary conduction pathway. Paresthesia is the perception of the action potentials generated by the electrical stimulations and propagated in the third direction to the brain. Sensed range of paresthesia can be used to select the optimal stimulating location for pain relieving. If the body parts feeling with paresthesia is exactly overlapped with the area of pain generating in a certain location, that location becomes the optimal electrical stimulation site. If the overlapped region is not satisfactory, the next location can be stimulated and compared again until to find the satisfactory stimulating location. Before permanently implanting the SCS device, optimal electrode location is sought in this way and verified through the temporary implantation period of a few days.

4.4 Stimulating Electrodes

Two types of electrodes, catheter-type electrodes and plate-type electrodes, are used for spinal cord stimulation (Fig. 15.15). Both types of electrodes are placed on the back of the spinal cord and do not penetrate any of the meninges of the spinal cord. They are placed out of the dura mater of the spinal cord. Electrodes are typically placed in the lower thoracic region for treating back and leg pain, and in the cervical region for treating neck and arm pain.

The catheter-type electrode is commonly used for trial stimulation or for permanent implantation. It has 4~16 cylindrical contacts at the tip part of the catheter. Catheter-type electrode is used to place the electrode percutaneously with a minimal incision at the back of the patient and is more appropriate for performing a trial to assess the candidacy for a permanent electrode implant. The catheter-type electrode is inserted longitudinally over several segments in the epidural space filled with adipose tissues within the bony structure. These electrodes are advanced over several segments in the epidural space, allowing testing of several spinal cord levels to assess for optimal electrode position under fluoroscopic guidance. The number of electrodes depend on how many segments of the spinal cord are to be covered. Generally, electrodes with larger spacing provide broader coverage, and electrodes with closer spacing allow better steering and electric field shaping. Percutaneous electrode tends to migrate due to their inherent flexibility, which is required for their insertion with a needle-type guidance tool. As electrical current is distributed circumferentially around the electrode, catheter-type electrode is less energy efficient than plate-type electrode. The slim-line electrode is a plate-type electrode with a narrow width to be implanted percutaneously. It has contacting surfaces only on the front side and the backside is insulated. Slimmer and plate-type profile also provides less migration after implantation against the movement and compression. Multiple catheter-type electrodes can be implanted side-by-side to form a two-dimensional array. Activating electrodes can be selected from this electrode configuration by refining the stimulating

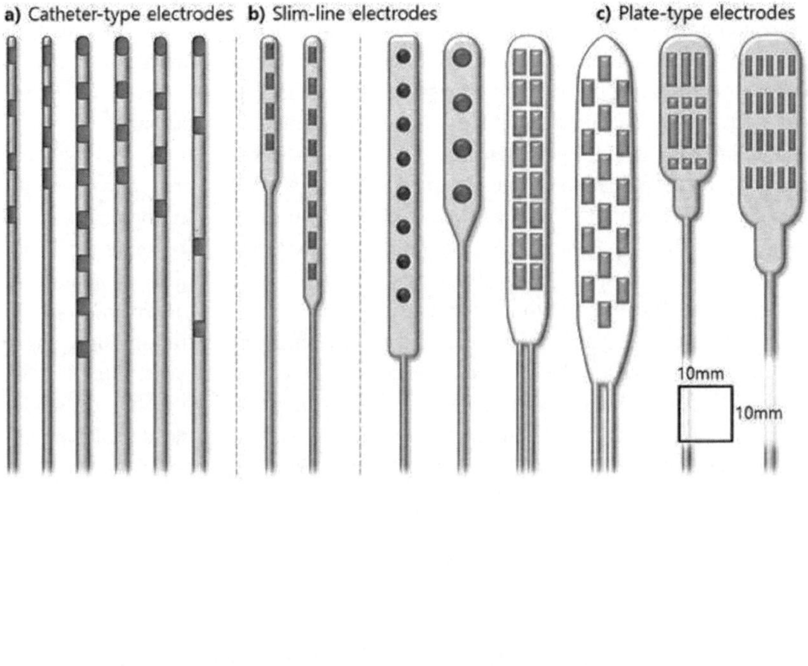

Fig. 15.15 Various shapes of electrodes for spinal cord stimulation. (**a**) Catheter-type electrodes implanted percutaneously, (**b**) slim-line plate-type electrodes for percutaneous implantation, (**c**) plate-type electrodes implanted with laminectomy

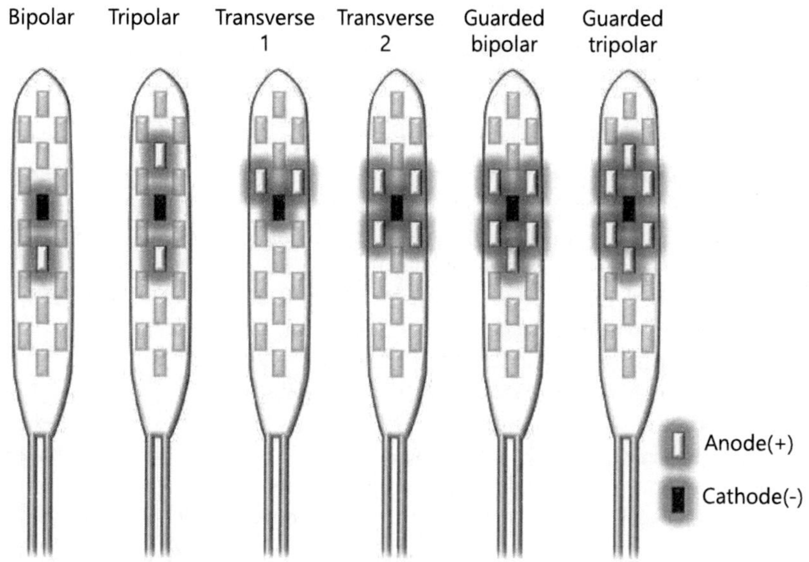

Fig. 15.16 Various electrode stimulating patterns in plate-type electrodes to optimize the region affected by stimulation

region and minimizing the side effects. By using the lateral electrodes as guard electrodes, recruitment causing unintended side effects can be mitigated. However, appropriate spacing between the electrodes is difficult to achieve and maintain.

To eliminate the problems occurring when the multiple catheter-type electrodes are used, a plate-type electrode, which has the electrodes in multiple columns of contacting points on the plate, can be used. However, plate-electrode cannot be inserted through a guiding needle into the epidural space. To prepare placement space for the plate of the electrode under direct vision, the surgical procedure of laminectomy is performed, which removes part of the vertebral bone to expose enough epidural space. The plate electrode has fixed spaces among the contacting electrodes and the backside of the plate is insulated to protect against unintended excessive stimulation. This electrode resides in an implanted place with inherent stability and a lesser propensity to migrate. It also provides a broader stimulation pattern with a lower level of stimulation current (Fig. 15.16). Among the various stimulation patterns, including the transverse stimulating and guided stimulating patterns, the optimal pattern can be selected. Even though the plate-type electrode is more invasive than the catheter-type electrode, the patterns of paresthesia induced by the plate electrodes are superior to ones produced by the catheter-type electrodes and the performance of the plate electrodes significantly exceeds that of the catheter-type electrodes.

An implantable pulse generator (IPG) is implanted into the buttock, abdomen, or subclavicular area. Implantation in the buttock is most common as this location makes it easier to connect the electrode leads to IPG in the same prone surgical position. In addition to stimulation circuits, IPG includes the battery and the logic circuits to program the optimal stimulating pattern. Stimulating parameters are modified by the transcutaneous wireless transmission of radiofrequency signals from the external unit. Batteries in the IPG can be a permanent or rechargeable one. Even though a rechargeable battery is not convenient with regular recharge, it has the advantage of smaller volume and can be used for an extended time without surgery for battery replacement.

4.5 Stimulating Parameters

Electrical current for SCS can be delivered in a number of ways and it diversifies the efficacy of SCS treatment. The basic control parameters of stimulating pulses are amplitude, frequency, pulse width, and bursting pattern.

Tonic stimulation is the basic stimulation pattern clearly demonstrated by the clinical efficacy of SCS and is most widely used and studied for the electrical treatment of pains. In tonic stimulation, stimulating pulses are continuously applied with a fixed frequency of 40~120 Hz, a fixed pulse width of 300~600 μs, and fixed current amplitude of 3.6~10 mA. This continuous electrical stimulation induces a noticeable sensation of paresthesia and the sensed paresthesia is used to optimally superimpose the region over the patient-specific pain region. According to the strength-duration curve principle of nerve stimulation, an increase in pulse amplitude and pulse width multiplicatively increases the number of fibers recruited. The stimulation level is optimized within the therapeutic window under the discomfort threshold.

Stimulation-induced paranesthesia is not pleasant for the patients, and sometimes the area of paresthesia does not overlap satisfactorily with the painful area of the patient.

This led to the search for new stimulating patterns that can relieve the pain without inducing paresthesia. New stimulating modes for paresthesia-free SCS include high-frequency stimulation, burst stimulation, and sub-perception stimulation.

High-frequency (HF) stimulation is a kind of tonic stimulation but with a higher frequency of 5~10 KHz compared to the traditional low-frequency tonic stimulation of 40~120 Hz. High-frequency stimulation with a frequency of 10 KHz, a pulse width of 30 μs, and a current amplitude of 1~5 mA has shown equivalent or better efficacy in the treatment of some pains without inducing any paresthesia (Ranjan et al., 2020). In burst stimulation, a burst has five pulses of 500~1 KHz frequency and the stimulation is applied with the bursting frequency of 40 Hz. Sub-perception SCS is the stimulating strategy using the multiple combinations of stimulating pulse parameters to deliver maximal amounts of energy to the nerves in the dorsal column spinal structures without discomfort or even a perceptible sensation. Sub-perception stimulation can be achieved utilizing stimulation frequencies in the range of 1~5 KHz.

4.6 Current Trends and Future Directions

Spinal cord stimulation (SCS) has been widely used for the treatment of intractable chronic pain syndromes of various origins. Recently, SCS is extending applicable indications other than pain relief including neurological, cardiac, and gastrointestinal disorders. While SCS is used as a treatment modality for pain associated with these new indications, SCS is also being used to treat primary nonpain symptoms of a specified disorder including major depressive disorder, hypertension, coma/decreased level of consciousness, and gait disturbances in Parkinson's disease (Samotus et al., 2018). SCS also can be used to improve the motor function of patients with spinal cord injury (SCI). SCS applied in pair with robot-assisted rehabilitation has helped paralyzed patients walk again (Wagner et al., 2018).

Many studies are focused on the settings of stimulating parameters to optimize the therapeutic outcomes for patients and reduce the adverse effects including paresthesia. Beyond parameter optimization, other cutting-edge SCS research is going on including the use of wireless stimulation devices and closed-loop systems (Harmsen et al., 2021).

5 Transcutaneous Electrical Nerve Stimulation (TENS)

Transcutaneous electrical nerve stimulation (TENS) is a non-invasive treatment stimulating peripheral nerves using the electrical current applied over the skin. TENS is mainly used for symptomatic management of acute and nonmalignant chronic pain by delivering current across the intact surface of the skin to excite underlying nerves. TENS pulses can be self-administered with inexpensive devices at home as there is no potential for overdose and there are few side effects. While pain relieving with electrical stimulation has a long history, modern TENS has emerged as a testing technique for the implantable spinal cord stimulation in 1974. During testing the tolerance to electrical stimulation, some subjects have reported that testing current itself relieved the pain.

5.1 Types of Transcutaneous Electrical Nerve Stimulation

TENS can be delivered using different stimulation parameters. The effect of TENS can be differentiated based on the parameters of electrical stimulating current and the location of electrode placement. There are three different types of TENS that have different stimulation parameter sets and activate different groups of nerve fibers. These are conventional TENS, acupuncture-like TENS (AL-TENS), and intense TENS. As the nerve fibers have different stimulating thresholds depending on their types, different types of TENS selectively stimulate different groups of nerve fibers by controlling the parameters of the stimulating current.

Figure 15.17 shows the strength-duration curve of different types of nerve fibers and a denervated muscle. Aβ nerve fibers have the lowest threshold and they are excited with low-intensity stimulation. Conventional TENS is targeting to stimulate the Aβ nerve fiber type only with a low level of current amplitude without stimulating other types of nerve fibers. Motor fibers have an increased threshold than Aβ nerve fibers but lower than Aδ nerve fibers. Increased stimulating current amplitude can excite the motor nerve fibers in addition to the Aβ nerve fibers. AL-TENS are targeting to stimulate this type of motor nerve to induce the extrasegmental analgesic effect without directly stimulating Aδ nerve fibers. When the strength of the stimulating current is increased more, Aδ nerve fibers are also recruited by the stimulation. Impulses generated in the Aδ nerve fibers are transmitted antidromically and block the pulses carrying the pain information. Intense TENS is targeting to stimulate Aδ nerve fibers and induce an analgesic effect by blocking the pain signals at the periphery. The strength of the stimulation is dependent on the delivered charge to the nerves, which increases multiplicatively with current amplitude and pulse width. So, the strength of the current can be selected by the combination of current amplitude and pulse width according to the strength-duration curve of the nerves. With the same stimulating current amplitude, different types of nerve fibers

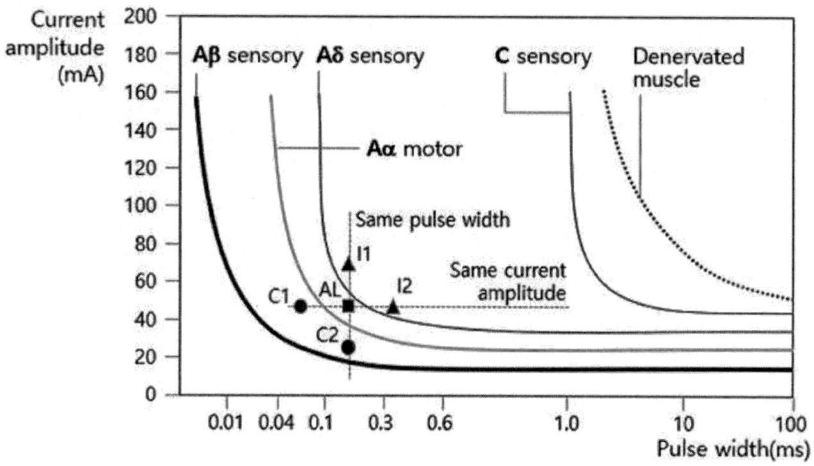

Fig. 15.17 Strength-duration curves of different types of nerve fibers and a denervated muscle. C1 & C2: stimulating parameter sets for conventional TENS, AL: stimulating parameter set for AL-TENS, I1 & I2: stimulating parameter sets for intense TENS

can be recruited by controlling the pulse width. And with the same pulse width, different types of nerve fibers can be recruited by controlling the amplitude of the current pulse. Figure 15.17 shows some example parameter sets for different types of TENS.

Conventional TENS is the most commonly used TENS type (Fig. 15.18a). Stimulating pulses are set at a high frequency of 10~250 Hz, a pulse width of 100~200 μs, and the intensity of the current is usually controlled below the pain threshold to produce a strong comfortable, and nonpainful paresthesia. This type of TENS is characterized by low-intensity and high-frequency stimulation. Conventional TENS can selectively excite the large diameter, low-threshold non-noxious afferent nerve (Aβ) fibers in a pain-related area without concurrently activating pain-transmitting small-diameter (Aδ and C) fibers and efferent motor fibers. As fibers of large diameter have shorter refractory times, they can generate nerve impulses in higher frequencies in response to high-frequency stimulation. Then, based on the gate control theory of the pain, excited Aβ fibers inhibit the activation of the second afferent neuron-transmitting nociceptive information from the smaller Aδ or C-fibers. The effect of conventional TENS is maximal when the stimulating electrode pads are placed around the painful area.

In acupuncture-like TENS (AL-TENS), the frequency of stimulating current is set in the lower frequency range of 2~4 Hz, and the stimulation is delivered with high intensity over the pain-sensing threshold and reaching the pain tolerance threshold (Fig. 15.18b). This type of TENS is characterized by the high-intensity and low-frequency stimulation inducing forceful but nonpainful phasic muscle contractions. Acupuncture-like TENS delivered over motor points stimulates Aα efferent fibers in addition to Aβ afferent fibers. Activated Aα efferent fibers by stimulation generate phasic muscle twitches and the discomfort of the muscle is sensed by the ergoreceptors within the muscle, which detect chemical by-products of skeletal muscle. Signals from the ergoreceptors are transmitted via small-diameter (Aδ and C) afferent fibers and these fibers subsequently activate extrasegmental descending pain inhibitory pathways to produce a spatially diffuse analgesic effect (Kalra et al., 2001). This analgesic effect is more related to the diffuse noxious inhibitory control (DNIC) phenomenon, a strong noxious input causes the release of endogenous opioids in the pain-regulating center in the brainstem, which in turn results in a diffuse descending inhibition of nociception (Mokhtari et al., 2020). In this way, the analgesic effect produced by AL-TENS is extrasegmental which is similar to the analgesic effect occurring with acupuncture treatment. Instead of paresthesia, AL-TENS induces a painful but tolerable sensation under the TENS electrodes. Unlikely conventional TENS, placement of a stimulating pad at the painful area is not required. A stimulating pad can be placed away from the painful area, as the extrasegmental analgesia effect much less influences the area where the TENS stimulation is applied. AL-TENS can be used in patients not responding to conventional TENS and it is recommended to be less frequently used than conventional TENS.

Intense TENS applies current stimulation with a high frequency in the range of less than 200 Hz and with a high amplitude current below the pain tolerance threshold (Fig. 15.18c). Intense TENS is characterized by high-intensity and high-frequency stimulation. Similar to AL-TENS, intense TENS also induces segmental analgesia by the stimulation of afferent Aβ fibers and extrasegmental analgesia by the stimulation of efferent Aα fibers. In addition, intense TENS directly stimulates the afferent fibers of small diameter (Aδ and C) and relieves pain by peripheral mechanism. Stimulated Aδ and C afferent fibers generate nerve

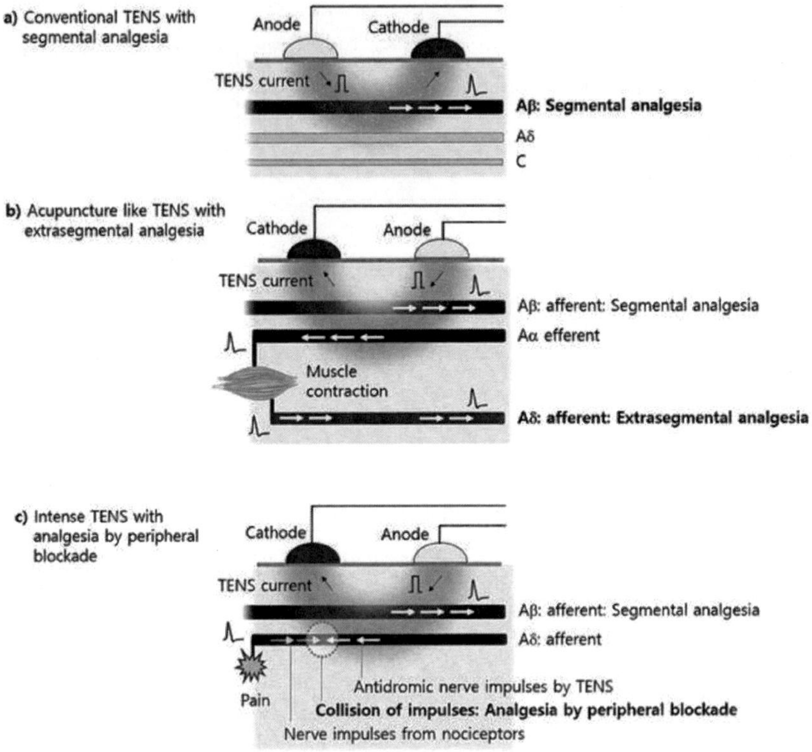

Fig. 15.18 Types of TENS and their mechanism of action. (**a**) Conventional TENS with segmental analgesia, (**b**) acupuncture-like TENS with extrasegmental analgesia, (**c**) intense TENS with analgesia by peripheral blockade

pulses in both directions, in antidromic as well as in orthodromic directions. Nerve impulses traveling antidromically toward the periphery collide with and extinguish afferent impulses arising from nociceptors. With these mechanisms, this technique can produce a spatially diffuse analgesic effect. For the sake of safety, intense TENS is suggested to be applied for a short period of less than a few minutes (Tashani & Johnson, 2009).

5.2 Indication of TENS

TENS is commonly used for the treatment of acute and chronic pain (Fig. 15.19). FDA approval for the use of TENS includes symptomatic relief of chronic intractable pain, relief of acute postsurgical and post-traumatic pain, and relief of pain associated with arthritis. However, based on potential benefits of TENS treatment including the safety, low cost, self-administering, and availability over the counter without a prescription, the use of TENS is widely accepted in the treatment for relieving pains of various origins. TENS is most commonly used for musculoskeletal pain. TENS is also applicable to alleviate central and peripheral neuropathic pains. Treatable neuropathic pains include pains related to diabetes, cancer, postherpetic neuralgia, spinal cord injury, stroke, and multiple sclerosis. A recent study reports the efficacy of TENS in relieving visceral pain and dysmenorrhea (Elboim-Gabyzon & Kalichman, 2020).

6 Sacral Nerve Stimulation

Sacral nerves are a collection of five pairs of nerves that exit the spinal column at its lowest point, the sacral vertebral area. Sacral nerves receive sensory information from and provide motor control to most of the pelvis and legs. Sacral nerve stimulation (SNS), which is also called sacral neuromodulation (SNM), is a treatment for subjects who experience bladder and bowel problems. The main indication of SNS is urinary incontinence, urgency, frequency, and urinary retention. It can also be the treatment for fecal incontinence. SNS is considered when the conservative options of behavioral and pharmacologic therapy are not successful.

Sacral neuromodulation therapy was developed in 1982 by Tanagho and Schmidt and received its first CE Mark in 1994 and FDA approval in 1997. Globally, about 300,000 patients have been treated with sacral nerve stimulation (De Wachter et al., 2020).

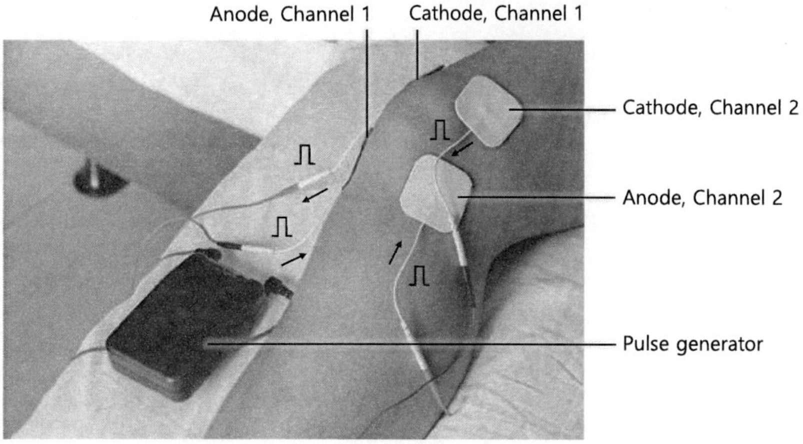

Fig. 15.19 Example of TENS for the treatment of knee pain with two pairs of stimulating electrode pads. (https://commons.wikimedia.org/wiki/File:Elettrostimolazione_con_TENS.jpg)

6.1 Micturition Reflex

Micturition or urination is the process by which the urinary bladder empties when it becomes full. The bladder normally accommodates up to 300–400 ml in adults. When the bladder is distended, it is sensed and the sensed signal is reported to the brain. With the perception of "full bladder" sensation, the micturition reflex starts and progressively increases its intensity as the volume increases. Basically, it is a visceral function that is under the control of the autonomic nervous system. However, micturition reflexes can be voluntarily suppressed up to the urine volume of 700 ml.

Micturition has two phases, filling and emptying the bladder. During the filling phase, the bladder is progressively filled below the threshold and the micturition reflex for emptying is unfavorable. However, the increase in bladder pressure is monitored by the stretch receptors and the afferent fibers. When the bladder volume reaches the threshold, it is sensed by increased bladder pressure. Then, emptying becomes favorable and the micturition reflex starts for emptying.

Filling and emptying of the bladder are controlled by the muscles of the bladder, detrusor and sphincter muscles. The detrusor muscle is a smooth and involuntary muscle found in the wall of the bladder. It is innervated sympathetically by the hypogastric nerve and parasympathetically by pelvic nerves. During the sympathetic dominance, it remains relaxed and does not contract to allow the bladder to fill. During parasympathetic dominance, the detrusor muscle contracts and decreases bladder pressure by releasing the urine. There are two sphincter muscles, the internal urethral sphincter and external urethral sphincter muscles. The internal urethral sphincter muscle is a smooth and involuntary muscle. It is innervated similarly to the detrusor muscle and functions in accordance with the detrusor muscle for filling and emptying. During the sympathetic dominance, it is contracted to close the urethra and allows the filling of the bladder. And, during the parasympathetic dominance for emptying, it relaxes and the urethra is open for urination. The external urethral sphincter muscle is a striated muscle that is innervated by the pudendal nerve and can be voluntarily controlled. Even though the bladder is filled with urine over the threshold volume, the micturition reflex for emptying can be suppressed by the extended contraction of the external urethral sphincter muscle. If the decision for urination is decided by the brain, an exciting signal to the external urethral sphincter muscle is blocked and the micturition reflex starts for emptying.

The micturition reflex center is located in the pons of the brainstem. It is activated via afferent signals from the urinary bladder as it is filling. Hypogastric nerve includes afferent fibers signaling the fullness and the pain of the bladder as well as sympathetic efferent fibers to detrusor muscle and internal sphincter muscle. The pelvic nerve includes afferent fibers reporting the degree of stretch of posturethra in addition to parasympathetic efferent fibers to detrusor muscle and internal sphincter muscle. The pudendal nerve also includes afferent fibers signaling the sensation from the external genitalia, anus, and perineum in addition to efferent fibers to the external urethral sphincter muscle. Based on the signals from these afferent fibers, the micturition reflex center recognizes the fullness of the bladder and activates parasympathetic tone to empty the bladder. Among these nerves, the pelvic nerve and pudendal nerves enter the spine via the foramina in the sacrum. Among the sacral foramina, the third sacral foramen (S3) is used to access these nerves and to place the stimulating electrodes.

6.2 Mechanism of Sacral Nerve Stimulation

The mechanism of action of sacral nerve stimulation is not completely understood. However, the SNS seems to function via afferent signaling rather than direct stimulation of the

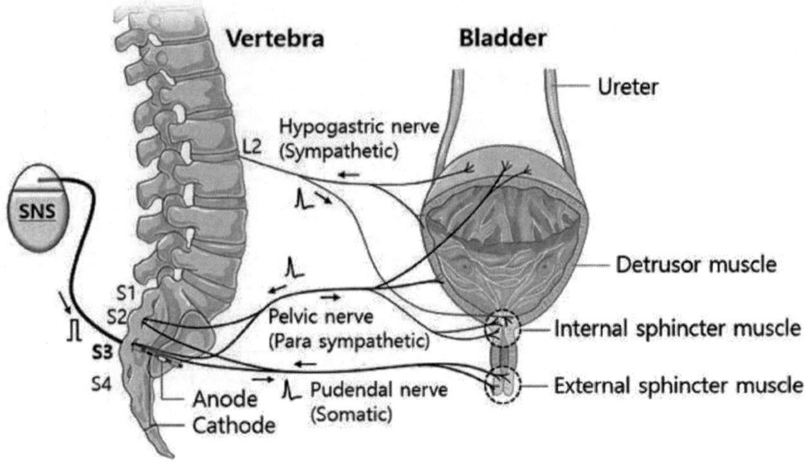

Fig. 15.20 Sacral nerve stimulation for the overactive bladder

detrusor or sphincter muscles. Afferent activation can modulate the forebrain structures involved in awareness and alertness. As C-fibers mostly respond to bladder distension and activate voiding reflexes, SNS is thought to block the overactive signal activities over C-fibers. In subjects with an overactive bladder, this modulation decreases the firing rate of the micturition center in the pontine, restores normal urinary continence, and results in clinical relief of urinary frequency and urgency. In subjects with an underactive bladder, who cannot sense bladder filling effectively, this modulation increases the awareness of bladder filling and pelvic floor contraction state. This awareness makes the micturition switch turn on at the appropriate time and the sphincter muscles relax effectively (Blok, 2018).

The mechanism of SNS action for fecal incontinence is also similarly understood. SNS blocks the overactive signal activities over C-fibers from the anal sphincter, rectum, and pelvic floor to the defecation reflex center in the pontine. This afferent modulation decreases the firing rates in the defecation center and restores normal fecal continence.

SNS is a minimally invasive surgical approach performed in outpatient procedures. As the sacral nerves include fibers connected to the lower extremity in addition to the bladder in the pelvic floor, their stimulation leads to the contraction of both the pelvic floor and various leg and foot muscles. Stimulating lead is placed through the third sacral foramen (S3) which provides maximal electrode exposure to nerves on the pelvic floor and minimal exposure to nerves in the lower extremity. After the lead is placed in the optimal position, patients undergo a trial period of stimulation for 1~2 weeks with an external pulse generator. When the incontinence score is improved by more than 50%, the pulse generator is permanently implanted.

6.3 Sacral Nerve Stimulation Device

SNS device is composed of the tined lead and implantable pulse generator (IPG) (Fig. 15.20). A lead has four stimulating electrodes and tines for anchoring the lead subcutaneously. While several stimulating modes are possible with these four electrodes, bipolar-stimulating mode, which uses the middle two electrodes as cathode and anode, is most commonly used. The amplitude of the stimulating current is delivered at sub sensory level, as low as 50% of the sensory threshold, which can prolong battery life as well as reduce the adverse effects associated with stimulation. The average amplitude of the stimulating current is 1.8 mA and in many cases amplitude of stimulation is controlled with stimulating voltages in the range of 0.75~1.5 V. While some studies report increased efficacy with a higher frequency of 31 Hz, a traditional stimulation frequency of 10~14 Hz is commonly used. Pulse width is almost fixed to 210 μs. Stimulation can be applied in cycling mode, which alternates turn-on and turn-off phases regularly. Stimulation in cycling mode can reduce the risk for minor neural damage and accommodation to the electrical stimulation as well as extend the battery life of implanted IPG.

References

Aaronson, S. T., Sears, P., Ruvuna, F., Bunkeret, M., et al. (2017). A 5-year observational study of patients with treatment-resistant depression treated with vagus nerve stimulation or treatment as usual: Comparison of response, remission, and suicidality. *The American Journal of Psychiatry, 174*(7), 640–648. https://doi.org/10.1176/appi.ajp.2017.16010034

Blok, B. F. M. (2018). Sacral neuromodulation for the treatment of urinary bladder dysfunction: Mechanism of action and future directions. *Bioelectronics in Medicine, 1*(1), 85–94. https://doi.org/10.2217/bem-2017-0003

Bodin, C., Aubert, S., Daquin, G., et al. (2015). Responders to vagus nerve stimulation (VNS) in refractory epilepsy have reduced interictal cortical synchronicity on scalp EEG. *Epilepsy Research, 113*, 98–113. https://doi.org/10.1016/j.eplepsyres.2015.03.018

Caylor, J., Reddy, R., et al. (2019). Spinal cord stimulation in chronic pain: Evidence and theory for mechanisms of action. *Bioelectronic Medicine, 5*(12). https://doi.org/10.1186/s42234-019-0023-1

Cheng, L. K., Nagahawatte, N. D., et al. (2021). Strategies to refine gastric stimulation and pacing protocols: Experimental and modeling approaches. *Frontiers in Neuroscience, 15*, 645472. https://doi.org/10.3389/fnins.2021.645472

Cimpianu, C. L., Strube, W., Falkai, P., et al. (2017). Vagus nerve stimulation in psychiatry: A systematic review of the available evidence. *Journal of Neural Transmission, 124*, 145–148. https://doi.org/10.1007/s00702-016-1642-2

Costantino, A., Rinaldi, V., Moffa, A., et al. (2020). Hypoglossal nerve stimulation long-term clinical outcomes: A systematic review and meta-analysis. *Sleep & Breathing, 24*, 399–411. https://doi.org/10.1007/s11325-019-01923-2

Daly, J. J., & Wolpaw, J. R. (2008). Brain–computer interfaces in neurological rehabilitation. *Lancet Neurology, 7*, 1032–1043.

De Wachter, S., Knowles, C. H., et al. (2020). New technologies and applications in sacral neuromodulation: An update. *Advances in Therapy, 37*(2), 637–643.

DiMarco, A. F. (1999). Diaphragm pacing in patients with spinal cord injury. *Topics in Spinal Cord Injury Rehabilitation, 5*, 6–20.

Elboim-Gabyzon, M., & Kalichman, L. (2020). Transcutaneous electrical nerve stimulation (TENS) for primary dysmenorrhea: An overview. *International Journal of Women's Health, 8*(12), 1–10. https://doi.org/10.2147/IJWH.S220523

Enoka, R. M., Amiridi, I. G., & Duchateau, J. (2020). Electrical stimulation of muscle: Electrophysiology and rehabilitation. *Physiology, 35*, 40–56. https://doi.org/10.1152/physiol.00015.2019

Gu, X. Y., Ren, S., et al. (2020). Restoration methods of respiratory function for spinal cord injury. *Mathematical Problems in Engineering, 2020*, 7398789. https://doi.org/10.1155/2020/7398789

Harden, C. L., Pulver, M. C., Ravdin, L. D., Nikolov, B., Halper, J. P., & Labar, D. R. (2000). A pilot study of mood in epilepsy patients treated with vagus nerve stimulation. *Epilepsy & Behavior, 1*(2), 93–99.

Harmsen, I. E., Hasanova, D., et al. (2021). Trends in clinical trials for spinal cord stimulation. *Stereotactic and Functional Neurosurgery, 99*, 123–134. https://doi.org/10.1159/000510775

Henssen, D. J. H. A., Derks, B., van Doorn, M., et al. (2019). Vagus nerve stimulation for primary headache disorders: An anatomical review to explain a clinical phenomenon. *Cephalalgia, 39*(9), 1180–1194. https://doi.org/10.1177/0333102419833076

Horn, C. C., Ardell, J. L., & Fisher, L. E. (2019). Electroceutical targeting of the autonomic nervous system. *Physiology, 34*(2), 150–162. https://doi.org/10.1152/physiol.00030.2018

Jiao, Y., Guo, X., et al. (2020). Effect of transcutaneous vagus nerve stimulation at auricular concha for insomnia: A randomized clinical trial. *Evidence-based Complementary and Alternative Medicine, 2020*, 6049891. https://doi.org/10.1155/2020/6049891

Kalra, A., Urban, M. O., & Sluka, K. A. (2001). Blockade of opioid receptors in rostral ventral medulla prevents antihyperalgesia produced by transcutaneous electrical nerve stimulation (TENS). *The Journal of Pharmacology and Experimental Therapeutics, 298*(1), 257–263.

Kaniusas, E., Szeles, J. C., et al. (2020). Non-invasive auricular vagus nerve stimulation as a potential treatment for covid19-originated acute respiratory distress syndrome. *Frontiers in Physiology, 11*, 890. https://doi.org/10.3389/fphys.2020.00890

Kok, B. E., Coffey, K. A., et al. (2013). How positive emotions build physical health: Perceived positive social connections account for the upward spiral between positive emotions and vagal tone. *Psychological Science, 24*(7), 1123–1132.

Koopman, F. A., Chavan, S. S., et al. (2016). Vagus nerve stimulation inhibits cytokine production and attenuates disease severity in rheumatoid arthriti. *Proceedings of the National Academy of Sciences, 113*, 8284–8289. https://doi.org/10.1073/pnas.1605635113

Krahl, S. E. (2012). Vagus nerve stimulation for epilepsy: A review of the peripheral mechanisms. *Surgical Neurology International, 3*(Suppl 1), S47–S52. https://doi.org/10.4103/2152-7806.91610

Kutlu, N., Özden, A. V., Alptekin, H. K., & Alptekin, J. O. (2020). The impact of auricular vagus nerve stimulation on pain and life quality in patients with fibromyalgia syndrome. *Biomedical Research International, 2020*, 8656218. https://doi.org/10.1155/2020/8656218

Malbert, C. H. (2017). Could vagus nerve stimulation have a role in the treatment of diabetes? *Bioelectronics in Medicine, 1*(1), 8. https://doi.org/10.2217/bem-2017-0008

Marquez-Chin, C., & Popovic, M. R. (2020). Functional electrical stimulation therapy for restoration of motor function after spinal cord injury and stroke: A review. *Biomedical Engineering Online, 19*, 34. https://doi.org/10.1186/s12938-020-00773-4

Miner, J. R., Lewis, L. M., Mosnaim, G. S., Varon, J., Theodor, D., & Hoffmann, T. J. (2012). Feasibility of percutaneous vagus nerve stimulation for the treatment of acute asthma exacerbations. *Academic Emergency Medicine, 19*, 421–429.

Mion, F., Pellissier, S., et al. (2020). Transcutaneous auricular vagus nerve stimulation for the treatment of irritable bowel syndrome: A pilot, open-label study. *Bioelectronics in Medicine, 3*(1), 4. https://doi.org/10.2217/bem-2020-0004

Moffitt, M. A., Lee, D. C., & Bradley, K. (2009). *Spinal cord stimulation: Engineering approaches to clinical and physiological challenges. Implantable neural prosthesis* (pp. 155–194). Springer. https://doi.org/10.1007/978-0-387-77261-5_5

Mokhtari, T., Ren, Q., et al. (2020). Transcutaneous electrical nerve stimulation in relieving neuropathic pain: Basic mechanisms and clinical applications. *Current Pain and Headache Reports, 24*, 14. https://doi.org/10.1007/s11916-020-0846-1

Nagai, M. K., Marquez-Chin, C., & Popovic, M. R. (2016). Why is functional electrical stimulation therapy capable of restoring motor function following severe injury to the central nervous system? In *Translational neuroscience* (pp. 479–498). Springer.

North, R. B. (2008). Neural interface devices: Spinal cord stimulation technology. *Proceedings of the IEEE, 96*(7), 1108–1119. https://doi.org/10.1109/JPROC.2008.922558

Pavlov, V. A. (2021). The evolving obesity challenge: Targeting the vagus nerve and the inflammatory reflex in the response. *Pharmacology & Therapeutics, 222*, 107794. https://doi.org/10.1016/j.pharmthera.2020.107794

Penry, J. K., & Dean, J. C. (1990). Prevention of intractable partial seizures by intermittent vagal stimulation in humans: Preliminary results. *Epilepsia, 31*, 40–43.

Pérez-Carbonell, L., Faulkner, H., Higgins, S., et al. (2020). Vagus nerve stimulation for drug-resistant epilepsy. *Practical Neurology, 20*, 189–198.

Ranjan, M., Kumar, P., Konrad, P., & Rezai, A. R. (2020). Finding optimal neuromodulation for chronic pain: Waves, bursts, and beyond. *Neurology India, 68*(8), 218–223. https://doi.org/10.4103/0028-3886.302465

Reddymasu, S. C., Sarosiek, I., & Mccallum, R. W. (2010). Severe gastroparesis: Medical therapy or gastric electrical stimulation. *Clinical Gastroenterology and Hepatology, 8*, 117–124.

Samotus, O., Parrent, A., & Jog, M. (2018). Spinal cord stimulation therapy for gait dysfunction in advanced Parkinson's disease patients. *Movement Disorders, 33*(5), 783–792.

Sinniger, V., & Pellissier, S. (2020). A 12-month pilot study outcomes of vagus nerve stimulation in Crohn's disease. *Neurogastroenterology and Motility, 32*, e13911. https://doi.org/10.1111/nmo.13911

Stakenborg, N., Pedro, J., Gomez-Pinilla, P. J., et al. (2020). Comparison between the cervical and abdominal vagus nerves in mice, pigs, and human. *Neurogastroenterology and Motility, 32*, e13889. https://doi.org/10.1111/nmo.13889

Stavrakis, S., Stoner, J. A., et al. (2020). TREAT AF (Transcutaneous electrical vagus nerve stimulation to suppress atrial fibrillation) A randomized clinical trial. *Journal of the American College of Cardiology EP, 6*(3), 282–291.

Steffen, A., Jeschke, S., et al. (2021). Impulse configuration in hypoglossal nerve stimulation in obstructive sleep apnea: The effect of modifying pulse width and frequency. *Neuromodulation, 2021*, 1–5. https://doi.org/10.1111/ner.13490

Steidel, K., Krause, K., et al. (2021). Transcutaneous auricular vagus nerve stimulation influences gastric motility: A randomized, double-blind trial in healthy individuals. *Brain Stimulation, 14*(5), 1126–1132. https://doi.org/10.1016/j.brs.2021.06.006

Tashani, O., & Johnson, M. I. (2009). Transcutaneous electrical nerve stimulation (TENS) a possible aid for pain relief in developing countries? *Libyan Journal of Medicine, 4*(2), 62–65.

Tatum, W. O., Johnson, K. D., Goff, S., Ferreira, J. A., & Vale, F. L. (2001). Vagus nerve stimulation and drug reduction. *Neurology, 56*, 561–563.

Thrasher, T. A., Flett, H. M., & Popovic, M. R. (2006). Gait training regimen for incomplete spinal cord injury using functional electrical stimulation. *Spinal Cord, 44*, 357–361.

Wagner, F. B., Mignardot, J. B., et al. (2018). Targeted neurotechnology restores walking in humans with spinal cord injury. *Nature, 563*(7729), 65–71.

Yap, J. Y. Y., Keatch, C., Lambert, E., Woods, W., Stoddart, P. R., & Kameneva, T. (2020). Critical review of transcutaneous vagus nerve stimulation: Challenges for translation to clinical practice. *Frontiers in Neuroscience, 14*, 284. https://doi.org/10.3389/fnins.2020.00284

Yuan, H., & Silberstein, S. D. (2016). Vagus nerve and vagus nerve stimulation, a comprehensive review: Part II. *Headache, 56*(2), 259–266. https://doi.org/10.1111/head.12650

Biological Effects of Electricity and Electromagnetic Field

16

In the many applications using electricity for therapeutic purposes, it is applied with precisely determined and verified protocols within the safety limit not to cause any harmful effects. However, there are chances that some amount of electricity is applied unintentionally by some faults. Depending on the strength of applied electricity, it may cause some harmful effects on the human body. The electromagnetic field ever presenting around our environment, even though we are not noticed it, is continuously applying electromagnetic energy to the human body. Depending on its intensity and duration of exposure, EMF can cause some biological and health-related effects.

1 Physiological Effect of Current

When the electricity is applied to the body, it can cause various physiological effects depending on its strength. To make current flow into the body, a current loop should be formed including the external electric source and the body. That is, there should be at least two contacting points between the external electric source and the body. When the nerves and the muscles are intentionally stimulated, contacting points are used to be the electrodes. Current flows via one of the electrodes and returns to the external unit via the other electrode. However, there can be unintended contacts to the external electricity by faults or by chances which used to result in electric injuries and hazards. As a person is touching the earth ground most of the time, simply touching an electric source at one more point can build a current loop through the body and earth ground. As the electricity is used all over our environments, accidents caused by unintended contact with the electricity can occur in any place including power plants, construction industries, streets, hospitals, and homes during daily living. In the USA only, there are approximately 1000 deaths per year as a result of electrical injuries and also at least 30,000 shock incidents per year that are nonfatal.

Among the variables describing the electricity applied to the body, the physiological effects of electricity are best explained in terms of current. As electrical behaviors of the nervous system are determined by the flow of charges, charged ions, and electrons, current can accurately describe the delivered amount of charges. So, it is the current that determines the physiological effects. When the electricity is applied in voltages, it should be calculated into the current with the resistances or impedances spread over the current loop.

The body has resistance to current flow. The major portion of the body's resistance to electric current flow is concentrated at contact points of the skin (Fig. 16.1). The resistance of dry hand may have more than 100 KΩ due to the thick outer layer of dead cells in the epidermis. However, internal resistance is about 300 Ω, due to the wet, relatively salty tissues beneath the skin. When the 100 V of external voltage is applied to the body through two contacting points over the skins having a resistance of 100 KΩ, there only flows a minute current of 0.499 mA, which is almost below the perception threshold.

$$\text{Current within the body} = \frac{100\ \text{V}}{100\ \text{K}\Omega + 100\ \text{K}\Omega + 300\ \Omega} = 0.499\ \text{mA} \quad (16.1)$$

If the skin resistance is reduced or bypassed by some reasons, like a cut, a deep abrasion, or wetness, the current flowing through the body tissue can increase dramatically. If the electricity is contacted through wet hands having decreased resistance of 1 KΩ, the current flowing through the internal tissue increases 87 times to 43.48 mA, which can contract muscles involuntarily and result in respiration paralysis.

$$\text{Current within the body} = \frac{100\ V}{1\ \text{K}\Omega + 1\ \text{K}\Omega + 300\ \Omega} = 43.48\ \text{mA} \quad (16.2)$$

K. S. Park, *Humans and Electricity*, https://doi.org/10.1007/978-3-031-20784-6_16

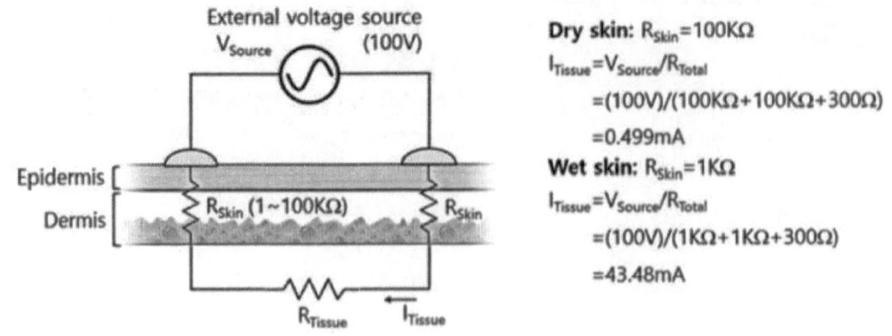

Fig. 16.1 Effect of skin-contacting impedance on the internal current flowing through the tissue

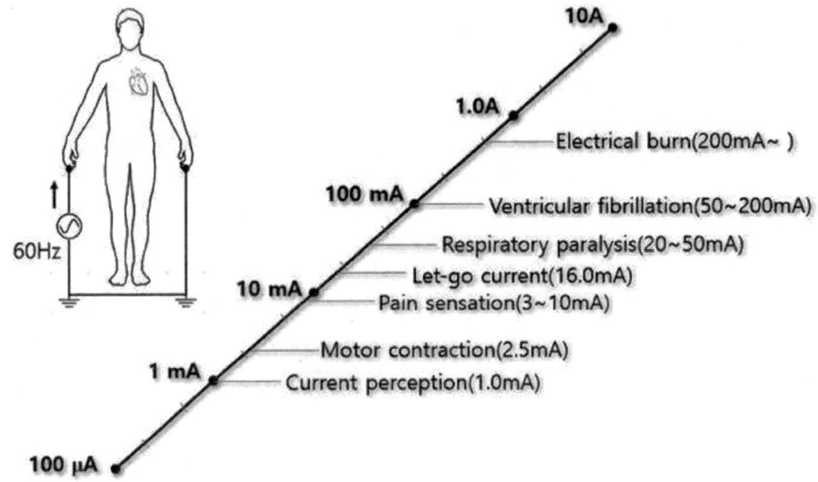

Fig. 16.2 Threshold of physiological effects of 60 Hz alternating current flowing through the body

When the external source of the same voltage is touched, the current flowing in the internal tissue is determined by the resistance of the contacting points. This is the reason that touching the power line with wet hands is much more dangerous. Skin does a critical role in protecting the body from electrical injuries at the first front to electrical exposure. Substantial reduction of skin conductance by moisture can convert what ordinary might be minor superficial skin damage into a life-threatening injury by allowing excessive current to flow into the deep internal part of the body including the heart. For electrical safety, most importantly, the subject should not be part of the current loop. If the chance of unintended contact with electricity is expected, insulation with gloves and shoes is compulsory to increase contacting resistance for safety.

When the electrical stimulation is applied to the nerves and muscle, controlling the stimulation with a constant current source is more stable and produces reproducible effects. However, it can cause a skin burn when the resistance of the skin is not lowered sufficiently. For example, if 10 mA of stimulating current is applied over the dry skin of 100 KΩ resistance, it generates a voltage of 1000 V across the skin which can easily make a skin burn. With this concern, in some stimulators, stimulation is controlled by the constant voltage source. Even though the skin resistance is not lowered sufficiently, the voltage over the skin does not rise over the preset constant voltage and does not generate a problem like skin burn. However, as the current flowing through internal tissue is varying depending on the unknown contacting resistance, an accurate amount of the internal current is not guaranteed and the performance of the treatment is less reproducible. When electric current flows through the body, it can cause various physiological effects depending on its intensity and frequency. Figure 16.2 summarizes the physiological effects of 60Hz alternating current flowing through the body.

1.1 Current Perception Threshold

When the current flows through the body tissue, it is sensed by the nerves first. While it does not excite the nerve or muscle, it is not sensed. After nerves are excited by the increased potential across the membrane, the action potential is generated and transmitted to the brain. Then, electricity is perceived.

Voltage is applied at two electrodes or entry points. The anode is the electrode where the positive voltage is connected and the cathode is the electrode where the negative voltage is connected. Theoretically, the direction of current flow indicates the flow of positive charges from the anode to the cathode (Fig. 16.3). However, in practice, negative charges of electrons flow in the reverse direction from the cathode to the anode. At the time of stimulation onset, there is an increase of electrons beneath the cathode and the positive ions beneath the anode. It increases the number of negative charges beneath the cathode which is also the outside of the underlying nerve fibers. An increase of negative charges outside of the underlying nerve fiber also can be thought caused by the rapid movement of positive ions to the cathode surface. As the potential of the outer side of the membrane is decreased, the potential across the membrane becomes increased. The resting potential of −70 mV referenced to the extracellular potential of 0 mV becomes −50 mV if the extracellular potential decreases to −20 mV due to the accumulated negative charges. When the membrane potential is depolarized across the threshold, the action potential is generated in the nerve fiber beneath the cathode.

Generated action potentials are propagating in both directions. They can be transmitted like intrinsic action potentials over the axonal nerve fiber by exciting the membrane of neighboring sites in one direction. However, it is difficult to be transmitted in the same manner toward another direction where the anode is located. Like the negative extracellular potential is developed in the nerve fibers underlying the cathode, the extracellular potential is increased in the nerve fibers underlying the anode. This can be caused by the rapid movement of negatively charged particles to the anode surface. Increased extracellular potential develops the potential across the membrane to a more negative direction and makes a fiber difficult to generate an action potential. Even though the action potential is propagated from the direction of the cathode, it is not sufficient to depolarize the membrane which is already hyperpolarized and the transmission of the action potential is less effectively transmitted or blocked in this direction.

The current perception threshold is the lowest current amplitude that a subject can detect. Sensitivity of the current including the perception is different among the subjects depending on the various factors including the sex, age, body weight, and chemical composition of the body. It is also greatly dependent on the frequency of applied electricity.

The current perception threshold is different between alternating current (AC) and direct current (DC). Membranes of nerve fibers are more easily stimulated with alternating current as the resulting extracellular voltages are alternating in both directions. With enough amount of alternating current strength, stimulation can be effective in every cycle up to the refractory limit of the nerve fiber. So, sensation or shock is continued during the contact is made to AC current. The perception threshold varies considerably among the subjects and is dependent on the measuring conditions. Generally, 1.0 mA is considered the perception threshold of 60 Hz AC current. Usually, subjects feel the AC current with a tingling sensation. In the experiment with 115 men measured with a hand-to-hand current loop, Dalziel derived the current perception threshold with a mean of 1.072 mA and with a range of 0.44~1.92 mA (Dalziel & Mansfield, 1950). Current lower than 0.4 mA was taken as the current level likely not perceived by the men, which is corresponding to the level that 99.5% of subjects did not perceive the current.

In contrast to AC current, humans are less sensitive to DC current. As there is no alternating of current directions, one type of charge is used to be built up in the extracellular space of the nerve fiber. DC current is sensed with slight warmth in the palm instead of the tingling sensation of AC current. Due to the additional capacitive component of the skin impedance, there occurs an abrupt capacitive current when the DC current is connected as well as it is disconnected. So, the perception of DC current is most sensitive at ON-time and OFF-time of current contact. In the same experiment, Dalziel derived the perception threshold for DC current with a mean of 5.2 mA and with a range of 2.1~12.6 mA (Dalziel & Mansfield, 1950). Similarly, DC current lower than 1.0 mA was taken as the current level likely not perceived by the men. The perception threshold for 60 Hz AC current is significantly lower in women compared to men, 0.85 mA

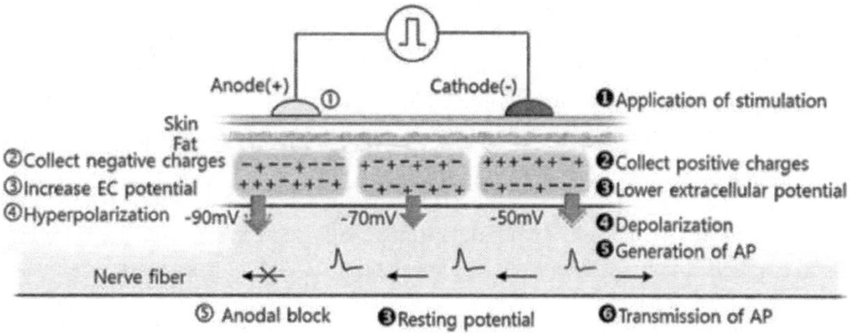

Fig. 16.3 Excitation of a nerve fiber by electrical stimulation

compared to 1.05 mA in men. This difference is derived from the characteristic difference between the sexes in body fat or body water percentages (Seno et al., 2019).

As the frequency increases, the perception threshold also increases. For AC current of 5 KHz, the mean perception threshold increases about sixfolds to 7.0 mA compared to 60 Hz AC current. For currents of higher frequencies above 100 KHz, the sensation of current changes from tingling to heat (Dalziel, 1972).

1.2 Threshold of Reaction Current

Reaction current is the current which can stimulate the motor nerves and cause an unexpected involuntary contraction of muscles. Even though a simple twitch of the muscle may not be hazardous by itself, it can cause secondary effects resulting in accidents. A housewife can drop a hot pot and a worker can fall from a ladder.

When a nerve trunk, which generally includes sensory and motor fibers of all types from Aα to C, is stimulated transcutaneously, as shown in Fig. 15.17 of Chap. 15, the sensory nerve fibers of Aβ are stimulated with the lowest threshold. Motor nerve fibers of the Aα type can be stimulated with an additional increase in current strength or duration. It can be easily confirmed during transcutaneous electrical nerve stimulation (TENS). As the strength or pulse width of TENS is increased slowly, tingling perception comes first and then the twitches of related muscles are followed.

Since motor nerve fibers of the Aα type have a larger diameter and higher conduction speed than Aβ sensory nerve fibers, it is strange that fiber with a larger diameter has a higher excitation threshold. It is caused by the difference between the sensory and motor nerve fibers. To be effective in their different roles, they are different in their physiological characteristics. Differences in membrane conductances including the difference in sodium channels between sensory and motor axons have been demonstrated in human peripheral nerves. There is a significant difference in the accommodation to hyperpolarization between sensory and motor axons. So, sensory axons in the human peripheral nerve are excited electrically at a lower threshold than motor axons (Burke et al., 2017).

When the sensory and motor thresholds were measured with increasing stimulating frequency in the range 1~35 KHz, the threshold ratio of motor nerves to sensory nerves was about 2.5 (Ward & Robertson, 1998). If this ratio is extended to an AC current of 60 Hz in men with hand-to-hand contacts, the reaction current becomes about 2.5 mA on average.

1.3 Threshold for Pain Sensation

Pain is a subjective matter, and there is no objective criterion for measuring the actual sensation of pain. Thus, although the pain threshold of a subject to electrical current stimulation may be constant, it is difficult to anticipate that this same stimulation level may produce a similar response in other subjects. So, it is difficult to determine a generally applicable pain threshold.

As the pain is sensed by the electrical stimulation of pain fibers of Aδ and C types, which is smaller in diameter than other sensory nerve fibers of Aβ type, a higher threshold is generally expected than the perception threshold of current and the threshold of reaction current.

Pain threshold is variant on the measurement conditions, and the generally accepted pain threshold is between 3 and 10 mA for 60 Hz AC current. Although males have a greater pain tolerance threshold than females, their threshold of pain perception is only slightly higher than females.

Pain threshold is used as one of the biomarkers for quantitative sensory testing (QST) to diagnose and evaluate the progress of neurological diseases including various chronic pains. In normal subjects, the pain thresholds of the limbs do not significantly differ from each other. However, in subjects suffering from chronic myofascial pain, the pain threshold of the limb on the affected side differs from normal subjects and it is higher or lower than the pain threshold on the opposite side.

1.4 Let-Go Current

When the strength of the current is increased, currents are sensed with increased tingling and warmth, muscles begin to contract involuntarily, and pains also can be sensed. Up to certain strength of the current, subjects can voluntarily move their muscles and withdraw from contacting the electricity. However, over certain strength of the current, subjects freeze and cannot let themselves go away from the contacting electricity. This level of current is called the let-go current, which describes the maximum current level that subjects can voluntarily disconnect themselves from the electricity. Above the let-go current, subjects involuntarily continue the grip on the conductor of the current. Due to the increased contacting pressure, contacting resistance is reduced and the flowing current can be increased. With this level of current, current causes the contraction of all the muscles of the body region. It contracts flexion muscles as well as extension muscles. However, the muscles of flexion to grip the objects are stronger than the extension muscles to release the objects. And, it makes the person unable to voluntarily let go. Mostly, the

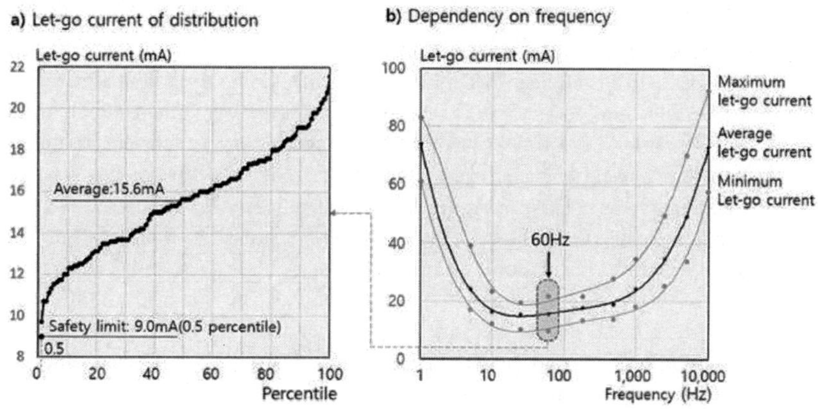

Fig. 16.4 Let-go current. (**a**) Distribution of let-go current for men at 60 Hz with contact in both hands, (**b**) variation of let-go current depending on the frequency of current. (Dalziel et al., 1941)

inability to let go occurs with alternating current. As alternating current repetitively stimulates nerves and muscles, it results in a sustained tetanic contraction which lasts as long as the contact is continued (Fish & Geddes, 2019).

The average let-go current for men is 16.0 mA and 9.0 mA, below which only 0.5% of subjects cannot release the contact, is considered as the reasonably safe current limit that normal men can withdraw voluntarily (Fig. 16.4a). For women, the average let-go current is much lowered to 10.5 mA, and the safety limit is also lowered to 6.0 mA. Let-go current also increases with the frequency of the electrical current (Fig. 16.4b). At 5 KHz, it increases about threefolds to 48.0 mA compared to the 60 Hz AC current. During the sustained contact due to the inability to let go, subjects feel electric shocks with alternating currents. With direct current, in contrast to AC current, subjects can feel the unpleasant electric shocks only when the contact is made or released. During the period that the contact is maintained, there is no sensation of shock. Below 300 mA DC, there is no let-go phenomenon because the hand is not involuntarily clamped.

Inability to let go results in more current flowing in a longer period of time. This will increase the pain and the damage due to the heating of muscle and nerves. With an increased level of current, it can be followed by severe incidences of respiratory paralysis or cardiac arrest. Unfortunately, at the frequency used in commercial power lines, subjects are most difficult to release voluntarily from contacting the electricity and easy to be electrically injured.

1.5 Respiration Paralysis

When the current higher than the leg-go current flows continuously, there occur involuntary muscle contractions in the hands and arms and it paralyzes or freezes the body movements. A subject is no longer able to release a tool, wire, or object in the hands. Actually, objects in the hands may be held even more tightly, and result in extended exposure to the hazardous current. The level of current becomes dangerous by causing respiratory paralysis and making a subject stop breathing. During the experiment for the determination of the let-go current, Dalziel observed two cases of respiratory paralysis at the current level of 18 mA and 22 mA, which is higher than the average let-go current of the men on 60 Hz AC current (Dalziel, 1972). On average, current in the range of 20~50 mA can cause respiratory paralysis. However, this respiratory paralysis can be discontinued by the interruption of current exposure and can be recovered to a normal state if the respiratory paralysis lasted only for a short duration. If the exposure to electricity is extended, it can cause lung injuries that accompany bilateral pulmonary infiltration and hemoptysis (Truong et al., 2017).

1.6 Ventricular Fibrillation

When the current flows through the heart, it can modulate the electrical rhythm of the heart and interrupt its function. Current can damage the heart in two ways, by neurologically disrupting the normal rhythm and by directly damaging heart muscles. As vascular axes of the heart are the part of the lowest resistance, current easily flows through this part and neighboring nerve conduction path. Currents flowing in the horizontal path of hand-to-hand and vertical path of hand-to-foot include the heart in the loop and current through the heart easily affects the electrical signals transmitting over the conduction pathway resulting in cardiac injuries. It has been hypothesized that ion channels in the sinoatrial (SA) node and atrioventricular (AV) node are the easiest to be disrupted and that ischemia and infarction in the right coronary artery distribution, which runs closest to the chest surface and supplies both nodes, make these nodes more vulnerable to electrical current (Spies & Trohman, 2006).

Arrhythmias are the most common complications of the electric current flowing through the heart. If the electrical activation in a part of the heart occurs at an improper time, the heart muscle cannot contract in a controlled manner by inducing arrhythmias. Electrical signals randomly propagate throughout the muscle and can cause the heart to quiver. They include sinus tachycardia, premature ventricular contractions, atrial fibrillation, ventricular tachycardia (VT), and ventricular fibrillation (VF). While most of these arrhythmias can be disappeared with the breaking of the current loop and subjects can restore normal heartbeats, ventricular fibrillation cannot be eliminated by simply stopping the current flows and can result in sudden cardiac death. In addition to the rhythmic disturbance by electrical stimulation, heart muscles can be damaged by the direct effects of the current like electrothermal conversion and electroporation.

Sudden cardiac death by ventricular fibrillation can be caused by the low voltage of alternating current as well as by the high voltage like lightning or direct current. In a canine experiment, ventricular fibrillation occurred nine times more often with the alternating current than with direct current shocks (Lown et al., 1962). The frequency of 50~60 Hz of the electrical power line, which is most widely used in home and industry, increases the likelihood of ventricular fibrillation with a high probability of electric shock during the vulnerable T-period. In another experiment, the current required to produce fibrillation in dogs at 3000 Hz was 22~28 times higher compared to the threshold at 60 Hz (Geddes & Baker, 1971). It implies that operating in a higher frequency current is much safer in the interface with subjects in medical and dental instruments.

During the different phases of a cycle of the heart beating rhythm, the first part of the T-wave is especially vulnerable to electrical stimulation. Electrical stimulation of sufficient magnitude applied during this period may induce ventricular fibrillation. This phase is corresponding to the timing of the end of the ventricular contraction. Electrical stimulation during this phase makes refractory areas of cardiac muscle spread among nonrefractory areas. Therefore, the probability of current stimulation during the approximately 100 ms duration of the T-wave is critical. If the stimulation is applied through the whole period of the heart cycle, some of them certainly fall on the vulnerable period of the T-wave. If the stimulation duration is less than 1 s and applied only during the part of the heart cycle, the risk of VF is determined by the chance of its coincidence with the T-wave.

When a single current pulse of 10 ms width is applied during this phase, ventricular fibrillation has occurred with a current amplitude of 15 mA in an experiment with dogs. Stimulation of 60 Hz current also induces ventricular fibrillation (VF) and the threshold of stimulating current inducing the VF decreases as the stimulated duration increases. VF occurred in a dog with a 60 Hz current of 24 mA when the stimulation is applied for a single heartbeat. However, the threshold for VF is reduced to 0.4 mA when the stimulation is applied for the duration of 5 heartbeats. This notifies that a very weak amount of AC current can result in VF as the exposure duration is extended (Sugimoto et al., 1967).

A *macroshock* is the situation where an electric shock is caused by the current applied at the body surface far from the organ of interest, usually the heart (Fig. 16.5a). Entered current spreads out more or less uniformly and only a small fraction of the current reaches the heart. Entry points of the current and the distance from the heart are important and determine the magnitude of the critical current. Usually, large skin resistance does the role of blocking or reducing the amount of current flowing through the heart. However, in the medical environment, there is a high chance that the skin resistance is lowered or bypassed by attaching moisturized electrode or by skin abrasion. To attain a critical level in the heart resulting in ventricular fibrillation, a rather large

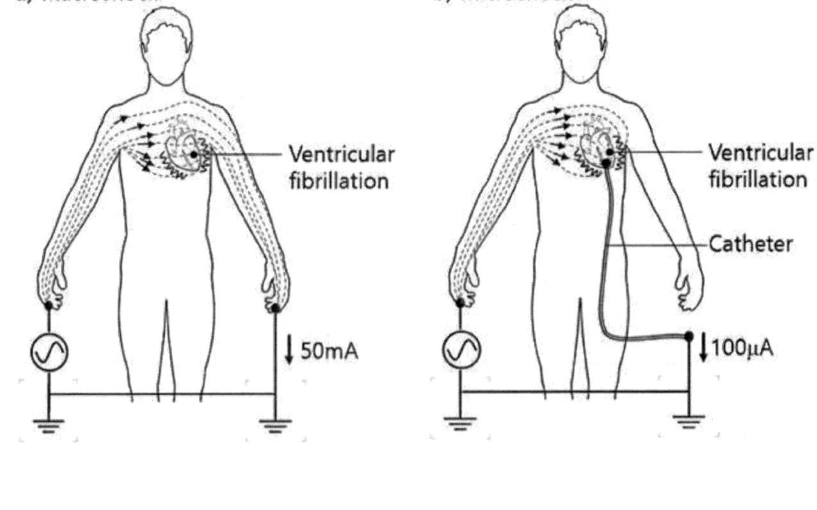

Fig. 16.5 Types of electrical shocks. (**a**) Macroshock, (**b**) Microshock

amount of current, usually higher than 50 mA at 60 Hz, is required from the external source. If skin resistances are bypassed or lowered to less than 1 KΩ, a macroshock can be developed with lower-level voltages less than 100 V in ordinary environments like households or hospitals. A macroshock can be occurred by touching the poorly designed or malfunctioning electrical equipment, damaged power cords, downed power lines as well as lighting. Most electrocutions, death by electric shock, are caused by the macroshock.

Instead of current being conducted through the tissue from the other part of the body as in the case of a macroshock, current can directly be applied to the critical organ like the heart and brainstem by bypassing the skin resistance. Many medical devices are implanted within the body, and surgical devices are contacting these organs directly during the surgeries. Pacemaker electrodes, catheter electrodes, and current-carrying fluid-filled cardiac catheters are some examples. Most of these devices are based on electrical circuits or conductive materials to carry the current. There often occurs accidents and an unintended current loop is developed by faults or chances through these devices. In these cases, a high density of current can easily flow through the contacting point into the critical organ. Even with a small amount of current flowing through, severe damage like ventricular fibrillation can be induced as the whole current is concentrated on the entering point in the heart (Fig. 16.5b). This is a *microschok*. Current in the range of 80~600 μA can cause microshock in humans. The current causing the macroshock is stronger than the current inducing the microshock around 100~1000 folds. The widely accepted safety current limit at 60 Hz for the applied part to the heart is 10 μA in normal mode which can prevent microshock.

1.7 Electrical Burns

Electrical burns result from the heating effect of the current. As the current flows through the resistance of the tissue, it generates heat like the current flows through the resistance of the heater. When the level of current is small at a lower level, the generated heat can be dissipated by the normal cooling system including radiation through the skin and vascular circulating. However, if the generated heat within a short duration of current shock is greater than the physiological cooling capacity, excessive heat starts to damage the tissues and occurs a burn.

Electrical burns produced by contact with an electrical source of fewer than 500 volts are classified as low-voltage burn. At this range, damage of electrical burn is concentrated on the contacting points, usually the skin. As the resistance of the skin takes the major part of the circuit resistance, it limits the current flowing in the internal tissue and the current is not enough to cause tissue damage along the internal path. This type of electrical burn may be mild and superficial if the contact is not sustained, but damage can be severe as the contacting time is extended. If the voltage of the contacting electrical source goes higher than 500 V, a high-voltage burn, in which damages spread throughout the body along the path of the current, can occur. In this type of electrical burn, damages in the deeper tissue and organs become more severe than the damage in the superficial area.

1.8 Stun Gun and Taser Gun

Stun guns and tasers are called "electroshock weapons." They are convenient self-defense weapons and fit easily into most bags and pockets. They are based on the use of electrical stimulations to stun an attacker with painful shock and to discourage further attack. These weapons are less likely to cause death or injury than other weapons, such as handguns. Electrical stimulation is applied by sharp metal probes to an attacker. When the electrical stimulation is applied, it temporarily causes partial paralysis by uncontrollable muscular contractions and makes a person usually fall to the ground.

The stun gun is a close-range device administering an electric shock through direct contact. In contrast, a taser gun administers the electrical shock at a distance through thin flexible wires connected to two probes that are fired into an attacking subject. Even though they do not cause death, they can cause serious injuries and medical problems, like irregular heartbeats and electrical burns.

Electrical shocks are administered in a series of high-amplitude short pulses of 10~100 μs width. Pulses are continuously administered with a frequency in the range of 15~40 Hz while the contact is sustained with a maximal limit of 5 s. The instantaneous peak output voltage may rise up to 6000 V, but the average voltage is in the range of 20~70 V. The amount of administered current is in the range of 10~200 mA with peak current rising to 2~20 A. The energy delivered per pulse is in the range of 0.05~1.0 J (McDaniel et al., 2009).

1.9 Lethality of Electricity

A high dosage of electric current is dangerous and lethal to life resulting in electrocution. The lethality of electricity is dependent on various parameters.

The first critical parameter is the magnitude of the current flowing through the body tissue and critical organs like the heart and brain. Even though the same voltage of electrical sources is contacted, the flowing current through the body greatly variable depending on the contacting resistance of

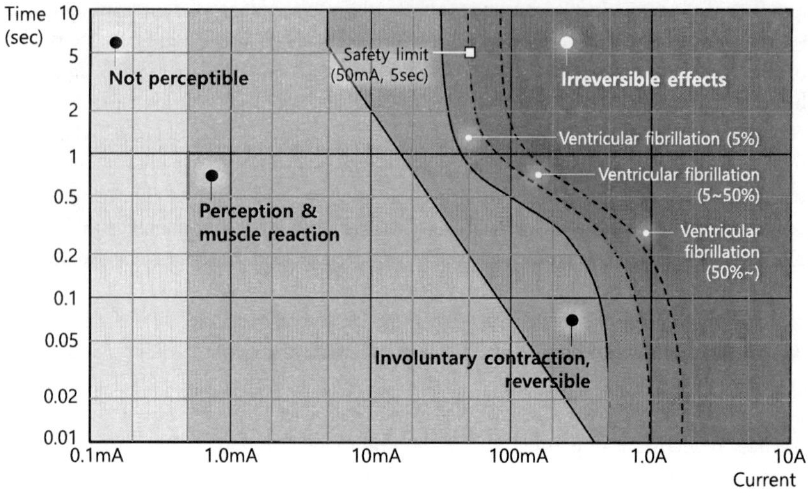

Fig. 16.6 Dependency of biological effects on current amplitude and contacting duration

entry points. As the resistance of the wet skin can be reduced to one-hundredth compared to the dry skin, the internally flowing current can be multiplied by almost a hundred folds with the same source voltage. The second critical parameter is the contacting duration of electricity. Even though the magnitude of the current is large enough, it cannot be lethal if it were applied only for a very short period of time. A longer duration increases the lethality. Generally, the amplitude of current and the contacting time multiplicatively strengthen the biological effects as the increased amount of electrical charges practically determines the voltage variations in the nerves and muscles. So, the first remedy for electrical accidents is to break the current loop and shorten the exposure time to electricity. Figure 16.6 shows the relationship, which is developed by the International Electrotechnical Commission (IEC), between the amplitude of current and the contacting time on the biological effect including the ventricular fibrillation for the alternating current of frequency in the range of 15~100 Hz.

The third critical parameter is the frequency of the electrical current. Unfortunately, the frequency of the power line is in the most susceptible range. While biological effects mostly occur only when the contact is made or broken in DC, biological effects are sustained during the contacting period in AC. Accumulation of charges repeatedly stimulates the nerves and muscles during the rising phase of the current before it is compensated by the falling phase of the current. An alternating current of very high frequency is less critical than the lower frequency AC. AC current of very high frequency cannot accumulate enough amount of charge in a single rising phase and accumulated charges are compensated in the falling phase before it activates the nerves or muscles. As the frequency of current increases, it is rather difficult to flow deep into the tissues and critical organs. It is more likely to cause a heating effect which can be led to burns. Lethality of AC was an issue from the earliest stage in the determination of frequency for commercial electric power transmission systems. Using AC was verified more dangerous than using DC. However, DC did not win the "War of the current" due to its inability and inefficiency in transmitting the long distance.

The fourth critical parameter is the entry points. As described in the microshock, if the skin resistance is bypassed and the critical organ is directly contacted by the electrical source, a very small amount of current, which even cannot be perceived, can cause lethal damage like ventricular fibrillation. In medical settings, accidents of a microshock may occur during the surgery where the body is open and metallic surgical devices and electrical instruments are touching the internal organs directly. When the heart is catheterized for monitoring or therapeutic purposes, the heart can be connected to an electrical source directly via the inserted catheter with a chance of a microshock. Points of entry are also important even when the current flows through the skin. If the current path includes a heart with a high probability, those entry points are more dangerous. Current paths between hands and between a hand and a foot are much more dangerous than the current path between a leg and the ground as they are likely to traverse the heart.

The voltage of the contacting electrical source affects when it is very high over about 600 V. In addition to high current flowing due to high voltage, it may cause dielectric breakdown in the skin. If the skin resistance is eliminated, the breakdown of the skin increases the current greatly and the lethality. Physiological characteristics of the subject also affect lethality. Female is more susceptible to electricity and body weight increases the thresholds for the biological effect of the electricity. If a subject has an implanted medical unit like a cardiac pacemaker and deep brain stimulator, it makes a subject very sensitive even to a small amount of current.

2 Biological Effects of Electromagnetic Field

With the progress of communication technologies and ever-advancing technologies based on electricity, exposure to man-made electromagnetic fields (EMF) is steadily on the increase. Everyone is easily exposed to a variety of electromagnetic fields across a wide range of frequency bands, including from the sources like mobile phones, home appliances, telecommunications, and broadcasting installations.

In the modern environment, people can be exposed to different types of EMF exposures. When subjects use any electric devices, they are *personally* exposed to EMF. EMF can be generated in most electric units personally used at home or office, which include mobile phones, cordless telephones, TV monitors, wireless internet routers, microwave ovens, personal computers, vacuum cleaners, and refrigerators. As these personal devices are used at close distances, the small strength of EMF might cause some biological effects. Even though subjects are not using any personal electrical devices, they are continuously exposed to the *environmental* EMF. Whether subjects are staying indoors or walking outdoors, they are not escaped from the access of EMF generated by many communication and broadcasting towers. Since environmental EMF exposure is ever presenting and its effect is accumulating with time, the strength of the environmental EMF needs to be maintained at a minimal level and its biological and potential hazardous health effects should be carefully evaluated. Some subjects like workers in power plants and broadcasting stations are *occupationally* exposed to higher intensity of EMF. As the biological effect of the EMF increases with the EMF intensity, their working hours in regard to EMF strength should be guided strictly not to cause any hazardous health effects.

2.1 Electromagnetic Field

The electromagnetic field (EMF) is a characteristic space produced by the movement of the electric charges. A static charge produces a stationary electric field in the surrounding space. If a charge is moving, which is the current, the electric field becomes time-varying and the magnetic field is produced additionally. Reversely, an electric field can be produced by the changing magnetic field. These mutual interactions of electric field and magnetic field produce electromagnetic field combinedly. The ways in which charges and currents interact with the electromagnetic field are described by Maxwell's equations. This electromagnetic field can be described as electromagnetic waves transmitting electromagnetic energy as well as electrical signals. The electromagnetic wave propagates with the speed of light and interacts with charges and currents within the space.

The electromagnetic wave is especially characterized by the frequency, which is the rate of the moving of current producing EMF (Table 16.1). The entire distribution of electromagnetic waves according to the frequency is called the electromagnetic spectrum. The electromagnetic spectrum includes a vast range of electromagnetic wave frequencies of $0{\sim}10^{25}$ Hz. In the higher frequency range, electromagnetic waves are more conveniently described by the particles. As electromagnetic waves have both characteristics of waves and particles, their interactions are understood sometimes by the waves and other times by the particles. Based on the different characteristics according to the frequency, they are widely used in various applications including communication, broadcasting, cooking, vision, diagnosis, and treatment of diseases.

Electromagnetic fields are present everywhere around our environment. But they are not visible to the human eyes. Electromagnetic fields can be produced naturally by the local build-up of electric charges in the atmosphere during a thunderstorm. The earth's magnetic field is also the natural field that is used in navigation by humans as well as fishes and birds. In addition to the natural EMF, there are many artificial sources producing EMF for their basic applications. Electricity from the power socket is associated with low-frequency electromagnetic fields. Various kinds of electrical units using motors, transformers, radio waves, and electric circuits, like a TV, radio, microwave oven, hair drier, vacuum cleaner, computer, and refrigerator, are also producing EMF.

In recent days, the rapid development of wireless communication using radiofrequency has increased the concern about the biological and health effect of exposure to electromagnetic fields (EMF) in the daily environment. As an electromagnetic field interacts with electrical charges and current within the space, humans having their own electricity within the body cannot escape from this interaction under the EMF. The effects of this interaction can be adverse or beneficial. The intensity of the electromagnetic field is most critical. In many cases, EMF of small magnitude can produce some biological effects, but they do not cause any health degrading effects due to the innate physiological compensatory system of the human. However, under a large magnitude of EMF, the effects of EMF may result in hazardous health effects.

Table 16.1 Electromagnetic wave spectrum and its types of biological effect

Type	Wavelength (m) Frequency (Hz) Energy (eV)	Production	Applications	Biological effect
Low frequency	10 Km~ **~30 KHz** ~124 peV	Accelerating charges	Navigation	Induced current
Radio frequency	1.0 m~10 Km **30 KHz~300 MHz** 124 peV~1.24 μeV		Communications, remote controls, MRI	Heating
Microwave	1.0 mm~1.0 m **300 MHz~300 GHz** 1.24 μeV~1.24 meV	Accelerating charges & thermal agitation	Communications, micro ovens, radar, cell phone use	
Infrared	**700 nm~1.0 mm** 300 GHz~430 THz 1.24 meV~1.7 eV	Thermal agitation & electronic transitions	Heating, imaging	Photo-chemical
Visible light	**400 nm~700 nm** 430 THz~790 THz 1.7 eV~3.3 eV		Vision, photosynthesis	
Ultraviolet	**10 nm~400 nm** 790 THz~30 PHz 3.3 eV~124 eV		Sterilization, vitamin D	Photo-chemical, ionization
X-ray	0.01 nm~10 nm 30 PHz~30 EHz **124 eV~124 KeV**	Electron to atom collision	X-ray imaging, cancer therapy	Ionization
Gamma ray	~0.01 nm 30 EHz~ **124 KeV~**	Nuclear decay	Nuclear medicine, cancer therapy	

p: pico (10^{-12}), n: nano (10^{-9}), μ: micro (10^{-6}), m: milli (10^{-3}), K: Kilo (10^{3}), M: Mega (10^{6}), G: Giga (10^{9}), T: Tera (10^{12}), P: Peta (10^{15}), E: Exa (10^{18})

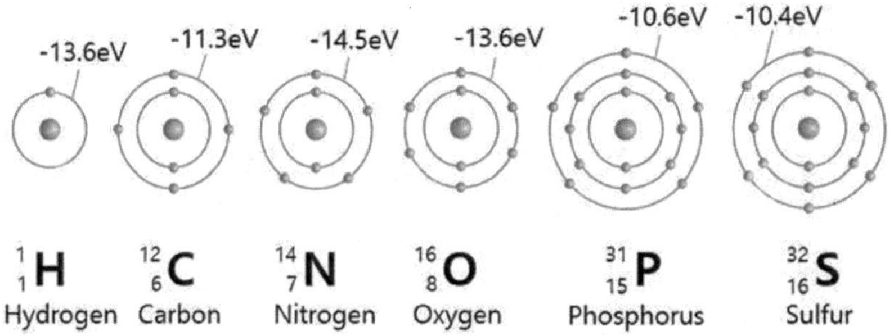

Fig. 16.7 Atoms constituting the body and their lowest binding energy susceptible to ionizing the atom

2.2 Ionization Effect in Ionizing EM Waves

There can occur various biological effects under electromagnetic (EM) wave exposure and they are depending on the frequency-specific characteristics of the electromagnetic wave.

Among the biological effects of the EMF, the most well-known adverse effect is ionization. Ionization does not occur in a lower frequency of electromagnetic (EM) waves and occurs only when the photon energy of the EM wave is greater than the energy required to ionize the atoms within the body. As the lowest binding energy of outermost electrons of atoms constituting the body is 10~15 eV, EM waves whose photon energy is greater than this level can detach an electron from an atom and make it a positive ion (Fig. 16.7). Ionization can cause damage to the cells leading to death and to DNA leading to cancer and heritable effects. EM wave having photon energy of 10 eV is ultraviolet. X-ray and gamma-ray also have photon energies higher than this level and can cause ionization. X-ray and gamma-ray are actively used in imaging opaque objects to see the inner contents and structure in the diagnostic examinations for medical, security, and safety purposes. With this ionization effect leading to the death of the cells, high doses of X-ray and gamma-ray are also used in cancer treatment by focusing the EM field on the target cancer cells. Like extended exposure to X-ray and gamma-ray can cause a mutation and cancer by the ionization, overexposure to ultraviolet on a hot summer day can lead to damage to the skin and to a malignant cancer of melanoma by the ionization.

In contrast to ionizing EM waves, EM waves whose photon energy is lower than the minimum ionizing energy level are called nonionizing EM waves.

2.3 Current-Inducing Effect in Low-Frequency EM Waves

Even though the nonionizing EM wave does not produce any hazardous ionization effect, it interacts with the body in different ways and produces some biological effects. EM wave in low frequency (LF) less than 30 KHz mostly interacts with the human body by inducing the current within the body. EM waves in the higher radiofrequency and microwave range up to 300 GHz mostly interact with the human body by producing the heating effects. Nonionizing EM waves higher than 300 GHz, which include infrared, visible light, and ultraviolet of lower energy, can cause a photochemical effect in addition to the heating effect.

A power supply of 50 or 60 Hz is the main source of EMF in low frequency. In these ranges, the wavelength of EM waves is greater than 10 km, which is much larger than the dimension of the human body. Also, the changing rate of the electric field or magnetic field is too slow to generate enough amount of accompanying magnetic or electric fields. A static electric field does not produce an accompanying magnetic field as there is no time-varying change. Likewise, a static magnetic field does not produce an accompanying electric field. When they are changing with low frequency, produced accompanying fields are too small and negligible. So, in this LF range, interaction with the human body is explained by an electric field or magnetic field separately instead of the combined electromagnetic field (Fig. 16.8).

Electric fields exist whenever a positive or negative charge is present. They interact with other charges within the field and affect forces on them. This interaction is described by voltage and the strength of the electric field is described in Volts per meter (V/m). The electric field becomes stronger as the voltage increases and the distance decreases. And it exists even though there is no current flowing as long as the voltage exists across the distance. The electric field is not transparent to many common materials like wood, water, and walls of the building. Burying the power lines in the ground reduces the electric fields at the surface to a level hardly detectable. When an electric field is applied to the conductive object, it interacts with the charges in the object and influences the distribution of electric charges at its surface. Ionic charges and tiny electrical currents flowing in the nervous system within the human body can be influenced by the external electric field. The external electric field usually causes linear current making charges flow through the body along the direction of the electric field to the ground.

The magnetic field arises between the magnetic poles and from the motion of electric charges. When the magnetic field is produced by the magnetic poles, it is measured in micro Tesla (μT). When the magnetic field is produced by the electric current, it can also be described in Amperes per meter (A/m) as well as micro Tesla. Higher current increases the strength of the produced magnetic field and its strength decreases as the distance increases from the current flowing site. The magnetic field exists only during the current is flowing. The magnetic field is transparent to many objects including the human body and not blocked by common materials like the walls of the building. Burying power lines

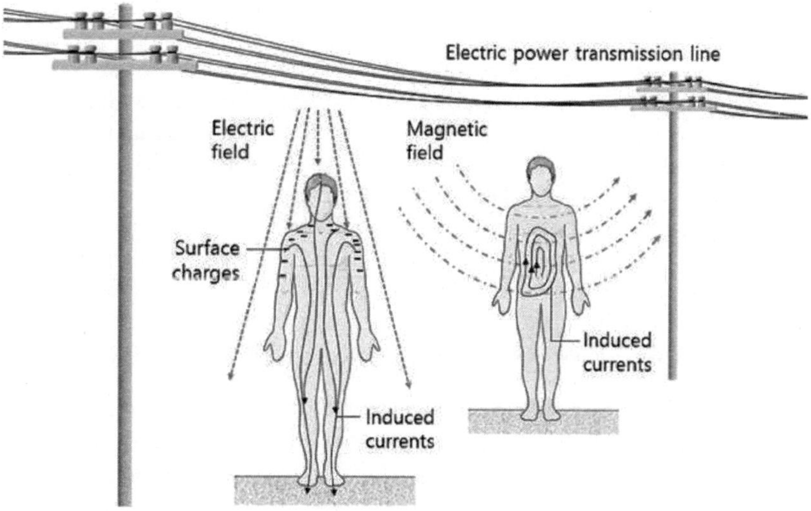

Fig. 16.8 Induced currents by the electric and magnetic field of low-frequency EM waves

does not reduce magnetic fields effectively as it does for electric fields. When the LF magnetic field is applied to the human, it induces circulating currents within the human body in the plane perpendicular to the direction of the magnetic field. The strength of the induced currents depends on the intensity of the applied external magnetic field. If it is sufficiently strong, induced currents may cause stimulation of excitable tissues like nerves and muscles affecting the ordinary neuronal behaviors. Both electric and magnetic fields can induce the disturbance in the voltages and currents in the body, but generally, disturbance by the induced currents is negligibly small in an ordinary daily environment.

2.4 Thermal Heating Effect in Radiofrequency and Microwave EM Waves

Radiofrequency and microwave EM waves cover the frequency range of 30 K~300 GHz and their wavelengths are 1 mm~10 Km, which can be comparable with the dimension of the body. As these EM waves radiate away from sources like light, they are used to transmit information over long distances in telecommunications as well as radio and TV broadcasting. The microwave used in microwave ovens has a higher frequency than radiofrequency EM waves. In the medical domain, magnetic resonance imaging (MRI) uses radiofrequency EM waves in the range of 1 M~300 MHz to excite and induce the resonance of the body tissue under the strong magnetic field of 3.0 Tesla.

In these range EM waves, an electric field accompanies magnetic fields as their changing rate is sufficiently high to produce an accompanying field. So, at radiofrequency and microwave ranges, electric and magnetic fields are closely interrelated and their strength is measured as power densities in Watts per square meter (W/m^2). The power density (W/m^2) indicates how much EMF passes through a certain surface and is thus appropriate for describing external exposure. For example, high EMF on the order of 1 KW/m^2, in which microwave ovens use to cook foods, can clearly result in heating of biological tissue and an increase in body temperature. EMF leaking from an oven is restricted to be lower than 50 W/m^2 at approximately 2 inches from the oven surface. Health damage in humans can occur if the exposure to EMF generates excessive heat that is beyond the human physiological capability to dissipate. For the description of internal exposure of EM wave power absorbed by the body, a specific absorption rate (SAR) in Watt per kilogram(W/Kg) is used as a reference unit. It indicates the power absorbed per kilogram of body tissue.

In radiofrequency and microwave EM waves, thermal heating is the major biological effect. A microwave oven, which is used to warm up the food, is a representative example of the heating effect of EM waves of this range. The heating effect occurs when radiofrequency EM waves are radiated on the body surface. Electronic and magnetic fields displace electrically charged ions and dipole molecules within the body by exerting force on them. Under the radiofrequency and microwave EMF, these particles move very fast with the frequency of the EMF. They hit and slide each other and heat is produced by the friction of the particles. Overall, body tissues function as a resistive load to induced currents, and the heat is generated by the currents flowing through resistance as in the electrical heater. When a whole body is exposed to radiofrequency EMF with an average SAR value of 4 W/Kg, the temperature rises approximately 1 °C in human subjects. This is comparable to normal muscle activity corresponding to 3~5 W/Kg.

The thermal heating effect varies with the absorption property of the tissues depending on the frequency of EMF and the constituents of tissues, like water, fat, protein, and salt. Tissue composition and distribution of water molecules, ions, and other molecules determine the electrical conducting properties of the tissue. The conductivity of the tissue influences the absorption level of EMF among different tissue types. The tissue of higher conductivity by the higher water or salt content increases the interaction with EMF and thus the thermal heating effect. There are areas in the body called hot spots which show higher local heating and areas which show lower heating relative to the surrounding tissues. For example, bone and fatty tissue are heated less than other tissues compared to the brain and muscles due to their lower water content. If the absorption is less effective, EMF can penetrate into deeper parts of the body. In bones and fatty tissues, EMF penetrates into the deeper regions than muscle, brain, and kidney.

As the frequency of EMF is increased, the depth of penetration is decreased and the absorption of EM energy as well as the heat conversion occur mostly at the body surface. EMF in the MHz range which is used for broadcasting can penetrate into the depth of the body about 10~30 cm. EMF around 1 GHz used in mobile communications can penetrate the body into the depth of a few cm. The penetration depth of EMF above 10 GHz used in radar equipment is below 1 mm. At still higher frequencies, the heating effect of EMF occurs effectively only at the skin surface.

Resonance is another interaction that can be occurred in the EM wave in the radiofrequency and microwave range. Resonance is also called whole-body absorption as it occurs when the body acts as an antenna to receive the transmitted RF energy. In this context, the body size is critical to act as an antenna. If the body size is approximately half of the EM wavelength, the body can act as an antenna and increase the energy absorption leading to resonance. The wavelength of

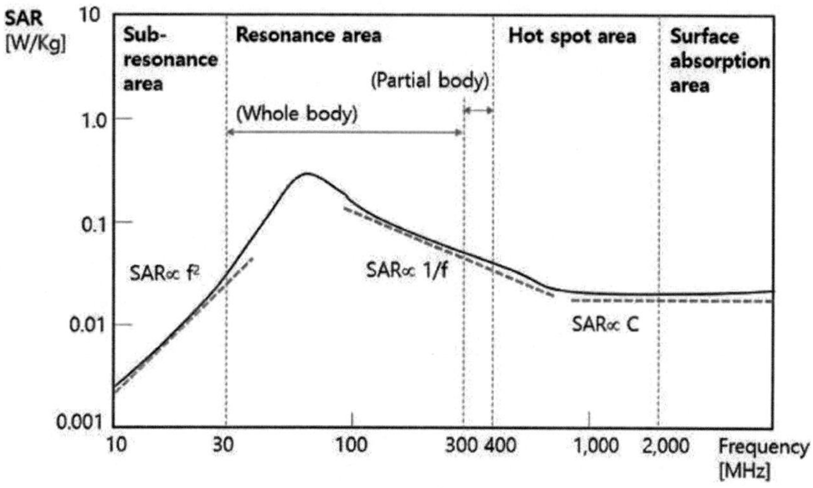

Fig. 16.9 Whole-body absorption characteristic of the human body under 10 W/m^2 plane EM wave. (Hagmann et al., 1979)

80 MHz radiofrequency is 3.75 m and half of this wavelength is 1.88 m, which is close to the height of adults. For the children, the resonance effect occurs in a higher frequency of a shorter wavelength around 100 MHz. So, the radiofrequency EM waves in the range of 30 M~150 MHz, which is the very high frequency (VHF) band used for FM and TV broadcasting, can increase the EMF absorption and heating effect additionally. Similar to whole-body absorption, partial body absorption can occur when the part of the body acts as an antenna. When people use a mobile phone, a head that is at a close distance to the EMF source can act as an antenna and increase the absorption of the EMF. The half-wavelength of radiofrequency around 300 MHz is 50 cm and it approximately matches the size of the head and other body parts.

Figure 16.9 shows the characteristic of the whole-body SAR caused by a plane EM with a power density of 10 W/m^2. From a frequency of 10 MHz, SAR is increased approximately in proportion to the square of the frequency before reaching a maximum of around 80 MHz, where the SAR has maximal value due to the whole-body resonance effect. Then, SAR starts to decrease approximately in proportion to the inverse of the frequency up to the frequency of 1000 MHz. In higher frequencies than 1000 MHz, SAR is almost constant and absorption occurs mostly on the skin surface. For an EMF power density of 10 W/m^2, a maximal SAR of 0.3 W/Kg occurs around 70 MHz.

2.5 Specific Absorption Rate (SAR)

Specific absorption rate (SAR) is a measurement of the amount of radiofrequency/microwave energy absorbed per unit mass of the body from the source of concern. SAR is averaged over the whole body or over a part of the body. For each type of body tissue, basically, SAR is determined by the density and the conductivity of the tissue under electric field or power density of EMF as follows.

$$\text{Specific absorption rate (SAR)} = \frac{\sigma}{\rho}|E|^2 = \frac{\sigma}{\rho}Z_0P_D \quad (16.3)$$

$$\text{Power density} = P_D = \frac{1}{Z_0}|E|^2 = \frac{1}{377}|E|^2 \quad (16.4)$$

where σ is the conductivity of the tissue (S/m), ρ is the mass density of the tissue (kg/m^3), $|E|$ is RMS electric field strength (V/m), and Z_0 is the characteristic impedance of free space.

Absorbed EM power is used to increase the body temperature globally as well as locally. Excessive whole-body heating occurs only in rare conditions and local heating occurs more relevantly. Local temperature rise can be approximated with Pennes bio-heat equation. In the brain tissue, for the given SAR, the temperature rises with the time constant as follows.

$$\Delta T = \frac{\tau \cdot SAR}{C_p}\left(1 - e^{-t/\tau}\right) \quad (16.5)$$

where τ is the thermal time constant of tissue, which is typically 2.5~6 min, C_p is the specific heat capacity of the tissue.

For the thermal time constant of 3 minutes, the temperature reaches 90% of the maximal temperature within 6 min.

For the thermal heating effect of the EMF to be harmful, the tissue must be heated by at least 1 °C. The international standardization organizations use this criterion as a foundation for setting the SAR safety limit. The body will no longer be able to cope with the additional heat if the temperature rises by more than 1 °C. Various physiological responses to heat effects appear at this point. Particularly, the eyes and the

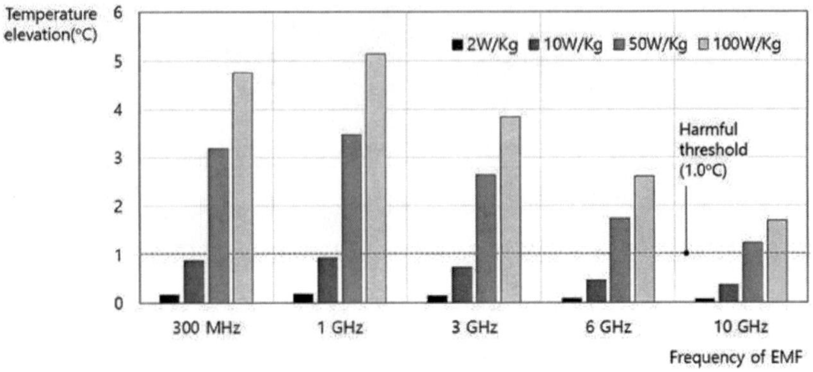

Fig. 16.10 Temperature elevation in the human brain and skin with thermoregulation during the exposure of radiofrequency energy depending on the frequency and the strength of the absorption. (Kodera et al., 2018)

Table 16.2 SAR standards of international organizations

International organization	Frequency range	Normal users (W/Kg) Whole body	Head & trunk	Limbs	Occupational users (W/Kg) Whole body	Head & trunk	Limbs
CENELEC	1 M~100 GHz	0.08	2.0	4.0	0.4	10.0	20.0
ICNIRP	100 K~10 GHz	0.08	2.0	4.0	0.4	10.0	20.0
ANSI/IEEE	100 K~6 GHz	0.08	1.6	4.0	0.4	8.0	20.0

CENELEC European Committee for Electrotechnical Standardization, *ICNIRP* International Commission on Non-Ionizing Radiation Protection, *ANSI/IEEE* American National Standards Institute/Institute of Electrical and Electronics Engineers

Table 16.3 Examples of measured SAR values of routinely used devices

Devices	SAR value	Measurement condition
WLAN antenna ($d = 3$ cm)	8.17 W/kg	2.4 GHz
Microwave oven ($d = 0$ cm)	5.7~8.0 W/kg	2450 MHz, 50 W/m^2
Microwave oven ($d = 1$ cm)	1.46 W/kg	2450 MHz, 50 W/m^2
Smartphones	0.88 ~1.4 W/kg	Available in 2020
Tablet PC	0.35~1.20 W/kg	
LTE/WiFi router	0.15~0.92 W/kg	23.4 dBm, 826.5 MHz
Microwave oven ($d = 5$ cm)	0.26 W/kg	2450 MHz, 50 W/m^2
Bluetooth headset	0.24 W/kg	
Cordless phone	0.01~0.05 W/kg	
WLAN antenna ($d = 34$ cm)	0.0004 W/kg	2.4 GHz

testes are known to be vulnerable to the heating effect of EMF, because they relatively lack available blood flows to dissipate the excessive heat generated. Cataracts and temporary sterility may be developed when a subject is exposed to a short-term of 0.5~1 hour and a high-level EMF of 1~2 KW/m^2. Figure 16.10 shows the simulated temperature elevation in the human brain and skin according to the strength of SAR and the EMF frequency.

To guide the safe applications of electrical devices and EM sources, the international organization developed the SAR standards. SAR for the environmental levels of EMF routinely encountered by the general public should be far below necessary to produce significant heating and to increase body temperature. These standards are developed separately for the general users and occupational users. Table 16.2 shows the SAR standards of international organizations for normal users and occupational users. Table 16.3 shows the measured SAR values for some routinely used electrical devices. Most of their measured SAR values are below the standard safety level in normally used distances. Even in the stronger case of 8.17 W/Kg of WLAN antenna at a distance of 3 cm, the elevated temperature is estimated less than 1.0 °C according to Fig. 16.10.

2.6 Complications with Increased Body Temperature

The generated heat can increase the body temperature locally or globally. As the thermal heating effect in EM waves is explicit, safety guideline of the EMF is developed basically based on this heating effect. When the subjects are exposed only for a short period under the safety limit, EM waves do not cause any significant rise in body temperature in normal subjects. The generated heat is cooled down by the compensatory thermoregulatory mechanism of the body. If the temperature is increased locally, additional heat is transferred away and dissipated by the blood flow. If the temperature is increased globally, the body increases the blood supply to the skin surface and heat is intensively dissipated through the skin by thermal radiation and evaporation via sweating.

When the EMF produces a thermal effect of heating the body tissues, the human body is under the control of physiological thermoregulation, which is a process that allows the body to maintain its normal core internal temperature. Even though the exposed EMF can produce the heating effect, normal core body temperature can be maintained by this thermoregulatory system to a certain degree. However, produced heat is more than that can be dissipated with the enhanced thermoregulatory feedback, core body temperature cannot but increase, and abnormal complications can occur depending on the degree of temperature rise (Table 16.4).

Normal range of core body temperature is 36.5 °C~37.5 °C. When heat is generated by the EMF exposure, it warms the blood circulating the tissue. Increased blood temperature higher than the set point of the thermoregulatory center in the hypothalamus activates the processes to lower the core body temperature. Vessels to the skin are dilated and the warmed blood circulates close to the skin surface and dissipates the heat by radiation. Sweat glands are also activated to secrete perspirations, which are evaporated by body heat and help to cool down the body. Even though the core temperature is in the normal range, the generated heat can induce locally in the body. Heat cramp, which is the mildest heat-related illness, is painful, involuntary muscle spasms and usually occurs during heavy exercise in hot environments. Heat cramps occur most often in muscles of the calves, arms, abdominal wall, and back. The spasms of the heat cramp may be more intense and prolonged than typical nighttime leg cramps. Fluid and electrolyte loss may facilitate heat cramps. Heat cramps can be remedied by cooling the body and hydration with rest.

If the body temperature is lowered, past the temperature range of feeling the cold, less than 35.0 °C, subjects are in *hypothermia*. Hypothermia is a state of a medical emergency in which the body loses heat faster than it gains. With the decreased temperature, body organs including the heart and the nervous system no longer work normally. Hypothermia can lead to complete failure of the heart and respiratory system leading to death. Hypothermia is often caused by exposure to cold weather or immersion in cold water and is not closely related to EMF exposure. The basic treatment for hypothermia is to warm the body back to a normal temperature.

If the thermoregulatory system fails to manage the core body temperate within the normal range, body temperature can rise into the state of *hyperthermia*. Hyperthermia is a state of abnormally high core body temperature above 38 °C. It occurs when the body absorbs or generates more heat than dissipates. *heat cramps* and *heat strokes* can occur in the subject with hyperthermia as well as heat cramps. Heat exhaustion is a body state with the symptoms of heavy sweating and a rapid pulse as a result of increased body core temperature. Heat exhaustion is a more severer illness than heat cramp and its symptoms develop suddenly or over time. Heat exhaustion is caused by the depletion of water, fluids, and electrolytes in the body. Signs or symptoms of heat exhaustion include cool and moist skin with goosebumps while in the heat, heavy sweating, faintness, dizziness, fatigue, weak and rapid pulse, low blood pressure upon standing, muscle cramps, nausea, and headache. If the signs of heat exhaustion are noticed, immediate cooling is required with medical attention. Heatstroke is the most serious heat-related illness. Heatstroke occurs when the body is unable to control its own temperature. With a failure of the thermoregulatory system, the sweating mechanism is unable

Table 16.4 Classification of body temperature and its accompanying symptoms or complications

Class	Temperature	Symptoms or complications
	<35.0 °C	Intense shivering, numbness, feeling very tired, confusion, fumbling hands, memory loss, slurred speech, drowsiness.
Feeling cold	36 °C	Vasoconstriction, piloerection, body curling up, shivering.
Normal	36.5~37.5 °C	Physiological thermoregulation maintains the normal body temperature under heat or cold stress. **Heat cramps** may occur.
Feeling hot	38 °C	Sweating, vasodilation, feeling thirsty, feeling uncomfortable.
Hypothermia	39 °C	Severe sweating, flushed, fast heart rate, breathlessness. **Heat exhaustions** can occur.
	40 °C	Profuse sweating, dehydration, fainting, weakness, vomiting, headache, breathlessness, and dizziness. **Heat strokes** can occur. Starts to be life-threatening with 0.5~2.0% death rate
Hyperpyrexia	41 °C	Fainting, vomiting, severe headache, dizziness, confusion, hallucinations, delirium, drowsiness, palpitations, and breathlessness. Life-threatening with 5.0~20% death rate.
	42 °C	Turn pale or remain flushed, comatose, severe delirium, vomiting, convulsions. Life-threatening with death rate reaching 50%.
	43 °C	Serious brain damage, continuous convulsions, shock, cardio-respiratory collapse. Highly life-threatening with death rate of 90%.
	>44 °C	Almost dead with rate 99.9%.

to cool down the body and temperature can increase rapidly. With a heat stroke, the core body temperature can rise to 40 °C or higher within 10~15 min. Symptoms of heatstroke include confusion, altered mental status, slurred speech, loss of consciousness, hot and dry skin, lack of sweating, and seizure. Heatstroke can cause death or permanent disability if emergent treatment is not given.

Hyperpyrexia is a state where the body temperature rises above 41.5 °C due to the change in the thermoregulatory center in the hypothalamus. In hyperpyrexia, the brain sends commands to raise its baseline temperature above normal as in most other cases of fever. The body responds to the brain's messages and raises its temperature to a new baseline. This reaction normally happens as a result of an infection or trauma. This is in contrast with hyperthermia where the source of temperature rise is external. Heat strokes are not developed with hyperpyrexia. Hyperpyrexia is a life-threatening emergency that demands urgent medical attention. Symptoms of hyperpyrexia include extreme sweating, thirst, dizziness, muscle cramps, fatigue, nausea, and lightheadedness. Without prompt proper treatment, hyperpyrexia can lead to long-term complications and death.

2.7 Photochemical Effect of Light

EM waves in the range of light including infra-red (IR), visible light, and ultra-violet (UV) produce a photochemical effect in addition to thermal heating effects. The frequency range of this EM wave is from 300 GHz to 1000 THz and the wavelength is from 300 nm to 1 mm, which is much smaller than the body size. The photochemical effect is a result of a chemical reaction initiated by the absorption of EM wave energy in the range of light. By absorbing the energy from light, molecules become transiently excited and their chemical and physical properties also change greatly from the original molecules. Excited molecules can fall apart, change to new structures, and combine with other molecules. They also can transfer electrons, hydrogen atoms, protons, or their electronic excitation energy to other molecules. The amount of energy absorbed is depending upon the wavelength. Excitation may occur at rotational, vibrational, or electronic levels depending on the energy absorbed. Absorption of energy in the far infra-red region causes mostly rotational excitation. Vibrational excitation requiring more energy occurs in the infra-red region of the EM waves. Excitation of electronic energy levels, on the other hand, may take place when a molecule absorbs energy in the visible or ultraviolet region of EM waves. The excited molecules become reactive transiently with excess energy and tend to return to the ground state level within a very short time of less than 1 ms (Mursyidi, 2002).

The photochemical effect mainly occurs in the ultraviolet EM wave and to a lesser degree in visible light. The photochemical effect has both positive and negative effects. The positive effect of the photochemical reaction is a photochemical reaction for vision and vitamin D synthesis in the human body. Human vision starts with a photochemical reaction in the retina, where molecules of rhodopsin change their shapes, after absorbing light. When exposed to UV EM waves, the epidermal layer of human skin synthesizes vitamin D, which is essential for normal bone and teeth development and kidney function. The photochemical effect of light is used in phototherapy. In prematurely born babies with neonatal jaundice, their blood bilirubin concentration is at a very high level and the white part of the eyes and skin turn yellowish. By the photochemical reaction under the exposure to UV or blue light, bilirubin is photooxidized into a nontoxic compound that is easily excreted. Photosynthesis and photodegradation of pesticides are examples of the photochemical effect in the environment which are beneficial to humans.

Negative effects of the photochemical reaction occur mostly on the skin because the EM waves in this range cannot penetrate into the deeper parts of the body. Exposure to UV light, either from the sun or from artificial sources like tanning beds, can cause sunburn. It also can cause premature aging of the skin and signs of sun damage such as wrinkles, liver spots, and leathery skin. In the eyes, UV exposure can also cause problems. There is persuasive evidence that the main types of skin cancer, basal cell carcinoma (BCC), squamous cell carcinoma (SCC), and melanoma, are caused by UV light exposure (Armstrong & Kricker, 2001). These cancers are induced by a series of photochemical reactions of the UV light directly damaging the DNA of the skin cells. If the damage to DNA builds up over time, it can be enough to cause cells to start growing out of control resulting in skin cancer. The colored skin is less sensitive compared to that of the colorless skin because the colored skin has enough melanin to absorb UV light. The cornea can be inflamed or burned. Vision can be impaired by the formation of cataracts, which clouds the lens of the eyes, and by tissue growth on the surface of the eye. Exposure to UV light can also weaken the immune system so that the body has a harder time in fighting against infections. Herpes can be triggered and reactivated by exposure to the sun or other sources of UV light. It can also make vaccines less effective.

2.8 Biological Effect and Health Effect of EMF

The biological effect is different from the health effect. The biological effect is the measurable physiological response when the subjects are exposed to EMF. An adverse health effect is any detectable impairment of the mental or physical health leading to a disorder or a disease in subjects exposed to EMF or in their offspring. EMF can cause various biological effects on DNAs, cells, tissues, organs, and systems. These biological effects can be beneficial or adverse to health. Even though a biological effect causes an adverse effect like temperature elevation, this effect is not necessarily harmful to health. For example, exposure to sunlight can increase the body temperature by the heating effect. Even though the heating effect is considered one of the adverse biological effects on health, generally, exposure to sunlight does not result in any hazardous health effects. The human body has sophisticated mechanisms controlling and reacting to various kinds of disturbances caused by internal and external stimulations. Ongoing variation of physiological variables and converging them to normal levels are the basic characteristics of the living system. Other environmental sources like air temperature, atmosphere pressure, humidity, wind speed, and many chemicals around can produce various biological effects like EMF. However, most of the produced biological effects do not result in health effects leading to disorders and diseases. The adverse biological effect is different from the adverse health effects.

However, the compensatory mechanism of the body may not be effective for certain kinds of stimulation and stimulations may be out of controllable range if they are too strong. If the stimulation is of a strange type, of high strength, or it lasts for a long period of time, it can cause irreversible damage to the cells leading to a hazardous health effect. Even though its strength is not strong to cause any acute adverse effects, EMF is a kind of strange stimulation and people can be exposed for a longer period of time. Continuing evaluation is required to verify that each of the biological effects does or does not result in any hazardous health effects.

3 Nonthermal Biological Effects of Nonionizing EMF

In addition to the univocal conclusion on the hazardous biological effects due to heating and ionization, many scientific studies report nonthermal biological effects due to nonionizing EMF. The intensive and rapid development of wireless communication technologies in our daily activities made people exposed to EMF even though they are not actively using the devices for communication. Applications and usage of wireless technology are expanding rapidly due to constant public demand for an increased speed, availability, and higher quality of transferred information. Bandwidth for wireless communication covers a wide radiofrequency range. For mobile phones, the frequency range of 900~1800 MHz is used for 3G standard, 2~8 GHz for 4G, and up to 60 GHz carrier frequency is used for 5G communication. Wi-Fi channels use frequencies ranging from 2.4 GHz up to 60 GHz of the "Industrial, scientific, and medical" (ISM) band. In addition to the direct radiation from individual devices they are using, humans are also exposed to the EMF from basic radio, TV, electric, and broadcasting stations. The growth of wireless networking technology in living environments like Wi-Fi at home, routers in the office, and fully covered Wi-Fi zones around, can also result in excessive exposure to radiofrequency EMF. In urban areas, the level of EMF from multisources including radiofrequency EM waves is high globally. All these technological progress and proximity to the EM sources inevitability extended the worries of nonthermal biological effects, especially by the low-level nonionizing EMF when the subjects are exposed continuously throughout the days and years.

The human nervous system functions with a minute level of endogenous ionic currents and electrical fields generated in the neuronal cells and they can be vulnerable to any oscillatory disturbance of an external electromagnetic field. Fundamentally, there is a strong chance that EMF may cause a disturbance and it may lead to a detrimental effect on cell differentiation and proliferation in adult tissues as well as young tissues in the developing stage.

3.1 Cancers and Brain Tumors

Among the nonthermal biological effect of nonionizing EMF, the association between EMF and cancers like leukemia and brain tumors has most collected the concern and been evaluated over the last three decades. There are many studies reporting the link between leukemia and EMF. These publications are mostly epidemiological studies that analyzed the potential correlation between the rate of blood cancers and exposure to different types of EMF.

Studies reported the twofold increase in the risk of childhood leukemia as a result of exposures to low-frequency EMF, along with the consistency of studies across different countries, types of study design, methods of exposure assessment, and systems of power transmission and distribution (Schuz, 2011). The incidence rate of acute myeloblastic leukemia has increased as the subjects live at closer distances to thermoelectric power plants (TPP) and high-power-lines (HPL) (Rodriguez-Garcia & Ramo, 2012). In a population-based case-control study of the link between residential power line exposure and childhood leukemia, a small excess risk has been reported in the highest exposure group (Kheifets et al., 2017). In a cross-sectional case-control study investigating the risk of childhood acute lymphoblastic leukemia, prenatal and postnatal exposure to high-voltage power lines has been indicated as the most important environmental risk factor (Tabrizi & Hossein, 2015).

In contrast to population bases studies, analytic studies have failed to verify the causal relationship between blood cancer and the exposure to nonionizing EMF. In a study on gene expression to investigate the possible association between EMF and childhood leukemia in volunteers, no gene response correlated to repetitive exposure to ELF-EMF was found (Kirschenlohr et al., 2012). Another case-control study among children aged less than 15 years was carried out on the risk of developing childhood acute lymphoblastic leukemia as a result of parental occupational exposure to low-frequency EMF (Reid et al., 2011). The authors did not find an association between maternal or paternal exposure to low-frequency EMF any time before the birth and risk of childhood acute lymphoblastic leukemia. In a study to investigate the association between regular mobile phone use and the risk of leukemia, no significant correlation was found. Analyses of risk in relation to years since first use, lifetime years of use, the cumulative number of calls, and cumulative hours of use produced no significantly raised risks, and there was no evidence of any trends. Only, a nonsignificantly raised risk of the odd rate of 1.87 was found in people who first used a phone 15 or more years ago (Cooke et al., 2010).

Despite the abovementioned findings and available information, evidence for the carcinogenic potential of low frequency and radiofrequency EMF is still insufficient to provide irrevocable proof of its carcinogenic and mutagenic potential for humans. The growing evidence in epidemiological studies must be verified with comprehensive analytic studies verifying any causal relationships. When an analytic study for claiming a potential threat of EMF is conducted either using cell culture or animal models, it is problematic to extrapolate the results directly to humans. At this point, blood cancer is considered as multifactorial, and EMF can be one of the environmental factors (Saliev et al., 2019).

With the rapidly increasing use of wireless phones, the brain became the mainly affected target of the nonthermal biological effect of radiofrequency EMF, and risks of brain tumors due to long-term use have become the major concern of the harmful effect of wireless phones. In a series of case-control studies, long-term use of wireless phones shows a significantly increased odds ratio in the brain tumors, glioma, meningioma, and acoustic neuroma. Glioma is the most common malignant brain tumor representing about 60% of all central nervous system (CNS) tumors. Among the subtypes, glioblastoma multiforme is the most common type of glioma with a peak incidence at the age of 45~75 years and median survival of less than one year. Meta-analysis for the longest cumulative user group longer than 1640 hours or 864 hours combinedly showed a significantly increased odds ratio of 1.9 for the glioma in all cases and an odds ratio of 2.54 for ipsilateral use. Meningioma is a tumor that arises from the meninges, the membranes that surround the brain and spinal cord. It is the most common type of tumor that forms in the head and accounts for about 30% of intracranial neoplasms. It is slowly growing and may give neurological symptoms by compressing or squeezing the adjacent brain, nerves, and vessels. Headaches and seizures can occur as its symptom. The results of the meta-analysis for cumulative exposure in the highest category show increased risk for meningioma with an odds ratio of 1.27 in all cases and with an odds ratio of 1.49 in ipsilateral users. Acoustic neuroma, which is also called vestibular schwannoma, is a benign tumor that develops on the eighth cranial nerve composed of vestibular and auditory nerves leading from the inner ears to the brain. While it grows slowly, narrowed anatomical space may compress vital brain stem structures. Usually, its symptoms begin with tinnitus and hearing problems. The results of the meta-analysis for cumulative exposure in the highest category show a significantly increased risk for acoustic neuroma with an odds ratio of 1.73 in all cases and with an odds ratio of 2.71 in ipsilateral users (Belpomme et al., 2018) (Fig. 16.11). These results of increased risks for brain tumors are supported by the results from animal studies showing tumor-promoting effects in similar types, glioma, and acoustic neuroma as in human studies, under radiofrequency EMF exposure (Falcioni et al., 2018). The incidence of brain tumors in children and adolescents is increasing globally and the use of wireless phones has been suggested as the cause of increased rates of brain tumors in Sweden (Hardell & Carlberg, 2017).

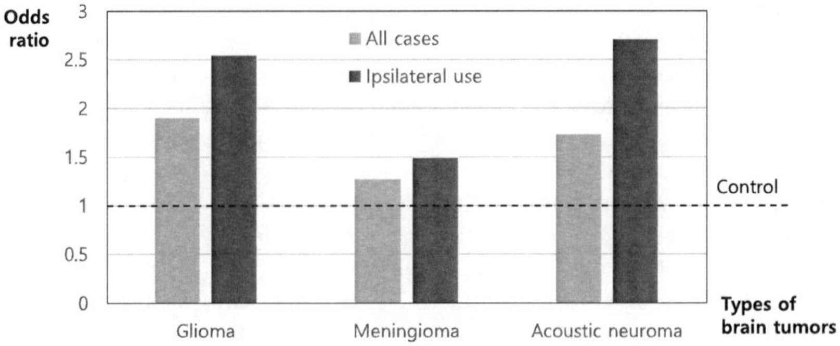

Fig. 16.11 Odds ratios of different types of brain tumors, glioma, meningioma, and acoustic neuroma, in the highest category of cumulative hours of mobile phone use. (Belpomme et al., 2018)

3.2 IARC Classification

International agency for research on cancer (IARC) has classified radio-frequency EMF and magnetic fields in low frequency as, Group 2B, which is the group of agents possibly carcinogenic to humans (Table 16.5). The IARC is a part of the world health organization (WHO), specially organized to classify different chemical agents, mixtures, or exposures, into one of four groups depending on the evidence for their cancer-causing potential, or carcinogenicity. IARC classification is not the direct assessment of the carcinogenic risk level of particular agents. This classification is the simple classification of the quality of the evidence supporting cancer-causing.

Biological effects caused by the ionizing radiation, like UV radiation, X-ray, and gamma-ray have classified into Group 1, which contains the agents described as carcinogenic to humans. Radiofrequency EMF and magnetic fields in low frequency are classified into Group 2B, which contains the agents possibly carcinogenic to humans. About 322 agents are grouped into this group including aloe vera, pickled vegetables, gasoline, carpentry, and many chemical agents. In contrast to radiofrequency EMF, static electric field, static magnetic field, and low-frequency electric field are classified into the rather safer Group 3, where the evidence for carcinogenicity is inadequate to classify.

The classification of radiofrequency EMFs as a Group 2B of "Possible carcinogenic to humans" was based primarily on evidence of elevated risk of developing brain cancer in the long-term users of mobile phones close to the head. However, the lack of clear evidence leading to cancer from animal studies prevented its classification into a more definitive Group of 1 or 2A. In recent animal studies of long-term exposure to mobile EMF, results showed a statistically significant increase in brain gliomas and schwannoma, which are the same type of cancers found in subjects after long-term cell phone use (Falcioni et al., 2018). With this new evidence for cancers in animals in response to prolonged exposure to RF-EMF, an opinion is enforcing that the IARC rating of long-term exposure to RF-EMF should be raised at least to Group 2A of "Probable carcinogenic to humans" (Belpomme et al., 2018) (Table 16.5).

3.3 Reproduction and Fertility

Another concern of long-term exposure to EMF is the possibility of risk to the human reproductive system, particularly on testes. Many studies have investigated the effects of EMF exposure mostly by analyzing the motility of the sperm under and after EMF exposures. When ejaculated human semen was analyzed under EMF by the mobile phone in talking mode, sperm motility and viability are decreased, and molecular damages were indicated by the augmented level of reactive oxygen species (ROS) and a decrease in total antioxidant capacity (TAC) (Agarwal et al., 2009). Also, the motility ratio showed a progressive decrease in subjects with prolonged cell phone daily usage of more than 4 hours/day (Hagras et al., 2016). Compared to the moderate user of mobile phones (<1 hour/day), active users of mobile phones (>1 hour/day) show a statistical difference in the frequencies of sperm concentration, volume, viscosity, liquefaction time, immotile sperms, and abnormal morphology. In addition, mean percentages of normal morphology have decreased significantly with time spent for watching television and using wireless phones, and a decrease in the sperm volume was significantly associated with the distance from the telecommunication tower. Holding an active wireless laptop in a man's lap, having an active mobile phone in a trouser pocket or on their belt, and time spent on sending or receiving messages were significantly associated with a decrease in sperm count and an increased number of immotile sperms (Al-Bayyari, 2017). In the study for the female, an association between spontaneous abortion and nonthermal EMF

Table 16.5 IARC classifications of carcinogenic agents

Group	Description	Definition	Number of agents	Example agents
1	Carcinogenic to humans.	Sufficient evidences in humans. Causal relationship has established	121	Tobacco, alcohol, **ionizing radiation**, **UV exposure**, air pollution.
2A	Probably carcinogenic to human.	Limited evidences in humans. Sufficient evidences in animals.	90	Steroid, red meat, night shift work, insecticides, hot beverages above 65 °C, hairdresser.
2B	Possibly carcinogenic to humans.	Limited evidences in humans. Insufficient evidences in animals.	322	**Radiofrequency EMF**, **magnetic fields in low frequency**, aloe vera, pickled vegetables, gasoline, carpentry.
3	Carcinogenicity not classifiable	Inadequate evidences in humans. Inadequate evidences in animals.	498	Coffee, tea, fluorescent light, **static magnetic field**, **static electric fields**, **electric fields in low frequency**.

exposure was reported in several case-control studies (Li et al., 2017).

Animal studies have intensively been utilized to investigate the effect of EMF exposure on reproductive organs, especially at the cellular level. In the morphological and microscopic analysis of male rats for the EMF exposure to the mobile phones, results have been observed indicating the reaction of testes to the EMF exposure through morphological re-organization. EMF exposure has resulted in a growth in the thickness of membrane propria and the amount of collagen fiber with an increased number of electron-dense mitochondria and cellular structures (Celik et al., 2012). Investigations on the female reproductive system were performed mostly in animal studies. Long-term EMF exposure to mobile phones in mice produces harmful effects on female fertility by increasing oxidative stress (Shahin et al., 2017).

3.4 Neurological Problems in Children

The age groups who are in their developing stage are more susceptible to EMF exposure than adults. These groups include fetuses, children, and adolescents. As their cells are rapidly dividing and their organs are not fully developed, any disturbance of their cellular function in the early life stage can induce abnormalities in the developing organs and systemic functions that last throughout the life. There are many studies reporting the adverse effects of EMF exposure on the cognition and neurobehavioral functions of children and adolescents. Innate vulnerability of the nervous system in the developing stage, higher conductivity due to higher moisture and ionic content than adults, smaller head size, and thinner skull, which is less than half of the adult until 10 years, increase the susceptibility of the young brain to EMF exposure.

Maternal mobile phone use and postnatal exposure to cell phones are associated with the emotional, hyperactivity, and inattention problems of young children (Birks et al., 2017). The use of portable screen-based wireless devices like mobile phones and tablets in children showed statistically significant associations with inadequate sleep quality and excessive daytime sleepiness (Carter et al., 2016). However, the association of EMF exposure to children comes with some inconsistent results. Ten-year-old children living in areas with higher RF exposure did not show any statistically significant effects in most of the cognitive parameters, even though they did show lower verbal scores and higher internalizing and total problems (Calvente et al., 2016). The prenatal or postnatal mobile phone exposure showed an association with the increased neurobehavioral effects combined with elevated lead levels in children (Byun et al., 2017). Similarly, EMF exposure may have synergistic effects with other environmental contaminants known to cause a reduction in intelligence quotient (IQ) and attention, like polychlorinated biphenyls, methyl mercury, environmental tobacco smoke, and probably others (Carpenter, 2006).

Similar effects of low-intensity EMF exposure have been confirmed in various animal studies. Mice exposed in utero were hyperactive and had impaired memory (Aldad et al., 2012). In utero exposure to microwave throughout gestation showed less movement and gender-dependent effects in learning and memory (Zhang et al., 2015). Exposure of young rats to nonthermal EMF intensities impairs learning and spatial memory and induces markers of oxidative stress and inflammation in the brain (Dasdag et al., 2015; Megha et al., 2015).

Even though the results show the adverse effect of EMF exposure on children, WHO still concludes that direct evidence that verifies the increased vulnerability of children to EMF exposure is not sufficient. However, considering the reported effects of EMF on children, WHO recommends minimizing EMF exposure in schools, kindergartens, and any locations where children remain for a substantial part of the day (Kheifets et al., 2005).

3.5 Electromagnetic Hypersensitivity

Electromagnetic hypersensitivity (EHS) is defined as "A phenomenon where individuals experience adverse health effects while using or being in the vicinity of devices emanating electric, magnetic or electromagnetic fields." While the normal subjects do not feel any senses, subjects with EHS report related symptoms under the EMF below the permitted standards. EHS is associated with a variety of EMF emitting devices including mobile phone base stations, mobile handsets, Wi-Fi routers, cordless phones, household appliances, compact fluorescent and halogen light bulbs, power lines, power transformers, and smart meters. EHS is characterized as syndromes with a broad spectrum of non-specific multiple organ symptoms including both acute and chronic inflammatory processes located mainly in the skin and nervous systems, as well as in respiratory, cardiovascular, and musculoskeletal systems (Kaszuba-Zwoińska et al., 2015). Sufferers of EHS predominantly report sleep disorders, asthenia, headaches, memory and concentration difficulties, dizziness, musculoskeletal pain, skin conditions, and mood disorders (Fig. 16.12). Sufferers sensitive to EMF try to reduce their exposure to electromagnetic sources by avoiding sources of exposure, disconnecting or removing electrical devices, and shielding or screening of self or residence. Some EHS sufferers have relocated them in seeking a relief place like "The United States National Radio Quiet

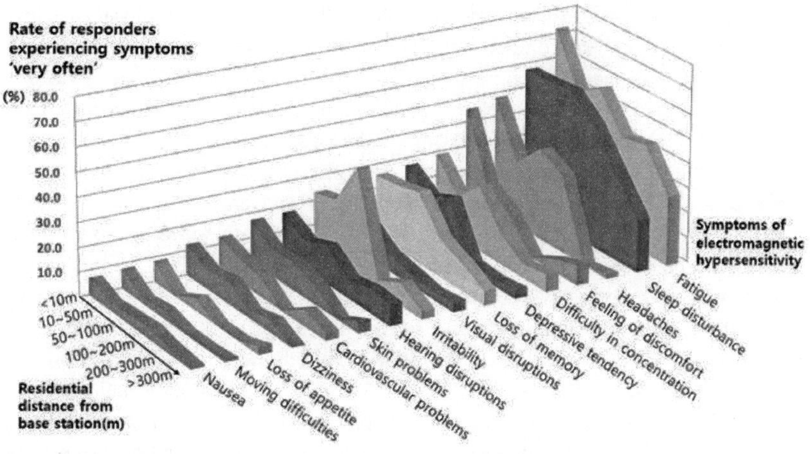

Fig. 16.12 Rate of reported electromagnetic hypersensitivity symptoms from the people in the vicinity of the base station. Rate is the percentage of the people experiencing the complications "very often" in relation to the distance from the base station. (Santini et al., 2002)

Zone," where wireless signals are restricted for scientific research purposes. Available epidemiological data show a relatively high prevalence of perceived EMF sensitivity in the general population ranging from 1.6% to 10.3% across the countries (Dieudonne, 2020).

Three explanatory models have been suggested for the EHS symptoms came to light. The first model is based on the electromagnetic hypothesis, which attributes EHS to EMF exposure. The second model is based on the cognitive hypothesis, which assumes that EHS results from false beliefs about EMF harmfulness and is promoted nocebo responses to perceived exposure. The nocebo effect is the opposite of the placebo effect. The nocebo effect describes a situation where a negative outcome occurs due to a belief that the intervention will cause harm. The third model is based on the attributive hypothesis, which conceives EHS as a coping strategy for explaining the symptoms they are suffering (Dieudonne, 2020).

The first model of the electromagnetic hypothesis is most straightforward. Many arguments support the electromagnetic hypothesis. There are too many people claiming they are EHS sufferers and their symptoms are not clearly diagnosed with any other causes. As EMF is a kind of energy it can be sensed, there exists a possibility that the human body can sense or respond to this type of energy. Light is another band of EM waves that humans sense and respond to the slightest intensity below the level causing a harmful biological effect. The increase of the EHS syndrome is paralleled with the recent massive increase of EMF exposure in the radiofrequency bands of 3 M~30 GHz. This is in contrast to the EMF exposures in the low-frequency bands of 3 Hz~300 KHz. Even though this band was widely associated with the production, transportation, and consumption of electricity long before the onset of EHS, there is no indication of a similar phenomenon. Many studies report the correlation between EMF exposure and the EHS, while many of them are disputed by the lack of completeness of the studies.

However, this electromagnetic hypothesis is not sufficiently supported by the biophysical relationships. The EHS symptoms are quite different from the known biological effect of EMF and the levels of EMF exposure of the EHS sufferers are far below the level to induce thermal effects. In experimental studies for EHS which intentionally expose EHS sufferers to EMF in order to observe their reactions, they failed to establish a link between EMF exposure and EHS sufferers' subjective assessments like the feeling of being exposed, number and intensity of reported symptoms (Rubin et al., 2010) as well as EHS sufferers' physiological responses like heart-rate variability, skin temperature and conductance, blood chemistry, and sleep EEG (Rubin et al., 2011).

The second cognitive hypothesis is raised by the insufficiency of the electromagnetic hypothesis in explicitly verifying EHS experimentally and biophysically. The cognitive model attributes that many EHS sufferers are probably wrong and EHS symptoms are caused through nocebo responses to recognized EMF exposure. Experimental studies support that the reactions of EHS sufferers are determined by their recognition of EMF exposure rather than biophysical sensation. In environmental studies of EHS, nonspecific physical symptoms in the general population and the sufferers' complaints appear mostly independent of estimated or measured EMF exposure levels. But they show a significant correlation with sufferers' recognized EMF exposure level. The more EHS sufferers believe to be EMF exposed, the sicker they are irrespective of their practical EMF exposure level. The cognitive hypothesis is also supported by the brain functioning model that the brain is not just a passive receiver of information. Rather, the brain has a prior model of the world to generate responses to expected sensory inputs. In EHS sufferers, the prior model can be built by a strong belief in the harmfulness of EMF, and this model leads the brain to pay extra attention to environmental cues related to EMF and easily produces expected responses.

However, the cognitive model also has limitations. Even though experimental studies failed to support the electromagnetic hypothesis and made the cognitive hypothesis arise, those studies did not cover various EMF sources that are supposed to cause EHS. In most cases, EHS symptoms do not begin with the reception of information on EMF health effects. With the onset of symptoms, EHS eventually appears as an explanation. Most EHS sufferers even did not recognize the existence of EMF until after they got symptoms and learned about EHS.

The third model of the attributive hypothesis partly parallels the cognitive hypothesis. The cognitive hypothesis can explain how EHS symptoms develop once negative expectations toward EMF have been acquired, but the initial step of the self-reinforcing process supposedly leading to EHS is not accounted for properly. The attributive hypothesis was proposed to explain the onset of EHS symptoms, rather than their persistence of aggravation. The attributive hypothesis regards EHS as a coping strategy of sufferers for their symptoms and this attribution to EMF exposure makes them easier to cope with. In most cases, EHS sufferers do not attribute spontaneously their symptoms to EMF exposure. They usually try to find medical diagnoses and treatment as the first step for their symptoms. However, they failed to get acceptable solutions and convincing diagnoses leading to effective treatments. The medically unexplainable nature of EHS sufferers' symptoms despite multiple clinical investigations makes them easy to accept that they may have EHS once they heard about it. Their commitment toward EHS is developed based on this attribution being the proper solution they found to understand and manage their symptoms.

However, the attributive hypothesis rests mostly on qualitative interview data, which suffer from significant weaknesses like selection bias. Only specific subjects may consent to an interview and qualitative interview data have limits in directly showing the unconscious aspects of cognitive activity. As there are many medically unexplained symptoms and EHS sufferers can attribute their symptoms to any of these, it is difficult to explain how they come to diagnose their symptoms are caused by EHS.

None of the three hypotheses on EHS proves totally acceptable. Lack of biophysical understanding, which is at the heart of the EHS symptom, is a major source of uncertainty. WHO does not consider the EHS as a disease defined on the basis of medical diagnosis and symptoms associated with any known syndrome. In the case of long-term symptoms of people suffering from EHS, it is appropriate for physicians to adapt therapeutic treatment to individual needs, and not just focus on the reduction of EMF intensity. The majority of reported EHS symptoms are still considered by scientists to be psychogenic subjective feelings (Levallois, 2002).

3.6 Beneficial Effect of Electromagnetic Field

EMF exposure also shows beneficial effects. As induced voltage under the EMF exposure can be in a similar range to the level produced normally during dynamic mechanical deformation of connective tissues, there is a chance that time-varying EMF can function beneficially in a process for the treatment of neurological and musculoskeletal disorders. The majority of beneficial effects of EMF are related to the applications of low-frequency-pulsed electromagnetic field (PEMF) in low intensity.

EMF exposure has shown a high therapeutic potential for joint and bone problems treatment. The pulsed electromagnetic field (PEMF) stimulates osteogenic differentiation and maturation of osteoblast and can be used for the prevention and treatment of osteoporosis (Xie et al., 2016). PEMF bone growth stimulator, which is noninvasive without risk, discomfort, and high costs of the operative procedure, is clinically used as a safe and effective treatment of bones. PEMF significantly improves both diabetic and normal wound healing in mice through upregulation of fibroblast growth factor (FGF)-2 and is able to prevent tissue necrosis in diabetic tissue after an ischemic insult (Callaghan et al., 2008). There is clinical evidence that very low and safe levels of amplitude-modulated electromagnetic fields administered may elicit therapeutic responses in patients with cancer (Zimmerman et al., 2012). Low-frequency-pulsed electromagnetic field (LF-PEMF) exposure modulates the expression of miRNAs that regulate the brain signaling and stimulates both tissue regeneration and brain signaling that are dysregulated in Alzheimer's disease (Capelli et al., 2017).

3.7 Cause-Effect Relation

Even though there are a vast amount of studies on nonthermal biological effects of low-intensity EMF exposure, most mainstream public health, and societal bodies claim accumulated data are not sufficient to prove that the clinical symptoms experienced and reported by EHS patients are caused by EMF exposure. This is mostly due to the inconsistency among the studies. Many of them are epidemiological studies based on the subject's self-reported interpretation. Even if there is an association between a symptom and the EMF exposure, it does not mean that a symptom is caused directly by the EMF exposure. Both a symptom and the EMF exposure might be associated commonly with the third variable. Also, the lack of biophysical and experimental evidence explaining the mechanism is an argument used in denying the causal-effect relationship of nonthermal biological effects to the EMF exposure. However, there are a number of mechanistic bases explaining the recognized nonthermal effects of low-intensity EMF exposures. These include induction of oxidative stress, DNA damage, epigenetic changes, altered gene expression, and induction including inhibition of DNA repair and changes in intracellular calcium metabolism.

Currently, the safety of the EMF is controlled by the specific absorption rate (SAR). It is basically based on the EMF energy absorption standardized using an adult male phantom to protect people from the thermal effect of EMF exposure. This safety level does not consider the vulnerability of the children and ignores the nonthermal biological effects reported in low-intensity EMF exposure for an extended period. Same responses used to be observed with lower power density but prolonged exposure as at higher power density and shorter exposure (Nordenson et al., 1994). EMF safety guidance simultaneously considered the duration and intensity of EMF exposure are anticipated to prevent and prepare for any hazardous effects of the nonthermal biological effects of the long-term low-intensity EMF exposure.

References

Agarwal, A., Desai, N. R., et al. (2009). Effects of radiofrequency electromagnetic waves (RF-EMW) from cellular phones on human ejaculated semen: An in vitro pilot study. *Fertility and Sterility, 92*, 1318–1325.

Al-Bayyari, N. (2017). The effect of cell phone usage on semen quality and fertility among Jordanian males. *Middle East Fertility Society Journal, 22*, 178–182.

Aldad, T. S., Gan, G., Gao, X. B., & Taylor, H. S. (2012). Fetal radiofrequency radiation exposure from 800-1900 MHz-rated cellular telephones affects neurodevelopment and behavior in mice. *Scientific Reports, 2*, 312.

Armstrong, B. K., & Kricker, A. (2001). The epidemiology of UV induced skin cancer. *Journal of Photochemistry and Photobiology B: Biology, 63*, 8–18.

Belpomme, D., Hardell, L., et al. (2018). Thermal and non-thermal health effects of low intensity non ionizing radiation: An international perspective. *Environmental Pollution, 242*, 643–658.

Birks, L., Guxens, M., et al. (2017). Maternal cell phone use during pregnancy and child behavioral problems in five birth cohorts. *Environment International, 104*, 122–131.

Burke, D., Howells, J., & Kiernan, M. C. (2017). Sensory and motor axons are different: Implications for neurological disease. *Annals of Clinical Neurophysiology, 19*, 3–12. https://doi.org/10.14253/acn.2017.19.1.3

Byun, Y. H., Ha, M., et al. (2017). Mobile phone use, blood lead levels, and attention deficit hyperactivity symptoms in children: A longitudinal study. *PLoS One, 8*, e59742.

Callaghan, M. J., Chang, E. I., et al. (2008). Pulsed electromagnetic fields accelerate normal and diabetic wound healing by increasing endogenous FGF-2 release. *Plastic and Reconstructive Surgery, 121*, 130–141.

Calvente, I., Pérez-Lobato, R., et al. (2016). Does exposure to environmental radiofrequency electromagnetic fields cause cognitive and behavioral effects in 10-year-old boys? *Bioelectromagnetics, 37*, 25–36.

Capelli, E., Torrisi, F., et al. (2017). Low-frequency pulsed electromagnetic field is able to modulate miRNAs in an experimental cell model of Alzheimer's disease. *Journal of Healthcare Engineering, 10*, 2530270.

Carpenter, D. O. (2006). Environmental contaminants and learning and memory. *International Congress, 1287*, 185–189.

Carter, B., Rees, P., Hale, L., Bhattacharjee, D., & Paradkar, M. S. (2016). Association between portable screen-based media device access or use and sleep outcomes: A systematic review and meta-analysis. *JAMA Pediatrics, 170*, 1202–1208.

Celik, S., Aridogan, I. A., et al. (2012). An evaluation of the effects of long-term cell phone use on the testes via light and electron microscope analysis. *Urology, 79*, 346–350.

Cooke, R., Laing, S., & Swerdlow, A. J. (2010). A case-control study of risk of leukemia in relation to mobile phone use. *British Journal of Cancer, 103*, 1729–1735.

Dalziel, C. F. (1972). Electric shock hazard. *IEEE Spectrum, 9*(2), 41–50. https://doi.org/10.1109/MSPEC.1972.5218692

Dalziel, C. F., & Mansfield, T. H. (1950). Perception of electric currents. *Electrical Engineering, 69*(9), 794–800. https://doi.org/10.1109/EE.1950.6434052

Dalziel, C. F., Lagen, J. B., & Thurston, J. I. (1941). Electric shock. *IEEE Transactions, 60*, 1073–1078.

Dasdag, S., Akdag, M. Z., et al. (2015). Effects of 2.4 GHz radiofrequency radiation emitted from Wi-fi equipment on microRNA expression in brain tissue. *International Journal of Radiation Biology, 91*, 555–561.

Dieudonne, M. (2020). Electromagnetic hypersensitivity: A critical review of explanatory hypotheses. *Environmental Health, 19*, 48. https://doi.org/10.1186/s12940-020-00602-0

Falcioni, L., Bua, L., Tibaldi, M., et al. (2018). Report of final results regarding brain and heart tumors in Sparague-Dawley rats exposed from prenatal life until natural death to mobile phone radiofrequency field representative of a 1.8 GHz GSM base station. *Environmental Research, 65*, 496–503.

Fish, R. M., & Geddes, L. A. (2019). Conduction of electrical current to and through the human body: A review. *Eplasty, 12*(9), e44.

Geddes, L. A., & Baker, L. E. (1971). Response to passage of electric current through the body. *Journal Association for the Advancement of Medical Instrumentation, 2*, 13–18.

Hagmann, M. J., Gandhi, O. P., & Durney, C. H. (1979). Numerical calculation of electromagnetic energy deposition for a realistic model of man. *IEEE Transactions on Microwave Theory and Techniques, 27*(9), 804–809.

Hagras, A. M., Toraih, E. A., & Fawzy, M. S. (2016). Mobile phones electromagnetic radiation and NADþ-dependent isocitrate dehydrogenase as a mitochondrial marker in asthenozoospermia. *Biochimie Open, 3*, 19–25.

Hardell, L., & Carlberg, M. (2017). Mobile phones, cordless phones and rates of brain tumors in different age groups in the Swedish National Inpatient Register and the Swedish cancer register during 1998–2015. *PLoS One, 12*(10), e0185461.

Kaszuba-Zwoińska, J., Gremba, J., Gałdzińska-calik, B., et al. (2015). Electromagnetic field induced biological effects in humans. *Przegląd Lekarski, 72*(11), 636–641.

Kheifets, L., Repacholi, M., Saunders, R., & van Deventer, E. (2005). The sensitivity of children to electromagnetic fields. *Pediatrics, 116*, e303–e313.

Kheifets, L., Crespi, C. M., et al. (2017). Residential magnetic fields exposure and childhood leukemia: A population-based case-control study in California. *Cancer Causes & Control, 28*, 1117–1123.

Kirschenlohr, H., Ellis, P., Hesketh, R., & Metcalfe, J. (2012). Gene expression profiles in white blood cells of volunteers exposed to a 50 Hz electromagnetic field. *Radiation Research, 178*, 138–149.

Kodera, S., Gomez-Tames, J., & Hirata, A. (2018). Temperature elevation in the human brain and skin with thermoregulation during exposure to RF energy. *Biomedical Engineering Online, 17*(1), 1. https://doi.org/10.1186/s12938-017-0432-x

Levallois, P. (2002). Hypersensitivity of human subjects to environmental exposure: A review of the literature. *Environmental Health Perspectives, 110*, 613–618.

Li, D. K., Chen, H., Ferber, J. R., Odouli, R., & Quesenberry, C. (2017). Exposure to magnetic field non-ionizing radiation and the risk of miscarriage: A prospective cohort study. *Scientific Reports, 7*, 17541.

Lown, B., Neuman, J., Amarasingham, R., & Berkovits, B. V. (1962). Comparison of alternating current with direct electroshock across the closed chest. *The American Journal of Cardiology, 10*, 223–233.

McDaniel, W. C., Benwell, A., & Kovaleski, S. (2009). Electrical parameters of projectile stun guns. *IEEE EMBS Proceeding, 2009*, 3184–3187. https://doi.org/10.1109/IEMBS.2009.5334541

Megha, K., Deshmukh, P. S., et al. (2015). Low intensity microwave radiation induced oxidative stress, inflammatory response and DNA damage in rat brain. *Neurotoxicology, 51*, 158–165.

Mursyidi, A. (2002). General aspects of photochemistry to human. *Majalah Farmasi Indonesia, 13*(3), 149–155.

Nordenson, I., Mild, K. H., et al. (1994). Chromosomal aberrations in human amniotic cells after intermittent exposure to fifty hertz magnetic fields. *Bioelectromagnetics, 15*, 293–301.

Reid, A., Glass, D. C., et al. (2011). Risk of childhood acute lymphoblastic leukemia following parental occupational exposure to extremely low-frequency electromagnetic fields. *British Journal of Cancer, 105*, 1409–1413.

Rodriguez-Garcia, J. A., & Ramo, F. (2012). High incidence of acute leukemia in the proximity of some industrial facilities in el Bierzo, Northwestern Spain. *International Journal of Occupational Medicine and Environmental Health, 25*, 22–30.

Rubin, G. J., Nieto-Hernandez, R., & Wessely, S. (2010). Idiopathic environmental intolerance attributed to electromagnetic fields (formerly 'electromagnetic hypersensitivity'): An updated systematic review of provocation studies. *Bioelectromagnetics, 31*, 1–11.

Rubin, G. J., Hillert, L., Nieto-Hernandez, R., van Rongen, E., & Oftedal, G. (2011). Do people with idiopathic environmental intolerance attributed to electromagnetic fields display physiological effects when exposed to electromagnetic fields? A systematic review of provocation studies. *Bioelectromagnetics, 32*, 593–609.

Saliev, T., Begimbetova, D., Masoud, A. R., & Matkarimov, B. (2019). Biological effects of non-ionizing electromagnetic fields: Two sides of a coin. *Progress in Biophysics and Molecular Biology, 141*, 25–36.

Santini, R., Santini, P., & Danze, J. M. (2002). Study of the health of people living in the vicinity of mobile phone base stations: Influence of distance and sex. *Pathologie et Biologie, 50*, S369–S373.

Schuz, J. (2011). Exposure to extremely low-frequency magnetic fields and the risk of childhood cancer: Update of the epidemiological evidence. *Progress in Biophysics and Molecular Biology, 107*, 339–342.

Seno, S. I., Shimazu, H., et al. (2019). Factors affecting and adjustments for sex differences in current perception threshold with transcutaneous electrical stimulation in healthy subjects. *Neuromodulation: Technology at the Neural Interface, 22*(5), 573–579.

Shahin, S., Singh, S. P., & Chaturvedi, C. M. (2017). Mobile phone (1800MHz) radiation impairs female reproduction in mice, Mus musculus, through stress-induced inhibition of ovarian and uterine activity. *Reproductive Toxicology, 73*, 41–60.

Spies, C., & Trohman, R. G. (2006). Narrative review: Electrocution and life-threatening electrical injuries. *Annals of Internal Medicine, 145*, 531–537.

Sugimoto, T., Schaal, S. F., & Wallace, A. G. (1967). Factors determining vulnerability to ventricular fibrillation induced by 60-cps alternating current. *Circulation Research, 21*, 601–608.

Tabrizi, M. M., & Hossein, S. A. (2015). Role of electromagnetic field exposure in childhood acute lymphoblastic leukemia and No impact of urinary alpha- amylase a case-control study in Tehran, Iran. *Asian Pacific Journal of Cancer Prevention, 16*, 7613–7618.

Truong, T., Le, T. V., Smith, D. L., Kantrow, S. P., & Tran, V. N. (2017). Low-voltage electricity-induced lung injury. *Respirology Case Reports, 6(2)*, e00292. https://doi.org/10.1002/rcr2.292. PMID: 29321936.

Ward, A. R., & Robertson, V. J. (1998). Sensory, motor, and pain thresholds for stimulation with medium frequency alternating current. *Archives of Physical Medicine and Rehabilitation, 79*(3), 273–278.

Xie, Y. F., Si, W. G., et al. (2016). Pulsed electromagnetic fields stimulate osteogenic differentiation and maturation of osteoblasts by upregulating the expression of BMPRII localized at the base of primary cilium. *Bone, 93*, 22–32.

Zhang, Y., Li, Z., Gao, Y., & Zhang, C. (2015). Effects of fetal microwave radiation exposure on offspring behavior in mice. *Journal of Radiation Research, 56*, 261–268.

Zimmerman, J. W., Pennison, M. J., et al. (2012). Cancer cell proliferation is inhibited by specific modulation frequencies. *British Journal of Cancer, 106*, 307–313.

Index

A
Absolute refractory period, 155
Acceptability, 251
Acetylcholine, 113
Actin filament, 202
Action potential, 18, 57
Active transport, 68
Adaptation, 127
Adrenaline, 113
Adrian, E., 14
Affecting factor, 68
Ag/AgCl electrodes, 180
Age-related macular degeneration (AMD), 345
Aldini, G., 18
All-or-none, 56
All-or-nothing, 14
Alpha values, 170
Alpha wave, 180
Amber, 5–6
Ambulation, 353–354
Ampere, 21
Ampere-Maxwell's Law, 4
Ampere's law, 4
Ampulla, 133
Amyotrophic lateral sclerosis (ALS), 226
Analog to digital conversion, 56
Analogy of neuron, 55–57
Anatomical biometric items, 252
Animal current, 11, 12, 18
Animal magnetism, 19
Annulospiral receptor, 133
Anterior horn, 45
Aortic arch, 130, 139
Approximate entropy, 169
Artificial electrical organ, 11
Artificial hearing, 325–326
Artificial vision, 326–327
Astatic, 17
Asynchronous BCI, 228
Asynchronous pacing, 287–288
Atmospheric electricity, 5
Atrioventricular (AV) node, 153
Attributive hypothesis, 398
Auditory conduction system, 331
Auditory evoked potentials, 186
Auditory P300, 234
Auditory pathway, 187
Auditory receptor, 126, 131–132
Augmentative and alternative communication (AAC), 226
Augmented limb leads, 159–160
Authentications, 249
Automated external defibrillator (AED), 297
Autonomic nervous system (ANS), 35, 100
Axons, 79, 111

B
Bakken, E., 279
Bardeen, J., 17
Barker, A.T., 300
Barlow, H.B., 20
Baroreceptors, 126, 130
Baroreflex, 130
Bartholow, R., 19
Basal ganglia, 39
Battery, 285, 310–311
BCI spellers, 232
Behavioral biometric items, 252
Benabid, A.L., 300
Beneficial effect, 398
Berger, H., 173
Bernstein, J., 13
Beta wave, 181
Bidirectional transmission, 115
Bimetallic arc, 10
Binary system, 75
Biological current, 12
Biological effects, 384, 386
Biometric authentication, 250
Bipolar, 55
Bipolar leads, 285
Bipolar pacing, 284
Birbaumer, N., 224
Biventricular pacing, 281
Blood pressure, 130
Blood vessel, 29
Body structure, 27–29
Body temperature, 390–392
Bowditch, H., 14
Bradycardia, 163, 278
Brain, 99
Brain-computer interface (BCI), 223, 225
Brain electrical stimulation, 299–301
BrainGate, 229
Brain-machine interface (BMI), 223
Brainstem, 43–44
Brain tumor, 393–394
Brattain, 17
Bremen speller, 239
Bundle branches, 153
Burst coding, 127

K. S. Park, *Humans and Electricity*, https://doi.org/10.1007/978-3-031-20784-6

C

Cable model, 118
Caloric test, 212
Cancers, 392–394
Capacitive recording, 162
Capillary electrometer, 16
Cardiac arrest, 298
Cardiac muscle, 154–155, 199
Cardiac pacemakers, 18, 154, 277
Cardiac pacing, 280
Cardiac vector, 149, 155–156
Cardiorespiratory resuscitation (CPR), 18
Cardiovascular system, 29
Carotid arteries, 139
Carotid sinus, 130
Catatonia, 313
Caton, R., 173
Cause-effect relation, 398–399
Central, 179
Central motor conduction time, 319
Central nervous system (CNS), 34, 36–45, 100
Central terminal, 158–159
Cephalic phase, 30
Cerebellum, 40–41
Cerebral cortex, 37, 38
Cerebral lobe, 37
Cerebrum, 36–40
Characteristic length constant, 101
Chemical receptor, 126, 135
Chemoreceptors, 126, 139, 140
Chloride, 15
Chronology, 23
Circumvention, 251, 260
Clamping voltage, 85
Classification, 245
Closed-loop DBS, 309
Cochlea, 131, 331
Cochlear implants, 334
Cochlear spiral, 332
Coding strategy, 56–57
Cognitive behavioral therapy (CBT), 363
Cognitive hypothesis, 397
Cohen, D., 175
Cold receptor, 126
Collectability, 251
Collecting electricity, 7–8
Communication pathways, 224
Complication, 390–392
Compound muscle action potential (CMAP), 208
Conductance of sodium, 82
Conduction, 153–154
Conduction current, 3
Conduction pathways, 153
Conduction speeds, 109, 110
Cone, 142–143, 342
Conjunctival electrodes, 217
Connexons, 114
Contact lens electrode, 216
Contingent negative variation, 232
Continuous interleaved sampling (CIS), 335
Continuous wavelet transform (CWT), 264
Contralaterally controlled FES, 355
Coordination, 41
Correlation, 265
Correlation dimension, 170
Correlation integral plot, 171
Cortical motor threshold, 318
Cortical silent period, 319
Cranial nerves, 46
Cupula, 133

D

Dalibard, 8
Dark current, 144
d'Arsonval, J., 19
Decibel, 328
Deep brain stimulation (DBS), 23, 301, 304, 323
Deep learning (DL), 245, 265
Defibrillators, 18, 291
de Forest, L., 17
Delgado, J., 300
Delor, 8
Delta wave, 181
Demyelination, 110
Dendrite, 53
Dendrite computation, 120–121
Dendrite potential, 120
Dendritic summation, 55–56
Depolarization, 75
Depolarization phase, 155
Depression, 313
Dermatome, 48
Detrended functional analysis (DFA), 169
Diaphragm pacing, 357
Dielectric constant, 59
Diencephalon, 42–43
Diffusion, 61
Diffusion coefficient, 61
Diffusion gradient, 60–61
Digestive system, 30
Digital pulse, 56–57
Dimension reduction, 263–264
Displacement voltage, 85
Distributed electrical equivalent model, 69–70
Dopamine, 113
Dorsal horn, 45
Double-layer, 57
Drift flow, 61
Dual chamber pacing, 281, 286
du Bois-Reymond, E., 13
du Fay, 6
Dynamic time warping (DTW), 265
Dyskinesia, 304
Dystonia, 305

E

Eardrum, 330
Early receptor potential (ERP), 213
Early repolarization, 155
Eccrine sweat gland, 219
ECG electrodes, 161–162
ECG mechanism, 150–151
ECG waveform, 157–158
Effective refractory, 155
Eigenfaces, 255
Einthoven, 17, 18, 150
Electric acoustic stimulation, 338
Electrical burn, 383
Electrical circuit model, 64–66
Electrical gap junctions, 114

Electrical muscle stimulation (EMS), 356
Electrical shock pulse, 294–296
Electrical stimulation, 18
Electrical transmission, 114–116
Electric charge, 1
Electric current field, 115
Electric fields, 1–2, 387
Electric fish, 5
Electrocardiogram (ECG), 18, 23, 149, 162, 200, 258–267
Electroceuticals, 364
Electroconvulsive therapy (ECT), 312, 323
Electrocorticogram (ECoG), 177
Electrode placement system, 178–179
Electrodermal activity (EDA), 200, 218
Electrodermal response (EDR), 218
Electroencephalogram (EEG), 18, 23, 174, 182, 195, 200
Electrogastrogram (EGG), 18, 200, 217, 218
Electromagnetic field (EMF), 385, 388, 395
Electromagnetic hypersensitivity (EHS), 396, 397
Electromagnetic hypothesis, 397
Electromagnetic signal, 227
Electromagnetic therapy, 20
Electromagnetic wave, 385
Electromagnetism, 19
Electromotive force (EMF), 2
Electromyogram (EMG), 18, 182, 199, 200, 320
Electroneurogram (ENG), 200, 207
Electronystagmogram, 212
Electrooculogram (EOG), 18, 182, 200, 209
Electrophosphene, 20
Electrophysiology, 8, 12–15
Electroreceptor, 124
Electroretinogram (ERG), 18, 200, 213
Electroscopes, 7
Electrostatic generator, 6–7
Encapsulated ending, 124
Endocrine system, 31
End plate potential, 202
Energy delivery, 29
Enteric nervous system (ENS), 35
Entry point, 384
Epicardial pacing, 281
Epilepsy, 195, 306
Epinephrine, 113
Epiretinal prosthesis, 344
Epithalamus, 43
Equal error rate (EER), 266, 271
Equations, 95
Equivalent circuit, 64, 69
Essential tremor (ET), 305
E-tran, 227
Euclidean distance, 265
Event-related desynchronization (ERD), 241
Event-related potential (ERP), 154, 188–191
Event-related synchronization (ERS), 240
Evoked potential (EP), 183, 184
Evolutionary success, 109–110, 116
Excitatory, 113
Excitatory postsynaptic potential (EPSP), 117
Exercise ECG, 162
External auditory canal, 330
Exteroceptor, 124
Extracellular concentration, 60
Extracellular recording, 77
Extrafusal muscle fiber, 133
EyeLink, 226
Eyes-closed, 269
Eyes-open, 269

F
Fake replica, 254
False alarm rate (FAR), 266
False rejection rate (FRR), 254, 266
Familiar faces, 236
Faraday, M., 18, 21, 22
Faraday's law, 2
Fasciculation, 207
Fast-fatigable motor unit, 203
Fast fatigue – resistant motor unit, 203
Fasting period, 218
Feature extraction, 244–245
Features selection, 270
Fertility, 395–396
FES bike, 355
Fetz, E., 223
Fibrillation, 207
Fiducial points, 261
Figure-of-eight coil, 321
Fingerprint, 253, 272
First motor, 21
1st derivative of ECG (FDECG), 262
Flashing light stimulation, 184
Fleming, J.A., 17
Flower spray receptor, 133
Foot drops, 352
Four-chamber heart, 152
Fovea, 143
Frankenstein, 18
Franklin, B., 8
Free nerve ending, 124, 129
Frequency distribution, 332
Frequency-domain parameters, 167
Frequency modulation (FM), 124
Frog's leg, 10
Frontal, 179
Full-field ERG (ffERG), 214
Functional classification, 33–34
Functional electrical stimulation (FES), 351, 354
Functional electrical stimulation therapy (FEST), 354–356

G
Gait, 257
Galvani, L., 8, 10, 12, 18
Galvani's Achievement, 11–12
Galvanic skin potential (GSP), 218
Galvanic skin response (GSR), 218
Gamma-aminobutyric acid (GABA), 113
Gamma-ray, 386
Gamma wave, 181
Ganglionic neurons, 49
Gas receptor, 126
Gastric electrical stimulation (GES), 358
Gastric phase, 30
Gate control theory, 367
Gauss's law, 1–3
Generating distance, 103
GeoSpell, 237
Gilbert, W., 6
Glass capillary, 80
Glomeruli, 138

Glutamate, 113
Goal selecting, 228
Goldman potential, 63–66
Goldman–Hodgkin–Katz equation, 66
G-protein, 113
G-protein-coupled receptor (GPCR), 136
Graphical effects, 235
g-ratio, 106
Gray matter, 37
Green, T., 18

H
Hair cells, 132
Hair follicle, 129
Hand blood vessels, 256
Hand geometry, 255
Hand grasping, 354
Handwritten signature, 257
Hauksbee, F., 7
Health effects, 385, 392–393
Hearing losses, 328, 333–334
Heart rate, 163
Heart rate variability (HRV), 165, 171
Heart rhythm, 163
Heat cramp, 391
Heat exhaustion, 391
Heat strokes, 391
Heaviside, O., 22
Hemichannels, 114
Hemisphere, 37
Hermann, L., 13
Hertz, H., 22
Heterotypic, 114
Hex – O – Speller, 237
Higher cognitive functions, 39
High-frequency (HF), 168
His bundle, 153
His bundle pacing (HBP), 281
Histamine, 113
History, 18, 325–327
Hodgkin, 14
Hodgkin–Huxley model, 14–15, 77–83
Holter monitoring, 162
Homotypic, 114
House, 325
HRV parameters, 165–167
HRV triangular index (HTI), 167
Human body, 27–33
Human heart, 151–153
Human visual system, 339
Huxley, 14
Hybrid BCI, 243–244
Hyperpolarization, 76
Hyperpyrexia, 391
Hypnogram, 182
Hypoglossal nerve stimulation (HNS), 359
Hypopolarization, 76
Hypothalamus, 43, 142
Hypothermia, 391
Hypoxia, 139, 292

I
IARC classification, 395
Identification, 249
Illuminance, 340
Implantable cardioverter-defibrillator (ICD), 292, 296
Implantable pacemaker, 283
Induced current, 19, 387
Induction law, 2
Inflammatory pain, 138
Information processing, 37
Information transfer rate (ITR), 234
Infrared, 386
Inhibitory, 113
Inhibitory post synaptic potential (IPSP), 117
Inion, 178
Inner ear, 131, 331
Inner plexiform layer, 342
Innocuous coolness, 141
Innocuous warmth receptor, 141
Integumentary system, 28
Interbeat interval, 164
Internal electrode, 80
Interneuron, 54
Interoceptor, 124
Intestine phase, 30
Intracellular recording, 77
Intracochlear electrode, 336
Intracortical electrodes, 228–230
Intrafusal muscle fibers, 133
Invasive BCIs, 228–231
Ion channel, 15, 58
Ion concentration, 60
Ion conductance, 83–90
Ion distribution, 60–61
Ionic current, 66–68
Ionization effect, 386–387
Ion pump, 67–68
Iris, 254–255, 260
Isometric contraction, 133, 205
Isotonic contraction, 133, 205

J
Joint capsule, 134

K
Key points, 96–97
Keystroke, 258
Kinesiology, 207
Kinesthetic receptor, 126, 134
Kite experiment, 8
k-nearest neighborhood (KNN), 265
Knowledge-based Authentication, 249
Kouwenhoven, W., 292

L
Labelled line principle, 125
Lateral geniculate nucleus (LGN), 339
Lateral horn, 45
Lateral single character (LSC), 236
Late receptor potential (LRP), 213
Lead vectors, 160
Leadless pacemaker, 290
Leadless pacing, 282
Leiden jars, 7–8, 17
Let-go current, 380
Lethality, 383

Levodopa, 304
LF/HF, 168
Ligament, 134
Light receptor, 142–144
Lighting, 8
Lightning, 5
Lightning rod, 8
Linear discriminant analysis (LDA), 263
Lippmann, G., 16
Local current, 13–14
Locked-in syndrome (LIS), 226
Local field potential (LFP), 177, 310
Loudness control, 337
Low-frequency (LF), 167, 386
Lucas, 14
Luminance, 340
Luminous flux, 340
Lymphatic system, 32

M
Macroshock, 383
Macula, 132
Magnet, 2–3
Magnetic field, 2–4, 387
Magnetic stimulation, 19
Magnetism, 2–3
Magnetoencephalogram (MEG), 18, 23, 175, 192, 193, 195, 225
Magnetocardiogram (MCG), 18
Magnetophosphene, 20
Maintaining, 27–29
Major depressive disorder (MDD), 363
Mania, 313
Manual defibrillator, 296
Matching, 265–266, 270–271
Matteucci, C., 12
Maximum voluntary contraction (MVC), 204
Maximum voluntary isometric contraction (MVIC), 207
Maxwell equations, 5
Maxwell, J., 22
Measured waveform, 92
Measurement, 13, 16–17
Mechanical receptor, 127
Mechanoreceptors, 124, 126
Medulla oblongata, 44
Meissner's corpuscle, 128
Mel-frequency cepstral, 257
Membrane, 57–60, 95
Membrane capacitance, 58, 70
Membrane potential, 61–66
Membrane resistance, 59, 70
Mental activities, 269
Merkel cell, 128
Metabolic signal, 228
Metabotropic receptor, 113
Metal plated electrode, 161
Metallic current, 10–12
mHR, 166
Microelectrode array (MEA), 229, 347
Microelectromechanical systems (MEMS), 229
Microshock, 383
Microwave, 386, 388–389
Micturition reflex, 373
Midbrain, 43
Middle ear, 131, 330
Mismatch negativity (MMN), 189
mNN, 165
Mobility, 61
Modeling ECG, 262–263
Montage, 179
Moore's law, 56
Morphological distortion, 163
Motor imagery (MI), 224, 225, 240–243
Motor nervous system, 35
Motor unit action potential (MUAP), 203
Motor unit recruitment, 204–205
Motor units, 202–205
Movement disorders, 40, 304
Multifactor authentication, 251, 272
Multifocal ERG (mfERG), 215
Multiple sleep latency test (MSLT), 196
Multiplikator, 16
Multipolar, 55
Muscle atonia, 207
Muscle contraction, 201–202
Muscle spindle, 126, 133
Muscle twitch, 203
Muscular system, 27
Myelinated axons, 37, 105–109
Myelinated fiber, 99
Myofibril, 201
Myosin filament, 202

N
Na-K pump, 15
Nasion, 178
Negative feedback, 95
Neher, E., 15
Nernst potentials, 61–63
Nervous system, 31
Neuralink, 230
Neurological disorders, 304
Neurological problems, 396
Neuromuscular junctions, 114
Neuron, 53–57
Neuropathic pain, 139
Neurotransmitters, 112, 113
nHF, 168
nLF, 168
NN50, 166
Nobili, L., 16
Nociceptive pain, 138
Nociceptors, 126, 138, 367
Node of Ranvier, 99
Noninvasive BCIs, 231–244
Nonionizing EMF, 393–399
Nonlinear computation, 120
Nonthermal, 393–399
Norepinephrine, 113
Noxious coldness, 141
Noxious heat, 141
N100 wave, 189
N200 wave, 189–190
N400 wave, 191
Nystagmus, 211

O
Occipital, 179
Oculomotor evaluation, 212
Oddball paradigm, 190, 233

Oersted, C., 16, 20
Ohm, G., 21
Olfactory bulb, 138
Olfactory receptor, 126, 137–138
Operational amplifier (Op-Amp), 56
Optically pumped magnetometer (OPM), 194
Optic nerve, 343
Optic radiation, 343
Optic tract, 343
Organ of Corti, 131
Otoconia, 132
Otolithic organ, 132
Outer ear, 131, 330
Outer plexiform layer, 342
Ovaries, 33

P
Pacemaker leads, 284
Pacemaker pacing mode, 287
Pacemaker potential, 154
Pacinian corpuscle, 128
Pain, 306
Pain receptor, 126, 138–139
Pain relieving, 366–367
Pain sensation, 380
Pain threshold, 380
Papillae, 135
Parasympathetic, 35, 51
Parasympathetic division, 50
Parietal, 179
Parkinson's disease (PD), 303, 306, 317
Pass-thought, 268, 270
Patch-clamp, 15
Patterned stimulation, 184
Pattern theory, 142
Perception threshold, 378–380
Percussive pacing, 280
Performance, 251
Peripheral nerve fibers, 109
Peripheral nervous system (PNS), 34, 45–51
Permanence, 251
Permeability, 63, 68
Phantom limb, 125
Phase coding, 127
Phasic component, 220
Phospholipid, 57
Photochemical effect, 392
Photopic vision, 143
Photopsin, 143
Photoreceptor, 342
Phototransduction, 143, 144
Phrenic nerve stimulation (PNS), 34, 357–358
Physical sensing element, 123
Physiological biometric items, 252
Pick-up coils, 192
Polarization, 75
Polymodality, 141
Polynomial equations, 262
Polysomnogram, 183
Pons, 44
Population coding, 56
Positioning and positional testing, 212
Positive feedback, 95
Possession-based authentication, 250
Posterior horn, 45
Postprandial period, 218
Postsynaptic potentials (PSPs), 53, 55, 100
Potassium, 15
Potassium ions, 62
Potential gradient, 61
Potential variation, 76
Power spectrum, 268
Precordial leads, 159
Pre-frontal, 179
Preganglionic, 49
Prepotential, 154
Preprocessing, 244
Primary coil, 19
Primary visual cortex, 343
Principal component analysis (PCA), 263
PR interval, 157
Procedural memory, 41
Processing, 34
Processing cycle, 100
Process selecting, 228
Propagation, 99–102, 105–109, 117–119
Prosthetic vision, 347–348
PR segment, 157
Psychiatric disorders, 304, 306
Psychogalvanic reflex (PGR), 218
Pulmonary circulation, 29
Pulse generator, 282
Pure tone average (PTA), 338
Purkinje fiber, 154
P-wave, 157
P50 wave, 189
P200 wave, 189
P300, 190–191, 224, 233–237
P600 wave, 191
Pyramidal cell, 176

Q
QRS complex, 157
QT interval, 158
Quantitative measurement, 12
Quantitative modeling, 83–90

R
Radio frequency, 386, 388–389
Rapid eye movement (REM), 212
Rapid serial visual presentation (RSVP), 237
Rate-adaptive pacing, 287, 289
Rate coding, 56, 125
Reaction current, 380
Readiness potential (RP), 232
Receptor potential, 123
Rectified transmission, 115
Reentry circuit, 294
Refractory period, 92–93
Regeneration, 99–102
Relative refractory period (RRP), 155
Renal system, 30
Repetitive transcranial magnetic stimulation (rTMS), 318, 321
Repolarization, 75, 155
Reproduction, 395–396
Reproductive system, 33
Resistance, 59–60
Resistivity, 60
Resonance effect, 389

Respiration paralysis, 381
Respiratory system, 30
Response, 34
Resting phase, 155
Resting potentials, 13, 66, 83
Resting state, 57
Retina, 341–343
Retinal pigment epithelium (RPE), 209, 343
Retinal prosthesis, 344–345
Retinal scan, 256
Retinitis pigmentosa, 345
RF transmission link, 335
Rhodopsin, 143
RMSSD, 167
Rod, 142–143, 342
Ruffini corpuscles, 129

S

Saccadic movement, 210–211
Sacral nerve stimulation (SNS), 372–374
Sakmann, B., 15
Sample entropy (SampEn), 169
Schizophrenia, 313
Schwann cell, 99
Schweigger, J., 16
Scotopic vison, 143
SD1, 168
SD2, 168
SDANN, 166
SDNN, 166
2nd derivative of ECG (SDECG), 262
Secondary coil, 19
Second messenger, 113
Semicircular canals, 132
Semipermeable, 61
Sensation, 33
Sensorimotor rhythm, 224, 240, 242
Sensory nerve action potential (SNAP), 209
Sensory nervous system, 35
Sensory receptors, 123–127
Serotonin, 113
Shockley, 17
Short-time Fourier transform (STFT), 264
Signal-averaging, 185–186
Signal-to-noise ratio (SNR), 186
Silver – silver chloride (Ag-AgCl), 162
Simulated waveforms, 92
Single-chamber pacing, 281, 288
Sinoatrial (SA) node, 153
Skeletal muscle, 199
Skeletal system, 27
Skin-contacting impedance, 378
Skin electrode, 217
Skin resistance, 377
Sleep-related disorders, 195
Slow cortical potential (SCP), 41, 224, 225, 231
Slow motor unit, 203
Smooth muscle, 199
Sodium, 15
Sodium conductances, 88
Sodium ion conductance, 87–89
Sodium–potassium pump, 67
Solenoid cage, 19
Soma, 53, 55
Somatic nervous system (SNS), 35
Somatosensory evoked potential (SEP), 188
Somatosensory P300, 234
Something you are, 250
Something you have, 250
Something you know, 249
Sound intensity, 328
Sound pressure level (SPL), 328
Sound processing, 331–333
Sound processor, 335–336
Sound transmission, 330–331
Space clamp, 78
Spatial recruitment, 204
Spatial summation, 119, 176–177
Specialized receptor, 124
Specific absorption rate (SAR), 388, 389
Specificity theory, 142
Spectral bands, 181
Speed of propagation, 13
Spinal cord, 44–45
Spinal cord stimulation (SCS), 366
Spinal nerves, 47
Spinal plexus, 47
Spinal reflex, 48
Stage N1, 182
Stage N2, 182
Stage N3, 183
Stage REM, 183
Stage W, 182
Standard 12-lead, 160–161
Standard limb leads, 156–157
Static electricity, 5–8
Steady-state potential, 70–72
Steady-state visual evoked potential (SSVEP), 185, 224, 225, 238, 269
Step current stimulation, 72
Stimulating current, 70–72
Stimulating electrodes, 368–369
Stimulating modes, 307
Stimulating pulse, 308–310, 315, 322
Stimulation dose, 315
Strength-duration curve, 371
String galvanometer, 17
Structural classification, 34–36
Structure of authentication, 260–261
ST segment, 158
Stun gun, 383
Subcutaneous implantable cardioverter-defibrillator (S-ICD), 297
Submatrix, 236
Subretinal prosthesis, 345
Subthreshold fluctuation, 57
Subthreshold stimulation, 69–72
Suction electrode, 161
Superconducting Quantum Interference Device (SQUID), 192–194
Suprachoroidal prostheses, 345
Surface EMG, 206
Survival, 32–33
Sweat gland, 218
Sympathetic, 35, 51
Sympathetic division, 50
Synaptic connection, 113
Synaptic transmission, 111–116
Synchronous BCI, 228
Systemic circulation, 29

T
Target placement, 311
Taser gun, 383
Taste bud, 136
Taste receptors, 126, 135
Temperature elevation, 390
Temperature receptor, 126
Temporal, 179
Temporal coding, 56, 126
Temporal recruitment, 205
Temporal summation, 119
Tendon organ, 126, 133
10/20 system, 178
Testes, 33
Tetraethylammonium (TEA), 83
Tetrodotoxin (TTX), 83
Thalamus, 42
Thales, 5
Theoretical progress, 20–22
Thermal heating effects, 388–389, 392
Thermal receptor, 140–142
Thermoreceptors, 124, 126, 140
Thermoregulatory, 390
Thermoregulatory feedback, 142
Theta wave, 181
3D virtual matrix, 236
Threshold, 96
Time course, 13, 76
Time-varying, 1–2
Tonic component, 220
Topographical mapping, 38
Torsional movement, 212
Touch receptor, 126, 128
Tourette's syndrome (TS), 306
Transcranial direct current stimulation (tDCS), 301, 316, 323
Transcranial magnetic stimulation (TMS), 2, 318
Transcutaneous electrical nerve stimulation (TENS), 370
Transcutaneous pacing, 280
Transcutaneous vagus nerve stimulation (tVNS), 363
Transducer, 123–124
Transepithelial potential (TEP), 209
Transfer coefficient, 85
Transient receptor potential (TRP), 140
Transvenous cardiac pacing, 281
Treatment-resistant depression (TRD), 363
12-Lead ECG, 151

U
Ultrasound, 327
Ultraviolet, 386
Unipolar, 55
Unipolar lead, 158–159, 284
Unipolar limb leads, 158
Uniqueness, 251
Universality, 251
Unmyelinated axon, 102, 109
Unmyelinated fiber, 99
Urinary system, 30
U wave, 158

V
Vagal tone, 360
Vagus nerve, 360–361
Vagus nerve stimulation (VNS), 360
van Musschenbroek, P., 7
Variable conversion element, 123
Varying electric field, 4
Ventilation, 139–140
Ventilation reflex, 140
Ventral horn, 45
Ventricular fibrillation (VF), 18, 158, 159, 293, 381–383
Ventricular tachycardia (VT), 293
Versorium, 7
Vertebra, 45
Very-low-frequency (VLF), 167
Vestibular receptors, 126, 132
Vestibule, 132
Vidal, J., 224
Video nystagmography (VNG), 212
Visible light, 340, 386
Visual P300, 233
Visual pathway, 184, 340, 343
Visual prosthesis, 344–348
Visual receptor, 126
Vocalization, 30
Voice, 257
Volta, 10, 12
Voltage clamp, 79–80
Voltage drop, 101
Voltage-gated ion channels, 93–95
Voltaic pile, 11
Voluntary movement, 41
von Békésy, G., 326
von Guericke, O., 6
von Helmholtz, H., 13
von Kleist, E.G., 7

W
Waller, A., 149
Warmth receptor, 126
Weakest audible sound (WAS), 328
White matter, 37
Whole-body absorption, 389
Wilson, F., 151

X
X-ray, 386

Z
Zimmerman, J.W., 175
Zoll, P., 279